GENETIC BREEDING AND AQUACULTURE OF SHELLFISH

GENETIC BREEDING AND AQUACULTURE OF SHELLFISH

Edited by

SHI WANG

College of Marine Life Sciences, Ocean University of China, Qingdao, China

ZHENMIN BAO

College of Marine Life Sciences, Ocean University of China, Qingdao, China

ELSEVIER

ACADEMIC PRESS

An imprint of Elsevier

elsevier.com/books-and-journals

Academic Press is an imprint of Elsevier

125 London Wall, London EC2Y 5AS, United Kingdom
50 Hampshire Street, 5th Floor, Cambridge, MA 02139, United States

ISBN: 978-0-443-33606-5

For Information on all Academic Press publications visit our website at https://www.elsevier.com/books-and-journals

Publisher: Mara Conner
Acquisitions Editor: Vaishali Singh
Editorial Project Manager: Ritika Dagar
Production Project Manager: M. Balakrishnan
Cover Designer: Gopalakrishnan Venkatraman

Typeset by Aptara, New Delhi, India

Contents

Contributors

Serean L Adams Cawthron Institute, Nelson, New Zealand

Antonio Agüera Benthic Ecology Research Group, Institute of Marine Research, Bergen, Norway

Isabelle Arzul ASIM Adaptation et Santé des Invertébrés Marins, Institut Francais de Recherche pour l'Exploitation de la Mer, La Tremblade, France

Zhiyi Bai Key Laboratory of Freshwater Aquatic Genetic Resources, Ministry of Agriculture and Rural Affairs, Shanghai Ocean University, Shanghai, China

Lisui Bao Institute of Evolution & Marine Biodiversity, Ocean University of China, Qingdao, China; College of Marine Life Sciences, Ocean University of China, Qingdao, China

Yongbo Bao College of Biological & Environmental Sciences, Zhejiang Wanli University, Ningbo, China

Zhenmin Bao College of Marine Life Sciences, Ocean University of China, Qingdao, China

Peter Cook University of Western Australia, Albany, Australia

Sergey Dobretsov Center of Excellence in Marine Biotechnology, Department of Marine Science and Fisheries, College of Agricultural and Marine Sciences, Sultan Qaboos University, Muscat, Oman

Yunwei Dong Fisheries College, Ocean University of China, Qingdao, China

Yinghui Dong College of Advanced Agricultural Science, Zhejiang Wanli University, Ningbo, China

Jianguang Fang Yellow Sea Fisheries Research Institute, Chinese Academy of Fishery Sciences, Qingdao, China

Ian G. Gleadall AiCeph LLC, Sendai, Japan

Ximing Guo Haskin Shellfish Research Laboratory, Department of Marine and Coastal Sciences, Rutgers University, Port Norris, New Jersey, United States

Wentao Han College of Marine Life Sciences, Ocean University of China, Qingdao, China

Lin He College of Advanced Agricultural Science, Zhejiang Wanli University, Ningbo, China

Xiaoli Hu College of Marine Life Sciences, Ocean University of China, Qingdao, China

Xiaoting Huang College of Marine Life Sciences, Ocean University of China, Qingdao, China

Zhongming Huo College of Fisheries and Life Science, Dalian Ocean University, Dalian, China

Dean Jerry Tropical Futures Institute, James Cook University, Singapore

Gaowei Jiang College of Fisheries, Ocean University of China, Qingdao, China

Zengjie Jiang Yellow Sea Fisheries Research Institute, Chinese Academy of Fishery Sciences, Qingdao, China

David Jones Centre for Sustainable Tropical Fisheries and Aquaculture, College of Science and Engineering, James Cook University, Townsville, Australia

Caihuan Ke College of Ocean and Earth Sciences, Xiamen University, Xiamen, China

Jiale Li Key Laboratory of Freshwater Aquatic Genetic Resources, Ministry of Agriculture and Rural Affairs, Shanghai Ocean University, Shanghai, China

Qi Li College of Fisheries, Ocean University of China, Qingdao, China

Xiaoxu Li Aquatic Sciences Centre, South Australian Research and Development Institute, Adelaide, Australia

Yuanning Li Institute of Marine Science and Technology, Shandong University, Qingdao, China

Yuli Li College of Marine Life Sciences, Ocean University of China, Qingdao, China

Shanshan Lian College of Marine Life Sciences, Ocean University of China, Qingdao, China

Xiao Liang Key Laboratory of Exploration and Utilization of Aquatic Genetic Resources, Ministry of Education, Shanghai Ocean University, Shanghai, China

Mingling Liao Fisheries College, Ocean University of China, Qingdao, China

Fan Lin Yellow Sea Fisheries Research Institute, Chinese Academy of Fishery Sciences, Qingdao, China

Zhihua Lin College of Advanced Agricultural Science, Zhejiang Wanli University, Ningbo, China; Ninghai Institute of Mariculture Breeding and Seed Industry, Zhejiang Wanli University, Ningbo, China

Fuyun Liu College of Marine Life Sciences, Ocean University of China, Qingdao, China

Pingping Liu College of Marine Life Sciences, Ocean University of China, Qingdao, China

Qigen Liu Key Laboratory of Freshwater Aquatic Genetic Resources, Ministry of Agriculture and Rural Affairs, Shanghai Ocean University, Shanghai, China

Shikai Liu Ocean University of China, Fisheries College, Qingdao, China

Yibing Liu Key Laboratory of Mariculture (Ministry of Education), Fisheries College, Ocean University of China, Qingdao, China

Yuan Lu College of Physics and Optoelectronic Engineering, Ocean University of China, Qingdao, China

Xuan Luo College of Ocean and Earth Sciences, Xiamen University, Xiamen, China

Konstantin Lutaenko Russian Academy of Sciences, A. V. Zhirmunsky National Scientific Center of Marine Biology, Moscow, Russia

Jia Lv College of Marine Life Sciences, Ocean University of China, Qingdao, China

Cecile Massault Tropical Futures Institute, James Cook University, Singapore

Jie Meng College of Marine Life Sciences, Ocean University of China, Qingdao, China

Jaruwat Nabhitabhata Centre for Marine and Coastal Studies, University of Sains Malaysia, Penang, Malaysia; Centre for Biodiversity of Peninsular Thailand, Prince of Songkla University, Hat Yai, Thailand

Deborah M. Power Centro de Ciências do Mar, University of the Algarve, Faro, Portugal

Tristan Renault Département Ressources Biologiques et Environnement, Institut Francais de Recherche pour l'Exploitation de la Mer, Nantes, France

Jin Sun Key Laboratory of Evolution & Marine Biodiversity (Ministry of Education) and Institute of Evolution & Marine Biodiversity, Ocean University of China, Qingdao, China; Laboratory for Marine Biology and Biotechnology, Qingdao Marine Science and Technology Center, Laoshan Laboratory, Qingdao, China

Karsoon Tan College of Marine Science, Beibu Gulf University, Qinzhou, Guangxi, China

Vengatesen Thiyagarajan School of Biological Sciences, The University of Hong Kong, Hong Kong SAR, China

Chunde Wang Qingdao Agricultural University, Qingdao, China

Huizhen Wang College of Marine Life Sciences, Ocean University of China, Qingdao, China

Jing Wang College of Marine Life Sciences, Ocean University of China, Qingdao, China

Shi Wang College of Marine Life Sciences, Ocean University of China, Qingdao, China

Yangfan Wang College of Marine Life Sciences, Ocean University of China, Qingdao, China

Zhiyan Wang Key Laboratory of Freshwater Aquatic Genetic Resources, Ministry of Agriculture and Rural Affairs, Shanghai Ocean University, Shanghai, China

Huilan Wei College of Marine Life Sciences, Ocean University of China, Qingdao, China

Xiaolin Wu University of Wisconsin-Madison, Madison, WI, United States

Qiang Xing College of Marine Life Sciences, Ocean University of China, Qingdao, China

Jinzhi Xu Key Laboratory of Evolution & Marine Biodiversity (Ministry of Education) and Institute of Evolution & Marine Biodiversity, Ocean University of China, Qingdao, China; Laboratory for Marine Biology and Biotechnology, Qingdao Marine Science and Technology Center, Laoshan Laboratory, Qingdao, China

Qinggang Xue Ninghai Institute of Mariculture Breeding and Seed Industry, Zhejiang Wanli University, Ningbo, China

Xiwu Yan College of Fisheries and Life Science, Dalian Ocean University, Dalian, China

Huiping Yang School of Forest, Fisheries, and Geomatics Sciences, Institute of Food and Agricultural Sciences, University of Florida, Gainesville, FL, United States

Zujing Yang College of Marine Life Sciences, Ocean University of China, Qingdao, China

Jinlong Yang Key Laboratory of Exploration and Utilization of Aquatic Genetic Resources, Ministry of Education, Shanghai Ocean University, Shanghai, China

Xiaoshen Yin College of Marine Life Sciences, Ocean University of China, Qingdao, China

Weiwei You College of Ocean and Earth Sciences, Xiamen University, Xiamen, China

Xijiang Yu Breeding, Genetics and Food Production Systems, Norwegian University of Life Sciences, Aas, Norway

Qifan Zeng College of Marine Life Sciences, Ocean University of China, Qingdao, China

Kyall Zenger Tropical Futures Institute, James Cook University, Singapore

Haiyan Zhang Ninghai Institute of Mariculture Breeding and Seed Industry, Zhejiang Wanli University, Ningbo, China

Junlong Zhang Institute of Oceanology, Chinese Academy of Sciences, Qingdao, China

Lingling Zhang College of Marine Life Sciences, Ocean University of China, Qingdao, China

Xiangchao Zhang College of Marine Life Sciences, Ocean University of China, Qingdao, China

Yuchuan Zhang South China Sea Institute of Oceanology, Chinese Academy of Sciences, Guangzhou, China

Jian Zheng Fisheries College, Ocean University of China, Qingdao, China

Xiafei Zheng Ninghai Institute of Mariculture Breeding and Seed Industry, Zhejiang Wanli University, Ningbo, China

Xiaodong Zheng Fisheries College, Ocean University of China, Qingdao, China

Preface

We stand at a critical moment for global aquaculture, an era where shellfish occupy a position of unusual strategic significance. The world faces the challenge of ending hunger and improving nutrition for a growing population, a task made even more complex by rapid climate change. Aquatic products, often called "blue food," are essential to solving this problem. Looking ahead to the year 2050, the demand for sustainable food from aquaculture will grow immensely, with researchers projecting a dramatic increase in the share of shellfish farming. Beyond providing high-quality protein, filter-feeding shellfish offer a unique environmental benefit: Their farming systems naturally capture and store carbon. Expanding shellfish farming not only increases our food supply but also directly supports global goals for carbon neutrality and marine conservation, uniting biological production with ecosystem stewardship.

At the same time, the science of genetic breeding has clearly entered a new era. For many years, progress in the sector depended largely on accumulated husbandry experience, conventional selection, and the gradual refinement of hatchery and grow-out techniques. Those foundations remain indispensable. Yet the scientific landscape has changed profoundly. Genomics, phenomics, genomic selection, and data-intensive tools are reshaping what is possible in shellfish improvement, driving a rapid transition from traditional farming to modern, technology-driven operations. However, this explosion of knowledge presents a significant challenge. Traditionally, shellfish breeding and aquaculture are presented as adjacent topics rather than as interconnected parts of a single system. Researchers, practitioners, and policymakers struggle to navigate these breakthroughs because existing reference materials no longer reflect this convergence. The central challenge is no longer simply how to grow shellfish more efficiently, but how to breed and farm them more intelligently, predictively, and sustainably.

This convergence of opportunity and pressure compelled us to assemble this book. Our goal is to build a bridge between fundamental biology, breeding science, and everyday farming operations, recognizing that a breeding objective is meaningful only if it can be expressed under real farming conditions. To make that integration clear, the book begins with chapters on the basic biology and ecology of shellfish. These chapters are not included merely as introductory background. Shellfish are biologically distinctive organisms, and their anatomy, life histories, reproductive modes, developmental pathways, environmental sensitivities, and ecological roles shape every downstream question in breeding and aquaculture. On this foundation, the book is organized into two closely connected sections that purposefully move from organism to system, and from trait to production.

Section One focuses on genetic breeding. It examines the conceptual and technological transition from conventional improvement to contemporary molecular and genomic breeding. We explore the enabling infrastructures of the discipline, including the mapping of genomes, advanced phenotyping systems, and modern methods such as genomic selection. This section aims to give readers a clear understanding of the cutting-edge tools that are reshaping how we breed shellfish today.

Section Two deliberately turns toward aquaculture application. Scientific progress in breeding achieves its full significance only when translated into hatchery reliability, seed quality, and farm performance. Therefore this section details how to achieve efficient, industrialized farming for major cultivated species, including oysters, scallops, clams, mussels, pearl oysters, abalone, and octopus. We extend the discussion beyond the farm itself to address broader requirements for responsible development, including managing diseases and protecting the environment by assessing carrying capacity. The goal is to cover critical aspects from land-based hatcheries to coastal grow-out strategies.

A book of this scale requires a global perspective. Shellfish aquaculture is global in distribution, yet highly regional in species composition and ecological context. We are deeply honored to collaborate with a distinguished international team of experts spanning Asia, North America, Europe, and Oceania. We invited the leading scientists from China, the United States, France, Norway, Australia, and New Zealand to share their latest research findings and discuss future trends. Together, they share their latest research findings alongside voices from the industry offering practical guidance on mature and novel farming models.

It is our sincere hope that this book will serve as a durable point of reference. By connecting scientific innovation with practical application, we aim to help scientists and technicians navigate the progress of shellfish aquaculture in the modern era. Ultimately, if the 21st century is to be an era of genuinely sustainable blue food development, we hope this unified science of improvement, cultivation, and stewardship will help secure our global food future.

Shi Wang
Zhenmin Bao

SECTION I

Introduction

CHAPTER

1

Shellfish basic biology and ecology

Jin Sun[a,b], *Jinzhi Xu*[a,b], *Konstantin Lutaenko*[c], *Yuanning Li*[d] *and Junlong Zhang*[e]

[a]Key Laboratory of Evolution & Marine Biodiversity (Ministry of Education) and Institute of Evolution & Marine Biodiversity, Ocean University of China, Qingdao, China [b]Laboratory for Marine Biology and Biotechnology, Qingdao Marine Science and Technology Center, Laoshan Laboratory, Qingdao, China [c]Russian Academy of Sciences, A. V. Zhirmunsky National Scientific Center of Marine Biology, Moscow, Russia [d]Institute of Marine Science and Technology, Shandong University, Qingdao, China [e]Institute of Oceanology, Chinese Academy of Sciences, Qingdao, China

1.1 Introduction: The molluscan phenomenon

The term "mollusc" originates from the 18th-century Latin "molluscus," with "mollis" denoting softness. Mollusks exemplify one of the animal kingdom's most remarkable evolutionary success narratives. As a branch of the Lophotrochozoa superphylum, Mollusca ranks as the second largest phylum, surpassed only by Arthropoda. Boasting an estimated 100,000 extant species across marine, freshwater, and terrestrial ecosystems, they demonstrate unparalleled morphological innovation, ecological adaptability, and evolutionary persistence.

This group maintains evolutionary prominence since the Cambrian, displays extraordinary size variation—from 20-m giant squids to millimeter-scale microspecies, and lifespans range from *Arctica islandica*'s four centuries to *Octopus wolfi*'s approximately 6-month existence. Mollusks have successfully colonized marine, freshwater, and terrestrial environments through diverse locomotion strategies, including burrowing, crawling, substrate cementation, floating, free swimming, and parasitism. Functioning as crucial trophic intermediaries, they connect pelagic and benthic ecosystems through varied feeding modes: Sediment processing, water column filtration, and direct predation.

Mollusks maintain ecological significance across Earth's extremes, inhabiting environments from 10,000-m oceanic trenches to 3000-m mountain elevations. While predominantly marine, bivalves and gastropods include freshwater representatives, with gastropods uniquely achieving full terrestrial adaptation through thousands of land-dwelling species by at least 30 independent terrestrializations (Vermeij & Watson-Zink, 2022). Intertidal species survive extreme desiccation and temperature fluctuations via sediment burrowing or substrate

Genetic Breeding and Aquaculture of Shellfish: A Collaborative Approach to a Sustainable Future
DOI: https://doi.org/10.1016/B978-0-443-33606-5.00015-1

adhesion. Deep-sea mollusks harbor symbionts which exhibit chemosynthetic capabilities utilizing hydrothermal vent emissions or cold seep chemicals, while select shallow-water species, such as *Elysia* and *Tridacna*, engage in photosynthesis also through symbiotic partnerships. The value of global bivalve aquaculture reached 23.9 billion USD annually, with broader ecosystem services provided by bivalve farming estimated at $6.47 billion per year in 2020 (Van der Schatte Olivier et al., 2020). China's annual shellfish production has now reached 16 million metric tons, supplying more than 70% of the global market.

1.2 Defining the molluscan blueprint

1.2.1 The soft tissues

Morphologically, mollusks display unique traits that distinguish them from other animal groups, which usually include a head, a specialized foot, mantle, and other tissues (Fig. 1.1).

Locomotory adaptations center on the muscular foot, comprising interwoven myocyte bundles with ventral ciliation for substrate adhesion. In burrowing species (e.g., razor clams), it develops into an axe-shaped structure for sediment penetration, whereas in clinging species (e.g., abalone), it forms a flattened sole with strong adhesive capacity suited for rocky reef cultivation. The suprapedal gland secretes mucus to facilitate movement. The condition of the foot serves as a critical health indicator—for instance, abalone may detach from substrates due to deteriorating water quality. Larger and more nutrient-rich feet in gastropods or bivalves have long been a key breeding target in aquaculture.

The mantle is a fold of the body wall composed of muscle and connective tissue covered by an integument. It adheres closely to the visceral mass near the shell's apex while diverging toward the opening to produce the mantle cavity as a protective barrier. The mantle demonstrates dual functionality as both a shell-forming organ and a respiratory chamber. Its epithelial edge continuously deposits calcium carbonate and conchiolin layers. In pearl oysters, this process can be artificially manipulated through nucleus implantation to induce pearl formation. In cowries, the mantle in turn extends over the shell, keeping the shell shiny and new. In certain species, the mantle extends into an inhalant siphon, sometimes accompanied by a shell siphonal canal, which serves to irrigate the gills with water. Anteriorly, at the entrance to the mantle cavity, there is an organ called osphradium, which regulates the chemical composition of the environment and sediment, allowing the animal to close the mantle cavity fully under hazardous conditions.

The visceral mass, including the gonads, digestive system, heart, and excretory organs, is located within the mantle cavity. The digestive system of mollusks is highly adapted to diet. In aquaculture, filter-feeding bivalves (e.g., *Crassostrea* spp., *Mytilus* spp.) rely entirely on gill filtration, requiring phytoplankton-rich water to sustain growth with supplemental *Spirulina* or *Chlorella* cultivation when necessary. In contrast, carnivorous species such as abalone demand targeted feeding regimens utilizing macroalgae or formulated feeds. After ingestion, a long esophagus leads food to the stomach. The hepatopancreas produces digestive enzymes, and digestion also occurs intracellularly in the hepatopancreas. Meanwhile, the stomach often owns a cuticularized shield to grind food. In some groups, a specialized crop serves as a physical processing area where sand grains further abrade food particles. The intestine only serves to transport feces rather than absorb nutrients, with the part coursing through the mantle as the rectum.

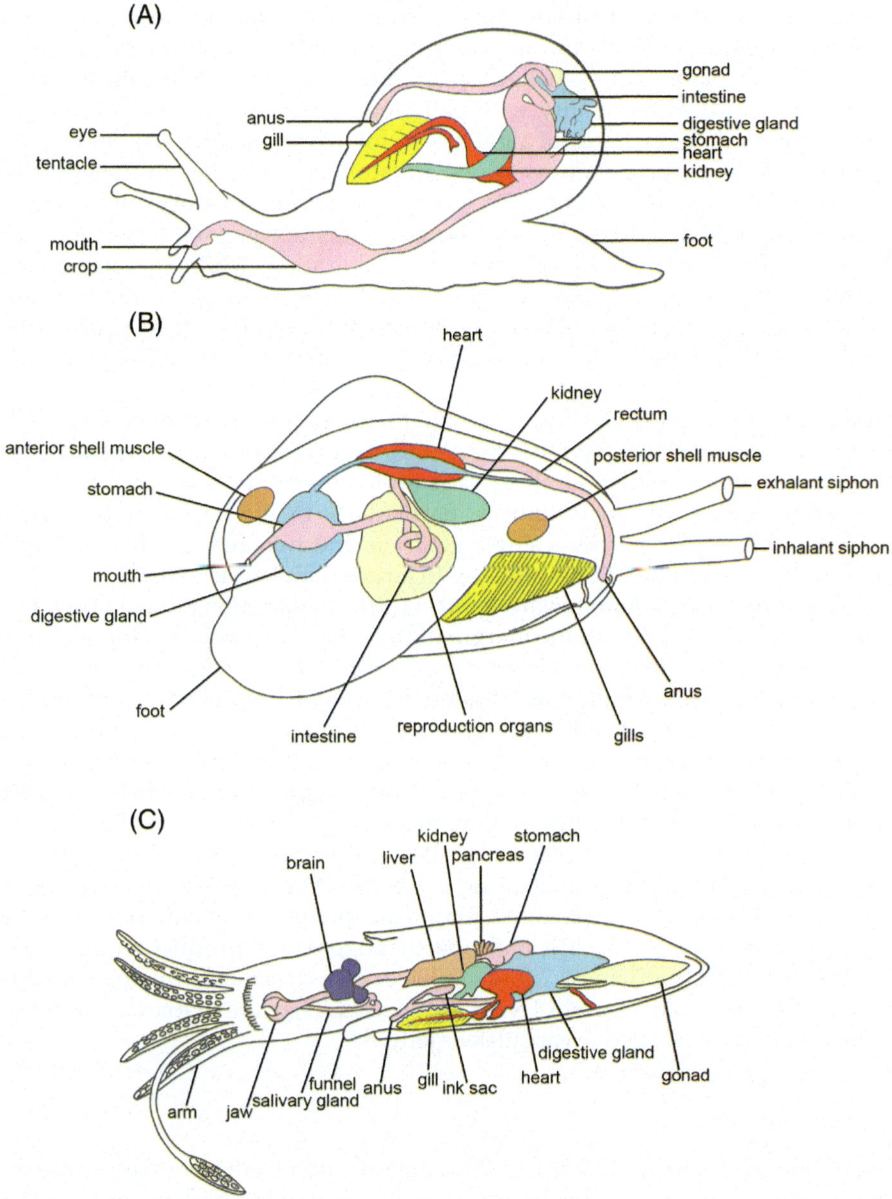

FIGURE 1.1 Schematic diagrams of Gastropoda (A), Bivalvia (B), and Cephalopoda (C). Depicted are schematic diagrams of three representative lineages of Mollusca: Gastropoda (A), Bivalvia (B), and Cephalopoda (C). The anatomy of different adult mollusks' taxa varies significantly across different organs. The head of mollusks is well-defined with concentrated sensory and feeding structures. The mouth opens into the buccal cavity with the muscular odontophore that bears the radula. The chitinous radula in the mouth is secreted by the radula sac, the diversity of which has led to great differentiation of mollusks. The esophagus connects the mouth to the rest of the digestive system. As for sensory organs, tentacles exhibit ecological adaptations: aquatic species typically bear basal eyes on tentacles, while some terrestrial groups develop terminal ocular structures.

The respiratory system of mollusks exhibits considerable variation. Most species utilize ctenidial gills with filament morphologies ranging from triangular lamellae to plumose structures, where water flow facilitates gas exchange across the gill surfaces. Gill condition serves as a sensitive bioindicator for water quality, with pathological manifestations observed in response to heavy metal contamination, harmful algal bloom toxins, or bacterial infections. The bivalves utilize the gill for food particle capture: cilia generate water currents to transport suspended particles to the labial palps for ingestion. Terrestrial and semiterrestrial groups have evolutionarily modified their mantle cavities into vascularized lungs, accompanied by ctenidial reduction or loss. Nudibranchs primarily employ cutaneous respiration, while Siphonaria demonstrate extreme adaptation through complete respiratory organ loss, relying on direct oxygen diffusion from seawater retained in their mantle cavity. Moreover, despite mollusks primarily utilizing hemocyanin, certain Planorbid possess hemoglobin to enhance the efficiency of oxygen transport.

Unlike other mollusks, the excretory and circulatory systems of gastropods are well modified due to torsion in the early stage. Asymmetric nephridial organization concentrates nitrogenous waste excretion in the left kidney. While arteries and veins are present, the circulatory system is open, featuring significant blood sinuses, particularly in the head, viscera, and foot. The chambered heart is characterized by a single ventricle and two auricles that correspond to the presence of either two ctenidia or one, along with a rudimentary structure.

The reproductive systems of mollusks exhibit significant variation among taxonomic groups. Typically, the gonad integrates with the margin of the digestive gland, coiling within the shell's spire. Genital ducts extend from the visceral mass to the mantle edge, running parallel to the rectum and terminating at a genital pore located on the right cephalic region. These ducts are accompanied by testes or ovaries that undergo substantial enlargement during reproductive seasons. Most marine mollusks are dioecious, enabling internal fertilization with an enlarged right tentacle or specific penis serving as a copulatory organ. In aquaculture, spawning can be induced through environmental manipulation, including temperature and light control. Effective spawning induction includes thermal stimulation with 2–3° C temperature elevation, aerial exposure stress, or hormonal treatment using serotonin. Meanwhile, excessive sperm density beyond 10^6/mL during oyster fertilization may lead to egg asphyxiation. Groups, including Pulmonates, are monoecious, which enables parthenogenetic reproduction for snails isolated from each other within a population. Most mollusks experience multiple reproductive cycles, though semelparity also occurs within 1-year life species. They are oviparous or ovoviviparous with eggs surrounded by protective capsules or jelly.

1.2.2 The hard parts

Building on the framework of Mollusca, biomineralization represents the optimal combination of conservatism and innovation. This process involves mineral deposition within organic matrices to form structures, such as shells and radula, playing a pivotal role in understanding molluscan evolution. Biologically controlled by diverse molecules, including proteins, lipids, and carbohydrates, among mollusks, calcium carbonates—particularly aragonite and calcite crystals—dominate biomineral composition.

The shell stands as the quintessential molluscan hard structure. Characteristic spicules or shells covering soft bodies define two subphyla: Aculifera (bearing spicules in Polyplacophora and Aplacophora) and Conchifera (shell-bearing groups). The Aculifera subphylum, as its name suggests, includes species whose bodies are covered with spicules embedded in a cuticular layer, such as members of the classes Polyplacophora and Aplacophora. The Conchifera subphylum is renowned for species that possess unique shells. Some, such as chitons, have both dorsal plates and spicules.

The secretion of the larval shell is deeply conserved at cellular and morphogenic levels of extant shelled mollusks. Such shell formation may use the same developmental program inherited from a common ancestor, an innovation dating back to the Cambrian. Although limpet-like originally, the most common shell form now is coiled, whether in gastropods, nautiloids, or ammonites. The mantle secretes material at the aperture, forming a new layer of the shell. As the aperture expands, the mollusk continuously coils around this conical structure, winding in circles. However, shell morphology is far more diverse than this coiled structure. From the coiled structures of snails to the flattened shapes of limpets in gastropods, the two hinged shells in bivalves, and the eight overlapping calcareous plates in chitons, each shelled species is unique in its own form. Some species have internalized or even degenerated shells, while slugs and squids are most typical. This morphological variety, spanning shape, color, and material composition, has been globally exploited for ornamental purposes throughout human history.

The formation of biomineralization structures may involve the synergistic action of multiple organs. Taking the shell as an example, the mantle serves as the primary organ responsible for shell formation. Calcium is absorbed by both the mantle and the gills in the form of divalent calcium ions, which are then transported to the extrapallial fluid through both passive and active transport mechanisms (Bleher & Machado, 2004; Neff, 1972; Sillanpää et al., 2018). Subsequently, these calcium ions continuously deposit, with different regions of the mantle producing different layers of the shell. A typical shell of bivalves consists of three main layers: the periostracum, which forms the outermost protective layer resistant to corrosion; the prismatic layer, which provides hardness and mechanical support as the middle layer; and the nacreous layer, the innermost layer with high strength and toughness, capable of forming pearls in pearl oysters. Calcium carbonate crystals in the shell are embedded in an organic matrix comprising proteins, acidic polysaccharides, and chitin. These proteins, although accounting for only 0.01% to 5% of the total dry shell weight, are critical factors in crystal nucleation events and growth. Transcriptomic analyses of the mantle have shown evolution with lineage-restricted biomineralization proteins and unique combinations of co-opted ancient genes resulting from domain shuffling, which laid the foundation for the rapid diversification of shells.

Molluscan shells have long been regarded as lifeless mineral deposits. In contrast, vertebrate bones with internal marrow play vital roles in lifelong blood formation and immune regulation, making them a central hub of vertebrate physiology. For decades, skeletal hematopoiesis has been considered a vertebrate-specific innovation during the water-to-land transition. However, this long-standing view has been challenged by a recent breakthrough discovery of abundant hematopoietic stem cells (HSCs) in adult mollusk shells, an invertebrate hematopoietic niche resembling vertebrate bone marrow (Lian et al., 2025). This study revealed the developmental origin of HSCs during larval shell formation and their participation in hemocyte-mediated shell regeneration and soft-body blood supply. Further analysis implicates the widespread presence of skeletal HSC niche in diverse invertebrate groups, boosting a new fundamental paradigm for hematopoiesis and stem cell research in invertebrates and redefining current understanding of vertebrate bone marrow evolution.

In addition to the shell, the radula of mollusks also possesses high diversity. The "radula," derived from the Latin word "little scraper," was introduced in the mid-19th century. Radula is the beginning part of the digestive system, secreted by the radula sac. It typically consists of several rows of chitinous teeth arranged horizontally and fixed to a basal membrane, with both the basal membrane and the teeth enveloped by a radula sheath. Radular teeth undergo continuous renewal, similar to shark teeth. The number of teeth covering a single radula ranges from a few to several hundred or even thousands, with each species exhibiting a unique arrangement of teeth and varied feeding strategies—from cone snails' harpoon-like predation to herbivorous scraping. The radula is primarily composed of dense chitinous fibers and various nonchitinous macromolecules, with some even containing metal ions, such as iron, in certain species, such as limpets and chitons. Most research has focused on the mineral composition and microstructure of the radula, particularly in chiton, which is among the hardest biological materials.

Both the radula and shell are common mineralized structures in mollusks, characterized by their formation on a chitinous matrix with novel orchestration of ancient mineralization gene toolkit and evolution of novel genes (Aguilera et al., 2017; Hilgers et al., 2018). These mineralized structures hold significant aquacultural importance. The growth lines on shells can serve as indicators of environmental changes such as water temperature and salinity, providing valuable data for growth assessment.

1.3 Life history strategies

Mollusks primarily exhibit an indirect developmental pattern, although some undergo direct development. Their life cycle mostly commences with reproduction via sexual fertilization. Following this, embryonic development progresses from the fertilized egg to the trochophore larva, which is a conserved feature among lophotrochozoans. In aquaculture, it requires specific water quality conditions, including salinity of 25‰–32‰, pH 7.8–8.2, and avoidance of heavy metal contamination. The subsequent planktonic larval phase involves a transition in which, among most mollusks such as marine gastropods and bivalves, the trochophore evolves into a veliger larva. This latter stage is distinguished by ciliated velar lobes that serve dual functions in filter-feeding and hydrodynamic propulsion. During this phase, appropriate feeding protocols should be implemented using unicellular algae (such as *Isochrysis* or *Chaetoceros*) at densities of 20,000–50,000 cells/mL, coupled with daily water exchange of 30%–50% to prevent ammonia accumulation (≤0.5 mg/L).

The larval stage is followed by settlement and metamorphosis, where the pelagic larva transitions to benthic life. During metamorphosis, these transient larval structures undergo systematic reorganization: the velum regresses while adult organs (including the foot and shell gland) progressively differentiate, culminating in the juvenile body plan. This biphasic ontogenetic pathway underscores both molluscan evolutionary relationships and their ecological strategies for planktonic dispersal. In aquaculture practices, substrate selection should be species-specific: Polyethylene ropes or shell strings for oysters, corrugated plates or PVC pipes for abalone, and net cages for scallops. In addition, application of γ-aminobutyric acid (GABA) or L-DOPA can improve larval settlement rates (Mesías-Gansbiller et al., 2008). Post-settlement, juvenile growth occurs as the spat (young shellfish) matures into the adult morphology. Finally, upon reaching sexual maturity and engaging in reproduction, the life cycle is completed (Fig. 1.2).

Moreover, hermaphroditism is also common in certain groups, particularly among hermaphroditic bivalves and prosobranch gastropods that keep both male and female gonadal

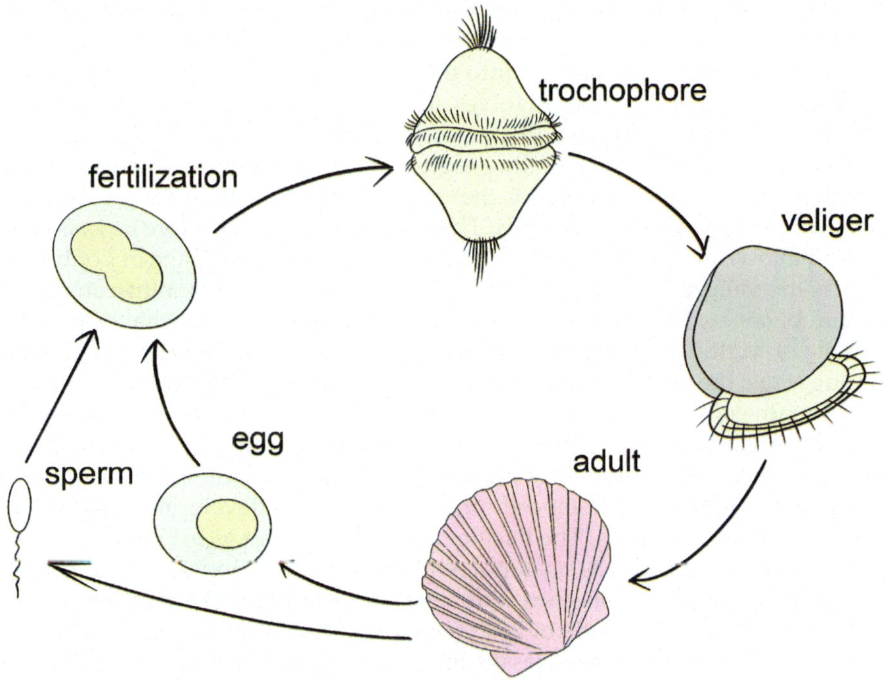

FIGURE 1.2 Life cycle of typical mollusks. Illustration of the life cycle of typical mollusks. They develop from egg to trochophore stage, which is common among lophotrochozoans. They progress to mollusks' unique veliger larval stage before ultimately maturing into adults. Specifically, mollusks exhibit reproductive and developmental strategies that align with their biological diversity. Most species are dioecious, with sexual reproduction remaining the primary mode through the release of gametes into the mantle cavity for internal or external fertilization. In some groups, eggs and sperm are released directly into water for external fertilization. Cephalopods and certain gastropods demonstrate more specialized mechanisms, typically involving sperm capsule transfer. Cephalopods employ a modified arm called the hectocotylus for this purpose, while gastropods often utilize various penis structures that may be accompanied by copulatory stylets or darts. In Solenogasters, copulation is frequently facilitated by a specialized genital cone.

functions. Oysters, for instance, initiate their reproductive life as males, producing sperm, and later switch to females, producing eggs. Such a sex change can occur multiple times throughout their lifespan. Some factors also involve an individual's sex. In sessile species like slipper-shell snails, the sex of an individual could be determined by its nearest neighbor. If isolated or placed near another male, the individual will change into a female. In addition, Parthenogenesis, which involves the development of embryos from unfertilized eggs, also occurs in certain terrestrial gastropods, including *Paludestrina jenkinsi, Potamopyrgus antipodarum*, and *Melanoides tuberculate*. Such a reproductive strategy contributes to their invasive potential in various ecosystems.

Most mollusks produce small eggs that develop into free-swimming larvae, which can be either non-feeding (lecithotrophic) or feeding (planktotrophic). These larvae typically pass through the trochophore stage and the veliger stage before settling down and undergoing metamorphosis. For the fertilized egg of mollusks, the initial cleavage is spiral and holoblastic, with the first two divisions being nearly meridional, resulting in the formation of the polar lobe.

Subsequent cleavages produce smaller micromeres, forming the characteristic spiral pattern of the developing embryo. Following cleavage, the embryo develops into a blastula, a hollow sphere of cells. Cells begin to reorganize into the three germ layers: ectoderm, mesoderm, and endoderm. This reorganization also involves the inward movement of cells, leading to the formation of the archenteron, which will eventually become the gut.

After gastrulation, the embryo typically progresses into larval form. There exist two primary larval stages in many mollusks. The first is the trochophore larva, which is pear-shaped and features a ciliary ring known as the prototroch. The pear-shaped trochophore larva is characterized by a ciliary ring known as the prototroch, which is a common larval stage in Lophophorata. The second stage is the veliger larva, a distinctive phase found in many molluscan groups such as gastropods and bivalves. This stage is characterized by a ciliated velum that extends beyond the shell as a single or multi-lobed structure, enabling swimming and food collection. During this free-swimming stage, the development of the foot and shell begins. In addition, some freshwater bivalves and certain cephalopods have evolved secondary larval forms. Eventually, the veliger larva undergoes metamorphosis, settling onto a substrate and transforming into the adult form. In contrast, direct development without a larval stage occurs in cephalopods and some terrestrial mollusks. Significant interspecific variation exists in developmental timelines among commercially important mollusks. The Pacific oyster (*Crassostrea gigas*) requires 1–2 years from fertilization to market size, whereas scallops complete larval development within 2–3 weeks but need 1.5–2 years of grow-out. Abalone exhibit remarkably brief planktonic phases (3–5 days) yet require extended cultivation periods of 2–3 years to reach commercial size, necessitating size-grading practices (typically 3–4 size classes in scallop aquaculture) to minimize intraspecific competition. Reproductive maturation timelines similarly vary, with oysters reaching sexual maturity at 6–12 months, scallops at 1–2 years, and abalone at 2–3 years.

The practical application of molluscan developmental biology manifests in aquaculture strategies. Hatchery production requires precise environmental control (temperature, salinity, and algal feed regimes) to optimize larval survival. Culture systems must be species-specific, employing either suspended culture (oysters, scallops) or bottom-sowing methods (clams, abalone), with polyculture systems (fish–bivalve–algae integration) providing ecological benefits. The proliferation of biological resources can be achieved through artificial release of larvae or the protection of spawning grounds, such as estuarine sandy-mud substrates.

In a nutshell, molluscan life history research provides fundamental guidance for aquaculture optimization and resource management. Larval phase management directly improves hatchery yields, growth stage knowledge informs feeding and grading protocols, and reproductive ecology studies enable sustainable wild stock utilization. The strategic application of developmental biology principles significantly enhances both economic returns and ecological sustainability in molluscan aquaculture operations.

1.4 Evolutionary origins and phylogeny

The extant mollusk lineages diversified rapidly during the early Cambrian (Song et al., 2023; Vinther et al., 2017), leading to the ongoing debate regarding the definition of strict mollusks and the instability of the mollusk phylogenetic tree or evolutionary history. The oldest shell fossils, known as "small shelly fossils," could date back to around 540 million years ago in the early Cambrian. These diverse small shells mark a significant point in the evolution of biomineralization. Among some bilaterally symmetrical shells, there are a few that show a partially coiled structure that quickly opens into a cap-like shape. This, with some other types, resembles mollusks we know today and is considered phylogenetically related to Mollusca. In the 1960s, the discovery of *Kimberella* fossils, dated to approximately 555 million years ago, added a new dimension to the understanding of early molluscan evolution. Initially, *Kimberella* was identified as a siphonophore or cubozoan. *Kimberella* was later reinterpreted as a benthic, likely motile organism with a complex body structure akin to triploblastic animals and possibly similar to shelled mollusks. Their elongated, slender foot is surrounded by a mantle, separated by a circumpedal mantle cavity, resembling the modern polyplacophoran (Vinther & Smith, 2015). No evidence for a mineralized exoskeleton has been found; Instead, some point-like structures in the fossils have been suggested as remains of sclerites (Fedonkin & Waggoner, 1997) like that of annelid chaetae. In addition, some brush—like imprints found in association with *Kimberella* specimens are interpreted as trace fossils of a scraping-type feeding apparatus similar to molluscan radula, although this remains disputed (Wanninger & Wollesen, 2019). Some locate *Kimberella* stem-group mollusk, supporting the hypothesis that mollusks may have originated in the Precambrian (Fedonkin & Waggoner, 1997).

Whether it is the aforementioned small-shelled organisms, *Kimberella*, or *Acaenoplax*, *Wiwaxia*, *Halkieria*, *Calvapilosa*, and *Shishania* (Wanninger & Wollesen, 2019; Zhang et al., 2024), these potential early stem-line mollusks provide us with a glimpse into the possible evolution of mollusks. These ancient forms lay the foundation for the organ systems that characterize modern mollusks. However, the evolution of key morphological features—including the muscular foot, radula, mantle cavity, and the parallel development of sensory or respiratory systems—reveals a complex interplay of adaptive mechanisms, which often obscure the phylogeny.

Molecular phylogenetic studies or recent phylogenomic research for class-level relationships within Mollusca. Most studies support the basal dichotomy that separates Mollusca into two lineages: Aculifera (Solenogasters, Caudofoveata, and Polyplacophora) and Conchifera (Gastropoda, Bivalvia, Cephalopoda, Monoplacophora, and Scaphopoda). Within Conchifera, Monoplacophora is typically placed as the basal group, although a hypothesis positing Monoplacophora as a sister group to Cephalopoda also exists (Han et al., 2024). The relationships among Bivalvia, Scaphopoda, and Gastropoda have been especially contentious. The clade Scaphopoda + Gastropoda as a sister group to Bivalvia is consistent with some hypotheses (Kocot et al., 2020; Smith et al., 2011), but this contradicts the morphological clade Diasoma (Scaphopoda + Bivalvia) supported by phylogenomic analysis (Chen et al., 2025; Song et al., 2023) (Fig. 1.3). Including the two additional genome sequences of two Monoplacophora together with an additional 11 new mollusk genomes sequenced recently, the molecular phylogeny of eight classes has been fixed, which is also supported by both morphology and most fossil classifications (Chen et al., 2025).

The study of molluscan biodiversity and phylogenetic relationships provides crucial theoretical foundations for aquaculture development. Systematic phylogenetic analyses guide germplasm innovation through targeted crossbreeding programs, while understanding ecological characteristics of different taxa enables optimized culture system design. Furthermore, evolutionary history informs effective resource management strategies for conservation purposes. Moving forward, priority research areas should emphasize: Comprehensive pan-genome studies of economically important species, the development of climate-resilient varieties through selective breeding, and continuous monitoring of genetic exchange between wild and cultured populations.

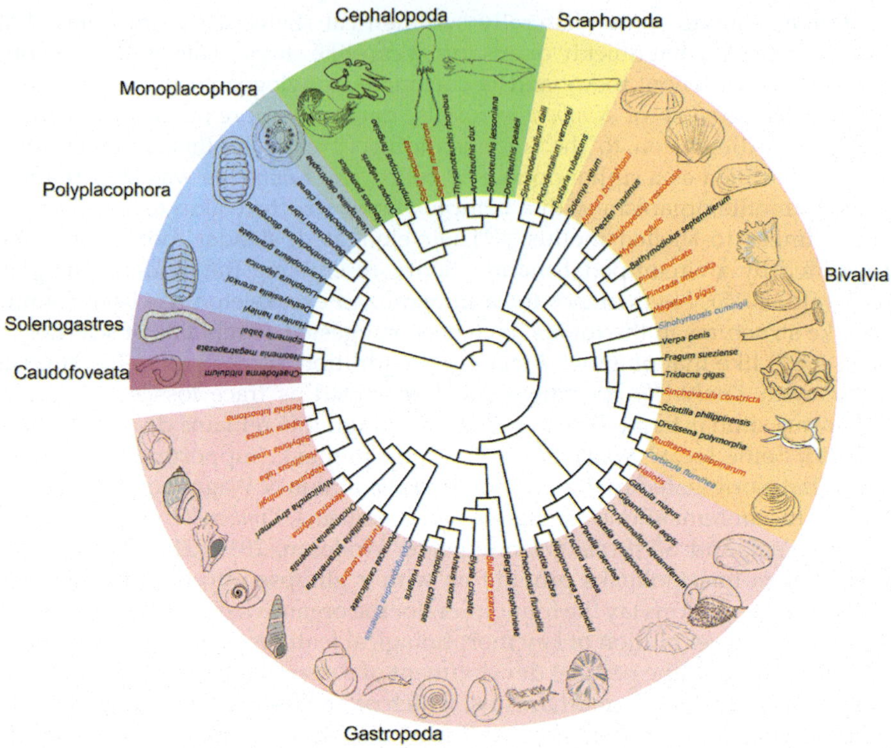

FIGURE 1.3 **Phylogenetic tree of major groups within Mollusca.** Exhibit the relationship of major groups within Mollusca (Chen et al., 2025), among cultured species are marked in red (marine species) and blue (freshwater species). Cultured species include species in Cephalopoda, Gastropoda, and Bivalvia.

1.5 Class-by-class adaptive radiations

Taxonomic understanding of mollusks has profoundly shaped aquaculture practices. However, the molluscan classification historically perplexed naturalists due to their extreme morphological diversity, leading early systems, such as the Compendium of Materia Medica, to erroneously group terrestrial snails with arthropods and aquatic mollusks with turtles. Modern taxonomy now recognizes eight extant classes: Gastropoda (~70% molluscan diversity, including shelled snails and shell-less slugs with global terrestrial/marine distribution), Bivalvia (dual-valved filter feeders such as clams/oysters), Cephalopoda (intelligent predators: Squids, octopuses, and nautiluses), Scaphopoda (tusk-shaped burrowers), Polyplacophora (eight-plated chitons), Monoplacophora (limpet-like with seriated organs), and vermiform Solenogasters/Caudofoveata.

1.5.1 Gastropoda

Gastropods, mollusks' most speciose class, exhibit extraordinary ecological and morphological diversity across marine, freshwater, and terrestrial habitats. Characterized by ventral muscular locomotion (via mucus-aided foot gliding in land snails) and protective spiral shells (absent in slugs), their defining evolutionary innovation is embryonic torsion—a 90–180° visceral rotation creating anatomical asymmetry from bilaterally symmetric ancestors. Ranging from 0.4 mm microgastropods to 700 mm marine giants *Syrinx aruanus*, they occupy virtually all trophic niches: predatory species employ radula drills or venomous harpoons; herbivores use scraping radula; chemosynthetic symbionts in either gills or esophageal gland. Adaptive radiation produced specialized swimmers (pteropods), sessile reef-dwellers, and even holothurian parasites. Notably, gastropods represent mollusks' sole terrestrial colonizers through at least 30 times of independent transitions (Vermeij & Watson-Zink, 2022). With approximately 74,000 extant species (~40,000 marine), gastropod diversity spans six major clades: Caenogastropoda, Heterobranchia, Neritimorpha, Patellogastropoda, Vetigastropoda, and Neomphaliones. In aquaculture, the global annual production of abalone exceeds 200,000 tons, with China accounting for 90% of the output. Meanwhile, groups such as *Babylonia* and *Rapana* are also cultured species with economical value. However, the moon snail poses a threat as the major predator in bivalve aquaculture.

In extant gastropods, the most diverse subclass is Caenogastropoda with approximately 35,000 species in 157 extant families recognized. Characterized by having a single gill and a single auricle in the heart, they exhibit nearly every conceivable shell form, including well-known taxa such as periwinkles, cowries, whelks, cone shells, volutes, miters, limpets, worm snails, and slipper shells. As the dominant group of marine gastropods, they inhabit virtually all marine. Meanwhile, estuarine, freshwater, and terrestrial habitats have been independently colonized by several groups. Species such as *Turritella terebra, Hemifusus tuba, Neptunea cumingi,* along with some freshwater snails from the family Viviparidae, are cultivated groups. In addition, the freshwater species *Pomacea canaliculata* is listed as one of the world's top 100 invasive alien species.

Heterobranchia, named for their varied gill structures (traditional, secondary, or lung-like mantle adaptations), dominate gastropod diversity across terrestrial and marine realms. This clade historically included Lower Heterobranchia (marine shell-bearers retaining opercula, except freshwater Valvatidae), Opisthobranchia, and Pulmonata. Notably, invasive Stylommatophorans (e.g., *Lissachatina fulica*) exemplify their ecological plasticity. Such slug-like shapes evolved in several groups independently (Gosliner, 1981; Medina et al., 2011). Those slugs always exhibit alternative chemical defense against predators or diseases directly from food or de novo synthesis (Avila & Angulo-Preckler, 2020; Gavagnin et al., 2019; Papu et al., 2020).

Typical Neritimorpha are marine grazers, feeding on algal spores, diatoms, and detritus, while this group's habitats range from the deep sea to freshwater and terrestrial. Two groups, Helicinoidea and Hydrocenoidea, have independently colonized the land (Bourne, 1908; Haszprunar, 1988), possessing lungs instead of gills. Certain phenacolepadids, such as *Bathynerita* and *Shinkailepas*, are chemosynthetic organisms found in cold seeps and hydrothermal vents. They possess special sense organs, including elongated cephalic and mantle edge tentacles, to cope with the absence of light and red blood containing hemoglobin within erythrocytes to adapt to low oxygen.

Patellogastropoda, also known as "true limpets," are all characterized by their cap-shaped shells with the apex situated at the center of the shell or slightly anterior. They are likely the earliest-diverging subclass in Gastropoda, which is widely recognized as a monophyletic group from both morphological and molecular evidence(Andrews, 1985; Haszprunar, 1985; Künz & Haszprunar, 2001). Patellogastropods are all marine creatures mostly inhabiting the intertidal and shallow subtidal regions, where they form a prominent component of the littoral marine ecosystems. However, some lineages, such as *Pectinodonta* and *Bathyacmaea*, can survive in chemosynthetic systems, such as hydrothermal vents and hydrocarbon seeps, at depths surpassing 4000 m. This adaptative radiation of deep water may be the result of a single colonization of the deep water around the early Jurassic (Qi et al., 2024).

Vetigastropoda is derived from the Latin word "vetus," meaning "old," with Pleurotomariidae among the oldest living lineages within Mollusca. They originated at the end of the Cambrian and have been common since the Palaeozoic (Frýda et al., 2008). It encompasses a diverse array of marine gastropods, including normal and keyhole limpets, abalones, slit snails, and turban snails. Among the Pleurotomarioidea or slit snails, they are one of the most primitive groups of marine gastropods whose fossil evidence dates to 550 million years ago. They inhabit various marine benthic habitats, ranging from the intertidal zone to the deep sea in tropical, temperate, and polar regions. However, they have not successfully colonized the uppermost littoral or supralittoral zones. Abalone, which belongs to Haliotidae, is indeed the most commonly cultured gastropods with China's annual production exceeds 200,000 tons.

Neomphaliones were truly separated in 2017 (Bouchet et al., 2017), encompassing Neomphalida and Cocculinida. Even within this small subclass, a diversity of shell types could be observed, ranging from limpets such as *Neomphalus* and *Echinopelta* to small, simple skeneiforms such as *Retiskenea* and *Depressigyra* or trochoid-like forms such as *Melanodrymia* (Kaim, 2022). In Neomphaliones, unique adaptations to extreme environments led to remarkable morphology. The most striking example, the scaly-foot snail *Chrysomallon squamiferum*, is covered by hundreds of chitinous sclerites on its dorsal foot. It is the only known gastropod with dermal sclerites. When first discovered, the outer layer of the scales was found to be a mixture of pyrite (FeS_2) and greigite (Fe_3S_4), making the scaly foot snail the only known animal to utilize iron as part of its exoskeleton (Warén et al., 2003).

1.5.2 Bivalvia

Bivalves are a group of headless marine or freshwater mollusks with two hinged shells, or valves. The "Bivalvia" originates from the Latin word "bis," meaning "two," and "valvae," which was first introduced by Carl Linnaeus in Systema Naturae in 1758. Comprising oysters, mussels, scallops, and clams, etc., the Bivalvia is the second largest group of mollusks with approximately 10,000 extant species, dominating broad rocky marine substrates, from intertidal zones to abyssal. Most bivalves live in the thickness of the bottom sediments, burying themselves in it with the help of a thick and powerful foot, and belong to the infauna, such as Tellinidae, Veneridae, Myidae, and many others. A small part bore dense clays or hard rocks, such as sandstones, and lived in burrows or holes, saving themselves from predators and the impact of waves. Such species belong to the endolithofauna, including some mytilids (e.g, *Adula*) and many pholadids (Pholadidae).

Bivalves are abundant in both freshwater and marine ecosystems. They are dominant filter feeders of phytoplankton and other suspended particles, contributing to increased water clarity and light penetration, which, in turn, promotes benthic and littoral primary production. Their feeding activity bridges pelagic and benthic zones, profoundly influencing energy transfer and nutrient cycling. Through nutrient excretion and biodeposition, they sustain aquatic productivity across multiple trophic levels.

For humans, bivalve aquaculture plays a significant role in the marine aquaculture industry (Fig. 1.3). Many species of Bivalve, including oysters, clams, scallops, and mussels, which are directly consumable by humans, are widely cultivated. The cultured species include the following groups: Clams from Arcidae, including *Scapharca subcrenata, Anadara broughtonii* and *Tegillarca granosa*; scallops from Pectinidae, including *Chlamys farreri, Mizuhopecten yessoensis, Argopecten irradians* and *Mimachlamys nobilis*; razor clams from Solenoidea, including *Sinonovacula constricta, Solen grandis* and *Cultellus attenuatus*; clams from Veneridae, including *Meretrix lusoria, Ruditapes philippinarum, Venerupis decussata, Paphia undulata, Mactra chinensis* and *Mercenaria mercenaria* among others; oysters from Ostreidae, including *Crassostrea gigas, Magallana gigas* and *Ostrea denselamellosa*; as well as pearl oysters from Pterioidea, including *Pinctada imbricata, Pteria penguin* and *Pinctada maxima*. As for freshwater cultured bivalves, there are river mussels *Corbicula fluminea* and river clams from Unionidae, including *Sinanodonta woodiana, Sinohyriopsis cumingii, Hyriopsis cumingii, Hyriopsis schlegelii*, and *Cristaria plicata*.

Bivalve aquaculture supports an industry with a production capacity of more than 100 billion tons per year, contributing not only to human consumption as food but also to medicinal and decorative purposes. China plays a dominant role in this sector, supplying over 70% of the world's farmed shellfish production.

Bivalves exhibit bilateral symmetry externally, though internal structures, such as hinge umbos, show asymmetry. Their defining adductor muscles control valve closure via contraction, with most species possessing anterior-posterior pairs (reduced to a single in some taxa). Notably, scallops achieve jet-propelled locomotion through rapid valve cycles. The mantle attaches dorsally along the pallial line, its margins often bearing sensory tentacles or photoreceptors. Within the mantle cavity, the dorsally positioned visceral mass contains core organs, whereas the ventral foot houses the byssal gland. Water circulation occurs through paired siphons formed by fused mantle margins—an inhalant anterior current and exhalant posterior flow. Flanking the visceral mass are two pairs of flaps at its anterior end, as well as leaf-like developed gills adorning each side. They primarily feed on phytoplankton. As water passes through the gills, which are covered with cilia, organic particles are trapped and coated in a mucilaginous substance secreted by the branchial epithelium, conveying toward the labial palps. Larger, coarser particles accumulate in the posterior region and are expelled as pseudofeces through the contraction of adductor muscles.

Bivalve shells comprise dorsally hinged calcareous valves formed through marginal accretion. Mantle-secreted valves consist of calcium carbonate crystals within a protein matrix, with structural variations across taxa in layer composition (aragonite/calcite), arrangement, and quantity. Standard shell architecture includes: (1) periostracum—a conchiolin protein outer layer; (2) prismatic aragonite midlayer; and (3) nacreous inner layer of alternating aragonite/conchiolin sheets. Notably, nacre combines vertebrate-like collagen and invertebrate-type chitin matrices, suggesting conserved biomineralization mechanisms across metazoans. Overall, bivalve shells exhibit more than 10 distinct morphological types, including simple prismatic, nacreous, foliated, composite prismatic, crossed, lamellar, and homogeneous structures (Kobayashi & Samata, 2006). In some species, the valves are not uniform in size or shape, leading to an "inequivalve condition." This disparity is pronounced as exemplified by *Corbula*.

Pearls, derived from Latin "pirula" (teardrop), form when foreign particles infiltrate bivalve mantle tissue, triggering layered secretion of aragonite and conchiolin to create protective nacre. While gastropods like conchs produce nonnacreous variants, true pearly specimens originate exclusively from pearl oysters and freshwater mussels. As *Zhuangzi* metaphorically noted, pearls' inherent value mirrors purposeful creation rather than posthumous possession. This biomineralization process epitomizes nature's defensive artistry through sustained material deposition.

Efforts at bivalve classification and phylogenetic analysis have traditionally been based on morphological features, including gill, stomach, hinge teeth, adductor scar, and siphons. Primarily, bivalve is classified into two major subclasses: Protobranchia and Autobranchia, with differentiation around 500 million years ago during the Cambrian epoch (Plazzi & Passamonti, 2010). The Protobranchia are considered "basal" than other groups, containing three orders, Nuculanida, Nuculida, and Solemyida, with about 740 extant species. It has a wide sole on the foot, the edges of which stretch and serve as an anchor when burying. As the name implies, the gills of Protobranchs are primarily used for respiration rather than feeding. Primarily comprises small, shallow-burrowing marine bivalves, Protobranchs are abundant in deep sea, with some inhabiting shallow coastal or estuarine environments. Protobranchs retain ancestral bivalve traits: Papillate foot with multi-paired pedal muscles, deposit-feeding proboscidate palps, and chitin-girdled stomach with sorting chambers—features reduced in derived lineages such as Solemyida. Total Autobranch diversity approaches 9200 extant species. Autobranchia diverges through filter-adapted W-shaped ctenidia (lost in septibranchs) and simplified palps without sensory organs. Comprising Heteroconchia (>7000 species with diverse feeding/hinge morphologies) and Pteriomorphia (economically vital oysters, mussels, etc.), this clade exhibits narrow/absent feet (except *Neotrigonia*), of which the foot is compressed from the sides and has the shape of an axe.

1.5.3 Cephalopoda

Cephalopods have been a star group since ancient times. As apex predators, they significantly impact marine ecosystems by consuming fish, crustaceans, and other mollusks. They were believed to have evolved from a limpet-like monoplacophoran ancestor during the late Cambrian epoch (House, 1988). During the Ordovician-Silurian periods, cephalopods emerged as dominant marine predators, with orthoceratoids occupying nearly all swimming ecological niches. Ancestral nautiloids developed coiled shell morphology approximately 440 million years ago, undergoing early evolutionary divergence. Modern Cephalopoda comprises Coleoidea (including octopuses, squids, and cuttlefish) and Nautiloidea (chambered nautilus). With approximately 800 living species, cephalopods form a relatively small group. However, they play crucial ecological roles as key food sources for fish, seabirds, and marine mammals, including whales. Cephalopod fisheries account for 2%–5% of global marine species landings compared to fish (Arkhipkin et al., 2021). Compared with the massive scale of fisheries catch, artificial breeding technologies for cephalopods remain underdeveloped. Current production is still largely dependent on wild-caught juveniles and stock enhancement programs. This technological bottleneck limits the expansion of sustainable cephalopod farming operations worldwide.

Cephalopods occupy all oceanic zones from shallow benthic regions to abyssal depths. The deepest recorded cephalopod observation occurred at 6957 m in the Java Trench. Many shallow-water species perform diurnal vertical migrations, descending to deeper waters by day and ascending at night to feed. While most species inhabit shallow benthic areas, some specialize in extreme environments, including brackish water, hydrothermal vents, and cold seeps.

Deep-sea adaptations include gigantism (e.g., giant squids), neutrally buoyant shells (*Spirula*), and photophores. Deep-water adaptations also drove the nautili's unique development of prey-capturing cirri and visual specialization through pinhole camera optics, accompanied by progressive reduction of ocular photoreceptive structures.

Cephalopods are known for many unique morphological novelties, such as a beak, arms, funnel, and highly intelligent neural systems. Specialized for active predation, cephalopods feature a chitinous beak with arms/tentacles derived evolutionarily from the foot, reflected in their name meaning "head-foot." In development, the head is completely merged with the foot, from which arms, tentacles, and funnel are derived, forming a head complex. Decabrachians possess eight sucker/hook-bearing arms and two retractable tentacles, while octopods exclusively retain eight arms.

Devoid of rigid skeletal components, the arms are composed of a sophisticated three-dimensional network of muscle fibers. By selectively activating the muscle groups, flexible movements such as elongation, shortening, bending, torsion, or stiffening could be produced. In most coleoids, arms bear suckers, and some decabrachians have hooks. The structural distinctions between the suckers of decabrachian and octobrachian cephalopods. Within the rim of decabrachian suckers, a horny ring is present, frequently developing into teeth, and in various squid families, these teeth further evolve into hooks. Conversely, octopod suckers feature an outer cup termed the infundibulum and an inner chamber referred to as the acetabulum.

Shallow-water coleoids exhibit semelparous reproduction with post-spawning demise, though some engage in multiple spawning bouts during brief terminal phases. Cuttlefish and coastal squids anchor eggs to benthic substrates, occasionally attempting camouflage, followed by maternal mortality after nonguarded egg deposition. Pelagic and deep-sea species frequently deviate from this strategy. With the adaptation to a low-oxygen environment, where predators are fewer, species such as vampire squid may benefit from longer life cycles. This survival advantage facilitates the possibility of engaging in numerous reproductive cycles while incurring minimal reproductive costs. Another typical example is the Argonaut, whose eggs are laid and carried inside its shell. Such a strategy enables foraging behavior, preventing individuals from becoming exhausted from caring for the eggs. This strategy looks so much like ammonite, considering that even the size and number of eggs are like those of ammonites.

Cephalopods exhibit advanced neural systems, with molluscan ganglia differentiating into cerebral (sensory), pedal (motor), and visceral clusters. Their cerebral ganglia account for 40% of neural mass, distributed through arms in squids, whose fin placement drives giant axon evolution. Octopuses demonstrate ganglionic fusion into centralized processing centers, supporting exceptional behavioral complexity exceeding most invertebrates and rivaling vertebrates. Contrary to Aristotle's "curious but simple" assessment, studies confirm sophisticated adaptations, including visual navigation systems and chromatophore-based communication/camouflage mechanisms. Such camouflage pattern matching relies on a visually driven statistical estimation and generation of the appearance, rather than a precise replication of the substrate's appearance. Meanwhile, the skin pattern is high-dimensional, and pattern matching is not a ballistic motion toward a memorized target (Woo et al., 2023). Thus the creation of a skin pattern requires the intricate coordination of motor and visual nerves. The intelligence of cephalopods can be manifested in a more intuitive form. They possess abilities such as the recognition of human faces and individuals, social learning, and episodic memory. As social animals, these cephalopods are even sophisticated enough for humans to easily discern individual personalities.

1.5.4 Polyplacophora (chitons)

Polyplacophorans, commonly known as chitons, typically adhere tightly to intertidal rocks. As their name, "bearer of many plates," suggests, these bluntly rounded mollusks possess eight overlapping dorsal shell plates arranged in parallel rows, surrounded by a flexible marginal girdle enabling them to curl defensively. When dislodged, their segmented plates reveal a limpet-like fleshy foot, confirming their molluscan identity. Ranging from 5 to 50 mm in length, they show remarkable morphological stability since the Carboniferous period (Sirenko, 2006), with exceptions such as *Cryptochiton stelleri* reaching 36 cm.

Most chitons are herbivores, using iron-mineralized radula to scrape algae and invertebrates included in their diets. Their teeth in radula are one of the hardest biomaterials in nature. Some are ambush predators or feed on sunken wood. These diets include sponges, bryozoans, barnacles, diatoms, foraminiferans, newly settled larvae, and algal crusts scraped from the substratum with the radula. Some selectively nip off small portions of leaf, whereas some deep-water species feed on detritus. Others, such as several species of *Leptochiton* and *Ferreiraella*, feed directly on decaying wood or overlying bacterial biofilm (Duperron et al., 2013). A few species are associated with hydrothermal vents (Saito & Okutani, 1990). They are prey for predatory animals, including fish, birds, starfish, and crabs, as well as other mollusks such as carnivorous whelks and octopuses. Larger species are sometimes eaten by early humans.

Chitons, known as "living fossils," are proposed to be one of the earliest molluscan differentiation, conserving body plans more than 300 million years with Upper Cambrian origin (Puchalski et al., 2008). They are distinguished by their eight-arranged shell plates, which contrast with most other mollusks that typically possess one or two shells. The eight-plate structure allows chitons to flex their bodies while maintaining protection or conforming to uneven surfaces. The plates are composed of two distinct layers. The outer layer is rich in organic content and sometimes brightly colored, which overlays an inner aragonite layer. In a few species, such as *Cryptochiton* and *Cryptoconchus*, the shell plates are significantly reduced or even internal. Some species in *Cryptoplax* have an elongated, somewhat worm-like body with plates that do not overlap.

Beyond protection, the shell plates also serve important sensory functions. All chitons have primitive sensory organs termed aesthetes, which are embedded in their shell plates and can detect light, chemical, and mechanical cues. Chitons exhibit at least three types of light-sensing systems: aesthetes alone, aesthetes with eyespots, and aesthetes with shell eyes (Chappell & Speiser, 2023). Simply put, species with only aesthetes do not show evidence of spatial resolution, while species with eyespots use spatial information about light to influence their locomotory behaviors. In addition to these shell plates, polyplacophorans could also develop spicules in a chitinous girdle, providing additional protection while maintaining flexibility. The spicules are composed of composite material that includes a fibrous organic matrix encased in calcium carbonate.

1.5.5 Scaphopoda, Monoplacophora, and Aplacophora: Unique adaptations in minor molluscan classes

These three classes—Scaphopoda (tusk shells), Monoplacophora (single-shelled "living fossils"), and Aplacophora (spicule-covered worm-like mollusks)—represent early-diverging lineages within Mollusca. Although less diverse than gastropods or bivalves, they exhibit striking morphological and ecological specializations that illuminate the phylum's evolutionary flexibility.

Scaphopoda are characterized by their tubular, tusk-shaped shells that are open at both ends, with a posterior aperture that regulates water flow for respiration. Inhabiting from intertidal to abyssal depths exceeding 6000 m, Scaphopoda have evolved to a significant reduction in most sensory organs, with the notable exception of their highly developed chemosensory captacula (Sigwart et al., 2017). Captacula involves their feeding mechanism as specialized tentacles, which extend from the mantle cavity to detect and transport microprey, such as foraminiferans, through mucus-coated ciliary action. The captacula possess terminal sensory ganglia that facilitate prey discrimination. In terms of their burrowing strategy, the ventral foot anchors in the sediment, allowing the shell to align vertically with the posterior apex exposed.

Monoplacophora retain ancestral metamerism, possessing 3–6 pairs of gills, 2–7 pairs of dorsoventral muscles, and multiple nephridia—traits lost in most Conchifera. Distinguished by a single limpet-like shell and broad muscular foot for adhesion, they lack the segmented plates of Polyplacophora. Rediscovered in 1952 (*Neopilina galatheae*) (Lemche & Wingstrand, 1959), these "living fossils" challenge assumptions about molluscan primitiveness through their Cambrian-like body plan. Their microphagous diet involves grazing deep-sea detritus with a radula structure intermediate between gastropods and chitons.

Aplacophora comprises two divergent lineages: Solenogasters and Caudofoveata. Solenogasters are carnivorous, characterized by a ventral foot groove and reduced radula, preying on cnidarians via secretory proboscises. Caudofoveata are headless burrowing detritivores lacking a foot but possessing a hardened oral shield for sediment penetration. Both lineages feature bodies covered with aragonitic sclerites (spicules) secreted by the mantle epithelium, varying morphologically from hair-like to scale-like for locomotion and defense. Solenogasters are hermaphroditic with internal fertilization, while Caudofoveata are dioecious with external gamete release.

These groups provide us with more perspectives on understanding mollusks. Serving as key infaunal predators in deep-sea ecosystems, Scaphopoda demonstrate convergent evolution with polychaete tentacles through captacula specialization. Monoplacophora provide critical insights into early molluscan body plans, with their organ seriality suggesting a link to annelid-like ancestors. Aplacophora highlight the transition from shelled to vermiform mollusks, where spicule-based defense replaces mineralized shells, enhancing their success in abyssal and interstitial habitats. Phylogenetically, Scaphopoda and Bivalvia constitute the clade Diasoma through shared burrowing and sediment-feeding adaptations. Monoplacophora are now considered sister to Polyplacophora (chitons) within Aculifera, united by spicule-like shell microstructures. Aplacophora diverged early from the main molluscan lineage, with Solenogasters and Caudofoveata splitting approximately 500 million years ago, reflecting distinct evolutionary pathways in body plan reduction.

Overall, these classes exemplify the interplay of constraint and innovation in evolution. Scaphopoda refine ancestral filter-feeding into precision micropredation, Monoplacophora preserve metameric traits lost in most mollusks, and Aplacophora reinvent body armor through biomineralized spicules. Their persistence in extreme environments underscores the remarkable adaptive breadth of the phylum Mollusca.

1.6 Ecological engineers of global systems

Throughout history, mollusks have subtly shaped the Earth's environment while also being influenced by it. Historically, bivalves were among the first reef builders, a role they still fulfill today. Rudist clams, for example, constructed reefs in shallow, equatorial waters during the Jurassic to Cretaceous periods. Modern shellfish reefs (oyster reefs, mussel beds) similarly enhance biodiversity, with Chesapeake Bay oyster reefs alone supporting over 200 associated species, including fish and crustaceans. The calcium carbonate provided by shells formed stable frameworks for colonization by other organisms. These structures provide coastal protection with a mere 1-m-wide oyster reef capable of dissipating 50%–90% of wave energy (Chowdhury et al., 2019). As reef-building rudists, their mineralization processes play a crucial role in calcium cycling, becoming an essential carbon sequestration component. Global shellfish aquaculture captures an estimated 1–2 million tons of carbon annually, equivalent to the carbon sequestration capacity of 50,000 ha of forest (Gu et al., 2022).

As efficient filter feeders, the major cultured bivalve groups, such as oysters and mussels, process 20–50 L of water daily per individual. Large populations can significantly improve water clarity; for example, a hectare of oyster reef filters approximately 10 m^3 of water annually. Meanwhile, their feeding activity helps control algal blooms by reducing phytoplankton densities while also mitigating eutrophication through nutrient removal—each ton of oysters absorbs 0.3 kg of nitrogen and 0.04 kg of phosphorus. Furthermore, their biodeposits act as a biological pump, enhancing benthic environments by promoting organic mineralization and boosting biodiversity, particularly for polychaetes and microcrustaceans. Through plankton filtration, bivalves serve as environmental monitors because of the accumulation of filtered materials in their soft tissues and shells. Especially, the direct injury of microplastics on bivalves in the physiology, food sources, and sedimentary habitats has been witnessed (Ding et al., 2021; Li et al., 2015; Zhang et al., 2020). These bioindicators detect pollutants, pathogens, and acidification effects, valued for their broad distribution, easy accessibility, and fixation.

Mollusk–human interactions hold significant ecological and economic importance. Modern mollusk aquaculture represents an environmentally friendly production model with distinct ecological advantages. As filter feeders, bivalves such as oysters and scallops primarily rely on natural planktonic food sources, significantly reducing dependence on artificial feeds. This trophic characteristic results in a substantially lower carbon footprint compared to finfish aquaculture—producing 1 kg of shellfish generates only 0.5–1 kg CO_2 emissions versus 2–5 kg CO_2 for finfish. Large-scale shellfish farming demonstrates remarkable water purification capacity, with total nitrogen (TN) and phosphorus (TP) levels typically 20%–40% lower in cultivation areas than in nonfarming zones. China's Sanggou Bay provides a successful example where combining the cultivation of shellfish, seaweed, and fish has achieved national Class I water quality standards.

The development of ecological aquaculture systems has focused on optimizing production while minimizing environmental impacts. Integrated multi-trophic aquaculture (IMTA) systems create synergistic relationships across water column layers: Macroalgae (e.g., kelp) absorb dissolved nutrients in surface waters; suspension-feeding bivalves (e.g., oysters) filter particulate organic matter at mid-depths; and deposit feeders (e.g., sea cucumbers) utilize settled detritus on the bottom. This integrated approach improves system efficiency, increasing nitrogen utilization by 30%–50% while reducing waste outputs. Artificial reef enhancement using oyster shells and scallop cages has also proven effective for wild stock restoration, exemplified by Chesapeake Bay's oyster reef rehabilitation program, which tripled local fishery resources.

However, mollusk aquaculture faces several ecological challenges requiring sustainable management solutions. High-density cultivation can cause localized hypoxia, with some scallop farming areas experiencing dissolved oxygen levels below 3 mg/L during summer months due to respiratory demand and organic decomposition. Genetic concerns include potential impacts on wild populations from farm escapees, as seen with Pacific oysters (*Crassostrea gigas*) hybridizing with native European species. Climate change presents additional pressures—while some populations demonstrate adaptation to acidified conditions, studies show a 10%–15% average reduction in shell thickness. Warming waters drive tropical species poleward, necessitating adjustments to farming zones, as observed with the northward expansion of *Perna viridis* into the Yellow Sea.

To address these challenges, the industry is implementing innovative management strategies. Carrying capacity models help determine optimal stocking densities to prevent environmental degradation. Rotational aquaculture systems periodically alternate cultivation areas to allow benthic recovery. Emphasizing native species cultivation, such as promoting the Fujian oyster (*Crassostrea angulata*) in southern China, reduces ecological risks compared to introduced species. These approaches aim to balance production needs with ecosystem health, ensuring the long-term sustainability of mollusk aquaculture's valuable ecological services.

Mollusk–human interactions hold significant ecological and economic importance. From prehistoric civilizations to the modern technological era, this relationship has been interwoven with societal development and ecological understanding. Aquaculture relies heavily on species such as oysters and abalone for food production. Biomedical innovations draw inspiration from nacre-based materials and biological adhesives.

1.7 Conclusion: Lessons from molluscan evolution and diversity

Mollusks exemplify evolutionary innovation through their unparalleled diversification from a conserved body plan. Their success—spanning deep-sea vents to Qinghai-Tibet Plateau—lies in balancing ancestral traits (e.g., mantle-driven biomineralization) with radical adaptations (e.g., cephalopod intelligence and shell loss in slugs). As ecosystem engineers, they purify waters, build reefs, and serve as bioindicators of environmental change. Through management practices, shellfish farming demonstrates remarkable potential to harmonize economic viability with ecological benefits, offering sustainable solutions for global aquaculture development. With global production projected to exceed 205 million tons by 2032, molluscan aquaculture stands at a convergence of food security and ecological restoration. By translating evolutionary innovations—from nacre's fracture resistance to scallops' jet propulsion—into cultivation paradigms, we position these ancient survivors as architects of tomorrow's sustainable ocean economies.

References

Aguilera, F., McDougall, C., Degnan, B. M., & Irwin, D. (2017). Co-option and *de novo* gene evolution underlie molluscan shell diversity. *Molecular Biology and Evolution, 34*(4), 779–792. doi:10.1093/molbev/msw294.

Andrews, E. B. (1985). Structure and function in the excretory system of archaeogastropods and their significance in the evolution of gastropods. *Philosophical Transactions of the Royal Society of London. Series B, Biological Sciences, 310*(1145), 383–406. doi:10.1098/rstb.1985.0124.

Arkhipkin, A. I., Hendrickson, L. C., Payá, I., Pierce, G. J., Roa-Ureta, R. H., Robin, J. P., & Winter, A. (2021). Stock assessment and management of cephalopods: Advances and challenges for short-lived fishery resources. *ICES Journal of Marine Science, 78*(2), 714–730. doi:10.1093/icesjms/fsaa038.

Avila, C., & Angulo-Preckler, C. (2020). Bioactive compounds from marine heterobranchs. *Marine Drugs, 18*(12), 657. doi:10.3390/md18120657.

Bleher, R., & Machado, J. (2004). Paracellular pathway in the shell epithelium of *Anodonta cygnea*. *Journal of Experimental Zoology Part A: Comparative Experimental Biology, 301*(5), 419–427. doi:10.1002/jez.a.20065.

Bouchet, P., Rocroi, J. P., Hausdorf, B., Kaim, A., Kano, Y., Nützel, A., Parkhaev, P., Schrödl, M., & Strong, E. E. (2017). Revised classification, nomenclator and typification of gastropod and monoplacophoran families. *Malacologia, 61*(1-2), 1–526. doi:10.4002/040.061.0201.

Bourne, G. C. (1908). Contributions to the morphology of the group neritacea of aspidobranch gastropods. Part I. The Neritidœ. *Proceedings of the Zoological Society of London, 78*(4), 810–887. doi:10.1111/j.1469-7998.1908.00810.x.

Chappell, D. R., & Speiser, D. I. (2023). Polarization sensitivity and decentralized visual processing in an animal with a distributed visual system. *Journal of Experimental Biology, 226*(4), jeb244710. doi:10.1242/jeb.244710.

Chen, Z., Baeza, J. A., Chen, C., Gonzalez, M. T, González, V. L, Greve, C., Kocot, K. M., Arbizu, P. M, Moles, J., Schell, T., Schwabe, E., Sun, J., Wong, N. L. W. S., Yap-Chiongco, M., & Sigwart, J. D. (2025). A genome-based phylogeny for Mollusca is concordant with fossils and morphology. *Science, 387*(6737), 1001–1007. doi:10.1126/science.ads0215.

Chowdhury, M. S. N., Walles, B., Sharifuzzaman, S., Shahadat Hossain, M., Ysebaert, T., & Smaal, A. C. (2019). Oyster breakwater reefs promote adjacent mudflat stability and salt marsh growth in a monsoon dominated subtropical coast. *Scientific Reports, 9*(1), 8549. doi:10.1038/s41598-019-44925-6.

Ding, J., Sun, C., He, C., Li, J., Ju, P., & Li, F. (2021). Microplastics in four bivalve species and basis for using bivalves as bioindicators of microplastic pollution. *Science of the Total Environment, 782*, 146830. doi:10.1016/j.scitotenv.2021.146830.

Duperron, S., Pottier, M. A., Léger, N., Gaudron, S. M., Puillandre, N., Le Prieur, S., Sigwart, J. D., Ravaux, J., & Zbinden, M. (2013). A tale of two chitons: Is habitat specialisation linked to distinct associated bacterial communities? *FEMS Microbiology Ecology, 83*(3), 552–567. doi:10.1111/1574-6941.12014.

Fedonkin, M. A., & Waggoner, B. M. (1997). The late Precambrian fossil *Kimberella* is a mollusc-like bilaterian organism. *Nature, 388*(6645), 868–871. doi:10.1038/42242.

Frýda, J. iří, Nützel, A., & Wagner, P. J. (2008). Paleozoic Gastropoda. *University of California Press*, 238–270. doi:10.1525/california/9780520250925.003.0010.

Gavagnin, M., Carbone, M., Ciavatta, M. L., & Mollo, E. (2019). Natural products from marine heterobranchs: An overview of recent results. *Chemistry Journal of Moldova, 14*(2), 9–31. doi:10.19261/cjm.2019.617.

Gosliner, T. M. (1981). Origins and relationships of primitive members of the Opisthobranchia (Mollusca: Gastropoda). *Biological Journal of the Linnean Society, 16*(3), 197–225. doi:10.1111/j.1095-8312.1981.tb01848.x.

Gu, Y., Lyu, S., Wang, L., Chen, Z., & Wang, X. (2022). Assessing the carbon sink capacity of coastal mariculture shellfish resources in China from 1981–2020. *Frontiers in Marine Science, 9*, 981569. doi:10.3389/fmars.2022.981569.

Han, X., Sun, S., Wang, Y., Liu, M., Shui, B., & Han, Z. (2024). Out of the ocean: The timescale of molluscan evolution based on phylogenomics revealed the ages of mollusks' evolutionary transitions into the novel environment. *Frontiers in Ecology and Evolution, 12*, 1327007. doi:10.3389/fevo.2024.1327007.

Haszprunar, G. (1985). The fine morphology of the osphradial sense organs of the Mollusca. 1. Gastropoda, Prosobranchia. *Philosophical Transactions of the Royal Society of London. Series B, Biological Sciences, 307*(1133), 457–496. doi:10.1098/rstb.1985.0003.

Haszprunar, G. (1988). On the origin and evolution of major gastropod groups, with special reference to the Streptoneura. *Journal of Molluscan Studies, 54*(4), 367–441. doi:10.1093/mollus/54.4.367.

Hilgers, L., Hartmann, S., & Hofreiter, M. (2018). Thomas von Rintelen, Iñaki Ruiz-Trillo, Novel genes, ancient genes, and gene co-option contributed to the genetic basis of the radula, a molluscan innovation. *Molecular Biology and Evolution, 35*(7), 1638–1652. doi:10.1093/molbev/msy052.

House, M. (1988). Major features of cephalopod evolution. In *Cephalopods: Present and past* (pp. 1–16). E. Schweizerbart.

Kaim, A. (2022). A review of gastropods at ancient hydrocarbon seeps. In Kiel S. (Ed.), *Ancient hydrocarbon seeps* (pp. 323–374). Springer. doi:10.1007/978-3-031-05623-9_11.

Kobayashi, I., & Samata, T. (2006). Bivalve shell structure and organic matrix. *Materials Science and Engineering: C, 26*(4), 692–698. doi:10.1016/j.msec.2005.09.101.

Kocot, K. M., Poustka, A. J., Stöger, I., Halanych, K. M., & Schrödl, M. (2020). New data from Monoplacophora and a carefully-curated dataset resolve molluscan relationships. *Scientific Reports, 10*(1), 101. doi:10.1038/s41598-019-56728-w.

Künz, E., & Haszprunar, G. (2001). Comparative ultrastructure of gastropod cephalic tentacles: Patellogastropoda, Neritaemorphi and Vetigastropoda. *Zoologischer Anzeiger - A Journal of Comparative Zoology, 240*(2), 137–165. doi:10.1078/0044-5231-00017.

Lemche, H., & Wingstrand, K. G. (1959). The anatomy of *Neopilina galatheae* Lemche, 1957. *Galathea Report, 3*, 9–71.

Li, J., Yang, D., Li, L., Jabeen, K., & Shi, H. (2015). Microplastics in commercial bivalves from China. *Environmental Pollution, 207*, 190–195. doi:10.1016/j.envpol.2015.09.018.

Lian, S., Hu, N., Chen, X., Dai, X., Zhu, X., Qiao, R., ... Wang, S. (2025). Widespread presence of bone marrow–like hematopoietic stem cell niche in invertebrate skeletons. *Science Advances, 11*(26), eadw0958. doi:10.1126/sciadv.adw0958.

Medina, M., Lal, S., Vallès, Y., Takaoka, T. L., Dayrat, B. A., Boore, J. L., & Gosliner, T. (2011). Crawling through time: Transition of snails to slugs dating back to the Paleozoic, based on mitochondrial phylogenomics. *Marine Genomics, 4*(1), 51–59. doi:10.1016/j.margen.2010.12.006.

Mesías-Gansbiller, C., Bendimerad, M. E. A., Román, G., Pazos, A. J., Sánchez, J. L., & Pérez-Parallé, M. L. (2008). Settlement behavior of black scallop larvae (*Chlamys varia*, L.) in response to GABA, epinephrine and IBMX. *Journal of Shellfish Research, 27*(2), 261–264. doi:10.2983/0730-8000(2008)27[261:SBOBSL]2.0.CO;2.

Neff, J. M. (1972). Ultrastructure of the outer epithelium of the mantle in the clam *Mercenaria mercenaria* in relation to calcification of the shell. *Tissue and Cell, 4*(4), 591–600. doi:10.1016/S0040-8166(72)80032-6.

Papu, A., Undap, N., Martinez, N. A., Segre, M. R., Datang, I. G., Kuada, R. R., ... Wägele, H. (2020). First study on marine heterobranchia (Gastropoda, Mollusca) in Bangka Archipelago, North Sulawesi, Indonesia. *Diversity, 12*(2), 52. doi:10.3390/d12020052.

Plazzi, F., & Passamonti, M. (2010). Towards a molecular phylogeny of Mollusks: Bivalves' early evolution as revealed by mitochondrial genes. *Molecular Phylogenetics and Evolution, 57*(2), 641–657. doi:10.1016/j.ympev.2010.08.032.

Puchalski, S. S., Eernisse, D. J., & Johnson, C. C. (2008). The effect of sampling bias on the fossil record of chitons (Mollusca, Polyplacophora).. *American Malacological Bulletin, 25*(1–2), 87–95. doi:10.4003/0740-2783-25.1.87.

Qi, Y., Zhong, Z., Liu, X., He, X., Zhou, Y., Zhang, L., Chen, C., Linse, K., Qiu, J.-W., & Sun, J. (2024). Phylogenomic analyses reveal a single deep-water colonisation in Patellogastropoda. *Molecular Phylogenetics and Evolution, 190*, 107968. doi:10.1016/j.ympev.2023.107968.

Saito, H., & Okutani, T. (1990). Two new chitons (Mollusca: Polyplacophora) from a hydrothermal vent site of the Iheya Small Ridge, Okinawa Trough, East China Sea. *Venus (Japanese Journal of Malacology), 49*(3), 165–179. doi:10.18941/venusjjm.49.3_165.

Sigwart, J. D., Sumner-Rooney, L. H., Dickey, J., & Carey, N. (2017). The scaphopod foot is ventral: More evidence from the anatomy of *Rhabdus rectius* (Carpenter, 1864) (Dentaliida: Rhabdidae). *Molluscan Research, 37*(2), 79–87. doi:10.1080/13235818.2016.1257970.

Sillanpää, J. K., Sundh, H., & Sundell, K. S. (2018). Calcium transfer across the outer mantle epithelium in the Pacific oyster, *Crassostrea gigas*. *Proceedings of the Royal Society B: Biological Sciences, 285*(1891), 20181676. doi:10.1098/rspb.2018.1676.

Sirenko, B. (2006). New outlook on the system of chitons (Mollusca: Polyplacophora) (Special Number: The 2nd International Chiton Symposium). *Venus (Journal of the Malacological Society of Japan), 65*(1–2), 27–49. doi:10.18941/venus.65.1-2_27.

Smith, S. A., Wilson, N. G., Goetz, F. E., Feehery, C., Andrade, S. C. S., Rouse, G. W., Giribet, G., & Dunn, C. W. (2011). Resolving the evolutionary relationships of molluscs with phylogenomic tools. *Nature, 480*(7377), 364–367. doi:10.1038/nature10526.

Song, H., Wang, Y., Shao, H., Li, Z., Hu, P., Yap-Chiongco, M. K., ... Kocot, K. M. (2023). Scaphopoda is the sister taxon to Bivalvia: Evidence of ancient incomplete lineage sorting. *Proceedings of the National Academy of Sciences, 120*(40), e2302361120. doi:10.1073/pnas.2302361120.

Van der Schatte Olivier, A., Jones, L., Vay, L. L., Christie, M., Wilson, J., & Malham, S. K. (2020). A global review of the ecosystem services provided by bivalve aquaculture. *Reviews in Aquaculture, 12*(1), 3–25. doi:10.1111/raq.12301.

Vermeij, G. J., & Watson-Zink, V. M. (2022). Terrestrialization in gastropods: Lineages, ecological constraints and comparisons with other animals. *Biological Journal of the Linnean Society, 136*(3), 393–404. doi:10.1093/biolinnean/blac053.

Vinther, J., Parry, L., Briggs, D. E. G., & Van Roy, P. (2017). Ancestral morphology of crown-group molluscs revealed by a new Ordovician stem aculiferan. *Nature, 542*(7642), 471–474. doi:10.1038/nature21055.

Vinther, J., & Smith, A. (2015). The origins of molluscs. *Palaeontology, 58*(1), 19–34. doi:10.1111/pala.12140.

Wanninger, A., & Wollesen, T. (2019). The evolution of molluscs. *Biological Reviews, 94*(1), 102–115. doi:10.1111/brv.12439.

Warén, A., Bengtson, S., Goffredi, S. K., & Van Dover, C. L. (2003). A hot-vent gastropod with iron sulfide dermal sclerites. *Science, 302*(5647), 1007. doi:10.1126/science.1087696.

Woo, T., Liang, X., Evans, D. A., Fernandez, O., Kretschmer, F., Reiter, S., & Laurent, G. (2023). The dynamics of pattern matching in camouflaging cuttlefish. *Nature, 619*(7968), 122–128. doi:10.1038/s41586-023-06259-2.

Zhang, F., Man, Y. B., Mo, W. Y., Man, K. Y., & Wong, M. H. (2020). Direct and indirect effects of microplastics on bivalves, with a focus on edible species: A mini-review. *Critical Reviews in Environmental Science and Technology, 50*(20), 2109–2143. doi:10.1080/10643389.2019.1700752.

Zhang, G., Parry, L. A., Vinther, J., & Ma, X. (2024). A Cambrian spiny stem mollusk and the deep homology of lophotrochozoan scleritomes. *Science, 385*(6708), 528–532. doi:10.1126/science.ado0059.

CHAPTER

2

Overview of molluscan aquaculture and breeding

Ximing Guo

Haskin Shellfish Research Laboratory, Department of Marine and Coastal Sciences, Rutgers University, Port Norris, New Jersey, United States

2.1 Introduction

Molluscan shellfish are important aquaculture species, accounting for 20% of total animal aquaculture production. While molluscan aquaculture has a long history, most cultured mollusks have not been domesticated like livestock. The development of hatchery production during the first half of the 20th century has paved the way for genetic improvement of molluscan shellfish. During the past decades, considerable progress has been made in the genetics and breeding of mollusks, contributing significantly to molluscan aquaculture. Advances in various species groups and subject areas are covered in depth in other chapters of this book, particularly for China, which dominates molluscan aquaculture. This chapter does not intend to comprehensively review the academic literature but provides an overview of molluscan aquaculture and breeding, focuing on major historical and global developments that have shaped the field. Challenges and future directions are also discussed.

2.2 Molluscan aquaculture

Molluscan shellfish have been exploited by humans for food and materials throughout the history of civilization. Archaeological evidence for human exploitation of molluscan shellfish dates to the early Holocene, with extensive shell middens found along the coasts of Europe, Asia, and the Americas. Analyses of shell remains from prehistoric sites have revealed signs of selective harvesting, indicating a form of protoaquaculture in transition to farming (Malindine, 2019). Archaeological studies show that indigenous people on the Northwest Coast of North America constructed intertidal clam gardens 3500 years ago to enhance production (Smith et al., 2019), which may represent the oldest system of mariculture. The earliest documented molluscan aquaculture dates back ~2100 years. Fishponds were constructed in ancient Rome to keep

Genetic Breeding and Aquaculture of Shellfish: A Collaborative Approach to a Sustainable Future
DOI: https://doi.org/10.1016/B978-0-443-33606-5.00022-9

TABLE 2.1 Global aquaculture production by major species groups in 2023. Data are from FAO FishStat (FAO, 2025).

Species groups	Metric tons	Fraction
Finfish	63,841,545	0.469
Mollusks	19,521,534	0.143
Crustaceans	13,688,028	0.101
Amphibians and reptiles	858,987	0.006
Other animals	1,465,867	0.011
Algae and aquatic plants	37,648,792	0.276
Total	136,165,767	1.000

and grow fish and shellfish, including oysters and snails (Bannon, 2014; Higginbotham, 1997; Nash, 2010). Sergius Orata is credited with inventing oyster farming in the 1st century BCE, which involved the construction of specially engineered intertidal ponds, collecting spat using twigs and ceramic tiles, and relocating oysters for fattening. Oyster farming was active in ancient Rome from the 1st century BCE to CE ~100 and declined during the CE 2nd and 3rd centuries. Oyster farming in China likely emerged around the same time (Guo et al., 1999; Hishamunda & Subasinghe, 2003), although no reliable records are available until around CE 1000. Historical documents described managed harvesting of marine pearl oysters around CE 900, the use of bamboo and stones to cultivate oysters around CE 1030, and the culturing of freshwater pearl mussels around CE 1082. Molluscan aquaculture became a significant economic activity in China by the Ming dynasty (16th century), with numerous records of oyster farming and the cultivation of cockles and razor clams in intertidal farming plots. Similar methods for oyster and clam farming are still being practiced in China today.

Molluscan aquaculture is now a significant sector of global aquatic food production, producing 19.5 million metric tons in 2023 and accounting for 14.3% of all aquaculture production (Table 2.1). Excluding algae and aquatic plants, mollusks account for 20% of all animals produced from aquaculture. Aquaculture production is reported for over 80 species of mollusks, including many species of oysters (38%), clams (30%), scallops (11%), mussels (10%), and gastropods (3%) (Table 2.2). Most molluscan aquaculture is from marine bivalves, and freshwater mollusks account for only about 1% of the production.

Molluscan aquaculture is dominated by Asian countries, with China alone accounting for 85.3% of the world production in 2023. Countries differ significantly in the molluscan species cultured (Table 2.2). In China, molluscan culture is primarily based on oysters (40%), clams (34%), scallops (11%), mussels (5%), and gastropods (4%) (Table 2.2). Oyster aquaculture includes three major species: the Pacific oyster (*Crassostrea gigas*) in the north, the Hong Kong oyster (*Crassostrea hongkongensis*) in the south, and the Portuguese or Fujian oyster (*Crassostrea angulata*) in between (but see below for the use of hybrids). Cultured clams include the Manila clam (*Ruditapes philippinarum*), blood cockles, and razor clams, and several other clam species (Guo et al., 1999). Scallop production primarily comes from bay (*Argopecten irradians*), Japanese (*Mizuhopecten yessoensis*), and Chinese (*Chlamys farreri*) scallops. Mussel aquaculture is based on the Mediterranean or blue mussel (*Mytilus galloprovincialis*), the thick-shell mussel (*Mytilus*

TABLE 2.2 The top 10 countries in molluscan aquaculture in 2023 by live weight and the major species cultured in each country (% of total production). Data are from FAO FishStat (FAO, 2025).

Country	Production	Major species
China	16,659,768	Oysters (40%), clams (34%), scallops (11%), mussels (5%), gastropods (4%), freshwater (1%)
South Korea	424,767	Oysters (73%), mussels (13%), gastropods (6%)
Vietnam	415,261	Clams and oysters (67%)
Chile	397,513	Mussels (98%), scallops (1%)
Japan	298,012	Scallops (51%), oysters (49%)
United States	202,356	Oysters (77%), clams (21%), mussels (2%)
Spain	158,231	Mussels (98%), oysters and clams (2%)
France	146,321	Oysters (62%), mussels (37%)
New Zealand	94,558	Mussels (98%), oysters (2%)
Thailand	89,428	Mussels (55%), clams (34%), oysters (11%)
World total	19,521,534	Oysters (38%), clams (30%), scallops (11%), mussels (10%), gastropods (3%), freshwater (1%)

coruscus), and the green mussel (*Perna viridis*), with *M. galloprovincialis* accounting for 55% of total production (Liu et al., 2021). Although abalone aquaculture is relatively small in quantity, it is significant in value, as abalone command higher prices. In other countries, the main molluscan species under culture include the Pacific oyster and blue mussel in South Korea, Japanese scallop and Pacific oyster in Japan, oysters and clams in the United States, and oysters and mussels in France (Table 2.2). Mussels are the primary species cultured in Chile, Spain, and New Zealand, accounting for 98% of the molluscan production in those countries.

Molluscan aquaculture relied on naturally available wild seed through much of its long history. Artificial fertilization was first achieved in the eastern oyster around 1880, and artificially produced embryos and larvae were used to stock ponds for oyster seed production with limited success (Brooks, 1996; Kennedy, 2014). In 1920, William F. Wells successfully produced eastern oysters in a hatchery for the first time (Wells, 1920). Between 1920 and the 1970s, techniques for controlled spawning, larval rearing, microalgal production, and reliable spat settlement of oysters were developed, optimized, and integrated, leading to reproducible hatchery production of bivalve mollusks (Breese & Malouf, 1975; Dupuy & Rivkin, 1972; Loosanoff & Davis, 1963). Commercial hatcheries have expanded significantly since the 1970s and have become a crucial component of molluscan aquaculture. Currently, more than 30 species of mollusks are produced in hatcheries and subjected to some degree of genetic improvement (Table 2.3). They include many species of oysters, scallops, mussels, clams, and abalone. The development of hatchery production has not only provided a reliable supply of seed to power aquaculture but also enabled genetic improvement of molluscan shellfish.

TABLE 2.3 Molluscan shellfish that are produced in hatcheries and subjected to genetic improvement (Based on Guo, 2021).

Species	Location	Breeding and improvement
Argopecten irradians	China, USA	Growth, survival, shell color, hybridization
Argopecten purpuratus	Chile, Peru	Hybridization for growth and survival
Argopecten ventricosus	Mexico	Shell width, body weight
Chlamys farreri	China	Growth, shell color
Crassostrea angulata	China, EU, Vietnam	Triploids in commercial production, tetraploids
Crassostrea belcheri	Southeast Asia	Induced triploidy
Crassostrea gasar	Brazil	Growth, shell color
Crassostrea gigas	Worldwide	Growth, disease resistance, polyploid, color, genomic selection
Crassostrea iredalei	Southeast Asia	Induced triploidy
Crassostrea virginica	USA, Canada	Disease resistance, growth, polyploids, genomic selection
Cyclina sinensis	China	Growth, intraspecific hybridization
Haliotis discus hannai	China, Japan, Korea	Survival, growth, intraspecific hybridization
Haliotis diversicolor	China	Growth, survival
Haliotis iris	New Zealand	Growth
Haliotis midae	South Africa	Growth
Haliotis rubra and laevigata	Australia	Growth, behavior, hybridization
Haliotis rufescens	USA, Chile	Heritability of growth and disease resistance
Hyriopsis cumingii	China	Growth, nacre color, pearl production, hybrids
Mercenaria mercenaria	USA, China	Body size, shell color pattern, disease resistance
Meretrix meretrix	China	Growth, shell color, intraspecific hybrids
Mimachlamys nobilis	China	Shell and meat color, growth
Mizuhopecten yessoensis	China, Japan	Growth, color, introduced to China
Mytilus chilensis	Chile	Live weight, shell length
Mytilus edulis	EU	Resistance to spring mortality, growth
Mytilus galloprovincialis	Australia	Growth
Panopea generosa	USA, Canada	Population genetics and induced triploidy
Pecten maximus	EU	Genetic diversity, triploidy
Perna calaniculus	New Zealand	Shell length, meat weight
Pinctada fucata martensii	China, Japan	Shell growth and color, survival

(*continued on next page*)

TABLE 2.3 Molluscan shellfish that are produced in hatcheries and subjected to genetic improvement (Based on Guo, 2021)—cont'd

Species	Location	Breeding and improvement
Ruditapes philippinarum	China, EU, USA	Growth, shell color patterns
Saccostrea glomerata	Australia	Resistance to QX and *Bonamia roughleyi*
Scapharca broughtonii	China	Growth, intraspecific hybridization
Sinonovacula constricta	China	Growth, survival, and intraspecific hybridization
Tegillarca granosa	China	Growth

Most of molluscan aquaculture (95%+) is from marine bivalves, which are filter feeders that do not require artificial feeding. Therefore, unlike finfish and crustacean farming, molluscan culture does not cause pollution or eutrophication. In fact, cultured bivalves help to clean the water, and the culture gear provides habitats for other organisms, thereby increasing biodiversity. In areas where the wild population is depleted, bivalve farming may also be considered restorative aquaculture because it supports recruitment into the wild population. Thus molluscan aquaculture is recognized as an environmentally friendly form of food production and has a bright outlook. The continued expansion and modernization of molluscan aquaculture are crucial to food security, economic prosperity, and the sustainable utilization of marine resources.

2.3 Molluscan genetics and breeding

2.3.1 Selective breeding

Traditional selective breeding is effective and has been used to improve many molluscan shellfish. The earliest record of molluscan breeding is the selection of eastern oysters for disease resistance at Rutgers University in the USA. The program began in 1960, soon after the outbreak of MSX (multinucleated sphere X disease, caused by the protist *Haplosporidium nelsoni*) in Delaware Bay, and has continued to this day (Guo, 2021; Haskin & Ford, 1987), making it the longest sustained shellfish breeding program. The program used mass selection for MSX resistance, selecting survivors from field exposure, since MSX cannot be artificially transmitted in the laboratory. Response to selection was rapid, and strong MSX resistance was obtained after five generations of selection, with selected lines surviving 8.8 times better than the susceptible control (Fig. 2.1). This result indicates that MSX resistance has strong heritability and may be controlled by major-effect genes. Dermo disease (caused by the protist *Perkinsus marinus*) appeared in Delaware Bay in 1990 and caused heavy mortality in the MSX-resistant lines. To restore genetic variability and breed for dermo resistance, two synthetic strains, NEH and DBH, were produced by crossing MSX-resistant lines originating from the Long Island Sound and Delaware Bay, respectively (Allen, 1993). In 1998, fast growth and shell shape were added to the breeding program by selecting the largest 20% of oysters with the desired shell attributes (a deep cup and straight hinge) that survived 3 years of field exposure to MSX and Dermo (Guo, 2021). New genetic materials were introduced, and sublines were created and managed through progressive rotational crossing. In repeated evaluations, Rutgers strains demonstrated

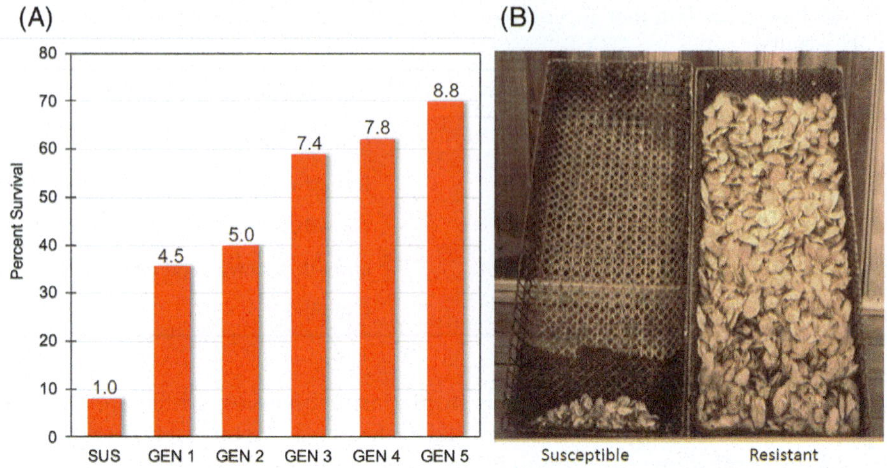

FIGURE 2.1 **Improvement of eastern oysters in MSX resistance after five generations of mass selection.** (A) Progressive improvement in survival under MSX standardized with the susceptible control (SUS) (Guo et al., 2008) and (B) survived oysters in susceptible control and MSX-resistant lines after exposure to MSX (Photo by S.E. Ford).

strong resistance to MSX and Dermo (Guo, 2021; Guo et al., 2003; Proestou et al., 2016), although the development of dermo resistance was relatively slow, indicating the trait is polygenic with low heritability. These disease-resistant strains are developed in a highly stressful intertidal environment in Delaware Bay, where oysters experience 3–4 h of air exposure twice a day, air temperature may reach 40°C in summer, and MSX and dermo are endemic (Guo, 2021). Besides disease resistance, hardiness may be an important contributor to the improved performance of these strains. Because of its strong disease resistance and hardiness, the Rutgers strains are widely used by oyster farmers in New Jersey and the Northeastern US, accounting for over 90% of oyster seed produced in Maine.

Several other breeding programs exist in the United States to improve the performance of the eastern oyster. At the Virginia Institute of Marine Science (VIMS), mass and family selection have produced strains that show resistance to MSX and Dermo and improved growth across different salinities (Allen et al., 2021; Calvo et al., 2003; Frank-Lawale et al., 2014). The VIMS strains are widely used by oyster farmers in Chesapeake Bay. Breeding at the University of Maine and a NY company has led to rapid and strong resistance to Roseovarius oyster disease (ROD) of juvenile eastern oysters, caused by the α-proteobacterium *Roseovarius crassostreae*, in 1–3 generations (Barber et al., 1998; Farley et al., 1998). Although not well-documented, breeding programs also exist in other Atlantic and Gulf states for the eastern oyster. The eastern oyster is the most important mariculture species in the United States. Nearly all hatchery production of eastern oyster seed is from stocks that have gone through some degree of selective breeding, especially for disease resistance. Aquaculture production of the eastern oyster has greatly expanded over the past two decades (Fig. 2.2), which is driven by economic forces as well as the availability of genetically improved stocks. The development of disease-resistant strains and triploids from these strains has contributed significantly to eastern oyster culture in the United States.

Selective breeding has also been successful in improving other mollusks. In the Pacific oyster, examples included the development of families resistant to summer mortalities after

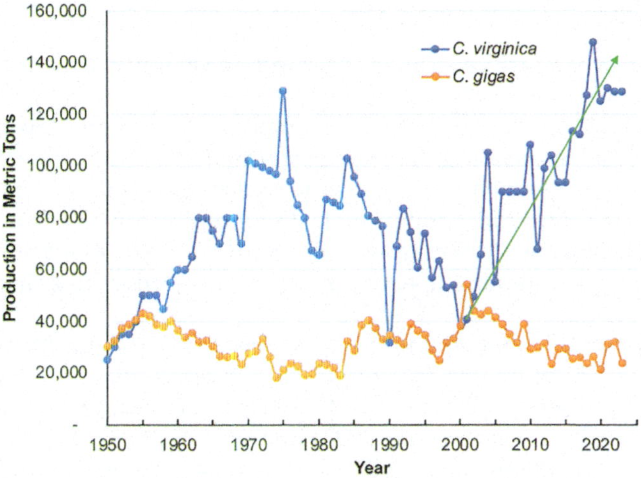

FIGURE 2.2 Oyster aquaculture production in the US since 1950 (FAO, 2025). The *green arrow* highlights the rapid increase in *C. virginica* production since 2010, which is supported by the release of diploid and triploid disease-resistant oysters.

three generations of selection (Hershberger et al., 1984), resistance to the Ostreid herpesvirus 1 μVar (OsHV-1) after four generations of mass selection in France (Dégremont et al., 2015) and similar progress in New Zealand (Camara et al., 2017), Australia (Kube et al., 2018) and the US (Divilov et al., 2023, 2019), improvement in growth through family-based (de Melo et al., 2016) and mass selection (Zhang et al., 2019). Selective breeding also improved resistance of the European flat oyster to bonamiasis (caused by *Bonamia ostrea*) after three generations of mass selection (Naciri-Graven et al., 1998) and resistance of the Sydney rock oyster to winter mortality caused by *Bonamia roughleyi* and QX (Queensland unknown) disease caused by *Marteilia sydneyi* after 3–4 generations of mass selection (Dove et al., 2013; Nell & Perkins, 2006). Disease resistance is the most important trait for aquaculture production. Resistance to several diseases, such as MSX, ROD, and OsHV-1, is achieved within a few generations, suggesting that resistance to these diseases has high heritability and may be under the control of some major-effect genes. The development of dermo resistance has been slow, suggesting that it may have low heritability or be under the control of many small-effect genes. Numerous studies have also demonstrated improvements in growth and survival through selective breeding in mollusks, although sustained breeding programs contributing to aquaculture production are relatively few.

Mollusks have colorful shells that are diverse and can be used for marketing and managing selected strains. Shell color is easier to breed or purify as it is usually controlled by one or a few genes following Mendelian inheritance. Selective breeding has led to the development of many shell color varieties, including red strains of Chinese and Japanese scallops, red and purple strains of bay scallops, golden and black Pacific oysters, golden noble scallops, red *Meretrix* clam, and "zebra" Manila clams in China. These color varieties are often associated with improved growth and survival, although some may also show signs of inbreeding. For hard clams cultured in the US, the *notata* shell color pattern has become a recognized marker for hatchery-produced clams, which are not always welcomed by farmers.

2.3.2 Managing genetic diversity

Demonstrating genetic improvement over a few generations of selection is straightforward. The challenge is to sustain the breeding program for the long term. Once the reproductive cycle is closed, selective breeding inevitably leads to a loss of genetic variation or inbreeding. Rapid inbreeding can lead to inbreeding depression, resulting in poor performance and the loss of inbred lines. Severe loss of genetic variation will limit response to future selection on other traits. Therefore managing genetic diversity is one of the most important tasks for molluscan breeding. A positive correlation between heterozygosity and growth rate is commonly observed in marine bivalves, underscoring the importance of maintaining genetic diversity in selective breeding programs.

Molluscan genomes are highly polymorphic and contain high genetic load, making them particularly sensitive to inbreeding (Guo, 2026; Launey & Hedgecock, 2001) . Inbreeding depression has been documented in many cultured mollusks, affecting larval and adult growth, development, and survival (Evans et al., 2004; Deng et al., 2005; Ibarra et al., 1995; Plough & Hedgecock, 2011; Wada & Komaru, 1994; Zheng et al., 2008, 2007). Most of those studies estimated inbreeding depression in families created by selfing or sister-brother matings. Slow or controlled inbreeding coupled with purifying selection may purge deleterious recessive alleles over time and reduce genetic load or the negative effects of inbreeding.

Inbreeding can be managed through family-based breeding, where pedigree information can be used to avoid mating between close relatives. Family-based selection is also advantageous for traits with low heritability or traits that cannot be measured in the individuals being selected. Family-based selection requires the production and evaluation of hundreds of families and proper control for family-bag confoundment. Mass selection, which is simple and strong, can lead to rapid inbreeding if not carefully managed, as mollusks have high fecundity and high variation in reproductive success. To avoid inbreeding in mass selection, the following measures are commonly taken: (1) using a large number (>50) of breeders per line, (2) balancing sex ratio and parental contribution with equalized factorial crossing, (3) maintaining multiple lines, (4) rotational crossing between the lines, and (5) monitoring inbreeding with genetic markers (Guo, 2021).

Hybridization between inbred or pure lines is a powerful approach to genetic improvement as it can restore genetic diversity, produce heterosis, and enhance uniformity of culture stocks (Guo, 2021). Hybrid crosses are widely used in crop improvement, where pure lines are developed over many generations and used in single-, double-, and three-way crosses. In mollusks, heterosis has been observed between inbred lines of oysters and scallops, although maintaining pure or inbred lines is difficult in mollusks because of their sensitivity to inbreeding. A modified scheme of progressive rotational crossbreeding that does not rely on pure or highly inbred lines has been used for oyster breeding (Guo, 2021). In progressive rotational crossbreeding, a minimum of three independent core lines is established from wild populations through strong, progressive mass selection and crossed in rotation with a production line to maintain genetic diversity and heterosis. With three or four core lines, the production line can consistently maintain heterozygosity levels of 86% or 93%, respectively, regardless of inbreeding level in the core lines (Guo, 2021). To prevent the collapse of core lines due to inbreeding, a bridging line is produced by crossing each core line with the source population and used to introduce new genetic material after vetting through in-crossing and backcrossing. In such a scheme, selection within core lines utilizes additive genetic variation, while rotational crossing restores heterozygosity and creates heterosis in the production line by capturing nonadditive genetic variation. The production line typically exhibits improved performance than the core lines and unselected controls. In the eastern oyster, it is also common for commercial hatcheries to cross their own locally adapted stock with a disease-resistant line from a breeding center and use the hybrid seed for production. In addition to heterosis, hybrids between pure lines can provide uniformity, a desirable trait for aquaculture production.

Intraspecific hybridization between non-inbred or geographic populations can also produce heterosis and improve the aquaculture performance of mollusks. In Pacific abalone (*Haliotis discus hannai*), for example, hybridization between Chinese and Japanese populations greatly improved survival and growth and had a major impact on the development of abalone aquaculture in China (Guo, 2009).

2.3.3 Interspecific hybridization and nonnative species

Interspecific hybridization can occur naturally and/or be artificially induced among many closely related mollusks, and in some cases, the hybrids have improved the performance of cultured stocks. Interspecific hybridization between abalone species is particularly fruitful (Lafarga de la Cruz & Gallardo-Escárate, 2011). Hybrids between the black-lip (*Haliotis rubra*) female and green-lip (*Haliotis laevigata*) male abalone show improved growth, behavior, and stress tolerance, and are used by farmers in Australia (Alter et al., 2017; Hamilton et al., 2009). Hybrids between *H. discus hannai* and *Haliotis fulgens* exhibited heterosis in growth, survival, and high-temperature tolerance and have become popular stocks for aquaculture in China (You et al., 2015). Hybrids between *Haliotis rufescens* and three other sister species, *Haliotis sorenseni, H. fulgens*, and *Haliotis corrugata*, showed improvements in growth and survival (Leighton & Lewis, 1982). Interspecific hybrids between the weathervane scallop (*Patinopecten caurinus*) and the Japanese scallop (*Patinopecten yessoensis*) exhibit fast growth and improved survival and have been cultured in British Columbia, Canada. Hybrids between the scallops *A. irradians* and *A. purpuratus* demonstrated heterosis and significant improvements in growth and adductor muscle size (Wang et al., 2011), although large-scale production of the hybrids is difficult because of hermaphroditism in both species.

Hybrids among several *Crassostrea* oyster species have been tested, yielding variable results (Jiang et al., 2024), and most have not led to commercial production. Hybrids between *C. gigas* and *C. angulata*, which are considered subspecies by some (Wang et al., 2010), are promising, but their use in large-scale aquaculture raises concerns about altering the genetics of natural populations, as the hybrids are fertile. Triploid hybrids between the two species have been widely used in aquaculture in China (Section 2.3.4.1). The performance of interspecific hybrids is unpredictable and depends on the degree of divergence and compatibility between the two genomes. Distantly related species may experience pre- and postgametic barriers to hybridization that affect fertilization, early development, growth, and gametogenesis. For example, hybrids between the eastern and Pacific oysters, two species that are separated by about 83 million years, are not viable, despite apparently normal fertilization and development to the D-stage. Hybrids between some distant mollusks may be sterile or partially sterile, such as the hybrids between *A. irradians* and *A. purpuratus*, which produce eggs but no or few sperm.

2.3.4 Polyploid breeding

2.3.4.1 *Triploids and their benefits*

Triploids have three sets of chromosomes, and because of the extra set of chromosomes, triploid mollusks exhibit several traits desirable in aquaculture, including sterility, rapid growth, and improved meat quality. Sterility is important for molluscan aquaculture, as mollusks are broadcast spawners cultured in open water, and uncontrolled reproduction of cultured stocks may alter the genetics of wild populations. While triploids are not completely sterile, the reproductive potential of triploids is extremely low, or about 0.0008% of diploids in the Pacific oyster (Guo and Allen, 1994b). The use of triploids can greatly reduce overreproduction in cultured stocks and the risk of impacting wild populations. This is especially important for nonnatives used in molluscan aquaculture. Nonnative species are a crucial part of molluscan aquaculture, and many mollusks are introduced and cultured as nonnatives, including major species such as Pacific oyster, Fujian oyster, European flat oyster, Kumamoto oyster, bay scallop, Japanese scallop, sea scallop, Manila clam, hard clam, and abalones (Guo, 2009).

Triploids have been produced and tested in dozens of mollusks, and in most species studied so far, triploids grew significantly faster than diploids (Guo et al., 2009). For example, triploid eastern oysters grew 32%–50% faster and survived 26%–38% better than diploids, yielding 88%–190% more than diploids in Delaware Bay (Guo et al., 2008a; Guo, 2021). Three hypotheses have been proposed to explain the superior growth of triploid mollusks, attributing it to sterility, increased heterozygosity, and/or increased cell size (Guo and Allen, 1994c). Sterility or the reallocation of energy from reproduction to somatic growth may contribute to the fast growth of triploids after sexual reproduction, but it cannot explain why triploids grow faster than diploids during larval and juvenile stages. Triploids are more heterozygous than diploids because of the extra set of chromosomes, and heterozygosity is positively correlated with growth (Wang et al., 2002). Triploid cells are significantly larger than diploid cells, and large cells may lead to large organs and organisms if cell number is not reduced under mosaic development in mollusks. Triploid scallops have significantly larger adductor muscles than diploids, supporting the cell-size hypothesis and arguing for the development of triploid scallops for aquaculture.

For oysters, another benefit of triploids is improved meat quality during the spawning season. Diploid oysters allocate over 50% of their body weight to gamete production. Mature or spawned out oysters cannot be marketed during the spawning season, especially for raw consumption. Because of their sterility or limited gonadal development, triploids can maintain high meat quality during the spawning season and provide a high-quality product year-round.

Whether triploid mollusks are more resistant to disease is uncertain and may depend on the species, disease, and environmental conditions. Triploids may show higher survival against some diseases but lower survival against others, and in most cases, triploids do not differ from diploids in disease resistance (Guo, 2021). Fast growth in triploid oysters shortens the time to refuge or market sizes and reduces their exposure to and losses from some diseases (Dégremont et al., 2012; Guo et al., 2008a). Triploid mollusks do not deplete their energy reserve for reproduction and may enjoy improved health and survival during the reproductive season. This advantage can be offset by abnormal gametogenesis and the newly acquired triploid genome, which has not been optimized by natural selection. Abnormal mortality has been reported in triploid oysters cultured in some environments, whereas most studies have demonstrated increased survival and yield from triploids. It should be noted that triploids are not created equal. The performance of triploids depends on their genetics and the environment, and triploids can differ, depending on their breeding history.

2.3.4.2 *Development of triploids and tetraploids*

Triploid oysters were first induced by inhibiting meiosis with cytochalasin B (Stanley et al., 1981). Triploids can also be induced by inhibiting meiosis II with heat shock or 6-dimethylaminopurine (Desrosiers et al., 1993). Triploid induction by inhibiting meiosis is highly variable and rarely 100% effective, hindering the commercial production of triploids. The ideal way to produce triploids is by mating diploids with tetraploids that have four sets of chromosomes and produce diploid gametes. In mollusks, tetraploidy can be induced by several methods, including inhibiting meiosis I, mitosis I, gynogenesis, or cell fusion. All these methods produced tetraploid embryos, at rates ranging from 28% to 95%, but these embryos did not develop normally and survive beyond the larval stage (Guo, 1991). It appeared that the eggs were too small to support the large tetraploid nuclei, leading to a deficiency in cell number or to incorrect segregation of morphogenic determinants that are crucial for embryogenesis in mollusks with mosaic development. Guo (Guo, 1991) hypothesized that using larger eggs from triploids, followed by blocking meiosis I, could produce viable tetraploids (Fig. 2.3A). Guo and Allen (Guo and Allen, 1994a) tested the hypothesis and, for the first time in 1993, succeeded in producing viable tetraploid Pacific oysters.

The successful production of viable tetraploids from large triploid eggs supports the cell-number deficiency hypothesis or, alternatively, the genome adaptation hypothesis, which suggests that the increased survival of tetraploids results from the triploid genome having become adapted to polyploidy (Guo et al., 2009). Although viable tetraploids have been produced from eggs of diploids in several species including blue mussel, pearl oyster, tropical oysters, Chinese scallop, bay scallop, Catarina scallop, Manila clam and dwarf surfclam (McCombie et al., 2005; Peruzzi & Guo, 2002; Scarpa et al., 1993; Yang et al., 2000), survival was extremely low, and large numbers of tetraploids could not be reliably obtained to create breeding populations of tetraploids. Blocking meiosis I in eggs of triploids (Fig. 2.3A) remains the most effective method for producing tetraploids. In addition to reported successes with Pacific and eastern oysters, this method has also led to the successful development of tetraploid populations in Fujian, Hong Kong, and Kumamoto oysters at the Qingdao Frontier Ocean Seed Company (QFOS, personal communication).

Tetraploid oysters are fertile and can produce triploids when mated with diploids (Guo et al., 1996). The mated triploids are not only highly pure (99%–100%) and free of chemical treatment, but also genetically superior to induced triploids, as they suffer no inbreeding from meiosis inhibition.

2.3.4.3 *Improvement of triploids and tetraploids*

While sterile triploids cannot be directly improved, their performance can be improved by enhancing that of their diploid and tetraploid parents. As a newly induced state that has undergone little natural selection, the tetraploid genome is relatively unstable and particularly in need of improvement. Tetraploids may lose chromosomes in somatic tissues, although this does not affect the germline cells and triploid production (Guo, 2021). Nevertheless, improving tetraploid growth and survival is critical for triploid performance, as tetraploids contribute two of the three sets of chromosomes.

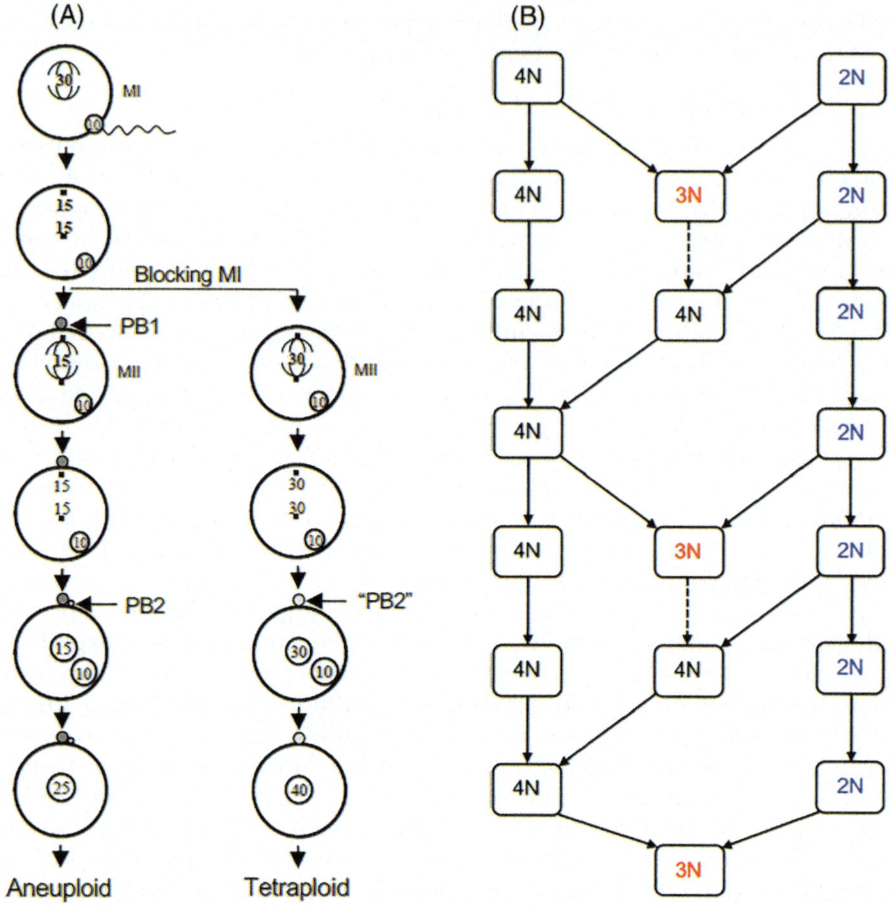

FIGURE 2.3 **Production and improvement of tetraploid oysters.** (A) Production of tetraploids by blocking meiosis I in eggs from triploids fertilized with haploid sperm from diploids (Guo, 1991). (B) Improvement of tetraploids and triploids by coalesced interploidy breeding (Modified from Guo (2021)), where parallel selections of tetraploids and diploids are connected by selected triploids and de novo tetraploids. *Dash lines* in (B) represent the production of de novo tetraploids as depicted in (A).

Tetraploid and triploid oysters can be improved through coalesced or systematic interploidy breeding, where selections on the diploids, triploids, and tetraploids are brought together (Guo, 2021). In coalesced interploidy breeding, diploid and tetraploid lines are improved through parallel, strong selection, and triploids are produced using the best-performing diploids and tetraploids (Fig. 2.3B). Every three generations, de novo tetraploids are produced with the best-performing triploids and diploids and crossed into the tetraploid line. By selecting and crossing the best-performing diploids, triploids, and tetraploids, coalesced interploidy breeding not only maintains genetic diversity in tetraploid lines but also brings genetic improvements from diploid, triploid, and tetraploid phases together into the new tetraploids and triploids.

Coalesced interplay breeding has significantly improved the performance of triploid and tetraploid eastern oysters (Guo, 2021).

It has been reported that tetraploids produced from diploid eggs may have a more stable genome and produce fewer mosaics (Benabdelmouna & Ledu, 2015). Repeated attempts to produce tetraploids with diploid eggs have failed in the Pacific and eastern oysters. Somatic chromosome loss does not affect the germline cells and the performance of triploid and tetraploid oysters, and most aneuploid embryos are eliminated by natural selection during the larval stage. More studies are needed to evaluate different tetraploid lines, and the ultimate measure is not somatic mosaics and aneuploidy, but the growth and survival of their triploid progeny.

2.3.4.4 Aquaculture production of triploids

The development of tetraploid Pacific oysters in 1993 broke the technical barrier for triploid production. Commercial production of triploid Pacific oysters began in the United States and France in the late 1990s. Tetraploid eastern oysters were developed from disease-resistant strains in 2001, and the combination of disease resistance and triploid advantages led to rapid adoption by US farmers, which contributed to the expansion of eastern oyster farming (Fig. 2.2). Now, triploid oysters have become an important part of oyster farming in many countries, including major producers such as the United States, France, China, Korea, Australia, and Chile. In the United States, triploids now account for 30%–40% of Pacific oysters cultured on the West Coast and 30%–90% of eastern oysters cultured along the Atlantic and Gulf Coasts. In France, hatcheries provide 30%–50% of the oyster seed, and about 90% of the hatchery-produced seed are triploids (Iitembu et al., 2023).

In China, commercial-scale production of triploids began in 2009, and it took five years for triploid oysters to gain wide market acceptance. Triploid production has expanded rapidly since 2015, following the establishment of large tetraploid populations at QFOS (Fig. 2.4). Triploid seed production at QFOS, which supplies ~70% of the nation's triploid seed, increased from 0.2 billion in 2015 to 41 billion in 2022. The rapid increase in triploid seed production corresponded to a higher rate of increase in oyster production. In the 10 years after 2015, oyster production in China increased by 66%, compared to an increase of 27% for the 10 years before 2015 (Fig. 2.4). In 2024, of the 7.3 million tons of oysters produced in China (China Ministry of Agriculture and Rural Affairs; China Society of Fisheries, 2025), an estimated 3.9 million tons, or 54%, were triploids (Table 2.4). The triploids consist of triploid Pacific oysters, triploid Fujian oysters, and different types of hybrid triploids between the two species, produced using tetraploid Pacific oyster, tetraploid Fujian oyster, and hybrid tetraploids. Hybrid triploids have become popular stocks due to their improved growth and heat tolerance, and about 50%–60% of the triploids in southern China and 30% in northern China are hybrid (QFOS, personal communication).

Traditionally, oyster farming in China is defined by Pacific oyster farming in the north (Liaoning to Jiangsu), Hong Kong oyster farming in the south (Guangdong and Guangxi), and Fujian oyster farming in between (Fujian and Zhejiang). The use of triploids and hybrid triploids has altered the species composition in different regions. In Guangdong, triploid Fujian oysters have been introduced and now account for ~47% of total oyster production, and virtually all (~98%) of Fujian oysters cultured in the province are triploids (Zhao et al., 2025). Currently, nearly all oysters (90%–95%) cultured in northern China are hatchery-produced and triploid Pacific oyster or triploid hybrids, and ~40% of the oysters produced in Fujian are triploid Fujian oyster or triploid hybrids (L. Li and QFOS, personal communication). Because of the hybrids,

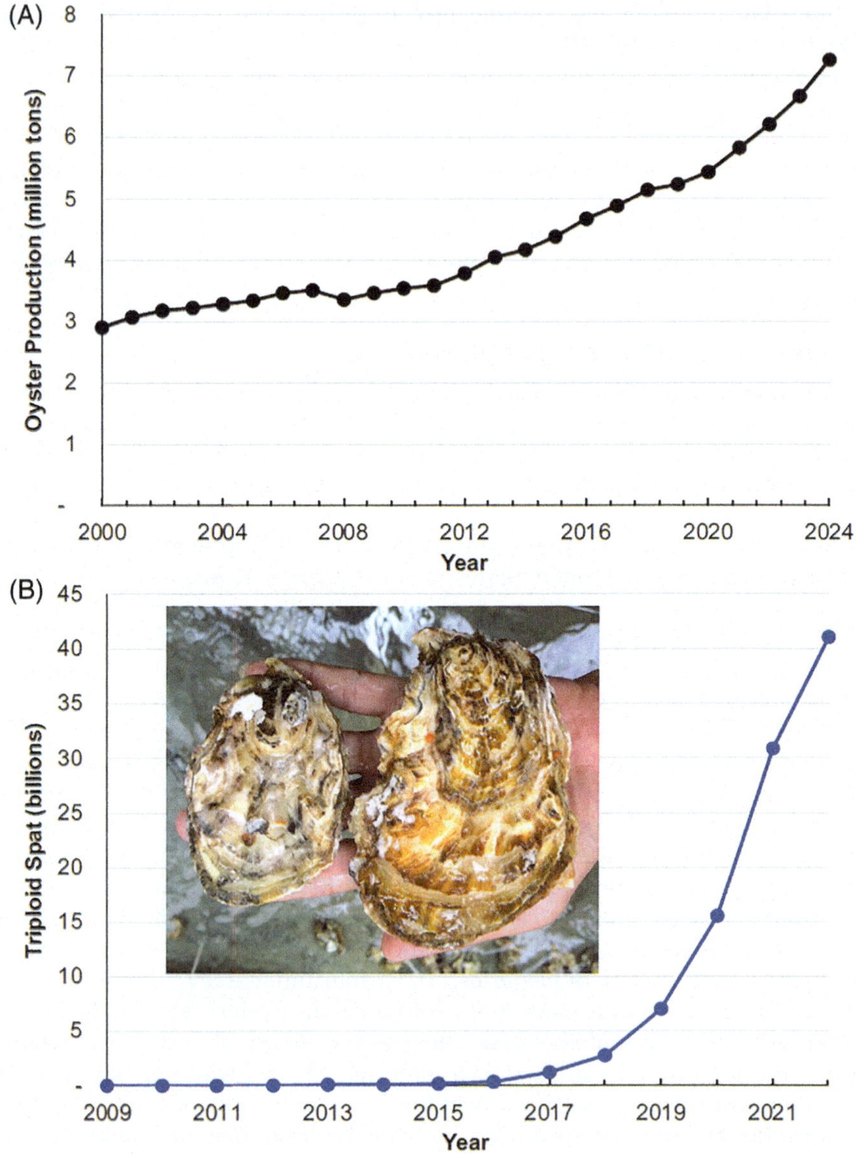

FIGURE 2.4 Oyster production of China from 2000 to 2024 (A) and triploid oyster seed production at Qingdao Frontier Ocean Seed Company (B), which supplies ~70% of the triploid oyster seed in China. Almost all production is spat on shells that are either cultured on ropes to market size or separated as juveniles and cultured in lantern nets as single oysters. Inside photo: Diploid (*left*) and triploid (*right*) Pacific oysters at ~8 months.

TABLE 2.4 Oyster production in metric tons (MT) from coastal provinces of China in 2024 (China Ministry of Agriculture and Rural Affairs; China Society of Fisheries, 2025) and the estimated production of triploid (3*n*) oysters.

Province	Production (MT)	3n (%)	3n (MT)	Notes
Liaoning	557,437	92	512,842	3n: Cgi, hybrid; 2n: Cgi
Hebei	17,634	92	16,223	3n: Cgi, hybrid; 2n: Cgi
Shandong	1,657,092	92	1,524,525	3n: Cgi, hybrid; 2n: Cgi
Jiangsu	225,071	92	207,065	3n: Cgi, hybrid; 2n: Cgi
Zhejiang	405,787	10	40,579	3n: Can, hybrid; 2n: Can, Csi, Cho
Fujian	2,356,598	40	942,639	3n: Can, hybrid; 2n: Can, Cho, Csi
Guangdong	1,288,665	47	600,000	3n: Can; 2n: Cho, Can, Csi
Guangxi	741,702	10	74,170	3n: Can; 2n: Cho, Can, Csi
Hainan	2508	5	125	3n: Can; 2n: Cho, Can, Csi
Total	7,252,494	0.54	3,918,169	All: Can/Cgi 78%, Cho 20%, Csi 2%

Species: Can, ***C. angulata***; Cgi, ***C. gigas***; Cho, ***C. hongkongensis***; Csi, ***C. sikamea***; hybrids, Can x Cgi and reverse. Production percentages are estimated based on communication with local industry experts.

it is no longer possible to separate Pacific and Fujian oyster production. Regardless of ploidy, oyster production in China can be roughly classified as 78% Pacific/Fujian oysters, 20% Hong Kong oysters, and ~2% Kumamoto oysters. Most of the Hong Kong oysters are produced in Guangdong (~688,665 tons) and Guangxi (630,447 tons), and most of the Kumamoto oysters (~100,000 tons) are produced in Zhejiang (Q. Xue, personal communication). Kumamoto oysters are also produced in Fujian, Guangdong, Guangxi, and Hainan, mostly as a by-product because they are present in wild seed.

2.3.5 Application of molecular genetics and genomics in shellfish breeding

Early studies on the genetics of mollusks included Mendelian inheritance of allozymes, inheritance of shell color and sex, population genetics, estimation of heritability, development of genetic markers and maps, identification of quantitative trait loci (QTL) and functional genes (Guo et al., 2008b; Hollenbeck & Johnston, 2018; Newkirk, 1980). In the past two decades, considerable progress has been made in molluscan genomics. Transcriptomic studies have been widely used to investigate molluscan development, reproduction, and responses to infections and environmental stress. Genomes of more than 50 mollusks have been sequenced and assembled to the chromosome level (Liu et al., 2024), providing unprecedented insights into molluscan biology and evolution. Advances in molluscan genetics and genomics have not only improved our basic understanding of mollusks but also provided knowledge and tools for molluscan breeding. The development of genetic markers has facilitated species identification, pedigree analysis, and monitoring of inbreeding. Large numbers of candidate genes and markers associated with economically important traits have been identified, including examples for growth (Ning et al., 2019; Zeng & Guo, 2022), disease resistance (He et al., 2012), and sex determination

(Li et al., 2018; Zhang et al., 2014). With refinement and validation, some of those genes and markers can potentially be used for marker-assisted selection or targeted gene editing. Marker-assisted selection has not been widely used in molluscan breeding. One successful case is the identification of a QTL explaining 13% of the variance for OsHV-1 resistance in the Pacific oyster, and marker-assisted selection yielded 47% improvement in survival (Divilov et al., 2023).

Marker-assisted selection is effective for traits controlled by major-effect genes, and most production traits are polygenic traits and controlled by many small-effect genes. Genomic selection uses all genetic variation across the genome, regardless of trait association, and is effective for polygenic traits. Genomic selection is more effective than family-based selection and especially powerful for traits that cannot be measured directly and quickly. Genomic selection has been widely used for genetic improvement of crops and livestock, but is relatively new to shellfish breeding. Recently, high-throughput genotyping by sequencing assays and SNP arrays have been developed for several molluscan species, including oysters, scallops, the hard clam, and abalone (Grouzdev et al., 2024; Guo et al., 2023; Liu et al., 2022; Lv et al., 2018; Qi et al., 2017). Several studies tested the accuracy of genomic prediction for different traits in mollusks (Dou et al., 2016; Gutierrez et al., 2018, 2020; Jourdan et al., 2023; McCarty et al., 2021; Vu et al., 2021; Wang et al., 2018; Zhu et al., 2021), demonstrating greater accuracy than family-based prediction, but did not actually perform genomic selection. Genomic selection has been conducted and led to improvements of vibriosis resistance in the Pacific oyster (Yang et al., 2024) and dermo resistance in the eastern oyster (Wang et al., 2025). All studies so far suggest that genomic selection has great potential, but further efforts are needed to reduce genotyping costs, increase sample sizes, test AI models, and implement it in sustained breeding of more species.

Transgenics have been used to improve many agricultural crops, and transgenic salmon has been approved for aquaculture production in the United States. While producing transgenic mollusks is technically possible (Lu et al., 1996), it has not been explored for shellfish aquaculture because of concerns about genetically modified organisms. Gene editing using CRISPR/Cas-9 is a new technology that allows modifying genes already present in a species, a process such as natural mutation. As it does not involve introducing new genes from another species, gene editing may pose less risk and face less regulation than transgenics. Gene editing has been achieved in the apple snail, producing mutant lines for a gene regulating eye development (Accorsi et al., 2025). In other mollusks, many studies have reported evidence of gene editing at the embryonic and larval stages, but none have reported survival to the adult stage. Most bivalves have small eggs and normally experience heavy larval mortality. The challenge is to develop a high-throughput system, such as electrophoresis or lipofection, capable of generating large numbers of gene-edited embryos that can survive to the adult stage.

2.4 Summary and perspectives

Molluscan aquaculture is a vibrant industry with a long history and a bright outlook. It accounts for 20% of animal aquaculture production and is rapidly growing. As most cultured mollusks are filter-feeding bivalves and ecoservice providers, molluscan aquaculture is environmentally friendly and favored for further expansion. Molluscan aquaculture and its sustainable development face some challenges. First, molluscan aquaculture mostly occurs in open water, where husbandry is labor-intensive, and working conditions are harsh. The development of mechanized equipment and AI-powered robots would greatly propel the industry. Second, molluscan culture encompasses more than 80 species, most of which have no history of domestication and for which the need for genetic improvement is great. Even for major aquaculture species, the level of genetic improvement is uneven, with some under genomic selection and others lacking any sustained breeding. Biologically, mollusks have a typical type III survivorship and are prone to mass mortalities. They are cultured in uncontrolled environments and are most vulnerable to disease and stress intensified by climate change. The sustainable development of molluscan aquaculture is dependent on genetic improvement.

Selective breeding is effective in improving disease resistance, growth, stress tolerance, shell color, and other key traits in cultured mollusks. The development of disease-resistant strains has contributed significantly to eastern oyster aquaculture in the United States. Intraspecific and interspecific hybridization, as well as the use of non-native species, have played an important role in molluscan aquaculture. The development of tetraploid-triploid oysters provided one of the best examples of a breakthrough in genetic breeding, transforming a major industry. Considerable progress has been made in molluscan genetics and genomics, paving the way for the advanced genetic improvement of these organisms. High-throughput genotyping tools have been developed, and genomic selection has demonstrated great potential to enhance selection accuracy and efficiency. Most exciting is the successful establishment of gene-edited lines of apple snails, which, if replicated in cultured mollusks, could open a new avenue for genetic improvement. It is also encouraging that private companies are now driving key advances in molluscan aquaculture and breeding, such as the development of triploid oysters in China.

Despite impressive advances in molluscan genetics, genomics, and breeding, major gaps exist in several areas. Although many studies reported response to selective breeding over a few generations, sustained breeding is lacking for many important species. Part of the problem is that inbreeding often leads to the collapse or deterioration of breeding populations without careful management. As family-based selection is demanding, using multiple lines and hybrid crosses provides a relatively simple and effective approach. As mollusks have high polymorphism and possibly high genetic load, hybrid crosses between divergent lines and populations are expected to generate heterosis. Another challenge for molluscan breeding is the difficulty of accurately measuring phenotypes, as we have little control over the environment, culture density, and biofouling. Improved culturing systems and AI-assisted phenotyping platforms would be instrumental. Genomic and transcriptomic studies have identified numerous candidate genes associated with production traits, but most of these associations are weak and explain only a small fraction of the variance. Greater efforts are needed to identify causal loci controlling critical phenotypes that can be targeted by gene editing. Model systems such as the bivalve model described in Chapter 12 are vital to functional genomics. Despite these challenges, continued advances in genomics, AI, and gene-editing technology are expected to spawn a new area of genetic improvement and greatly contribute to molluscan aquaculture.

References

Accorsi, A., Pardo, B., Ross, E., Corbin, T. J., McClain, M., Weaver, K., ... Alvarado, A. S. (2025). A genetically tractable non-vertebrate system to study complete camera-type eye regeneration. *Nature Communications, 16*(1), 6698. doi:10.1038/s41467-025-61681-6.

Allen, S. K (1993). Development of high survival resistant lines in oysters using MSX-resistant strains. *Journal of Shellfish Research, 12*(1), 103.

Allen, S. K, Small, J. M, & Kube, P. D (2021). Genetic parameters for *Crassostrea virginica* and their application to family-based breeding in the mid-Atlantic, USA. *Aquaculture, 538*, 736578. doi:10.1016/j.aquaculture.2021.736578.

Alter, K., Andrewartha, S. J., Morash, A. J., Clark, T. D., Hellicar, A. D., León, R. I., & Elliott, N. G. (2017). Hybrid abalone are more robust to multi-stressor environments than pure parental species. *Aquaculture, 478*, 25–34. doi:10.1016/j.aquaculture.2017.04.035.

Bannon, C. (2014). Sergius Orata and the Rhetoric of Fishponds. *The Classical Quarterly, 64*(1), 166–182. doi:10.1017/s0009838813000621.

Barber, B. J., Davis, C. V., & Crosby, M. A. (1998). Cultured oysters, *Crassostrea virginica*, genetically selected for fast growth in the Damariscotta River, Maine, are resistant to mortality caused by juvenile oyster disease (JOD). *Journal of Shellfish Research, 17*(4), 1171–1176.

Benabdelmouna, A., & Ledu, C. (2015). Autotetraploid Pacific oysters (*Crassostrea gigas*) obtained using normal diploid eggs: Induction and impact on cytogenetic stability. *Genome, 58*(7), 333–348. doi:10.1139/gen-2015-0014.

Breese, W. P, & Malouf R. E. (1975) Hatchery manual for the Pacific oyster (Publication no. ORESU-H-75002). Oregon State University Sea Grant College Program.

Brooks, W. K (1996). *The oyster*. JHU Press.

Calvo, L. M. R., Calvo, G. W, & Burreson, E. M (2003). Dual disease resistance in a selectively bred eastern oyster, *Crassostrea virginica*, strain tested in Chesapeake Bay. *Aquaculture, 220*, 69–87.

Camara, M. D., Yen, S., Kaspar, H. F., Kesarcodi-Watson, A., King, N., Jeffs, A. G., & Tremblay, L. A. (2017). Assessment of heat shock and laboratory virus challenges to selectively breed for ostreid herpesvirus 1 (OsHV-1) resistance in the Pacific oyster, *Crassostrea gigas*. *Aquaculture, 469*, 50–58. doi:10.1016/j.aquaculture.2016.11.031.

China Society of Fisheries. (2025). *China fishery statistical yearbook 2025*. Beijing: China Agriculture Press.

Dégremont, L., Garcia, C., Frank-Lawale, A., & Allen, S. K. (2012). Triploid oysters in the Chesapeake Bay: Comparison of diploid and triploid *Crassostrea virginica*. *Journal of Shellfish Research, 31*(1), 21–31.

Dégremont, L., Nourry, M., & Maurouard, E. (2015). Mass selection for survival and resistance to OsHV-1 infection in *Crassostrea gigas* spat in field conditions: Response to selection after four generations. *Aquaculture, 446*, 111–121. doi:10.1016/j.aquaculture.2015.04.029.

de Melo, C. M. R., Durland, E., & Langdon, C. (2016). Improvements in desirable traits of the Pacific oyster, *Crassostrea gigas*, as a result of five generations of selection on the West Coast, USA. *Aquaculture, 460*, 105–115. doi:10.1016/j.aquaculture.2016.04.017.

Deng, Y., Liu, X., Zhang, G., & Guo, X. (2005). Inbreeding depression and maternal effects on early performance of Pacific abalone. *North American Journal of Aquaculture, 67*(3), 231–236. doi:10.1577/a04-021.1.

Desrosiers, R. R., Gérard, A., Peignon, J. M., Naciri, Y., Dufresne, L., Morasse, J., Ledu, C., Phelipot, P., Guerrier, P., & Dubé, F. C. (1993). A novel method to produce triploids in bivalve molluscs by the use of 6-dimethylaminopurine. *Journal of Experimental Marine Biology and Ecology, 170*, 29–43.

Divilov, K., Merz, N., Schoolfield, B., Green, T. J., & Langdon, C. (2023). Marker-assisted selection in a Pacific oyster population for an antiviral QTL conferring increased survival to OsHV-1 mortality events in Tomales Bay. *Aquaculture, 567*, 739291. doi:10.1016/j.aquaculture.2023.739291.

Divilov, K., Schoolfield, B., Morga, B., Dégremont, L., Burge, C. A., Cortez, D. M., ... Langdon, C. (2019). First evaluation of resistance to both a California OsHV-1 variant and a French OsHV-1 microvariant in Pacific oysters. *BMC Genetics, 20*(1), 96. doi:10.1186/s12863-019-0791-3.

Dou, J., Li, X., Fu, Q., Jiao, W., Li, Y., Li, T., ... Bao, Z. (2016). Evaluation of the 2b-RAD method for genomic selection in scallop breeding. *Scientific Reports, 6*(1), 19244. doi:10.1038/srep19244.

Dove, M. C., Nell, J. A., & O'Connor, W. A. (2013). Evaluation of the progeny of the fourth-generation Sydney rock oyster *Saccostrea glomerata* (Gould, 1850) breeding lines for resistance to QX disease (Marteilia sydneyi) and winter mortality (Bonamia roughleyi). *Aquaculture Research, 44*(11), 1791–1800. doi:10.1111/are.12012.

Dupuy, J. L., & Rivkin, S. (1972). The development of laboratory techniques for the production of cultch-free spat of the oyster, *Crassostrea virginica*. *Chesapeake Science, 13*(1), 45–52. doi:10.2307/1350550.

Evans, F., Matson, S., Brake, J., & Langdon, C. (2004). The effects of inbreeding on performance traits of adult Pacific oysters (*Crassostrea gigas*). *Aquaculture, 230*(1), 89–98. doi:10.1016/j.aquaculture.2003.09.023.

Farley, C., Lewis, E., Relyea, D., Zahtila, J., & Rivara, G. (1998). Resistance studies for juvenile oyster disease (JOD) 1997: Some early insights into genetic aspects. *Journal of Shellfish Research, 17*(1), 352–353.

Food and Agriculture Organization of the United Nations, FishStatJ – software for fishery statistical time series. (2025). http://www.fao.org/fishery/en/statistics/software/fishstatj.

Frank-Lawale, A., Allen, S. K., & Dégremont, L. (2014). Breeding and domestication of eastern oyster (*Crassostrea virginica*) lines for culture in the Mid-Atlantic, USA: Line development and mass selection for disease resistance. *Journal of Shellfish Research, 33*(1), 153–165.

Grouzdev, D., Farhat, S., Guo, X., Espinosa, E. P., Reece, K., McDowell, J., … Allam, B. (2024). Development and validation of a 66K SNP array for the hard clam (*Mercenaria mercenaria*). *BMC Genomics, 25*(1). doi:10.1186/s12864-024-10756-7.

Guo, X. (1991). Studies on tetraploid induction in the Pacific oyster, *Crassostrea gigas* (Thunberg).

Guo, X. (2009). Use and exchange of genetic resources in molluscan aquaculture. *Reviews in Aquaculture, 1*(3-4), 251–259. doi:10.1111/j.1753-5131.2009.01014.x.

Guo, X. (2021). Genetics in shellfish culture. In SE. Shumway (Ed.), *Molluscan shellfish aquaculture: A practical guide*. 5m Books Ltd.

Guo, X. (2026). Genetics and Epigenetics in Life History and Reproduction: Bivalve Molluscs. In M. K. kinner (Ed.), *Encyclopedia of Reproduction: 6* (pp. 946–972). US: Elsevier.

Guo, X., & Allen, S. K (1994a). Viable tetraploids in the Pacific oyster (*Crassostrea gigas Thunberg*) produced by inhibiting polar body 1 in eggs. *Molecular Marine Biology and Biotechnology, 1994*(1), 42–50.

Guo, X., & Allen, S. K. (1994b). Reproductive potential and genetics of triploid Pacific Oysters, *Crassostrea gigas (Thunberg). The Biological Bulletin, 187*(3), 309–318. doi:10.2307/1542288.

Guo, X., & Allen, S. K. (1994c). Sex determination and polyploid gigantism in the dwarf surfclam *(Mulinia lateralis say). Genetics, 138*(4), 1199–1206.

Guo, X., DeBrosse, G., & Allen, S. (1996). All-triploid Pacific oysters (*Crassostrea gigas* Thunberg) produced by mating tetraploids and diploids. *Aquaculture, 142*(3–4), 149–161. doi:10.1016/0044-8486(95)01243-5.

Guo, X., Ford, S., DeBrosse, G., Smolowitz, R., & Sunila, I. (2003). Breeding and evaluation of eastern oyster strains selected for MSX, Dermo and JOD resistance. *Journal of Shellfish Research, 22*, 353–354.

Guo, X., Ford, S. E, & Zhang, F. (1999). Molluscan aquaculture in China. *Journal of Shellfish Research, 18*(1), 19–31.

Guo, X., Puritz, J. B, Wang, Z., Proestou, D., Allen, S., Small, J., … Lotterhos, K. E. (2023). Development and evaluation of high-density SNP arrays for the eastern oyster *Crassostrea virginica*. *Marine Biotechnology, 25*(1), 174–191. doi:10.1007/s10126-022-10191-3.

Guo, X., Wang, Y., DeBrosse, G. A, Bushek, D., & Ford, S. E (2008a). Building a superior oyster for aquaculture. *Jersey Shoreline, 25*(1), 7–9.

Guo, X., Wang, Y., Wang, Y., & Lee, J.-H. (2008b). Oysters. In TD. Kocher, & C. Kole (Eds.), *Genome mapping and genomics in fishes and aquatic animals*. Springer.

Guo, X., Wang, Y., Xu, Z., & Yang, H. (2009). Chromosome set manipulation in shellfish. In G. Burnell, & G. Allan (Eds.), *New technologies in aquaculture: Improving production efficiency, quality and environmental management*. Woodhead Publishing.

Gutierrez, A. P, Matika, O., Bean, T. P, & Houston, R. D (2018). Genomic selection for growth traits in Pacific Oyster (*Crassostrea gigas*): Potential of low-density marker panels for breeding value prediction. *Frontiers in Genetics, 9*, 391. doi:10.3389/fgene.2018.00391.

Gutierrez, A. P., Symonds, J., King, N., Steiner, K., Bean, T. P., & Houston, R. D. (2020). Potential of genomic selection for improvement of resistance to ostreid herpesvirus in Pacific oyster (*Crassostrea gigas*). *Animal Genetics, 51*(2), 249–257. doi:10.1111/age.12909.

Hamilton, M., Kube, P., Elliott, N., McPherson, L., & Krsinich, A. (2009). Development of a breeding strategy for hybrid abalone. *Proceedings of the Association for the Advancement of Animal Breeding and Genetics, 18*, 350–353.

Haskin, H. H., & Ford, S. E. (1987). Breeding for resistance in molluscs. In *Proceedings of the World Symposium on Selection, Hybridization, and Genetic Engineering in Aquaculture* (pp. 432–441). Heenemann.

He, Y., Yu, H., Bao, Z., Zhang, Q., & Guo, X. (2012). Mutation in promoter region of a serine protease inhibitor confers Perkinsus marinus resistance in the eastern oyster (*Crassostrea virginica*). *Fish and Shellfish Immunology, 33*(2), 411–417. doi:10.1016/j.fsi.2012.05.028.

Hershberger, W. K., Perdue, J. A., & Beattie, J. H. (1984). Genetic selection and systematic breeding in Pacific oyster culture. *Aquaculture, 39*(1-4), 237–245. doi:10.1016/0044-8486(84)90269-2.

Higginbotham, J. A. (1997). *Piscinae: Artificial fishponds in Roman Italy*. UNC Press Books.

Hishamunda, N., & Subasinghe, R. P. (2003). *Aquaculture development in China: The role of public sector policies. FAO Fisheries Technical Paper. Food & Agriculture Organization*. Rome Food & Agriculture Organization.

Hollenbeck, C. M., & Johnston, I. A. (2018). Genomic tools and selective breeding in molluscs. *Frontiers in Genetics, 9*, 253. doi:10.3389/fgene.2018.00253.

Ibarra, A. M., Cruz, P., & Romero, B. A. (1995). Effects of inbreeding on growth and survival of self-fertilized catarina scallop larvae, *Argopecten circularis*. *Aquaculture, 134*(1), 37–47. doi:10.1016/0044-8486(95)00022-T.

Iitembu, J. A., Fitzgerald, D., Altintzoglou, T., Boudry, P., Britz, P., Byron, C. J., … Strand, Å. (2023). Comparative description and analysis of oyster aquaculture in selected Atlantic regions: Production, market dynamics, and consumption patterns. *Fishes, 8*(12), 584.

Jiang, K., Chen, C., Jiang, G., Chi, Y., Xu, C., Kong, L., Yu, H., Liu, S., & Li, Q. (2024). Genetic improvement of oysters: Current status, challenges, and prospects. *Reviews in Aquaculture, 16*(2), 796–817. doi:10.1111/raq.12868.

Jourdan, A., Morvezen, R., Enez, F., Haffray, P., Lange, A., Vétois, E., … Boudry, P. (2023). Potential of genomic selection for growth, meat content and colour traits in mixed-family breeding designs for the Pacific oyster *Crassostrea gigas*. *Aquaculture, 576*, 739878. doi:10.1016/j.aquaculture.2023.739878.

Kennedy, V. S. (2014). Technological constraints during the first 40 years of eastern oyster *Crassostrea virginica* aquaculture. *Reviews in Fisheries Science and Aquaculture, 22*(1), 55–72. doi:10.1080/10641262.2013.822464.

Kube, P., Dove, M., Cunningham, M., Kirkland, P., Gu, X., & Hick, P. (2018). *Genetic selection for resistance to Pacific oyster mortality syndrome. CSIRO Marine and Atmospheric Research, FRDC and Seafood CRC Project 2012/760*. CSIRO Agriculture and Food; NSW Department of Primary Industries.

Lafarga de la Cruz, F., & Gallardo-Escárate, C. (2011). Intraspecies and interspecies hybrids in Haliotis: Natural and experimental evidence and its impact on abalone aquaculture. *Reviews in Aquaculture, 3*(2), 74–99. doi:10.1111/j.1753-5131.2011.01045.x.

Launey, S., & Hedgecock, D. (2001). High genetic load in the Pacific Oyster *Crassostrea gigas*. *Genetics, 159*, 255–265. doi:10.1093/genetics/159.1.255.

Leighton, D. L., & Lewis, C. A. (1982). Experimental hybridization in abalones. *International Journal of Invertebrate Reproduction, 5*(5), 273–282. doi:10.1080/01651269.1982.10553479.

Li, R., Zhang, L., Li, W., Zhang, Y., Li, Y., Zhang, M., … Bao, Z. (2018). FOXL2 and DMRT1L are Yin and Yang genes for determining timing of sex differentiation in the Bivalve Mollusk Patinopecten yessoensis. *Frontiers in Physiology, 9*, 1166. doi:10.3389/fphys.2018.01166.

Liu, F., Cai, B., Lian, S., Chang, X., Chen, D., Pu, Z., Bao, L., Wang, J., Lv, J., Zheng, H., Bao, Z., Zhang, L., Wang, S., & Li, Y. (2024). MolluscDB 2.0: A comprehensive functional and evolutionary genomics database for over 1400 molluscan species. *Nucleic Acids Research, 53*(D1), D1075. doi:10.1093/nar/gkae1026.

Liu, J., Peng, W., Yu, F., Lin, W., Shen, Y., Yu, W., … Ke, C. (2022). Development and validation of a 40-K multiple-SNP array for Pacific abalone (*Haliotis discus hannai*). *Aquaculture, 558*. doi:10.1016/j.aquaculture.2022.738393.

Liu, M., Yang, L., Que, H., Li, L., & Zhang, G. (2021). Evaluation of translocation impacts on genetic patterns in farmed and naturalized populations of *Mytilus galloprovincialis* along the China coast: Clues from mitochondrial cytochrome c oxidase I sequences. *Israeli Journal of Aquaculture - Bamidgeh, 73*, 1–17. doi:10.46989/001C.28258.

Loosanoff, V. L, & Davis, H. C (1963). Rearing of bivalve mollusks. In FS. Russell (Ed.), *Advances in marine biology*. Academic Press.

Lu, J. K., Chen, T. T., Allen, S. K., Matsubara, T., & Burns, J. C. (1996). Production of transgenic dwarf surfclams, *Mulinia lateralis*, with pantropic retroviral vectors. *Proceedings of the National Academy of Sciences, 93*, 3482–3486. doi:10.1073/pnas.93.8.3482.

Lv, J., Jiao, W., Guo, H., Liu, P., Wang, R., Zhang, L., Zeng, Q., Hu, X., Bao, Z., & Wang, S. (2018). HD-Marker: A highly multiplexed and flexible approach for targeted genotyping of more than 10,000 genes in a single-tube assay. *Genome Research, 28*(12), 1919–1930.

Malindine, J. (2019). Prehistoric aquaculture: Origins, implications, and an argument for inclusion. *Culture, Agriculture, Food and Environment, 41*(1), 66–70. doi:10.1111/cuag.12226.

McCarty, A. J., Allen, S. K., & Plough, L. V. (2021). Genome-wide analysis of acute low salinity tolerance in the eastern oyster *Crassostrea virginica* and potential of genomic selection for trait improvement. *G3: Genes, Genomes, Genetics, 12*(1), jkab368. doi:10.1093/G3JOURNAL/JKAB368.

McCombie, H., Ledu, C., Phelipot, P., Lapègue, S., Boudry, P., & Gérard, A. (2005). A complementary method for production of tetraploid *Crassostrea gigas* using crosses between diploids and tetraploids with Cytochalasin B treatments. *Marine Biotechnology, 7*(4), 318–330.

Naciri-Graven, Y., Martin, A.-G., Baud, J.-P., Renault, T., & Gérard, A. (1998). Selecting the flat oyster *Ostrea edulis* (L.) for survival when infected with the parasite *Bonamia ostreae*. *Journal of Experimental Marine Biology and Ecology, 224*(1), 91–107. doi:10.1016/s0022-0981(97)00171-8.

Nash, C. (2010). *The history of aquaculture*. John Wiley & Sons.

Nell, J. A., & Perkins, B. (2006). Evaluation of the progeny of third-generation Sydney rock oyster *Saccostrea glomerata* (Gould, 1850) breeding lines for resistance to QX disease *Marteilia sydneyi* and winter mortality *Bonamia roughleyi*. *Aquaculture Research, 37*, 693–700. doi:10.1111/j.1365-2109.2006.01482.x.

Newkirk, G. F. (1980). Review of the genetics and the potential for selective breeding of commercially important bivalves. *Aquaculture, 19*(3), 209–228. doi:10.1016/0044-8486(80)90045-9.

Ning, X., Li, X., Wang, J., Zhang, X., Kong, L., Meng, D., Wang, H., Li, Y., Zhang, L., Wang, S., Hu, X., & Bao, Z. (2019). Genome-wide association study reveals E2F3 as the candidate gene for scallop growth. *Aquaculture, 511*, 734216. doi:10.1016/j.aquaculture.2019.734216.

Peruzzi, S., & Guo, X. (2002). Tetraploid induction by meiosis inhibition with cytochalasin B in the dwarf surfclam, *Mulinia lateralis* Say: Effects of temperature. *Journal of Shellfish Research, 21*(2), 677–684.

Plough, L. V., & Hedgecock, D. (2011). Quantitative trait locus analysis of stage-specific inbreeding depression in the pacific oyster *Crassostrea gigas*. *Genetics, 189*, 1473. doi:10.1534/genetics.111.131854.

Proestou, D. A., Vinyard, B. T., Corbett, R. J., Piesz, J., Allen, S. K., Small, J. M., ... Gómez-Chiarri, M. (2016). Performance of selectively-bred lines of eastern oyster, *Crassostrea virginica*, across eastern US estuaries. *Aquaculture, 464*, 17–27.

Qi, H., Song, K., Li, C., Wang, W., Li, B., Li, L., ... Xu, P. (2017). Construction and evaluation of a high-density SNP array for the Pacific oyster (*Crassostrea gigas*). *PLOS ONE, 12*(3), e0174007. doi:10.1371/journal.pone.0174007.

Scarpa, J., Wada, K. T., & Komaru, A. (1993). Induction of tetraploidy in mussels by suppression of polar body formation. *Nippon Suisan Gakkaishi (Japanese Edition), 59*(12), 2017–2023. doi:10.2331/suisan.59.2017.

Smith, N. F., Lepofsky, D., Toniello, G., Holmes, K., Wilson, L., Neudorf, C. M., & Roberts, C. (2019). 3500 years of shellfish mariculture on the Northwest Coast of North America. *PLoS ONE, 14*(2), e0211194. doi:10.1371/journal.pone.0211194.

Stanley, J. G., Allen, S. K., & Hidu, H. (1981). Polyploidy induced in the American oyster, *Crassostrea virginica*, with cytochalasin B. *Aquaculture, 23*, 1–10.

Vu, S. V., Gondro, C., Nguyen, N. T. H., Gilmour, A. R., Tearle, R., Knibb, W., ... O'Connor, W. (2021). Prediction accuracies of genomic selection for nine commercially important traits in the portuguese oyster (*Crassostrea angulata*) using dart-seq technology. *Genes, 12*(2), 1.

Wada, K. T., & Komaru, A. (1994). Effect of selection for shell coloration on growth rate and mortality in the Japanese pearl oyster, *Pinctada fucata martensii*. . *Aquaculture, 125*(1), 59–65. doi:10.1016/0044-8486(94)90282-8.

Wang, C., Liu, B., Li, J., Liu, S., Li, J., Hu, L., Fan, X., Du, H., & Fang, H. (2011). Introduction of the Peruvian scallop and its hybridization with the bay scallop in China. *Aquaculture, 310*(3), 380–387. doi:10.1016/j.aquaculture.2010.11.014.

Wang, H., Qian, L., Liu, X., Zhang, G., & Guo, X. (2010). Classification of a common cupped oyster from Southern China. *Journal of Shellfish Research, 29*(4), 857–866. doi:10.2983/035.029.0420.

Wang, Y., Sun, G., Zeng, Q., Chen, Z., Hu, X., Li, H., ... Bao, Z. (2018). Predicting growth traits with genomic selection methods in Zhikong scallop (*Chlamys farreri*). *Marine Biotechnology, 20*(6), 769–779. doi:10.1007/s10126-018-9847-z.

Wang, Z., Casas, S., Peyre, J. L., Rikard, S., Williams, M. L., Tarnecki, A., ... Guo, X. (2025). Genomic selection for dermo resistance in the eastern oyster *Crassostrea virginica*: Production and laboratory testing of F1 generation. *Journal of Shellfish Research, 44*(1), 75–87. doi:10.2983/035.044.0108.

Wang, Z., Guo, X., Allen, S. K., & Wang, R. (2002). Heterozygosity and body size in triploid Pacific oysters, *Crassostrea gigas* Thunberg, produced from meiosis II inhibition and tetraploids. *Aquaculture, 204*(3–4), 337–348. doi:10.1016/S0044-8486(01)00845-6.

Wells, W. (1920). Growing oysters artificially. *Conservationist, 3*, 151–154.

Yang, B., Zhi, C., Li, P., Xu, C., Li, Q., & Liu, S. (2024). Genomic selection accelerates genetic improvement of resistance to Vibriosis in the Pacific oyster, *Crassostrea gigas*. *Aquaculture, 584*. doi:10.1016/j.aquaculture.2024.740679.

Yang, H., Zhang, F., & Guo, X. (2000). Triploid and tetraploid Zhikong scallop, *Chlamys farreri* Jones & Preston, produced by inhibiting polar body I. *Marine Biotechnology, 2*(5), 466–475.

You, W., Guo, Q., Fan, F., Ren, P., Luo, X., & Ke, C. (2015). Experimental hybridization and genetic identification of Pacific abalone *Haliotis discus hannai* and green abalone H. fulgens. *Aquaculture, 448*, 243–249. doi:10.1016/j.aquaculture.2015.05.043.

Zeng, D., & Guo, X. (2022). Mantle transcriptome provides insights into biomineralization and growth regulation in the Eastern Oyster (*Crassostrea virginica*). *Marine Biotechnology, 24*(1), 82–96. doi:10.1007/s10126-021-10088-7.

Zhang, J., Li, Q., Xu, C., & Han, Z. (2019). Response to selection for growth in three selected strains of the Pacific oyster *Crassostrea gigas*. *Aquaculture, 503*, 34–39. doi:10.1016/j.aquaculture.2018.12.076.

Zhang, N., Xu, F., & Guo, X. (2014). Genomic analysis of the pacific oyster (*Crassostrea gigas*) reveals possible conservation of vertebrate sex determination in a mollusc. *G3: Genes, Genomes, Genetics, 4*(11), 2207–2217. doi:10.1534/g3.114.013904.

Zhao, L., He, Z., Wu, Y., Zhang, Y., Liu, Z., & Chan, J. (2025). *South Guangdong Oyster Boom: Oyster aquaculture and industry development in Guangdong*. Ocean University of China Press.

Zheng, H., Zhang, G., Guo, X., & Liu, X. (2008). Inbreeding depression for various traits in two cultured populations of the American bay scallop, *Argopecten irradians irradians* Lamarck (1819) introduced into China. *Journal of Experimental Marine Biology and Ecology, 364*(1), 42–47. doi:10.1016/j.jembe.2008.06.027.

Zheng, H., Zhang, G., Liu, X., & Guo, X. (2007). Inbreeding depression for various traits in two stocks of the bay scallop, *Argopecten irradians irradians* Lamarck (1819). *Aquaculture, 272*, S320. doi:10.1016/j.aquaculture.2007.07.208.

Zhu, X., Ni, P., Xing, Q., Wang, Y., Huang, X., Hu, X., Hu, J., Wu, X., & Bao, Z. (2021). Genomic prediction of growth traits in scallops using convolutional neural networks. *Aquaculture, 545*. doi:10.1016/j.aquaculture.2021.737171.

SECTION II

Shellfish Genetic Breeding

CHAPTER

3

Multiomic resources and MolluscDB database

Yuli Li, Jing Wang, Jia Lv, Fuyun Liu and Shi Wang

College of Marine Life Sciences, Ocean University of China, Qingdao, China

3.1 Molluscan genomics

Mollusca is the second most speciose phylum in the animal kingdom, encompassing over 100,000 described extant species with a broad distribution across marine, freshwater, and terrestrial ecosystems (Haszprunar & Wanninger, 2012). Mollusks hold a prominent position in the aquatic industry, not only as globally important food sources but also for their economic and nutritional value. They are prized for their delicate flavor, rich protein content, and essential minerals, making them a cornerstone of global seafood markets. The genetic makeup of mollusks, encoded within their complex genomes, is a treasure trove of potential for advancing genetic breeding programs. Understanding the intricate details of these genomes can unlock traits related to growth rate, disease resistance, environmental adaptability, and the evolution of specific traits, all of which are crucial for sustainable aquaculture practices.

3.1.1 Summary of molluscan genome sequencing

The rapid advance in sequencing technologies, assembly strategies, and computational tools has continuously improved the assembly accuracy of mollusk genomes. By the end of 2025, genomes of 467 mollusk species had been sequenced according to NCBI Database, covering 7 out of the all 8 mollusk classes, including 165 in Bivalvia, 3 in Scaphopoda, 225 in Gastropoda, 47 in Cephalopoda, 21 in Polyplacophora, 2 in Caudofoveata and 2 in Solenogastres, 2 in Monoplacophora (Fig. 3.1). The long reads from third-generation sequencing technologies such as PacBio or Nanopore have significantly enhanced genome assembly quality by increasing Contig N50 while reducing scaffold numbers. The emergence of high-resolution chromosome conformation capture (Hi-C) technology has enabled chromosome-level genome assemblies, with the scaffold N50 of several newest molluscan genome assemblies reaching up to mega levels and the number of scaffolds reduced to less than or around one thousand. There are

Genetic Breeding and Aquaculture of Shellfish: A Collaborative Approach to a Sustainable Future
DOI: https://doi.org/10.1016/B978-0-443-33606-5.00006-0

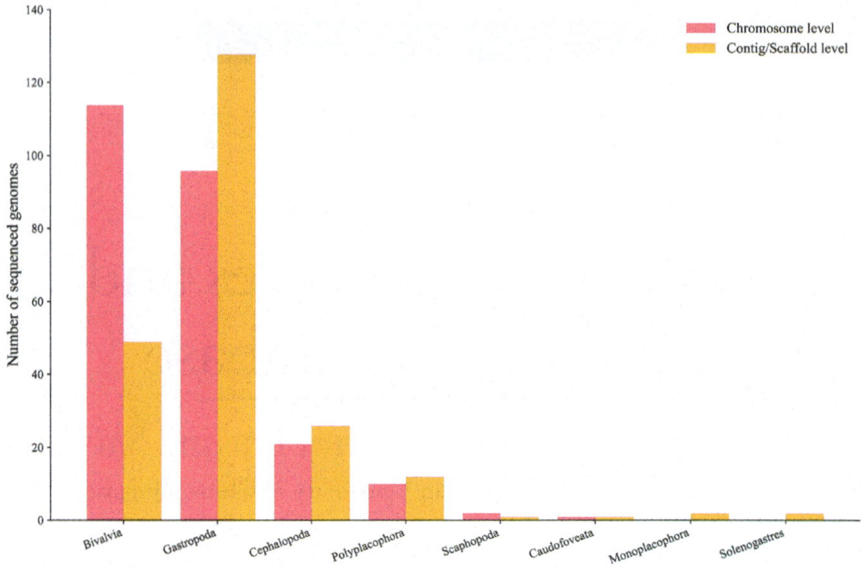

FIGURE 3.1 Summary of sequenced molluscan genomes.

now 236 high-quality genomes reaching chromosome-level assembly. The availability of these molluscan genome sequences, particularly high-quality reference genomes, has driven various biological studies in mollusks, especially functional genomics research. To date, thousands of molluscan functional genomic datasets, including bulk transcriptomes, single-cell transcriptomes, proteomes, epigenomes, and metabolomes, have been released for mollusks (Liu et al., 2025), which deepens our understanding of molluscan genomic regulation.

3.1.2 Sequencing platform and assembly strategy

One of the most striking features of molluscan genomes is their high level of heterozygosity (Hollenbeck & Johnston, 2018; Takeuchi, 2017). To date, the molluscan genome with the highest recorded heterozygosity is that of the sea slug *Elysia chlorotica*, reaching 3.66% (Cai et al., 2019). In stark contrast, the human genome typically exhibits heterozygosity levels below 0.1%. Highly heterozygous individuals posed significant challenges for sequencing. Early sequencing efforts conquered the high-heterozygosity problem of mollusks by using highly inbred individuals and/or advanced assembly approaches (Takeuchi et al., 2012; Wang et al., 2017; Zhang et al., 2012). Recent advancements in long-read sequencing technologies (e.g., PacBio HiFi and Oxford Nanopore) have significantly improved allele-resolved genome assemblies in heterozygous species, as demonstrated in the Pacific oyster *Crassostrea gigas* and *Crassostrea angulata* (Qi et al., 2023) and the pearl oyster *Pinctada fucata* (Takeuchi et al., 2022), allowing researchers to distinguish between paralogs and allelic variants with higher confidence. However, this genetic variation is also a crucial factor in the adaptability of mollusks, enabling them to thrive in a wide range of environmental conditions (Harrang et al., 2013; Li et al., 2018; Song et al., 2019). Genomic analysis of the invasive Mediterranean mussel *Mytilus galloprovincialis* showed that heterozygous loci are enriched in genes related to immune response and xenobiotic detoxification, facilitating colonization of polluted habitats (Figueras et al., 2019). These findings underscore the complex interplay between heterozygosity, genome architecture, and evolutionary innovation in mollusks. Multiple members of all eight classes show widespread genomic variation in mollusks. The flexibility of the molluscan genome likely explains both historic challenges with their genomes and their evolutionary success (Chen et al., 2025).

3.1.3 Molluscan genomic characteristics

3.1.3.1 *Gene and repeat content*

Most molluscan genomes sequenced to date contain approximately 20,000–30,000 genes (Liu et al., 2025). Whole-genome duplication (WGD) events were less reported in molluscan genomes, since comparative genome analysis revealed that the gene repertoires of mollusks are generally well conserved (Simakov et al., 2013; Wang et al., 2017). However, recent genome sequencing studies of hyperdiverse venomous gastropods have supported the occurrence of WGD events in molluscan lineages (Farhat et al., 2023). Most sequenced molluscan genomes exhibit a single, nonduplicated Hox gene cluster (Albertin et al., 2015; Biscotti et al., 2014; Wang et al., 2017; Zhang et al., 2021); however, venomous gastropods uniquely possess two Hox gene clusters, directly validating the occurrence of WGD in these gastropod lineages. The genome sizes of mollusks are primarily influenced by highly dynamic repeat content, varying from ~10% to 70% (Yang et al., 2020). Although transposable elements predominate in many molluscan genomes, they are less active in scallops (Li et al., 2017; Wang et al., 2017), potentially contributing to their conserved genome structure (Wang et al., 2017).

3.1.3.2 *Gene family expansion*

Mollusks exhibit a particularly striking expansion of lineage-specific gene families, which play a key role in their environmental adaptation (Takeuchi, 2017; Zhang et al., 2016). In the Pacific oyster (*C. gigas*), the expansion of heat shock protein 70 (HSP70) and apoptosis inhibitor genes enables survival in the highly stressful intertidal zone (Zhang et al., 2012). Expansion of C1q domain gene family in *C. gigas* and *Ruditapes philippinarum* may account for immunological defense mechanisms against various biotic and abiotic factors in the marine habitats (Gerdol et al., 2015; Mun et al., 2017). In cephalopods, substantial expansion of protocadherins and the C_2H_2 superfamily of zinc-finger transcription factors played a critical role in the evolution of cephalopod morphological innovations, including their large and complex nervous systems (Albertin et al., 2015). Functional diversification through gene duplication is further exemplified by species-specific innovations. The scallop *Chlamys farreri* employs expanded retroposition-derived intronless opsins for broad spectral sensitivity, expanded tyrosinases for rapid byssal secretion, and sulfotransferases for neurotoxin production (Li et al., 2017). The deep-sea mussel *Bathymodiolus* (*Gigantidas*) *platifrons* exhibits expanded immune recognition and apoptosis-related genes in its gills to manage chemoautotrophic symbionts (Sun et al., 2017), whereas the freshwater snail *Biomphalaria glabrata* counters parasitic schistosomes through expanded fibrinogen-related proteins (FREPs) and immunoglobulin superfamily members (Pennance et al., 2024). Collectively, these findings underscore gene family expansion as a cornerstone of molluscan evolutionary innovation, enabling resilience in dynamic and challenging environments (Takeuchi, 2017; Zhang et al., 2016).

3.1.4 Genomic dark matter

The concept of "dark matter" was initially proposed in human genome studies, referring to the non-coding regions that constitute approximately 99% of the genome (Chi, 2016; The ENCODE Project Consortium, 2012). This concept has since been extended, particularly in nonmodel organisms such as mollusks and other marine invertebrates, where it now describes both coding and noncoding regions with undetermined functions (Edsinger & Moroz, 2024). This conceptual expansion arises from recent research findings revealing that nonmodel organisms harbor vast numbers of unannotated genes—those that cannot be functionally characterized through sequence homology alignment (David & Halanych, 2023). As one of the major phyla of protostomes, the Mollusca phylum suffers from extremely scarce genomic resources (accounting for only about 5% according to NCBI statistics), relatively lagging genome sequencing progress, and a lack of model organism platforms within the phylum. Consequently, gene functional annotation relies solely on homologous sequence alignment, leaving approximately 25% or more genes unannotated on average in their genomes (Edsinger & Moroz, 2024). Therefore compared with model organisms, mollusks not only exhibit a higher proportion of "absolutely unknown genes" awaiting annotation, but their pleiotropic gene annotation also faces greater technical challenges due to phylogenetic particularities.

In recent years, the paradigm integration of deep learning and systems biology has made it possible to construct systems biology networks and comprehensively understand gene pleiotropy function (Ma et al., 2018; Nilsson et al., 2022; Novakovsky et al., 2023). Heterogeneous graph neural networks can simultaneously process multidimensional data from genomic variations, transcriptional regulation, and metabolomics, achieving systematic integration of multidimensional omics data (Eicher et al., 2020; Erfanian et al., 2023; Gao et al., 2023; Subramanian et al., 2020). By constructing cross-omics dimensional systems biology networks, it is possible not only to circumvent the species-dependent limitations of homology inference (Barabási et al., 2023) but also to enable functional deduction from lower-dimensional data through overlapping functional modules (Blumenthal et al., 2024). Systems biology approaches have achieved initial progress in gene function prediction for model organisms (Ma et al., 2024; Pazos Obregón et al., 2022). Applying this framework to nonmodel organisms for deciphering gene function presents both challenges and opportunities: (1) The omics data resources for nonmodel organisms are relatively limited in volume, dimensionality, and functional omics data, currently insufficient for de novo training of large language models; (2) Because of distinct biological characteristics across taxa, large models built on model organism omics data cannot be directly applied to nonmodel organisms; and (3) integrating gene function research from nonmodel organisms with knowledge from model organisms to further refine the cognitive framework holds promise for accelerating the understanding of dark matter of molluscan genomes.

3.1.5 Pangenome and telomere-to-telomere assembly

Pangenomics enables us to trace the evolutionary history of species and offers new perspectives on sources of genomic variation and adaptation of organisms. A pangenome aggregates genomic sequences from multiple individuals within a species and therefore represents the full genetic repertoire of a species. It consists of "core" genes that are present in all species and "accessory" genes found in only a subset of species. The progress of pangenome research in humans, plants, and domestic animals has proved that the missing genetic components and the identification of large structural variants (SVs) can be explored through pangenomic studies, and many individual-specific sequences have been shown to be related to biological adaptability, phenotype, and important economic traits (Bozan et al., 2023; Liu et al., 2023; Wang et al., 2021; Xin et al., 2024). It was reported that the Mediterranean mussel (*M. galloprovincialis*) pangenome harbors an ultra-long accessory genomic region (~580 Mb), exceeding by orders of magnitude those observed in other metazoans, which correlates with hemizygous regions marked by SV (Gerdol et al., 2020). This distinctive genome architecture in the *M. galloprovincialis* genome, characterized by a preponderance of hemizygous segments, is enriched for genes implicated in stress resilience and immune response, potentially underpinning the species' evolutionary success and global distribution in temperate coastal marine habitats. Such extensive presence–absence variation may also characterize other cosmopolitan marine invertebrates, particularly broadcast-spawning bivalves with large effective populations and analogous environmental stressors, where high heterozygosity has been documented. The maturity of techniques such as third-generation sequencing, telomere-to-telomere (T2T) genome, graphical genome, and reference-free assembly will further promote the development of the pangenome (Gong et al., 2023). In the future, pan-genomes combined with long-read data and multiomics will help to analyze large structural variations and their relationship with major economic traits, thereby aiding molecular breeding and seed industry advancement of mollusks.

T2T assembly is a cutting-edge strategy designed to construct gap-free genome sequences encompassing all chromosomal regions—including telomeres, centromeres, and complex repetitive elements—thereby overcoming limitations inherent to traditional sequencing approaches. While particularly challenging for mollusks characterized by extensive tandem repeats and structural variations, advances in long-read sequencing (e.g., PacBio HiFi and Oxford Nanopore) now enable T2T assembly even for these recalcitrant systems. Some research teams are making every effort to advance the T2T genome project for mollusks. For instance, our team have recently obtained a preliminary T2T genome assembly of the Yesso scallop with the genome size of 1.28 Gb and all 19 T2T-level chromosomes, which will be released publicly soon. The revolutionary progress in decoding complex genomic regions by the T2T genome will empower precise discovery of functional genes and structural variations and thus pave the way for accelerated molecular breeding of mollusks.

3.2 Molluscan transcriptomics

Transcriptomics is a large-scale study of all RNA transcripts present in a cell or tissue at a specific moment. From early Sanger sequencing and microarray chips to high-throughput RNA-seq, and further to single-cell/spatial transcriptomics, transcriptomic technologies have achieved a leap from low throughput to high resolution. This progress has accelerated understanding of mollusk developmental biology, adaptive evolution, and immune mechanisms, while facilitating the integration of multi-omics data to unravel complex trait regulation (Fig. 3.2). In aquaculture, it has enhanced the identification of key genetic determinants for economic traits, promoting the development of precision breeding strategies.

3.2.1 Early molluscan transcriptomics

The advent of microarray technology marked the beginning of high-throughput gene expression analysis, which gained prominence in the late 1990s and early 2000s (Schena et al., 1995). During the first 15 years of the 21st century, the emergence of early high-throughput

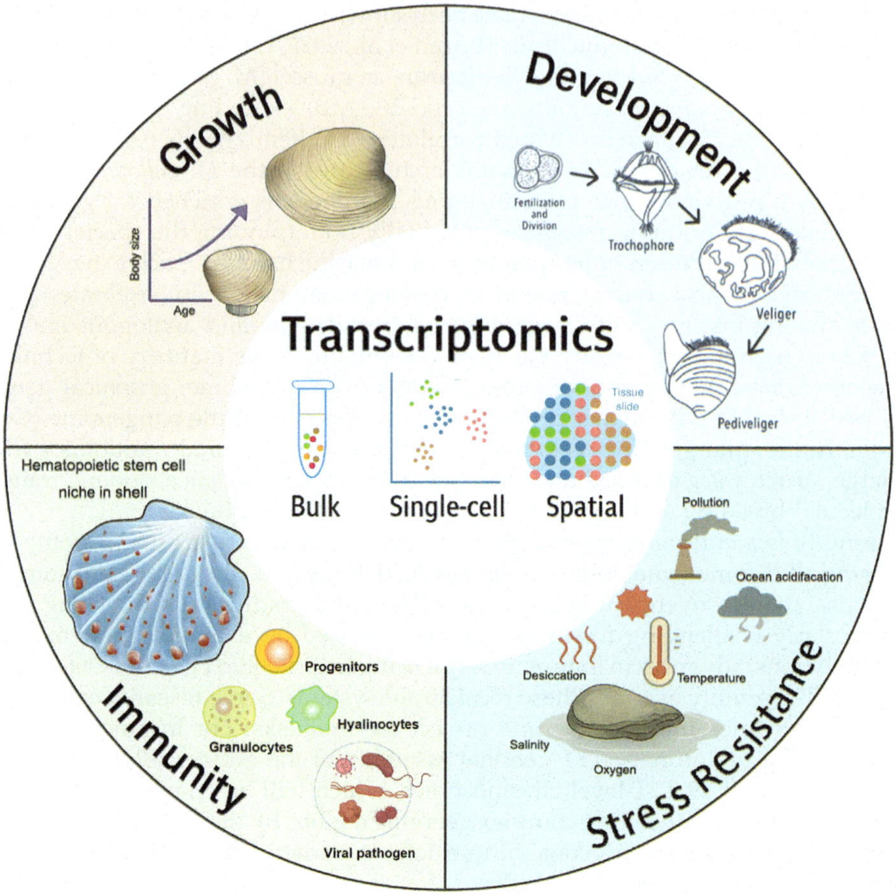

FIGURE 3.2 The application of transcriptome in mollusks.

platforms, represented by 454 GS FLX and deep RNA sequencing combined with EST analysis and microarray, reduced per-base costs and deeper coverage, has enabled scientists to identify a series of functional genes and achieved significant progress across multiple molluscan species. Key outcomes include identifying shell formation-related genes (e.g., novel genes in nacreous and prismatic layer tissues of pearl oysters) through de novo transcriptome assembly and deep EST sequencing (Hou et al., 2011; Kinoshita et al., 2011), and revealing the parallel evolutionary mechanism of nacre formation gene sets (Jackson et al., 2010); developing microarray tools to support population genetics and environmental monitoring (Milan et al., 2011); and analyzing the central nervous system transcriptome of freshwater snail to facilitate neurofunctional research (Sadamoto et al., 2012). These studies have advanced the mining of mollusk functional genes, the development of molecular markers, and the understanding of environmental adaptation mechanisms, laying a foundation for subsequent functional genomics and evolutionary research.

3.2.2 High-throughput transcriptomics

The development of next-generation sequencing (NGS) technologies, particularly RNA-seq, has revolutionized transcriptomics by overcoming many of these limitations (Wang et al., 2009). In aquaculture, RNA-seq has been established as an effective tool for identifying genes and pathways responsible for key aquaculture traits such as growth, development, and environmental stress tolerance. By capturing genome-wide expression profiles, RNA-seq enables the identification of key pathways involved in growth regulation, including growth hormone signaling, cell proliferation, and nutrient metabolism (Mokrani et al., 2025). For example, studies in the Pacific oyster (*C. gigas*) have linked rapid growth to upregulated genes in the insulin-like growth factor (IGF) and energy metabolism pathways (Kim & Choi, 2019; Li et al., 2021). RNA-seq analyses in scallops (*P. yessoensis*) and clam (*M. lateralis*) have identified the conserved function of somatostatin receptor (SSTR) genes in growth regulation (Zhang et al., 2024). Scallops have a remarkably large adductor muscle compared to most bivalves, such as oysters, mussels, and clams. RNA-seq revealed heightened energy metabolism gene expression (e.g., glycolysis and TCA cycle) in the striated muscle of the scallop *C. farreri*. Comparative transcriptomics highlighted conserved expression of ancestral bilaterian muscle components, supporting the hypothesis that scallop muscles retain plesiomorphic traits shared with other invertebrates (Li et al., 2017). Furthermore, RNA-seq has identified potential long noncoding RNAs (lncRNAs) and microRNAs (miRNAs) implicated in posttranscriptional control of growth pathways (Sun et al., 2022; Li et al., 2023).

Transcriptomic approaches, particularly RNA-seq, have profoundly advanced our understanding of the molecular mechanisms underpinning environmental adaptation in economically important mollusks. Studies across species such as oysters, scallops, mussels, and clams have delineated how these organisms respond to diverse stressors, including temperature fluctuations, ocean acidification, hypoxia, pollution, and pathogen challenge (Gleason, 2019). For instance, in the Pacific oyster (*C. gigas*), transcriptome analyses have revealed extensive gene family expansions related to heat shock proteins (*HSP70*) and inhibitor of apoptosis proteins (*IAP*), facilitating survival in intertidal zones through enhanced stress tolerance and phenotypic plasticity (Zhang et al., 2012). Remarkably, recent work on the octopus (*Octopus bimaculoides*) demonstrated extensive temperature-dependent RNA editing, particularly in neural transcripts, enabling rapid proteomic reconfiguration in response to thermal shifts—a novel mechanism underlying acute environmental acclimation (Anania, 2023; Birk et al., 2023). Investigations into the blood clam (*Tegillarca granosa*) uncovered a mollusk-specific expansion of hemoglobin genes, correlating with its adaptation to hypoxic conditions (Bao et al., 2021). Under hypoxic stress, transcriptomic profiling in the pearl oyster (*P. fucata*) revealed activation of antioxidant defenses, metabolic suppression, and cytoskeletal remodeling, illustrating a coordinated molecular strategy to maintain homeostasis (Chen et al., 2023). In response to ocean acidification, RNA-seq has identified differentially expressed genes involved in cell cycle regulation, oxidative stress response, and biomineralization, suggesting potential pathways for resilience in oyster populations under high CO_2 conditions (Goncalves et al., 2017; Zhang et al., 2012). Pollutant adaptation studies have uncovered unique detoxification and defense strategies. For example, scallops tolerate neurotoxin exposure through mutations in voltage-gated sodium channels and enhanced toxin transformation via sulfotransferases (Li et al., 2017). Collectively, these findings underscore the power of RNA-seq to not only enrich fundamental knowledge of molluscan biology but also provide critical insights for aquaculture breeding programs aimed at enhancing stress resilience and disease resistance in commercial species.

3.2.3 Single-cell transcriptomics

Single-cell RNA sequencing (scRNA-seq) has revolutionized molluscan research by providing unprecedented resolution into cellular heterogeneity, developmental processes, and evolutionary innovations across this diverse phylum (Bump & Lubeck, 2023). One of the most striking applications of scRNA-seq in bivalves has been characterizing the immune cells (hemocytes). This technology has uncovered intricate immune cell diversity in oysters, revealing specialized hemocyte subtypes responsible for phagocytosis, reactive oxygen species production, and antimicrobial peptide expression (De La Forest Divonne et al., 2025; Meng & Wang, 2022). Lian et al. found the presence of bone marrow-like hematopoietic stem cell (HSC) niche in mollusk shells, with an extraordinarily high concentration of HSC-like cells (~60%). This discovery challenged the textbook view that bone marrow is a vertebrate-specific innovation and suggesting an ancient origin of skeletal hematopoiesis (Lian et al., 2025). Beyond immunity, scRNA-seq has illuminated germline development and gonadal niche interactions, delineating germ cell trajectories and somatic cell communication via pathways such as NOTCH and BMP in oysters and scallops (Shu et al., 2025; Wang et al., 2024), and elucidating spermatogenesis dynamics in *Mulinia lateralis* (Li et al., 2025). Larval cell atlases in bivalves such as *Dreissena rostriformis* have molecularly defined trochophore structures, enabling tests of homology across spiralians (Salamanca-Díaz et al., 2022). More recently, single-cell transcriptomics was proposed as a new source of evidence to assess whether spiralian larvae are homologous by cataloguing their cell-type complement and comparing their transcriptomic signatures (Piovani & Marlétaz, 2023; Salamanca-Díaz et al., 2022). In addition, cellular diversity in scallop adductor muscles has been categorized, highlighting distinct neuronal and muscle fiber subtypes (Sun et al., 2021). In cephalopods, scRNA-seq has decoded the complexity of neural and visual systems, revealing both conserved and novel cell types in Octopus and squid species (Styfhals et al., 2022). In summary, despite challenges such as the need for high-quality genomes and optimized protocols for marine taxa, scRNA-seq bridges traditional morphology with molecular genetics, offering transformative insights into molluscan biology, aquaculture, and evolutionary history.

3.2.4 Single-cell spatial transcriptomics

Single-cell spatial transcriptomics aims to understand the spatial organization and localization of gene expression within tissues. This technology combines histology with RNA sequencing, enabling the mapping of gene expression patterns to their exact spatial locations. Spatial transcriptomics is typically performed using methods such as in situ sequencing or slide-based approaches, where RNA is extracted and sequenced directly from tissue sections. Notable platforms for spatial transcriptomics include the spatial transcriptomics method developed by 10x Genomics (Zheng et al., 2017), the GeoMx Digital Spatial Profiling (DSP) instrument by Nanostring (Bonnett et al., 2023; Merritt et al., 2020), and the Stereo-seq by BGI (Wei et al., 2022). Currently, no publicly accessible single-cell spatial transcriptomic data are available for mollusks. In a recent study as part of the Heart Evolution and Development program led by the Fang Zongxi Center for Marine Evo-Devo at Ocean University of China, a spatially resolved single-cell atlas of the adult scallop heart was constructed using the Stereo-seq platform. This atlas identified 10 distinct cell clusters, including cardiomyocytes, endothelial cells, epithelial cells, and fibroblasts. The spatial distribution and gene expression profiles of these cell types are accessible via an open-access database (https://www.bgiocean.com/EvoHeart_Atlas/index.html). It can be anticipated that this technology will bring breakthrough discoveries to mollusk research, uncovering molecular mechanisms underlying development, physiology, and adaptation, and enhance functional genomics research and precise breeding strategies.

3.2.5 Transcriptome analysis methods

RNA-seq data analysis represents a sophisticated and intricate process that is essential for unraveling the complexities of the transcriptome. It starts with expression quantification, using metrics such as transcripts per million (TPM) and fragments per kilobase of transcript per million mapped reads (FPKM) to measure transcript abundance. Differential expression analysis follows to identify differential expressed genes (DEGs) between experimental groups. Weighted gene correlation network analysis (WGCNA) constructs weighted coexpression networks, linking gene modules to phenotypic traits. Transcriptome age index (TAI) evaluates evolutionary conservation via phylostratigraphy, inferring expressed gene age and divergence. Cross-species transcriptome similarity is assessed through orthologous gene profiling, phylogenetic comparisons, and expression pattern correlation, revealing conserved regulatory networks or species-specific adaptations, collectively decoding transcriptomic dynamics, functional interactions, and evolutionary relationships.

Gene expression quantification is foundational for comparing transcript abundance across samples, but raw read counts require normalization to account for sequencing depth variations and technical biases. Key normalization methods, such as TPM and FPKM, adjust for transcript length and sequencing depth, enabling accurate cross-sample comparisons of expression levels. Following quantification, differential expression analysis emerges as a pivotal step, employing statistical tools to identify genes with significant expression changes between experimental conditions. A classic application includes comparing tissue-specific expression profiles, which facilitates the discovery of biomarkers or hub regulatory genes underlying phenotypic differences. The comprehensive genomic database MolluscDB integrates expression profiles of 1413 transcriptomes from 61 mollusk species currently available in NCBI, providing a critical resource for systematic research on mollusk gene expression through comprehensive compilation (Liu et al., 2025).

Beyond individual gene expression levels, coexpression network analysis illuminates regulatory relationships among genes by constructing networks where nodes represent genes and edges denote coexpression strength. This approach enables the identification of functionally coordinated gene clusters involved in biological pathways or processes (Fu et al., 2014). In developmental studies, such analyses have yielded critical insights. For instance, WGCNA of *Patinopecten yessoensis* (Yesso scallop) across developmental stages revealed modules associated with body axis formation, neurogenesis, and myogenesis in larvae (Jing et al., 2024; Wang et al., 2020). Similarly, transcriptomic profiling of Pacific oyster metamorphosis identified upregulated genes linked to nervous system development and shell formation, with distinct temporal expression patterns of transcription factors (e.g., homeobox, bHLH, and nuclear receptor genes) suggesting a complex regulatory network during this transition (Xu et al., 2023). WGCNA further pinpointed *slc4a8* and *rnf41* as hub genes potentially governing biomineralization and larval shell formation (Guo et al., 2023). Complementary comparative transcriptome analyses have extended these findings: studies on larval shells identified numerous shell matrix proteins (SMPs) and hypothesized that developmentally unique larval shell transcription represents a conserved trait in Conchifera (Cavallo et al., 2022; Zhao et al., 2018).

The TAI analysis provides critical evolutionary insights into transcriptomes, as established in foundational research (Domazet-Lošo & Tautz, 2010). By quantifying transcriptional complexity, TAI links gene expression patterns to evolutionary history and functional significance, complementing gene age estimation techniques that distinguish ancient versus recently evolved genes. A study on scallop transcriptomes across developmental stages employed TAI to trace lifecycle evolution, revealing that the trochophore stage harbors the youngest transcriptome among developmental phases. This finding supports the "adult-first" evolutionary model, challenging the traditional "larva-first" paradigm (Wang et al., 2020). Wang et al. modified Domazet-Lošo's formula to calculate the relative TAI contribution of a given phylostratum to the "ancient" transcriptome, and traced the first appearance of proto-larva back to the divergence of Ctenophora from the metazoan ancestor (Wang et al., 2020). Extending TAI analysis to cell-type resolution of lophotrochozoan larvae showed that the lowest TAI occurs in proliferative cells and hemocytes, which are potentially homologous across all Bilateria (Piovani et al., 2023).

3.3 Molluscan metabolomics

The concept of metabolomics, proposed by Jeremy Nicholson and Oliver Fiehn in the late 20th century (Fiehn et al., 2000; Nicholson et al., 1999), aims to qualitatively and quantitatively analyze all low molecular weight compounds within organisms, tissues, or cells under specific conditions (pathological, physiological, or genotypic) (Fiehn, 2002). In molluscan aquaculture, metabolite patterns have been effectively used to address various critical issues, including nutrition and flavor (Li et al., 2017; Luo et al., 2025; Tan et al., 2024; Zhang et al., 2023), disease and immunology (Dong et al., 2022; Sun et al., 2021), and molecular analysis of important traits (Zhang et al., 2024). These applications demonstrate the importance of metabolomics in optimizing bivalve production, enhancing health management, improving nutritional strategies, and ensuring the quality and sustainability of bivalve products in aquaculture. A metabolomics study generally involves several key steps: (1) collecting and preparing samples, (2) performing analytical measurements and acquiring data, (3) bioinformatics for data quality control and metabolite identification, (4) statistical analysis, and (5) biological interpretation and/or biomarker validation. At present, the core methods of metabolomics detection mainly include techniques based on nuclear magnetic resonance spectroscopy and techniques combining mass spectrometry (MS) and chromatography (Young & Alfaro, 2018) (Fig. 3.3).

3.3.1 Nuclear magnetic resonance technology

Nuclear magnetic resonance (NMR) is one of the fundamental methods in metabolomics research. The basic principle of NMR is that nuclei whose spin quantum numbers are not 0, such as 1H, 1C, and 1F, undergo Zeeman energy level splitting in the presence of an external magnetic field. When a radiofrequency (RF) field is applied perpendicular to the external magnetic field, and the frequency of the RF field matches the transition frequency between adjacent energy levels, the phenomenon of nuclear spin energy level transition occurs, thereby generating

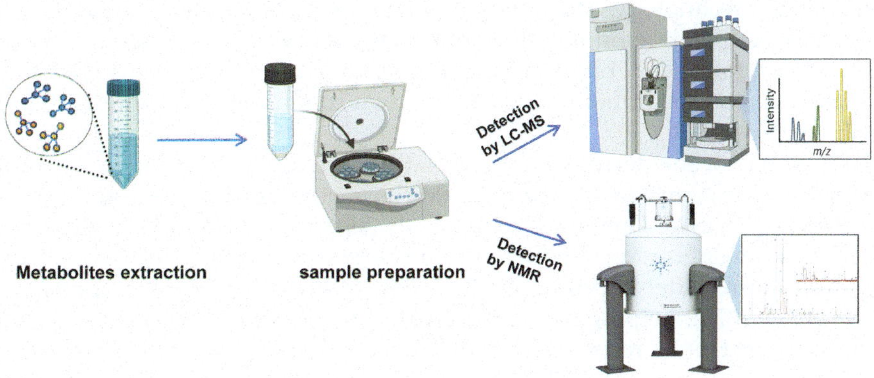

FIGURE 3.3 The workflow of metabolomics.

NMR signals (Beckonert et al., 2007). Through different ways, such as Fourier transform, phase adjustment, chemical shift calibration, integration, and time domain experiment, the NMR signal created can be measured and processed to yield a certain NMR spectrum for the nucleus concerned (Cao et al., 2021). The technical parameters of NMR mainly include resonance frequency, spectral width, relaxation time, acquisition time, sampling points, peak integral, and scanning frequency. By analyzing the chemical shift, coupling constant, peak integral, and other information obtained from NMR, it can be used to identify the functional groups, structure, molecular weight, and other information of compounds. As a nondestructive and noninvasive technique, NMR could retain liquid or solid samples for future studies. Furthermore, samples for NMR detection require minimal preparation (either dissolved or diluted in a deuterium solvent), whereas other methods usually require a separation and purification step prior to analysis. At present, NMR has been applied to obtain the metabolic fingerprint across various tissues in fish or bivalve, as well as to distinguish between wild and cultured aquatic species (Frizzo et al., 2021; Mannina et al., 2008). For example, Frizzo et al. (2021) used NMR-based metabolomics to analyze the hemolymph metabolome of *M. galloprovincialis* in response to *Vibrio splendidus* injection, revealing metabolic changes associated with osmoregulation, aerobic respiration, and protein turnover during stress. Götze et al. (2020) investigated the metabolic responses of king scallops to warming, hypercapnia, and hypoxia using NMR spectroscopy, revealing significant alterations in mitochondrial energy metabolism and a reduction in thermal tolerance, primarily due to an imbalance between oxygen demand and supply. However, the challenge of metabolite identification has limited the broader application of NMR in aquaculture research. Metabolomic analysis of individuals with different breeds or traits can not only provide an in-depth understanding of the metabolic regulatory mechanism behind important phenotypes, but also reveal how organisms adapt to their environment.

3.3.2 Mass spectrometry technology

MS technology is another widely used technique in metabolomics research. Its basic principle is to ionize the sample in an ion source, detect the mass to charge ratio (m/z) and other information of ions, and finally obtain a mass spectrum containing m/z and content information.

Compared to NMR technology, MS technology has the advantages of higher sensitivity and greater specificity. Depending on the metabolites, different ion sources are selected to achieve better detection performance. Liquid chromatography-mass spectrometry (LC-MS) and gas chromatography-mass spectrometry (GC-MS) are two commonly used chromatographic-mass spectrometric techniques.

LC-MS is suitable for the detection of nonvolatile, thermally unstable, polar substances and high molecular weight compounds. It uses liquid chromatography as the separation system and MS as the detection system. Unlike GC-MS, LC-MS separates substances at room temperature through liquid chromatography without requiring derivatization. In recent years, metabolomics techniques have been widely applied in the study of growth and development (Tan et al., 2024; Zhang et al., 2024), helping to identify chemical signatures that reflect specific cellular activities, such as metabolic responses to environmental stimuli or stress in various bivalves (Dong et al., 2022; Pérez-López et al., 2023; Sun et al., 2021). For example, Zhang et al. (2024) used LC-MS to perform metabolic profiles of multiple tissues of *C. farreri* with different growth stages and identified growth-related metabolic pathways that contributed to growth. Tan et al. conducted integrative metabolome and proteome analysis of two Pacific oyster (*C. gigas*) strains with different growth phenotypes, showing that the fast-growing strain utilizes energy-efficient mechanisms of ATP generation (Tan et al., 2024).

The basic principle of GC-MS is to first separate volatile and thermally stable substances using chromatography, and then ionizing the separated compounds in the mass spectrometer (MSD). GC-MS has the advantages of high sensitivity and good reproducibility for volatile and thermally stable substances (Chan et al., 2011). Volatile compounds play a crucial role in the flavor profile of food, with their cooperative interactions contributing to the complexity and diversity of flavors. Because of its sensitivity and accuracy, GC-MS is extensively employed in the evaluation of flavor characteristics in bivalve species (such as oysters, scallops, and clams) (Liu et al., 2024; Luo et al., 2025; Zhang et al., 2023), enabling a detailed analysis of the volatile compounds present. This technique is valuable for investigating the nutritional composition of aquaculture species, facilitating a better understanding of their dietary contributions and how they vary across different species or processing methods.

3.3.3 The principles of spatial metabolomics

Spatial metabolomics offers substantial potential, particularly in the fields of biomedical research and agricultural sciences, providing a new visual perspective and multidimensional information depth for traditional metabolomics analysis. Traditional metabolomics focuses on identifying and quantifying the overall metabolite content and variations within tissues or cells (Hu et al., 2023). However, as research has progressed, it has become increasingly clear that the lack of spatial resolution in conventional metabolomics methods represents a significant limitation. Currently, spatial metabolomics, a technique combining mass spectrometry imaging (MSI) with metabolomics techniques, has been able to accurately determine the type, content, and spatial distribution of metabolites within tissues and cells. By integrating MS imaging with metabolomics, this technology enables the generation of detailed metabolic maps at both the tissue and cellular levels, thereby facilitating a deeper understanding of metabolic processes and disease mechanisms in living organisms. Furthermore, the incorporation of artificial intelligence and big data analytics will significantly expand the scope of metabolomics applications, provide more comprehensive biological insights, and enable precise predictive modeling. Recently our team have successfully employed Raman imaging technology to characterize the spatial distribution patterns of metabolites in the adductor muscle of Yesso scallop, supporting the strong potential of spatial metabolomics for quality trait assessment and breeding applications in bivalves. As costs decrease and user-friendliness improves, spatial metabolomics is expected to become increasingly accessible across various laboratories and research fields, thus advancing the field of aquaculture research (Alexandrov, 2023).

3.4 Molluscan metagenomics

Microorganisms refer to the small and medium-sized organisms in nature that are invisible to the naked eye, including bacteria, archaea, algae, fungi, and microanimals (Berg et al., 2020). Microorganisms are commonly found in many aquatic ecosystems as they contribute to essential ecosystem functions, such as nutrient cycling and climate regulation. The importance of the intestinal microbiota in aquaculture species has also been highlighted, and it plays a crucial role in shaping the aquaculture species' phenotypic traits, immune mechanisms, and animal fitness in response to climate change by the dynamic interplay with the host (Diwan et al., 2022; Zhang et al., 2025).

With the continuous development of microbial molecular sequencing technology, the concept of the metagenome was first proposed in 1998, which takes the total DNA in a specific environment as the research object. With the revolutionary breakthrough of high-throughput sequencing and the rapid development of bioinformatics, the research methods of microbiome have been constantly innovated, which has also greatly promoted the research process of microbiome. Presently, amplicon sequencing and metagenomic sequencing are the most mainstream DNA sequencing technologies for microbiome research.

3.4.1 Amplicon sequencing in metagenomics

Amplicon sequencing typically targets the 16S rRNA gene for bacteria and archaea, and the internal transcribed spacer (ITS) region for fungi, as the 16S rRNA gene contains several conserved regions and nine variable regions (V1–V9). Universal PCR primer was designed to amplify the conserved and variable regions of the 16S rRNA gene, serving as a taxonomic marker to achieve the taxonomic profiles. Given that 16S rRNA sequencing remains more cost-effective than whole metagenome shotgun (WMS) sequencing, it is well-suited for monitoring the diversity of specific taxonomic groups and tracking microbiome composition changes across various aquaculture systems. Recent microbial research in aquaculture have focus on exploring the symbiotic and antagonistic interactions between microbes and their hosts, such as oysters, clam and scallops (Ma et al., 2024; Paillard et al., 2022; Wilkins et al., 2019). In this context, metagenomics can offer a more comprehensive understanding of these relationships by sequencing microbial DNA extracted from specific niches within host organisms (Gignoux-Wolfsohn et al., 2024; Harke et al., 2016; Paillard et al., 2022; Pérez-Cobas et al., 2020; Scanes et al., 2021). In the case of 16S rRNA gene amplicon sequencing, the taxa identification also has several limitations: The primer choice will introduce a bias for or against certain taxonomic groups; furthermore, this method lack of species-level taxonomic resolution required for microbiota studies.

3.4.2 Whole-metagenome shotgun sequencing

WMS sequencing aims to sequence the entire set of genomes present in an environmental sample, providing a comprehensive characterization of microorganisms in terms of whole genomes, genes, and genetic traits. This approach, known as shotgun metagenomics, involves fragmenting total DNA from the sample, preparing sequencing libraries, and subsequently performing sequencing, assembly, and annotation. It allows researchers to access the complete gene content of the microbial community. As a result, WGS metagenomics not only facilitates species- or strain-level identification but also uncovers potential and novel functions within the community. With the continued decrease in NGS costs, whole-metagenome sequencing has become an attractive alternative to traditional 16S rDNA methods. Metagenomics has overcome the limitations of amplification-based methods, offering more detailed insights into functional characteristics, not only detects pathogens but also identifies microbial genes related to antibiotic resistance, virulence, and metabolic functions, facilitating better disease management. In addition, it aids in understanding the complex interactions between the microbiomes of aquaculture species and their environments. This is particularly useful for optimizing health management practices and monitoring environmental impacts. The ability to analyze functional genes involved in nutrient cycling, waste degradation, and stress tolerance makes WMS a crucial tool for improving the sustainability of aquaculture systems (Perry et al., 2020). With advancements in sequencing technologies, the cost of WMS has significantly decreased, making it a scalable solution for large-scale studies. Consequently, WMS is poised to play an increasingly important role in bivalves, from disease prevention and environmental monitoring to enhancing production efficiency and sustainability (Diwan et al., 2022, 2024). For example, Pathak et al. (2024) used WMS to reveal the microbial community characteristics of diploid and triploid oysters and found that *Psychrobacter*-dominated microbiomes drive denitrification and nitrogen removal in eastern oysters.

3.4.3 2bRAD-M

The current metagenomic sequencing technologies, such as 16S rRNA sequencing and WGS sequencing, are widely used methods for studying microbiomes. However, they still face several limitations in practical applications. 16S rRNA sequencing is susceptible to issues such as amplification bias and low resolution, and it can only detect bacteria and is difficult to detect microbes at the species level. Although WGS sequencing addresses some of these challenges, it requires high-quality DNA samples and is often constrained by cost, particularly for large-scale studies. In addition, WGS sequencing is not ideal for samples with high host DNA content or degraded DNA. The 2bRAD-M method (Sun et al., 2022), a recent innovation, enables high-resolution, cost-effective microbiome characterization at the species level. It is particularly useful for challenging samples, requiring as little as 1 pg of total DNA, even when dealing with high host DNA contamination or degraded DNA. This technology has proven effective for clinical samples with low biomass or highly degraded DNA, such as paraffin samples. Based on the Type IIB restriction system of the 2bRAD-M, an advancing metagenomic profiler method called MAP2B, a metagenomic profiler based on the Type IIB restriction site) has been proposed to enhance the accuracy of taxonomic profiles derived from WMS data by significantly reducing false positives (Sun et al., 2023). MAP2B improves precision and recall in species identification, optimizing the decoding of taxonomic structures in microbiome studies and improving the resolution for differential abundance analysis, biomarker detection, phenotype classification, and disease prediction. Recently, a simple and efficient technique for dynamic hologenomic analysis, known as holo-2bRAD, was introduced (Ma et al., 2024). This approach enables the collection of hologenomic data, which includes both the genetic information of the microbiome and its associated host, all from a single 2bRAD sequencing library, making it a highly efficient tool for investigating host–microbe interactions in bivalves. For example, Ma et al. (2024) adopted this method to explore the host–microbe associations underlying mass mortality events in bivalve larvae and identified core microbial species that changed with developmental stages and two potentially associated host SNPs. Holo-2bRAD provides an efficient tool for investigating host–microbe coevolution, immune regulation, microbial symbiosis, and disease susceptibility, particularly in bivalves. Looking ahead, as the technology continues to evolve and be applied, it is expected to deepen our understanding of the mechanisms underlying host–microbe interactions in molluscan species.

3.5 Integrated genomic database of mollusks

The development of the molluscan aquaculture industry has undergone significant stages, transitioning from artificial breeding to molecular breeding and further to genomic selection. With advancements in technology and increasing demands, the future of molluscan aquaculture lies in digital breeding, a process that requires vast amounts of multidimensional omics data for trait analysis and germplasm evaluation. To efficiently integrate and mine these data to guide breeding efforts, the establishment of a comprehensive genomic database is particularly crucial. By continuously optimizing the functions and services of the database and enhancing the efficiency of data integration and mining, it will provide strong support for the sustainable development of the molluscan aquaculture industry. This will facilitate the transition from genomic selection to digital breeding in aquaculture, accelerating the cultivation and promotion of new molluscan varieties.

To address the rapid expansion of massive, complex, and high-dimensional omics resources, the first comprehensive molluscan genomic database, MolluscDB, was constructed, integrating approximately 1000 molluscan genomic and transcriptomic data, and providing multiple genomic comparative analysis tools (Liu et al., 2021). Widely recognized in the community, it amassed over 20,000 accesses from more than 70 countries. To meet future research trends, the database was recently upgraded to MolluscDB 2.0 (http://mgbase.qnlm.ac), providing a systematic view of molluscan multi-omics information (Liu et al., 2025). MolluscDB 2.0 integrates 4200 multiomics datasets—including genomes, transcriptomes, proteomes, epigenomes, and metagenomes, derived from 1450 species from all eight molluscan classes and approximately 92% of total orders. The database features 34 customized modules, involved in the analysis of comparative genomes, gene expression network and regulatory genomes, assisting in the location and regulation of functional genes in genetic breeding (Fig. 3.4).

Taxonomy, paleontological records, and multiomics data: Mollusks exhibit vast taxonomic diversity, with rich paleontological records and extensive multiomics data from living species. The "Taxonomy" module in MolluscDB 2.0 integrates omics data across all eight molluscan classes, offering taxonomic classifications linked to NCBI and World Register of Marine Species (WoRMS) databases, and providing genome assembly statistics for species with high-quality genomes. The "Paleobiology" module aggregates 56,704 fossil records from the Paleobiology

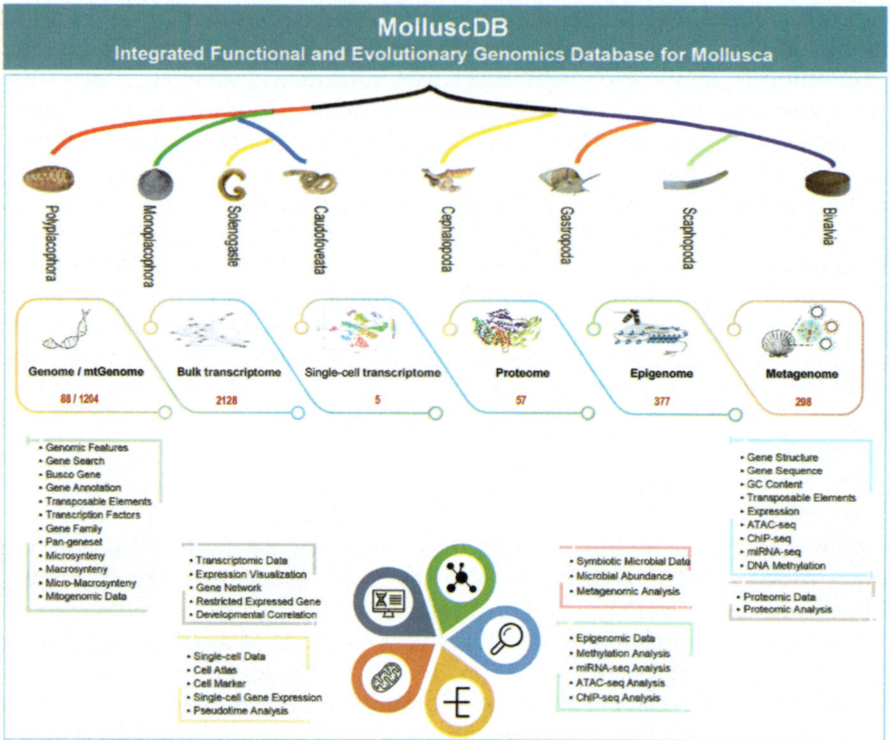

FIGURE 3.4 Overview of MolluscDB database architecture and function. The database integrates 4200 multiomics datasets—including genomes, transcriptomes, proteomes, epigenomes, and metagenomes, derived from all eight molluscan classes.

Database (PBDB). In addition, the "Genomic Data" module presents 88 high-quality genomes, including 52 chromosome-level assemblies, marking significant advancements in genome assembly. The transcriptomic data summary includes 265 developmental stage datasets and 1078 adult tissue datasets from 61 species. 552 key datasets for single-cell, proteomic, epigenomic, and metagenomic studies are also included.

Genomic modules: The genomic modules enhance research on genome function and evolution through cross-species comparisons and panevolutionary analysis. The "Gene Search" module integrates data from over 2.6 million protein-coding genes across 88 species. The "Transposable Elements" module catalogs 139 million TEs linked to more than 1 million genes. The "Mitogenomic Data" module offers phylogenetic insights with 1204 mitochondrial genomes. Functional annotation is provided by the "Gene Annotation" module using six major databases. The BUSCO Gene module aids phylogenetic studies by identifying conserved genes, while the "Transcription Factors" module catalogs 45,712 TF genes crucial for regulation. The "Gene Family" module identifies 939,599 gene families, linking to "Gene Search" and PANTHER for comparative analysis. The Pan-geneset module explores evolutionary changes in gene families. Genomic synteny analysis is available through the "Microsynteny" and "Macrosynteny" modules using the PanSyn toolkit (Yu et al., 2024).

Bulk and single-cell transcriptomic modules: Understanding the transcriptome is essential for identifying all transcribed genomic elements and studying gene functions during development and evolution. The transcriptomic modules provide comprehensive gene expression data across developmental stages and adult tissues. The "Expression Visualization" module allows exploration of gene expression profiles for 61 molluscan species. The "Restricted Expressed Gene" module identifies genes specifically expressed in particular tissues or developmental stages, including 134 developmental stages and 452 adult tissues/organs. The "Gene Network" module constructs coexpression networks, featuring more than 223 million highly correlated gene pairs depicting developmental stages of 265 species and adult tissues/organs of 1078 species. The "Developmental Correlation" module performs comparative transcriptomic analysis for 120 species pairs, revealing gene expression patterns during development across species. Single-cell transcriptomic data are available in the "Cell Atlas" module, identifying 206 cell clusters in five molluscan species. The "Cell Marker" and "Single-cell Gene Expression" modules visualize gene expression at the single-cell level, whereas the "Pseudotime Analysis" module studies developmental trajectories by reconstructing cells' temporal progression in pseudotime.

Proteomic, epigenomic, and metagenomic modules: The proteomic modules advance molluscan proteomics and support materials science. The MS data from 15 molluscan species cover biomineralization, environmental adaptation, and biomaterial synthesis. The epigenomic modules explore genomic modifications regulating gene function. The "Methylation Analysis" module compiles 153 million methylation sites across 12 species, offering insights into DNA methylation patterns by genomic location, motif type, or gene ID. The "miRNAseq Analysis" module includes 36,905 miRNAs from 17 species for cross-species comparisons. Limited ATAC-seq and ChIP-seq datasets provide preliminary insights into chromatin accessibility and protein-DNA interactions. The metagenomic modules catalog microbial composition and abundance in mollusks. The "Microbial Abundance Analysis" module visualizes dominant microorganisms by relative abundance, whereas the "Symbiotic Microbial Analysis" module provides detailed annotations of microbial genomes and functional characteristics, enhancing understanding of host adaptation and symbiotic interactions.

Integration of functional and evolutionary genomics: The abundance of molluscan multiomics data enables deeper insights into functional and evolutionary genomics. The MolluscDB 2.0 integrates diverse multiomics modules—including gene structure, sequence, GC content, TEs, bulk/single-cell expression, and epigenomics (ATAC-seq, ChIP-seq, and DNA methylation)—via a customized genome browser. Moreover, MolluscDB 2.0 also achieves integrated pan-evolutionary analysis at both the micro- and macroevolutionary scale through the PanSyn toolkit: Profiling gene clusters in extant species to deduce ancestral conserved clusters (bottom-up synteny) and tracing how ancestral macroevolution events shape extant microsynteny (top-down perspective). This platform highlights integrated analysis advantages, encouraging customized research for detailed functional and evolutionary investigations.

3.6 Future directions

Building upon the foundational multiomic resources currently released in Molluscan community, the future direction of molluscan research must pivot toward systematic data integration and high-resolution functional genomics. As sequencing technologies continue to mature, beyond standard whole-genome and transcriptome profiles, the sequencing of high-dimensional datasets, such as single-cell RNA sequencing, spatial transcriptomics, and epigenomic landscapes would be speed up. This expansion will be crucial for deciphering the cellular heterogeneities and precise regulatory networks underlying molluscan phenotypic traits. Ongoing projects such as the Earth BioGenome Project (Lewin et al., 2018) and molluscan M10K genome project will enable the expansion of molluscan multi-omics data of more molluscan species. New layers of multiomics data, advanced multiomics integration strategies and the comprehensive MolluscDB database (Ballard et al., 2024; Picard et al., 2021; Wang et al., 2025), will further enhance functional genomics research and accelerate the cultivation and promotion of new molluscan varieties.

References

Albertin, C. B., Simakov, O., Mitros, T., Wang, Z. Y., Pungor, J. R., Edsinger-Gonzales, E., Brenner, S., Ragsdale, C. W., & Rokhsar, D. S. (2015). The octopus genome and the evolution of cephalopod neural and morphological novelties. *Nature, 524*(7564), 220–224. doi:10.1038/nature14668.

Alexandrov, T. (2023). Spatial metabolomics: From a niche field towards a driver of innovation. *Nature Metabolism, 5*(9), 1443–1445. doi:10.1038/s42255-023-00881-0.

Anania, C. (2023). Temperature-induced RNA recoding in octopus. *Nature Genetics, 55*(7), 1082. doi:10.1038/s41588-023-01456-1.

Ballard, J. L., Wang, Z., Li, W., Shen, L., & Long, Q. (2024). Deep learning-based approaches for multi-omics data integration and analysis. *BioData Mining, 17*(1), 38. doi:10.1186/s13040-024-00391-z.

Bao, Y., Zeng, Q., Wang, J., Zhang, Z., Zhang, Y., Wang, S., … Lin, Z. (2021). Genomic insights into the origin and evolution of molluscan red-bloodedness in the blood clam *Tegillarca granosa*. *Molecular Biology and Evolution, 38*(6), 2351–2365. doi:10.1093/molbev/msab030.

Barabási, D. L., Beynon, T., Katona, Á., & Perez-Nieves, N. (2023). Complex computation from developmental priors. *Nature Communications, 14*(1), 2226. doi:10.1038/s41467-023-37980-1.

Beckonert, O., Keun, H. C., Ebbels, T. M. D., Bundy, J., Holmes, E., Lindon, J. C., & Nicholson, J. K. (2007). Metabolic profiling, metabolomic and metabonomic procedures for NMR spectroscopy of urine, plasma, serum and tissue extracts. *Nature Protocols, 2*(11), 2692–2703. doi:10.1038/nprot.2007.376.

Berg, G., Rybakova, D., Fischer, D., Cernava, T., Vergès, M. C. C., Charles, T., … Schloter, M. (2020). Microbiome definition re-visited: Old concepts and new challenges. *Microbiome, 8*(1). doi:10.1186/s40168-020-00875-0.

Birk, M. A., Liscovitch-Brauer, N., Dominguez, M. J., McNeme, S., Yue, Y., Hoff, J. D., Twersky, I., Verhey, K. J., Sutton, R. B., Eisenberg, E., & Rosenthal, J. J. C. (2023). Temperature-dependent RNA editing in octopus extensively recodes the neural proteome. *Cell, 186*(12), 2544. doi:10.1016/j.cell.2023.05.004.

Biscotti, M. A, Canapa, A., Forconi, M., & Barucca, M. (2014). Hox and ParaHox genes: A review on molluscs. *Genesis, 52*(12), 935–945. doi:10.1002/dvg.22839.

Blumenthal, D. B., Lucchetta, M., Kleist, L., Fekete, S. P., List, M., & Schaefer, M. H. (2024). Emergence of power law distributions in protein-protein interaction networks through study bias. *eLife, 13*, e99951. doi:10.7554/elife.99951.

Bonnett, S. A., Rosenbloom, A. B., Ong, G. T., Conner, M., Rininger, A. B. E., Newhouse, D., New, F., Phan, C. Q., Ilcisin, S., Sato, H., Lyssand, J. S., Geiss, G., & Beechem, J. M. (2023). Ultra high-plex spatial proteogenomic investigation of giant cell glioblastoma multiforme immune infiltrates reveals distinct protein and RNA expression profiles. *Cancer Research Communications, 3*(5), 763–779. doi:10.1158/2767-9764.CRC-22-0396.

Bozan, I., Achakkagari, S. R., Anglin, N. L., Ellis, D., Tai, H. H., & Strömvik, M. V. (2023). Pangenome analyses reveal impact of transposable elements and ploidy on the evolution of potato species. *Proceedings of the National Academy of Sciences of the United States of America, 120*(31), e2211117120. doi:10.1073/pnas.2211117120.

Bump, P., & Lubeck, L. (2023). Marine invertebrates one cell at a time: Insights from single-cell analysis. *Integrative and Comparative Biology, 63*(5), 999–1009. doi:10.1093/icb/icad034.

Cai, H., Li, Q., Fang, X., Li, J., Curtis, N. E., Altenburger, A., … Wang, J. (2019). A draft genome assembly of the solar-powered sea slug *Elysia chlorotica*. *Scientific Data, 6*(1), 190022. doi:10.1038/sdata.2019.22.

Cao, R., Liu, X., Liu, Y., Zhai, X., Cao, T., Wang, A., & Qiu, J. (2021). Applications of nuclear magnetic resonance spectroscopy to the evaluation of complex food constituents. *Food Chemistry, 342*, 128258. doi:10.1016/j.foodchem.2020.128258.

Cavallo, A., Clark, M. S., Peck, L. S., Harper, E. M., & Sleight, V. A. (2022). Evolutionary conservation and divergence of the transcriptional regulation of bivalve shell secretion across life history stages. *Royal Society open science, 9*(12), 221022. doi:10.1098/rsos.221022.

Chan, E. C. Y., Pasikanti, K. K., & Nicholson, J. K. (2011). Global urinary metabolic profiling procedures using gas chromatography-mass spectrometry. *Nature Protocols, 6*(10), 1483–1499. doi:10.1038/nprot.2011.375.

Chen, J., Qiu, J., Yang, C., Liao, Y., He, M., Mkuye, R., … Du, X. (2023). Integrated transcriptomic and metabolomic analysis sheds new light on adaptation of *Pinctada fucata martensii* to short-term hypoxic stress. *Marine Pollution Bulletin, 187*, 114534. doi:10.1016/j.marpolbul.2022.114534.

Chen, Z., Baeza, J. A., Chen, C., Gonzalez, M. T, González, V. L, Greve, C., Kocot, K. M., Arbizu, P. M, Moles, J., Schell, T., Schwabe, E., Sun, J., Wong, N. L. W. S., Yap-Chiongco, M., & Sigwart, J. D. (2025). A genome-based phylogeny for Mollusca is concordant with fossils and morphology. *Science, 387*(6737), 1001–1007. doi:10.1126/science.ads0215.

Chi, K. R. (2016). The dark side of the human genome. *Nature, 538*(7624), 275–277. doi:10.1038/538275a.

David, K. T., & Halanych, K. M. (2023). Unsupervised deep learning can identify protein functional groups from unaligned sequences. *Genome Biology and Evolution, 15*(5), evad084. doi:10.1093/gbe/evad084.

De La Forest Divonne, S., Pouzadoux, J., Romatif, O., Montagnani, C., Mitta, G., Destoumieux-Garzón, D., … Vignal, E. (2025). Diversity and functional specialization of oyster immune cells uncovered by integrative single-cell level investigations. *eLife, 13*, RP102622. doi:10.7554/elife.102622.

Diwan, A. D., Harke, S. N., & Gopalkrishna, A. N. P. (2022). Aquaculture industry prospective from gut microbiome of fish and shellfish: An overview. *Journal of Animal Physiology and Animal Nutrition, 106*(2), 441–469. doi:10.1111/jpn.13619.

Diwan, A. D., Harke, S. N., & Panche, A. N. (2024). Studies on exploring the potentials of gut microbiomes to mitigate the bacterial and viral diseases of fish and shellfish in aquaculture farming. *The Microbe, 2*, 100031. doi:10.1016/j.microb.2023.100031.

Domazet-Lošo, T., & Tautz, D. (2010). A phylogenetically based transcriptome age index mirrors ontogenetic divergence patterns. *Nature, 468*(7325), 815–818. doi:10.1038/nature09632.

Dong, X., Yang, Z., Liu, Z., Wang, X., Yu, H., Peng, C., … Bao, Z. (2022). Metabonomic analysis provides new insights into the response of Zhikong Scallop (*Chlamys farreri*) to heat stress by improving energy metabolism and antioxidant capacity. *Antioxidants, 11*(6), 1084. doi:10.3390/antiox11061084.

Edsinger, E., & Moroz, L. L. (2024). Genomic hotspots: Localized chromosome gene expansions identify lineage-specific innovations as targets for functional biodiversity and predictions of stress resilience. *Frontiers in Marine Science, 11*, 1434130. doi:10.3389/fmars.2024.1434130.

Eicher, T., Kinnebrew, G., Patt, A., Spencer, K., Ying, K., Ma, Q., Machiraju, R., & Mathé, E. A. (2020). Metabolomics and multi-omics integration: A survey of computational methods and resources. *Metabolites, 10*(5), 202. doi:10.3390/metabo10050202.

Erfanian, N., Heydari, A. A., Feriz, A. M, Iañez, P., Derakhshani, A., Ghasemigol, M., Farahpour, M., Razavi, S. M, Nasseri, S., Safarpour, H., & Sahebkar, A. (2023). Deep learning applications in single-cell genomics and transcriptomics data analysis. *Biomedicine & Pharmacotherapy, 165*, 115077. doi:10.1016/j.biopha.2023.115077.

Farhat, S., Modica, M. V., & Puillandre, N. (2023). Whole genome duplication and gene evolution in the hyperdiverse venomous gastropods. *Molecular Biology and Evolution, 40*(8), msad171. doi:10.1093/molbev/msad171.

Fiehn, O. (2002). Metabolomics—The link between genotypes and phenotypes. *Plant Molecular Biology, 48*(1-2), 155–171. doi:10.1023/A:1013713905833.

Fiehn, O., Kopka, J., Dörmann, P., Altmann, T., Trethewey, R. N., & Willmitzer, L. (2000). Metabolite profiling for plant functional genomics. *Nature Biotechnology, 18*(11), 1157–1161. doi:10.1038/81137.

Figueras, A., Moreira, R., Sendra, M., & Novoa, B. (2019). Genomics and immunity of the Mediterranean mussel *Mytilus galloprovincialis* in a changing environment. *Fish & Shellfish Immunology, 90*, 440–445. doi:10.1016/j.fsi.2019.04.064.

Frizzo, R., Bortoletto, E., Riello, T., Leanza, L., Schievano, E., Venier, P., & Mammi, S. (2021). NMR metabolite profiles of the bivalve mollusc *Mytilus galloprovincialis* before and after immune stimulation with *Vibrio splendidus*. *Frontiers in Molecular Biosciences, 8*, 686770. doi:10.3389/fmolb.2021.686770.

Fu, X., Sun, Y., Wang, J., Xing, Q., Zou, J., Li, R., … Bao, Z. (2014). Sequencing-based gene network analysis provides a core set of gene resource for understanding thermal adaptation in Zhikong scallop *Chlamys farreri*. *Molecular Ecology Resources, 14*(1), 184–198. doi:10.1111/1755-0998.12169.

Gao, H., Zhang, B., Liu, L., Li, S., Gao, X., & Yu, B. (2023). A universal framework for single-cell multi-omics data integration with graph convolutional networks. *Briefings in Bioinformatics, 24*(3), bbad081. doi:10.1093/bib/bbad081.

Gerdol, M., Moreira, R., Cruz, F., Gómez-Garrido, J., Vlasova, A., Rosani, U., Venier, P., Naranjo-Ortiz, M. A., Murgarella, M., Greco, S., Balseiro, P., Corvelo, A., Frias, L., Gut, M., Gabaldón, T., Pallavicini, A., Canchaya, C., Novoa, B., Alioto, T. S., Posada, D., & Figueras, A. (2020). Massive gene presence-absence variation shapes an open pan-genome in the Mediterranean mussel. *Genome Biology, 21*(1), 275. doi:10.1186/s13059-020-02180-3.

Gerdol, M., Venier, P., & Pallavicini, A. (2015). The genome of the Pacific oyster *Crassostrea gigas* brings new insights on the massive expansion of the C1q gene family in Bivalvia. *Developmental & Comparative Immunology, 49*(1), 59–71. doi:10.1016/j.dci.2014.11.007.

Gignoux-Wolfsohn, S., Garcia Ruiz, M., Barron, D. P, Ruiz, G., & Lohan, K. (2024). Bivalve microbiomes are shaped by host species, size, parasite infection, and environment. *PeerJ, 12*(10), e18082. doi:10.7717/peerj.18082.

Gleason, L. U. (2019). Applications and future directions for population transcriptomics in marine invertebrates. *Current Molecular Biology Reports, 5*(3), 116–127. doi:10.1007/s40610-019-00121-z.

Goncalves, P., Jones, D. B., Thompson, E. L., Parker, L. M., Ross, P. M., & Raftos, D. A. (2017). Transcriptomic profiling of adaptive responses to ocean acidification. *Molecular Ecology, 26*(21), 5974–5988. doi:10.1111/mec.14333.

Gong, Y., Li, Y., Liu, X., Ma, Y., & Jiang, L. (2023). A review of the pangenome: How it affects our understanding of genomic variation, selection and breeding in domestic animals? *Journal of Animal Science and Biotechnology, 14*(1), 73. doi:10.1186/s40104-023-00860-1.

Götze, S., Bock, C., Eymann, C., Lannig, G., Steffen, J. B. M., & Pörtner, H.-O. (2020). Single and combined effects of the "Deadly trio" hypoxia, hypercapnia and warming on the cellular metabolism of the great scallop *Pecten maximus*. *Comparative Biochemistry and Physiology Part B: Biochemistry and Molecular Biology, 243-244*, 110438. doi:10.1016/j.cbpb.2020.110438.

Guo, X., Li, X., Zhao, F., Liu, D., Yang, Z., Li, M., ... Bao, Z. (2023). Full-length transcriptome analysis provides insights into larval shell formation in *Mulinia lateralis*. *Frontiers in Marine Science, 9*. doi:10.3389/fmars.2022.1111241.

Harke, M. J., Steffen, M. M., Gobler, C. J., Otten, T. G., Wilhelm, S. W., Wood, S. A., & Paerl, H. W. (2016). A review of the global ecology, genomics, and biogeography of the toxic cyanobacterium, *Microcystis spp. Harmful Algae, 54*, 4–20. doi:10.1016/j.hal.2015.12.007.

Harrang, E., Lapègue, S., Morga, B., & Bierne, N. (2013). A high load of non-neutral amino-acid polymorphisms explains high protein diversity despite moderate effective population size in a marine bivalve with sweepstakes reproduction. *G3 Genes|Genomes|Genetics, 3*(2), 333–341. doi:10.1534/g3.112.005181.

Haszprunar, G., & Wanninger, A. (2012). Molluscs. *Current Biology, 22*(13), R510–R514. doi:10.1016/j.cub.2012.05.039.

Hollenbeck, C. M., & Johnston, I. A. (2018). Genomic tools and selective breeding in molluscs. *Frontiers in Genetics, 9*, 253. doi:10.3389/fgene.2018.00253.

Hou, R., Bao, Z., Wang, S., Su, H., Li, Y., Du, H., ... Hu, X. (2011). Transcriptome sequencing and de novo analysis for Yesso Scallop (*Patinopecten yessoensis*) using 454 GS FLX. *PLoS ONE, 6*(6), e21560. doi:10.1371/journal.pone.0021560.

Hu, T., Allam, M., Cai, S., Henderson, W., Yueh, B., Garipcan, A., ... Coskun, A. F. (2023). Single-cell spatial metabolomics with cell-type specific protein profiling for tissue systems biology. *Nature Communications, 14*(1), 8260. doi:10.1038/s41467-023-43917-5.

Jackson, D. J., McDougall, C., Woodcroft, B., Moase, P., Rose, R. A., Kube, M., Reinhardt, R., Rokhsar, D. S., Montagnani, C., Joubert, C., Piquemal, D., & Degnan, B. M. (2010). Parallel evolution of nacre building gene sets in molluscs. *Molecular Biology and Evolution, 27*(3), 591–608. doi:10.1093/molbev/msp278.

Jing, H., Zhou, L., Zheng, Y., Yu, T., Wu, B., Liu, Z., ... Sun, X. (2024). Comparative transcriptome analysis reveals the developmental program in Yesso scallop *Patinopecten yessoensis*. *Aquaculture Reports, 38*(3), 102357. doi:10.1016/j.aqrep.2024.102357.

Kim, E. Y., & Choi, Y. H. (2019). Regulation of adductor muscle growth by the IGF-1/AKT pathway in the triploid Pacific oyster, *Crassostrea gigas*. *Fisheries and Aquatic Sciences., 22*(1), 19. doi:10.1186/s41240-019-0134-3.

Kinoshita, S., Wang, N., Inoue, H., Maeyama, K., Okamoto, K., Nagai, K., ... Watabe, S. (2011). Deep sequencing of ESTs from nacreous and prismatic layer producing tissues and a screen for novel shell formation-related genes in the pearl oyster. *PLoS ONE, 6*(6), e21238. doi:10.1371/journal.pone.0021238.

Lewin, H. A., Robinson, G. E., Kress, W. J., Baker, W. J., Coddington, J., Crandall, K. A., ... Zhang, G. (2018). Earth BioGenome Project: Sequencing life for the future of life. *Proceedings of the National Academy of Sciences of the United States of America, 115*(17), 4325–4333. doi:10.1073/pnas.1720115115.

Li, B., Song, K., Meng, J., Li, L., & Zhang, G. (2017a). Integrated application of transcriptomics and metabolomics provides insights into glycogen content regulation in the Pacific oyster *Crassostrea gigas*. *BMC Genomics*, *18*(1), 713. doi:10.1186/s12864-017-4069-8.

Li, C., Liu, X., Liu, B., Ma, B., Liu, F., Liu, G., … Wang, C. (2018). Draft genome of the Peruvian scallop *Argopecten purpuratus*. *GigaScience*, *7*(4), giy031. doi:10.1093/gigascience/giy031.

Li, Y., Fu, H., Zhang, F., Ren, L., Tian, J., Li, Q., & Liu, S. (2021). Insulin-like peptide receptor-mediated signaling pathways orchestrate regulation of growth in the Pacific Oyster (*Crassostrea gigas*), as revealed by gene expression profiles. *International Journal of Molecular Sciences*, *22*(10), 5259. doi:10.3390/ijms22105259.

Li, Y., Sun, X., Hu, X., Xun, X., Zhang, J., Guo, X., Jiao, W., Zhang, L., Liu, W., Wang, J., Li, J., Sun, Y., Miao, Y., Zhang, X., Cheng, T., Xu, G., Fu, X., Wang, Y., Yu, X., Huang, X., … Bao, Z. (2017b). Scallop genome reveals molecular adaptations to semi-sessile life and neurotoxins. *Nature Communications*, *8*(1), 1721. doi:10.1038/s41467-017-01927-0.

Li, Y., Wei, H., Dai, X., Zhang, L., Liu, L., Chen, X., … Zhang, L. (2025). Insights from the single-cell level: Lineage trajectory and somatic-germline interactions during spermatogenesis in dwarf surfclam *Mulinia lateralis*. *BMC Genomics*, *26*(1), 69. doi:10.1186/s12864-025-11266-w.

Li, Y., Yang, B., Shi, C., Tan, Y., Ren, L., Mokrani, A., … Liu, S. (2023). Integrated analysis of mRNAs and lncRNAs reveals candidate marker genes and potential hub lncRNAs associated with growth regulation of the Pacific Oyster, *Crassostrea gigas*. *BMC Genomics*, *24*(1), 453. doi:10.1186/s12864-023-09543-7.

Lian, S., Hu, N., Chen, X., Dai, X., Zhu, X., Qiao, R., … Wang, S. (2025). Widespread presence of bone marrow–like hematopoietic stem cell niche in invertebrate skeletons. *Science Advances*, *11*(26), eadw0958. doi:10.1126/sciadv.adw0958.

Liu, F., Cai, B., Lian, S., Chang, X., Chen, D., Pu, Z., Bao, L., Wang, J., Lv, J., Zheng, H., Bao, Z., Zhang, L., Wang, S., & Li, Y. (2025). MolluscDB 2.0: A comprehensive functional and evolutionary genomics database for over 1400 molluscan species. *Nucleic Acids Research*, *53*(D1), D1075. doi:10.1093/nar/gkae1026.

Liu, F., Li, Y., Yu, H., Zhang, L., Hu, J., Bao, Z., & Shi, W. (2021). MolluscDB: An integrated functional and evolutionary genomics database for the hyper-diverse animal phylum Mollusca. *Nucleic Acids Research*, *49*(D1), D988. doi:10.1093/nar/gkaa918.

Liu, J., Yu, M., Wang, X., Qi, M., & Wu, M. (2024). Dynamic changes in microbiome and metabolome during sun-drying of oysters (*Crassostrea gigas*), a traditional procedure in South China. *Journal of Microbiology and Biotechnology*, *34*(12), 2555–2564. doi:10.4014/jmb.2312.12033.

Liu, X., Liu, W., Lenstra, J. A., Zheng, Z., Wu, X., Yang, J., … Liu, J. (2023). Evolutionary origin of genomic structural variations in domestic yaks. *Nature Communications*, *14*(1), 5617. doi:10.1038/s41467-023-41220-x.

Luo, J., Frank, D., Peterson, A. L., Nijagal, B., & Arcot, J. (2025). Metabolomic characterisation and flavour profiles of prawn, scallop, squid, barramundi, salmon, snapper, and tuna. *Food Chemistry: X*, *26*, 102284. doi:10.1016/j.fochx.2025.102284.

Ma, C., Xu, C., Zhang, T., Mu, Q., Lv, J., Xing, Q., Yang, Z., Xu, Z., Guan, Y., Chen, C., Ni, K., Dai, X., Ding, W., Hu, J., Bao, Z., Wang, S., & Liu, P. (2024a). Tracking the hologenome dynamics in aquatic invertebrates by the holo-2bRAD approach. *Communications Biology*, *7*(1), 827. doi:10.1038/s42003-024-06509-7.

Ma, J., Yu, M. K., Fong, S., Ono, K., Sage, E., Demchak, B., Sharan, R., & Ideker, T. (2018). Using deep learning to model the hierarchical structure and function of a cell. *Nature Methods*, *15*(4), 290–298. doi:10.1038/nmeth.4627.

Ma, W., Bi, X., Jiang, H., Wei, Z., & Zhang, S. (2024b). Annotating protein functions via fusing multiple biological modalities. *Communications Biology*, *7*(1), 1705. doi:10.1038/s42003-024-07411-y.

Mannina, L., Sobolev, A. P., Capitani, D., Iaffaldano, N., Rosato, M. P., Ragni, P., Reale, A., Sorrentino, E., D'Amico, I., & Coppola, R. (2008). NMR metabolic profiling of organic and aqueous sea bass extracts: Implications in the discrimination of wild and cultured sea bass. *Talanta*, *77*(1), 433–444. doi:10.1016/j.talanta.2008.07.006.

Meng, J., & Wang, W.-X. (2022). Highly sensitive and specific responses of oyster hemocytes to copper exposure: Single-cell transcriptomic analysis of different cell populations. *Environmental Science & Technology*, *56*(4), 2497–2510. doi:10.1021/acs.est.1c07510.

Merritt, C. R., Ong, G. T., Church, S. E., Barker, K., Danaher, P., Geiss, G., … Beechem, D. L. (2020). Multiplex digital spatial profiling of proteins and RNA in fixed tissue. *Nature Biotechnology*, *38*(5), 586–599. doi:10.1038/s41587-020-0472-9.

Milan, M., Coppe, A., Reinhardt, R., Cancela, L. M., Leite, R. B., Saavedra, C., … Bargelloni, L. (2011). Transcriptome sequencing and microarray development for the *Manila clam, Ruditapes philippinarum*: Genomic tools for environmental monitoring. *BMC Genomics*, *12*, 234. doi:10.1186/1471-2164-12-234.

Mokrani, A., Li, J., Li, Q., & Liu, S. (2025). Toward understanding mechanistic regulation of body size and growth control in bivalve mollusks. *Reviews in Aquaculture, 17*(1), e12962. doi:10.1111/raq.12962.

Mun, S., Kim, Y. J., Markkandan, K., Shin, W., Oh, S., Woo, J., … Han, K. (2017). The whole-genome and transcriptome of the Manila clam (*Ruditapes philippinarum*). *Genome Biology and Evolution, 9*(6), 1487–1498. doi:10.1093/gbe/evx096.

Nicholson, J. K., Lindon, J. C., & Holmes, E. (1999). Metabonomics': Understanding the metabolic responses of living systems to pathophysiological stimuli via multivariate statistical analysis of biological NMR spectroscopic data. *Xenobiotica, 29*(11), 1181–1189. doi:10.1080/004982599238047.

Nilsson, A., Peters, J. M., Meimetis, N., Bryson, B., & Lauffenburger, D. A. (2022). Artificial neural networks enable genome-scale simulations of intracellular signaling. *Nature Communications, 13*(1), 3069. doi:10.1038/s41467-022-30684-y.

Novakovsky, G., Dexter, N., Libbrecht, M. W., Wasserman, W. W., & Mostafavi, S. (2023). Obtaining genetics insights from deep learning via explainable artificial intelligence. *Nature Reviews Genetics, 24*, 125–137 2. doi:10.1038/s41576-022-00532-2.

Paillard, C., Gueguen, Y., Wegner, K. M., Bass, D., Pallavicini, A., Vezzulli, L., & Arzul, I. (2022). Recent advances in bivalve-microbiota interactions for disease prevention in aquaculture. *Current Opinion in Biotechnology, 73*, 225–232. doi:10.1016/j.copbio.2021.07.026.

Pathak, A., Marquez, M., Stothard, P., Chukwujindu, C., Su, J.-Q., Zhou, Y., Zhou, X.-Y., Jagoe, C. H., & Chauhan, A. (2024). A seasonal study on the microbiomes of diploid vs. triploid eastern oysters and their denitrification potential. *iScience, 27*(7), 110193. doi:10.1016/j.isci.2024.110193.

Pazos Obregón, F., Silvera, D., Soto, P., Yankilevich, P., Guerberoff, G., & Cantera, R. (2022). Gene function prediction in five model eukaryotes exclusively based on gene relative location through machine learning. *Scientific Reports, 12*(1), 11655. doi:10.1038/s41598-022-15329-w.

Pennance, T., Calvelo, J., Tennessen, J. A., Burd, R., Cayton, J., Bollmann, S. R., … Steinauer, M. L. (2024). The genome and transcriptome of the snail *Biomphalaria sudanica s.l.*: Immune gene diversification and highly polymorphic genomic regions in an important *African vector* of *Schistosoma mansoni*. *BMC Genomics, 25*(1), 192. doi:10.1186/s12864-024-10103-w.

Pérez-Cobas, A. E., Gomez-Valero, L., & Buchrieser, C. (2020). Metagenomic approaches in microbial ecology: An update on whole-genome and marker gene sequencing analyses. *Microbial Genomics, 6*(8), 1–22. doi:10.1099/mgen.0.000409.

Pérez-López, C., Rodríguez-Mozaz, S., Serra-Compte, A., Alvarez-Muñoz, D., Ginebreda, A., Barceló, D., & Tauler, R. (2023). Effects of sulfamethoxazole exposure on mussels (*Mytilus galloprovincialis*) metabolome using retrospective non-target high-resolution mass spectrometry and chemometric tools. *Talanta, 252*, 123804. doi:10.1016/j.talanta.2022.123804.

Perry, W. B., Lindsay, E., Payne, C. J., Brodie, C., & Kazlauskaite, R. (2020). The role of the gut microbiome in sustainable teleost aquaculture. *Proceedings of the Royal Society B: Biological Sciences, 287*(1926), 20200184. doi:10.1098/rspb.2020.0184.

Picard, M., Scott-Boyer, M.-P., Bodein, A., Périn, O., & Droit, A. (2021). Integration strategies of multi-omics data for machine learning analysis. *Computational and Structural Biotechnology Journal, 19*, 3735–3746. doi:10.1016/j.csbj.2021.06.030.

Piovani, L., Leite, D. J., Guerra, L. A. Y., Simpson, F., Musser, J. M., Salvador-Martínez, I., … Telford, M. J. (2023). Single-cell atlases of two lophotrochozoan larvae highlight their complex evolutionary histories. *Science Advances, 9*(31), eadg6034. doi:10.1126/sciadv.adg6034.

Piovani, L., & Marlétaz, F. (2023). Single-cell transcriptomics refuels the exploration of spiralian biology. *Briefings in Functional Genomics, 22*(6), 517–524. doi:10.1093/bfgp/elad038.

Qi, H., Cong, R., Wang, Y., Li, L., & Zhang, G. (2023). Construction and analysis of the chromosome-level haplotype-resolved genomes of two Crassostrea oyster congeners: *Crassostrea angulata* and *Crassostrea gigas*. *GigaScience, 12*, giad077. doi:10.1093/gigascience/giad077.

Sadamoto, H., Takahashi, H., Okada, T., Kenmoku, H., Toyota, M., & Asakawa, Y. (2012). De novo sequencing and transcriptome analysis of the central nervous system of mollusc *Lymnaea stagnalis* by deep RNA sequencing. *PLoS ONE, 7*(8), e42546. doi:10.1371/journal.pone.0042546.

Salamanca-Díaz, D. A., Ritschard, E. A., Schmidbaur, H., & Wanninger, A. (2022a). Comparative single-cell transcriptomics reveals novel genes involved in bivalve embryonic shell formation and questions ontogenetic homology of molluscan shell types. *Frontiers in Cell and Developmental Biology, 10*, 883755. doi:10.3389/fcell.2022.883755.

Salamanca-Díaz, D. A., Schulreich, S. M., Cole, A. G., & Wanninger, A. (2022b). Single-cell RNA sequencing atlas from a bivalve larva enhances classical cell lineage studies. *Frontiers in Ecology and Evolution, 9*, 783984. doi:10.3389/fevo.2021.783984.

Scanes, E., Parker, L. M., Seymour, J. R., Siboni, N., King, W. L., Danckert, N. P., … Ross, P. M. (2021). Climate change alters the haemolymph microbiome of oysters. *Marine Pollution Bulletin, 164*, 111911. doi:10.1016/j.marpolbul.2021.111991.

Schena, M., Shalon, D., Davis, R. W., & Brown, P. O. (1995). Quantitative monitoring of gene expression patterns with a complementary DNA microarray. *Science, 270*(5235), 467–470. doi:10.1126/science.270.5235.467.

Shu, Y., Li, Y., Wei, H., Liu, L., Liu, T., Zhang, L., … Zhang, L. (2025). Single-cell RNA sequencing reveals cellular heterogeneity and gonadal niche interactions in the hermaphroditic scallop *Argopecten irradians*. *Water Biology and Security, 5*(1), 100405. doi:10.1016/j.watbs.2025.100405.

Simakov, O., Marletaz, F., Cho, S. J., Edsinger-Gonzales, E., Havlak, P., Hellsten, U., … Rokhsar, D. S. (2013). Insights into bilaterian evolution from three spiralian genomes. *Nature, 493*(7433), 526–531. doi:10.1038/nature11696.

Song, K., Wen, S., & Zhang, G. (2019). Adaptive evolution patterns in the Pacific oyster *Crassostrea gigas*. *Marine Biotechnology, 21*(5), 614–622. doi:10.1007/s10126-019-09906-w.

Styfhals, R., Zolotarov, G., Hulselmans, G., Spanier, K. I., Poovathingal, S., Elagoz, A. M., … Seuntjens, E. (2022). Cell type diversity in a developing octopus brain. *Nature communications, 13*(1), 7392. doi:10.1101/2022.01.24.477459.

Subramanian, I., Verma, S., Kumar, S., Jere, A., & Anamika, K. (2020). Multi-omics data integration, interpretation, and its application. *Bioinformatics and Biology Insights, 14*, 1177932219899051. doi:10.1177/1177932219899051.

Sun, J., Zhang, Y., Xu, T., Zhang, Y., Mu, H., Zhang, Y., … Qian, P. Y. (2017). Adaptation to deep-sea chemosynthetic environments as revealed by mussel genomes. *Nature Ecology and Evolution, 1*(5), 0121. doi:10.1038/s41559-017-0121.

Sun, X., Li, L., Wu, B., Ge, J., Zheng, Y., Yu, T., Zhou, L., Zhang, T., Yang, A., & Liu, Z. (2021a). Cell type diversity in scallop adductor muscles revealed by single-cell RNA-Seq. *Genomics, 113*(6), 3582–3598. doi:10.1016/j.ygeno.2021.08.015.

Sun, X., Tu, K., Li, L., Wu, B., Wu, L., Liu, Z., Zhou, L., Tian, J., & Yang, A. (2021b). Integrated transcriptome and metabolome analysis reveals molecular responses of the clams to acute hypoxia. *Marine Environmental Research, 168*, 105317. doi:10.1016/j.marenvres.2021.105317.

Sun, X., Zhang, T., Li, L., Tu, K., Yu, T., Wu, B., Zhou, L., Tian, J., & Liu, Z. (2022a). MicroRNA expression signature in the striated and smooth adductor muscles of Yesso scallop *Patinopecten yessoensis*. *Genomics, 114*(4), 110409. doi:10.1016/j.ygeno.2022.110409.

Sun, Z., Huang, S., Zhu, P., Tzehau, L., Zhao, H., Lv, J., Zhang, R., Zhou, L., Niu, Q., Wang, X., Zhang, M., Jing, G., Bao, Z., Liu, J., Wang, S., & Xu, J. (2022b). Species-resolved sequencing of low-biomass or degraded microbiomes using 2bRAD-M. *Genome Biology, 23*(1), 36. doi:10.1186/s13059-021-02576-9.

Sun, Z., Liu, J., Zhang, M., Wang, T., Huang, S., Weiss, S. T., & Liu, Y. Y. (2023). Removal of false positives in metagenomics-based taxonomy profiling via targeting Type IIB restriction sites. *Nature Communications, 14*(1), 5321. doi:10.1038/s41467-023-41099-8.

Takeuchi, T. (2017). Molluscan genomics: Implications for biology and aquaculture. *Current Molecular Biology Reports, 3*(4), 297–305. doi:10.1007/s40610-017-0077-3.

Takeuchi, T., Kawashima, T., Koyanagi, R., Gyoja, F., Tanaka, M., Ikuta, T., … Satoh, N. (2012). Draft genome of the pearl oyster *Pinctada fucata*: A platform for understanding bivalve biology. *DNA Research, 19*(2), 117–130. doi:10.1093/dnares/dss005.

Takeuchi, T., Suzuki, Y., Watabe, S., Nagai, K., Masaoka, T., Fujie, M., … Myers, E. W (2022). A high-quality, haplotype-phased genome reconstruction reveals unexpected haplotype diversity in a pearl oyster. *DNA Research, 29*(6), dsac035. doi:10.1093/dnares/dsac035.

Tan, Y., Li, Y., Ren, L., Fu, H., Li, Q., & Liu, S. (2024). Integrative proteome and metabolome analyses reveal molecular basis underlying growth and nutrient composition in the Pacific oyster, *Crassostrea gigas*. *Journal of Proteomics, 290*, 105021. doi:10.1016/j.jprot.2023.105021.

The ENCODE Project Consortium. (2012). An integrated encyclopedia of DNA elements in the human genome. *Nature, 489*(7414), 57–74. doi:10.1038/nature11247.

Wang, H., Yu, H., & Li, Q. (2024). Integrative analysis of single-nucleus RNA-seq and bulk RNA-seq reveals germline cells development dynamics and niches in the Pacific oyster gonad. *iScience, 27*(4), 109499. doi:10.1016/j.isci.2024.109499.

Wang, J., Zhang, L., Lian, S., Qin, Z., Zhu, X., Dai, X., … Wang, S. (2020). Evolutionary transcriptomics of metazoan biphasic life cycle supports a single intercalation origin of metazoan larvae. *Nature Ecology & Evolution, 4*(5), 725–736. doi:10.1038/s41559-020-1138-1.

Wang, K., Hu, H., Tian, Y., Li, J., Scheben, A., Zhang, C., … Kang, X. (2021). The Chicken pan-genome reveals gene content variation and a promoter region deletion in IGF2BP1 affecting body size. *Molecular Biology and Evolution, 38*(11), 5066–5081. doi:10.1093/molbev/msab231.

Wang, S., Zhang, J., Jiao, W., Li, J., Xun, X., Sun, Y., … Bao, Z. (2017). Scallop genome provides insights into evolution of bilaterian karyotype and development. *Nature Ecology & Evolution, 1*(5), 0120. doi:10.1038/s41559-017-0120.

Wang, Y., Wang, Z., Yu, X., Wang, X., Song, J., Yu, D. J., & Ge, F. (2025). MORE: A multi-omics data-driven hypergraph integration network for biomedical data classification and biomarker identification. *Briefings in Bioinformatics, 26*(1), bbae658. doi:10.1093/bib/bbae658.

Wang, Z., Gerstein, M., & Snyder, M. (2009). RNA-Seq: A revolutionary tool for transcriptomics. *Nature Reviews Genetics, 10*(1), 57–63. doi:10.1038/nrg2484.

Wei, X., Fu, S., Li, H., Liu, Y., Wang, S., Feng, W., … Gu, Y. (2022). Single-cell stereo-seq reveals induced progenitor cells involved in axolotl brain regeneration. *Science, 377*(6610), eabp9444. doi:10.1126/science.abp9444.

Wilkins, L. G. E., Leray, M., O'Dea, A., Yuen, B., Peixoto, R. S., Pereira, T. J., … Eisen, J. A. (2019). Host-associated microbiomes drive structure and function of marine ecosystems. *PLOS Biology, 17*(11), e3000533. doi:10.1371/journal.pbio.3000533.

Xin, L., Guan, D., Wei, N., Zhang, X., Deng, W., Li, X., & Song, J. (2024). Genomic analysis reveals novel genes and adaptive mechanisms for artificial diet utilization in the silkworm strain Guican No.5. *Insects, 15*(12), 1010. doi:10.3390/insects15121010.

Xu, F., Deng, S., Gavriouchkina, D., & Zhang, G. (2023). Transcriptional regulation analysis reveals the complexity of metamorphosis in the Pacific oyster (*Crassostrea gigas*). *Marine Life Science & Technology, 5*(4), 467–477. doi:10.1007/s42995-023-00204-y.

Yang, Z., Zhang, L., Hu, J., Wang, J., Bao, Z., & Shi, W. (2020). The evo-devo of molluscs: Insights from a genomic perspective. *Evolution & Development, 22*(6), 409–424. doi:10.1111/ede.12336.

Young, T., & Alfaro, A. C. (2018). Metabolomic strategies for aquaculture research: A primer. *Reviews in Aquaculture, 10*(1), 26–56. doi:10.1111/raq.12146.

Yu, H., Li, Y., Han, W., Bao, L., Liu, F., Ma, Y., Pu, Z., Zeng, Q., Zhang, L., Bao, Z., & Shi, W. (2024). Pan-evolutionary and regulatory genome architecture delineated by an integrated macro- and microsynteny approach. *Nature Protocols, 19*(6), 1623–1678. doi:10.1038/s41596-024-00966-4.

Zhang, G., Fang, X., Guo, X., Li, L., Luo, R., Xu, F., … Wang, J. (2012). The oyster genome reveals stress adaptation and complexity of shell formation. *Nature, 490*(7418), 49–54. doi:10.1038/nature11413.

Zhang, G., Li, L., Meng, J., Qi, H., Qu, T., Xu, F., & Zhang, L. (2016). Molecular basis for adaptation of oysters to stressful marine intertidal environments. *Annual Review of Animal Biosciences, 4*, 357–381. doi:10.1146/annurev-animal-022114-110903.

Zhang, X., Niu, Y., Gao, C., Kong, L., Yang, Z., Chang, L., Kong, X., Bao, Z., & Hu, X. (2024a). Somatostatin receptor gene functions in growth regulation in bivalve scallop and clam. *International Journal of Molecular Sciences, 25*(9), 4813. doi:10.3390/ijms25094813.

Zhang, Y., Mao, F., Mu, H., Huang, M., Bao, Y., Wang, L., … Yu, Z. (2021). The genome of Nautilus pompilius illuminates eye evolution and biomineralization. *Nature Ecology & Evolution, 5*(7), 927–938. doi:10.1038/s41559-021-01448-6.

Zhang, Y., Wang, H., Liu, S., Kong, X., Chang, L., Zhao, L., Bao, Z., & Hu, X. (2024b). Multi-tissue metabolomic profiling reveals the crucial metabolites and pathways associated with scallop growth. *BMC Genomics, 25*(1), 1091. doi:10.1186/s12864-024-11016-4.

Zhang, Z., Chen, S., Chen, A., Xu, Y., Zhang, Y., Yu, W., … Wu, Y. (2023). Comparison of the flavor qualities between two varieties of *Mercenaria mercenaria*. *Scientific Reports, 13*(1), 13047. doi:10.1038/s41598-023-39757-4.

Zhang, Z., Yang, Q., Liu, H., Jin, J., Yang, Y., Zhu, X., … Xie, S. (2025). Potential functions of the gut microbiome and modulation strategies for improving aquatic animal growth. *Reviews in Aquaculture, 17*(1), e12959. doi:10.1111/raq.12959.

Zhao, R., Takeuchi, T., Luo, Y. J., Ishikawa, A., Kobayashi, T., Koyanagi, R., Villar-Briones, A., Yamada, L., Sawada, H., Iwanaga, S., Nagai, K., Satoh, N., & Endo, K. (2018). Dual gene repertoires for larval and adult shells reveal molecules essential for molluscan shell formation. *Molecular Biology and Evolution*, *35*(11), 2751–2761. doi:10.1093/molbev/msy172.

Zheng, G. X. Y., Terry, J. M., Belgrader, P., Ryvkin, P., Bent, Z. W., Wilson, R., . . . Bharadwaj, R. (2017). Massively parallel digital transcriptional profiling of single cells. *Nature Communications*, *8*, 14049. doi:10.1038/ncomms14049.

CHAPTER

4

Genotyping and molecular breeding platform

Jia Lv, Qifan Zeng, Pingping Liu, Zhenmin Bao and Shi Wang

College of Marine Life Sciences, Ocean University of China, Qingdao, China

4.1 Introduction

4.1.1 The concept of molecular markers

Molecular or DNA markers are indispensable tools in modern genetics and genomics, defined as specific DNA sequences that enable the precise identification of particular genomic regions (Terzin et al., 2024; Wenne, 2023b). Acting as landmarks or signposts, they allow researchers to trace the inheritance of genes and genetic traits across generations. Molecular markers provide several notable advantages. Since they are DNA-based, they remain stable and are not affected by tissue type, developmental stage, season, or environmental conditions, nor do they suffer from issues related to gene expression. They are numerous, distributed across the entire genome, and exhibit high polymorphism, with extensive allelic variation in natural populations. Many are codominant, allowing the differentiation of homozygous and heterozygous genotypes, which is especially useful in the selection of recessive agronomic traits (Bhatia & Bajwa, 2022; Khatei et al., 2022). Importantly, they are generally neutral, exerting no influence on trait expression and avoiding undesirable genetic linkages. In addition, molecular markers can be detected through relatively simple and rapid methods, often compatible with high-throughput analysis, making them powerful and versatile tools for genetic research and breeding.

4.1.2 Brief history and development in genotyping

The development of genotyping has undergone substantial evolution over the past decades. Early approaches relied primarily on hybridization-based marker systems, such as restriction fragment length polymorphism (RFLP), which provided valuable insights into genetic variation but were constrained by limited scale and efficiency (Ota et al., 2007). The advent of polymerase chain reaction (PCR) established the technical foundation for amplification-based marker

Genetic Breeding and Aquaculture of Shellfish: A Collaborative Approach to a Sustainable Future
DOI: https://doi.org/10.1016/B978-0-443-33606-5.00004-7

systems, including randomly amplified polymorphic DNA (RAPD), amplified fragment length polymorphism (AFLP), and simple sequence repeat (SSR). PCR-based markers generally employ primer-dependent amplification and/or DNA hybridization followed by electrophoresis, with polymorphisms detected through the presence or absence of amplicons or differences in band size and mobility. RAPD amplifies random DNA segments using short, arbitrary primers under low-stringency conditions, producing banding patterns that can be compared across individuals. They require only small amounts of DNA and are relatively easy to generate, making them attractive for preliminary applications, such as phylogenetic and taxonomic studies, genetic diversity assessment, and species identification. However, their dominant inheritance (inability to distinguish homozygotes from heterozygotes) and poor reproducibility across experiments and laboratories have limited their popularity in aquaculture genetics (Liu & Cordes, 2004). AFLP, in contrast, combine restriction digestion with PCR amplification to generate highly reproducible, multilocus genomic fingerprints (Sheeja et al., 2021). This method does not require prior sequence knowledge and is valued for its high sensitivity, reproducibility, and ability to generate numerous polymorphic markers. AFLP was instrumental in aquaculture genetics, particularly in linkage mapping, as many of the first genetic linkage maps for aquatic species were constructed primarily using AFLP markers (Rogers et al., 2001). Despite their technical complexity, AFLP represented a powerful intermediate step between early low-throughput markers and the modern high-resolution systems now in use.

Early molecular markers such as RFLP, RAPD, and AFLP thus played a pivotal role in the initial development of molecular genetics. RFLP provided the first marker framework but were limited by technical complexity; RAPD offered simplicity and speed, yet suffered from poor reproducibility; and AFLP combined reproducibility with high polymorphism but required considerable expertise. Despite their limitations, these earlier systems established the conceptual and methodological foundation for subsequent advances. Building on this groundwork, more refined and versatile markers such as SSR and single-nucleotide polymorphisms (SNP) have emerged, offering higher reproducibility, efficiency, and scalability (Gopikrishna, 2023; Maqsood & Ahmad, 2017). Today, SSR and SNP dominate aquaculture genetics research, illustrating how the evolution from earlier markers to modern platforms has progressively advanced the field and enabled powerful tools for genetic improvement. In the following section, we will focus on these widely used marker systems, providing an overview of their principles, characteristics, and applications in molecular breeding and aquaculture genetics.

4.2 The types and characteristics of modern genetic markers

4.2.1 Simple sequence repeats

SSRs, also known as microsatellites, are highly polymorphic DNA markers characterized by short, repeated sequences of 1–6 base pairs. These sequences are widely distributed across the genome and exhibit significant variability in the number of repeat units among individuals, making them highly informative for genetic studies (Duran et al., 2009; Guichoux et al., 2011). Compared to other molecular markers, SSR markers are amenable to relatively high-throughput detection; for example, microsatellite loci can be genotyped by analyzing amplicon length using capillary electrophoresis (Vieira et al., 2016). Their high level of polymorphism, reproducibility, and codominant inheritance has made SSR particularly valuable for distinguishing between homozygous and heterozygous individuals. The markers have become widely used in molluscan species for diverse applications, including species and varieties identification, parentage analyses, and population genetics studies, as demonstrated in *Mizuhopecten yessoensis* (Khamatova et al., 2025; Xu et al., 2008), *Crassostrea gigas* (Kochmann et al., 2012; Liang et al., 2023), *Ruditapes philippinarum* (Tan et al., 2020, Tan et al., 2024), and *Meretrix petechialis* (Xu et al., 2020). Although newer marker systems such as SNP are increasingly favored for large-scale studies, SSR continue to be indispensable in many applied and experimental contexts (Wenne, 2023a; Yuan et al., 2021). For example, SSR was identified from the draft genomes of eight *Lamprotula* species, enabling the development of molecular markers for species identification and advancing population genetics research in this genus (Jiang et al., 2025).

4.2.2 Single-nucleotide polymorphisms

SNP represent the most prevalent form of genetic variation, consisting of single-base pair changes in the DNA sequence. Because of their abundant presence and relatively uniform distribution across the genome, SNP have become a cornerstone of modern genotyping technologies, particularly in aquaculture (Hollenbeck & Johnston, 2018; Seeb et al., 2011; Yu et al., 2025). They offer significant advantages by providing numerous markers dispersed evenly throughout the genome, facilitating high-resolution genetic analysis. The automation of SNP detection, combined with its integration into high-throughput genotyping platforms, has significantly advanced the ability to conduct large-scale genetic studies involving thousands of markers and individuals (Kockum et al., 2023; Robledo et al., 2018). This is particularly beneficial for aquaculture species with complex genomes, where traditional markers might fall short in providing comprehensive genetic insight. The discovery of SNP can be achieved through a variety of techniques, ranging from smaller-scale methods such as high-resolution melting (HRM), PCR-based assays, and TaqMan genotyping, to more advanced, high-throughput sequencing technologies, such as reduced-representation sequencing (RRS) and target genotyping methods (Kockum et al., 2023; Lawrie & Massey, 2023). HRM was once considered a highly effective technique for rapid and accurate profiling of genetic variation within PCR amplicons. The first large-scale gene-associated marker development and evaluation for the bivalve mollusk, *Chlamys farreri*, was performed using this high-resolution melting genotyping platform(Jiao et al., 2014). Nowadays, next-generation sequencing-based methods, which have become more affordable and precise in recent years, are enabling the rapid identification of SNP variants across large populations. As sequencing costs continue to decline, these technologies are progressively replacing older nonsequencing–based genotyping approaches, allowing for deeper and more expansive genetic analysis. In mollusks, SNP is predominantly utilized in breeding programs for a wide array of applications, including the construction of genetic linkage maps, the identification of quantitative trait loci (QTL) associated with key traits such as growth, disease resistance, and survival, and for genomic selection (GS) programs (Robledo et al., 2018). The advent of SNP-based tools has revolutionized the precision and efficiency of aquaculture genetics, paving the way for more sustainable and profitable breeding strategies.

4.2.3 Structural variants and copy number variants

Structural variant (SV) is now recognized as major components of genetic polymorphism, involving genomic segments larger than 50 bp where the structural arrangement differs between genomes. These include deletions, duplications, insertions, inversions, and translocations (Sudmant et al., 2015). Unlike SNP, which typically involve single-base pair changes, SV can lead to extensive modifications in cisregulatory regions, thereby exerting more pronounced effects on gene expression and phenotypic traits. They can also alter gene dosage by modifying the copy number of specific genes. Structural variant identification and comparative genomic analyses in oysters revealed widespread haplotype divergent sequences (HDSs), which are largely driven by transposable elements and contribute to genetic innovation and variation in gene expression (Liu et al., 2025). Despite their importance, the reliable detection of SVs remains technically challenging, particularly when using short-read sequencing technologies. Recent advances in long-read sequencing platforms, such as Oxford nanopore technologies (ONT), now provide powerful opportunities to characterize SVs at the population level, offering new insights into their role in aquaculture genetics (Smolka et al., 2024).

Copy number variant (CNV) represent a specific subset of SV that involve changes in the number of copies of a given genomic region (Ho et al., 2020). Although less frequent than SNP, their larger size often results in substantial effects on gene expression, with potential contributions to phenotypic diversity, disease susceptibility, and adaptive evolution. In aquaculture species, CNV has recently gained attention due to their relevance for economically important traits such as growth, stress tolerance, and disease resistance. CNV is prevalent in oysters and plays an important role in shaping genomic variation and adaptive potential (Modak et al., 2021). Future research directions will likely focus on elucidating the functional consequences of CNV, developing more accurate and cost-effective detection platforms, and incorporating CNV into genomics-assisted breeding programs.

4.2.4 Mitochondrial markers

Mitochondrial markers are derived from mitochondrial DNA (mtDNA), a small circular genome that is maternally inherited and independent of the nuclear genome. MtDNA is a small, circular molecule that encodes a few essential genes involved in energy production, as well as the ribosomal and transfer RNAs needed for mitochondrial protein synthesis (West & Shadel, 2017). Because of its high mutation rate and lack of recombination, mtDNA is highly useful for studying maternal lineages, phylogenetics, and population genetics. The relatively small size and high copy number of mtDNA within cells make it easier to extract and sequence, even from degraded samples. Nevertheless, mitochondrial markers provide only a maternal perspective of genetic inheritance and represent a small fraction of the total genetic information of an individual. Furthermore, their nonrecombining nature restricts their ability to capture the genetic reshuffling that occurs in the nuclear genome. Despite these limitations, mitochondrial markers remain widely used in molluscan genetics and evolutionary studies, offering valuable insights into species identification, evolutionary history across diverse molluscan species, such as mussels, (Lee et al., 2021; Noorullah et al., 2023).

4.3 High-throughput genotyping technologies

4.3.1 Whole-genome sequencing

Genome analysis relies heavily on whole-genome sequencing (WGS), which has become the most comprehensive approach for characterizing genetic variation. Unlike reduced-representation or targeted genotyping methods, WGS provides an unbiased survey of the entire genome, allowing simultaneous detection of SNP, insertion and deletion (InDel), SV, CNV, and repetitive elements (Wang et al., 2017; Xu et al., 2024). Sequencing reads are generally aligned to reference genomes using algorithms such as BWA-MEM, followed by variant discovery with tools such as GATK, SAMtools, or deep learning–based approaches such as DeepVariant. By sequencing each nucleotide position multiple times, typically at coverage depths of 30 × or higher, WGS ensures that each base is repeatedly sampled, thereby improving variant detection accuracy and reducing genotyping errors. High-quality reference genomes assembled for many molluscan species, which enabled the identification of SNP loci underlying traits of economic importance (Zheng et al., 2023), population differentiation among wild and cultured varieties (Lv et al., 2022; Wang et al., 2021), and divergence and adaptive potential of the mollusks (Li et al., 2018b). Though WGS can obtain the highest marker density with comprehensive and reliable genomic information, it is still largely cost-prohibitive for large sample sizes. RRS is an attractive genotyping tool since it sequences only a small fraction of the genome, thus reducing the cost of the analysis with respect to WGS.

4.3.2 Reduced-representation genome sequencing

RRS has emerged as a transformative approach in aquaculture genomics by significantly reducing both the cost and complexity of sequencing while still providing sufficient information for genome-wide analyses (Robledo et al., 2018). Unlike WGS, which generates complete genomic data but remains costly and computationally intensive for many aquaculture species with large or complex genomes, RRS achieves an optimal balance between cost and resolution by selectively sequencing only a representative fraction of the genome. Over the past decade, several cost-effective RRS methods have been developed, including restriction-site associated DNA sequencing (RAD-seq) (Baird et al., 2008), genotyping-by-sequencing (GBS) (Elshire et al., 2011), and the more recent 2b-RAD (Wang et al., 2012; Wang et al., 2016). Collectively, these approaches have underpinned a wide range of applications in aquatic species, from variant discovery and population genetics to linkage disequilibrium mapping, genome-wide association studies (GWAS), QTL mapping, genomic prediction, and even epigenetic profiling (Crawford & Oleksiak, 2016; Robledo et al., 2018; Wang et al., 2015). A notable advantage of RRS is its suitability for nonmodel aquaculture species, where reference genomes may be incomplete or entirely absent.

All RRS methods follow a broadly similar workflow: (1) genomic DNA fragment with restriction enzymes or sonication, (2) fragment selection and adaptor ligation, (3) library amplification and quality control, (4) high-throughput sequencing and bioinformatic analysis for polymorphism detection and genotyping (Fig. 4.1). Despite these shared steps, RRS approaches differ in several critical respects, including the complexity of library preparation, the proportion of the genome surveyed, sequencing depth requirements, per-sample cost, and the accuracy of genotyping. These methodological differences largely determine the suitability of each approach for specific research goals. For example, RAD-seq has been widely employed in population-level studies, GBS has become a preferred strategy for SNP discovery in large-scale sample sets, and 2b-RAD offers distinct advantages by generating uniform isolength fragments, thereby extending its utility beyond genotyping to multiomics applications such as epigenetic profiling and metagenomic analysis.

4.3.2.1 Restriction-site associated DNA

Since its introduction in 2008, restriction-site associated DNA sequencing (RAD-seq) has emerged as one of the most promising and widely adopted RRS methods in aquaculture genetics. By integrating restriction enzyme digestion with NGS, RAD-seq enables rapid discovery and genotyping of thousands of SNPs across the genome, even in species lacking

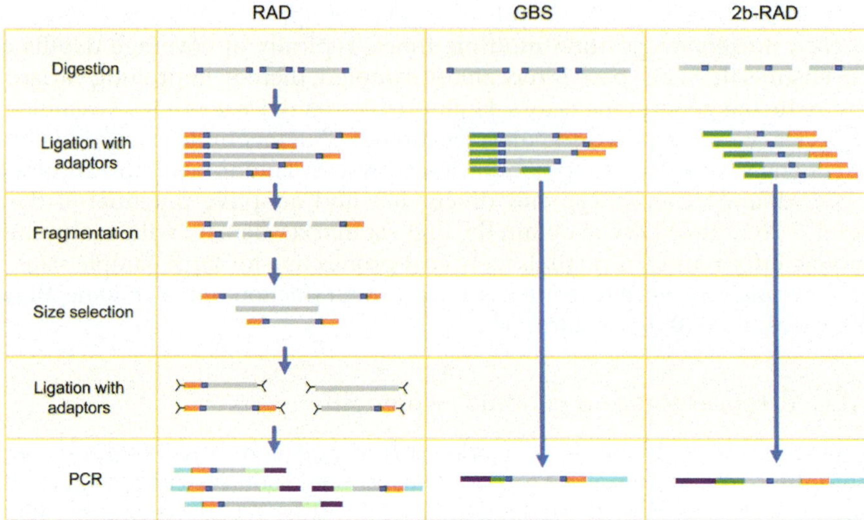

FIGURE 4.1 General workflow of reduced-representation genome sequencing.

a reference genome. This technology has substantially accelerated research on population genetics (Bucholz et al., 2022; Sekino et al., 2023), high-density linkage map construction and QTL mapping (Li et al., 2014), thereby providing novel insights into the genetic basis of economically important traits in mollusks such as scallops and oysters, as well as in crustaceans and other aquaculture species (Robledo et al., 2018). The standard RAD-seq workflow involves several key steps. First, genomic DNA is digested with a restriction enzyme, and a P1 adapter is ligated to the digested fragments. This adapter is complementary to the restriction site, incorporates a unique barcode for sample identification, and provides primer binding sites for subsequent amplification and sequencing. The fragments are then mechanically sheared, followed by size selection to ensure compatibility with sequencing requirements. A second adapter (P2) is ligated to the selected fragments, after which PCR amplification with P1 and P2 primers selectively enriches DNA fragments containing both adapters. Despite its efficiency, this protocol introduces certain technical challenges. The requirement for DNA shearing and precise size selection increases laboratory complexity, can lead to uneven genomic coverage, and may introduce strand bias, which in turn affects downstream data interpretation (Andrews et al., 2016; Robledo et al., 2018). To overcome these limitations, several modified RAD-based methods have been developed, including double-digest RAD (ddRAD), 2b-RAD, and ezRAD. These approaches simplify the workflow by eliminating the need for random shearing, reducing size-selection bias, and increasing reproducibility across laboratories. As a result, they have broadened the applicability of RAD-seq in aquaculture genetics, making it more robust for large-scale population studies, GWAS, and GS.

4.3.2.2 Genotyping-by-sequencing

GBS represents a streamlined alternative to RAD-seq, primarily because of the elimination of the DNA shearing and size-selection steps, which greatly simplifies library construction and reduces technical variability. In a typical GBS protocol, a two-enzyme digestion system is employed, often incorporating methylation-sensitive restriction enzymes. The latter are particularly effective in reducing repetitive sequences, thereby increasing the proportion of informative reads and improving the efficiency of downstream SNP discovery (Elshire et al., 2011). GBS has been widely applied across a wide range of organisms, but its adoption has been especially extensive in plant genomics, where it has become a core method for genetic diversity assessment, domestication research, germplasm characterization, and trait improvement (Kim et al., 2016). Nevertheless, one of the main challenges of GBS lies in its relatively high proportion of missing data compared to other genotyping approaches. This limitation arises because the efficiency of restriction enzyme digestion is not uniform across all individuals or genomic contexts, which may result in incomplete representation of certain genomic regions. As a consequence, datasets generated through GBS often contain substantial amounts of missing genotypes, potentially affecting downstream analyses such as GWAS or genomic prediction. GBS has been primarily utilized in various oyster species (Silliman, 2019; White et al., 2017; Yin et al., 2021) and seldom utilized in other mollusks. To address this issue, various imputation strategies have been developed and are routinely employed to infer missing genotypes, thereby improving data completeness and analytical power (Howie et al., 2009; Li et al., 2010).

4.3.2.3 2b-RAD

The 2b-RAD method represents a significant innovation in RRS. Unlike RAD-seq and GBS, which generate fragments of variable length, 2b-RAD employs type IIB restriction enzymes that cut at fixed distances on both sides of their recognition sites, producing uniform 33 bp tags across the genome (Wang et al., 2012). Moreover, it eliminates the need for fragment selection, enhances reproducibility, and circumvents the strand bias introduced by random shearing in RAD-seq, as well as simplifying library construction (Puritz et al., 2014). A distinctive feature of 2b-RAD is its ability to adjust the density of genomic tags by adding selective bases to the adapters, allowing researchers to develop SNP markers tailored to their research goals and needs, making it more cost-effective. Nowadays, 2b-RAD has been widely applied not only to nonmodel organisms but also to hundreds of animal and plant species. It is commonly used in wide range of applications for molluscks, including linkage map construction and QTL identification (Jiao, 2014; Mao et al., 2020; Shi et al., 2014), genetic analysis of important traits (Li et al., 2019; Zhao et al., 2017), population genetics (Xu et al., 2018; Yang et al., 2024) and GS (Dou et al., 2016; Wang et al., 2018). The first high-density genetic linkage map for scallops, comprising 3,806 markers, represents one of the densest in mollusk species and was constructed using the 2b-RAD method. Based on these high-density maps, several loci associated with key traits, such as sex and growth, have been identified (Jiao et al., 2014). Beyond its traditional use in genotyping, several methodological variants of 2b-RAD have been developed to extend its applications, including DNA methylation profiling, genome assembly, and microbiome characterization (Fig. 4.2).

MethylRAD is a variant of the 2bRAD method, specifically developed for DNA methylation analysis. By adopting Mrr-like enzymes, which specifically capture methylated DNA fragments, it enables efficient profiling of DNA methylation patterns in both model and non-model species (Wang et al., 2015). The technique can perform *de novo* (reference-free) methylation analysis, making it applicable to species with incomplete or no reference genomes (Matz, 2018). MethylRAD requires extremely low amounts of DNA (as little as 1 ng), making it highly versatile for working with precious or limited samples. Furthermore, the tag density in MethylRAD can be adjusted, offering flexibility to optimize for specific research needs. Currently, it is highly

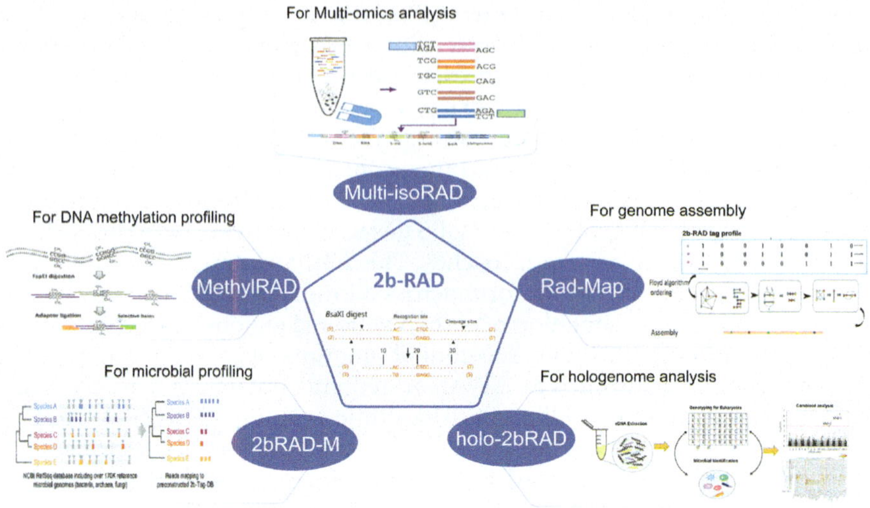

FIGURE 4.2 2b-RAD and its derived method.

regarded for its cost-effectiveness, scalability, and ability to perform reference-free DNA methylation analysis, making it a valuable tool for both model and non-model species in methylation studies (Matz, 2018). For example, Wang et al. (2015) employed methylRAD to identify genes potentially critical for carotenoid accumulation in scallop adductor muscle, demonstrating its capacity to efficiently capture DNA methylation patterns and facilitate the localization of key trait genes. By capturing DNA methylation patterns with high precision, MethylRAD plays a pivotal role in the identification of key trait genes, facilitating the discovery of gene variations linked to specific traits. Nowadays, MethylRAD has been widely applied in genetic research of both plants and animals, encompassing comparative DNA methylation analysis (Li et al., 2018a; Li et al., 2018c), evolutionary studies (Durante et al., 2019; Scheschonk et al., 2024), and even medical research (Li et al., 2020; Zheng et al., 2020).

The multi-isoRAD technique represents an innovative extension of reduced representation sequencing methods to multi-omics analysis (Wang et al., 2016). A key feature of this approach is the concatenation of multiple source tags, which enables the simultaneous detection of various omics layers, such as genotyping and DNA methylation. The configuration of five concatenated tags is highly flexible, allowing users to customize the combination of samples and restriction enzymes according to specific research requirements. This flexibility not only enhances the scope of genomic research but also significantly reduces costs, facilitating the broader application of 2b-RAD technology in larger populations and expanding its potential across diverse research fields.

The emergence of 2bRAD-M, a derivative of the 2b-RAD technique, represents a significant advancement in microbiome analysis, particularly for microbial species identification (Sun et al., 2022). Its key advantage lies in generating consistent, high-resolution taxonomic profiles for bacteria, archaea, and fungi, even from challenging samples with low microbial biomass, high host DNA contamination, or degraded DNA. By sequencing only approximately 1% of the metagenome, 2bRAD-M offers a highly cost-effective and efficient method for species-level profiling. This technique holds considerable promise for expanding our understanding of microbiomes in aquatic environments. Its ability to profile microbial communities in a variety of complex and degraded samples, such as environmental water samples, fish tissue, and aquaculture biofilms, makes it an invaluable tool for studying the microbial ecology of aquatic organisms. Furthermore, the development of integrative techniques such as holo-2bRAD enables the simultaneous acquisition of both host and microbial genomic data from a single sequencing library. For example, Ma et al. (2024) adopted this method to explore the host-microbe associations underlying mass mortality events in bivalve larvae and identified core microbial species that changed with developmental stages and two potentially associated host SNPs . This holistic approach enhances the study of host-microbe interactions, particularly in areas like co-evolution, immune regulation, and disease susceptibility in aquatic species. Overall, 2b-RAD and its derivatives offer a cost-effective, high-resolution, and scalable toolkit for genetic breeding, microbiome research, ecology, and conservation.

4.3.3 Targeted genotyping technology

Targeted genotyping represents a powerful approach for the detection of known genetic variations that are biologically or economically important. Unlike reduced-representation sequencing, which samples random genomic fractions, targeted approaches specifically enrich predefined DNA regions of interest and generate high sequencing coverage for those loci. This design enables accurate and efficient interrogation of candidate variants, particularly for applications such as GS, marker-assisted breeding, and functional validation of trait-associated markers.

Array-based genotyping platforms (e.g., Illumina Infinium and Axiom Affymetrix platforms) have been widely used in human genetics and model organisms. These systems offer high reproducibility and throughput but are generally less accessible for aquaculture species due to the absence of inexpensive standardized arrays and the high development cost of species-specific SNP chips. Nonetheless, some efforts have been made to establish chip-based genotyping platforms in aquaculture, particularly for commercially important fish and shellfish, using Illumina and Affymetrix technologies to generate species-specific SNP panels (Abdelrahman et al., 2017). A 190K SNP array for oysters was developed based on the Affymetrix Axiom genotyping platform, enabling high-throughput and accurate genotyping (Qi et al., 2017). Recent advances in next-generation sequencing have accelerated the development of several sequencing-based target enrichment methods, which can overcome many of the limitations of microarray platforms. Two major categories dominate targeted sequencing: PCR-based and hybrid capture-based approaches.

Hybrid capture-based methods (e.g., NimbleGen and SureSelect) utilize long DNA or RNA probes to hybridize and capture genomic regions of interest after DNA fragmentation. This approach typically allows enrichment of 250 Kb–5 Mb of genomic sequence, supporting genotyping of over 50,000 loci. While powerful for surveying broad genomic regions and discovering novel variants, hybrid capture tends to have modest target specificity (40%–60%) and is less efficient for interrogating small numbers of loci with high precision (Teer et al., 2010). The HD-Marker system was developed as a high-density in-solution probe array for targeted genotyping (Lv et al., 2018). HD-Marker uses locus-specific probes (LSPs) designed to flank target loci, followed by extension, ligation, and amplification to construct libraries for sequencing. This approach enables genotyping of more than 86,000 user-defined markers in a single-tube assay,

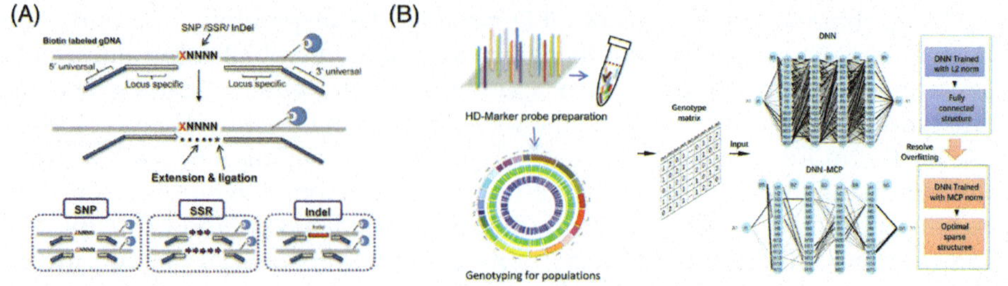

FIGURE 4.3 **Workflow of HD-Marker and its integration with neural networks for breeding applications.** (A) The workflow of HD-Marker technology. (B) The development of the Haishen 24k chip and its integration with neural networks in GS.

combining the specificity of GoldenGate technology with the cost-effectiveness of modern sequencing platforms (Liu et al., 2022). Unlike array-based platforms, HD-Marker does not require expensive instruments, making it suitable for easy-adoption in standard molecular laboratories. Importantly, HD-Marker has already been successfully applied in aquaculture species, including scallop, shrimp, and sea cucumber (Liu et al., 2022; Liu et al., 2025; Lv et al., 2018; Lv et al., 2022) (Fig. 4.3). Ultrahigh-Multiplex HD-Marker enables targeted genotyping of up to 86,000 loci, offering comprehensive coverage of the entire gene repertoire for Yesso scallop (Liu et al., 2022). By leveraging the HD-Marker array for large-scale genotyping data acquisition and applying deep neural networks for genomic prediction, this approach significantly enhances the accuracy of genetic prediction for complex traits, thereby advancing precision breeding (Lv et al., 2022). With its exceptional flexibility, cost-effectiveness, and scalability, HD-Marker represents a promising tool for broad applications in genetics, ecology, and evolutionary studies of nonmodel aquatic organisms.

4.4 Genotype imputation

Low-coverage whole-genome sequencing (lcWGS) has emerged as a cost-effective alternative to high-depth WGS by distributing the sequencing effort across more samples at reduced coverage per individual. Although this strategy sacrifices the genotype call rate and accuracy at the individual level, it enables genome-wide screening across entire populations while retaining enough individual information to draw meaningful conclusions. Importantly, when combined with robust genotype imputation strategies, lcWGS can recover much of the missing information and achieve accuracy levels comparable to high-coverage WGS (Lou et al., 2021). In the following section, we will first discuss the feasibility and conceptual framework of lcWGS, then evaluate how genotype imputation enhances its accuracy and utility, and finally highlight its practical applications in GS and GWAS in scallops and other aquaculture species.

4.4.1 Framework and feasibility of low-coverage whole-genome sequencing

The process of lcWGS begins with WGS of a reference population to identify genetic variants. Subsequently, high-density haplotype panels are constructed based on the reference data. For

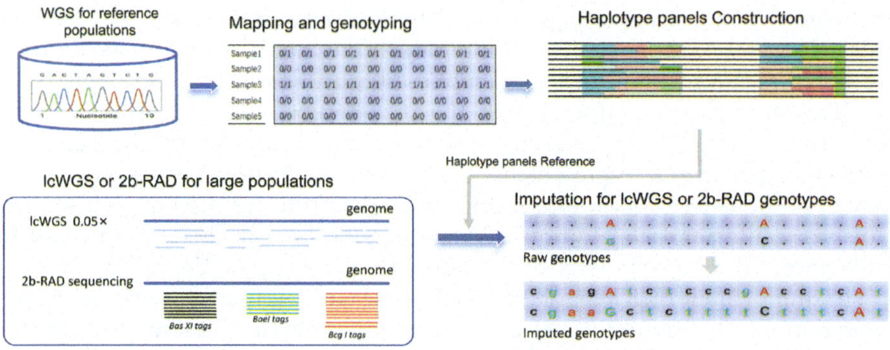

FIGURE 4.4 The framework of lcWGS combined with the genotype imputation. *lcWGS*, Low-coverage whole genome sequencing.

large populations, lcWGS is employed for a broad range of individuals at a reduced cost. Imputation algorithms are subsequently applied to predict missing genotypes within these low-coverage datasets, utilizing the haplotype panels derived from the reference population (Naj, 2019; Sun & Li, 2025). This process enables the accurate recovery of genotypes, positioning lcWGS imputation as a cost-effective and reliable approach for large-scale genetic studies and GS (Fig. 4.4). Although lcWGS is a promising technology, it faces several challenges in DNA extraction, sample preparation, genotype inference, and data analysis. To address these issues, the field has focused on optimizing extraction protocols and constructing high-quality reference panels to improve the accuracy of low-depth sequencing data imputation.

Obtaining sufficient DNA material from living animals in a rapid, cost-effective, and non-lethal manner for lcWGS is a key challenge in implementing GS in aquaculture. Liu et al. (2021) describe an integrated approach for rapid, cost-effective nonlethal gill DNA sampling and genome-scale genotyping in scallops. This nonlethal DNA sampling procedure can be completed within 20 min, bypassing many of the tedious and time-consuming steps involved in the traditional phenol–chloroform method, and does not require using any commercial kits. The method has also demonstrated broad applicability in other bivalve species, such as *Chlamys farreri*, *Mytilus edulis*, and *Ruditapes philippinarum* (Liu et al., 2021). Another bottleneck in lcWGS has traditionally been the preparation of individually indexed libraries for large numbers of samples, which was both labor-intensive and often more expensive than sequencing itself. Strategies such as reducing reaction volumes (Therkildsen and Palumbi, 2017) and employing automated extraction and library construction systems (e.g., the Agilent NGS Workstation) have greatly improved efficiency. With continuing advances in sequencing technologies, these costs are expected to decline further, enhancing the feasibility of lcWGS for population-scale applications in aquaculture genomics.

The quality control and read alignment process in lcWGS closely follows the standard procedures used in WGS. After trimming adapters and removing low-quality bases, the short reads are aligned to a reference genome for SNP calling. In cases where a reference genome is unavailable, reference transcriptome sequences or genomes from closely related species can be used as substitutes. However, using these alternatives may result in incomplete coverage of noncoding regions and species-specific sequences, potentially introducing biases in genotype imputation and downstream population genomic analyses (Therkildsen and Palumbi, 2017). With recent advancements in long-read sequencing, assembly algorithms, and scaffolding techniques, it is now possible to generate chromosome-scale assemblies at much lower costs and with greater speed than before. Therefore a high-quality, species-specific reference genome is preferable for starting a genotyping study using lcWGS, as it ensures more accurate and comprehensive results.

The major challenge in genotyping using lcWGS is that reduced coverage creates uncertainty about whether both alleles of a diploid individual have been sampled during sequencing and whether the base calling at each position is accurate (Nielsen et al., 2011). As a result, genotypes are typically inferred as likelihoods of all possible genotypes rather than as definite copy numbers of the minor allele (e.g., 0, 1, or 2), as is the case with high-coverage sequencing datasets. Currently, genotype likelihoods (GLs) are commonly estimated using probabilistic models implemented in GATK or SAMtools (Li et al., 2009; McKenna et al., 2010). These two models differ in their assumptions about the error distribution of base calling, which can sometimes affect the estimation of rare alleles (Korneliussen et al., 2014). Specialized analytical tools have been developed to account for the uncertainty in genotype calls and are effective in population structure inference. However, certain genetic studies, such as rare variant association analysis and demographic reconstruction, are highly sensitive to low-frequency alleles and require precise genotype calls. Because of the probabilistic nature of LcWGS data and the high missing rates, imputation is used to refine the GLs and fill in the gaps between the sparsely mapped reads by leveraging information from a reference panel of haplotypes. Therefore when applying lcWGS to nonmodel species, constructing a comprehensive reference panel using a subset of high-coverage samples is a promising strategy to improve genotype accuracy through imputation.

4.4.2 Genotype imputation in low-coverage whole-genome sequencing

Genotype imputation is a computational strategy used to infer missing or untyped genotypes by leveraging information from known genotypes and patterns of linkage disequilibrium (LD). Because alleles at nearby loci are often inherited together more frequently than expected by chance, LD structures allow imputation algorithms to predict unobserved variants using the observed haplotypes in surrounding genomic regions. A well-constructed reference panel, which contains fully genotyped individuals, is essential for accurate imputation. Such panels capture the underlying genetic diversity of the study population and reflect the historical signatures of recombination and mutation, thereby enabling precise inference at untyped loci.

Most imputation algorithms are built on the Li and Stephens model, introduced in 2003 (Li & Stephens, 2003), which conceptualizes a sample haplotype as a mosaic of short haplotype segments drawn from the reference panel. The imputation process involves hidden Markov models (HMMs), which estimate the probability that a study haplotype follows a particular path through the reference haplotypes. The model accounts for both recombination events (as transitions between haplotypes) and mutations or genotype errors (as differences between study and reference haplotypes), allowing the imputation algorithm to assign posterior probabilities to candidate genotypes at unobserved sites.

Early implementations of this framework included fastPHASE, designed for handling large-scale datasets (Scheet & Stephens, 2006). Subsequent developments, such as MaCH and IMPUTE, provided improved accuracy and computational efficiency (Howie et al., 2009; Li et al., 2010). More recently, Beagle introduced refinements to the Li and Stephens model, enabling direct construction of haplotype models from both reference and study samples, thereby enhancing computational efficiency while maintaining accuracy (Ayres et al., 2019). Although HMM-based approaches remain dominant, alternative strategies have been developed to further improve efficiency. The Positional Burrows-Wheeler Transformation (PBWT) is one such method, which uses positional matching between haplotypes to infer missing genotypes. Although highly efficient, PBWT sacrifices some accuracy compared to HMM-based methods, as it focuses on the longest matching haplotype segments and does not consider all possible mosaic configurations. Another approach is SNP-tagging, which identifies a small set of markers to serve as proxies for untyped variants. Though faster, SNP-tagging is less accurate because it does not capture the full complexity of recombination patterns across the chromosome.

As the volume of reference haplotype panels expands in scale, computational challenges have arisen. There is increasing demand for higher accuracy from lower sequencing depths, driving the development of imputation methods specifically tailored for lcWGS data. GLIMPSE is a notable example, bypassing the genotyping process by directly using initial sequencing data for imputation (Rubinacci et al., 2021). GLIMPSE employs a Gibbs sampling procedure to perform genotype imputation at all variable positions in the reference panel. This approach has delivered satisfactory results across various species, demonstrating both its effectiveness and versatility.

4.4.3 Development of Pan-genomic Identity-by-Descent (IBD) panels

GS has increasingly been adopted in aquaculture, offering substantial genetic gains compared to traditional breeding methods. Traditionally, genotype imputation has relied on SNP arrays but recent advancements have shifted toward using lcWGS or RRS for genotype imputation in large cohorts of species such as scallops, oysters, and shrimp (Garcia et al., 2022; Huang et al., 2025; Zhang et al., 2021). For example, studies on the Yesso scallop have demonstrated that genomic predictions using imputed SNPs from lcWGS at 0.5 × coverage can yield a saturated number of imputed SNPs comparable to WGS, with imputation accuracy reaching up to 94.76%. The combination of low-coverage sequencing with imputation in scallops (*Patinopecten yessoensis*) and oysters (*Crassostrea gigas*) has enabled the recovery of accurate genotypes for GWAS and GS, while significantly reducing costs compared to high-depth WGS (Huang et al., 2025; Sutherland et al., 2026). This combination proves both cost-effective and feasible, positioning lcWGS-based genotype imputation as a promising tool for advancing genetic selection in aquaculture. In addition, for 2b-RAD datasets, imputed moderate or high-density reference panels, with a GP threshold of 0.95, provided genome-wide SNP markers with an imputation accuracy of approximately 95%. Integrating RRS data with imputation could offer a cost-effective method for generating robust genome-wide SNP datasets for GWAS, enhancing the genetic analysis capabilities of mollusk species (Huang et al., 2025).

As more high-quality reference panels become available for aquaculture species, the accuracy and applicability of lcWGS-based imputation are expected to improve, further supporting cost-effective breeding programs and population genetic studies. In scallops (*Chlamys farreri*), the first haplotype reference panel using whole-genome resequencing from 174 individuals representing six major farming populations was established (Sui et al., 2024). Population-specific imputation accuracies exceeded 0.90, indicating a strong positive correlation between population kinship and imputation performance. Notably, accuracy improved with sequencing depth and plateaued at 0.5 × coverage, with accuracies of 0.91–0.93 observed in artificially bred populations. This growing body of research highlights the potential of lcWGS and imputation in making GS more accessible and efficient for mollusk species.

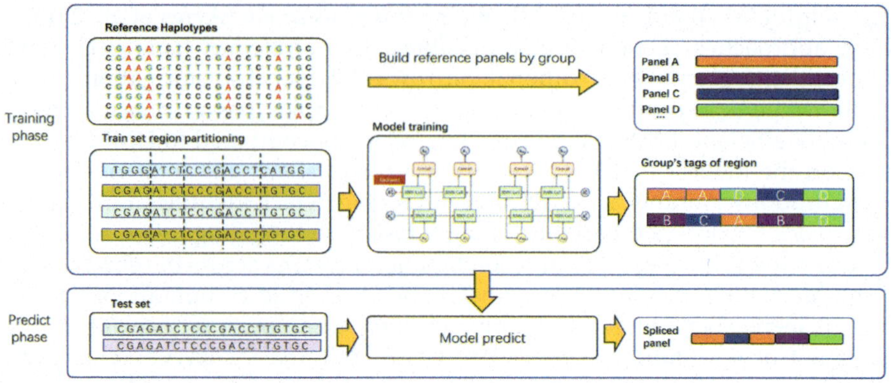

FIGURE 4.5 The specific panel reconstruction algorithm based on the GRU neural network. *GRU,* Gated recurrent unit.

Despite these successes, challenges remain in extending lcWGS-based imputation to diverse aquaculture populations. Most existing reference panels are constructed from limited populations, which can reduce imputation accuracy in underrepresented groups. Expanding panels to capture broader population diversity is therefore essential. Another limitation lies in the difficulty of imputing complex genomic variants such as structural variations and CNV, which are often overlooked by SNP-focused algorithms. Our team recently developed a deep learning–based pan-genome IBD cluster genotyping technique, which includes two parts: IBD cluster reconstruction using gated recurrent unit (GRU) neural networks with special panel reconfiguration (SPR) and genotype imputation through a convolutional denoising autoencoder (Fig. 4.5). The SPR process refines the reference haplotype panel by comparing genetic similarities within the IBD cluster. During the imputation phase, a pruning-mask module is introduced to prevent excessive variance in highly linked regions, improving the model's learning ability. The Pan-genomic IBD algorithm presents an innovative approach to enhancing genotype imputation accuracy in aquaculture species by integrating local and global genetic information to create a tailored reference panel.

4.5 Genomic and breeding database

Genomic tools have significantly transformed the field of selective breeding, particularly in aquaculture species. Existing breeding databases, however, have been primarily developed for livestock and plant species, integrating resources for genotype imputation, GWAS, and GS. Platforms such as Animal-ImputeDB and Plant-ImputeDB offer reference panels and tools optimized for these organisms, yet there is a noticeable gap in aquaculture species-specific resources (Gao et al., 2021; Yang et al., 2020) . While some platforms, such as the Michigan Imputation Server, provide robust imputation services, they lack specialized data and functionalities tailored to aquatic species' unique genetic characteristics. The absence of high-quality reference panels for aquatic species has limited the broader application of genotype imputation in aquaculture. This gap underscores the need for dedicated platforms that can streamline the genetic analysis and breeding strategies for economically significant aquaculture species.

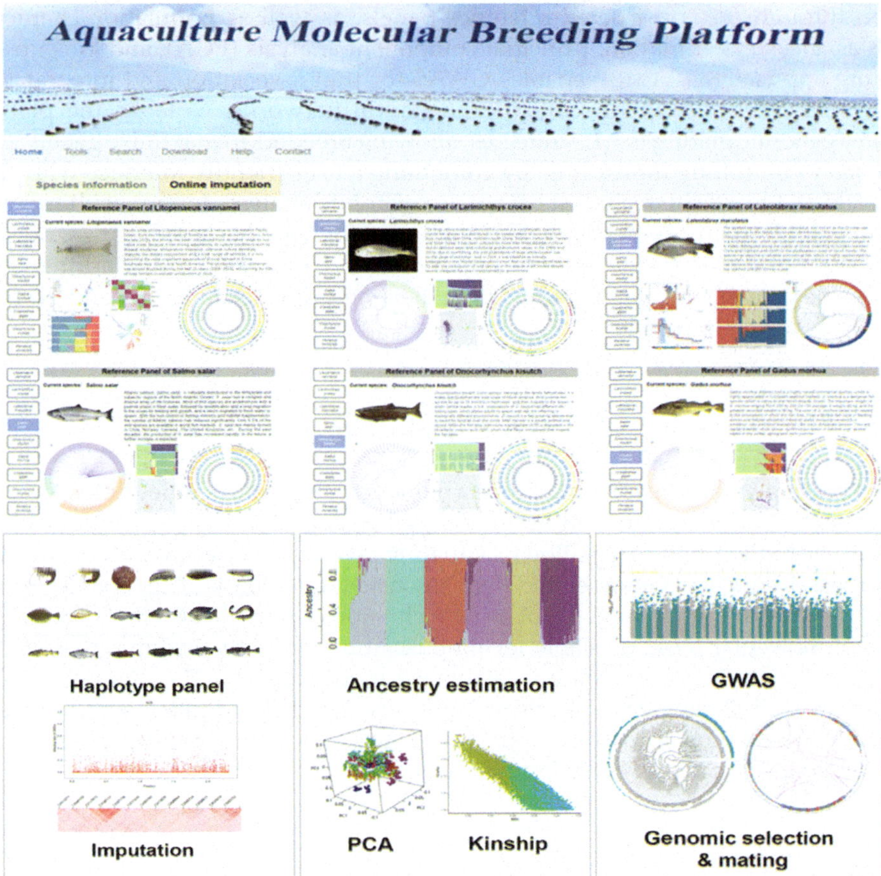

FIGURE 4.6 The architecture and main functions of the AMBP platform. *AMBP*, Aquaculture molecular breeding platform.

To address the limitations of existing tools, the aquaculture molecular breeding platform (AMBP) was developed to meet the unique requirements of aquaculture species (Fig. 4.6). AMBP integrates high-density haplotype reference panels covering 18 economically important aquaculture species, including crustaceans, mollusks, and various finfish (Zeng et al., 2022). The panels were constructed using whole-genome resequencing data from more than 2,000 individuals, followed by rigorous variant calling and filtering pipelines. Clean reads were aligned using the Burrows-Wheeler Aligner (BWA), and variant calling was performed using tools such as GATK to ensure high-quality SNP detection. Imputation pipelines were optimized with software such as Beagle and GLIMPSE, achieving high imputation accuracy, with concordance rates exceeding 92% across various species. By leveraging this robust database, AMBP enhances the accuracy of genotype imputation, thus providing a reliable foundation for downstream genetic analysis and breeding applications.

Beyond reference panel construction, AMBP is a comprehensive web-based platform designed to support the full spectrum of genetic analyses in aquaculture. It offers online genotype imputation through pre-constructed reference panels, as well as population characterization tools such as ancestry estimation, principal component analysis (PCA), and kinship inference. For breeding applications, AMBP enables GWAS for trait association and integrates genomic prediction models, including RR-GBLUP and neural networks. Notably, the platform also incorporates genomic mating (GM) strategies, allowing breeders to maximize genetic gain while simultaneously controlling inbreeding. By combining a user-friendly interface with automated pipelines, AMBP lowers the technical barriers to genomic analysis for aquaculture researchers and practitioners. It thus provides a reliable foundation for selective breeding programs, enhancing the accuracy and efficiency of genetic improvement. Ultimately, AMBP addresses a critical gap in available genomic resources, supporting the long-term sustainability and profitability of the global aquaculture industry.

4.6 Conclusion

In recent years, genotyping has undergone a paradigm shift from single-dimensional, marker-based analyses toward integrated, multidimensional strategies. Conceptual frameworks such as the hologenome highlight this transition by situating genotyping within a systems-level context that encompasses both host genomes and their associated microbiota. The rapid development of high-throughput and multilayered genotyping technologies underscores the increasing breadth and depth of molecular tools available for molluscan research. By enabling the integration of diverse genomic data, these approaches facilitate a more holistic and systematic investigation of mollusks, as well as the genetic basis of key traits and their adaptive significance, thereby laying a solid foundation for future applications in aquaculture improvement, biodiversity conservation, and sustainable resource management.

References

Abdelrahman, H., ElHady, M., Alcivar-Warren, A., Allen, S., Al-Tobasei, R., Bao, L., … Zhou, T. (2017). Aquaculture genomics, genetics and breeding in the United States: Current status, challenges, and priorities for future research. *BMC Genomics, 18*(1), 191. doi:10.1186/s12864-017-3557-1.

Andrews, K. R., Good, J. M., Miller, M. R., Luikart, G., & Hohenlohe, P. A. (2016). Harnessing the power of RADseq for ecological and evolutionary genomics. *Nature Reviews Genetics, 17*(2), 81–92. doi:10.1038/nrg.2015.28.

Ayres, D. L., Cummings, M. P., Baele, G., Darling, A. E., Lewis, P. O., Swofford, D. L., Huelsenbeck, J. P., Lemey, P., Rambaut, A., & Suchard, M. A. (2019). BEAGLE 3: Improved performance, scaling, and usability for a high-performance computing library for statistical phylogenetics. *Systematic Biology, 68*(6), 1052–1061. doi:10.1093/sysbio/syz020.

Baird, N. A., Etter, P. D., Atwood, T. S., Currey, M. C., Shiver, A. L., Lewis, Z. A., … Johnson, E. A. (2008). Rapid SNP discovery and genetic mapping using sequenced RAD markers. *PLoS One, 3*(10), e3376. doi:10.1371/journal.pone.0003376.

Bhatia, D., & Bajwa, G. S. (2022). *Molecular marker techniques and recent advancements genotyping by sequencing for crop improvement* (pp. 1–21). Wiley Blackwell. doi:10.1002/9781119745686.ch1.

Bucholz, J. R., Sard, N. M., VanTassel, N. M., Lozier, J. D., Morris, T. J., Paquet, A., & Zanatta, D. T. (2022). RAD-tag and mitochondrial DNA sequencing reveal the genetic structure of a widespread and regionally imperiled freshwater mussel, *Obovaria olivaria* (Bivalvia: Unionidae). *Ecology and Evolution, 12*(1), e8560. doi:10.1002/ece3.8560.

Crawford, D. L., & Oleksiak, M. F. (2016). Ecological population genomics in the marine environment. *Briefings in Functional Genomics, 15*(5), 342–351. doi:10.1093/bfgp/elw008.

Dou, J., Li, X., Fu, Q., Jiao, W., Li, Y., Li, T., … Bao, Z. (2016). Evaluation of the 2b-RAD method for genomic selection in scallop breeding. *Scientific Reports, 6*(1), 19244. doi:10.1038/srep19244.

Duran, C., Appleby, N., Edwards, D., & Batley, J. (2009). Molecular genetic markers: Discovery, applications, data storage and visualisation. *Current Bioinformatics, 4*(1), 16–27. doi:10.2174/157489309787158198.

Durante, M. K., Baums, I. B., Williams, D. E., Vohsen, S., & Kemp, D. W. (2019). What drives phenotypic divergence among coral clonemates of Acropora palmata? *Molecular Ecology, 28*(13), 3208–3224. doi:10.1111/mec.15140.

Elshire, R. J, Glaubitz, J. C, Sun, Q., Poland, J. A, Kawamoto, K., Buckler, E. S, & Mitchell, S. E. (2011). A robust, simple genotyping-by-sequencing (GBS) approach for high diversity species. *PLoS One, 6*, e19379. https://doi.org/10.1371/journal.pone.0019379.

Gao, Y., Yang, Z., Yang, W., Yang, Y., Gong, J., Yang, Q. Y., & Niu, X. (2021). Plant-ImputeDB: An integrated multiple plant reference panel database for genotype imputation. *Nucleic Acids Research, 49*(1), D1480. doi:10.1093/nar/gkaa953.

Garcia, B. F., Yoshida, G. M., Carvalheiro, R., & Yáñez, J. M. (2022). Accuracy of genotype imputation to whole genome sequencing level using different populations of Nile tilapia. *Aquaculture, 551*, 737947. doi:10.1016/j.aquaculture.2022.737947.

Gopikrishna, G. (2023). Application of molecular markers in aquaculture. *Elsevier BV*, 47–53. doi:10.1016/b978-0-323-91240-2.00014-2.

Guichoux, E., Lagache, L., Wagner, S., Chaumeil, P., Léger, P., Lepais, O., Lepoittevin, C., Malausa, T., Revardel, E., Salin, F., & Petit, R. J. (2011). Current trends in microsatellite genotyping. *Molecular Ecology Resources, 11*(4), 591–611. doi:10.1111/j.1755-0998.2011.03014.x.

Ho, S. S., Urban, A. E., & Mills, R. E. (2020). Structural variation in the sequencing era. *Nature Reviews Genetics, 21*(3), 171–189. doi:10.1038/s41576-019-0180-9.

Hollenbeck, C. M., & Johnston, I. A. (2018). Genomic tools and selective breeding in molluscs. *Frontiers in Genetics, 9*, 253. doi:10.3389/fgene.2018.00253.

Howie, B. N., Donnelly, P., Marchini, J., & Schork, N. J. (2009). A flexible and accurate genotype imputation method for the next generation of genome-wide association studies. *PLoS Genetics, 5*(6), e1000529. doi:10.1371/journal.pgen.1000529.

Huang, X., Duan, X., Ma, S., Sun, X., Geng, Q., Sun, X., Xing, Q., Zhang, L., Wang, S., & Lv, J. (2025). Optimizing genotype imputation strategies for low coverage whole genome and 2b-RAD sequencing in bivalve molluscs. *Aquaculture, 608*, 742713. doi:10.1016/j.aquaculture.2025.742713.

Jiang, M., Liu, Q., Jiang, C., Zhan, M., Wen, H., Shu, F., … Liu, K. (2025). Genome survey and evolutionary analysis of 8 Lamprotula species: SSR profiling, mitochondrial characterization, and population dynamics inference. *DNA Research, 32*(5), dsaf020. doi:10.1093/dnares/dsaf020.

Jiao, W., Fu, X., Dou, J., Li, H., Su, H., Mao, J., Yu, Q., Zhang, L., Hu, X., Huang, X., Wang, Y., Wang, S., & Bao, Z. (2014). High-resolution linkage and quantitative trait locus mapping aided by genome survey sequencing: Building up an integrative genomic framework for a bivalve mollusc. *DNA Research, 21*(1), 85–101. doi:10.1093/dnares/dst043.

Jiao, W., Fu, X., Li, J., Li, L., Feng, L., Lv, J., Zhang, L., Wang, X., Li, Y., Hou, R., Zhang, L., Hu, X., Wang, S., & Bao, Z. (2014). Large-scale development of gene-associated single-nucleotide polymorphism markers for molluscan population genomic, Comparative genomic, and genome-wide association studies. *DNA Research, 21*(2), 183–193. doi:10.1093/dnares/dst048.

Khamatova, A. Y., Malyar, V. V., Yagodina, V. D., & Bondar, E. I. (2025). Population genetic structure of Yesso scallop (*Mizuhopecten Yessoensis*) of the Russian Far East revealed by mitochondrial and microsatellite genetic markers. *Thalassas, 41*(2), 123. doi:10.1007/s41208-025-00881-9.

Khatei, A., Tripathy, P. S., & Parhi, J. (2022). *Molecular markers in aquaculture advances in fisheries biotechnology* (pp. 165–174). India: Springer Nature. doi:10.1007/978-981-16-3215-0_11.

Kim, C., Guo, H., Kong, W., Chandnani, R., Shuang, L. S., & Paterson, A. H. (2016). Application of genotyping by sequencing technology to a variety of crop breeding programs. *Plant Science, 242*, 14–22. doi:10.1016/j.plantsci.2015.04.016.

Kochmann, J., Carlsson, J., Crowe, T. P., & Mariani, S. (2012). Genetic evidence for the uncoupling of local aquaculture activities and a population of an invasive species-a case study of pacific oysters (*Crassostrea gigas*). *Journal of Heredity, 103*(5), 661–671. doi:10.1093/jhered/ess042.

Kockum, I., Huang, J., & Stridh, P. (2023). Overview of genotyping technologies and methods. *Current Protocols., 3*(4), e727. doi:10.1002/cpz1.727.

Korneliussen, T. S., Albrechtsen, A., & Nielsen, R. (2014). ANGSD: Analysis of next generation sequencing data. *BMC Bioinformatics, 15*(1), 356. doi:10.1186/s12859-014-0356-4.

Lawrie, R. D., & Massey, S. E. (2023). Agrigenomic diversity unleashed: Current single nucleotide polymorphism genotyping methods for the agricultural sciences. *Applied Biosciences, 2*(4), 565–585. doi:10.3390/applbiosci2040036.

Lee, Y., Ni, G., Shin, J., Kim, T., Kern, E. M. A., Kim, Y., ... Park, J.-K. (2021). Phylogeography of *Mytilisepta virgata* (Mytilidae: Bivalvia) in the northwestern Pacific: Cryptic mitochondrial lineages and mito-nuclear discordance. *Molecular Phylogenetics and Evolution, 157*, 107037. doi:10.1016/j.ympev.2020.107037.

Li, H., Handsaker, B., Wysoker, A., Fennell, T., Ruan, J., Homer, N., Marth, G., Abecasis, G., & Durbin, R. (2009). The sequence alignment/map format and SAMtools. *Bioinformatics, 25*(16), 2078–2079. doi:10.1093/bioinformatics/btp352.

Li, H., Yuan, J., Wu, M., Han, Z., Li, L., Jiang, H., Jia, Y., Han, X., Liu, M., Sun, D., Chen, C., Song, W., & Wang, C. (2018a). Transcriptome and DNA methylome reveal insights into yield heterosis in the curds of broccoli (Brassica oleracea L var. italic). *BMC Plant Biology, 18*(1). doi:10.1186/s12870-018-1384-4.

Li, L., Li, A., Song, K., Meng, J., Guo, X., Li, S., Li, C., De Wit, P., Que, H., Wu, F., Wang, W., Qi, H., Xu, F., Cong, R., Huang, B., Li, Y., Wang, T., Tang, X., Liu, S., ... Zhang, G. (2018b). Divergence and plasticity shape adaptive potential of the Pacific oyster. *Nature Ecology & Evolution, 2*(11), 1751–1760. doi:10.1038/s41559-018-0668-2.

Li, L., Yang, L., Fan, Z., Xue, W., Shen, Z., Yuan, Y., ... Zhang, Y. (2020). Hypoxia-induced GBE1 expression promotes tumor progression through metabolic reprogramming in lung adenocarcinoma. *Signal Transduction and Targeted Therapy, 5*(1), 54. doi:10.1038/s41392-020-0152-8.

Li, N., & Stephens, M. (2003). Modeling linkage disequilibrium and identifying recombination hotspots using single-nucleotide polymorphism data. *Genetics, 165*(4), 2213–2233. doi:10.1093/genetics/165.4.2213.

Li, X., Wang, S., Xun, X., Zhang, M., Wang, S., Li, H., Zhao, L., Fu, Q., Wang, H., Li, T., Lian, S., Xing, Q., Li, X., Wu, W., Zhang, L., Hu, X., & Bao, Z. (2019). A carotenoid oxygenase is responsible for muscle coloration in scallop. *Biochimica et Biophysica Acta (BBA) - Molecular and Cell Biology of Lipids, 1864*(7), 966–975. doi:10.1016/j.bbalip.2019.03.003.

Li, Y., Willer, C. J., Ding, J., Scheet, P., & Abecasis, G. R. (2010). MaCH: Using sequence and genotype data to estimate haplotypes and unobserved genotypes. *Genetic Epidemiology, 34*(8), 816–834. doi:10.1002/gepi.20533.

Li, Y., He, M., & Yue, G. H. (2014). Genetic mapping and QTL analysis of growth-related traits in Pinctada fucata using restriction-site associated DNA sequencing. *PLoS ONE, 9*(11), e111707. doi:10.1371/journal.pone.0111707.

Li, Y., Wang, R., Xun, X., Wang, J., Bao, L., Thimmappa, R., Ding, J., Jiang, J., Zhang, L., Li, T., Lv, J., Mu, C., Hu, X., Zhang, L., Liu, J., Li, Y., Yao, L., Jiao, W., Wang, Y., ... Shi, W. (2018c). Sea cucumber genome provides insights into saponin biosynthesis and aestivation regulation. *Cell Discovery, 4*(1). doi:10.1038/s41421-018-0030-5.

Liang, Y., Xu, C., & Li, Q. (2023). Heterosis and genetic diversity of intraspecific hybrids crosses between two selected lines of the Pacific oyster *Crassostrea gigas*. *Aquaculture, 569*, 739369. doi:10.1016/j.aquaculture.2023.739369.

Liu, P., Zhang, T., Lv, J., Ma, C., Yang, Z., Huang, X., Zhang, L., Bao, Z., & Shi, W. (2021). An efficient integrated approach for nonlethal DNA sampling and genome-wide genotyping in bivalve molluscs. *Aquaculture, 536*, 736489. doi:10.1016/j.aquaculture.2021.736489.

Liu, M., Yang, Z., Huang, X., Zeng, Q., Mu, Q., Xu, Z., ... Bao, Z. (2025). Development and evaluation of high-throughput HD-Marker arrays in pacific white shrimp (*Litopenaeus vannamei*). *Aquaculture, 599*, 742119. doi:10.1016/j.aquaculture.2025.742119.

Liu, P., Lv, J., Ma, C., Zhang, T., Huang, X., Yang, Z., Zhang, L., Hu, J., Wang, S., & Bao, Z. (2022). Targeted genotyping of a whole-gene repertoire by an ultrahigh-multiplex and flexible HD-marker approach. *Engineering, 13*, 186–196. doi:10.1016/j.eng.2021.07.027.

Liu, S., Shi, C., Chen, C., Tan, Y., Macqueen, D., & Li, Q. (2025). Haplotype-resolved genomes provide insights into the origins and functional significance of genome diversity in bivalves. *Cell Reports, 44*, 115697.

Liu, Z., & Cordes, J. (2004). DNA marker technologies and their applications in aquaculture genetics. *Aquaculture, 238*(1–4), 1–37. doi:10.1016/j.aquaculture.2004.05.027.

Lou, R. N., Jacobs, A., Wilder, A. P., & Therkildsen, N. O. (2021). A beginner's guide to low-coverage whole genome sequencing for population genomics. *Molecular Ecology, 22*, 1–14.

Lv, J., Jiao, W., Guo, H., Liu, P., Wang, R., Zhang, L., ... Wang, S. (2018). HD-Marker: A highly multiplexed and flexible approach for targeted genotyping of more than 10,000 genes in a single-tube assay. *Genome Research, 28*(12), 1919–1930. doi:10.1101/gr.235820.118.

Lv, J., Wang, Y., Ni, P., Lin, P., Hou, H., Ding, J., Chang, Y., Hu, J., Wang, S., & Bao, Z. (2022). Development of a high-throughput SNP array for sea cucumber (Apostichopus japonicus) and its application in genomic selection with MCP regularized deep neural networks. *Genomics, 114*(4), 110426. doi:10.1016/j.ygeno.2022.110426.

Lv, J., Cai, Y., Liu, P., Huang, X., Lv, Z., Zhang, R., … Hu, J. (2022). Genomic differentiation and selection signatures of two elite varieties of Yesso scallop *Mizuhopecten yessoensis*. *Aquaculture, 550*, 737842. doi:10.1016/j.aquaculture.2021.737842.

Ma, C., Xu, C., Zhang, T., Mu, Q., Lv, J., Xing, Q., … Liu, P. (2024). Tracking the hologenome dynamics in aquatic invertebrates by the holo-2bRAD approach. *Communications Biology, 7*(1), 827. doi:10.1038/s42003-024-06509-7.

Mao, J., Zeng, Q., Yang, Z., Pan, H., Yao, L., Bao, Z., … Wang, S. (2020). High-resolution linkage and quantitative trait locus mapping using an interspecific cross between *Argopecten irradians irradians* (♀) and *A. purpuratus* (♂). *Marine Life Science & Technology, 2*(2), 123–134. doi:10.1007/s42995-020-00029-z.

Maqsood, H. M., & Ahmad, S. M. (2017). Advances in molecular markers and their applications in aquaculture and fisheries. *Genetics of Aquatic Organisms, 1*(1), 27–41. doi:10.4194/2459-1831-v1_1_05.

Matz, M. V. (2018). Fantastic beasts and how to sequence them: Ecological genomics for obscure model organisms. *Trends in Genetics, 34*(2), 121–132. doi:10.1016/j.tig.2017.11.002.

McKenna, A., Hanna, M., Banks, E., Sivachenko, A., Cibulskis, K., Kernytsky, A., Garimella, K., Altshuler, D., Gabriel, S., Daly, M., & DePristo, M. A. (2010). The genome analysis toolkit: A MapReduce framework for analyzing next-generation DNA sequencing data. *Genome Research, 20*(9), 1297–1303. doi:10.1101/gr.107524.110.

Modak, T., Literman, R., Puritz, J., Johnson, K., Roberts, E., Proestou, D., … Schwartz, R. (2021). Extensive genome-wide duplications in the eastern oyster (*Crassostrea virginica*). *Philosophical Transactions of the Royal Society B: Biological Sciences, 376*, 20200164.

Naj, A. C. (2019). Genotype imputation in genome-wide association studies. *Current Protocols in Human Genetics, 102*(1), e84. doi:10.1002/cphg.84.

Nielsen, R., Paul, J. S., Albrechtsen, A., & Song, Y. S. (2011). Genotype and SNP calling from next-generation sequencing data. *Nature Reviews Genetics, 12*(6), 443–451. doi:10.1038/nrg2986.

Noorullah, M., Zuberi, A., Zaman, M., Younas, W., Hussain, S., & Kamran, M. (2023). Assessment of genetic diversity among wild and captive-bred Labeo rohita through microsatellite markers and mitochondrial DNA. *Fisheries and Aquatic Sciences, 26*(12), 752–761. doi:10.47853/fas.2023.e67.

Ota, M., Fukushima, H., Kulski, J. K., & Inoko, H. (2007). Single nucleotide polymorphism detection by polymerase chain reaction-restriction fragment length polymorphism. *Nature Protocols, 2*(11), 2857–2864. doi:10.1038/nprot.2007.407.

Puritz, J. B., Matz, M. V., Toonen, R. J., Weber, J. N., Bolnick, D. I., & Bird, C. E. (2014). Demystifying the RAD fad. *Molecular Ecology, 23*(24), 5937–5942. doi:10.1111/mec.12965.

Qi, H., Song, K., Li, C., Wang, W, Li, B., & Zhang, G. (2017). Construction and evaluation of a high-density SNP array for the Pacific oyster (*Crassostrea gigas*). *PLOS ONE, 12*(3), e0174007.

Robledo, D., Palaiokostas, C., Bargelloni, L., Martínez, P., & Houston, R. (2018). Applications of genotyping by sequencing in aquaculture breeding and genetics. *Reviews in Aquaculture, 10*(3), 670–682. doi:10.1111/raq.12193.

Rogers, S. M., Campbell, D., Baird, S. J. E., Danzmann, R. G., & Bernatchez, L. (2001). Combining the analyses of introgressive hybridisation and linkage mapping to investigate the genetic architecture of population divergence in the lake whitefish (*Coregonus clupeaformis*, Mitchill). *Genetica, 111*(1–3), 25–41. doi:10.1023/a:1013773600304.

Rubinacci, S., Ribeiro, D. M., Hofmeister, R. J., & Delaneau, O. (2021). Efficient phasing and imputation of low-coverage sequencing data using large reference panels. *Nature Genetics, 53*(1), 120–126. doi:10.1038/s41588-020-00756-0.

Scheet, P., & Stephens, M. (2006). A fast and flexible statistical model for large-scale population genotype data: Applications to inferring missing genotypes and haplotypic phase. *The American Journal of Human Genetics, 78*(4), 629–644. doi:10.1086/502802.

Scheschonk, L., Nilsen, A. M. L., Bischof, K., & Jueterbock, A. (2024). Chloroplast DNA methylation in the kelp *Saccharina latissima* is determined by origin and possibly influenced by cultivation. *Evolutionary Applications, 17*(7), e13744. doi:10.1111/eva.13744.

Seeb, J. E., Carvalho, G., Hauser, L., Naish, K., Roberts, S., & Seeb, L. W. (2011). Single-nucleotide polymorphism (SNP) discovery and applications of SNP genotyping in nonmodel organisms. *Molecular Ecology Resources, 11*(1), 1–8. doi:10.1111/j.1755-0998.2010.02979.x.

Sekino, M., Hashimoto, K., Nakamichi, R., Yamamoto, M., Fujinami, Y., & Sasaki, T. (2023). Introgressive hybridization in the west Pacific pen shells (genus *Atrina*): Restricted interspecies gene flow within the genome. *Molecular Ecology, 32*(11), 2945–2963. doi:10.1111/mec.16908.

Sheeja, T. E., Kumar, I. P. V., Giridhari, A., Minoo, D., Rajesh, M. K., & Babu, K. N. (2021). Amplified fragment length polymorphism: Applications and recent developments. *Methods in Molecular Biology, 2222*, 187–218. doi:10.1007/978-1-0716-0997-2_12.

Shi, Y., Wang, S., Gu, Z., Lv, J., Zhan, X., Yu, C., … Wang, A. (2014). High-density single nucleotide polymorphisms linkage and quantitative trait locus mapping of the pearl oyster, *Pinctada fucata martensii* Dunker. *Aquaculture., 434*, 376–384. doi:10.1016/j.aquaculture.2014.08.044.

Silliman, K. (2019). Population structure, genetic connectivity, and adaptation in the Olympia oyster (*Ostrea lurida*) along the west coast of North America. *Evolutionary Applications, 12*(5), 923–939. doi:10.1111/eva.12766.

Smolka, M., Paulin, L. F., Grochowski, C. M., Horner, D. W., Mahmoud, M., Behera, S., Kalef-Ezra, E., Gandhi, M., Hong, K., Pehlivan, D., Scholz, S. W., Carvalho, C. M. B., Proukakis, C., & Sedlazeck, F. J. (2024). Detection of mosaic and population-level structural variants with Sniffles. *Nature Biotechnology, 42*(10), 1571–1580. doi:10.1038/s41587-023-02024-y.

Sudmant, P. H., Rausch, T., Gardner, E. J., Handsaker, R. E., Abyzov, A., Huddleston, J., Zhang, Y., Ye, K., Jun, G., Fritz, M. H. Y., Konkel, M. K., Malhotra, A., Stütz, A. M., Shi, X., Casale, F. P., Chen, J., Hormozdiari, F., Dayama, G., Chen, K., … Korbel, J. O. (2015). An integrated map of structural variation in 2,504 human genomes. *Nature, 526*(7571), 75–81. doi:10.1038/nature15394.

Sui, M., Liu, Z., Huang, X., Yang, Z., Yu, H., Cui, C., … Huang, X. (2024). Development and evaluation of a haplotype reference panel of Zhikong scallop (*Chlamys farreri*) for genotype imputation. *Aquaculture, 582*, 740497. doi:10.1016/j.aquaculture.2023.740497.

Sun, Q., & Li, Y. (2025). Advances in haplotype phasing and genotype imputation. *Nature Reviews Genetics, 27*, 155–169. doi:10.1038/s41576-025-00895-2.

Sun, Z., Huang, S., Zhu, P., Tzehau, L., Zhao, H., Lv, J., … Xu, J. (2022). Species-resolved sequencing of low-biomass or degraded microbiomes using 2bRAD-M. *Genome Biology, 23*(1), 36. doi:10.1186/s13059-021-02576-9.

Sutherland, B., Divilov, K., Thompson, N., Delomas, T., Lunda, S., Langdon, C., & Green, T. (2026). Pedigree-based genome-wide imputation using a low-density amplicon panel for the highly polymorphic Pacific oyster *Crassostrea (Magallana) gigas*. *Aquaculture, 612*, 743096. doi:10.1016/j.aquaculture.2025.743096.

Tan, Y., Fang, L., Qiu, M., Huo, Z., & Yan, X. (2020). Population genetics of the Manila clam (*Ruditapes philippinarum*) in East Asia. *Scientific Reports., 10*(1), 21890. doi:10.1038/s41598-020-78923-w.

Tan, Y., Liu, L. J., Zhang, Y. X., Fang, L., Liu, Y., & Huo, Z. M. (2024). Geographic patterns and ecological adaptations of Manila clam (*Ruditapes philippinarum*) along the Chinese coast: Insights from migration dynamics and gene flow. *Global Ecology and Conservation, 54*, 21890. doi:10.1016/j.gecco.2024.e03192.

Teer, J. K., Bonnycastle, L. L., Chines, P. S., Hansen, N. F., Aoyama, N., Swift, A. J., Abaan, H. O., Albert, T. J., Margulies, E. H., Green, E. D., Collins, F. S., Mullikin, J. C., & Biesecker, L. G. (2010). Systematic comparison of three genomic enrichment methods for massively parallel DNA sequencing. *Genome Research, 20*(10), 1420–1431. doi:10.1101/gr.106716.110.

Terzin, M., Villamor, A., Marincich, L., Matterson, K., Paletta, M. G., Bertuccio, V., … Costantini, F. (2024). 2bRAD reveals fine-scale genetic structuring among populations within the Mediterranean zoanthid *Parazoanthus axinellae* (Schmidt, 1862). *Coral Reefs, 43*(2), 357–370. doi:10.1007/s00338-023-02456-7.

Wang, S., Lv, J., Zhang, L., Dou, J., Sun, Y., Li, X., Fu, X., Dou, H., Mao, J., Hu, X., & Bao, Z. (2015). MethylRAD: A simple and scalable method for genome-wide DNA methylation profiling using methylation-dependent restriction enzymes. *Open Biology, 5*(11), 150130. doi:10.1098/rsob.150130.

Therkildsen, N. O., & Palumbi, S. R. (2017). Practical low-coverage genomewide sequencing of hundreds of individually barcoded samples for population and evolutionary genomics in nonmodel species. *Molecular Ecology Resources, 17*(2), 194–208. doi:10.1111/1755-0998.12593.

Vieira, M. L. C., Santini, L., Diniz, A. L., & Munhoz, C. D. F. (2016). Microsatellite markers: What they mean and why they are so useful. *Genetics and Molecular Biology, 39*(3), 312–328. doi:10.1590/1678-4685-GMB-2016-0027.

Wang, H., Lv, J., Zeng, Q., Liu, Y., Xing, Q., Wang, S., … Bao, L. (2021). Genetic differentiation and selection signatures in two bay scallop (*Argopecten irradians*) breeds revealed by whole-genome resequencing analysis. *Aquaculture, 543*, 736944. doi:10.1016/j.aquaculture.2021.736944.

Wang, S., Liu, P., Lv, J., Li, Y., Cheng, T., Zhang, L., Xia, Y. u, Sun, H., Hu, X., & Bao, Z. (2016). Serial sequencing of isolength RAD tags for cost-efficient genome-wide profiling of genetic and epigenetic variations. *Nature Protocols, 11*(11), 2189–2200. doi:10.1038/nprot.2016.133.

Wang, S., Meyer, E., Mckay, J. K., & Matz, M. V. (2012). 2b-RAD: A simple and flexible method for genome-wide genotyping. *Nature Methods, 9*(8), 808–810. doi:10.1038/nmeth.2023.

Wang, S., Zhang, J., Jiao, W., Li, J., Xun, X., Sun, Y., … Bao, Z. (2017). Scallop genome provides insights into evolution of bilaterian karyotype and development. *Nature Ecology & Evolution, 1*(5), 0120. doi:10.1038/s41559-017-0120.

Wang, Y., Sun, G., Zeng, Q., Chen, Z., Hu, X., Li, H., … Bao, Z. (2018). Predicting growth traits with genomic selection methods in Zhikong Scallop (*Chlamys farreri*). *Marine Biotechnology, 20*(6), 769–779. doi:10.1007/s10126-018-9847-z.

Wenne, R. (2023). Microsatellites as molecular markers with applications in exploitation and conservation of aquatic animal populations. *Genes, 14*(4), 808. doi:10.3390/genes14040808.

Wenne, R. (2023). Single nucleotide polymorphism markers with applications in conservation and exploitation of aquatic natural populations. *Animals, 13*(6), 1089. doi:10.3390/ani13061089.

West, A. P., & Shadel, G. S. (2017). Mitochondrial DNA in innate immune responses and inflammatory pathology. *Nature Reviews Immunology, 17*(6), 363–375. doi:10.1038/nri.2017.21.

White, S., Vadopalas, B., Silliman, K., & Roberts, S. (2017). Genotoype-by-sequencing of three geographically distinct populations of Olympia oysters. *Scientific Data, 4*(1), 170130. doi:10.1038/sdata.2017.130.

Xu, K., Li, Q., Kong, L., & Yu, R. (2008). A first-generation genetic map of the Japanese scallop *Patinopecten yessoensis*-based AFLP and microsatellite markers. *Aquaculture Research, 40*(1), 35–43. doi:10.1111/j.1365-2109.2008.02058.x.

Xu, Q., Zheng, J., Yan, X., & Nie, H. (2020). Genetic diversity and differentiation of nine populations of the hard clam (*Meretrix petechialis*) assessed by EST-derived microsatellites. *Electronic Journal of Biotechnology, 48*, 23–28. doi:10.1016/j.ejbt.2020.09.003.

Xu, T., Sun, J., Watanabe, H. K., Chen, C., Nakamura, M., Ji, R., … Qiu, J.-W. (2018). Population genetic structure of the deep-sea mussel *Bathymodiolus platifrons* (Bivalvia: Mytilidae) in the Northwest Pacific. *Evolutionary Applications, 11*(10), 1915–1930. doi:10.1111/eva.12696.

Xu, W., Liu, Y., Li, M., Lu, S., & Chen, S. (2024). Advances in biotechnology and breeding innovations in China's marine aquaculture. *Advanced Biotechnology, 2*(4), 38. doi:10.1007/s44307-024-00043-7.

Yang, W., Yang, Y., Zhao, C., Yang, K., Wang, D., Yang, J., Niu, X., & Gong, J. (2020). Animal-ImputeDB: A comprehensive database with multiple animal reference panels for genotype imputation. *Nucleic Acids Research, 48*(D1), D659. doi:10.1093/nar/gkz854.

Yang, Z., Zhao, A., Teng, M., Li, M., Wang, H., Wang, X., … Huang, X. (2024). Signatures of selection in *Mulinia lateralis* underpinning its rapid adaptation to laboratory conditions. *Evolutionary Applications, 17*(2), e13657. doi:10.1111/eva.13657.

Yin, X., Hedgecock, D., & Yu, J. (2021). Overt and concealed genetic loads revealed by QTL mapping of genotype-dependent viability in the Pacific oyster *Crassostrea gigas*. *Genetics, 219*(4), iyab165. doi:10.1093/genetics/iyab165.

Yu, X., Faggion, S., Liu, Y., Wang, B., Zeng, Q., Lu, C., Hu, J., Bargelloni, L., Fang, L., & Bao, Z. (2025). Role of multi-omics in aquaculture genetics and breeding: current status and future perspective. *Science China Life Sciences, 68*(9), 2591–2604. doi:10.1007/s11427-024-2828-8.

Yuan, J., Zhang, X., Wang, M., Sun, Y., Liu, C., Li, S., … Li, F. (2021). Simple sequence repeats drive genome plasticity and promote adaptive evolution in penaeid shrimp. *Communications Biology, 4*(1), 186. doi:10.1038/s42003-021-01716-y.

Zeng, Q., Zhao, B., Wang, H., Wang, M., Teng, M., Hu, J., … Wang, Y. (2022). Aquaculture Molecular Breeding Platform (AMBP): A comprehensive web server for genotype imputation and genetic analysis in aquaculture. *Nucleic Acids Research, 50*(W1), W66–W74. doi:10.1093/nar/gkac424.

Zhang, W., Li, W., Liu, G., Gu, L., Ye, K., Zhang, Y., Li, W., Jiang, D., Wang, Z., & Fang, M. (2021). Evaluation for the effect of low-coverage sequencing on genomic selection in large yellow croaker. *Aquaculture, 534*, 736323. doi:10.1016/j.aquaculture.2020.736323.

Zhao, L., Li, Y., Li, Y., Yu, J., Liao, H., Wang, S., … Bao, Z., et al. (2017). A genome-wide association study identifies the genomic region associated with shell color in Yesso Scallop, *Patinopecten yessoensis*. *Marine Biotechnology, 19*(3), 301–309. doi:10.1007/s10126-017-9751-y.

Zheng, G., Zhang, Y., Wang, H., Ding, E., Qu, A., Su, P., Yang, Y., Zou, M., & Zhang, Y. (2020). Genome-wide DNA methylation analysis by MethylRad and the transcriptome profiles reveal the potential cancer-related lncRNAs in colon cancer. *Cancer Medicine, 9*(20), 7601–7612. doi:10.1002/cam4.3412.

Zheng, Z., Hao, R., Yang, C., Jiao, Y. u, Wang, Q., Huang, R., … Du, X. (2023). Genome-wide association study analysis to resolve the key regulatory mechanism of biomineralization in *Pinctada fucata martensii*. *Molecular Ecology Resources, 23*(3), 680–693. doi:10.1111/1755-0998.13743.

CHAPTER

5

Phenotyping and high-throughput phenomics

Xiaoting Huang[a], *Yunwei Dong*[b], *Yuan Lu*[c] *and Mingling Liao*[b]

[a]College of Marine Life Sciences, Ocean University of China, Qingdao, China [b]Fisheries College, Ocean University of China, Qingdao, China [c]College of Physics and Optoelectronic Engineering, Ocean University of China, Qingdao, China

5.1 Overview of molluscan biological and ecological traits

Phenotypes, which refer to the observable characteristics or traits of an organism, can be influenced by both genetic and environmental factors. The phylum Mollusca, recognized as the second-largest animal phylum, exhibits a remarkable diversity of biological and ecological traits, thereby offering significant implications for both scientific research and economic applications. Historically, phenotypes were mainly mentioned to macroscopic traits, such as size, shape, color, and growth rate. And traditional phenotyping typically relies on basic visual and morphological assessments. For example, early taxonomic studies of molluscan species were mainly based on the external shell features, such as shape, color, and texture (Brown & Neville, 2015). However, these conventional methodologies are constrained by their inherent subjectivity, susceptibility to error, and lack of quantification, which impede precise analysis and comparison of diverse molluscan phenotypes. With the advancements in molecular biology, phenotypes including transcript, protein, and metabolite abundances, as well as telomere length, and epigenetic features, have become accessible for genetic or evolutionary studies (Cook et al., 2016; Damerval et al., 1994; Johannes et al., 2008). Consequently, the fields of transcriptomes, proteomes, metabolomes, and epigenomes have emerged as invaluable resources of molecular phenotypes. In addition, with the advancements in automated and high-throughput phenotypic analysis techniques, phenomics has become an important interdisciplinary field that drives the development of life sciences after genomics, and is profoundly changing the technological innovation model in agriculture, livestock, and aquaculture (Hancock, 2014).

Genetic Breeding and Aquaculture of Shellfish: A Collaborative Approach to a Sustainable Future
DOI: https://doi.org/10.1016/B978-0-443-33606-5.00017-5

5.2 The pivotal roles of phenotyping

Phenotyping plays a crucial role in the study of mollusks, which can provide some critical insights into the diverse biological and evolutionary characteristics. It enables researchers to systematically categorize and analyze these variations, which is essential for understanding the adaptive strategies and ecological roles of various molluscan species.

Understanding the characteristics is crucial for developing comprehensive databases that can aid in the resource conservation and management of molluscan species, as well as in exploring their potential economic benefits (Costello et al., 2015). Phenotyping provides valuable insights into the interaction between genetic and environmental factors that affect the growth, development, and fitness of mollusks (Bayne, 2004). By tracking the performance of mollusks in diverse environments, researchers can evaluate the effects of water quality, feed, and stocking density on their growth and health. Understanding how mollusks perform under various environmental conditions aids in developing effective scientific breeding programs and enhancing production efficiency. Furthermore, by evaluating key traits such as the growth rate and disease resistance, breeders can select superior individuals for breeding and improve breeding efficiency. These analyses can effectively facilitate the selection of high-yield and disease-resistant mollusk varieties.

Deciphering phenotypes in mollusks allows researchers to identify genes or genomic regions associated with specific traits, thereby enhancing our comprehension of the relationship between genes and phenotypes. One of the primary methods used for this analysis is genome-wide association studies (GWAS), which evaluate the association between phenotypes and high-density markers distributed across the genome using statistical methodologies. This approach has successfully identified numerous key genes that influence agricultural economic traits. A comprehensive analysis of phenotypic traits of mollusks facilitates the collection of scientific data to support breeding practices, management strategies, and genetic research, which contributes to the sustainable development of aquaculture.

5.3 Phenotyping of molluscan aquaculture-related traits

5.3.1 Phenotyping of growth-related traits

In phenotypic studies of marine mollusks, growth-related traits constitute critical indicators for assessing growth performance and aquaculture potential. These traits typically encompass morphometric parameters, body weight parameters, as well as feed ingestion efficiency and nutrient assimilation capacity.

Morphometric traits are fundamental metrics for evaluating molluscan growth status. In bivalve mollusks, these traits generally include shell height (SH), shell length (SL), and shell width (SW) (Ziegler et al., 2010). SH is defined as the vertical distance from the shell apex to the ventral margin. For scallops, mussels, and clams, SH is frequently employed as the most direct growth indicator. SL refers to the anterior-posterior distance of the shell, with significant interspecific variations in SL/SH ratios observed among bivalve taxa. SW represents the maximum lateral distance between the left and right shell valves. Collectively, SH, SL, and SW provide essential three-dimensional descriptors of volumetric growth, offering comprehensive insights into individual growth patterns, thereby enhancing assessments of health status, growth potential, and environmental adaptability. In cephalopods, mantle length (ML)—the dorsal midline length of the mantle—serves as the primary morphometric index, extensively utilized in taxonomic classification, growth studies, and fisheries resource assessments.

Weight parameters are equally vital for evaluating molluscan growth and nutritional value. In marine bivalves, these include total weight (TW), soft tissue weight (STW), and muscle weight (MW). TW denotes the entire body mass, including shell, visceral organs, and muscular tissues. In aquaculture, TW is a critical parameter for assessing cultivation efficacy and individual health, with commercial value often determined by size and weight. STW refers to the mass of soft tissues (adductor muscles, mantle, gills, and digestive gland) after manual shell removal. MW quantifies muscular tissue mass, specifically the combined weight of adductor and foot muscles. For bivalves such as scallops and abalones, MW is a key determinant of edible value. In cephalopods, where the internalized shell contributes minimally to total mass, TW remains the sole practical weight metric. Accurate weight quantification necessitates high-precision instrumentation, including electronic balances and analytical weighing systems. Routine weight monitoring enables real-time growth tracking and facilitates analysis of environmental influences (e.g., water, temperature, salinity, and feed composition) on development. Such data inform both aquaculture management and ecological conservation strategies.

As filter feeders, mollusks primarily derive nutrition through filtration of algae and uptake of organic particulates. Their growth rates and health status are directly influenced by feeding efficiency. Common methodologies for investigating algal ingestion include: Examination of algal remnants in feces or ambient water allows indirect estimation of ingestion rates and dietary composition. This approach is particularly effective for species with specialized algal diets. Stable isotope analysis provides quantitative insights into nutrient assimilation pathways. By introducing isotopically labeled algae into aquatic systems, researchers analyzed tissue isotopic ratios to determine trophic positions and dietary dependencies (Zhang et al., 2024). Incorporation of fluorophore-tagged algal particles enables noninvasive, real-time tracking of ingestion dynamics via fluorescence microscopy or spectroscopy. This high-sensitivity technique excels in tracing microparticle distribution within organisms. Accompanying the incorporation of algae, changes in nutrient absorption can be reflected through the elemental composition of mollusks, and elemental analysis serves as an effective method to interpret this process. For example, strontium (Sr) is a reliable indicator of mollusks' physiological and metabolic activities at different developmental stages and has been found to correlate strongly with shell growth ring patterns (Li et al., 2018b). Nitrogen (N) has been confirmed as a marker for identifying organic layers within shells, offering evidence for the essential role of amino acids in mollusk development (Ren et al., 2024). Furthermore, shell elemental data provide valuable biogeochemical insights into habitat variations. The Mg/Ca ratio is a well-established proxy for reconstructing sea surface temperatures (SST) (Fig. 5.1) (Li et al., 2023b). Biogenic elements such as nitrogen (N) and phosphorus (P) in shells are closely associated with nutrient levels in seawater (Buer et al., 2020). In addition, heavy metals (e.g., Pb, Cd, As) present in shells serve as key indicators for assessing marine pollution and tracing pollution sources (Belcheva et al., 2006). These elemental indicators are not universally applicable across all mollusk species. The indicative elements may vary not only between different species but also among individuals of the same species in diverse regions. Therefore a comprehensive analysis must consider both the specific mollusk species and environmental conditions. Nonetheless, elemental analysis holds significant potential in studies of mollusk growth, environmental responses, and biogeochemical cycles.

FIGURE 5.1 The study of laser-induced breakdown spectroscopy for Mg/Ca investigation in cultured scallop shells. (A) The seashell (*Chlamys farreri*); (B) The seashell (*Patinopecten yessoensis*); (C) LIBS experimental setup for seashell analysis. *DM*: Dichroic mirror; *L1, L2*: lenses for light collection; *L3*: doublet lens for imaging; *LD*: laser diode; *OBJ*: objective lens. *From Li et al. (2023).*

5.3.2 Phenotyping of stress-resistant traits

Phenotyping stress-resistant traits in mollusks is a critical area of artificial breeding, especially in the context of increasing environmental stressors such as global warming, pollution, and habitat loss. Understanding these traits can provide insights into the resilience and adaptability of mollusk species, which are vital for developing stress-resistant genetic variants and identifying suitable habitats. Recent studies have focused on identifying suitable physiological markers that can be applied to assessing the impacts of environmental stressors.

Among these physiological markers, upper thermal physiological limits are some of the most well-developed markers. Thermal tolerance limits of mollusks are primarily shaped by their evolutionary thermal histories. Adaptation to higher temperatures induces genetic modifications that enhance heat tolerance across various biological levels, including the organism, organ, cellular, and molecular domains (Dong et al., 2022). Two commonly employed indices of thermal tolerance are the lethal temperature for 50% individuals (LT50) and critical thermal maximum (CTmax). The LT50 value is derived by exposing organisms to a range of temperatures in a controlled laboratory setting, with the temperature at which 50% of the study organisms experience lethal effects being used as an index of thermal tolerance. This temperature represents the threshold at which physiological systems suffer irreversible failure. In comparison, CTmax is characterized as the temperature at which an organism experiences a significant loss of critical functional capacities, such as maintaining posture or adhering to a substrate. This threshold is often identified as the point at which an organism transitions to a condition often referred to as "ecologically dead," despite potentially remaining physiologically alive. Furthermore, the CTmax is often associated with specific behavioral traits, which may differ among species, complicating cross-species comparisons (Dong et al., 2022).

Cellular physiological responses, particularly mitochondrial functions, are also essential for assessing how mollusks respond to environmental stress. Mitochondria in mollusks, such as mussels, must balance ATP production with stress-induced damage, such as reactive oxygen species (ROS) overproduction (Vakifahmetoglu-Norberg et al., 2017). To mitigate energy deficits during hypoxia or anoxia, many species shift to anaerobic glycolysis, which conserves ATP and minimizes the accumulation of cytotoxic metabolites (Sokolova et al., 2011). Enhanced glycolytic capacity, coupled with tissue buffering and efficient waste excretion, enables survival under

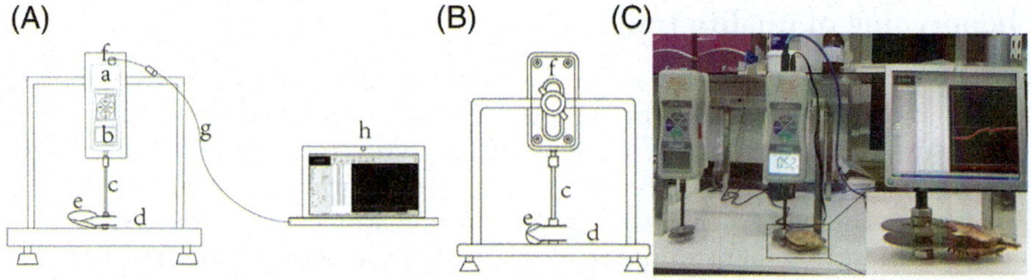

FIGURE 5.2 The self-developed device for measurement of shell-closing strength (Zeng et al., 2025). (A) Front view; (B) Rear view; (C) The real measured device. a. digital force gauge; b. display; c. force measurement axis; d. circular plates; e. scallop; f. vertical adjustment plate; g. data cable; h. computer. Shell-closing strength measurement of the scallop.

prolonged stress (Eberlee & Storey, 1984). Collectively, these adaptations highlight mitochondria as central hubs for stress sensing, energy homeostasis, and mitochondrial respiration, and ROS production are potentially important stress-related traits.

Cellar stress responses (CSR) are also important stress-related traits due to their crucial roles in maintaining protein structure and function. The CSR involves a multitiered hierarchy of processes, including protein repair, degradation of irreparably damaged proteins, and, if necessary, programmed cell death (PCD) (Kültz, 2020). Activation of the CSR begins with the detection of protein damage, which triggers the heat shock response (HSR), a highly conserved stress response across taxa (Feder & Hofmann, 1999). The HSR protects cellular proteostasis by synthesizing heat shock proteins (HSPs), which act as molecular chaperones to refold denatured proteins and prevent aggregation (Somero, 2020). As HSP levels rise, HSP70/90 triggers negative feedback regulation by rebinding to heat shock factor 1 (HSF1), which dissociates from DNA and returns to the cytoplasm, terminating the stress response. This self-limiting regulatory mechanism ensures that cells quickly return to homeostasis after clearing abnormal proteins (Hartl et al., 2011).

Shell-closing strength (SCS), the load value required to open the shell, was reported to be a useful selection indicator for breeding pearl oysters with a high survival rate in summer, because pearl oysters with strong SCS had lower mortality because of malnutrition, and SCS was a heritable character. Since pearl oysters with weak SCS had a low filtering ability and SCS is related to the muscular weight of pearl oysters and the glycogen content of adductor muscles, SCS was thought to reflect the overall health of pearl oysters. Consequently, SCS serves as a reliable indicator for evaluating the performance of the adductor muscle in bivalves. The force gauge, serving as the primary instrument for quantifying SCS, was both economical and simple. Moreover, the noninvasive measurement process preserves the integrity of bivalves, thereby highlighting the significant practical applications of SCS (Aoki et al., 2010). Recently, a rapid and noninvasive method was developed to quantify SCS with four indicators, including total force (TF), time of shell closing (TSC), average force (AF), and maximum force (MF), using a self-developed force gauge device (Fig. 5.2). Subsequent analysis revealed that the four SCS indicators of scallops were significantly associated with the adductor muscle weight (AMW) and the adductor muscle glycogen (AMG) content. Furthermore, scallops with stronger SCS were proven to have longer survival times under heat stress, with correlation coefficients ranging from 0.455 to 0.619 ($P < 0.01$). This finding indicated that SCS could serve as a reliable metric for assessing the heat resistance of scallops (Zeng et al., 2025).

5.3.3 Phenotyping of quality traits

Phenotyping of quality traits in mollusks is a critical area of research that focuses on understanding and improving the various characteristics that determine the quality of these aquatic organisms. For a long time, meat yield, growth, and mortality rate have been the primary selective traits pursued in breeding programs for marine economic species. However, in recent years, due to intensified competition, the marine aquaculture industry faces challenges in innovation and improvement in order to obtain unique products with guaranteed quality (Brake et al., 2003). Research focusing on quality traits has gradually become an important direction for breeding work. Quality traits, such as shell color, muscle quality, blood color, and shell morphology, directly affect the economic value and market competitiveness of marine mollusks. In addition, they reflect the genetic diversity and environmental adaptability of these species (Mizuta & Wikfors, 2019).

The shell color of mollusks exhibits considerable diversity and is often regarded as a typical qualitative trait, with variations such as black, white, orange, or partial coloration (Wang et al., 2023a). These color variations are usually controlled by a limited number of genes associated with major pigments, including melanin, carotenoids, and porphyrins (Hu et al., 2025). Shell color not only serves as an important basis for taxonomy but is also closely related to the environmental adaptability and market value of mollusks (Han & Li, 2020). As a result, the shell color of many mollusks, such as scallops, clams, and oysters, has been used in breeding programs to develop new strains (Evans et al., 2009).

During the breeding process of the scallop *Patinopecten yessoensis*, a rare variant with orange adductor muscle instead of the usual white one was discovered, occurring in approximately 0.2% of the natural population (Li et al., 2010). The strain named "Haida golden scallop" was selected based on the color of the adductor muscle (Fig. 5.3). Further analysis showed that the color of the orange adductor was because of the high carotenoid content, which is not only highly nutritious but also aligns with consumer demand for premium marine products (Li et al., 2022). This color variation is regulated by a limited number of key genes, such as those involved in carotenoid biosynthesis, and is unaffected by environmental factors, thus representing an important quality trait (Li et al., 2023).

In addition, the *Tegillarca granosa* (blood clam) is a bivalve mollusk belonging to the family Veneridae. Because of its unique red blood, it is commonly referred to as the "blood clam." The red blood phenotype is an important economic trait of *T. granosa*, reflecting the degree of redness in its blood. The red blood phenotype specifically refers to the pigmentation level of the hemolymph, with hemoglobin concentration (HGC) and total hemocyte count (THC) serving as key indicators of this phenotype (Jin et al., 2024). The red blood phenotype is an indicator of the health status of blood clams to some extent. Specifically, among a batch of blood clams, individuals with lighter blood color typically indicate poor health, which may reduce consumer interest in purchasing them (Fig. 5.4) (Yang et al., 2023). The blood color of *T. granosa* is also a typical qualitative trait. Some blood clams exhibit deep red or light red blood, and this color variation is related to the type and content of hemoglobin. Studies have shown that targeted knockdown of the Tg-TAL1 gene expression can significantly lighten the blood color (Jin et al., 2024). These quality traits are not only of biological significance but also play a crucial role in economic value and breeding applications.

In the phenotypic determination of qualitative traits, researchers have developed various phenotyping techniques. For the analysis of shell color and body color, common methods

FIGURE 5.3 **The selected scallop strain "Haida golden scallop" (Patinopecten yessoensis) with orange adductor muscle.**

include visual observation and image analysis techniques. Visual observation involves classifying shell color or body color by the naked eye, while image analysis techniques use digital image processing software to analyze color parameters. For instance, digital image analysis was used to calculate light density for determining shell pigmentation (Brake et al., 2004). More recently, a computer vision system (CVS) was used to assess traits related to shell color (Wan et al., 2017). In addition, spectral analysis is widely used for measuring shell color and body

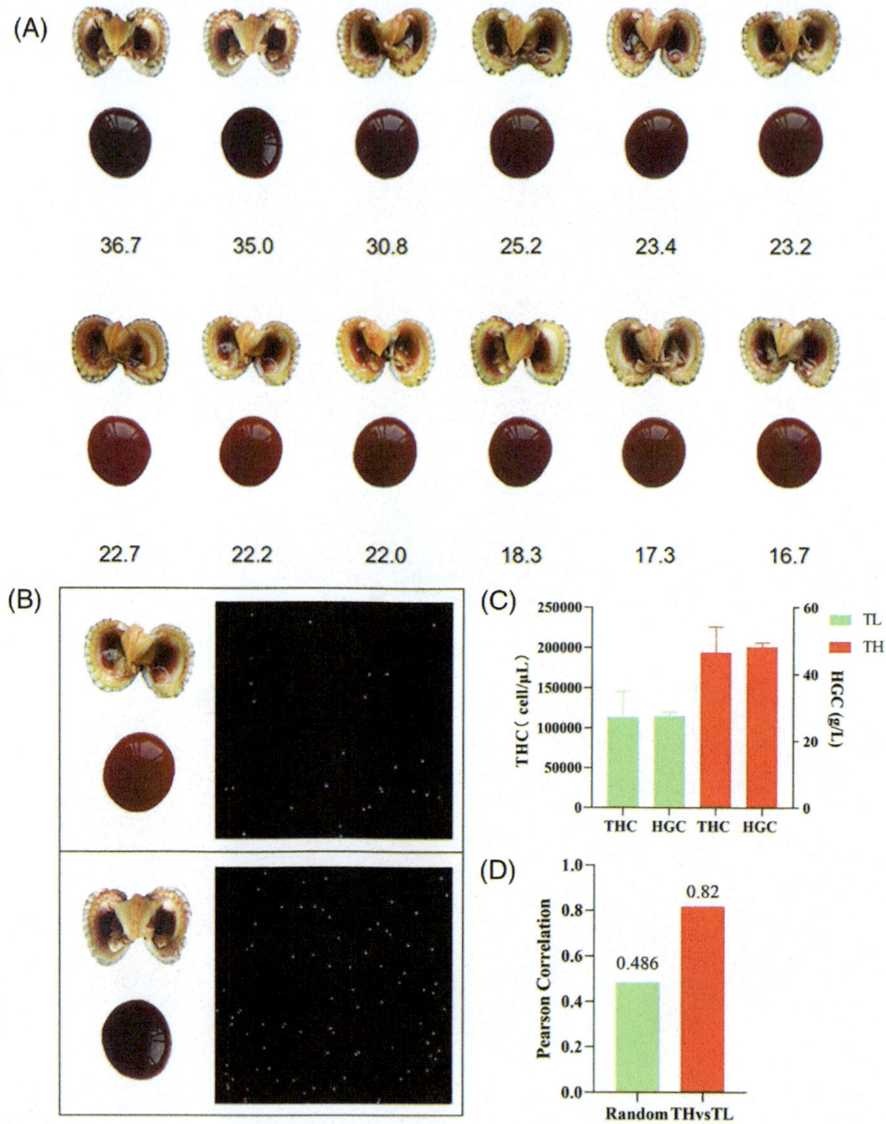

FIGURE 5.4 The red blood phenotype is an indicator of the health status of blood clams *Tegillarca granosa*. (A) Blood color and total hemocyte count (THC) (g/L). (B) Association between blood color shade and THC (stained by Hoechst). (C) Differences in THC and hemoglobin concentration (HGC) between the light-blood clams (HGC> 45.7 g/L, designated TL) and dark-blood clams (HGC < 31.3 g/L, designated TH). (D) Pearson correlation of HGC and THC in the TH and TL groups and of HGC and THC in random individuals. *From Yang et al. (2023).*

color by analyzing the pigment composition through reflectance spectroscopy. To determine the adductor muscle color in "Haida golden scallop," high-performance liquid chromatography (HPLC) technology was used to confirm that this color-related phenotype is associated with carotenoids (Li et al., 2022). Spectral analysis is a commonly used phenotyping technique that measures a substance's absorption, reflection, or emission characteristics at specific wavelengths of light to analyze its composition, structure, and properties. It is widely applied in both qualitative and quantitative analysis. For blood color determination, spectrophotometry and visual observation are the primary techniques. Spectrophotometry quantifies blood color by measuring the absorbance of hemoglobin at specific wavelengths. Moreover, blood cell counting and flow cytometry are also used to analyze blood cell quantity and type, providing valuable insights into the erythrocyte phenotype of blood clams (Jin et al., 2024).

Therefore research on quality traits can not only enhance the economic value of marine mollusks but also drive innovation in breeding technologies and contribute to the sustainable development of the industry (Mizuta & Wikfors, 2019).

5.3.4 Phenotyping of reproduction traits

Phenotypic analysis of reproductive traits in mollusks is crucial for understanding their reproductive biology and applications in aquaculture and conservation. Traditionally, gender identification within this analysis has relied on morphological or behavioral indicators. However, advancements in molecular biology have introduced molecular markers as reliable methods for gender identification. These markers, such as β-catenin, FOXL2, DMRT1, and SOXE, exhibit sex-specific expression patterns (Li et al., 2018a; Santerre et al., 2014). For instance, a gender identification method was established according to the expression of sex-determination genes FOXL2 and DMRT in *P. yessoensis*, utilizing the LOG_{10}(DMRT/FOXL2) ratio (Li et al., 2018a). The dynamic changes in gene expression ratios offer insights into the molecular mechanisms underlying sex differentiation and gonadal development. These discoveries have provided a high degree of accuracy and reproducibility in gender identification, significantly enhancing the precision of sex determination in mollusks.

Gonadal development is critical for ensuring effective reproduction, and histological methods are among the most commonly used techniques for studying this process. Paraffin sectioning, a classic method, categorizes gonadal development into different stages by observing the occurrence, morphological characteristics, and distribution of germ cells. The use of antibodies designed against specific developmental processes of germ cells, coupled with immunofluorescence techniques, allows for further investigation into the expression of certain genes within cells during gonadal development. Techniques such as transmission electron microscopy (TEM) and scanning electron microscopy (SEM) provide a more microscopic perspective on changes in germ cells during gonadal development. Using endoscopy, scallop gender and gonadal maturity can be rapidly identified (Kong et al., 2025). These observations, in conjunction with gene expression analysis, play a significant role in determining the sex fate of gonads.

Mature gametes are a necessary condition for reproduction, and assessing gamete viability is essential for both industrial production and for understanding mollusk reproduction. Sperm viability and motility are important characteristics for evaluating reproductive success. Various techniques, including computer-assisted sperm analysis (CASA) and fluorescence staining (Kong & Sokolova, 2024), are used to assess sperm quality. CASA provides a detailed analysis of sperm motility and viability, while fluorescence staining offers a visual assessment of sperm viability. Furthermore, dual fluorescence staining combined with flow cytometry allows for rapid evaluation of a large number of sperm. These methods are crucial for understanding the factors affecting sperm performance and the reproductive potential of mollusk populations.

5.3.5 Phenotyping of aging-related traits

Mollusks have become fascinating aging research models by virtue of their remarkable species-specific lifespan plasticity (1~500 years). The assessment system for the aging-related phenotypic traits of mollusks usually involves a combination of molecular, cellular, and physiological parameters. Enzyme activity assays, especially the determination of antioxidant enzymes, are widely used to evaluate the level of oxidative stress. Telomere length measurements provide insights into genomic stability and cellular aging. In addition, proteomic and transcriptomic analyses have become increasingly valuable tools for identifying age-related changes in protein expression and gene regulation. This section reviews the research progress on aging-related traits in mollusks. It also synthesizes phenotypic evaluation systems and research technologies focused on aging markers such as enzymatic activity and telomere length, while addressing species-specific variations and methodological advancements.

Enzymatic activities play a pivotal role in maintaining cellular homeostasis and regulating metabolic processes. In mollusks, changes in enzyme levels and activities have been linked to aging (Lian et al., 2019). Bivalve mollusks, which can be collected from diverse environmental contexts with documented chronological age data at the individual level, serve as a robust model for investigating the temporal dynamics of environmental variables and stress-related molecular mechanisms underlying organismal aging (Appeltans et al., 2012). Oysters, scallops, and abalones are among the most studied bivalve mollusks in aging research. Comparative analyses of marine bivalves have demonstrated age-dependent progressive attenuation in key enzymatic antioxidant defenses, particularly superoxide dismutase (SOD; EC 1.15.1.1) and catalase (CAT; EC 1.11.1.6) activities (Lian et al., 2019). The research highlighted that genes related to DNA damage response, cellular senescence, and stress resistance exhibit convergent evolution in long-lived species, suggesting a potential molecular framework for their extended lifespans (Appeltans et al., 2012). Enzymatic activity, particularly in antioxidant enzymes such as superoxide dismutase and catalase, plays a crucial role in mitigating oxidative stress, a key driver of aging in these organisms (Lian et al., 2019).

Telomeres, specialized nucleoprotein complexes located at chromosomal termini, undergo progressive attrition during successive cellular replication cycles, functioning as conserved biomarkers of organismal senescence (Zhu et al., 2019). In marine mollusks, empirical evidence from telomeric investigations demonstrates age-dependent diminution patterns across taxonomic groups. The shortening of telomeres in tissues of *A. irradians* may determine the short lifespan of this scallop (18~22 months), and note a correlation of telomere length with lifespan by comparing telomere length of *A.irradians* with that of the longer-lived *A. purpuratus* (7 years) (Estabrooks, 2007). This attrition mechanism appears evolutionarily conserved within Bivalvia; clams and mussels are particularly notable for their exceptionally long lifespans, with some individuals exceeding 500 years (Bodnar, 2009; Zhu et al., 2019). These species have been found to exhibit minimal signs of cellular damage, including reduced telomere shortening rates compared to shorter-lived counterparts (Bodnar, 2009). As corroborated by cross-species investigations documenting inverse proportionality between telomere length quantitation metrics and gerontological indices, thereby implicating telomeric dynamics in the regulation of invertebrate cellular senescence mechanisms.

While general trends in enzymatic activities and telomere length are observed across molluscan species, there are also notable species-specific variations. For example, certain cephalopod species, known for their rapid growth and short lifespans, display less pronounced changes in telomere length, possibly due to their distinctive biological characteristics. Cephalopods, such as the common cuttlefish (*Sepia officinalis*), offer unique insights into aging due to their relatively short lifespans and complex behaviors. Research on these species has focused on molecular mechanisms of aging, including changes in mitochondrial function and oxidative damage (Zielinski & Pörtner, 2000). Enzymatic activity assays have shown that cephalopods exhibit significant declines in antioxidant capacity with age, which may play a role in their rapid senescence (Zielinski & Pörtner, 2000).

In conclusion, mollusks provide valuable models for studying aging due to their diverse lifespans and unique biological characteristics. Research on enzymatic activity, telomere length, and other aging-related traits has significantly advanced our understanding of the molecular mechanisms underlying longevity in these organisms.

5.4 High-throughput phenomics

5.4.1 Morphological phenomics

Traditional morphometric analyses predominantly depend on manual methodologies, such as caliper-based measurements. These labor-intensive procedures are not only time-consuming but also susceptible to human error, which may introduce observational bias. These challenges are further exacerbated when handling large cohorts or removing specimens from aquatic environments for measurement purposes (Zenger et al., 2019). Consequently, high-throughput and precise phenotyping is imperative for molluscan breeding programs, particularly in the era of multiomics-assisted breeding that necessitates individual-level accuracy. High-throughput morphological phenomics emphasizes the comprehensive and systematic analysis of morphological phenotypes using high-throughput technologies, characterized by large-scale, rapid, and automated workflows. The objective is to efficiently and accurately collect extensive morphological phenotypic data from individuals or cells, thereby elucidating the diversity and complexity of morphological phenotypes and their interactions with genetic and environmental factors.

To acquire morphological phenotypic data efficiently, emerging technologies such as digital imaging, computer vision, remote sensing, and multisensor systems are increasingly applied to plants, animals, and microorganisms (Ziegler et al., 2018). For molluscan research, high-throughput morphological phenomics relies on a framework integrating three-dimensional (3D) imaging technologies and intelligent analytical algorithms, tailored to sample characteristics and scientific objectives, enabling multidimensional and multiscale investigations. Digital imaging technology has provided new possibilities for the study of malacology. It offers a method for rapidly obtaining complex anatomical features (Ziegler et al., 2010). This technology allows for the quick acquisition of digital data from live, fixed, and fossil specimens without causing damage, and in some cases, it is even completely non-invasive (Zenger et al., 2019). As a result, it enables large-scale comparative analyses both within and across taxonomic units. Time-of-flight (ToF) cameras, for instance, generate depth distribution maps within 2D images to quantify

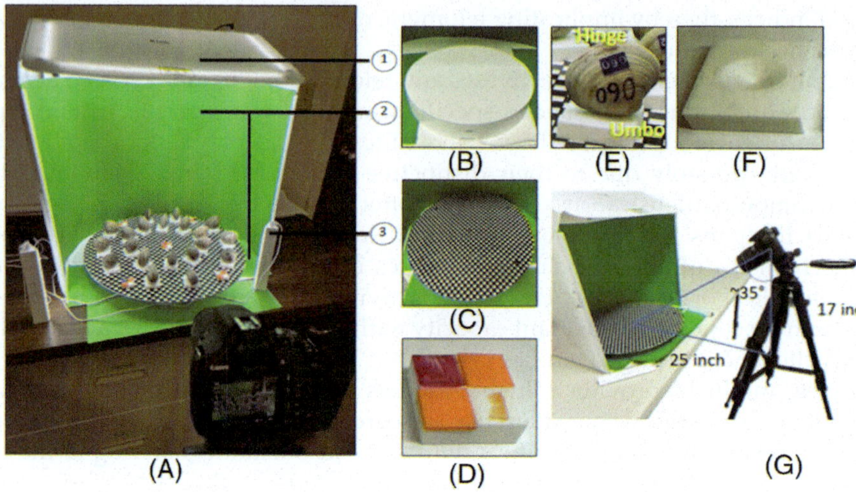

FIGURE 5.5 The system for capturing digital 3D replicas of hard clams. (A) Video collection with (1) the lightbox studio, (2) green background, (3) standalone light sources, (B) the turntable, (C) the plate covered by a round checkerboard, (D) the marker for spatial correction, (E) clam orientation, (F) the clam holder, and (G) camera and tripod setup. *From Shen et al. (2025).*

animal morphology (Ziegler et al., 2010, 2018, 2019). A rapid and nondestructive method for assessing scallop muscle traits was established by using X-ray technology. This technique presents a clear image of the shell and adductor muscle cross-section for calculating adductor muscle area (AMA) and shell area (SA) (Zhao et al., 2021). Similarly, laser scanning techniques rapidly capture 3D surface data by emitting and measuring reflected laser beams (e.g., oysters, scallops, and abalone) (Duan et al., 2025; Shen et al., 2025). For example, an automated system integrating 3D laser scanning with deep learning algorithms has been developed in northern quahogs, *Mercenaria mercenaria*, for dimensional trait measurement and weight prediction (Fig. 5.5) (Duan et al., 2025; Shen et al., 2025). The combination of magnetic resonance imaging (MRI) and computed tomography (CT) technology can cover the morphological characteristics of mollusks at different scales, ranging from the millimeter level to the meter level. MRI does not involve radiation and is suitable for the visualization of the soft tissues of mollusks, enabling the visualization of both the internal and external structures of mollusks (Pouvreau et al., 2006). CT is particularly suitable for the rapid 3D visualization of the morphology of the hard parts of mollusks, including spicules, tooth bases, and ligaments, especially the micro- and macrostructures of mollusk shells, which was used to study the morphology of the odontophoral cartilages of gastropods (Shen et al., 2025; Kong et al., 2025). Microcomputed tomography (μCT): μCT is a more advanced CT technology. It utilizes synchrotron radiation or microfocus X-ray sources and software algorithms designed for 3D image reconstruction to achieve micron- and nanometer-scale resolution. Existing research mainly focuses on the pore structure of shells, and the morphology of spicules and other related aspects (Alba-Tercedor & Sánchez-Tocino, 2011). In addition to the techniques mentioned above, MRI and CT have also undergone expansions and advances, including functional MRI (fMRI), diffusion tensor imaging (DTI), and synchrotron-based μCT (SRμCT).

(A)

(B)

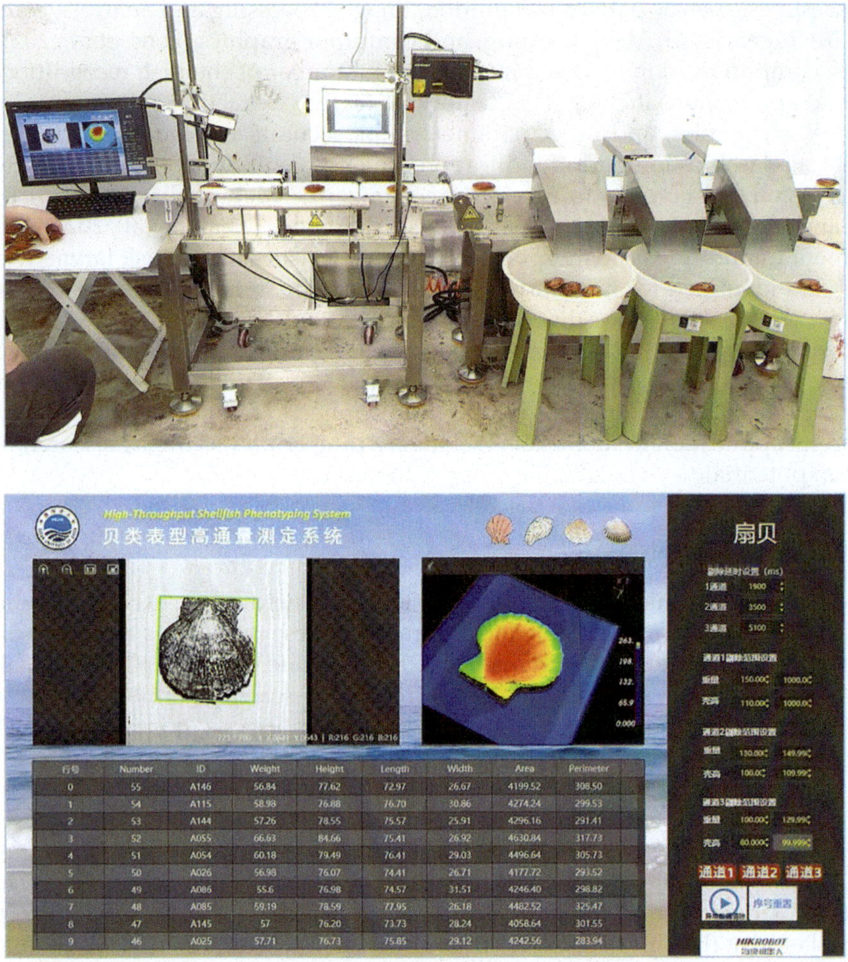

FIGURE 5.6 Automated phenotyping system for growth traits in bivalves (Kong et al., 2025). (A) High-throughput bivalve phenotyping system, and (B) intelligent management software. *From Kong et al. (2025).*

With the rapid development of machine learning, computer vision-related technologies have been applied to various fields and have begun to be applied in aquaculture. Through image recognition and machine learning algorithms, the automatic extraction and classification of mollusk phenotypic features can be realized. For example, the automated phenotyping system for growth traits in bivalves can accurately measure the body size parameters of SL, SW, and SH by using the contour recognition algorithm based on multi-view images (including side view and top view). Through the external rectangle fitting and feature point calibration technology, rapid screening of batch samples and low-cost, high-throughput phenotypic determination was achieved, which significantly improved the detection efficiency (Fig. 5.6) (Kong et al., 2025). The detection and recognition technology of mollusk morphological phenotype based on computer vision is essentially an image analysis, which needs to involve the theoretical knowledge of many disciplines, such as pattern recognition, image processing and analysis, computer vision, artificial intelligence (AI), deep learning, and computer graphics (Feng et al., 2021). In the past few years, computer vision has been involved in a number of shellfish recognition studies, and its potential applications include: (1) Phenotypic feature extraction and quantification, such as SH, SL, SW, shell color, and other parameters for accurate measurement (Feng et al., 2021). (2) Biological classification and identification. The convolution neural network technology in computer vision can be used to precisely recognize the high-similarity shellfish through the learning and training of a large number of shellfish images (Zhang et al., 2023). (3) Growth and development monitoring. The morphological characteristics of mollusk larvae were effectively extracted and identified for monitoring the early growth and development stages of mollusks (Wang et al., 2023b). (4) Disease diagnosis and prevention. Through the analysis of the abnormal characteristics of mollusk body surface, the potential signs of disease can be found in time, which is helpful to take preventive measures and reduce the impact of disease on the breeding industry (Fitzgerald et al., 2025). This technology has been applied in fish disease detection (Ahmed et al., 2022), and its application in mollusks is relatively rare, but it has broad research prospects and application potential.

5.4.2 Physiological phenomics

Physiological phenomics of mollusks is an emerging field that seeks to understand the complex interactions between the physiological traits of these organisms and their environments. Mollusks, a diverse group of invertebrates, exhibit a wide range of physiological adaptations that allow them to thrive in various habitats, from deep-sea environments to terrestrial ecosystems. These adaptations are often reflected across various organizational levels, including organ, cellular, and molecular levels, influenced by both genetic and environmental factors.

5.4.2.1 Cardiac thermal response

Cardiac function plays a crucial role in the thermal adaptation process of mollusks. Since there is a close association between cardiac performance and metabolism in ectotherms. It is essential for determining the optimal thermal temperature, thermal tolerance limits, and adaptive plasticity (Dong et al., 2022). In high-temperature environments, cardiac failure can lead to severe oxygen supply shortages, which is considered a primary determinant of the upper thermal tolerance limit in aquatic ectotherms (Pörtner & Farrell, 2008). In addition, impaired cardiac function caused by high temperatures can lead to a decrease in aerobic metabolism and, in certain cases, an increase in anaerobic metabolism (Liao et al., 2021). Therefore, the response of heart rate (HR) to temperature changes has become an important biological indicator for studying the impact of temperature on species distribution, with applications ranging from microhabitat levels to broad geographic scales (Dong et al., 2017).

The temperature at which cardiac activity reaches its maximum rate is a crucial parameter for characterizing cardiac performance. In studies of acute cardiac thermal responses in intertidal mollusks, five terms are commonly used to describe the temperature point at which the heart rate sharply declines after reaching its maximum: Critical temperature (Tcrit or Hcrit) (Moyen et al., 2019), Arrhenius Break Temperature (ABT) (Stenseng et al., 2005), breakpoint temperature (BPT) (Bjelde & Todgham, 2013), and the temperature of 50% of maximum heart rate (T1/2) (Stenseng et al., 2005). "Critical temperature" usually refers to Tcrit, Hcrit, ABT, or BPT unless a study specifies one of these terms. Although the critical temperature is not an acutely lethal temperature for many mollusks, prolonged exposure can still threaten survival (Dowd & Somero, 2013). For most aquatic ectotherm species, HR declines steadily, often sharply, as the temperature exceeds the critical temperature, eventually stopping at the FLT. Unlike the critical temperature, FLT is invariably lethal for the mollusks discussed here, typically within hours to days. So, it is worth noting that these intertidal mollusks exhibit a distinct relationship between critical temperature and thermal tolerance compared to some other invertebrate species (Moyen et al., 2019; Stenseng et al., 2005).

A non-invasive method based on an infrared (IR) sensor is popular for measuring the heartbeat of mollusks. This method consists of an infrared light-emitting diode and a phototransistor detector. HR signals can be detected by an IR sensor and transduced into changes in electrical current, which are then amplified by an amplifier and processed by software (Chelazzi et al., 1999; Depledge & Andersen, 1990). PowerLab (ADInstruments) and PicoScope (Pico Technology) are two widely used oscilloscopes. PowerLab is a multichannel oscilloscope that allows for multiple individuals to be monitored simultaneously, and PicoScope can be powered by a laptop and can be used for measuring HR in the field.

5.4.2.2 *Cellular response*

In addition to organ-level performance, cellular responses also demonstrate thermal sensitivity and adaptation in mollusks. In some instances, cellular-level thermal damage may occur noticeably before organ-level responses, such as cardiac critical temperature, become apparent (Moyen et al., 2019).

Mollusks serve as an ideal model for studying the effects of temperature on cellular responses. The hemocytes in mollusk hemolymph allow for the examination of temperature effects on cells, making it an easy-to-obtain model. For instance, hemolymph extracted from the adductor muscle of *Mytilus* spp. provides individual cells suitable for experimentation, and fluorescence-activated cell sorting (FACS) can be used for hemocyte analyses (Moyen et al., 2020). Both heat and cold stress significantly increased haemocyte mortality in a time- and temperature-dependent manner. Given that these mussels likely encounter comparable temperature ranges in the wild, the cost of maintaining haemocyte and other cell populations may represent a crucial component of sublethal thermal stress. Regardless, the ability to isolate specific hemocyte populations, such as hyalinocytes vs. granulocytes, opens the door to future downstream analyses of these cell types via FACS. For example, one could explore whether these cell populations respond differently to various stressors based on changes in their transcriptional or proteomic profiles.

In addition, cold stress similarly led to hematological cell death, a finding that has important implications in the context of species range expansion. Although many studies have shown that as global warming occurs, species migrate to higher latitudes to escape warming-induced heat stress, the limiting effect of low winter temperatures on further range expansion is often overlooked. For instance, research on *Nerita yoldii* along the Chinese coast indicates that low winter temperatures could limit the northward expansion of warm-adapted species (Wang et al., 2020). In the case of the southern species *N. yoldii*, for example, low winter temperatures resulted in high mortality rates (95%–100%) despite the expansion of its northern range in the last decade (Wang et al., 2018). Cold stress-induced decreases in haematological cell survival may reflect the cellular damage mechanism behind this cold-limited distribution expansion. Therefore not only do the effects of high temperatures need to be considered when predicting species range changes, but the potential limiting effects of low temperatures on cell and individual survival also need to be fully considered.

5.4.2.3 *Protein stability*

Protein structural stability has long been regarded as an indicator of high-level organizational traits such as thermal tolerance, growth performance, and geographic distribution. Highlighting the importance of protein stability across species and the mechanisms through which adaptive changes in stability occur via amino acid substitutions is essential. The thermal structural stability of cytosolic malate dehydrogenase (cMDH) in *Tegula* species is clearly related to their habitat temperature (Dong et al., 2022). This pattern is consistent with broader observations across 23 marine mollusk species, where cMDH thermal stabilities differ significantly, reflecting their varying upper lethal temperatures (Liao et al., 2019). Similar patterns have been observed for other enzymes, such as isocitrate dehydrogenases (IDHs) in *Mytilus* species (Lockwood & Somero, 2012). Notably, even a single amino acid substitution in a protein of over 333 residues can significantly affect thermal stability (Dong & Somero, 2009). These results further confirm that the structural stability of proteins can serve as an indicator of a species' thermal tolerance. A remarkably high fraction of the total proteome mass in all kingdoms is dedicated to protein homeostasis and folding, highlighting the biological challenge of maintaining protein structure in all branches of life (Müller et al., 2020). With advancements in in silico protein structure simulation techniques, protein structural stability will increasingly become an important stress-related trait for phenomics.

5.4.3 Molecular phenomics

Molecular phenomics in mollusks is a burgeoning field that integrates various omics technologies to unravel the complex biological processes and evolutionary pathways of these diverse organisms. The advent of high-throughput sequencing technologies has significantly advanced our understanding of molluscan genomics, transcriptomics, proteomics, metabolomics, and lipidomics, which are collectively termed molecular phenomics (Fig. 5.7) (Laudicella et al., 2020; Liu et al., 2025). These technologies enable researchers to explore the genetic and molecular foundations of molluscan biology, including development, adaptation, and symbiotic interactions.

Transcriptomics, as an omics field, plays a pivotal role in elucidating the various factors that can modify gene expression profiles. The transcriptome is dynamic and can adapt to both internal and external changes in the subject at any given time. Transcriptomics has significantly advanced our understanding of gene regulation by leveraging high-throughput sequencing technologies such as RNA-Seq, which allowed researchers to identify novel genes, splice isoforms, and RNA editing events, thereby offering insights into the complex regulatory networks that govern cellular functions. For example, transcriptomic analysis reveals genes and pathways involved in gonadal development, sex differentiation, and gametogenesis, providing a deeper understanding of the molecular basis of these processes in mollusks (Yu et al., 2018). Recently, one of the significant advancements in transcriptomics is the development of single-cell RNA sequencing (scRNA-seq), which has revolutionized our understanding of cellular heterogeneity and gene regulatory network plasticity (Iacono et al., 2019). Single-cell transcriptomics enables the analysis of gene expression at an unprecedented resolution, allowing for the discovery of novel cell types and the elucidation of cell-to-cell variability in gene expression (Lin et al., 2021). An increasing number of scRNA-Seq studies have shown that gametogenesis is a continuous and ongoing differentiation process (Green et al., 2018) and has been increasingly applied in

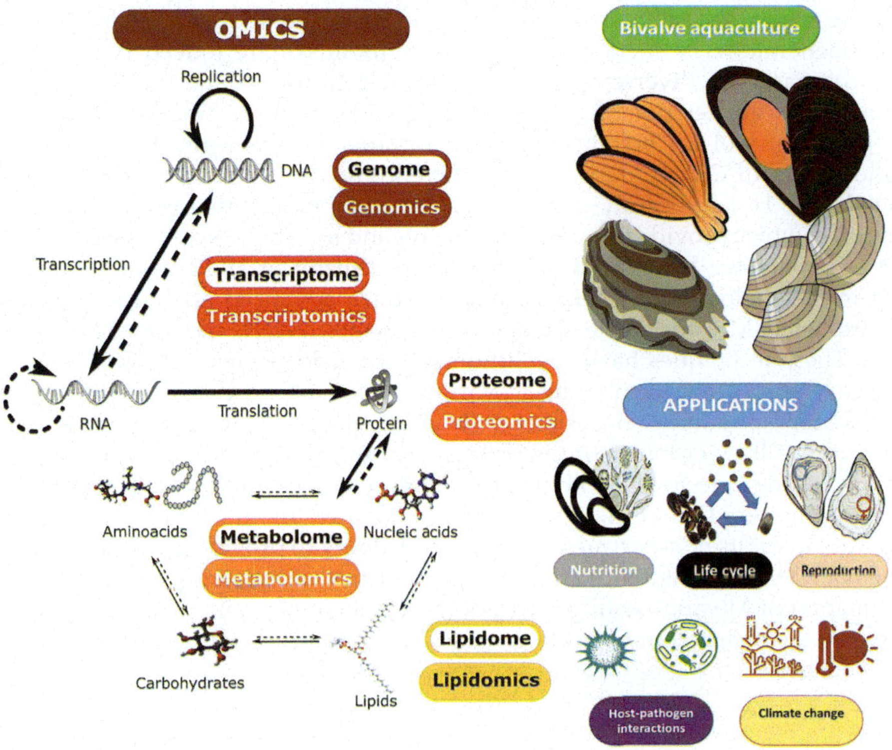

FIGURE 5.7 Schematic relation of molecular biology central dogma and "-omic" sciences adapted from (Barillot et al., 2012).

the study of gonadal development in mollusks. Spatial transcriptomics, another innovative approach, has enabled the high-resolution mapping of gene expression patterns within intact tissues, providing insights into the spatial organization of gene expression (Thrane et al., 2018). By integrating spatial information with gene expression data, researchers can better understand the microenvironmental dynamics and cellular interactions (Thrane et al., 2018).

Recent advancements in structural proteomics have been driven by the introduction of deep neural networks, such as AlphaFold3, ESMFold, and EVOLVEpro (Abramson et al., 2024; Jiang et al., 2025; Lin et al., 2023), which enable highly accurate protein structure prediction from sequence data. Building on this, integrative computational frameworks now combine coevolutionary analysis (e.g., direct coupling analysis), variational autoencoders (VAEs), and molecular dynamics (MD)simulations to map functional landscapes and reveal mechanisms of protein adaptation (Shukla et al., 2025). Proteome-wide 3D structural analyses of ancient bacteria and archaea from deep-sea hydrothermal vents reveal conserved metabolic modules, enabling the reconstruction of ancestral metabolism and tracing of origin (Zhao et al., 2022). These findings validate high-throughput protein structure prediction as a pivotal approach for future evolutionary exploration.

The Investigation of molluscan metabolism and its metabolites is a fascinating area within molecular phenomics. The metabolic reactions of mollusks are indeed complex and multifaceted, encompassing a diverse range of biochemical pathways and physiological processes. Metabolomics has greater potential in predicting complex traits such as growth and flavor, and is a powerful tool for breeding research in aquatic organisms. The metabolic profiles of multiple tissues of *Chlamys farreri* with different growth performance were systematically investigated, and 11 significantly different metabolites (SDMs) associated with growth traits were identified, which provide the crucial pathways and metabolites associated with the growth of *C. farreri* for bivalve breeding and aquaculture (Zhang et al., 2024). The methodologies employed in metabolomics detection primarily utilize advanced analytical techniques, with mass spectrometry (MS) and nuclear magnetic resonance (NMR) spectroscopy being the most prominent. These techniques have revolutionized the field of metabolomics by enabling the comprehensive analysis of metabolites in various biological samples. MS techniques, such as liquid chromatography-mass spectrometry (LC-MS), gas chromatography-mass spectrometry (GC-MS), and capillary electrophoresis-mass spectrometry (CE-MS), are extensively used for metabolite profiling due to their ability to handle complex mixtures and provide detailed molecular information (Lei et al., 2011). NMR is another fundamental tool in metabolomics detection, valued for its non-destructive nature and ability to provide quantitative data across a wide range of metabolites without the need for extensive sample preparation (Brennan, 2014). The combination of MS and NMR techniques in metabolomics will provide a comprehensive perspective on the metabolome, enabling researchers to capture both the breadth and depth of metabolic changes.

5.5 Future trends and challenges of phenotyping

With the completion of whole-genome sequencing for various marine mollusks, research in molluscan functional genomics has entered an era dominated by big data and high-throughput technologies. Nevertheless, the acquisition of large-scale phenotypic data has become a major bottleneck, hindering progress in both marine mollusk breeding and functional genomics research. Traditional methods of phenotypic analysis of mollusks are labor-intensive, time-consuming, subjective, and often destructive to organisms. Although recent advances in phenomics of the aquaculture field have improved the accuracy and reproducibility of growth-related trait measurements (Ventura et al., 2020), these methods still fall short of achieving a high level of intelligence and automation. In contrast, in the field of crop breeding, many automated and intelligent phenotyping technologies and equipment have been widely adopted, such as sensor technologies, machine vision technologies, and automation technologies (Yang et al., 2020). For example, high-throughput rice phenotyping systems integrate color imaging, X-ray CT, automated control, and image analysis pipelines, enabling the simultaneous monitoring of at least 15 agronomic traits in up to 1,920 rice plants (Yang et al., 2014). In addition, traits that are difficult to assess manually, such as leaf rolling and stay-green characteristics in drought-resistant rice, can also be quantitatively analyzed using these systems (Duan et al., 2018). However, in the field of aquaculture, phenomics platforms for key traits such as stress resistance and health status remain underdeveloped. To address the challenges in phenomics for aquaculture species, multidisciplinary collaboration is urgently needed to develop digital phenotyping, remote phenotyping technologies, and automated assessments of animal health and welfare at the whole-body, tissue, and cellular levels (Fu & Yuna, 2022).

In the future, phenotypic analysis of mollusks will undoubtedly move toward intelligence and data-driven approaches. Next-generation phenotyping technologies will integrate advanced physical, chemical, and biological sensors, graphic imaging technologies, AI, and the Internet of Things (IoT). In the field of crop phenomics, transforming raw data into biologically meaningful phenotypic traits has become a critical task. In recent years, the widespread application of computer vision algorithms, machine learning, and deep learning methods in phenotypic data analysis has significantly enhanced the classification, interpretation, and visualization of large-scale phenotypic data (Wen et al., 2024). Therefore in mollusk phenomics research, the development of intelligent and automated phenotyping platforms, along with the creation of new algorithms for intelligent phenotypic data analysis, is not only a key focus for the future but also an inevitable trend. With the continuous advancement of intelligent phenotyping technologies, mollusk phenomics will usher in new opportunities for development. In the future, the achievement of comprehensive, precise, and efficient analysis of mollusk phenotypes will be facilitated through multidisciplinary collaboration and technological innovation, thereby providing robust scientific support for the sustainable development of aquaculture. At the same time, the findings from phenomics research of mollusks will offer valuable insights into understanding the adaptive evolution, ecosystem functions, and the impacts of climate change, thereby driving overall progress in marine science.

References

Abramson, J., Adler, J., Dunger, J., Evans, R., Green, T., Pritzel, A., Ronneberger, O., Willmore, L., Ballard, A. J., Bambrick, J., Bodenstein, S. W., Evans, D. A., Hung, C.-C., O'Neill, M., Reiman, D., Tunyasuvunakool, K., Wu, Z., Žemgulytė, A., Arvaniti, E., … Jumper, J. M. (2024). Accurate structure prediction of biomolecular interactions with AlphaFold 3. *Nature, 630*(8016), 493–500. doi:10.1038/s41586-024-07487-w.

Ahmed, M. S., Aurpa, T. T., & Azad, M. A. K. (2022). Fish disease detection using image based machine learning technique in aquaculture. *Journal of King Saud University - Computer and Information Sciences, 34*(8), 5170–5182. doi:10.1016/j.jksuci.2021.05.003.

Alba-Tercedor, J., & Sánchez-Tocino, L. (2011). The use of the SkyScan 1172 high-resolution micro-CT to elucidate if the spicules of the "sea slugs" (Mollusca: Nudibranchia, Opisthobranchia) have a structural or a defensive function. *SkyScan Users Meeting, 2011.*

Aoki, H., Ishikawa, T., Fujiwara, T., Atsumi, T., Nishikawa, H., Okamoto, C., & Komaru, A. (2010). Utility of shell-closing strength as the indicator of good health in breeding and culture management of Japanese pearl oyster *Pinctada fucata. Aquaculture, 308*(1), S115–S118. doi:10.1016/j.aquaculture.2010.06.041.

Appeltans, W., Ahyong, S. T., Anderson, G., Angel, M. V., Artois, T., Bailly, N., Bamber, R., Barber, A., Bartsch, I., Berta, A., Błazewicz-Paszkowycz, M., Bock, P., Boxshall, G., Boyko, C. B., Brandão, S. N., Bray, R. A., Bruce, N. L., Cairns, S. D., Chan, T. Y., … Costello, M. J. (2012). The magnitude of global marine species diversity. *Current Biology, 22*(23), 2189–2202. doi:10.1016/j.cub.2012.09.036.

Barillot, E., Calzone, L., Hupe, P., Vert, J. P., & Zinovyev, A. (2012). *Computational Systems Biology of Cancer* (1st). CRC Press https://doi.org/10.1201/b12677.

Bayne, B. L. (2004). Phenotypic flexibility and physiological tradeoffs in the feeding and growth of marine bivalve molluscs. *Integrative and Comparative Biology, 44*(6), 425–432. doi:10.1093/icb/44.6.425.

Belcheva, N. N., Zakhartsev, M., Silina, A. V., Slinko, E. N., & Chelomin, V. P. (2006). Relationship between shell weight and cadmium content in whole digestive gland of the Japanese scallop *Patinopecten yessoensis* (Jay). *Marine Environmental Research, 61*(4), 396–409. doi:10.1016/j.marenvres.2005.12.001.

Bjelde, B. E., & Todgham, A. E. (2013). Thermal physiology of the fingered limpet *Lottia digitalis* under emersion and immersion. *Journal of Experimental Biology, 216*(15), 2858–2869. doi:10.1242/jeb.084178.

Bodnar, A. G. (2009). Marine invertebrates as models for aging research. *Experimental Gerontology, 44*(8), 477–484. doi:10.1016/j.exger.2009.05.001.

Brake, J., Evans, F., & Langdon, C. (2003). Is beauty in the eye of the beholder? Development of a simple method to describe desirable shell shape for the Pacific oyster industry. *Journal of Shellfish Research, 22*(3), 767–771.

Brake, J., Evans, F., & Langdon, C. (2004). Evidence for genetic control of pigmentation of shell and mantle edge in selected families of Pacific oysters, *Crassostrea gigas*. *Aquaculture, 229*(1–4), 89–98. doi:10.1016/s0044-8486(03)00325-9.

Brennan, L. (2014). NMR-based metabolomics: From sample preparation to applications in nutrition research. *Progress in Nuclear Magnetic Resonance Spectroscopy, 83*, 42–49. doi:10.1016/j.pnmrs.2014.09.001.

Brown, L. G., & Neville, B. D. (2015). Catalog of the recent taxa of the families Epitoniidae and Nystiellidae (Mollusca: Gastropoda) with a bibliography of the descriptive and systematic literature. *Zootaxa, 3907*(1), 1–188. doi:10.11646/zootaxa.3907.1.1.

Buer, A.-L., Taylor, D., Bergström, P., Ritzenhofen, L., & Klemmstein, A. (2020). Nitrogen and phosphorous content in Blue Mussels (*Mytilus spp.*) across the Baltic Sea. *Frontiers in Marine Science, 7*. doi:10.3389/fmars.2020.00705.

Chelazzi, G., Williams, G. A., & Gray, D. R. (1999). Field and laboratory measurement of heart rate in a tropical limpet, *Cellana grata*. *Journal of the Marine Biological Association of the United Kingdom, 79*(4), 749–751. doi:10.1017/S0025315498000915.

Cook, D. E., Zdraljevic, S., Tanny, R. E., Seo, B., Riccardi, D. D., Noble, L. M., ... Andersen, E. C. (2016). The genetic basis of natural variation in *Caenorhabditis elegans* telomere length. *Genetics, 204*(1), 371–383. doi:10.1534/genetics.116.191148.

Costello, M. J, Claus, S., Dekeyzer, S., Vandepitte, L., Tuama, É. Ó, Lear, D., & Tyler-Walters, H. (2015). Biological and ecological traits of marine species. *PeerJ, 3*(8), e1201. doi:10.7717/peerj.1201.

Damerval, C., Maurice, A., Josse, J. M, & de Vienne, D. (1994). Quantitative trait loci underlying gene product variation: a novel perspective for analyzing regulation of genome expression. *Genetics, 137*(1), 289–301. doi:10.1093/genetics/137.1.289.

Depledge, M. H., & Andersen, B. B. (1990). A computer-aided physiological monitoring system for continuous, long-term recording of cardiac activity in selected invertebrates. *Comparative Biochemistry and Physiology Part A: Physiology, 96*(4), 473–477. doi:10.1016/0300-9629(90)90664-e.

Dong, Y., Li, X. X., Choi, F. M. P., Williams, G. A., Somero, G. N., & Helmuth, B. (2017). Untangling the roles of microclimate, behaviour and physiological polymorphism in governing vulnerability of intertidal snails to heat stress. *Proceedings of the Royal Society B: Biological Sciences, 284*(1854), 20162367. doi:10.1098/rspb.2016.2367.

Dong, Y., & Somero, G. N. (2009). Temperature adaptation of cytosolic malate dehydrogenases of limpets (genus *Lottia*): Differences in stability and function due to minor changes in sequence correlate with biogeographic and vertical distributions. *Journal of Experimental Biology, 212*(2), 169–177. doi:10.1242/jeb.024505.

Dong, Y. W., Liao, M. L., Han, G. D., & Somero, G. N. (2022). An integrated, multi-level analysis of thermal effects on intertidal molluscs for understanding species distribution patterns. *Biological Reviews, 97*(2), 554–581. doi:10.1111/brv.12811.

Dowd, W. W., & Somero, G. N. (2013). Behavior and survival of *Mytilus* congeners following episodes of elevated body temperature in air and seawater. *Journal of Experimental Biology, 216*(3), 502–514. doi:10.1242/jeb.076620.

Duan, L., Han, J., Guo, Z., Tu, H., Yang, P., Zhang, D., ... Yang, W. (2018). Novel digital features discriminate between drought resistant and drought sensitive rice under controlled and field conditions. *Frontiers in Plant Science, 9*, 492. doi:10.3389/fpls.2018.00492.

Duan, S., Gao, Y., Hu, X., & Yin, X. (2025). A novel method for in vivo assessment of soft tissue weight in scallops. *Aquaculture Reports, 43*, 102959. doi:10.1016/j.aqrep.2025.102959.

Eberlee, J. C., & Storey, K. B. (1984). Buffering capacities of the tissues of marine molluscs. *Physiological Zoology, 57*(5), 567–572. doi:10.1086/physzool.57.5.30163950.

Estabrooks, S. L. (2007). The possible role of telomeres in the short life span of the Bay scallop, *Argopecten irradians irradians* (Lamarck 1819). *Journal of Shellfish Research, 26*(2), 307–313. doi:10.2983/0730-8000(2007)26[307:TPROTI]2.0.CO;2.

Evans, S., Camara, M. D., & Langdon, C. J. (2009). Heritability of shell pigmentation in the Pacific oyster, *Crassostrea gigas*. *Aquaculture, 286*(3–4), 211–216. doi:10.1016/j.aquaculture.2008.09.022.

Feder, M. E., & Hofmann, G. E. (1999). Heat-shock proteins, molecular chaperones, and the stress response: Evolutionary and ecological physiology. *Annual Review of Physiology, 61*, 243–282. doi:10.1146/annurev.physiol.61.1.243.

Feng, Y., Tao, X., & Lee, E. J. (2021). Classification of shellfish recognition based on improved faster R-CNN framework of deep learning. *Mathematical Problems in Engineering, 2021*, 1966848. doi:10.1155/2021/1966848.

Fitzgerald, A., Ioannou, C. C., Consuegra, S., Dowsey, A., & Garcia de Leaniz, C. (2025). Machine vision applications for welfare monitoring in aquaculture: Challenges and opportunities. *Aquaculture, Fish and Fisheries, 5*, e70036. doi:10.1002/aff2.70036.

Fu, G., & Yuna, Y. (2022). Phenotyping and phenomics in aquaculture breeding. *Aquaculture and Fisheries, 7*(2), 140–146. doi:10.1016/j.aaf.2021.07.001.

Green, C. D., Ma, Q., Manske, G. L., Shami, A. N., Zheng, X., Marini, S., ... Hammoud, S. S. (2018). A comprehensive roadmap of murine spermatogenesis defined by single-cell RNA-seq. *Developmental Cell, 46*(5), 651–667. doi:10.1016/j.devcel.2018.07.025.

Han, Z., & Li, Q. (2020). Mendelian inheritance of orange shell color in the Pacific oyster *Crassostrea gigas*. *Aquaculture, 516*, 734616. doi:10.1016/j.aquaculture.2019.734616.

Hancock, J. M. (2014). *Phenomics*. Boca Raton, FL: CRC Press.

Hartl, F. U., Bracher, A., & Hayer-Hartl, M. (2011). Molecular chaperones in protein folding and proteostasis. *Nature, 475*(7356), 324–332. doi:10.1038/nature10317.

Jin, H. Y., Liu, H. X., Wang, J. C., Zhang, W. W., & Bao, Y. B. (2024a). CBFA2T3 affects red blood phenotype in the blood clam *Tegillarca granosa* by regulating hemocyte proliferation. *Aquaculture, 581*, 740462.

Jin, H. Y., Liu, H. X., Wang, J. C., Zhang, W. W., Bao, Y. B., et al. (2024b). TAL1-mediated regulation of hemocyte proliferation influences red blood phenotype in the blood clam *Tegillarca granosa*. *Aquaculture, 586*, 740801. doi:10.1016/j.aquaculture.2024.740801.

Hu, B., Yu, H., Xu, C., Kong, L., Liu, S., & Li, Q. (2025). Shell colour diversity in marine molluscs: From current knowledge to future aquaculture applications. *Reviews in Aquaculture, 17*, e70038. doi:10.1111/raq.70038.

Iacono, G., Massoni-Badosa, R., & Heyn, H. (2019). Single-cell transcriptomics unveils gene regulatory network plasticity. *Genome Biology, 20*(1), 110. doi:10.1186/s13059-019-1713-4.

Jiang, K., Yan, Z., Di Bernardo, M., Sgrizzi, S. R., Villiger, L., Kayabolen, A., ... Abudayyeh, O. O. (2025). Rapid in silico directed evolution by a protein language model with EVOLVEpro. *Science, 387*, 6732. doi:10.1126/science.adr6006.

Johannes, F., Colot, V., & Jansen, R. C. (2008). Epigenome dynamics: A quantitative genetics perspective. *Nature Reviews Genetics, 9*(11), 883–890. doi:10.1038/nrg2467.

Kong, H., & Sokolova, I. M. (2024). Oxidative phosphorylation rather than glycolysis is the primary energy source for sperm motility in the mussels *Mytilus edulis*. *Comparative Biochemistry and Physiology Part B: Biochemistry and Molecular Biology, 270*, 110909. doi:10.1016/j.cbpb.2023.110909.

Kong, X., Huang, S., Wang, X., Liang, H., Hu, C., Wang, Y., Bao, Z., & Hu, X. (2025). Development of automated phenotyping system for growth traits in bivalves. *Aquaculture Reports, 44*, 103019. doi:10.1016/j.aqrep.2025.103019.

Kong, X., Liang, H., Huang, S., Bao, Z., & Hu, X. (2025). Rapid identification of scallop gender and gonadal maturity using endoscopy. *Journal of Fisheries of China, 49*(5), 05960. doi:10.11964/jfc.20240214366.

Kültz, D. (2020). Defining biological stress and stress responses based on principles of physics. *Journal of Experimental Zoology Part A: Ecological and Integrative Physiology, 333*(6), 350–358. doi:10.1002/jez.2340.

Laudicella, V. A., Whitfield, P. D., Carboni, S., Doherty, M. K., & Hughes, A. D. (2020). Application of lipidomics in bivalve aquaculture: A review. *Reviews in Aquaculture, 12*(2), 678–702. doi:10.1111/raq.12346.

Lei, Z., Huhman, D. V., & Sumner, L. W. (2011). Mass spectrometry strategies in metabolomics. *Journal of Biological Chemistry, 286*(29), 25435–25442. doi:10.1074/jbc.R111.238691.

Li, N., Hu, J., Wang, S., Cheng, J., Hu, X., Lu, Z., Lin, Z., Zhu, W., & Bao, Z. (2010). Isolation and identification of the main carotenoid pigment from the rare orange muscle of the Yesso scallop. *Food Chemistry, 118*(3), 616–619. doi:10.1016/j.foodchem.2009.05.043.

Li, R., Zhang, L., Li, W., Zhang, Y., Li, Y., Zhang, M., Zhao, L., Hu, X., Wang, S., & Bao, Z. (2018a). FOXL2 and DMRT1L are yin and yang genes for determining timing of sex differentiation in the bivalve mollusk *Patinopecten yessoensis*. *Frontiers in Physiology, 9*. doi:10.3389/fphys.2018.01166.

Li, T., Zhang, Y., Liu, S., Li, M., Yao, R., Niu, S., ... Hu, X. (2023). Transcriptome and network analyses reveal key pathways and genes involved in response to carotenoid deposition in scallop muscle. *Frontiers in Marine Science, 10*, 1158325. doi:10.3389/fmars.2023.1158325.

Li, X., Li, N., Zhao, L., Shi, J., Wang, S., Ning, X., ... Hu, X. (2022). Tissue distribution and seasonal accumulation of carotenoids in Yesso scallop (*Mizuhopecten yessoensis*) with orange adductor muscle. *Food Chemistry, 367*, 130701. doi:10.1016/j.foodchem.2021.130701.

Li, Y., Lu, Y., Lan, Y., Li, Y., Guo, J., & Zheng, R. (2018b). Sr/Ca ratio analysis of seashells using laser-induced breakdown spectroscopy under objective-lens focusing and single-lens focusing. *Applied Optics, 57*(13), 3539–3545. doi:10.1364/AO.57.003539.

Li, Y., Wang, Y., Wang, S., Zheng, R., Bao, Z., & Lu, Y. (2023). Preliminary study of laser-induced breakdown spectroscopy for Mg/Ca investigation in cultured scallop shells. *Journal of Laser Applications, 35*, 022018. doi:10.2351/7.0000937.

Lian, S., Wang, J., Zhang, L., Xing, Q., Hu, N., Liu, S., … Wang, S. (2019). Integration of biochemical, cellular, and genetic indicators for understanding the aging process in a bivalve mollusk *Chlamys farreri*. *Marine Biotechnology, 21*(5), 718–730. doi:10.1007/s10126-019-09917-7.

Liao, M. L., Dong, Y. W., & Somero, G. N. (2021). Thermal adaptation of mRNA secondary structure: Stability versus lability. *Proceedings of the National Academy of Sciences of the United States of America, 118*(45), e2113324118. doi:10.1073/pnas.2113324118.

Liao, M. L., Somero, G. N., & Dong, Y. (2019). Comparing mutagenesis and simulations as tools for identifying functionally important sequence changes for protein thermal adaptation. *Proceedings of the National Academy of Sciences of the United States of America, 116*(2), 679–688. doi:10.1073/pnas.1817455116.

Lin, S., Liu, Y., Zhang, M., Xu, X., Chen, Y., Zhang, H., & Yang, C. (2021). Microfluidic single-cell transcriptomics: moving towards multimodal and spatiotemporal omics. *Lab on a Chip, 21*(20), 3829–3849. doi:10.1039/d1lc00607j.

Lin, Z., Akin, H., Rao, R., Hie, B., Zhu, Z., Lu, W., Smetanin, N., Verkuil, R., Kabeli, O., Shmueli, Y., dos Santos Costa, A., Fazel-Zarandi, M., Sercu, T., Candido, S., & Rives, A. (2023). Evolutionary-scale prediction of atomic-level protein structure with a language model. *Science, 379*(6637), 1123–1130. doi:10.1126/science.ade2574.

Liu, F., Cai, B., Lian, S., Chang, X., Chen, D., Pu, Z., Bao, L., Wang, J., Lv, J., Zheng, H., Bao, Z., Zhang, L., Wang, S., & Li, Y. (2025). MolluscDB 2.0: A comprehensive functional and evolutionary genomics database for over 1400 molluscan species. *Nucleic Acids Research, 53*(1), D1075. doi:10.1093/nar/gkae1026.

Lockwood, B. L., & Somero, G. N. (2012). Functional determinants of temperature adaptation in enzymes of cold-versus warm-adapted mussels (genus *mytilus*). *Molecular Biology and Evolution, 29*(10), 3061–3070. doi:10.1093/molbev/mss111.

Mizuta, D. D., & Wikfors, G. H. (2019). Seeking the perfect oyster shell: a brief review of current knowledge. *Reviews in Aquaculture, 11*(3), 586–602. doi:10.1111/raq.12247.

Moyen, N. E., Bump, P. A., Somero, G. N., & Denny, M. W. (2020). Establishing typical values for hemocyte mortality in individual California mussels, *Mytilus californianus*. *Fish and Shellfish Immunology., 100*, 70–79. Available from http://www.elsevier.com/inca/publications/store/6/2/2/8/3/2/index.htt . doi:10.1016/j.fsi.2020.02.069.

Moyen, N. E., Somero, G. N., & Denny, M. W. (2019). Impact of heating rate on cardiac thermal tolerance in the California mussel, Mytilus californianus. *Journal of Experimental Biology, 222*(17), jeb203166. doi:10.1242/jeb.203166.

Müller, J. B., Geyer, P. E., Colaço, A. R., Treit, P. V., Strauss, M. T., Oroshi, M., Doll, S., Virreira Winter, S., Bader, J. M., Köhler, N., Theis, F., Santos, A., & Mann, M. (2020). The proteome landscape of the kingdoms of life. *Nature, 582*(7813), 592–596. doi:10.1038/s41586-020-2402-x.

Pörtner, H. O., & Farrell, A. P. (2008). Physiology and Climate Change. *Science, 322*(5902), 690–692. doi:10.1126/science.1163156.

Pouvreau, S., Rambeau, M., Cochard, J. C., & Robert, R. (2006). Investigation of marine bivalve morphology by in vivo MR imaging: First anatomical results of a promising technique. *Aquaculture, 259*(1-4), 415–423. doi:10.1016/j.aquaculture.2006.05.018.

Ren, L., Li, S., Ye, W., Lv, Q., Sun, Y., Zhou, X., Lian, S., Lv, J., Wang, S., Guo, J., Tian, Y., Zheng, R., & Lu, Y. (2024). Tracking organic matrix in the seashell by elemental mapping under laser-induced breakdown spectroscopy. *Talanta, 271*, 125658. doi:10.1016/j.talanta.2024.125658.

Santerre, C., Sourdaine, P., Adeline, B., & Martinez, A. S. (2014). Cg-SoxE and Cg-β-catenin, two new potential actors of the sex-determining pathway in a hermaphrodite lophotrochozoan, the Pacific oyster *Crassostrea gigas*. *Comparative Biochemistry and Physiology - A Molecular and Integrative Physiology, 167*, 68–76. doi:10.1016/j.cbpa.2013.09.018.

Shen, K., Wang, X., Gao, K., Yee, J., McDonald, P., & Yang, H. (2025). High-throughput 3D phenotyping in northern quahogs *Mercenaria mercenaria* for dimensional trait measurement and weight prediction. *Smart Agricultural Technology, 10*, 100840. doi:10.1016/j.atech.2025.100840.

Shukla, D., Martin, J., Morcos, F., & Potoyan, D. A. (2025). Thermal adaptation of cytosolic malate dehydrogenase revealed by deep learning and coevolutionary analysis. *Journal of Chemical Theory and Computation, 21*(6), 3277–3287. doi:10.1021/acs.jctc.4c01774.

Sokolova, I., Sukhotin, A., & Lannig, G. (2011). Stress Effects on Metabolism and Energy Budgets in Mollusks. *Oxidative Stress in Aquatic Ecosystems*. Wiley-Blackwell https://doi.org/10.1002/9781444345988.ch19.

Somero, G. N. (2020). The cellular stress response and temperature: function, regulation, and evolution. *Journal of Experimental Zoology Part A: Ecological and Integrative Physiology, 333*(6), 379–397. doi:10.1002/jez.2344.

Stenseng, E., Braby, C. E., & Somero, G. N. (2005). Evolutionary and acclimation-induced variation in the thermal limits of heart function in congeneric marine snails (Genus *Tegula*): Implications for vertical zonation. *Biological Bulletin, 208*(2), 138–144. doi:10.2307/3593122.

Thrane, K., Eriksson, H., Maaskola, J., Hansson, J., & Lundeberg, J. (2018). Spatially resolved transcriptomics enables dissection of genetic heterogeneity in stage III cutaneous malignant melanoma. *Cancer Research, 78*(20), 5970–5979. doi:10.1158/0008-5472.can-18-0747.

Vakifahmetoglu-Norberg, H., Ouchida, A. T., & Norberg, E. (2017). The role of mitochondria in metabolism and cell death. *Biochemical and Biophysical Research Communications, 482*(3), 426–431. doi:10.1016/j.bbrc.2016.11.088.

Ventura, R. V, Silva, F. F., Yáñez, J. M., & Brito, L. F (2020). Opportunities and challenges of phenomics applied to livestock and aquaculture breeding in South America. *Animal Frontiers, 10*(2), 45–52. doi:10.1093/af/vfaa008.

Wan, S., Li, Q., Liu, T., Yu, H., & Kong, L. (2017). Heritability estimates for shell color-related traits in the golden shell strain of Pacific oyster (*Crassostrea gigas*) using a molecular pedigree. *Aquaculture, 476*, 65–71. doi:10.1016/j.aquaculture.2017.04.012.

Wang, J., Yan, H. Y., Cheng, Z. Y., Huang, X. W., Wang, W., Ding, M. W., & Dong, Y. W. (2018). Recent northward range extension of *nerita yoldii* (gastropoda: neritidae) on artificial rocky shores in China. *Journal of Molluscan Studies, 84*(4), 345–353. doi:10.1093/mollus/eyy042.

Wang, W., Wang, J., Choi, F. M. P., Ding, P., Li, X. X., Han, G. D., Ding, M. W., Guo, M., Huang, X. W., Duan, W. X., Cheng, Z. Y., Chen, Z. Y., Hawkins, S. J., Jiang, Y., Helmuth, B., & Dong, Y. W. (2020). Global warming and artificial shorelines reshape seashore biogeography. *Global Ecology and Biogeography, 29*(2), 220–231. doi:10.1111/geb.13019.

Wang, X., Yang, Z., Jiang, L., Liu, Z., Dong, X., Sui, M., Yin, C., Shen, X., Zhao, A., Hu, J., Bao, Z., Hu, L., & Huang, X. (2023a). Assessment of germplasm resource and detection of genomic signature under artificial selection of Zhikong scallop *(Chlamys farreri). Aquaculture, 574*. doi:10.1016/j.aquaculture.2023.739730.

Wang, Y., Wu, H., Hua, X., Ren, D., Li, Y., Mu, Z., Xu, W., Wei, Y., Zhang, T., & Jiang, Y. (2023b). Biological characters identification for hard clam larva based on the improved YOLOX-s. *Computers and Electronics in Agriculture, 212*. doi:10.1016/j.compag.2023.108103.

Wen, W., Gu, S., Zhang, Y., Yang, W., & Guo, X. (2024). Standard framework construction of technology and equipment for big data in crop phenomics. *Engineering, 42*, 175–184. doi:10.1016/j.eng.2024.06.001.

Yang, W., Feng, H., Zhang, X., Zhang, J., Doonan, J. H., Batchelor, W. D., Xiong, L., & Yan, J. (2020). Crop phenomics and high-throughput phenotyping: Past decades, current challenges, and future perspectives. *Molecular Plant, 13*(2), 187–214. doi:10.1016/j.molp.2020.01.008.

Yang, W., Guo, Z., Huang, C., Duan, L., Chen, G., Jiang, N., … Xiong, L. (2014). Combining high-throughput phenotyping and genome-wide association studies to reveal natural genetic variation in rice. *Nature Communications, 5*, 5087. doi:10.1038/ncomms6087.

Yang, Z., He, X., Jin, H., Su, D., Lin, Z., Liu, H., & Bao, Y. (2023). Hemocyte proliferation is associated with blood color shade variation in the blood clam, *Tegillarca granosa. Aquaculture, 571*, 739447. doi:10.1016/j.aquaculture.2023.739447.

Yu, L., Xu, D., Ye, H., Yue, H., Ooka, S., Kondo, H., … Takeuchi, Y. (2018). Gonadal transcriptome analysis of Pacific abalone *Haliotis discus discus*: Identification of genes involved in germ cell development. *Marine Biotechnology, 20*(4), 467–480. doi:10.1007/s10126-018-9809-5.

Zeng, X., Yang, Z., Sui, M., Liu, Z., Xing, Q., Hu, L., … Huang, X. (2025). Rapid and noninvasive assessment of adductor muscle performance through shell-closing strength of Zhikong scallop *Chlamys farreri. Aquaculture Reports, 40*, 102589. doi:10.1016/j.aqrep.2024.102589.

Zenger, K. R., Khatkar, M. S., Jones, D. B., Khalilisamani, N., Jerry, D. R., & Raadsma, H. W. (2019). Genomic selection in aquaculture: Application, limitations and opportunities with special reference to marine shrimp and pearl oysters. *Frontiers in Genetics, 9*, 693. doi:10.3389/fgene.2018.00693.

Zhang, Y., Wang, H., Liu, S., Kong, X., Chang, L., Zhao, L., … Hu, X. (2024). Multi-tissue metabolomic profiling reveals the crucial metabolites and pathways associated with scallop growth. *BMC Genomics, 25*(1), 1091. doi:10.1186/s12864-024-11016-4.

Zhang, Y., Wang, L., Sun, X., Dong, J., Zhan, Q., & Zhang, X. (2024). Trophic niche partitioning of golden cuttlefish (*Sepia esculenta*) during ontogeny revealed by stable isotope analysis and DNA metabarcoding. *Fisheries Research, 271*, 106921. doi:10.1016/j.fishres.2023.106921.

Zhang, Y., Yue, J., Song, A., Jia, S., & Li, Z. (2023). A High-similarity shellfish recognition method based on convolutional neural network. *Information Processing in Agriculture, 10*(2), 149–163. doi:10.1016/j.inpa.2022.05.009.

Zhao, L., Lian, S., Ren, Q., Yang, Z., Guo, Z., Lou, J., … Hu, X. (2021). *In vivo* and rapid assessment of scallop muscle trait. *Aquaculture, 530*, 735817. doi:10.1016/j.aquaculture.2020.735817.

Zhao, W., Zhong, B., Zheng, L., Tan, P., Wang, Y., Leng, H., … Xiao, X. (2022). Proteome-wide 3D structure prediction provides insights into the ancestral metabolism of ancient archaea and bacteria. *Nature Communications, 13*(1), 7861. doi:10.1038/s41467-022-35523-8.

Zhu, Y., Liu, X., Ding, X., Wang, F., & Geng, X. (2019). Telomere and its role in the aging pathways: telomere shortening, cell senescence and mitochondria dysfunction. *Biogerontology, 20*(1), 1–16. doi:10.1007/s10522-018-9769-1.

Ziegler, A., Bock, C., Ketten, D. R., Mair, R. W., Mueller, S., Nagelmann, N., Pracht, E. D., & Schröder, L. (2018). Digital three-dimensional imaging techniques provide new analytical pathways for malacological research. *American Malacological Bulletin, 36*(2), 248–273. doi:10.4003/006.036.0205.

Ziegler, A., Ogurreck, M., Steinke, T., Beckmann, F., Prohaska, S., & Ziegler, A. (2010). Opportunities and challenges for digital morphology. *Biology Direct, 5*, 45. doi:10.1186/1745-6150-5-45.

Zielinski, S., & Pörtner, H. O. (2000). Oxidative stress and antioxidative defense in cephalopods: A function of metabolic rate or age? *Comparative Biochemistry and Physiology - B Biochemistry and Molecular Biology, 125*(2), 147–160. doi:10.1016/S0305-0491(99)00162-5.

CHAPTER

6

Hybridization and polyploidy breeding

Yuehuan Zhang[a], *Weiwei You*[b], *Chunde Wang*[c] *and Karsoon Tan*[d]

[a] **South China Sea Institute of Oceanology, Chinese Academy of Sciences, Guangzhou, China** [b] **Xiamen University, Xaimen, China** [c] **Qingdao Agricultural University, Qingdao, China** [d] **College of Marine Science, Beibu Gulf University, Qinzhou, Guangxi, China**

6.1 Introduction

The term "blue foods" typically refers to aquatic organisms such as fish, shellfish and algae. Among these, shellfish account for a significant yet often overlooked portion(Cao et al., 2023). In 2024, global shellfish production reached 20.50 million tons, showing an upward trend (FAO, 2025). China alone contributed 17.39 million tons, accounting for 84.83% of the world's total production (China Fishery Statistical Yearbook, 2025). Genetic improvement has played an important role in advancing shellfish breeding, with notable progress achieved through various methods, including selective breeding, family breeding, hybridization, and polyploid breeding (Hulata, 1995; Zhang et al., 2012).

Hybridization is defined as crossing between two different individuals or higher-level taxa, facilitating genome transfer, leading to phenotypic and genotypic changes in the progeny. Based on parental genetic distance, hybridization can be categorized into strain hybridization, population hybridization, interspecific hybridization, and intergenus hybridization (Zhang et al., 2012). Hybridization has been shown to enhance growth performance, improve meat quality, and strengthen disease resistance in shellfish (Hulata, 1995). Beyond generating heterosis, it serves as an effective strategy for expanding genetic diversity and developing new varieties (Bartley et al., 2000). However, not all interspecific hybridization results in heterosis. In some cases, genetic incompatibilities between parental lineages can lead to outbreeding depression or hybrid breakdown, reducing fitness. Therefore continuous hybridization experiments are necessary to identify and select hybrids with desirable characteristics.

Genetic Breeding and Aquaculture of Shellfish: A Collaborative Approach to a Sustainable Future
DOI: https://doi.org/10.1016/B978-0-443-33606-5.00007-2

Polyploid organisms possess additional chromosome sets beyond the typical diploid number. Triploids, organisms with three sets of homologous chromosomes, which arise spontaneously or through artificial induction, are particularly valuable in aquaculture due to their sterility, which often translates into enhanced growth by redirecting energy from reproduction to somatic development. Tetraploid shellfish play an important role in the global shellfish industry because crossing them with diploids can produce 100% triploid offspring (Arnold, 2004). Since the first successful induction of tetraploid *Crassostrea gigas*, viable tetraploid oysters have been reported in *C. virginica*, *C. ariakensis*, *C. angulata*, etc. In fact, hybridization and polyploidization are powerful tools for genetic enhancement in aquaculture. Despite the scarce reports of heterosis in shellfish, hybrids may combine advantageous traits from both parental species, such as faster growth, improved environmental tolerance, or overall hardiness. Triploid organisms are usually sterile, a characteristic that proves advantageous in aquaculture as it often results in superior growth performance (Zhang et al., 2017). When hybridization is combined with polyploidization to create triploid hybrids (allotriploids), the sterility effect is typically enhanced because of additional complications in gametogenesis. On the other hand, these triploid hybrids may combine desirable traits from both parental species, potentially exhibiting increased growth rates, enhanced environmental tolerance, or greater overall hardiness under culture conditions.

It can be seen that both hybridization of the two types of fusion and polyploid breeding with chromosome doubling have created significant advantages, playing an important role in the genetic improvement and industrial development of global aquatic seed stocks.

6.2 Hybridization

Shellfish hybridization, which originated with oysters, has a history of 140 years. The hybridization process can be divided into four stages: (1) From 1882 to 1950, scientists focused on determining whether oviparous and ovoviviparous mollusks could successfully hybridize; (2) From 1950 to 1993, scientists carried out distant hybridization experiments. However, because of the lack of genetic verification of hybrid offspring, the results were often compromised by gamete contamination, larval contamination, or gynogenesis, etc. Consequently, most findings from this era remained to be confirmed by molecular evidence. (3) From 1993 to 2009, due to the influence of the concept of inactivation of interspecific hybrids, research efforts successfully produced almost exclusively true hybrid offspring. (4) Since 2009, breakthroughs have been achieved in shellfish hybrid breeding, resulting in a number of superior varieties with significant advantages in species such as oysters, scallops, and abalones.

6.2.1 Natural hybridization

Natural hybridization serves as a fundamental evolutionary mechanism that facilitates gene introgression, the transfer of genetic material between hybridizing taxa through backcrossing, and then the formation of new varieties or even new species (Arnold, 2004). While natural hybridization is commonly observed in fish populations, it occurs less frequently in shellfish. Current reports of natural hybridization in shellfish have primarily focused on mussels, with only rare cases reported in clams and oysters. Specific natural hybridization events for shellfish are shown in Table 6.1.

TABLE 6.1 Natural hybridization of shellfish.

Order	Nature hybridization	Main results	References
1	*Crassostrea hongkongensis*♀ × *C. ariakensis*♂	Only a few hybrids derived from *C. hongkongensis* and *C. ariakensis* were found in Southern China.	Zhang et al. (2017)
2	*C. sikamea*♀ × *C. gigas*♂	Only one hybrid derived from *C. sikamea* and *C. gigas* was identified from Janap samples.	Camara et al. (2008)
3	*Ruditapes decussatus* × *R. philippinarum*	Only a small number of hybrids derived from *Ruditapes decussatus* × *R. philippinarum* was confirmed by molecular markers and their morphs in Spain and Italy.	Hurtado et al. (2011), Carugati et al. (2024)
4	*Mercenaria mercenaria* × *M. campechiensis*	Six hybrids were found from the Indian River of Florida.	Dillon and Manzi (1989)
5	*Mytilus trossulus* × *M. edulis*	A natural hybrid zone exists between two species, formed through crossing and backcrossing, in Atlantic Canadian waters, Newfoundland, and Greenland.	Comesaña et al. (1999), Toro et al. (2004), Wenne et al. (2016)
6	*M. galloprovincialis* × *M. trossulus*	Natural hybrids dfrom these two species were found at Gavoy in Canada.	Springer and Heath, 2007
7	*M. coruscus* × *M. galloprovincialis*	An identification method for these natural hybrids was constructed.	Zhang et al. (2020)
8	*Dreissena polymorpha* × *D. bugensis*	Natural hybrids were found on the Volga River in Russia.	Voroshilova et al. (2010)
9	*Lampsilis siliquoidea* × *L. radiate*	12 natural hybrids were found in the lower Laurentian Great Lakes and St. Lawrence River drainage.	Porto-Hannes et al. (2021)
10	*Haliotis corrugata* × *H. fulgens*	Hybrids were found in North America. They were described from abalone fisheries, based on morphological characteristics. Induced hybridization experiments at a later date confirmed the hybrids between the abalone species.	Leighton (2000), Li et al. (2025b)
11	*H. rufescens* × *H. fulgens*	Hybrids were found in North America. They were described from abalone fisheries, based on morphological characteristics. Induced hybridization experiments at a later date confirmed the hybrids between the abalone species.	Leighton (2000)

(*continued on next page*)

TABLE 6.1 Natural hybridization of shellfish—cont'd

Order	Nature hybridization	Main results	References
12	*H. rufescens* × *H. sorenseni*	Hybrids were found in North America. They were described from abalone fisheries, based on morphological characteristics. Induced hybridization experiments at a later date confirmed the hybrids between the abalone species.	Leighton (2000)
13	*H. discus hannai* × *H. discus discus*	Hybrids were found in Japan. Four categories of hybrids between *H. discus discus* and *H. discus hannai* were described, three categories of wild origin and another from artificial crossing, as characterized by the presence or absence of the three enzyme loci encoding.	Fujino et al. (1980)
14	*H. discus discus* × *H. gigantea*	Hybrids were found in Japan. Eight hybrid specimens from Awa Island, Niigata, exhibited enzyme activities at all four key enzyme loci without variation.	Fujino et al. (1980)
15	*H. discus hannai* × *H. gigantea*	Hybrids were found in Japan. Hybridization evidence was found based on the views of shells of the abalones.	Fujino et al. (1980)
16	*H. rubra* × *H. laevigata*	Hybrids were found in Australia. Protein electrophoresis revealed that rare, morphologically intermediate abalones are predominantly F1 hybrids.	Brown (1995)

6.2.2 Artificial hybridization

Artificial hybridization is widely used to improve specific traits in aquatic animals and stocks by using new genetic combinations in the offspring. In aquaculture, hybridization offers several benefits, including improved growth rates, survival, feed conversion efficiency, and stress resistance (to temperature and disease). Under commercial conditions, hybrids also enhance industrial diversification, creating opportunities for new products and markets (Wang et al., 2011; You et al., 2015).

6.2.2.1 Self-fertilization

Compared with typical hybridization, self-fertilization is a unique form of breeding in hermaphroditic species. For most hermaphroditic marine invertebrates, self-fertilization frequently occurs during reproduction. However, inbreeding depression, a reduction in biological fitness, is common, particularly in hermaphroditic shellfish (Qin et al., 2020). This phenomenon has been studied in species such as *Crassostrea virginica, Ostrea edulis, Pinctada fucata martensii,* and *C. gigas* (Launey & Hedgecock, 2001), and *Haliotis discus hannai* using methods such as family selection, successive mass selection, and population directive breeding. The effect of inbreeding on performance traits has also been analyzed in hermaphrodite scallops and giant clams, including *Argopecten circularis* (Ibarra et al., 1995), *A. purpuratus, A. irradians irradians, Patinopecten yessoensis, Tridacna derasa, T. squamosa*, and *T. crocea*, by evaluating the success of self-fertilization in progeny production. Interestingly, the severity of inbreeding depression varies dramatically depending on fitness traits, species, lineages, and even stages in the life cycle.

6.2.2.2 Strain hybridization

Strain hybridization refers to the crossing of artificially selected lines of the same species. Its primary goal is to combine the advantageous traits of two strains or to develop new varieties. Since no reproductive isolation barrier exists, this method is widely used in seed production. Several examples in shellfish breeding illustrate its applications: (1) Color recombination: Hybridizing blue-strain and brown-strain boring giant clams can produce a new line with unique color patterns. (2) Trait integration: Crossing disease-resistant and fast-growth lines of the American oyster can yield offspring with both desirable traits. (3) Enhanced resistance: Hybridizing two or more disease-resistant lines may produce a "super-resistant" strain capable of surviving frequent disease outbreaks that typically cause high mortality (Dégremont et al., 2015). In short, strain hybridization is an efficient, feasible, and widely used strategy in shellfish breeding.

6.2.2.3 Interpopulation hybridization

Population hybridization involves crossing distinct populations within a species to combine their desirable traits, serving as an effective method for genetic improvement in marine shellfish (Zhang et al., 2007). This approach has been successfully applied in scallops, pearl oysters, and boring giant clams. Beyond genetic background differences, several factors influence the successful development and growth of shellfish. The maternal effect is particularly significant, sometimes persisting into adulthood and even shaping certain phenotypic traits in mature individuals. In addition, nutrient availability and environmental conditions play key roles in regulating lipid accumulation. Moreover, survivability at the early stages is mainly affected by the combination of factors including egg origin and mating strategy (Zhang et al., 2012, 2007).

6.2.2.4 Interspecific hybridization

Interspecific hybridization is defined as crossing between different species, often within the same genus, which has been reported to increase growth performance, improve meat quality, and strengthen disease resistance in hybrid offspring (Hulata, 1995). Such hybridization not only produces heterosis but also serves as an effective method for expanding genetic variation and breeding new varieties (Bartley et al., 2000). However, successful heterosis is not guaranteed in all cases due to reproductive isolation barriers and hybrid incompatibility. Among marine shellfish, positive interspecific heterosis has been predominantly observed in scallops, abalones, and giant clams, with only limited successful cases reported in oysters (Zhang et al., 2022, 2014). The practice of interspecific hybridization in shellfish began with oysters, though early hybridization attempts lacked molecular genetic verification. Consequently, these early studies may have been affected by species misidentification, sperm contamination, and larvae pollution, raising questions about the authenticity of the reported hybrids. With the advent of

TABLE 6.2 Hybridization feasibility among eight *Crassostrea* oysters in China as indicated by fertilization success (√), failure (×), or unknown results (——).

Parents	H♂	R♂	G♂	A♂	S♂	N♂	D♂	I♂
H♀	HH	√	√	√	√	√	√	√
R♀	×	R R	√	√	×	——	——	——
G♀	×	√	GG	√	×	×	——	——
A♀	×	√	√	AA	×	×	——	——
S♀	√	√	√	√	SS	——	——	——
N♀	√	——	√	√	——	NN	——	——
D♀	√	——	——	——	——	——	DD	√
I♀	√	——	——	——	——	——	√	II

H = *C. hongkongensis*, R = *C. ariakensis*, G = *C. gigas*, A = *C. angulata*, S = *C. sikamea*, N = *C. nippon*, D = *C. dianbaigensis*, I = *C. iredalei*.

modern molecular techniques, genetic markers have become essential tools for verifying hybrid status, ensuring the accuracy of hybridization outcomes in contemporary studies (Gaffney & Allen, 1993; Zhang et al., 2012).

Interestingly, not all *Crassostrea* oysters exhibit reciprocal fertilization. In some cases, only unidirectional fertilization is possible because of differences in sperm-egg recognition proteins between parental species (Table 6.2). Moreover, even when fertilization occurs, hybrid viability is not guaranteed. High mortality rates during metamorphosis are often observed in the hybridization of certain species (Li et al., 2025a).

6.2.2.5 Intergeneric hybridization

Intergeneric hybridization is defined as crosses between species from different genera within the same family, which is rare in shellfish and often results in gamete incompatibility or hybrid lethality (Su et al., 2016). A case study involving *C. angulata* and *Saccostrea cucullata*, two oysters coexisting along the southern coasts of China, demonstrated this phenomenon. While reciprocal fertilization occurred, both hybrid combinations showed a low hatching rate. Although the resulting larvae were confirmed as true hybrids through genetic analysis, all offspring died within 6–8 days during the planktonic stage. These results demonstrate both the presence of postzygotic reproductive isolation between *C. angulata* and *S. cucullata* and the unsuccessful outcome of this intergeneric cross (Su et al., 2016).

6.2.3 Heterosis and performance

Heterosis is defined as the increase in performance displayed by hybrids compared with their inbred parents. As performance can be subjective (e.g., age of reproduction), a more precise definition involves non-additive inheritance where F1 hybrid traits exceed the range of both parental values (Comai, 2005). The degree of heterosis depends on both hybridization direction and parental genetic distance. In mollusks, including scallops, abalones, and giant clams, growth-related heterosis initially increases with genetic distance but declines when distance becomes excessive. Interspecific hybrids typically show the strongest heterosis effects. For oysters, optimal hybrid combinations must be empirically determined through systematic screening. Significant heterosis has been demonstrated in *C. gigas* × *C. angulata* (Tan et al., 2020), *C. hongkongensis*♀ × *C. ariakensis*♂ (Qin et al., 2020), and *C. dianbaigensis*♀ × *C. iredalei*♂(Li et al., 2025a).

6.2.4 Case description

To date, interspecific hybridization has become the best-developed and extensively studied breeding technique in shellfish aquaculture. Representative case studies from major commercial species are presented here, including documented examples from oysters (Table 6.3), ablones (Table 6.4), and scallops (Table 6.5).

6.2.5 Commercial production of shellfish hybrids in China

As the world's largest shellfish-producing country, China has achieved significant breakthroughs in hybrid seed production, particularly with oysters, scallops, and abalones.

Oysters: Three hybrid oysters are now being commercially produced and cultured in China. These include: (1) *C. gigas*♀ × *C. angulata*♂hybrids, which have fast growth and disease resistance traits and are mainly cultured in northern China, including Liaoning Province, Shandong Province, and Hebei Province; (2) *C. angulata*♀ × *C. gigas*♂hybrids, also with fast growth and disease resistance traits, mainly grown in Fujian Province in southern China; and (3) *C. hongkongensis*♀ × *C. ariakensis*♂ (Fig. 6.1), which display fast growth and high salinity resistance, mainly cultured in southern China, including Guangdong Province, and Guangxi Province.

Scallops: Hybrid scallops now commercially cultured in China were derived from the *Argopecten* and *Patinopecten* genera. In *Argopecten* scallops, three new strains were selected from F1 hybrids between the bay scallops and the Peruvian scallop (Fig. 6.2). The scallop Bohai Red has purplish-red shells and outperformed the bay scallop by 38% in whole weight (Wang et al., 2017). The scallop QN-2 has black and white shell coloration and outperformed the bay scallop by 43.6% in whole weight and 65.3% in adductor muscle weight (Xu et al., 2019). The third strain, QN Orange, has a golden shell color and orange adductor muscle and mantles because of a high content of carotenoids (Chen et al., 2020). The former two strains, Bohai Red and QN-2, are now the most cultured bay scallop varieties in northern China waters. Since 2019, backcrosses of Bohai Red or QN-2 with sperm from the Peruvian scallop, namely the Coffee scallop or Black Coffee scallop, have become increasingly popular in northern China, as they exhibited tremendous heterosis in growth comparable to that of F1 hybrids. However, because of the hermaphroditic nature of the broodstocks, the proportion of hybrids in the cohorts produced by mass spawning of two stocks still needs to be improved.

Abalone: In China, extensive studies on abalone hybridization have been conducted, and hybridization technologies have played an important role in increasing abalone production over the past two decades. For example, Japanese stocks of *H. discus hannai* and *H. diversicolor* were introduced into China, and intraspecific hybrids between the Chinese and Japanese stocks were developed. These hybrids exhibited significantly improved survival and growth as compared to wild-types, supporting the development of the Chinese abalone industry. A strong heterotic effect has also recently been observed in hybrids of *H. sieboldii* and *H. discus hannai*: Both reciprocal hybrids had significantly faster growth rates after 100 days of culture, and their

TABLE 6.3 Interspecific hybridization of ***Crassostrea*** oysters with molecular confirmation.

Order	Interspecific hybridization	Main results	References
1	*C. hongkongensis* × *C. gigas*	One-way fertilization, slow growth, and a high degree of sterility.	Zhang et al. (2012, 2014, 2017)
2	*C. hongkongensis* × *C. angulata*	One-way fertilization, fast growth, and complete fertility.	Zhang et al. (2016)
3	*C. hongkongensis* × *C. ariakensis*	One-way fertilization, fast growth, and complete fertility.	Huo et al. (2013), Qin et al. (2020, 2021)
4	*C. hongkongensis* × *C. sikamea*	Two-way fertilization, fast growth (Compared with *C. sikamea*), improvement in meat quality, and complete fertility.	Zhang et al. (2017)
5	*C. hongkongensis* × *C. dianbaigensis*	Two-way fertilization, adults were obtained from only one-way hybridization, showing fast growth and complete fertility.	Li et al. (2024)
6	*C. hongkongensis* × *C. iredalei*	Two-way fertilization, adults were obtained from only one-way hybridization, showing fast growth (Compared with *C. iredalei*) and complete fertility.	Li et al. (2025a)
7	*C. gigas* × *C. angulata*	Two-way fertilization, improvement in high temperature resistance, and complete fertility.	Tan et al. (2020), Jiang et al. (2021), Qin et al., 2023
8	*C. gigas* × *C. ariakensis*	Two-way fertilization, hardly any growth heterosis, and complete fertility.	Zhang et al. (2012), Li et al. (2022)
9	*C. gigas* × *C. sikamea*	One-way fertilization, fast growth (Compared with *C. sikamea*), and complete fertility.	Xu et al. (2019)
10	*C. gigas* × *C. nippon*	One-way fertilization, fast growth (Compared with *C. nippon*), and complete fertility.	Xu et al. (2019)

(*continued on next page*)

TABLE 6.3 Interspecific hybridization of ***Crassostrea*** oysters with molecular confirmation—cont'd

Order	Interspecific hybridization	Main results	References
11	*C. ariakensis* × *C. augulata*	Two-way fertilization, slow growth, and complete fertility.	Yao et al. (2015)
12	*C. ariakensis* × *C. sikamea*	One-way fertilization, slow growth, and complete fertility.	Xu et al. (2019), 2011, 2014
13	*C. angulata* × *C. sikamea*	Two-way fertilization, fast growth (Compared with *C. sikamea*), and complete fertility.	Yan et al. (2017, 2018), Xu et al. (2014)
14	*C. angulata* × *C. nippon*	One-way fertilization, fast growth (Compared with *C. nippon*), and complete fertility.	Li et al. (2023)
15	*C. dianbaigensis* × *C. iredalei*	Two-way fertilization, adults were obtained from only one-way hybridization, showing moderate heterosis in growth, improvement in high salinity resistance, and complete fertility.	Li et al. (2025b)

Most interspecific hybrids in oysters showed heterosis in survival till adulthood, with environmental factors playing a key role in the survival rate.

survival rates were similar to those of the parental species after 180 days of grow-out. The North American green abalone *H. fulgens* was also introduced into China and successfully hybridized with the Pacific abalone. The hybrids *H. discus hannai*♀ × *H. fulgens*♂ exhibited heterosis in growth and survival, and demonstrated increased tolerance to high temperatures (Fig. 6.3). Over the last two decades, four hybrids, *H. discus hannai* "Dalian-I", *H. diversicolor* "Dongyou-I", *H. sieboldii*♀ × *H. discus hannai*♂, and *H. discus hannai*♀ × *H. fulgens*♂, have received "new variety certificates" from the Chinese Ministry of Agriculture, and have been commercially produced to improve abalone farming efficiency.

6.3 Autopolyploids

As an important breeding method, in addition to hybridization, polyploid breeding has greatly promoted the development of aquaculture. Generally, most Metazoa are diploid, possessing a duplicated set of homologous chromosomes in somatic cells, a condition that evolved together with gametogenesis via meiotic cell division (Piferrer et al., 2009). Polyploids, defined as organisms that have multiple sets of chromosomes, are common in certain taxa and can be

TABLE 6.4 Interspecific hybridization of ***Haliotis*** abalones.

Order	Interspecific hybridization	Main results	References
1	*H. rufescens* × *H. fulgens*	Two-way fertilization, 1% survival of juveniles.	Lewis et al. (1982), Leighton (2000)
2	*H. rufescens* × *H. corrugata*	Two-way fertilization, 5% survival of juveniles.	Lewis et al. (1982), Leighton (2000)
3	*H. fulgens* × *H. corrugata*	One-way fertilization, 1% survival of juveniles.	Lewis et al. (1982), Leighton (2000)
4	*H. rufescens* × *H. sorenseni*	Two-way fertilization, 5% survival of juveniles.	Lewis et al. (1982), Leighton (2000)
5	*H. fulgens* × *H. sorenseni*	Two-way fertilization, one-way, 1% survival of juveniles, and the other way, 0.1% survival of post-larvae.	Lewis et al. (1982), Leighton (2000)
6	*H. rufescens* × *H. k.* *assimilis*	—	Lewis et al. (1982), Leighton (2000)
7	*H. rufescens* × *H. walallensis*	—	Lewis et al. (1982), Leighton (2000)
8	*H. rufescens* × *H. k. ramtschatkana*	—	Lewis et al. (1982), Leighton (2000)
9	*H. corrugata* × *H. sorenseni*	Specimens produced in the laboratory.	Lewis et al. (1982), Leighton (2000)
10	*H. corrugata* × *H. cracherodii*	Specimens produced in the laboratory.	Lewis et al. (1982), Leighton (2000)
11	*H. corrugata* × *H. walallensis*	Specimens produced in the laboratory.	Lewis et al. (1982), Leighton (2000)
12	*H. corrugata* × *H. k.* *assimilis*	—	Lewis et al. (1982), Leighton (2000)
13	*H. fulgens* × *H. cracherodii*	Specimens produced in the laboratory.	Lewis et al. (1982), Leighton (2000)
14	*H. sorenseni* × *H.k. Assimilis*	Specimens produced in the laboratory.	Lewis et al. (1982), Leighton (2000)
15	*H. sorenseni* × *H. walallensis*	—	Lewis et al. (1982), Leighton (2000)
16	*H. k.* *assimilis* × *H. walallensis*	—	Lewis et al. (1982), Leighton (2000)
17	*H.k.* *assimilis* × *H. k. ramtschatkana*	Specimens produced in the laboratory.	Lewis et al. (1982), Leighton (2000)

(*continued on next page*)

TABLE 6.4 Interspecific hybridization of ***Haliotis*** abalones—cont'd

Order	Interspecific hybridization	Main results	References
18	*H. discus discus* × *H. gigantea*	Two-way fertilization. Specimens produced in the laboratory with a lower fertilization rate (means of 1.3%–60.8%) than the parental species (34.3%–90%). The hybrid showed heterosis in growth and survival rates in summer in grow-out experiments.	Ahmed et al. (2008), Guo (2009)
19	*H. discus hannai* × *H. fulgens*	Two-way fertilization. Specimens produced in the laboratory with a fertilization rate between 12.2% and 56.1%. The hybrids exhibited heterosis for growth and survival, as well as a higher temperature tolerance.	You et al. (2015)
20	*H. rubra* × *H. laevigata*	Two-way fertilization. Specimens produced in the laboratory. Heterosis was present for total weight and meat yield.	Kube et al. (2024)
21	*H. rufescens* × *H. discus hannai*	One-way fertilization. Specimens produced in the laboratory.	Lafarga-de la Cruz et al. (2015)

Survival reported as the number of live young stages/number of larvae at settlement (%).

surprisingly stable (Comai, 2005). They are very common among plants, occur frequently in fish and amphibians, and are usually fit and well adapted. Regarding mollusks, since the first case of triploid oysters was cultivated, this approach has greatly advanced the triploid shellfish industry (Stanley et al., 1981). In particular, 100% triploid oyster seedlings were produced by mating male tetraploids with female diploids after the creation of tetraploid oysters, which greatly popularized the global triploid oyster industrialization (Guo et al., 1996).

In fact, the polyploid in shellfish originated with oysters. Its development history can be divided into the following three stages. (1) Early exploration stage (early 1980s): In 1981, American scientists first successfully produced triploid American oysters through chemical induction. This achievement marked the beginning of the artificial cultivation of polyploid shellfish (Stanley et al., 1981). (2) Technological breakthrough stage (1990s–early 2000s): Building on triploid breeding techniques, the core germplasm resources of tetraploid oysters were successfully created (Guo et al., 1996). (3) Industrial application stage (2005–present): Through diploid and tetraploid biological hybridization, 100% triploid superior varieties have been produced on a

TABLE 6.5 Interspecific hybridization in scallops with molecular confirmation.

Order	Interspecific hybridization	Main results	References
1	*Patinopecten caurinus* × *P. yessoensis*	Reciprocal hybrids were obtained between the two species.	Guilong et al. (2018), Xing et al. (2022)
2	*Argopecten irradians irradians* × *A. purpuratus*	Reciprocal male sterile F1 hybrids with significant heterosis in growth were obtained. The backcross of the F1 hybrids with *A. purpuratus* also exhibited high heterosis in growth. Three new strains were selected from the hybrids.	(Wang et al. (2009), 2011, Feng et al. (2012), Wang (2017), Chen et al. (2020)
3	*Argopecten irradians concentricus* × *A. purpuratus*	Male sterile hybrids with significant heterosis in growth were obtained.	Nan et al. (2012), Shen et al. (2024)
4	*Argopecten irradians concentricus* × *A.i. irradians*	Intraspecific hybrids were obtained by hybridizing between the two subspecies of the bay scallops.	Zheng et al. (2007), Wang et al. (2023)

large scale. These triploids have been widely adopted and promoted in the United States, China, France, Australia, and other countries (Guo et al., 1996).

6.3.1 Nature polyploids

Spontaneous polyploidy (nature polyploids) arises through the multiplication of chromosome sets, triggered by disruptions of the normal meiotic process that lead to polyploidization. Some of these disruptions have become evolutionarily fixed, enabling the development of polyploid taxa at different ploidy levels (Comai, 2005; Piferrer et al., 2009). So far, only diploids, triploids, and tetraploids of Asian clams have been reported, and no similar reports have been found in other shellfish. Interestingly, these clams reproduce through androgenesis by nonreductional spermatozoa, which is one of the few methods of reproduction way in shellfish. Unfortunately, the tetraploid Asian clams that were found in Anyue County, Sichuan Province, China, have disappeared due to freshwater pollution and river reconstruction.

6.3.2 Artificial polyploids

6.3.2.1 Triploids

Triploids, organisms with three sets of homologous chromosomes, are mainly cultured populations and can be easily induced in many commercially relevant species of shellfish. The major consequence of triploids is a certain degree of gonadal sterility, which is advantageous in the aquaculture of mollusks because it can result in superior growth and strong disease resistance

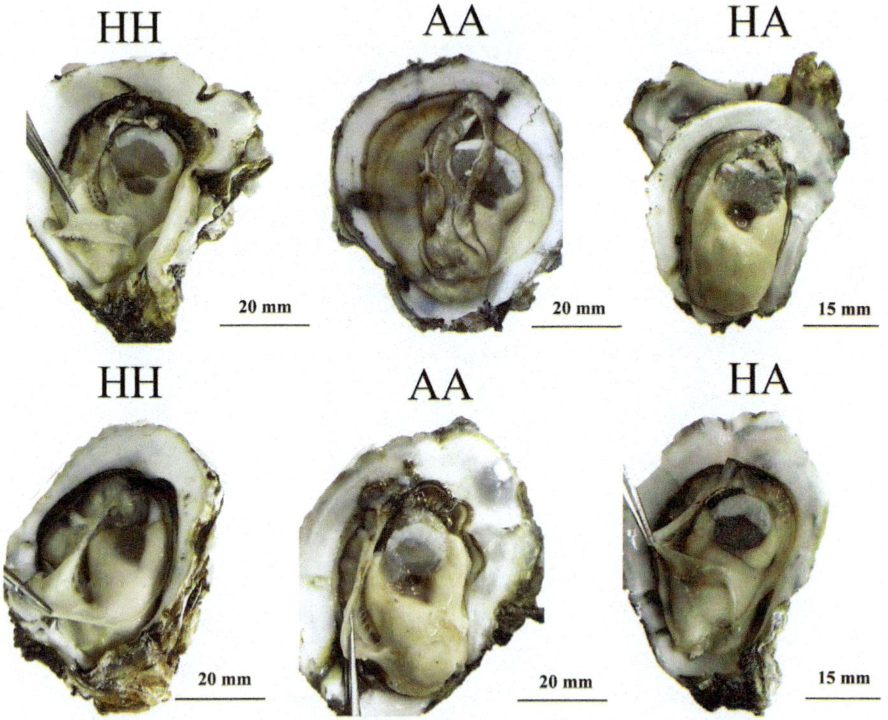

FIGURE 6.1 Morphological identification of *Crassostrea hongkongensis*, *Crassostrea ariakensis*, and their hybrids HH: *C. hongkongensis*; AA: *C. ariakensis*; HA: *C. hongkongensis*♀ × *C. ariakensis* ♂ hybrids. *From Qin et al., 2020*

(Piferrer et al., 2009). Triploids can be produced either by direct artificial induction methods or indirectly through mating strategies between tetraploids and diploids (Fig. 6.4) (Guo et al., 1996). Triploid shellfish include various oysters, scallops, mussels, clams, and others. Many of these have been developed, and some have been applied in aquaculture worldwide (Table 6.6). Typically, the performance traits of triploid shellfish produced through controlled mating were better than those of introduced triploids, especially in oyster species (Wadsworth et al., 2019).

6.3.2.2 Tetraploids

Tetraploids, organisms with four sets of homologous chromosomes, are derived from artificial induction, except for the naturally occurring tetraploids found in Asian clams (Piferrer et al., 2009). Although tetraploids have been reported in many species, truly successful cases have mainly focused on oysters (Zhang et al., 2022). Tetraploid oysters, as a core germplasm resource, play an important role in the global oyster industry because interploid matings between diploids and tetraploids can produce 100% triploids (Guo et al., 1996; Matt & Allen, 2014; Qin et al., 2022; Zhang et al., 2022). Since the first successful production of viable tetraploid *C. gigas*, a series of tetraploid oysters have been developed (Tan et al., 2020; Qin et al., 2021). However, truly viable adult tetraploids have only been reported in *C. virginica, C. ariakensis, C. angulata, C.*

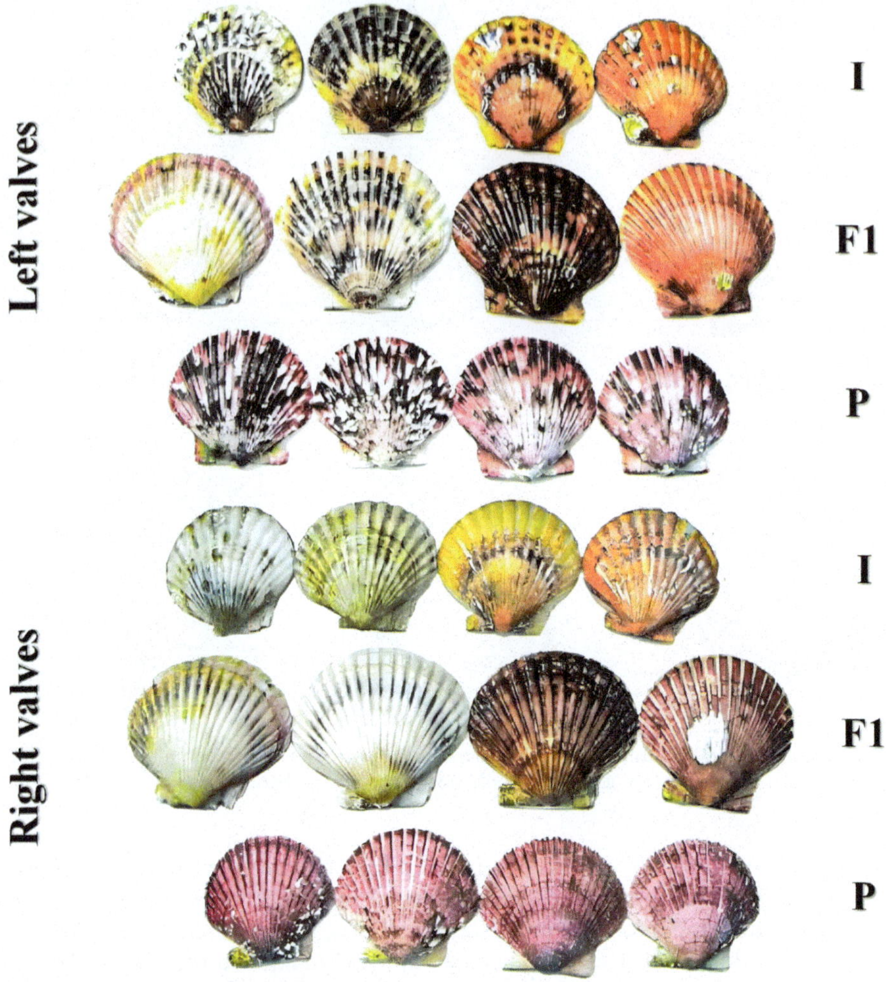

FIGURE 6.2 Shell colors of the brood stocks and offspring cohorts. The left valves are shown on the top pane, and the right valves are on the bottom pane. I: bay scallops; P: Peruvian scallops; and F1: hybrids. *From Wang et al. (2011)*

hongkongensis, etc. (Allen et al., 2005; Qin et al., 2024; Zhang et al., 2022) (Table 6.7). There are five methods for producing tetraploid shellfish: (1) Inhibiting the first polar body of the fertilized egg produced by diploid egg and diploid sperm; (2) Inhibiting the second polar body of the fertilized egg from diploid egg and tetraploid sperm (Fig. 6.5A); (3) Inhibiting the first polar body of the fertilized egg from triploid egg and diploid sperm (Fig. 6.5B); (4) Inhibiting the second polar body of the fertilized egg from triploid egg and diploid sperm (Fig. 6.5C); (5) tetraploid self-propagation (Li & Li, 2022; Matt & Allen, 2014; Qin et al., 2023,2022,2024; Yue et al., 2024; Zhang et al., 2022;). Generally, tetraploid shellfish showed reduced growth and lower survivability,

FIGURE 6.3 **Morphological features for the broodstock and hybrid.** (A) *H. discus hannai*. (B) *H. sieboldii*. (C) *H. sieboldii* ♀ × *H. discus hannai* ♂ hybrids.

which may be due to the larger cells consuming more energy (Guo et al., 1996; Piferrer et al., 2009).

6.3.2.3 *Polyploid advances and dis-advances*

As for shellfish, most of the triploids have obviously performance advances, such as fast growth, a high degree of sterility, strong disease resistance, and so on. Possible causes of these traits include: (1) high heterozygosity; (2) cell enlargement; (3) meiosis disorder; (4) high expression of insulin growth factor (Qin et al., 2019). However, most tetraploid oysters show growth weakness and low viability. Although scientists have genetically improved the tetraploid growth trait through successive generations of selection, tetraploids are still not as large as diploids (Qin et al., 2023; Yue et al., 2024; Zhang et al., 2022). The possible explanations include the disrupting effects of nuclear and cellular enlargement, the tendency of tetraploid mitosis and meiosis to produce aneuploid cells, which ultimately develop into tetraploid, aneuploid,

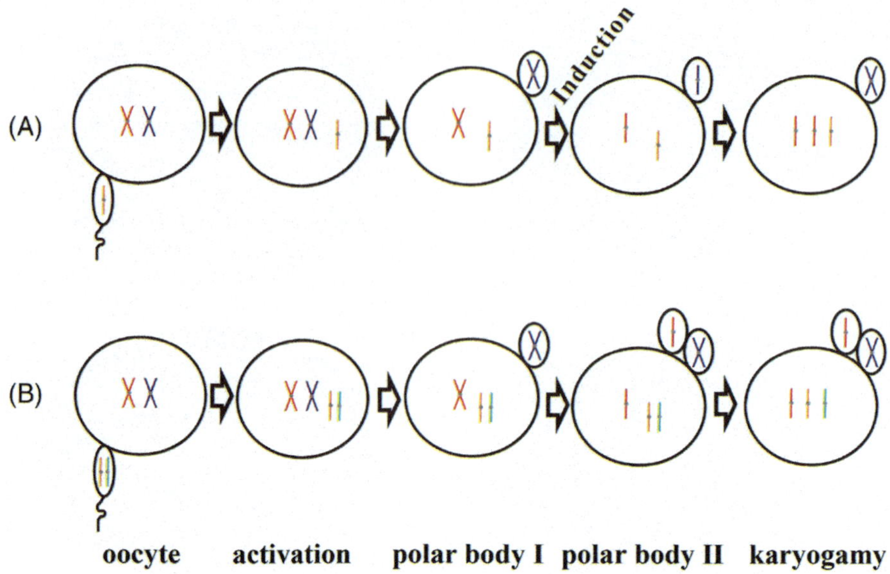

FIGURE 6.4 Two triploid production methods. (A) Introduced method. (B) Mating method (tetraploid males × diploid females). *From Qin et al. (2024)*

TABLE 6.6 Global triploid shellfish culture situation.

Order	Triploid species	Scale and culture area	References
1	*C. gigas*	Larger-scale culture. Area: China, Korea, USA, Australia, France, New Zealand, Africa, etc.	Guo et al. (1996), Nell (2002), Guo (2009), Qin et al. (2022)
2	*C. angulata*	Larger-scale culture in Southern China.	Zhang et al. (2022)
3	Allotriploid oysters from *C. gigas* and *C. angulata*	Larger-scale culture in Northern and Southern China.	Qin et al., 2023
4	*C. hongkongensis*	A small-scale culture in Southern China.	Zhang et al. (2017), Qin et al. (2019, 2024)
5	*C. virginica*	Larger-scale culture in the United States.	Guo et al. (2002), Matt and Allen (2014)
6	*Saccostrea glomerat*	A small-scale culture in Australia.	Nell (2002)
7	*Haliotis discus hannai*	A small-scale culture in Southern China.	Wang et al. (2021)
8	Allotriploid abalones from *H. discus hannai and H. fulgens*	A small-scale culture in Southern China.	Wang et al. (2021)

TABLE 6.7 Global generation of tetraploid shellfish.

Order	Tetraploid species	Main results	References
1	*C. gigas*	First case of artificial tetraploid shellfish.	Guo and Allen (1994), Guo et al. (1996), Wang et al. (2003), McCombie et al. (2005), Zhang et al. (2022), Li et al. (2022)
2	*C. angulata*	A closely related species of *C. gigas*.	Zhang et al. (2022)
3	*C. Hongkongensis*	An important brackish oyster species in Southeast Asia.	Zhang et al. (2017)
4	*C. virginica*	An important brackish oyster species in USA.	Guo et al. (2002)
5	*C. fluminea*	Nature tetraploids of shellfish.	Qiu et al. (2001)

triploid, and diploid individuals (Figs. 6.6 and 6.7). Another reason is epigenetic instability, which results in transgressive gene regulation. Over time, the process of depolyploidization eventually leads to stable diploidy (Comai, 2005; Piferrer et al., 2009).

6.3.2.4 Case description

At present, triploids of several species produced by mating tetraploids with diploids have been applied on a large scale, including *C. gigas*, *C. angulata*, *C. hongkongensis*, and *C. virginica* (Matt & Allen, 2014; Qin et al., 2024; Zhang et al., 2022). Meanwhile, triploids of several species produced by direct induction methods have also been applied on a, including *C. ariakensis* (Qin et al., 2019), *C. sikamea*, *Haliotis rubra* (Liu et al., 2009), *H. discus*, *H. laevigata* (Dunstan et al., 2007), *Argopecten irradians*, and *A. ventricosus*. Finally, many shellfish are still being tested with triploid breeding methods to evaluate their production performance, which can provide technical support for future development and application.

6.3.2.5 Polyploid industry

The shellfish polyploid breeding industry includes two main components: triploid commercial seed production and tetraploid core germplasm resources. The most successful example is oyster polyploid breeding. In particular, for the Pacific oyster, the market share of triploid seeds is approximately more than 80% in China, about 50% in the United States, about 40% in France, about 30% in Australia, and about 10% in Korea. Tetraploid Pacific oysters were successively developed in the United States, France (McCombie et al., 2005), and China (Li & Li, 2022), and can be supplied to hatcheries for commercial triploid seed production through cooperation or as globally traded commodities.

6.4 Allopolyploids

Allopolyploidy arises through the multiplication of chromosome sets resulting from interspecific hybridizations, and includes forms such as allotriploids and allotetraploids (Piferrer et al., 2009). Allopolyploids combine the advantages of autoployploid advances with

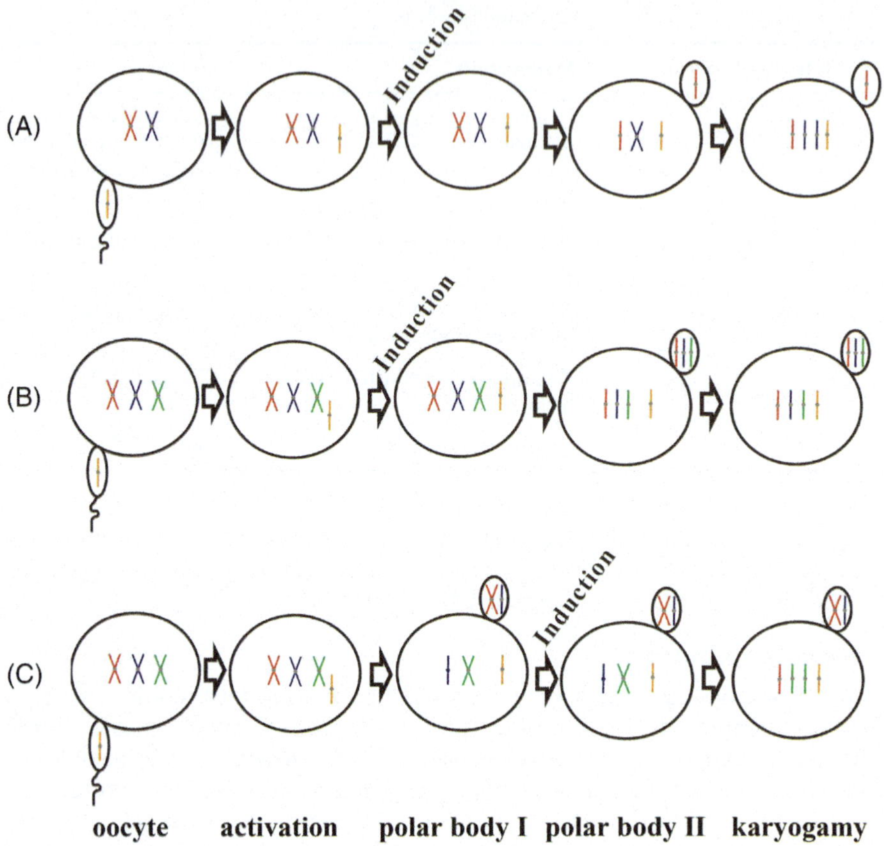

FIGURE 6.5 Three tetraploid production methods. (A) Inhibit the second polar body of the fertilized egg from diploid egg and tetraploid sperm. (B) Inhibit the first polar body of the fertilized egg from triploid egg and diploid sperm. (C) Inhibit the second polar body of the fertilized egg from triploid egg and diploid sperm. *From Qin et al. (2021)*

heterosis, often displaying significant improvements in performance and evolutionary potential (Comai, 2005). However, polyploid weaknesses still exist in some respects, as observed in allotetraploid Pacific oysters (Yue et al., 2024).

6.4.1 Allotriploids

Triploidy is usually associated with sterility, which is advantageous in aquaculture because it can result in superior growth (Piferrer et al., 2009). Hence, triploid hybrids (allotriploids), which combine hybridization and triploidization, are expected to be more sterile than diploids because of additional complications in gametogenesis. On the other hand, triploid hybrids can also integrate desirable traits from both parental species, leading to increased growth rates, increased environmental tolerance, and improved disease resistance under culture conditions (Zhang et al., 2014). Allotriploids of shellfish have been reported in oysters and abalones, such

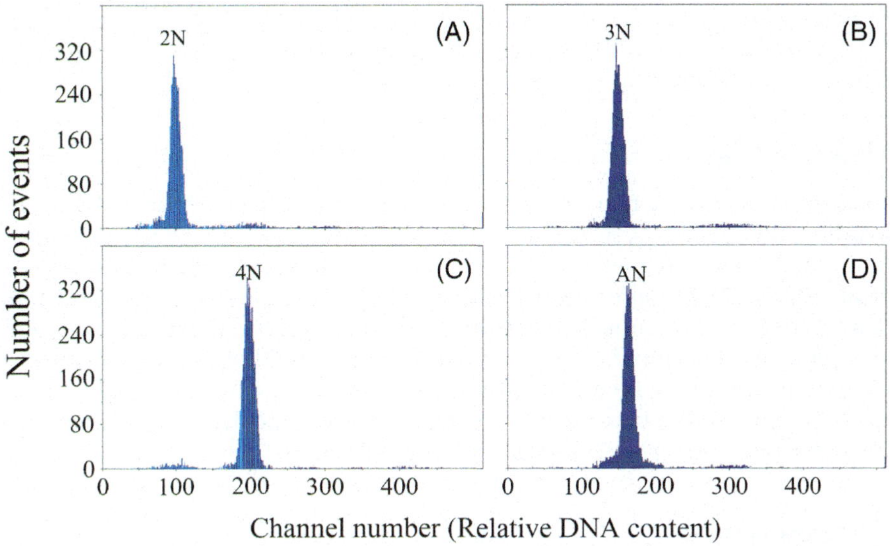

FIGURE 6.6 Tetraploid oyster transforms into diploid, triploid, tetraploid, and aneuploid by flow cytometer examination (A) Diploids. (B) Triploids. (C) Tetraploids. (D) Aneuploid. *From Zhang et al. (2022)*

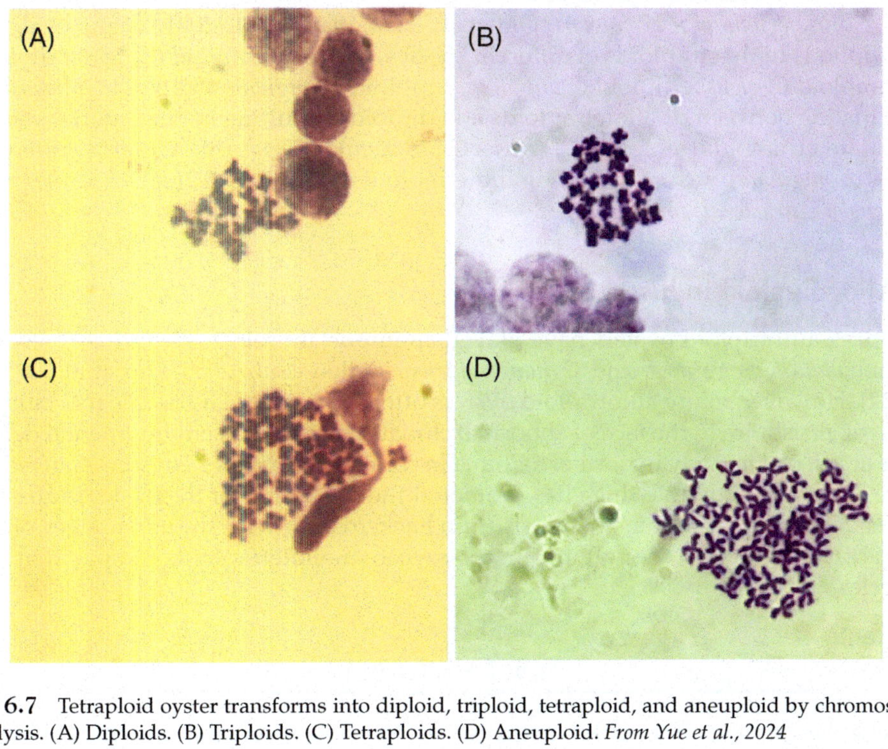

FIGURE 6.7 Tetraploid oyster transforms into diploid, triploid, tetraploid, and aneuploid by chromosome karyotype analysis. (A) Diploids. (B) Triploids. (C) Tetraploids. (D) Aneuploid. *From Yue et al., 2024*

as *C. gigas* × *C. ariakensis, C. gigas* × *C. hongkongensis* (Zhang et al., 2014), *C. hongkongensis* × *C. ariakensis* (Qin et al., 2020), *C. gigas* × *C.angulata* (Qin et al., 2023), *H. discus hannai* × *H. fulgens* (Wang et al., 2021), and so on.

6.4.2 Allotetraploids

Autotetraploids of several marine bivalves have been artificially induced over the past 30 years, including the Pacific oyster (Guo et al., 1996), Manila clam, Eastern oyster, Catarina scallop, Suminoe oyster (Allen et al., 2005), Sydney rock oyster (Nell, 2002), Hongkong oyster (Zhang et al., 2017), Portuguese oysters (Zhang et al., 2022), pearl oyster, and blue mussel. Unfortunately, there is only one relevant research on bivalve allotetraploids, derived from tetraploid *C. gigas* and tetraploid *C. angulata*. These reciprocal allotetraploids have shown good performance traits, including favorable growth and suitability (Yue et al., 2024). Allotetraploid oysters have the potential to become new genetic resources for the improvement of polyploid oysters through backcrossing or three-way crossing strategies.

6.4.3 Case description

Allotriploid shellfish breeding has mainly focused on oysters and abalones. There are three principal approaches: (1) directly induction from diploids to triploids by inhibiting the second polar body of the fertilized egg, such as in *C. gigas* × *C. hongkongensis, C. hongkongensis* × *C. ariakensis, H. discus hannai* × *H. fulgens* (Qin et al., 2020; Wang et al., 2021; Zhang et al., 2014). (2) crossing dipliods with tetraploids of different species, such as tetraploid *C. gigas* × diploid *C. ariakensis*, teraploid *C. gigas* × diploid *C. angulata*, teraploid *C. angulata* × diploid *C. gigas* (Qin et al., 2023). (3) mating between di-allotetraploids and diploids, resulting in new, special allotriploids, such as di-allotetraploids (teraploid *C. gigas* × *C. angulata*) × *C. gigas*, di-allotetraploids (teraploid *C. gigas* × *C. angulata*) × *C. angulata*, di-allotetraploids (teraploid *C. gigas* × *C. angulata*) × *C. hongkongensis, C.ariakensis, or C. sikamea*.

6.4.4 Allopolyploid industry

In the past, autotriploid oysters have played an important role in the industry, especially *C. gigas, C. angulata, C. virginica*, and *C. hongkongensis* (Allen *et al.*, 2005; Qin et al., 2024; Zhang et al., 2022). However, many autotriploid oysters often experience large-scale mortality because of their weak resistance in the face of abnormal climate changes and environmental degradation in culture areas, particularly in northern China. In contrast, allotriploid oysters showed superior resistance and rapid growth, which has promoted the prosperity of the oyster industry. Meanwhile, breeders hope to combine heterosis from backcrossing with the advantages of triploidy, and they are testing backcrossed allotriploids between di-allotetraploids and parental species or other species.

6.5 Conclusions

This chapter reviews the latest progress in hybridization and polyploid breeding of shellfish, describing in details the current status and application prospects of these two breeding strategies. In particular, the effective combination of the two breeding methods through allopolyploid to create reliable new seed resources shows great promise for industrial application. However, although both methods have achieved considerable success in a small number of species, further development is still needed for most other species. These methods can also be integrated with other breeding techniques, such as gene editing, to create superior varieties. Finally, the authors put forward constructive breeding suggestions to help the future shellfish industry achieve even greater success.

References

Ahmed, F., Koike, Y., Strüssmann, C.A., Yamasaki, I., Yokota, M., Watanabe, S., 2008. Genetic characterization and gonad development of artificially produced interspecific hybrids of the abalones, Reeve, Gmelin and Habe. Aquaculture Research 39, 532–541. doi:10.1111/j.1365-2109.2008.01911.x.

Allen, S.K., Erskine, A.J., Walker, E.J., DeBrosse, G.A., 2005. Production of tetraploid Suminoe oysters, *Crassostrea ariakensis*. Aquaculture 247 (1–4), 3.

Arnold, M.L., 2004. Natural hybridization and the evolution of domesticated, pest and disease organisms. Molecular Ecology 13 (5), 997–1007. doi:10.1111/j.1365-294X.2004.02145.x.

Bartley, D.M., Rana, K., Immink, A.J., 2000. The use of inter-specific hybrids in aquaculture and fisheries. Reviews in Fish Biology and Fisheries 10 (3), 325–337. doi:10.1023/a:1016691725361.

Brown, L.D., 1995. Genetic evidence for hybridisation between *Haliotis rubra* and *H. laevigata*. Marine Biology 123, 89–93. doi:10.1007/BF00350327.

Camara, M.D., Davis, J.P., Sekino, M., Hedgecock, D., Li, G., Langdon, C.J., Evans, S., 2008. The Kumamoto oyster *Crassostrea sikamea* is neither rare nor threatened by hybridization in the northern Ariake Sea, Japan. Journal of Shellfish Research 27, 313–322. doi:10.2983/0730-8000(2008)27[313:TKOCSI]2.0.CO;2.

Cao, L., Halpern, B., Troell, M., Short, R., Zeng, C., Jiang, Z., … Tigchelaar, et al., 2023. Vulnerability of blue foods to human-induced environmental change. Nature Sustainability 6, 1186–1198.

Carugati, L., Pinna, V., Demurtas, R., Cau, A., Cannas, R., 2024. Genetic diversity of the *Ruditapes decussatus* and evidence of its hybridization with the alien R. philippinarum in the Western Mediterranean Sea. Estuarine, Coastal and Shelf Science 15, 108903. doi:10.1016/j.ecss.2024.108903.

Chen, M., Liu, B., Ma, B., Liu, G., Cao, W., Liu, X., … Wang, C., 2020. Selection of a carotenoid-rich scallop strain, QN Orange, from the inter-specific hybrids between the bay scallop and the Peruvian scallop. Aquaculture 528, 735513. doi:10.1016/j.aquaculture.2020.735513.

Fishery and Fishery Administration Bureau of the Ministry of Agriculture and Rural Affairs, National Fisheries Technology Extension Center, China Society of Fisheries, 2025. China Fishery Statistical Yearbook. China Agriculture Press, Beijing, p. 27.

Comai, L., 2005. The advantages and disadvantages of being polyploid. Nature Reviews Genetics 6 (11), 836–846. doi:10.1038/nrg1711.

Comesaña, A.S., Toro, J.E., Innes, D.J., Thompson, R.J., 1999. A molecular approach to the ecology of a mussel (*Mytilus edulis-M. trossulus*) hybrid zone on the east coast of Newfoundland. Canada. Marine Biology 133, 213–221. doi:10.1007/s002270050460.

Dégremont, L., Garcia, C., Allen, S.K., 2015. Genetic improvement for disease resistance in oysters: A review. Journal of Invertebrate Pathology 131, 226–241. doi:10.1016/j.jip.2015.05.010.

Dillon, Jr.RT., Manzi, J.J., 1989. Genetics and shell morphology in a hybrid zone between the hard clams *Mercenaria mercenaria* and *M. campechiensis*. Marine Biology 100, 217–222. doi:10.1007/BF00391961.

Dunstan, G.A., Elliott, N.G., Appleyard, S.A., Holmes, B.H., Conod, N., Grubert, M.A., Cozens, M.A., 2007. Culture of triploid greenlip abalone (*Haliotis laevigata* Donovan) to market size: Commercial implications. Aquaculture 271 (1–4), 130–141. doi:10.1016/j.aquaculture.2007.06.023.

Food and Agriculture Organization of the United Nations(FAO). FishStat: Global Aquaculture Production [Dataset].

Feng, W., Li, J., Cai, D., Wang, C., Nan, Y., 2012. Comparison of development and growth of the F1 and F2 hybrids between the Peruvian and bay scallop. Chinese Agricultural Science Bulletin 28 (20), 118–124. doi:10.1007/s11783-011-0280-z.

Fujino, K., Sasaki, K., Wilkins, N.P., 1980. Differences in Electrophoretic Patterns between *Haliotis discus* Reeve and *H. discus hannai* Ino. Bulletin of the Japanese Society of Scientific Fisheries 46, 543–548. doi:10.2331/suisan.46.543.

Gaffney, P.M., Allen, S.K., 1993. Hybridization among *Crassostrea* species: A review. Aquaculture 116 (1), 1–13. doi:10.1016/0044-8486(93)90217-M.

Guilong, L, Jiliang, L., Bin, Y., Yin, C., Bin, M., Zhenlu, W., … Chunde, W., 2018. Preliminary study on interspecific hybridization between *Patinopecten yessoensis* and *Mizuhoecten (Patinopecten) caurinus*. Journal of Fisheries of China 42 (11), 1880–1888.

Guo, X., Allen, S.K., 1994. Viable tetraploids in the Pacific oyster (*Crassostrea gigas* Thunberg) produced by inhibiting polar body in eggs from triploids. Mol. Mar. Biol. Biotechnol. 3, 42–50.

Guo, X., DeBrosse, G.A., Allen, S.K., 1996. All-triploid pacific oysters (*Crassostrea gigas* thunberg) produced by mating tetraploids and diploids. Aquaculture 142 (3–4), 149–161. doi:10.1016/0044-8486(95)01243-5.

Guo, X., Wang, J., Landau, B.J., Li, L., DeBrosse, G.A., Krista, K.D., 2002. The successful production of tetraploid eastern oyster *Crassostrea virginica* Gmelin. J. Shellfish Res. 21, 380–381 (abstract).

Guo, X., 2009. Use and exchange of genetic resources in molluscan aquaculture. Reviews of Aquaculture 1, 251–259. doi:10.1111/j.1753-5131.2009.01014.x.

Hulata, G., 1995. A review of genetic improvement of the common carp (*Cyprinus carpio* L.) and other cyprinids by crossbreeding, hybridization and selection. Aquaculture 129 (1–4), 143–155. doi:10.1016/0044-8486(94)00244-I.

Huo, Z., Wang, Z., Yan, X., Gaffney, P.M., 2013. Fertilization, survival, and growth of *Crassostrea hongkongensis* ♀× *Crassostrea ariakensis* ♂ hybrids in northern China. Journal of Shellfish Research 32, 377–385. doi:10.2983/035.032.0217.

Hurtado, N.S., Pérez-García, C., Morán, P., Pasantes, J.J., 2011. Genetic and cytological evidence of hybridization between native *Ruditapes decussatus* and introduced *Ruditapes philippinarum* (Mollusca, Bivalvia, Veneridae) in NW Spain. Aquaculture 311, 123–128. doi:10.1016/j.aquaculture.2010.12.015.

Ibarra, A.M., Cruz, P., Romero, B.A., 1995. Effects of inbreeding on growth and survival of self-fertilized catarina scallop larvae, *Argopecten circularis*. Aquaculture 134 (1–2), 37–47. doi:10.1016/0044-8486(95)00022-T.

Jiang, G., Li, Q., Xu, C., Liu, S., Kong, L., Yu, H., 2021. Reciprocal hybrids derived from *Crassostrea gigas* and *C. angulata* exhibit high heterosis in growth, survival and thermotolerance in northern China. Aquaculture 545, 737173. doi:10.1016/j.aquaculture.2021.737173.

Kube, P.D., McPherson, L., Krsinich, A., King, H., Raidan, F.S.S., Reverter, A., Kijas, J., Elliott, N.G., 2024. Genetic variation and heterosis of the interspecific abalone hybrid of *Haliotis rubra* and *H. laevigata*. Aquaculture 580, 740275. doi:10.1016/j.aquaculture.2023.740275.

Lafarga-de la Cruz, F., Aguilar-Espinoza, A., Gallardo-Escárate, C., 2015. Parentage assignment in hybrid abalones (*Haliotis rufescens*× *Haliotis discus hannai*) based on microsatellite DNA markers. Aquaculture Research 46, 216–225. doi:10.1111/are.12169.

Launey, S., Hedgecock, D., 2001. High genetic load in the pacific oyster *Crassostrea gigas*. Genetics 159 (1), 255–265.

Leighton, L.R., 2000. Environmental distribution of spinose brachiopods from the Devonian of New York: Test of the soft-substrate hypothesis. Palaios 15, 184–193. doi:10.2307/3515641.

Leighton, D., 2000. The biology and culture of the California abalones. Dorrance Publishing Co., Inc, Pittsburgh, PA.

Lewis, C.A., Talbot, C.F., Vacquier, V.D., 1982. A protein from abalone sperm dissolves the egg vitelline layer by a nonenzymatic mechanism. Developmental Biology 92 (1), 227–239. doi:10.1016/0012-1606(82)90167-1.

Li, Y., Li, Q., 2022. The growth, survival and ploidy of diploid, triploid and tetraploid of the Pacific oyster (*Crassostrea gigas*) in larval and juvenile stages. Aquaculture 553, 738083. doi:10.1016/j.aquaculture.2022.738083.

Li, S., Qin, Y., Shi, G., Wan, W., Liao, Q., Ma, H., Li, J., Yu, Z., Pan, Y., Zhang, Y., 2023. Feasibility of interspecific hybridization between Iwagaki oyster, *Crassostrea nippona* and Portuguese oyster, *Crassostrea angulata*. Aquaculture 563, 739002. doi:10.1016/j.aquaculture.2022.739002.

Li, J., Qin, Y., Peng, Z., Wan, W., Yue, C., Wang, Z., Ma, H., Xing, Q., Li, J., Yu, Z., Pan, Y., Zhang, Y., 2024. Feasibility exploration and comparative analysis of interspecific hybridization between *Crassostrea hongkongensis* and *Crassostrea dianbaiensis*. Aquaculture 589, 740963. doi:10.1016/j.aquaculture.2024.740963.

Li, J., Pan, Y., Xing, Q., Ma, H., Yue, C., Wan, W., Wu, J., Wang, Z., Li, S., Li, J., Yu, Z., Qin, Y., Zhang, Y., 2025a. Hybridization and growth characteristics of interspecific crosses between *Crassostrea hongkongensis* and *Crassostrea iredalei*. Aquaculture 595. doi:10.1016/j.aquaculture.2024.741656, 741656.

Li, J., Qin, Y., Xing, Q., Wu, J., Ma, H., Wan, W., Yue, C., Wang, Z., Li, J., Yu, Z., Pan, Y., Zhang, Y., 2025b. Offspring performance of backcrossing between hybrid line of *Crassostrea dianbaiensis* ♀ × *Crassostrea iredalei* ♂ and their parental species. Aquaculture 599. doi:10.1016/j.aquaculture.2025.742140, 742140.

Liu, W., Heasman, M., Simpson, R., 2009. Growth and reproductive performance of triploid and diploid blacklip abalone, *Haliotis rubra* (Leach, 1814). Aquaculture Research 40 (2), 188–203. doi:10.1111/j.1365-2109.2008.02082.x.

Matt, J.L., Allen, S.K., 2014. Heteroploid mosaic tetraploids of *Crassostrea virginica* produce normal triploid larvae and juveniles as revealed by flow cytometry. Aquaculture 432, 336–345. doi:10.1016/j.aquaculture.2014.05.015.

McCombie, H., Lapègue, S., Cornette, F., Ledu, C., Boudry, P., 2005. France chromosome loss in bi-parental progenies of tetraploid Pacific oyster *Crassostrea gigas*. Aquaculture 247 (1–4), 97–105. doi:10.1016/j.aquaculture.2005.02.003.

Nan, L., Zhang, J., Feng, W., Liu, X., Li, Z., Wang, C., 2012. Inter-specific hybridization between *Argopecten purpuratus* and *Argopecten irradians concentricus*. Chin. Agric. Sci. Bull. 28(20), 131â€"135. doi:https://doi.org/10.3969/j.issn.1000-6850.2012.20.025.

Nell, J.A., 2002. Farming triploid oysters. Aquaculture 210 (1-4), 69–88. doi:10.1016/S0044-8486(01)00861-4.

Piferrer, F., Beaumont, A., Falguière, J.C., Flajšhans, M., Haffray, P., Colombo, L., 2009. Polyploid fish and shellfish: Production, biology and applications to aquaculture for performance improvement and genetic containment. Aquaculture 293 (3-4), 125–156. doi:10.1016/j.aquaculture.2009.04.036.

Porto-Hannes, I., Burlakova, L.E., Zanatta, D.T., Lasker, H.R., 2021. Boundaries and hybridization in a secondary contact zone between freshwater mussel species (Family: Unionidae). Heredity 126, 955–973. doi:10.1038/s41437-021-00424-x.

Qin, Y., Zhang, Y., Mo, R., Zhang, Y., Li, J., Zhou, Y., ... Yu, Z., 2019. Influence of ploidy and environment on grow-out traits of diploid and triploid Hong Kong oysters *Crassostrea hongkongensis* in southern China. Aquaculture 507, 108–118. doi:10.1016/j.aquaculture.2019.04.017.

Qin, Y., Li, X., Noor, Z., Li, J., Zhou, Z., Ma, H., ... Yu, Z., 2020. A comparative analysis of the growth, survival and reproduction of *Crassostrea hongkongensis*, *Crassostrea ariakensis*, and their diploid and triploid hybrids. Aquaculture 520, 734946. doi:10.1016/j.aquaculture.2020.734946

Qin, Y., Noor, Z., Li, X., Ma, H., Li, J., Zhou, Y., ... Yu, Z., 2021. Tetraploid induction of *Crassostrea hongkongensis* and *C. sikamea* by inhibiting the polar body 1 release in diploid fertilized eggs. Marine Life Science and Technology 3 (4), 463–473. doi:10.1007/s42995-021-00107-w.

Qin, Y., Zhang, Y., Yu, Z., 2022. Aquaculture performance comparison of reciprocal triploid *C. gigas* produced by mating tetraploids and diploids in China. Aquaculture 552, 738044. doi:10.1016/j.aquaculture.2022.738044.

Qin, Y., Shi, G., Wan, W., Li, S., Li, Y., Li, J., ... Yu, Z., 2023. Comparative analysis of growth, survival and sex proportion among tetraploid-based autotriploid (*Crassostrea gigas* and *C. angulata*) and their allotriploid oysters. Aquaculture 563, 739026. doi:10.1016/j.aquaculture.2022.739026.

Qin, Y., Wan, W., Li, J., Wang, Z., Ma, H., Li, J., ... Zhang, Y., 2024. Growth, survival and reproductive traits of chemical induced and tetraploid-derived triploid Hong Kong oysters (*Crassostrea hongkongensis*). Aquaculture 586, 740759. doi:10.1016/j.aquaculture.2024.740759.

Qiu, A., Shi, A., Komaru, A., 2001. Yellow and brown shell color morphs of *Corbicula fluminea* (Bivalve: Corbiculidae) from Sichuan Province, China, are triploids and tetraploids. Jounal of Shelfish Research 20, 323–328.

Shen, M., Wang, J., Chen, R., Tingjin, L.V., Chen, J., Lu, X., ... Wang, C., 2024. Heterosis, genetic diversity and fertility of reciprocal hybrids between *Argopecten irradians concentricus* and *A. purpuratus*. Aquaculture 586, 740765.

Springer, S.A., Heath, D.D., 2007. Environment-specific heterozygote deficiency and developmental instability in hybrid Mytilus. Marine Biology Research 3, 182–187. doi:10.1080/17451000701320564.

Stanley, J.G., Allen, S.K., Hidu, H., 1981. Polyploidy induced in the American oyster, *Crassostrea virginica*, with cytochalasin B. Aquaculture 23 (1–4), 1–10. doi:10.1016/0044-8486(81)90002-8.

Su, J., Wang, Z., Zhang, Y., Yan, X., Li, Q., Yu, R., 2016. Early embryo and larval development of inviable intergeneric hybrids derived from *Crassostrea angulata* and *Saccostrea cucullata*. Journal of Ocean University of China 15 (3), 515–522. doi:10.1007/s11802-016-2877-0.

Tan, K., Liu, H., Ye, T., Ma, H., Li, S., Zheng, H., 2020. Growth, survival and lipid composition of *Crassostrea gigas*, *C. angulata* and their reciprocal hybrids cultured in southern China. Aquaculture 516, 734524. doi:10.1016/j.aquaculture.2019.734524.

Toro, J., Innes, D.J., Thompson, R.J., 2004. Genetic variation among life-history stages of mussels in a *Mytilus edulis–M. trossulus* hybrid zone. Marine Biology 145, 713–725. doi:10.1007/s00227-004-1363-1.

Voroshilova, I.S., Artamonova, V.S., Makhrov, A.A., Slyn'ko, Y.V., 2010. Natural hybridization of two mussel species *Dreissena polymorpha* (Pallas, 1771) and *Dreissena bugensis* (Andrusov, 1897). Biology Bulletin 37, 542–547. doi:10.1134/S1062359010050158.

Wadsworth, P., Wilson, A.E., Walton, W.C., 2019. A meta-analysis of growth rate in diploid and triploid oysters. Aquaculture 499, 9–16. doi:10.1016/j.aquaculture.2018.09.018.

Wang, Z., Yu, R., Wang, R., Zheng, X., Li, K., Jiang, B. 2003. Method for inducting oyster tetraploid using 6-dimethylin purine. Chinese invention patent, No.: ZL200310114497.1.

Wang, C., Liu, B., Li, J., Liu, S., Li, J., Hu, L., Fan, X., Du, H., Fang, H., 2011. Introduction of the Peruvian scallop and its hybridization with the bay scallop in China. Aquaculture 310 (3-4), 380–387. doi:10.1016/j.aquaculture.2010.11.014.

Wang, C., Liu, B., Liu, X., Ma, B., Zhao, Y., Zhao, X., Liu, F., Liu, G., 2017. Selection of a new scallop strain, the Bohai Red, from the hybrid between the bay scallop and the Peruvian scallop. Aquaculture 479, 250–255. doi:10.1016/j.aquaculture.2017.05.045.

Wang, Y., Wang, X., Gan, Y., Huang, M., You, W., Luo, X., Ke, C., 2021. Autotriploid and allotriploid abalone larvae (*Haliotis discus hannai* and *H. discus hannai* ♀ × *Haliotis fulgens* ♂) produced by two chemical methods. Aquaculture Reports 20, 100769. doi:10.1016/j.aqrep.2021.100769.

Wang, J., Ning, J., Lu, X., Chen, M., Liu, G., Xu, X., … Wang, C., 2023. Production traits and fertility of reciprocal hybrids between *Argopecten irradians irradians* and *A. i. concentricus*. Journal of Oceanology and Limnology 42 (1), 304–315.

Wang, C., Liu, B., Li, J., Liu, S., 2009. Inter-specific hybridization between *Argopecten purpuratus* and *Argopecten irradians irradians*. Marine Sciences. 33(10), 84–75. doi:CNKI: SUN:HYKX.0.2009-10-016.

Wenne, R., Bach, L., Zbawicka, M., Strand, J., McDonald, J.H., 2016. A first report on coexistence and hybridization of *Mytilus trossulus* and *M. edulis* mussels in Greenland. Polar Biology 39, 343–355. doi:10.1007/s00300-015-1785-x.

Xing, Q., Yang, Z., Zhu, X., Liu, J., Huang, X., Hu, J., Bao, Z., 2022. Interspecific hybridization between *Patinopecten yessoensis* (♀) and *P. caurinus* (♂) with heterosis in growth and temperature tolerance. Aquaculture 547, 737489. doi:10.1016/j.aquaculture.2021.737489.

Xu, F., Guo, X., Li, L., Zhang, G., 2011. Effects of salinity on larvae of the oysters *Crassostrea ariakensis*, *C. sikamea* and the hybrid cross. Marine Biology Research 7, 796–803. doi:10.1080/17451000.2011.569555.

Xu, F., Li, L., Wang, J., Zhang, G., 2014. Use of high-resolution melting analysis for detecting hybrids between the oysters *Crassostrea sikamea* and *C.angulata* reveals bidirectional gametic compatibility. Journal of Molluscan Studies 80, 435–443. doi:10.1093/mollus/eyu040.

Xu, D., Wang, C., Liu, B., Chen, M., Ke K., Zhao, Y., Ma, B., Liu, G., Zhao, X., Zhang., J. 2019. Selection of a new scallop strain, QN-2, from the backcross of Peruvian scallop × bay scallop F1 hybrids with bay scallops. Aquaculture Research, 50(12), 3692-3699

Yan, L., Su, J., Wang, Z., Yan, X., Yu, R., Ma, P., Li, Y., Du, J., 2017. Transcriptomic analysis of *Crassostrea sikamea* × *Crassostrea angulata* hybrids in response to low salinity stress. PLoS One 12, e0171483. doi:10.1371/journal.pone.0171483.

Yan, L., Su, J., Wang, Z., Zhang, Y., Yan, X., Yu, R., 2018. Growth performance and biochemical composition of the oysters *Crassostrea sikamea*, *Crassostrea angulata* and their hybrids in southern China. Aquaculture Research 49, 1020–1028. doi:10.1111/are.13549.

Yao, T., Zhang, Y., Yan, X., Wang, Z., Li, C., Su, J., Yu, R., 2015. Interspecific hybridization between *Crassostrea angulata* and *C. ariakensis*. Journal of Ocean University of China 14, 710–716. doi:10.1007/s11802-015-2546-8.

You, W., Guo, Q., Fan, F., Ren, P., Luo, X., Ke, C., 2015. Experimental hybridization and genetic identification of Pacific abalone *Haliotis discus hannai* and green abalone *H. fulgens*. Aquaculture 448, 243–249. doi:10.1016/j.aquaculture.2015.05.043.

Yue, C., Qin, Y., Wan, W., Shi, G., Li, S., Li, J., … Zhang, Y., 2024. Phenotypic traits of reciprocal tetraploid hybrids derived from tetraploid *Crassostrea gigas* and tetraploid *Crassostrea angulata*. Aquaculture 582, 740495. doi:10.1016/j.aquaculture.2023.740495.

Zhang, H., Liu, X., Zhang, G., Wang, C., 2007. Growth and survival of reciprocal crosses between two bay scallops, *Argopecten irradians concentricus* Say and *Argopecten irradians irradians* Lamarck (1819). Aquaculture 272 (1), S88–S93. doi:10.1016/j.aquaculture.2007.08.008.

Zhang, Y., Wang, Z., Yan, X., Yu, R., Kong, J., Liu, J., … Guo, X., 2012. Laboratory hybridization between two oysters: *Crassostrea gigas* and *Crassostrea hongkongensis*. Journal of Shellfish Research 31 (3), 619–625. doi:10.2983/035.031.0304.

Zhang, Y., Zhang, Y., Wang, Z., Yan, X., Yu, Z., 2014. Phenotypic trait analysis of diploid and triploid hybrids from female *Crassostrea hongkongensis*×male *C. gigas*. Aquaculture 434, 307–314. doi:10.1016/j.aquaculture.2014.08.039.

Zhang, Y., Zhang, Y., Jun, L., Wang, Z., Yan, X., Yu, Z., 2016. Phenotypic trait of *Crassostrea hongkongensis* ♀× *C. angulata* ♂ hybrids in southern China. Aquaculture Research 47, 3399–3409. doi:10.1111/are.12787.

Zhang, Y., Li, J., Zhang, Y., Ma, H., Xiao, S., Xiang, Z., Yu, Z., 2017. Performance evaluation of reciprocal hybrids derived from the two brackish oysters, *Crassostrea hongkongensis* and *Crassostrea sikamea* in southern China. Aquaculture 473, 310–316. doi:10.1016/j.aquaculture.2017.02.031.

Zhang, W., Li, R., Chen, X., Wang, C., Gu, Z., Mu, C., Song, W., Zhan, P., Huang, J., 2020. Molecular identification reveals hybrids of *Mytilus coruscus* × *Mytilus galloprovincialis* in mussel hatcheries of China. Aquaculture International 28, 85–93. doi:10.1007/s10499-019-00445-8.

Zhang, Y., Qin, Y., Yu, Z., 2022. Comparative study of tetraploid-based reciprocal triploid Portuguese oysters, *Crassostrea angulata*, from seed to marketsize. Aquaculture 547, 737523. doi:10.1016/j.aquaculture.2021.737523.

CHAPTER

7

Pedigree-based and marker-assisted breeding

Xiaoli Hu[a], *Jie Meng*[a], *Jing Wang*[a], *Shikai Liu*[b], *Xiaoshen Yin*[a], *Huizhen Wang*[a] *and Xiangchao Zhang*[a]

[a]College of Marine Life Sciences, Ocean University of China, Qingdao, China [b]Ocean University of China, Fisheries College, Qingdao, China

7.1 Introduction

Shellfish aquaculture is a growing sector that provides high-quality protein and supports both economic and ecological sustainability. The rapid expansion in shellfish aquaculture necessitates genetic improvement of economically important traits such as growth rate, disease resistance, and environmental adaptability through effective breeding strategies. Traditional mass selection that focuses on traits such as body weight has been used in mollusks, but it often suffers from environmental influence on phenotypes and risks of inbreeding when broodstock are limited. In contrast, pedigree-based selective breeding and quantitative genetic methods have enabled more accurate evaluation of breeding value (BV), better control of inbreeding, and simultaneous improvement of multiple traits (e.g., growth, disease resistance, and shell quality). To exploit heterosis and nonadditive genetic effects, pedigree-based hybridization is used to improve production and resilience traits by crossing elite families/lines. Over the past two decades, advances in molecular genetics have introduced marker-assisted selection (MAS) and even genomic selection (GS) as powerful tools to complement traditional pedigree-based approaches. Together with rapidly declining genotyping costs and improved statistical and machine-learning methodologies, these tools support more precise selection decisions, facilitate earlier selection of candidates, and accelerate genetic gain while managing inbreeding. By identifying genetic markers linked to desirable alleles, breeders can enhance selection accuracy and accelerate genetic progress for economically important traits. In this chapter, we synthesize these developments to provide a practical overview of pedigree-based breeding and MAS in mollusks, and to highlight key considerations for implementation.

Genetic Breeding and Aquaculture of Shellfish: A Collaborative Approach to a Sustainable Future
DOI: https://doi.org/10.1016/B978-0-443-33606-5.00005-9

7.2 Pedigree-based selection in mollusks

Pedigree-based selection uses pedigree and family information to evaluate and select superior individuals, which enhances the accuracy of BVs estimation by evaluating full-sib or half-sib family performance rather than individual phenotypes. This approach is especially advantageous for complex traits such as growth rate, survival, and stress resistance that are influenced by many genes and environmental factors. By using family means to buffer environmental noise, breeders can achieve more reliable genetic evaluations and faster genetic gains. This is critical for long-term productivity and genetic diversity of molluscan stocks, which face dynamic environments and emerging pathogens in culture. An additional benefit is improved control of inbreeding: by maintaining pedigree records and optimally mating individuals from different families, breeders can minimize inbreeding depression and preserve genetic diversity over generations (Hill et al., 2008; Lind et al., 2012). Mollusks (e.g., oysters, clams, and scallops) often have high fecundity (producing millions of offspring per spawn), which allows large family sizes and high selection intensity (Guo et al., 2012; Houston et al., 2022; Wang et al., 2017). While tagging individual mollusks is difficult, early programs relied on separate family rearing or molecular parentage analysis. These characteristics make pedigree-based selection particularly well suited to molluscan breeding.

7.2.1 Genetic parameters estimation

7.2.1.1 Genetic parameters

Accurate estimation of genetic parameters is foundational to the design of efficient breeding programs, as these parameters determine both the expected response to selection and the optimal choice of breeding strategies. Genetic parameters are population- and environment specific; therefore estimates are most informative when derived from the target population under representative farming conditions and consistent phenotyping protocols. Two key parameters are heritability and genetic correlations (Boudry, 2009; Falconer, 1981). Narrow-sense heritability (h^2) quantifies the fraction of phenotypic variance attributable to additive genetic effects. Only the additive genetic variance (the cumulative effect of genes that is transmitted to offspring) directly contributes to response to selection. Heritability directly determines the efficiency of selection methods. Traits with high heritability can be effectively improved through individual phenotypic selection, whereas for traits with low heritability, reliance on family information or molecular markers is essential.

In mollusks, heritabilities have been estimated for traits such as shell growth, meat yield, survival, and pathogen resistance, using data from pedigree-designed experiments. These estimates are commonly obtained using a mixed-model framework, which helps separate additive genetic effects from shared environmental and family effects. Growth traits (such as growth rate, shell length, body weight) and disease resistance usually show moderate to high heritability (even higher than 0.7), making selection for these traits highly effective in mollusks (Hollenbeck & Johnston, 2018; Wang et al., 2010). For example, resistance to *Ostreid herpesvirus 1* (*OsHV-1*) in *Crassostrea gigas* has shown heritability estimates ranging from 0.22 to 0.81 under different experimental conditions, indicating a strong genetic component (Dégremont et al., 2015). Similarly, in selectively bred oyster families exposed to *Vibrio spp.*, survival h^2 was estimated at 0.535–0.541 on the latent liability scale, which represents an underlying continuous propensity for survival, and 0.113–0.114 on the observed binary scale, which corresponds to the recorded alive/dead outcome.

In contrast to growth and disease resistance, environmental stress tolerance traits often exhibit more complex genetic architectures. Moreover, environmental stress tolerance traits, such as tolerance to heat, hypoxia, and salinity, often show lower but still notable heritability values (Azéma et al., 2017; Chi et al., 2022; Liang et al., 2017). For example, in the dwarf surf clam (*Mulinia lateralis*), the heritability of heat tolerance (measured as survival under thermal stress) reached 0.375 ± 0.127, higher than values reported for *C. gigas* (0.16–0.36) and *Haliotis discus hannai* (0.35–0.42) (Wang et al., 2024). In *Chlamys farreri*, heat tolerance showed a heritability of 0.52 when treated as a binary survival trait, but only 0.24 when modeled as continuous survival time (Yu et al., 2023). These estimates support the feasibility of genetic improvement for stress tolerance traits. Despite evidence of heritable control, resistance traits in mollusks often exhibit significant phenotypic complexity due to interactions with environmental variables. Genotype-by-environment (G × E) interactions occur when the relative performance of genotypes differs across environmental contexts. These interactions complicate selection because individuals or families that perform well in one environment may perform poorly in another. Therefore multienvironment testing and explicit modeling of G × E (e.g., reaction norms or environment-specific genetic correlations) are often needed to support robust selection decisions. For example, resistance to *Vibrio splendidus*-induced mortality in mussels exhibited lower heritability estimates (0.12–0.37), alongside pronounced *G × E* effects (Liang et al., 2017). Genetic correlations of mortality between different test sites ranged from 0.05 to 0.43, suggesting weak correspondence in family performance across environments (Chi et al., 2022).

Genetic correlation refers to the genetic relationship between two different traits, which arises from pleiotropy or linkage disequilibrium among the genes controlling those traits. Genetic correlations between traits measure the extent to which the same genes affect multiple traits (Falconer, 1981). In molluscan breeding, understanding genetic correlations is important because improvement of one trait may cause changes in another, either positively or negatively (Jones et al., 2014; Kong et al., 2015; Kvingedal et al., 2010). Studies have reported positive genetic correlations among various growth metrics (e.g., shell length and body weight), meaning selecting for one will incidentally improve the others. However, there can sometimes be a trade-off between traits. In the Fujian oyster (*C. angulata*), growth traits exhibited a low genetic correlation with gonad development and with heat tolerance, suggesting these traits could be improved independently (Han et al., 2022; Zhai et al., 2021). Knowledge of these correlations allows the development of selection indices to balance multiple goals. For example, a selection index could assign weights to both growth rate and shell strength, aiming to maximize growth without compromising shell quality to avoid deteriorating shell quality while maximizing growth. It also helps identify situations when independent culling or separate selection lines may be needed for antagonistic traits.

Ultimately, the value of genetic parameter estimation is reflected in realized genetic progress. Regular assessment of genetic progress is essential. Genetic progress (gain) per generation is measured as the difference in average BV of offspring compared to the previous generation, for the target traits. This can be calculated from selection differentials and heritability, or observed directly in the results of a breeding program. In well-managed mollusk breeding programs, substantial genetic gains have been achieved. As noted, selection for oyster growth can yield

~10% improvement per generation, and for survival in disease challenges around ~15% per generation (Hollenbeck & Johnston, 2018).

7.2.1.2 Family design and statistical models

The precise and accurate estimate of genetic parameters is a prerequisite for making efficient selection strategies in breeding programs. However, accurate parameter estimation relies on rigorous experimental design and powerful statistical tools. Regarding mating design, the most prevalent is the nested design, where one male is mated with multiple females to create multiple full-sib families, which in turn are connected through the common sire to form half-sib families. This design effectively separates additive genetic variance from common environmental effects and is considered the gold standard for heritability estimation. An alternative is the factorial design, which involves all possible male–female crosses and provides the richest genetic information, but its operational scale increases geometrically with the number of parents (Barros et al., 2018; Liang et al., 2016). Nevertheless, it remains feasible in mollusks because of their high fecundity. A critical point is that families must be reared separately until individuals can be tagged to prevent environmental effects from confounding genetic influences, followed by common-garden testing to ensure phenotype differences are primarily of genetic origin.

For statistical modeling, the best linear unbiased prediction (BLUP) method represents the most widely used model (Farias et al., 2017). As a mixed linear model, BLUP combines pedigree relationships, trait measurements, and fixed environmental effects into a single model to estimate the BV of each individual or family. In practice, BLUP uses an animal model or family model that accounts for additive genetic effects (random) and known environmental or management effects (fixed). For example, a simplified mixed-model equation is:

$$y = X\beta + Zu + e$$

where y is the vector of observed phenotype, Xβ is the fixed effect part of the model, where X is the design matrix that accounts for fixed effects (e.g., environmental factors), and β is the vector of fixed effect parameters. Zu is the random effect part, where Z is the design matrix for random effects, and u is the vector of random effects. e is the residual error vector, representing the variation in phenotype that is not explained by the model.

The model is solved using maximum likelihood estimation (MLE) or restricted maximum likelihood estimation (REML), which allows for the optimal estimation of both fixed and random effects.

Assuming the random effects and residuals are jointly normally distributed with:

$$\begin{bmatrix} u \\ e \end{bmatrix} \sim N\left(\begin{bmatrix} 0 \\ 0 \end{bmatrix}, \begin{bmatrix} \sigma_u^2 A & 0 \\ 0 & \sigma_e^2 I \end{bmatrix}\right)$$

where A is the additive genetic relationship matrix (positive semi-definite) and I is an identity matrix. The mixed-model equations (MME) are derived by maximizing the joint likelihood function, yielding the following system:

$$\begin{bmatrix} X^T X & X^T Z \\ Z^T X & Z^T Z + A^{-1}\sigma_e^2/\sigma_u^2 \end{bmatrix} \begin{bmatrix} \beta \\ \hat{u} \end{bmatrix} = \begin{bmatrix} X^T y \\ Z^T y \end{bmatrix}$$

where β is the estimated value of the fixed effect, and $\hat{u}$ is the estimated value of the random effect (i.e., estimated BV values).

In aquaculture breeding, the effectiveness of BLUP depends on key factors such as pedigree depth and trait heritability (Gutierrez et al., 2018; Vu, 2023). Pedigree depth and, critically, connectedness among families determine how much information can be shared across the dataset, which directly affects prediction accuracy. A deeper, multigenerational pedigree provides more genetic information, improving the accuracy of BV estimates. The more complete the pedigree, the better we can understand genetic relationships and make reliable predictions. Trait heritability also matters—traits with high heritability (e.g., growth rate) are easier to improve and predict, since genetics plays a larger role. Traits with low heritability (e.g., disease resistance) are harder to predict because of strong environmental influences. Still, with enough pedigree and accurate trait data, BLUP can provide useful guidance even for low-heritability traits (Yang et al., 2024). In addition, careful phenotyping (consistent protocols, appropriate contemporary groups, and relevant covariates) is often as important as model choice for achieving reliable estimates.

7.2.2 Pedigree-based hybridization

Pedigree-based hybridization refers to a breeding strategy in which hybridization (crossing) is carried out between individuals from different families or genetic lines under controlled mating schemes. This approach aims to combine desirable traits from parental families and exploit heterosis (hybrid vigor) in offspring, thereby improving commercially important traits that may show limited response to within-family selection alone . In aquaculture, family-based hybridization is widely used to improve production and resilience traits by crossing elite families/lines while maintaining rigorous pedigree control (Hollenbeck & Johnston, 2018). A key advantage of pedigree-based hybridization is that it naturally interfaces with quantitative genetic evaluation. Specifically, it benefits from: (1) accurate pedigree tracking, enabling correct assignment of parents and crosses; (2) estimation of genetic parameters using family data; and (3) application of BLUP and mixed-model methodology to standardize comparisons across cohorts, environments, and unbalanced family sizes—common issues in aquaculture breeding datasets (Lynch & Walsh, 1998).

7.2.2.1 Heterosis analysis for family-based hybridization

To assess the relative merits of hybridization schemes, the magnitude of heterosis (hybrid vigor) must be quantified. Heterosis refers to the phenomenon that progeny of two genetically distinct parents perform better (e.g., greater biomass, higher fertility, and increased resistance to unfavorable conditions) than the better parent (Shull, 1948). Although "heterosis" and "hybrid vigor" are often used interchangeably, classical usage emphasizes heterosis as a genetically defined deviation of hybrids from parental expectations. Heterosis has been broadly exploited in major crops and livestock, and it is commonly observed as the recovery of performance reduced by inbreeding when genetically divergent parents are crossed (Crow, 1998). Heterosis can occur in either interspecific or intraspecific crosses. In applied breeding programs, intraspecific heterosis is typically constructed by crossing distinct geographic populations, varieties, or pedigreed lines within the same species. Quantification of heterosis is generally based on replicated family means (rather than single individuals), and is evaluated across multiple environments and production stages using standardized indices such as mid-parent heterosis (MPH) and better-parent heterosis (BPH) (Kaeppler, 2012). These measures provide a comparable basis for ranking cross combinations and identifying trait- and environment-specific hybrid advantages.

7.2.2.2 *Genetic basis of heterosis*

Historically, three main genetic explanations for heterosis have been proposed: dominance (complementation of deleterious recessive alleles by dominant alleles in hybrids), overdominance (heterozygote superiority), and epistasis (interaction among loci affecting fitness-related traits). With advances in molecular and multi-omics approaches, additional hypotheses have been discussed, including the gene-balance hypothesis (improved dosage balance for dosage-sensitive components), multigenic metabolic efficiency models, and genome-wide regulatory mechanisms involving chromatin state and DNA methylation (Kaeppler, 2011). At present, no single unifying theory explains heterosis across all traits and species (Schnable & Springer, 2013). From an applied perspective, the value of genetic dissection is that it clarifies which components of performance are primarily additive versus nonadditive, and whether hybrid advantage is expected to be stable across environments or life stages. Genetic mapping frameworks have been widely used to infer the number, genomic location, effect sizes, and modes of gene action underlying quantitative traits (Mackay et al., 2009). In the context of heterotic traits, distinguishing additive, dominance/overdominance, and epistatic contributions is critical because it informs whether improvement is best pursued via within-line selection, structured crossbreeding, or a combined strategy. Moreover, evidence that different life-stage components of performance can be governed by distinct genetic architectures highlights the importance of multistage testing: hybrids that perform well early may not necessarily maximize later performance if hidden trade-offs exist (Yin & Hedgecock, 2021). Collectively, these insights provide a mechanistic rationale for why family-based designs are particularly valuable in aquaculture: They allow breeders to quantify inbreeding depression and heterosis within a unified framework, to model genotype-by-environment interaction explicitly, and to design crossbreeding schemes that exploit both additive and nonadditive components of variation to reduce early mortality while enhancing later growth and robustness.

7.2.2.3 *Crossbreeding Strategies for family-based hybridization*

Identifying elite parent lines is a prerequisite for setting up a crossbreeding program to produce hybrid, high-performing seed, which requires diallel or factorial test crosses among candidate lines. Diallel crosses allow the estimation of causal components of variance in performance—general combining ability (GCA), specific combining ability (SCA), and reciprocal effects (R). Estimates of these components, typically derived from linear/mixed models for diallel designs, enable breeders to select superior parental lines and optimize hybrid combinations for commercial production (Schnable & Springer, 2013). For complete diallel crosses, both generalized linear/mixed models and Bayesian hierarchical models can partition variance in hybrid-family performance into interpretable genetic components. In practice, aquaculture crossing datasets are often incomplete or unbalanced due to variation in reproductive success and early survival. Under these conditions, Bayesian hierarchical models can provide shrinkage estimates that incorporate uncertainty from missing crosses and uneven sample sizes, improving the robustness of inference for GCA, SCA, and reciprocal effects (Gianola, 2013; Lenarcic et al., 2012). Although producing and maintaining inbred lines can be challenging in many aquaculture species, structured crossbreeding programs can still be implemented using well-pedigreed families or lines. Given the large diversity of farmed aquatic species and the limited infrastructure available for long-term domestication across all of them, family-based crossbreeding offers a pragmatic strategy to generate high-yielding, disease-resistant, and stress-tolerant seed through controlled mating and quantitative genetic evaluation. In this sense, crossbreeding programs provide an effective route to reconcile the high demand for improved seed with constrained breeding resources, while creating a bridge to future molecular breeding and GS (Hollenbeck & Johnston, 2018).

7.2.3 Application of pedigree-based breeding

Pedigree-based selective breeding has become a cornerstone of modern molluscan aquaculture. The Australian programme for Sydney rock oyster illustrates the advantages of this approach: After decades of mass selection for growth and disease resistance, breeders established a pedigree-based program that produced roughly eighty families per year and ultimately sought to double this to 160 families; this shift enabled them to assess disease resistance under biosecure conditions, implement multi-trait selection and obtain better estimates of genetic gain. Annual family production increased genetic progress and allowed accurate evaluation of QX disease (Queensland unknown disease) resistance within a single year, and the program reported low-to-moderate heritability for winter mortality resistance, suggesting it could respond to selection but at a slower rate than growth or QX resistance (Dove et al., 2020). In Vietnam, the first selective breeding program for the Portuguese oyster also used strip spawning to create paternal full- and half-sibling families and maintained full pedigree records to guide mate allocation (Vu et al., 2020). Analysis of more than 30, 000 progeny revealed low-to-moderate heritabilities for whole weight (0.13–0.24) and lower heritabilities for soft tissue weight and shell shape; however, selection for whole weight yielded a direct response of about 6% per generation and 8.8% in the most recent generation, with correlated improvements in meat yield and shell shape, demonstrating that family selection can effectively improve multiple traits in this species. The Aquaculture Genetics and Breeding Technology Center in the United States adopted family selection for the Eastern oyster because survival exhibited low heritability and strong genotype–environment interactions across the Chesapeake Bay salinity gradient. Their program shows that by testing separate families in high- and low-salinity environments, breeders could treat survival at each salinity as a distinct trait while capturing higher heritability for growth and meat yield (Allen et al., 2021). In China, researchers conducted two generations of family selection for summer survival using two fast-growing lines of Pacific oysters. The study found that heritability for this trait ranged from 0.12 to 0.28; even at these low-to-moderate levels, the realized genetic gain in survival was substantial, with a cumulative improvement of 24% across two generations and no adverse genetic correlations with growth (Chi et al., 2022). The research team led by Professor Zhenmin Bao at the Ocean University of China pioneered the world's first BLUP-based genetic evaluation system for shellfish and developed the first new scallop variety, "Penglai Red" (*C. farreri*), in 2005 (Guo, 2009), which showed improvements in both yield and survival rate compared to commonly used commercial stocks. Collectively, these examples show that pedigree-based selection schemes allow molluscan aquaculturists to handle traits with low heritability, incorporate multiple environments or disease challenges into selection, and achieve appreciable genetic gains for growth and survival. Family selection also provides a foundation for integrating molecular tools and GS strategies in the future. Crossbreeding and heterosis provide an additional, widely used route to genetic improvement that is naturally enabled by pedigree-based programs. In bivalve mollusks, early evidence for heterosis came from the eastern oyster *C. virginica*, where growth was positively associated with allozyme heterozygosity, prompting subsequent work linking individual heterozygosity to fitness-related traits such as growth rate and metabolism. More direct support has since been obtained from controlled intraspecific crosses among different geographic populations, varieties, and (where available) inbred lines for oysters, scallops, abalones, and clams. These crosses frequently improve performance in survival, growth, environmental stress tolerance, and immune response. In the Pacific oyster *C. gigas*, controlled crosses among inbred lines formally confirmed heterosis for larval growth and survival, and heterosis has also been repeatedly reported for growth in sessile juvenile and adult stages (Kong et al., 2017). Importantly, pedigree-based diallel or factorial crossing designs further allow breeders to quantify GCA, SCA, and reciprocal effects, thereby identifying elite parental lines and optimizing hybrid combinations for commercial seed production. Together, these examples indicate that family selection and structured crossbreeding are complementary strategies: the former provides robust evaluation of additive genetic merit, whereas the latter exploits both additive and nonadditive variation to reduce early mortality and enhance later growth—an especially pragmatic route for improving farmed oysters and other highly fecund aquaculture species.

7.3 Quantitative trait locus mapping

Selective breeding could greatly benefit from the application of molecular markers and the ability to map the genome regions influencing quantitative traits. The past two decades have seen a progression of molecular marker systems applied in mollusks, from early low-density markers, such as amplified fragment length polymorphisms (AFLPs) and microsatellites (simple sequence repeats, SSRs), to modern high-density single-nucleotide polymorphisms (SNPs). These markers have enabled the construction of genetic linkage maps and the mapping of quantitative trait locus (QTL) for a variety of traits (Gopikrishna, 2022; Maqsood & Ahmad, 2017). Together, these advances have progressively increased the resolution at which genotype–phenotype relationships can be interrogated in molluscan species.

7.3.1 Genetic linkage maps

Genetic linkage maps constitute the primary framework for detecting QTL and for anchoring molecular markers to chromosomes in mollusks. Early linkage mapping efforts in aquatic mollusks produced low-density frameworks using AFLP and microsatellite markers. For example, first-generation maps in oysters and scallops had on the order of 100–200 markers with large intervals of ~5–15 cM. Pioneering studies in the eastern oyster and Pacific oyster using AFLPs or microsatellites established ~10–12 linkage groups but covered only ~70%–80% of the genome (Hubert & Hedgecock, 2004; Yu & Guo, 2003). Similarly, initial maps for scallops and abalone based on AFLPs yielded 18–20 linkage groups with markers spaced ~10–15 cM apart (Wang et al., 2004; Li et al., 2007; Xu et al., 2008). The introduction of more polymorphic microsatellite markers improved map density and enabled preliminary trait mapping. In Zhikong scallop *C. farreri*, an integrated microsatellite map identified 19 linkage groups (matching the haploid chromosome number) with an average spacing of ~12 cM (Zhan et al., 2009). A consensus map in bay scallop (*Argopecten irradians*) used 161 SSR markers to achieve ~5.9 cM resolution (Li et al., 2012). The advent of SNP markers and high-throughput sequencing in the 2010s revolutionized molluscan linkage maps. High-density SNP maps were developed for pearl oysters and clams, dramatically increasing marker numbers into the thousands (Hedgecock et al., 2015; Jones et al., 2013; Nie et al., 2017; Shi et al., 2014). These maps typically encompass >95% of the genome with average marker intervals below 1 cM. For instance, a SNP-based map for the silver-lipped pearl oyster (*Pinctada maxima*) placed 887 markers in 14 linkage groups (covering 96% of the genome). The dramatic increase in marker density fundamentally changed the scale at which the genetic architecture of complex traits could be resolved. Genotyping-by-sequencing also boosted map resolution. A notable example is a *C. farreri* map constructed via 2b-RAD sequencing, which achieved 3806 markers and 0.41 cM average interval (Jiao et al., 2014). Likewise, genotyping by sequencing (GBS)-based maps in Pacific abalone and Manila clam reached marker densities of 3000–10,000 loci with ~0.5 cM spacing (Kho et al., 2021; Nie et al., 2017). Mao et al. (2020) constructed high-resolution SNP-based maps for the bay scallop and Peruvian scallop using a 2b-RAD approach in an interspecific hybrid family, illustrating the feasibility of dense map construction even in hybrid pedigrees (Fig. 7.1). The power of these dense maps is evidenced by consistent identification of multiple loci for growth, shell traits, and other aquaculture-related phenotypes (Kho et al., 2021; Shi et al., 2014). In the freshwater pearl mussel (*Hyriopsis cumingii*), an initial 506-locus microsatellite map was rapidly superseded by a high-density combined map (4,983 markers) using SNP genotyping, which enabled fine-mapping of pearl quality QTL (Bai et al., 2016, 2015). In summary, over the past two decades, molluscan linkage mapping has progressed from sparse AFLP/SSR maps to high-resolution SNP-based maps, greatly enhancing the precision of QTL mapping and providing valuable resources for comparative genomics and MAS in aquaculture (Table 7.1).

7.3.2 Quantitative trait locus identification in molluscan traits

QTL mapping has been applied across a range of traits in mollusks, from growth and size traits to survival, disease resistance, and product quality characteristics. Building on the availability of genetic linkage maps, QTL mapping provides a statistical framework to link phenotypic variation with specific genomic regions, thereby forming a direct bridge between molecular markers and breeding-relevant traits. The general approach is to create a mapping population (often by crossing two divergent parents or by using F_2 or backcross designs, or even using full-sib family analysis in outbred populations), genotype the offspring with genome-wide markers, and test for statistical associations between marker genotype and trait values. Significant associations indicate the presence of a QTL near that marker. In mollusks, practical constraints such as high heterozygosity, large family sizes, and the frequent use of outbred full-sib families have shaped the choice of QTL mapping designs and statistical models.

7.3.2.1 *Growth-related traits*

Growth performance represents one of the most economically important traits in molluscan aquaculture, directly determining yield, production efficiency, and farming profitability. Growth traits in mollusks are typically characterized by shell length, shell height, total weight, and soft tissue mass, all of which exhibit substantial variation among individuals and populations. Molluscan growth is a complex quantitative trait influenced by both genetic and environmental factors. Traditional approaches based solely on phenotypic selection are often inefficient, as growth traits are polygenic and late-expressed. These features make growth an ideal target

TABLE 7.1 Summary of key features of genetic linkage mapping studies in aquatic mollusks.

Species (Latin name)	Marker type	No. of markers	Avg. marker interval (cM)	No. of linkage groups
Argopecten irradians (Bay scallop) ♀× *A. purpuratus* ♂	2b-RAD-derived SNPs	2,994 (♀) & 1,585 (♂)	0.32~0.51	16
Chlamys farreri	2b-RAD-derived SNPs	3,806	0.41	19
Pinctada maxima	SNP array	887	~0.94	14
Haliotis discus hannai	GBS-derived SNPs	3,345	0.55	18
Crassostrea gigas	SSR markers	86 (♀) / 88 (♂)	~7 (♂), ~9 (♀)	10 (consensus)
Crassostrea gigas	SNP (exonic) + SSR markers	656 (consensus)	1.0	10
Chlamys farreri	AFLP markers	166 (♀) / 197 (♂)	10.2 (♀), 9.2 (♂)	19 (♀), 20 (♂)
Chlamys farreri	SSR markers	154	12.3	19
Crassostrea virginica	AFLP + SSR markers	114 (♂) / 84 (♀)	~5.7 (♂), ~10.9 (♀)	12 (♂), 12 (♀)

(*continued on next page*)

TABLE 7.1 Summary of key features of genetic linkage mapping studies in aquatic mollusks—cont'd

Species (Latin name)	Marker type	No. of markers	Avg. marker interval (cM)	No. of linkage groups
Pinctada fucata martensii	GBS-derived SNPs	3,117	0.39	14
Haliotis discus hannai	AFLP markers	174 (♀) / 195 (♂)	13.0 (♀), 12.9 (♂)	18 (♀), 19 (♂)
Argopecten irradians	SSR markers	162	5.9	16
Ruditapes philippinarum	GBS-derived SNPs	9,658	0.42	18
Hyriopsis cumingii	SSR markers	506	3.99	19
Hyriopsis cumingii	SLAF-seq-derived SNPs + SSR markers	4,983	1.81	19
*Patinopecten yessoensis*Yesso	AFLP + SSR markers	161 (♀) / 166 (♂)	14.7 (♀), 15.6 (♂)	20 (♀), 21 (♂)
*Patinopecten yessoensis*Yesso	SNP markers	7,489	0.26	19

AFLP, amplified fragment length polymorphism; SNP, single-nucleotide polymorphism; SSR, simple sequence repeat; GBS, genotyping-by-sequencing; SLAF-seq, specific-locus amplified fragement sequencing; 2b-RAD, type IIB restriction-site-associated DNA sequencing

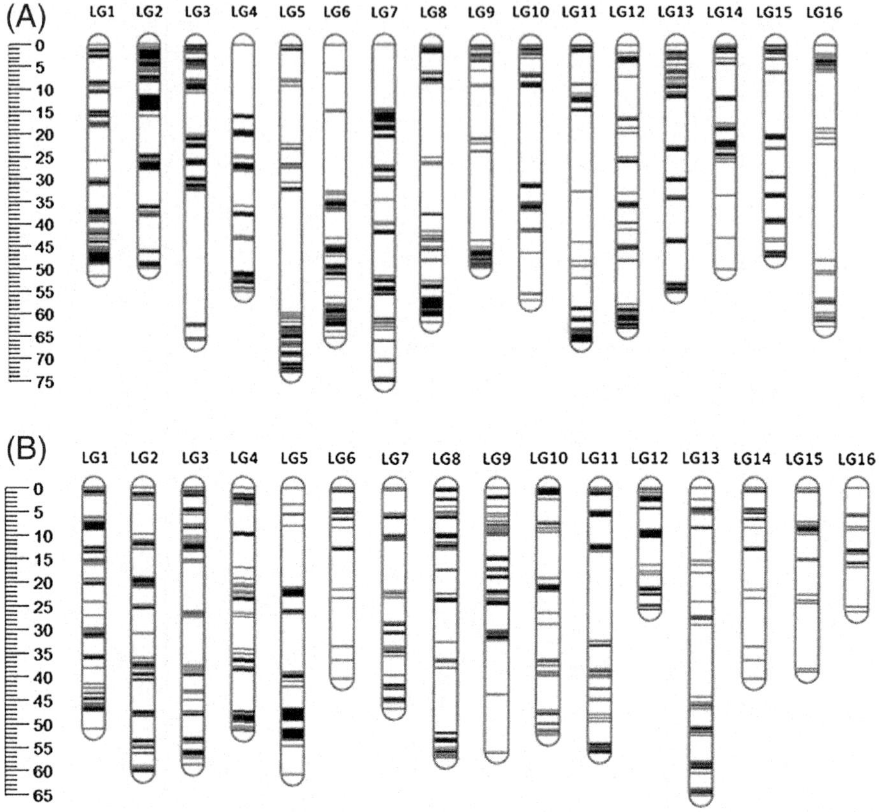

FIGURE 7.1 High-resolution genetic linkage maps. The high-resolution genetic linkage map for (A) *Argopecten irradians* containing 2,994 markers in 16 linkage groups (LGs), and (B) *A. purpuratus* containing 1,585 markers in 16 LGs. The white spaces are the gaps without markers mapped, and *black lines* represent the positions of markers in the linkage maps. *From Mao et al. (2020).*

for QTL analysis, as molecular markers can help uncover genetic effects that are obscured by environmental noise or delayed phenotypic expression.

Early molecular studies in mollusks primarily relied on low- to medium-density marker systems such as AFLPs and SSRs to identify growth-associated loci. Besides, several candidate genes were also used for marker identification. For example, Guo et al. (2012) performed QTL mapping in the Pacific oyster and identified several genomic regions linked to growth rate. One of their notable findings was a QTL on oyster linkage group 2 that had a moderate effect on shell height. Jiao et al. (2014) constructed a high-resolution linkage map for *C. farreri* and detected two QTLs for growth traits, including one on linkage group LG3 harboring the candidate gene *Prop1*, suggesting its potential role in growth regulation (Fig. 7.2). In the Yesso scallop (*P. yessoensis*), a genome-wide association study (GWAS) found two major QTL for shell growth, one of which harbored a strong candidate gene *E2F3*, a cell cycle regulator, and its function in growth regulation was further verified *via* RNA interference (Ning et al., 2019; Wang et al., 2023). Through identification of the gene *Tgfbr1* (transforming growth factor beta receptor type I) in

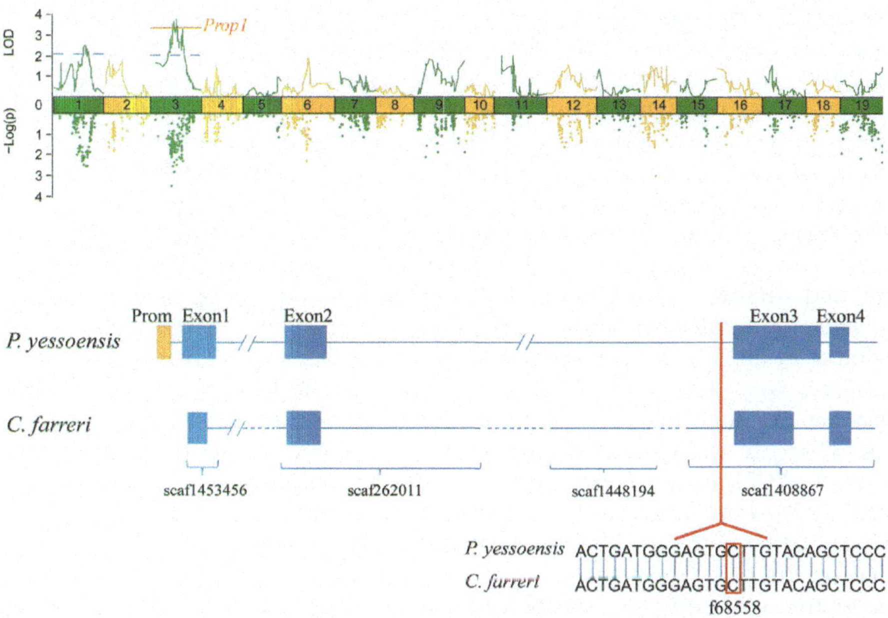

FIGURE 7.2 QTL (Quantitative Trait Locus) mapping and association analysis of growth traits in *Chlamys farreri*. The solid and dashed lines indicate the genome-wide and chromosome-wide significance thresholds, respectively. *From Jiao et al. (2014).*

scallops, certain allelic variants in *Tgfbr1* were found significantly associated with faster growth (Guo et al., 2012). Zheng et al. (2023) applied GWAS to *Pinctada fucata martensii* and identified regulatory genes, including *FoxO* and *MIPs*, associated with shell growth, revealing molecular pathways involved in somatic growth regulation. These genes and markers could be included in selective breeding programs for growth improvement in bivalves.

7.3.2.2 Stress and disease resistance

Disease outbreaks and environmental stressors are among the most critical constraints affecting the sustainability and productivity of molluscan aquaculture. Pathogens such as *Perkinsus marinus, Vibrio spp.*, and *OsHV-1*, together with abiotic challenges including temperature fluctuation, salinity stress, hypoxia, and ocean acidification, frequently cause mass mortality events and significant economic losses. Unlike growth traits, resistance-related traits are often expressed only under specific challenge conditions and are therefore difficult to evaluate accurately through routine phenotypic selection. As filter-feeding organisms inhabiting dynamic coastal ecosystems, mollusks are continuously exposed to multiple biotic and abiotic stressors, making stress and disease-resistance essential targets for genetic improvement. QTL mapping for stress and disease-resistance traits has been actively pursued. Early studies relied on dominant markers such as AFLPs, which helped identify genomic regions linked to *P. marinus* resistance in the eastern oyster (*C. virginica*). The subsequent adoption of co-dominant markers, especially SNPs, greatly enhanced the feasibility of MAS. The availability of SNP markers enabled direct genotyping of resistance-associated alleles in breeding populations, transforming

QTL discoveries into operational selection tools. For example, SNPs within the immune-related gene *CvSI-1* and a 25-bp promoter deletion were highly correlated with dermo disease resistance (He et al., 2012). This deletion was shown to affect gene regulation, providing a mechanistic basis for the observed resistance phenotype. In the Pacific oyster (*C. gigas*), specific haplotypes of HSP family genes were associated with bacterial resistance. Similar advances were reported in *Meretrix meretrix*, *A. irradians*, *Tegillarca granosa*, and *C. farreri*, where numerous stress- or pathogen-responsive loci were identified under experimental challenge.

With the advent of high-throughput sequencing, QTL mapping and GWAS have revealed multiple resistance loci associated with Roseovarius Oyster Disease (ROD), bonamiosis, *OsHV-1* infection, and summer mortality across several oyster and scallop species. QTL analysis in *Chlamys nobilis* × *A. irradians* hybrids identified 42 loci affecting low-temperature survival and growth (Zhang et al., 2024a). For salinity tolerance, QTLs and SNPs involved in osmoregulatory processes were detected in *C. virginica*, *C. gigas*, and *C. hongkongensis* (McCarty et al., 2022). Genome resequencing of *C. gigas* revealed more than 680 salinity-associated SNPs, while ddRAD-seq in the eastern oyster pinpointed major QTLs on chromosomes 1 and 7 (McCarty et al., 2022). In *A. irradians*, SNPs located in the promoter regions of *metallothionein 1*, *PC4*, and *HSP90* were linked to thermal tolerance, whereas in *M. lateralis*, genes within the *ETHR/EHF* signaling pathway were implicated via GWAS (Wang et al., 2024; Yang et al., 2013, 2015; Zhu et al., 2021).

Despite evidence of heritable control, resistance traits in mollusks often exhibit significant phenotypic complexity because of interactions with environmental variables. Genotype-by-environment (G × E) interactions occur when the relative performance of genotypes differs across environmental contexts. These interactions complicate selection because individuals or families that perform well in one environment may perform poorly in another. For example, resistance to *V. splendidus*-induced mortality in mussels exhibited lower heritability estimates (0.12–0.37), alongside pronounced G × E effects. Genetic correlations of mortality between different test sites ranged from 0.05 to 0.43, suggesting inconsistent resistance expression across environments. Such variability highlights the limitations of single-environment selection and underscores the importance of incorporating G × E interactions into statistical models and breeding design. This highlights the necessity of accounting for G × E interactions in breeding models to ensure the robustness and transferability of genetic improvement.

7.3.2.3 Quality traits

Mollusks are a vital component of global aquaculture, contributing significantly to food security and human nutrition. They contain appreciable quantities of high-quality proteins, essential amino acids, long-chain polyunsaturated fatty acids (such as eicosapentaenoic acid (EPA) and docosahexaenoic acid (DHA)), glycogen, carotenoids, vitamins (such as vitamins B and E), essential minerals (e.g., zinc (Zn), selenium (Se), iron (Fe)), and also other nutrients, which offer a variety of health benefits to the consumer. With the increasing demand for high-quality seafood and the need for sustainable aquaculture practices, there is a growing interest in improving the nutritional quality of mollusks through breeding work. Many quality traits are difficult, destructive, or expensive to measure phenotypically, which makes molecular-marker-based approaches especially valuable for their improvement.

Glycogen content, a major determinant of flavor and energy storage, has been a central focus in multiple molluscan species. GWAS in oysters has uncovered several significant SNP loci strongly associated with glycogen levels, with the *PPP1R3B* gene identified as a putative regulator of glycogen synthesis (Meng et al., 2019). Similar approaches revealed SNPs and candidate genes linked to glycogen variation in abalone (*H. discus hannai*), further supporting the conserved genetic basis of this trait across mollusks (Tang et al., 2024). Fatty acid composition, particularly omega-3 fatty acids such as EPA and DHA, is another important nutritional trait. In both oysters and scallops, integrative analyses combining GWAS and QTL mapping have pinpointed candidate genomic regions and genes influencing fatty acid profiles (Shi et al., 2020; Zhang et al., 2024c). These markers offer potential for selecting individuals with higher health-beneficial lipid content through MAS. Notably, studies in abalone have also revealed SNPs related to total protein and taurine content, with genes like *SLC6A7* implicated in taurine transport regulation (Yin et al., 2025). Trace elements such as zinc, magnesium, and selenium are critical micronutrients that also vary genetically among individuals. In oysters, QTL mapping and GWAS have identified SNPs linked to Zn and Se accumulation (Meng et al., 2020). These studies suggest that selective breeding for mineral-enriched mollusks is feasible through marker-based approaches.

Carotenoid accumulation, especially in the muscles, affects both the visual quality and nutritional value of mollusks. Detailed genetic analyses revealed that the orange muscle phenotype is inherited as a recessive Mendelian trait, with carotenoid deposition mainly comprising pectenolone and pectenoxanthin (Li et al., 2022; Zhang et al., 2014). This trait represents a rare case where a major-effect locus dominates phenotypic variation in a quality-related character. Fine-mapping positioned the causal locus on chromosome 8, in scallop *P. yessoensis*, where *PyBCO-like 1* was the only differentially expressed gene between orange and white muscles (FDR = 2.75×10^{-21}), showing an ~11-fold down-regulation in orange individuals (Fig. 7.3). Functional assays further confirmed that *PyBCO-like 1* catalyzes β-carotene degradation, and RNA interference of this gene in white-muscle scallops induced visible orange coloration with increased carotenoid deposition. This functional validation establishes a direct causal link between gene expression, metabolic activity, and the observed quality phenotype. Interestingly, the expression of *PyBCO-like 1* did not differ in the hepatopancreas between color morphs, suggesting that both phenotypes absorb carotenoids similarly, but degradation in the muscle differs (Li et al., 2019). A GWAS further dissected the quantitative variation in carotenoid content and pigmentation intensity. The study revealed multiple SNPs on chromosome 7 significantly associated with concentrations of pectenolone and pectenoxanthin. Among these, the most significant SNPs were located upstream of two tandemly arranged Poly(ADP-ribose) polymerase (*PARP9/14/15*) genes, whose expression levels were strongly correlated with carotenoid concentrations ($r > 0.7$, $P < 0.05$). Additional associated loci were identified near other PARP copies (*PARP9/14/15-3* and *-4*) and a Cytochrome P450 2J6 (*CYP2J6*) gene cluster, implying a regulatory network linking PARPs with carotenoid absorption and *CYP2J6*-mediated ketocarotenoid conversion (Wang et al., 2022). These findings suggest a multi-gene regulatory framework for carotenoid accumulation in scallops, where *PyBCO-like 1* primarily controls carotenoid degradation, whereas *PARP* and *CYP* gene families contribute to differential carotenoid modification and deposition. The identification of these genetic loci not only elucidates the molecular mechanism of muscle coloration but also provides reliable molecular markers for MAS aimed at developing scallop strains with enhanced carotenoid pigmentation and nutritional value.

7.3.3 Integrating multiomics for marker discovery

Because molluscan traits are complex, a multiomics approach has been increasingly adopted to validate and enrich marker-trait associations. Single-layer genetic association analyses often provide limited resolution for complex traits, especially when multiple genes and regulatory

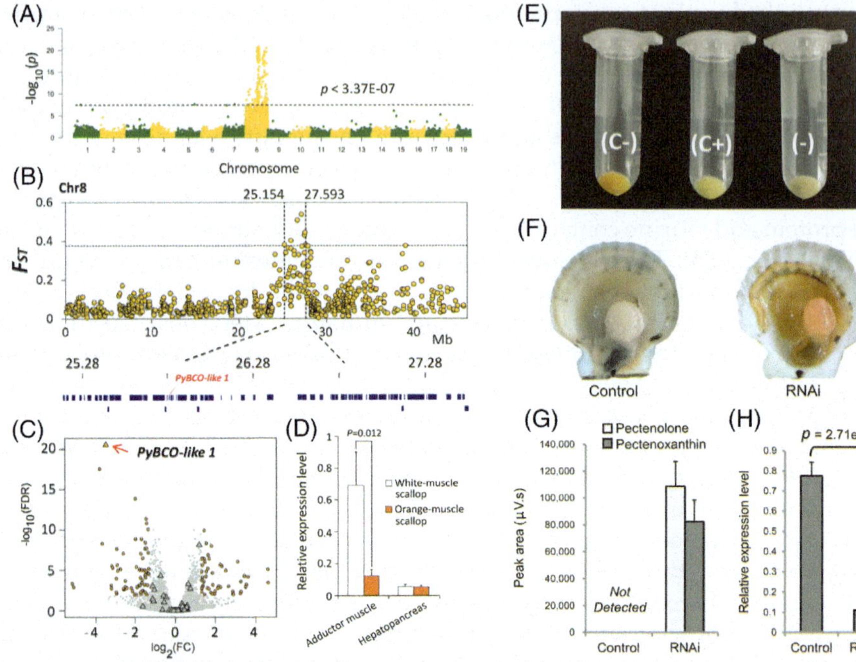

FIGURE 7.3 Identification of *PyBCO-like 1*, the causal gene responsible for carotenoid coloration in scallop muscle through multiomics analysis and gene function validation (Li et al. 2019). (A) GWAS in scallop families revealed that the genomic region associated with carotenoid coloration in scallop muscle was located on Chromosome 8 (Chr8, $P < 3.37E-07$). (B) fixation index (Fst) between orange- and white-muscle populations was assessed through genome resequencing. An ~2.5 Mb candidate region underlying muscle coloration was identified on Chr 8, which contains all the values above the 99th percentile (Fst > 0.376) (upper panel). Blue boxes in the lower panel represent the genes harbored in this region. (C) Volcano plot of gene expression comparison between white and orange muscle. Genes located in the 2.5 Mb candidate region that was identified by GWAS and Fst analysis are indicated by triangles, whereas dots represent genes located in the other region of the genome. Differentially expressed genes are marked in orange. *PyBCO-like 1* was the most significantly expressed gene (FDR: 2.75E-21) in the whole transcriptome analysis, and the only differentially expressed gene in the 2.5 Mb candidate region. (D) qRT-PCR analysis of *PyBCO-like 1* expression in the adductor muscle and hepatopancreas of orange- and white-muscle scallops. (E) *PyBCO-like 1* expressed in *E. coli* could degrade β-carotene. The *E. coli* cells in the tube marked by (C−) contain the pORANGE plasmid carrying β-carotene synthesis genes and exhibit orange color, and cells in (C+) tube contain both the pORANGE plasmid and the pET-28a (+) plasmid with the *PyBCO-like 1* gene. Decoloration occurred in (C+) tube after PyBCO-like 1 expression was induced. E. coli cells without any plasmid are marked by (-). (F) In vivo verification of *PyBCO-like 1* function by RNAi. White muscle shifted to orange after injection of *PyBCO-like 1* dsRNA. (G) Carotenoids pectenolone and pectenoxanthin were detected in the muscle injected with *PyBCO-like 1* dsRNA, while they were not detected in the control muscle injected with PBS. (H) *PyBCO-like 1* expression was inhibited by dsRNA, as revealed by qRT-PCR. GWAS, Genome-wide association study.

mechanisms are involved. Transcriptome (RNA-seq) studies have been especially useful to identify candidate genes and pathways underlying traits of interest, which can then guide marker discovery. For instance, in scallops, researchers compared gene expression profiles between fast-growing and slow-growing families: One outcome was the identification of a gene called *SSTR-1* (somatostatin receptor 1) that was consistently upregulated in fast growers. Functional assays (RNA interference knockdown) confirmed *SSTR-1* influences growth (silencing it caused a significant reduction in growth), highlighting it as a potential marker or even a direct target for genomic editing. This kind of analysis goes beyond correlation by demonstrating causation, thereby strengthening the case for using a marker in MAS (Zhang et al., 2024b). Metabolomic evidence further supports this multiomics framework: in scallops, multi-tissue UPLC-Q-TOF-MS profiling of fast- and slow-growing scallops identified growth-associated metabolites and pathways, including fatty acid biosynthesis, and ABC transporter. Together with transcriptomic evidence, these findings suggest that scallop growth is regulated by coordinated gene-expression and metabolic remodeling (Zhang et al., 2024c). Expression quantitative trait locus (eQTL) mapping further connected genetic variation to stress-induced transcriptional responses. In *C. ariakensis*, thermal and salinity stress eQTL analysis identified 1194 eSNPs, 85% of which exhibited genotype-by-environment (G × E) interactions, underscoring the complexity of regulatory mechanisms (Zhang et al., 2023). In *C. gigas* × *C. angulata* hybrids, 164 regulatory SNPs and 831 stress-related genes were mapped, including *HspA9*-associated loci, providing a foundation for cross-species comparative analysis. More recently, epigenetic regulation has been recognized as an additional layer influencing stress tolerance. Studies on *C. gigas* have shown that DNA methylation patterns interact with genetic variants to modulate gene expression during environmental stress, highlighting the need for integrated analyses combining SNP, transcriptomic, and epigenomic data (Wang et al., 2021). Together, these studies demonstrate that multiomics integration transforms marker discovery from a purely statistical exercise into a mechanistic framework, thereby increasing the robustness and transferability of markers for selective breeding.

7.4 Marker-assisted selection

The success of MAS in aquaculture marks a shift from selecting individuals to selecting genotypes. More fundamentally, it represents a transition from phenotype-driven inference to increasingly genotype-informed decision-making. The key driver of this transformation lies in the ability of molecular markers to directly reveal DNA-level variations, enabling precise genotype-based selection. It is important to emphasize that MAS does not replace traditional pedigree-based selection, but rather forms a powerful synergy with it. Pedigree-based selection provides essential phenotypic data and population foundation for MAS, whereas MAS pinpoints superior BVs to specific genomic regions, laying the foundation for precision breeding practices moving forward.

7.4.1 Algorithms and breeding design

In practice, MAS in molluscan breeding is not a single algorithm but a set of breeding designs that differ in how marker information is incorporated into selection decisions. In molluscan breeding, MAS can be implemented through two main approaches that integrate marker information into breeding decisions. The first strategy is MAS based on significant QTLs. Its core principle lies in using a small number of key molecular markers that are tightly linked to the target trait as the basis for selection. This strategy is suitable for traits with a relatively clear genetic architecture, controlled by a few major genes or large-effect QTLs, such as vertical resistance to specific pathogens or qualitative-quantitative traits such as specific color (Robledo et al., 2018; Divilov et al., 2023; Li et al., 2019; Zhao et al., 2017). Its technical implementation primarily relies on two core models: (1) Marker-assisted BLUP, which incorporates known QTL-linked markers into the animal/line evaluation model as fixed effects or covariates. By including the genotype at key loci directly in the BLUP analysis, one can account for the favorable allele effects when estimating BVs. The marker-assisted BLUP framework increases the accuracy of estimated breeding values (EBVs), especially in genomic regions harboring major genes, thereby improving the selection of candidates carrying desirable alleles (Guimarães et al., 2007); (2) Combined selection index, by constructing a composite index that merges conventional phenotypic EBVs with molecular marker scores for the trait. Each candidate receives a weighted index value that reflects both its performance (or pedigree-based selection value) and the presence of favorable marker alleles. This framework allows breeders to flexibly balance phenotypic performance and marker information according to trait importance and economic weights. By integrating direct genomic information into a traditional selection index, we can achieve a more efficient selection response on traits influenced by known QTL, balancing phenotypic merit with genotype data. However, MAS utility diminishes for complex traits controlled by many loci of small effect. In such polygenic cases, focusing on only a few markers captures only a fraction of the genetic variance. This limitation has led to the development of more comprehensive GS strategies, which can exploit information from across the entire genome (Meuwissen et al., 2001, 2013).

Conceptually, GS can be viewed as a natural extension of MAS when trait architecture becomes highly polygenic and no single QTL explains a large proportion of genetic variance. In a GS scheme, individuals are genotyped for tens of thousands of SNP markers covering the whole genome, and statistical models are used to estimate an effect for every marker. Summing up the contributions of all these markers yields the genomic estimated breeding value (GEBV) for each individual. GS can utilize markers in linkage disequilibrium with QTL across the genome to capture the total genetic merit of an individual without needing to pinpoint specific major genes. Genome-wide prediction harnesses essentially all the genetic variance, providing an unbiased and much more accurate evaluation of BV for complex traits. This makes GS a powerful tool for accelerating genetic improvement in molluscan aquaculture and other farmed species with complex trait targets (Hollenbeck & Johnston, 2018). To implement GS, a large reference population with both genotype data and phenotype records is first used to train a prediction model, after which the model can predict GEBVs of selection candidates from genomic data alone. Several classes of statistical models are commonly used for GS predictions, each with different assumptions about marker effect distributions. Genomic BLUP (G-BLUP) is an extension of the BLUP approach using a genomic relationship matrix in place of the traditional pedigree-based relationship matrix (Zenger et al., 2019). All genome-wide SNP markers contribute to the relationships among individuals, and GEBVs are predicted under the assumption that every marker captures an equal proportion of the genetic variance (i.e., all SNP effects are assumed to have the same prior variance). G-BLUP is conceptually simple and has been widely applied, effectively capturing the realized relationships and yielding higher prediction accuracy than pedigree-only BLUP. However, the equal-variance assumption may not hold when a few loci have much larger effects than others. A variety of Bayesian genomic prediction methods (e.g., BayesA, BayesB, BayesCπ, and Bayesian Lasso) have been developed to relax the equal-effect assumption of G-BLUP (Gianola, 2013). These models impose different prior distributions on marker effects, allowing for heterogeneity in the magnitude of effects across loci. By permitting some SNPs to explain disproportionately large portions of variance, these Bayesian approaches can more flexibly capture genetic architectures where both minor and major QTL coexist. In practice, models such as BayesA/B and Bayesian Lasso often yield improved prediction accuracy in the presence of major QTL, as they effectively upweight markers in strong linkage with those large-effect loci.

A comprehensive and efficient MAS/GS breeding program is structured as a rigorous, iterative cycle. In this framework, MAS and GS are not mutually exclusive, but are deployed according to trait architecture, data availability, and breeding stage. In the initial phase, clear breeding objectives are defined (e.g., faster growth, enhanced disease resistance, and improved product quality), and a broad-based founder population is established (Hayes et al., 2006). Ensuring wide genetic diversity in the base population is critical for sustaining long-term genetic gains. The next phase focuses on data generation, involving high-throughput genotyping and precise phenotypic evaluation. Genome-wide molecular markers (such as SNP arrays or reduced-representation sequencing, such as GBS/RAD-seq) are applied to obtain dense genotypic profiles for each individual. Simultaneously, standardized phenotyping protocols are used to collect high-quality trait data (e.g., growth metrics, survival under disease challenge) on these individuals (Duan et al., 2025; Kong et al., 2025; Robledo et al., 2018b; Tsai et al., 2015; Zhao et al., 2021). The integration of extensive SNP genotypes with reliable phenotypes yields a reference dataset suitable for downstream genomic prediction modeling. The cost-effective 2b-RAD sequencing has been successfully used to genotype scallops, demonstrating the feasibility of GS in mollusks (Wang et al., 2012). Such rich data resources form the foundation for building robust prediction models in the breeding program. In the model-building phase, the genotyped and phenotyped individuals serve as a training population to calibrate a genomic prediction model. Statistical algorithms, such as G-BLUP or Bayesian regression, are employed to estimate the effects of genome-wide markers on the trait of interest. Essentially, the model uses the reference population data to capture associations between SNP markers and phenotypic variation, thereby explaining the genetic architecture of the trait (including many small-effect loci). In practice, one can rapidly obtain GEBVs for a candidate population based solely on their DNA profiles (Song et al., 2023). The adoption of genomic prediction improved the accuracy of selection compared to traditional methods. The final phase is decision-making and mating execution. Candidates are ranked by their GEBVs for the target traits, and those with superior predicted BVs are selected as parents for the next generation. When devising crossing schemes, breeders consider the complementarity and additive combination of parental GEBVs to maximize favorable trait expression in progeny. In addition, genomic relationship information among candidates is utilized to manage kinship and minimize inbreeding accumulation. Unlike pedigree estimates, genome-based relationships more precisely measure genetic similarity between individuals, helping avoid the coselection of close relatives. By optimizing mate pairings with molecular kinship data, the program achieves genetic gain while keeping inbreeding at bay. This genome-informed mating strategy is particularly valuable in high-fecundity mollusks (oysters, scallops, etc.), where large full-sib families are common and controlling relatedness is essential to maintaining long-term population fitness. Overall, the cyclical process of selection and mating, grounded in genomic predictions, iteratively drives the population toward the breeding objectives with enhanced efficiency each generation.

7.4.2 Marker-assisted selection applications in mollusks: Case studies

Growth rate (often measured as shell length, shell height, live weight, or meat weight at a given age) is a primary target in virtually all aquaculture breeding programs, as faster growth shortens the production cycle and increases yield. Mollusks show substantial genetic variation for growth-related traits, and traditional selective breeding has already achieved moderate gains per generation in many species. However, despite its economic importance, growth in mollusks is genetically complex, which places inherent constraints on the effectiveness of MAS targeting individual loci. Although RNA interference and expression analyses have validated several growth-associated genes in mollusks, growth is generally not controlled by many loci with small effects. In practical breeding, integrating genome-wide marker information provides a more effective strategy for growth improvement, which is essentially the basis of GS. This distinction is important in practice because it defines the boundary between MAS as a targeted tool and GS as a genome-wide strategy. An example of MAS contributing to growth improvement is seen in hybrid oyster breeding in China. Researchers identified SNP markers associated with heterosis for growth in hybrid oysters (between *C. gigas* and *C. angulata*) (Wang et al., 2016). By selecting broodstock that had complementary genotypes at these loci (one parent homozygous for allele A, the other for allele B, maximizing offspring heterozygosity), they were able to produce F1 hybrids with superior growth. This is a form of marker-assisted mate selection to exploit heterosis. It demonstrates the utility of markers in designing crosses for growth. Another case study comes from abalone (*H. discus hannai*) breeding in Korea. Quantitative genetic studies had found that abalone growth was moderately heritable, and a QTL study identified some markers linked to juvenile growth. They created a MAS experiment where, among the offspring of a pair mating, they preferentially chose those with the favorable marker genotypes to be the next-generation's parents. Over two generations, they observed a larger increase in shell length in the MAS-treated line compared to a control line (selected only on phenotype). This indicates MAS gave an extra boost by capturing some genetic variance that was not evident from phenotype alone (perhaps due to environmental noise or because the marker predicted adult size from juvenile genotype). Together, these studies suggest that MAS can contribute incremental gains for growth-related traits when applied in well-defined contexts. However, growth traits in mollusks are highly polygenic, and the effect of any single marker is generally limited. Therefore, GS integrates genome-wide marker information, which is better suited for achieving sustained genetic gain in growth-related traits.

Improving disease resistance is a critical need in molluscan aquaculture, given the devastating outbreaks that have occurred (e.g., oyster herpesvirus, bacterial vibriosis, and protozoan parasites such as *Perkinsus*). Traditional selective breeding has achieved some success in increasing survival under disease pressure, but often it is slow and challenged by low heritability and variable pathogen strains. Compared with growth-related traits, disease resistance is often better suited for MAS because resistance phenotypes are difficult, costly, or risky to measure directly, and because specific genomic regions with relatively large effects can sometimes be identified. MAS offers a way to accelerate gains by directly selecting for genotypes known to confer resistance. One of the pioneering examples of MAS in aquaculture was in the Pacific oyster *C. gigas* for resistance to *OsHV-1*. In the Molluscan Broodstock Program (MBP) in the United States, a breeding population of oysters was exposed to *OsHV-1* in field trials, and families were evaluated. Researchers identified a QTL on chromosome 8 significantly associated with survival (oysters with certain marker alleles had much higher survival). The MBP implemented

MAS by genotyping broodstock for SNPs linked to this QTL and selecting individuals with the "resistance" allele. This approach allowed resistance to be selected without deliberately exposing all candidates to lethal viral challenge, highlighting a major practical advantage of MAS for disease traits. As a result, the next-generation families showed substantially improved survival in *OsHV-1* infected waters—their EBV for survival were 47% higher than those of families selected by pedigree information alone (Divilov et al., 2023). This is a landmark case demonstrating MAS can produce a measurable advantage. Another species where MAS has been applied is the eastern oyster (*C. virginica*) for resistance to dermo disease (caused by *P. marinus*). Researchers discovered a polymorphism in the promoter of a serine protease inhibitor gene (*cvSI-1*) that is associated with lower parasite burden and higher survival (He et al., 2012). In contrast to the OsHV-1 example, which relied on a QTL identified through family-based analysis, this case represents a candidate-gene-based MAS approach. Specifically, a 25-bp deletion in the *cvSI-1* promoter leads to upregulation of this immune-related gene and confers improved dermo resistance in oysters carrying the mutation. Breeders at Rutgers University incorporated this marker into their selection program by preferentially breeding from oysters with the favorable allele. Over successive generations, the frequency of that allele increased in the population. Field trials indeed showed an increase in dermo survival. Like growth, most disease-resistance traits are polygenic; known QTLs typically explain only a fraction of the total genetic variance, so selecting on a single marker will not capture all the resistance potential. Furthermore, pathogens can evolve: heavy reliance on one resistance allele could impose selection on the pathogen population (analogous to pests overcoming single-gene resistances in crops). A sustainable strategy is to breed for broad, durable resistance by accumulating multiple resistance alleles. MAS facilitates such gene pyramiding—for instance, one might combine an *OsHV-1* resistance QTL from one oyster line with a *Vibrio* resistance QTL from another line by crossing and then selecting offspring that inherited both favorable loci. Such marker-assisted introgression or stacking of QTLs can produce progeny with enhanced, multifaceted disease resistance. In summary, MAS has proven useful for accelerating gains in mollusk disease resistance (as shown by the *OsHV-1* QTL case), but it is best deployed as part of an integrated breeding strategy that also addresses polygenic variation and long-term pathogen coevolution.

Mollusks are valued not just for quantity (growth) and survival, but also for the quality of the product traits, such as meat yield, texture, taste, nutritional content (glycogen, fatty acids, etc.), and even aesthetic traits such as shell color or the presence of pearls in pearl oysters. Many of these quality traits are difficult to assess early in life, may require destructive sampling, or are expressed only at market size, making them particularly well suited to MAS. One quality trait that has seen successful genetic improvement is glycogen content in oysters. Glycogen (the carbohydrate stored in oyster tissues) contributes to the flavor and condition of oysters, with higher glycogen giving a sweeter taste and plumper appearance. In China, a breeding program aimed to develop a "high-glycogen" strain of Pacific oyster for the premium half-shell market. Through GWAS, researchers identified several SNPs strongly associated with glycogen content, including loci near genes such as *PPP1R3B* (involved in glycogen metabolism) (Liu et al., 2019). By selecting broodstock that had the favorable alleles at these loci, within a couple of generations, they achieved a new oyster line with ~30% higher glycogen content than the base population. One challenge when improving quality traits is the potential for trade-offs with growth or other performance traits. For instance, extremely fast-growing oysters might divert more energy to somatic growth and relatively less to storage reserves such as glycogen. Indeed, some studies have noted a slight negative genetic correlation between growth rate and certain quality indices (such as condition index or glycogen percentage) in tissues. This means that selecting purely for maximum growth could inadvertently reduce some aspects of product quality. MAS offers a more nuanced approach to navigate such trade-offs. Instead of selecting one trait at the expense of another, breeders can use markers to balance multiple traits in their selection index. This illustrates how MAS can be used not only to improve individual traits but also to manage genetic trade-offs in a more targeted and predictable manner. In conclusion, MAS is proving its worth not only for productivity traits but also for creating higher-value, differentiated products in molluscan aquaculture.

7.5 Conclusions

In summary, the combination of pedigree-based and marker-assisted approaches in molluscan breeding offers exciting prospects for the future of aquaculture. With continued advancements in genomics and statistical modeling, breeders can expect even more rapid genetic improvements in traits such as growth, disease resistance, and environmental adaptability. These developments will play a pivotal role in meeting the challenges of modern aquaculture, ensuring the long-term sustainability of the molluscan farming industry, and contributing to global food security. The integration of these advanced breeding methods, alongside careful attention to ecological and environmental considerations, will be essential for achieving a balance between productivity and sustainability in the rapidly growing field of molluscan aquaculture.

References

Allen, S.K., Small, J.M., Kube, P.D., 2021. Genetic parameters for *Crassostrea virginica* and their application to family-based breeding in the mid-Atlantic, USA. Aquaculture 538, 736578. doi:10.1016/j.aquaculture.2021.736578.

Azéma, P., Lamy, J.B., Boudry, P., Renault, T., Travers, M.A., Dégremont, L., 2017. Genetic parameters of resistance to *Vibrio aestuarianus*, and OsHV-1 infections in the Pacific oyster, *Crassostrea gigas*, at three different life stages. Genetics Selection Evolution 49 (1), 23. doi:10.1186/s12711-017-0297-2.

Bai, Z., Han, X., Luo, M., Lin, J., Wang, G., Li, J., 2015. Constructing a microsatellite-based linkage map and identifying QTL for pearl quality traits in triangle pearl mussel (*Hyriopsis cumingii*). Aquaculture 437, 102–110. doi:10.1016/j.aquaculture.2014.11.008.

Bai, Z.Y., Han, X.K., Liu, X.J., Li, Q.Q., Li, J.L., 2016. Construction of a high-density genetic map and QTL mapping for pearl quality-related traits in *Hyriopsis cumingii*. Scientific Reports 6, 32608. doi:10.1038/srep32608.

Barros, J., Velasco, L.A., Winkler, F.M., 2018. Heritability, genetic correlations and genotype by environment interactions in productive traits of the Caribbean scallop, *Argopecten nucleus* (Mollusca: Bivalvia). Aquaculture 488, 39–48. doi:10.1016/j.aquaculture.2018.01.011.

Boudry, P., 2009. Genetic variation and selective breeding in hatchery-propagated molluscan shellfish. Elsevier BV 87–108. doi:10.1533/9781845696474.1.87.

Chi, Y., Jiang, G., Liang, Y., Xu, C., Li, Q., 2022. Selective breeding for summer survival in Pacific oyster (*Crassostrea gigas*): Genetic parameters and response to selection. Aquaculture 556, 738271. doi:10.1016/j.aquaculture.2022.738271.

Crow, J.F., 1998. 90 years ago: The beginning of hybrid maize. Genetics 148 (3), 923–928.

Dégremont, L., Lamy, J.B., Pépin, J.F., Travers, M.A., Renault, T., 2015. New insight for the genetic evaluation of resistance to ostreid herpesvirus infection, a worldwide disease, in *Crassostrea gigas*. PLoS One 10 (6), e0127917. doi:10.1371/journal.pone.0127917.

Divilov, K., Merz, N., Schoolfield, B., Green, T.J., Langdon, C., 2023. Marker-assisted selection in a Pacific oyster population for an antiviral QTL conferring increased survival to OsHV-1 mortality events in Tomales Bay. Aquaculture 567, 739291. doi:10.1016/j.aquaculture.2023.739291.

Dove, M., Kube, P., Lind, C., Cumbo, V., Raftos, D., O'Connor, W., 2020. Accelerated Sydney Rock Oyster (SRO) Breeding Research. Department of Planning, Industry and Environment.

Duan, S., Gao, Y., Hu, X., Yin, X., 2025. A novel method for in vivo assessment of soft tissue weight in scallops. Aquaculture Reports 43, 102959. doi:10.1016/j.aqrep.2025.102959.

Falconer, D.S., 1981. Introduction to quantitative genetics. Longman Group Ltd, London, pp. 1–133.

Farias, T.F, César, J., Silva, L.P.D., 2017. Methods of selection using the quantitative genetics in aquaculture-A short review. Insights in Aquaculture and Biotechnology 1, 1.

Gianola, D., 2013. Priors in whole-genome regression: The Bayesian alphabet returns. Genetics 194 (3), 573–596. doi:10.1534/genetics.113.151753.

Gopikrishna, G., 2022. Application of molecular markers in aquaculture. In: Frontiers in Aquaculture Biotechnology. Elsevier, pp. 47–53.

Guimarães, E.P., Ruane, J., Scherf, B.D., Sonnino, A., Dargie, J.D., 2007. Marker-assisted selection: Current status and future perspectives in crops, livestock, forestry and fish. Food and Agriculture Organization of the United Nations.

Guo, H., Bao, Z., Li, J., Lian, S., Wang, S., He, Y., … Orban, L., 2012. Molecular characterization of TGF-β Type I receptor gene (*Tgfbr1*) in *Chlamys farreri*, and the association of allelic variants with growth traits. PLoS One 7 (11), e51005. doi:10.1371/journal.pone.0051005.

Guo, X., 2009. Use and exchange of genetic resources in molluscan aquaculture. Reviews in Aquaculture 1, 251–259. doi:10.1111/j.1753-5131.2009.01014.x.

Gutierrez, A.P., Matika, O., Bean, T.P., Houston, R.D., 2018. Genomic selection for growth traits in Pacific Oyster (*Crassostrea gigas*): Potential of low-density marker panels for breeding value prediction. Frontiers in Genetics 9, 391. doi:10.3389/fgene.2018.00391.

Han, Z., Guo, X., Lu, Z., Song, Y., Chen, R., Han, X., … Que, H., 2022. Heritability estimates for gonadal development traits and their genetic correlations with growth and heat tolerance traits in the Fujian Oyster *Crassostrea angulata*. Frontiers in Marine Science 9, 986441. doi:10.3389/fmars.2022.986441.

Hayes, B., He, J., Moen, T., Bennewitz, J., 2006. Use of molecular markers to maximise diversity of founder populations for aquaculture breeding programs. Aquaculture 255 (1-4), 573–578. doi:10.1016/j.aquaculture.2005.11.038.

He, Y., Yu, H., Bao, Z., Zhang, Q., Guo, X., 2012. , Mutation in promoter region of a serine protease inhibitor confers Perkinsus marinus resistance in the eastern oyster (*Crassostrea virginica*). Fish & Shellfish Immunology 33 (2), 411–417. doi:10.1016/j.fsi.2012.05.028.

Hedgecock, D., Shin, G., Gracey, A.Y., van den Berg, D., Samanta, M.P., 2015. Second-generation linkage maps for the pacific oyster *Crassostrea gigas* reveal errors in assembly of genome scaffolds. G3: Genes, Genomes, Genetics 5 (10), 2007–2019. doi:10.1534/g3.115.019570.

Hill, W.G., Goddard, M.E., Visscher, P.M., Mackay, T.F.C., 2008. Data and theory point to mainly additive genetic variance for complex traits. PLoS Genetics 4 (2), e1000008. doi:10.1371/journal.pgen.1000008.

Hollenbeck, C.M., Johnston, I.A., 2018. Genomic tools and selective breeding in molluscs. Frontiers in Genetics 9, 253. doi:10.3389/fgene.2018.00253.

Houston, R.D, Kriaridou, C., Robledo, D., 2022. Animal board invited review: Widespread adoption of genetic technologies is key to sustainable expansion of global aquaculture. Animal 16 (10), 100642. doi:10.1016/j.animal.2022.100642.

Hubert, S., Hedgecock, D., 2004. Linkage maps of microsatellite DNA markers for the pacific oyster *Crassostrea gigas*. Genetics 168 (1), 351–362. doi:10.1534/genetics.104.027342.

Jiao, W., Fu, X., Dou, J., Li, H., Su, H., Mao, J., Yu, Q., Zhang, L., Hu, X., Huang, X., Wang, Y., Wang, S., Bao, Z., 2014. High-resolution linkage and quantitative trait locus mapping aided by genome survey sequencing: Building up an integrative genomic framework for a bivalve mollusc. DNA Research 21 (1), 85–101. doi:10.1093/dnares/dst043.

Jones, D., Jerry, D.R., Khatkar, M.S., Moser, G., Raadsma, H.W., Taylor, J.J., Zenger, K.R., 2014. Quantitative trait loci and genetic association analysis reveals insights into complex pearl quality traits in donor silver-lipped pearl oysters. Aquaculture 434, 476–485. doi:10.1016/j.aquaculture.2014.08.038.

Jones, D.B., Jerry, D.R., Khatkar, M.S., Raadsma, H.W., Zenger, K.R., 2013. A high-density SNP genetic linkage map for the silver-lipped pearl oyster, *Pinctada maxima*: A valuable resource for gene localisation and marker-assisted selection. BMC Genomics 14 (1), 810. doi:10.1186/1471-2164-14-810Australia.

Kaeppler, S., 2011. Heterosis: One boat at a time, or a rising tide? New Phytologist 189 (4), 900–902.

Kaeppler, S., 2012. Heterosis: many genes, many mechanisms—end the search for an undiscovered unifying theory. In: International Scholarly Research Notices, 2012. International Scholarly Research Notices, pp. 1–12.

Kho, K.H., Sukhan, Z.P., Hossen, S., Cho, Y., Kim, S.C., Sharker, M.R., … Nou, I.S., 2021. Construction of a genetic linkage map based on SNP markers, QTL mapping and detection of candidate genes of growth-related traits in

Pacific abalone using genotyping-by-sequencing. Frontiers in Marine Science 8 (8), 713783. doi:10.3389/fmars.2021.713783.

Kong, N., Li, Q., Yu, H., Kong, L.F., 2015. Heritability estimates for growth-related traits in the Pacific oyster (*Crassostrea gigas*) using a molecular pedigree. Aquaculture Research 46 (2), 499–508. doi:10.1111/are.12205.

Kong, X., Huang, S., Wang, X., Liang, H., Hu, C., Wang, Y., Bao, Z., Hu, X., 2025. Development of automated phenotyping system for growth traits in bivalves. Aquaculture Reports 44, 103019. doi:10.1016/j.aqrep.2025.103019.

Kong, L.F, Song, S.L, Li, Q., 2017. The effect of interstrain hybridization on the production performance in the Pacific oyster *Crassostrea gigas*. Aquaculture 472, 44–49. doi:10.1016/j.aquaculture.2016.07.018.

Kvingedal, R., Evans, B.S., Lind, C.E., Taylor, J.J.U., Dupont-Nivet, M., Jerry, D.R., 2010. Population and family growth response to different rearing location, heritability estimates and genotype × environment interaction in the silver-lip pearl oyster (*Pinctada maxima*). Aquaculture 304 (1–4), 1–6. doi:10.1016/j.aquaculture.2010.02.035.

Lenarcic, A.B., Svenson, K.L., Churchill, G.A., Valdar, W., 2012. General Bayesian approach to analyzing diallel crosses of inbred strains. Genetics 190 (2), 413–435. doi:10.1534/genetics.111.132563.

Li, H., Liu, X., Zhang, G., Knight, M., 2012. A consensus microsatellite-based linkage map for the hermaphroditic Bay Scallop (*Argopecten irradians*) and its application in size-related QTL analysis. PLoS One 7 (10), e46926. doi:10.1371/journal.pone.0046926.

Li, X., Li, N., Zhao, L., Shi, J., Wang, S., Ning, X., … Hu, X., 2022. Tissue distribution and seasonal accumulation of carotenoids in Yesso scallop (*Mizuhopecten yessoensis*) with orange adductor muscle. Food Chemistry 367, 130701. doi:10.1016/j.foodchem.2021.130701.

Li, X., Wang, S., Xun, X., Zhang, M., Wang, S., Li, H., Zhao, L., Fu, Q., Wang, H., Li, T., Lian, S., Xing, Q., Li, X., Wu, W., Zhang, L., Hu, X., Bao, Z., 2019. A carotenoid oxygenase is responsible for muscle coloration in scallop. Biochimica et Biophysica Acta (BBA) - Molecular and Cell Biology of Lipids 1864 (7), 966–975. doi:10.1016/j.bbalip.2019.03.003.

Li, Q, Xu, Y, Yu, R., Akihiro, K., 2007. An AFLP genetic linkage map of pacific abalone (*Haliotis discus hannai*). Journal of Ocean University of China 6 (3), 259–267. doi:10.1007/s11802-007-0259-3.

Liang, B., Jiang, F., Zhang, S., Yue, X., Wang, H., Liu, B., 2017. Genetic variation in vibrio resistance in the clam *Meretrix petechialis* under the challenge of *Vibrio parahaemolyticus*. Aquaculture 468, 458–463. doi:10.1016/j.aquaculture.2016.10.037.

Liang, B., Yue, X., Wang, H., Liu, B., 2016. Influence of parental sample sizes on the estimating genetic parameters in cultured clam *Meretrix meretrix* based on factorial mating designs. Acta Oceanologica Sinica 35 (6), 42–49. doi:10.1007/s13131-016-0875-0.

Lind, C.E., Ponzoni, R.W., Nguyen, N.H., Khaw, H.L., 2012. Selective breeding in fish and conservation of genetic resources for aquaculture. Reproduction in Domestic Animals 47 (4), 255–263. doi:10.1111/j.1439-0531.2012.02084.x.

Liu, S., Li, L., Meng, J., Song, K., Huang, B., Wang, W., Zhang, G., 2019. Association and functional analyses revealed that PPP1R3B Plays an important role in the regulation of glycogen content in the Pacific Oyster *Crassostrea gigas*. Frontiers in Genetics 10, 106. doi:10.3389/fgene.2019.00106.

Lynch, M., Walsh, B., 1998. Genetics and analysis of quantitative traits. Sinauer Associates.

Mackay, T.F., Stone, E.A., Ayroles, J.F., 2009. The genetics of quantitative traits: challenges and prospects. Nature Reviews Genetics 10 (8), 565–577.

Mao, J., Zeng, Q., Yang, Z., Pan, H., Yao, L., Bao, Z., … Wang, S., 2020. High-resolution linkage and quantitative trait locus mapping using an interspecific cross between *Argopecten irradians irradians* (♀) and *A. purpuratus* (♂). Marine Life Science & Technology 2 (2), 123–134. doi:10.1007/s42995-020-00029-z.

Maqsood, H.M., Ahmad, S.M., 2017. Advances in molecular markers and their applications in aquaculture and fisheries. Central Fisheries Research Inst, India Genetics of Aquatic Organisms 1 (1), 27–41. doi:10.4194/2459-1831-v1_1_05.

McCarty, A.J, Allen, S.K, Plough, L.V, Houston, R., 2022. Genome-wide analysis of acute low salinity tolerance in the eastern oyster *Crassostrea virginica* and potential of genomic selection for trait improvement. G3 Genes | Genomes | Genetics 12 (1), jkab368. doi:10.1093/g3journal/jkab368.

Meng, J., Song, K., Li, C., Liu, S., Shi, R., Li, B., … Zhang, G., 2019. Genome-wide association analysis of nutrient traits in the oyster *Crassostrea gigas*: Genetic effect and interaction network. BMC Genomics 20 (1), 625. doi:10.1186/s12864-019-5971-z.

Meng, J., Wang, W., Shi, R., Song, K., Li, L., Que, H., Zhang, G., 2020. Identification of SNPs involved in Zn and Cu accumulation in the Pacific oyster (*Crassostrea gigas*) by genome-wide association analysis. Ecotoxicology and Environmental Safety 192, 110208. doi:10.1016/j.ecoenv.2020.110208.

Meuwissen, T.H., Hayes, B.J, Goddard, M.E, 2001. Prediction of total genetic value using genome-wide dense marker maps. Genetics 157 (4), 1819–1829. doi:10.1093/genetics/157.4.1819.

Meuwissen, T., Hayes, B., Goddard, M., 2013. Accelerating improvement of livestock with genomic selection. Annual Review of Animal Biosciences 1 (1), 221–237. doi:10.1146/annurev-animal-031412-103705.

Nie, H., Yan, X., Huo, Z., Jiang, L., Chen, P., Liu, H., … Yang, F., 2017. Construction of a high-density genetic map and quantitative trait locus mapping in the Manila clam *Ruditapes philippinarum*. Scientific Reports 7 (1), 11953. doi:10.1038/s41598-017-00246-0.

Ning, X., Li, X., Wang, J., Zhang, X., Kong, L., Meng, D., … Bao, Z., 2019. Genome-wide association study reveals *E2F3* as the candidate gene for scallop growth. Aquaculture 511, 734216. doi:10.1016/j.aquaculture.2019.734216.

Robledo, D., Matika, O., Hamilton, A., Houston, R.D., 2018a. Genome-wide association and genomic selection for resistance to amoebic gill disease in Atlantic salmon. G3: Genes, Genomes, Genetics 8 (4), 1195–1203. doi:10.1534/g3.118.200075.

Robledo, D., Palaiokostas, C., Bargelloni, L., Martínez, P., Houston, R., 2018b. Applications of genotyping by sequencing in aquaculture breeding and genetics. Reviews in Aquaculture 10 (3), 670–682. doi:10.1111/raq.12193.

Schnable, P.S., Springer, N.M., 2013. Progress toward understanding heterosis in crop plants. Annual Review of Plant Biology 64 (1), 71–88.

Shi, R., Li, C., Qi, H., Liu, S., Wang, W., Li, L., Zhang, G., 2020. Construction of a high-resolution genetic map of *Crassostrea gigas*: QTL mapping and GWAS applications revealed candidate genes controlling nutritional traits. Aquaculture 527, 735427. doi:10.1016/j.aquaculture.2020.735427.

Shi, Y., Wang, S., Gu, Z., Lv, J., Zhan, X., Yu, C., … Wang, A., 2014. High-density single nucleotide polymorphisms linkage and quantitative trait locus mapping of the pearl oyster, *Pinctada fucata martensii* Dunker. Aquaculture. 434, 376–384. doi:10.1016/j.aquaculture.2014.08.044.

Shull, G.H., 1948. What is "heterosis"? Genetics 33 (5), 439–446.

Song, H., Dong, T., Yan, X., Wang, W., Tian, Z., Sun, A., Dong, Y., Zhu, H., Hu, H., 2023. Genomic selection and its research progress in aquaculture breeding. Reviews in Aquaculture 15 (1), 274–291. doi:10.1111/raq.12716.

Tang, X., Li, S., Zhuo, L., Zhang, Q., Wu, F., Lin, S., 2024. Correlation study of SNP of glycogen degradation metabolism related genes and glycogen content in *Haliotis discus hannai*. Aquaculture 578, 740022. doi:10.1016/j.aquaculture.2023.740022.

Tsai, H.Y., Hamilton, A., Tinch, A.E., Guy, D.R., Gharbi, K., Stear, M.J., … Houston, R.D., 2015. Genome wide association and genomic prediction for growth traits in juvenile farmed Atlantic salmon using a high density SNP array. BMC Genomics 16 (1), 969. doi:10.1186/s12864-015-2117-9.

Vu, S.V., Knibb, W., Nguyen, N.T.H., Vu, I.V., O'Connor, W., Dove, M., Nguyen, N.H., 2020. First breeding program of the Portuguese oyster *Crassostrea angulata* demonstrated significant selection response in traits of economic importance. Aquaculture 518, 7. doi:10.1016/j.aquaculture.2019.734664.

Vu Van, S., 2023. Doctoral dissertation, University of the Sunshine Coast doi:10.25907/00729.

Wang, H., Yang, Z., Wang, S., Zhao, A., Wang, H., Liu, Z., … Huang, X., 2024. Genome-wide association analysis reveals the genetic basis of thermal tolerance in dwarf surf clam *Mulinia lateralis*. Genomics 116 (5), 110904. doi:10.1016/j.ygeno.2024.110904.

Wang, H., Du, X., Lü, W., Liu, Z., 2010. Estimating the heritability for growth-related traits in the pearl oyster, *Pinctada fucata martensii* (Dunker). Aquaculture Research 42 (1), 57–64. doi:10.1111/j.1365-2109.2010.02552.x.

Wang, J., Li, L., Zhang, G., 2016. A high-density SNP genetic linkage map and QTL analysis of growth-related traits in a hybrid family of oysters (*Crassostrea gigas* × *Crassostrea angulata*) using genotyping-by-sequencing. G3: Genes, Genomes, Genetics 6 (5), 1417–1426. doi:10.1534/g3.116.026971.

Wang, S., Meyer, E., Mckay, J.K., Matz, M.V., 2012. 2b-RAD: A simple and flexible method for genome-wide genotyping. Nature Methods 9 (8), 808–810. doi:10.1038/nmeth.2023.

Wang, S., Wang, H., Zhao, L., Zhang, Y., Li, T., Liu, S., … Hu, X., 2022. Identification of genes associated with carotenoids accumulation in scallop (*Patinopecten yessoensis*). Aquaculture 550, 737850. doi:10.1016/j.aquaculture.2021.737850.

Wang, S., Bao, Z., Pan, J., Zhang, L., Yao, B., Zhan, A., … Zhang, Q., 2004. AFLP linkage map of an intraspecific cross in *Chlamys farreri*. Journal of Shellfish Research 23 (2), 491–499.

Wang, S., Zhang, J., Jiao, W., Li, J., Xun, X., Sun, Y., ... Bao, Z., 2017. Scallop genome provides insights into evolution of bilaterian karyotype and development. Nature Ecology & Evolution 1 (5), 120. doi:10.1038/s41559-017-0120.

Wang, X., Li, A., Wang, W., Que, H., Zhang, G., Li, L., 2021. DNA methylation mediates differentiation in thermal responses of Pacific oyster (*Crassostrea gigas*) derived from different tidal levels. Heredity 126 (1), 10–22. doi:10.1038/s41437-020-0351-7.

Wang, C., Kong, L., Lian, S., Yang, Z., Meng, D., Li, M., ... Hu, X., 2023. Verify the function of a potential growth-regulating gene in marine bivalve using a candidate model organism *Mulinia lateralis*. Journal of Ocean University of China 22 (4), 1012–1022. doi:10.1007/s11802-023-5382-2.

Xu, K., Li, Q.I., Kong, L., Yu, R., 2008. A first-generation genetic map of the Japanese scallop *Patinopecten yessoensis*-based AFLP and microsatellite markers. Aquaculture Research 40 (1), 35–43. doi:10.1111/j.1365-2109.2008.02058.x.

Yang, B., Zhi, C., Li, P., Xu, C., Li, Q.I., Liu, S., 2024. Genomic selection accelerates genetic improvement of resistance to *Vibriosis* in the Pacific oyster, *Crassostrea gigas*. Aquaculture 584, 740679. doi:10.1016/j.aquaculture.2024.740679.

Yang, C., Wang, L., Jiang, Q., Wang, J., Yue, F., Zhang, H., ... Song, L., 2013. The polymorphism in the promoter region of metallothionein 1 is associated with heat tolerance of scallop *Argopecten irradians*. Gene 526 (2), 429–436. doi:10.1016/j.gene.2013.05.033.

Yang, C., Wang, L., Liu, C., Zhou, Z., Zhao, X., Song, L., 2015. The polymorphisms in the promoter of HSP90 gene and their association with heat tolerance of bay scallop. Cell Stress and Chaperones 20 (2), 297–308. doi:10.1007/s12192-014-0546-z.

Yin, Z., Liu, J., Yu, W., Shen, Y., Gan, Y., Chen, Y., ... You, W., 2025. Seasonal variation in two culture systems and genome-wide association analysis of taurine content in the foot muscles of the Pacific abalone (*Haliotis discus hannai*). Aquaculture 604, 742431. doi:10.1016/j.aquaculture.2025.742431.

Yin, X., Hedgecock, D., 2021. Overt and concealed genetic loads revealed by QTL mapping of genotype-dependent viability in the Pacific oyster *Crassostrea gigas*. Genetics 219 (4), iyab165. doi:10.1093/genetics/iyab165.

Yu, H., Sui, M., Yang, Z., Cui, C., Hou, X., Liu, Z., ... Bao, Z., 2023. Deciphering the genetic basis and prediction genomic estimated breeding values of heat tolerance in Zhikong scallop *Chlamys farreri*. Aquaculture 565, 739090. doi:10.1016/j.aquaculture.2022.739090.

Yu, Z., Guo, X., 2003. Genetic linkage map of the Eastern Oyster *Crassostrea virginica* Gmelin. The Biological Bulletin 204 (3), 327–338. doi:10.2307/1543603.

Zenger, K.R., Khatkar, M.S., Jones, D.B., Khalilisamani, N., Jerry, D.R., Raadsma, H.W., 2019. Genomic selection in aquaculture: Application, limitations and opportunities with special reference to marine shrimp and pearl oysters. Frontiers in Genetics 10, 693. doi:10.3389/fgene.2018.00693.

Zhai, S., Yang, B., Zhang, F., Li, Q.I., Liu, S., 2021. Estimation of genetic parameters for resistance to Vibrio alginolyticus infection in the Pacific oyster (*Crassostrea gigas*). Aquaculture 538, 736545. doi:10.1016/j.aquaculture.2021.736545.

Zhan, A., Hu, J., Hu, X., Hui, M., Wang, M., Peng, W., ... Bao, Z., 2009. Construction of microsatellite-based linkage maps and identification of size-related quantitative trait loci for Zhikong scallop (*Chlamys farreri*). Animal Genetics 40 (6), 821–831. doi:10.1111/j.1365-2052.2009.01920.x.

Zhang, C., He, C., Sun, Y., Zhang, H., Ye, T., Liu, F., Ma, H., Li, S., Zheng, H., 2024a. Selective breeding in the noble scallop (*Chlamys nobilis*) for low-temperature resistance to reduce overwintering losses. Aquaculture 586, 740737. doi:10.1016/j.aquaculture.2024.740737.

Zhang, K., Yang, Q., Du, M., Zhang, Z., Wang, W., Zhang, G., ... Li, L., 2023. Genome-wide mapping of regulatory variants for temperature- and salinity-adaptive genes reveals genetic basis of genotype-by-environment interaction in *Crassostrea ariakensis*. Environmental Research 236, 116614. doi:10.1016/j.envres.2023.116614.

Zhang, X., Niu, Y., Gao, C., Kong, L., Yang, Z., Chang, L., Kong, X., Bao, Z., Hu, X., 2024b. Somatostatin receptor gene functions in growth regulation in bivalve scallop and clam. International Journal of Molecular Sciences 25 (9), 4813. doi:10.3390/ijms25094813.

Zhang, Y., Zhang, L., Sun, J., Qiu, J., Hu, X., Hu, J., Bao, Z., 2014. Proteomic analysis identifies proteins related to carotenoid accumulation in Yesso scallop (*Patinopecten yessoensis*). Food Chemistry 147, 111–116. doi:10.1016/j.foodchem.2013.09.078.

Zhang, Y., Wang, H., Liu, S., Kong, X., Chang, L., Zhao, L., Bao, Z., Hu, X., 2024c. Multi-tissue metabolomic profiling reveals the crucial metabolites and pathways associated with scallop growth. BMC Genomics 25 (1), 1091. doi:10.1186/s12864-024-11016-4.

Zhao, L., Li, Y., Li, Y., Yu, J., Liao, H., Wang, S., ... Bao, Z., 2017. A Genome-wide association study identifies the genomic region associated with shell color in Yesso scallop, *Patinopecten yessoensis*. Marine Biotechnology 19 (3), 301–309. doi:10.1007/s10126-017-9751-y.

Zhao, L., Lian, S., Ren, Q., Yang, Z., Guo, Z., Lou, J., Kong, X., Li, M., Bao, Z., Hu, X., 2021. In vivo and rapid assessment of scallop muscle trait. Aquaculture 530, 735817. doi:10.1016/j.aquaculture.2020.735817.

Zheng, Z., Hao, R., Yang, C., Jiao, Y.U., Wang, Q., Huang, R., ... Du, X., 2023. Genome-wide association study analysis to resolve the key regulatory mechanism of biomineralization in *Pinctada fucata martensii*. Molecular Ecology Resources 23 (3), 680–693. doi:10.1111/1755-0998.13743.

Zhu, X., Liu, P., Hou, X., Zhang, J., Lv, J., Lu, W., ... Bao, Z., 2021. Genome-wide association study reveals PC4 as the candidate gene for thermal tolerance in Bay scallop (*Argopecten irradians irradians*). Frontiers in Genetics 12, 650045. doi:10.3389/fgene.2021.650045.

CHAPTER

8

Reproduction and sex-controlled breeding

Jing Wang[a], *Huilan Wei*[a], *Wentao Han*[a], *Lisui Bao*[b] *and Lingling Zhang*[a]

[a]College of Marine Life Sciences, Ocean University of China, Qingdao, China
[b]Institute of Evolution & Marine Biodiversity, Ocean University of China, Qingdao, China

8.1 Introduction

Mollusks exhibit a fascinating variety of reproductive strategies that align with their highly adaptable nature to different environmental and ecological conditions. This diversity spans sexual systems (from gonochorism to various forms of hermaphroditism), rare instances of asexual reproduction, and different fertilization modes.

8.1.1 Sexual reproduction

The vast majority of mollusks are sexually reproducing, with distinct gametes and fertilization processes (Collin, 2013; Pandian, 2018) (Fig. 8.1). Gonochorism is the most common system (~75%), which is found across nearly all cephalopods (all squids, cuttlefish, octopuses), many gastropods (e.g., abalone), and bivalves (e.g., most oysters, clams, and scallops). For example, an adult Yesso scallop (*Patinopecten yessoensis*) is either male or female in a given spawning season, releasing millions of gametes into the water for external fertilization. Hermaphroditism, however, is also widespread in mollusks, particularly in gastropods (many snails and slugs) and in certain bivalves (e.g., *Argopecten irradians*), possessing functional male and female gonads at the same time. *A. irradians* can produce both eggs and sperm, either self-fertilizing or outcrossing with others. Simultaneous hermaphroditism allows reproductive assurance in low-density populations. However, in aquaculture, excessive self-fertilization in hermaphrodites can lead to inbreeding depression, so broodstock are typically kept in groups to promote outcrossing.

Genetic Breeding and Aquaculture of Shellfish: A Collaborative Approach to a Sustainable Future
DOI: https://doi.org/10.1016/B978-0-443-33606-5.00011-4

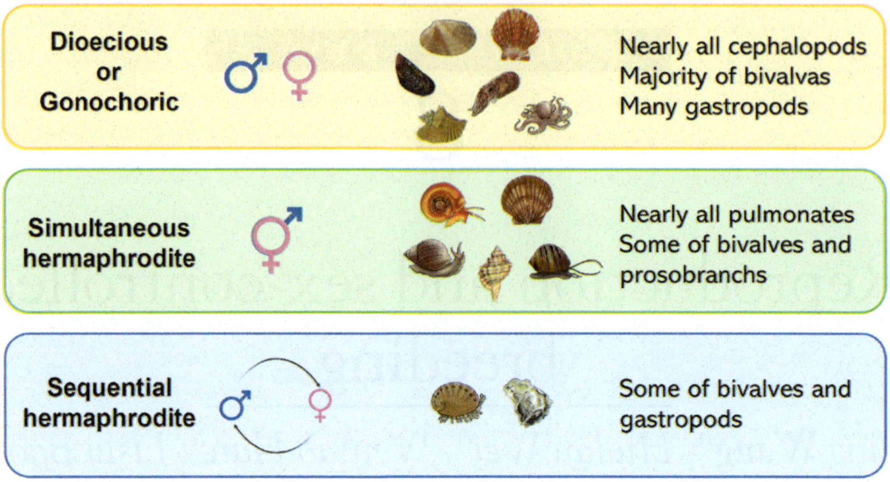

FIGURE 8.1 Reproductive traits in Mollusca. *From Rato et al. (2025).*

Moreover, sequential hermaphroditism (i.e., organisms that change sex during their lifetime) has also been reported in some gastropods (e.g., slipper limpet *Crepidula* and abalone *Haliotis*) and certain bivalves (e.g., *Crassostrea gigas*). Protandry (from male to female) is dominant in mollusks. For example, most young oysters start as males and later transition to females as they grow, a strategy that maximizes reproductive output since larger, older individuals can produce more eggs. Environmental factors can influence this sex change; higher rearing temperatures have been shown to skew *C. gigas* toward maleness. In sessile species like slipper-shell snails, the sex of an individual could be determined by its nearest neighbor. If isolated or placed near another male, the individual will change into a female. Some mollusks can even undergo repeated sex reversals (serial hermaphroditism). A striking case is the oyster *Ostrea puelchana*, in which individuals function as males in their first year, then as females in the next, and can continue alternating sex in subsequent breeding seasons. Such rhythmic sex changes, while rare, demonstrate the plasticity of molluscan reproductive systems (Guo et al., 1998; Hoving et al., 2006; Kim et al., 2021).

8.1.2 Asexual reproduction

True asexual reproduction is exceedingly rare in mollusks (Komaru et al., 1997; McElroy et al., 2022). Parthenogenesis (development of embryos from unfertilized eggs) occurs in a handful of mollusks, usually under unusual circumstances, including *Potamopyrgus antipodarum*, and *Melanoides tuberculata*. In these parthenogenetic mollusks, females produce diploid eggs that develop without sperm, often because of endosymbiotic or environmental influences that eliminate males. For example, certain freshwater snails can form all-female lineages that reproduce parthenogenetically. This can lead to rapid population growth, but at the cost of low genetic diversity in the offspring. In aquaculture or invasive species contexts, such parthenogenetic snails (e.g., the New Zealand mud snail *P. antipodarum* in some cases) can quickly become pests since a single individual can establish a population.

FIGURE 8.2 Fertilization mechanisms in Mollusca. (A) Broadcast spawning in Bivalvia: External fertilization (*Mulinia lateralis*, Photo by Zujing Yang); (B) Copulation in Gastropoda: Internal fertilization (*Polymita picta*, Photo by Joris M Koene (Reyes-Tur et al., 2015); (C) Spermatophore transfer in Cephalopoda: Internal fertilization. Taking *Doryteuthis plei* as an example, large consort males (green) deposit spermatophores within the female (red) mantle cavity near the oviduct opening (green star) through "male-parallel" mating. Small sneaker males (blue) employ furtive mating and deposit spermatophores near the female buccal seminal receptacle (blue star) through "head-to-head" mating (Marian et al., 2019).

8.1.3 Fertilization mechanisms

Molluscan fertilization strategies range from broadcast spawning in open water to internal fertilization with copulation (Fig. 8.2). External fertilization is typified by many marine bivalves and some marine gastropods. In these species, males and females release sperm and eggs into the water column, where gametes meet by chance. For instance, oysters, clams, and scallops are classic broadcast spawners—a single female oyster can shed millions of eggs per spawning, and neighboring males simultaneously release sperm, resulting in a fertilization plume in the sea. This strategy requires synchronization and high gamete output to compensate for dilution; natural fertilization rates in open water can be under 10%. In hatcheries, techniques such as thermal or chemical spawning induction are used to ensure males and females spawn at the same time, thereby boosting fertilization success. External fertilizers usually have simple reproductive anatomies, as gametes exit directly via gonoducts or siphons into the environment (Rato et al., 2025). By contrast, internal fertilization has evolved in many gastropods and cephalopods. Sperm are introduced into the female's body (or onto her eggs) via mating behaviors or specialized structures. Many simultaneous hermaphrodite snails engage in reciprocal copulation, exchanging sperm packets. Terrestrial snails (e.g. *Helix* species) famously use love-darts and copulatory organs to stimulate and transfer sperm internally (Staikou & Koemtzopoulos, 2019). In such species, females (or hermaphrodites) often possess sperm storage organs (seminal receptacles or spermathecae) to keep viable sperm for extended periods. Sperm storage allows a mollusk to fertilize multiple batches of eggs over time from one mating and can also facilitate multiple paternity within a single clutch, which is documented in some gastropods and bivalves. Cephalopods have highly developed internal fertilization: Males produce spermatophores (encased sperm packets) and use a modified arm (hectocotylus) or a long, extensible penis to implant these spermatophores into the female's mantle cavity during mating. The female later utilizes the stored sperm to fertilize her eggs as she lays them in protected locations. This is seen in octopuses, squids, and cuttlefish—e.g., the octopus *Argonauta argo* (paper nautilus) where the male's hectocotylus autonomously transfers to the female, and in loliginid squids where large "consort" males place spermatophores inside the female's oviduct while small "sneaker" males stick them near the female's buccal area (Ibáñez et al., 2019). In internally fertilizing mollusks, sperm storage is often crucial; octopus females, for instance, can retain spermatophores until their eggs are ready to be laid, and some snails can delay fertilization to better timing or conditions. Thus mollusks exhibit both extremes of fertilization, with corresponding anatomical adaptations (from simple gonadal ducts to complex intromittent organs and storage glands) to ensure successful reproduction.

8.2 Reproductive systems and gonadal development

8.2.1 Reproductive system anatomy

Bivalves: The gonads are typically diffuse, consisting of follicles or acini interwoven through the visceral mass and surrounding organs such as the digestive gland. During the breeding season, these gonadal follicles expand and fill with gametes, often causing visible gonad enlargement in mature individuals. Lacking elaborate accessory reproductive ducts or copulatory organs, bivalves simply release their gametes from the gonoducts into the mantle cavity.

Gastropods: Gastropod mollusks show more varied reproductive anatomies. Many marine gastropods (e.g., abalone) are gonochoric with relatively simple reproductive tracts (gonads and ducts). In abalone (*Haliotis*), individuals have a single gonad (ovary or testis) located in a hepatogonadal complex—a conical organ encasing the digestive gland (hepatopancreas) internally and the gonad as an outer layer. The gonad opens to the exterior via a gonoduct. Abalone males and females release gametes through a pore near the head/epipodium region, relying on external fertilization. There are typically no penes or specialized copulatory organs. By contrast, many land snails are simultaneous hermaphrodites with complex reproductive systems (Fig. 8.3). Pulmonate snails (terrestrial or freshwater) are hermaphroditic, possessing an ovotestis that produces both eggs and sperm. Their reproductive system includes an ovotestis embedded in the digestive gland, leading to a common hermaphroditic duct, and separate male and female tracts with accessory glands (e.g., albumen gland, prostate, bursa copulatrix, and

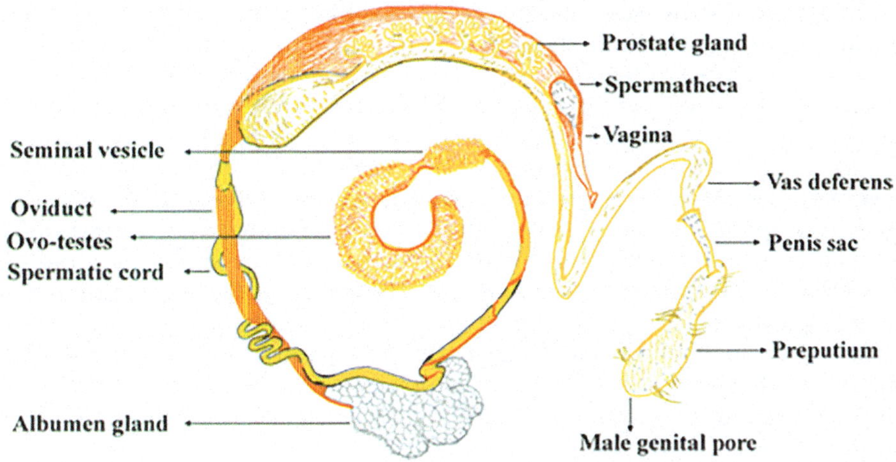

FIGURE 8.3 Reproductive system of a snail: Male in yellow and female in orange (Pathak et al., 2024).

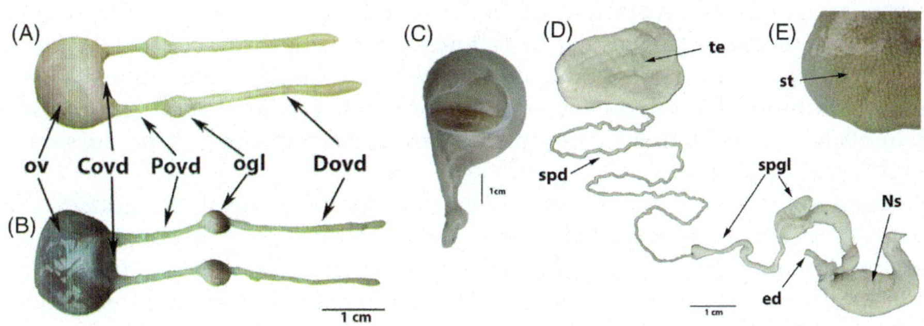

FIGURE 8.4 External anatomy of the reproductive system of *Octopus hubbsorum* García-Flores et al. (2019). (A & B) Ripening stage and postspawning stage of female individual. (C) Male reproductive system enveloped by the membrane. (D) Expanded reproductive system of male individual. (E) Detail of testis showing many seminiferous tubules. *Covd*: Common oviduct; *Dovd*: distal oviduct; *ogl*: oviducal gland; *ov*: ovary; *Povd*: proximal oviduct; *ed*: excurrent duct or terminal organ; *Ns*: Needham's sac or spermatophoric sac; *spd*: spermatophoric duct; *spgl*: spermatophoric gland; *st*: seminiferous tubules; *te*: testis.

dart sac in land snails) for internal fertilization (Pathak et al., 2024). In snail species that copulate, a penis and vas deferens are present for sperm transfer, and a vaginal tract and sperm storage organs (spermathecae) are present for receiving sperm. Notably, structures such as the albumen gland (for nourishing eggs), the spermoviduct (combined oviduct/sperm duct) with internal fertilization pouches, and love dart apparatus (in terrestrial snails) have no analogs in bivalves or cephalopods.

Cephalopods: Cephalopods (squid, cuttlefish, and octopus) are exclusively gonochoric and have the most specialized reproductive systems among mollusks (Fig. 8.4). Males and females have distinctly different structures adapted for internal fertilization. Female cephalopods typically possess a single ovary located in the posterior of the mantle cavity, and associated oviducts and glands. The ovary is often large and sac-like, containing oocytes in all stages of development simultaneously. From the ovary, mature eggs travel through an oviduct (usually one functional oviduct, except *Nautilus*, which has two) to an oviducal gland where jelly coats or egg capsules are added. Most female decapod cephalopods (squid and cuttlefish) also have large nidamental glands (paired in squid/cuttlefish) that secrete additional egg masses or coverings, and often accessory nidamental glands housing symbiotic bacteria thought to help protect eggs. A seminal receptacle may be present for storing sperm received from males (for instance, some squid have sperm storage sites near the buccal membrane). Male cephalopods have a single testis in the mantle cavity, which produces sperm that are transported through a long vas deferens to a set of accessory glands collectively known as the spermatophoric complex. This complex typically includes: A spermatophoric organ (Needham's sac) that stores packaged spermatophores, spermatophoric glands that add coatings to form the spermatophore, and a penis (appendage) that everts to transfer spermatophores to the female. Many male cephalopods also modify one arm into a hectocotylus specialized for delivering spermatophores to the female (e.g. the octopus's hectocotylus arm).

8.2.2 Gametogenesis: Spermatogenesis and oogenesis in mollusks

Molluscan gametogenesis encompasses the formation and maturation of sperm (spermatogenesis) and eggs/oocytes (oogenesis). It follows a fundamentally conserved process across bivalves, gastropods, and cephalopods, albeit with class-specific adaptations. The process of gametogenesis in mollusks is typically seasonal and synchronized with environmental cues, resulting in cyclical development of germ cells from undifferentiated precursors to mature gametes ready for spawning.

Spermatogenesis typically occurs within the follicles or acini of the gonad, exhibiting a distinct spatial organization where development proceeds from the periphery toward the lumen. Spermatogonia, the diploid stem cells located near the outer follicle wall, undergo mitotic division. These give rise to primary and secondary spermatocytes, which undergo meiosis, resulting in haploid spermatids that subsequently differentiate into mature spermatozoa through spermiogenesis. The mature sperm of broadcast-spawning bivalves, such as scallops, are typically primitive, flagellated aquasperm with a curved nucleus, a conical acrosome, and a midpiece containing a few mitochondria to power motility for external fertilization. Somatic accessory or follicle cells, reminiscent of Sertoli cells, are crucial in bivalve and gastropod testes, providing structural support, nutrients, and phagocytosing residual cytoplasm. Gastropod sperm structure can be more variable, with some terrestrial forms producing complex sperm stored in spermatophores. In cephalopods, spermatogenesis occurs in seminiferous tubules rather than spherical follicles, but the cellular stages from spermatogonia to spermatozoa are similar. The resulting sperm are highly elongated and are uniquely packaged into complex, elaborate spermatophores within the male's reproductive tract before transfer to the female.

Oogenesis similarly involves a conserved sequence of stages: Oogonia, primary oocyte, secondary oocyte, and ovum. Oogonia at the follicle periphery proliferate mitotically and enter meiosis, becoming primary oocytes that arrest in prophase I during a prolonged growth phase. A critical event is vitellogenesis, where the oocyte accumulates yolk reserves to support future embryonic development. This occurs through a combination of autosynthesis, where the oocyte's own organelles produce yolk, and heterosynthesis, where yolk protein precursors, such as vitellogenin, are synthesized externally in organs like the hepatopancreas or by ovarian follicle/auxiliary cells adjacent to the oocyte, and taken up by the oocyte via endocytosis. In bivalves such as oysters and scallops, developing oocytes are often attached to the follicle wall by a stalk and are closely associated with follicle cells that facilitate nutrient transfer during early vitellogenesis; these same cells often take on a phagocytic role postspawning to recycle residual nutrients. The ovary contains a heterogeneous mix of oocyte stages, and mature oocytes are large and yolk-rich. Gastropod oogenesis follows a similar pattern, though in simultaneous hermaphrodites, the ovotestis contains both spermatogenic cysts and growing oocytes within the same organ, often in distinct lobes to prevent self-fertilization. An albumen gland adds nutritive layers to the eggs before laying. Cephalopod oogenesis is distinguished by the production of a finite batch of very large, yolk-rich eggs for semelparous reproduction. Their ovaries lack a continuous follicular epithelium; instead, oocytes grow attached to an ovarian stroma before being released into the coelom to enter the oviduct, where oviducal glands add egg capsule layers.

8.2.3 Gonadal developmental regulation

Molluscan gonads cycle through recognizable stages: Resting (undeveloped) when gonads are small and mostly occupied by spent tissue or early germ cells; developing (active gametogenesis) as germ cells proliferate and grow; mature (ripe) when gonads are filled with gametes (sperm or mature oocytes); and spawning or spent stage when gametes are released and the gonad tissue regresses. Histological criteria clearly define these stages: A developing gonad is characterized by clusters of spermatocytes or previtellogenic oocytes; a ripe gonad displays follicles completely filled with sperm or tightly packed oocytes; and a spent gonad reveals residual gametes alongside phagocytes that are consuming atretic material. Similar staging exists in abalone: "Developing" where the gonad begins to cover the digestive gland with early gametes, "mature," where it forms a thick layer full of gametes, "spawning," with some gamete release evident, and "resting," with a thin gonad layer and many atretic oocytes or residual sperm. These stages are important for aquaculture monitoring of broodstock condition (Fig. 8.5).

Molluscan gonadal development is orchestrated by a complex interplay of environmental cues and intrinsic molecular mechanisms. As ectotherms, mollusks (bivalves, gastropods, and cephalopods) rely on external factors like temperature, photoperiod, salinity, and nutrition to time their reproductive cycles (El-Sokkary et al., 2023). Seasonal temperature shifts are especially influential. For many species, sustained warming serves as a trigger for gametogenesis and spawning, while cooler periods correspond with gonadal quiescence. However, the effect of temperature on earlier developmental stages can be complex and species-specific, particularly among cold-adapted taxa. For instance, in the cold-water Yesso scallop (*P. yessoensis*), lower temperatures at the onset of gonadal differentiation markedly accelerate the process, whereas higher temperatures can delay it—underscoring temperature's precise and stage-dependent role in modulating gonad development (Liu et al., 2022). Photoperiod (day length) also modulates reproduction—short-day conditions can stimulate reproductive activity under otherwise permissive temperatures, as demonstrated in the marine gastropod *Aplysia californica* (Wayne & Block, 1992), where individuals on short-day light cycles laid eggs more frequently than those on long-day cycles (given the same warm temperature). Likewise, ample food availability is crucial; abundant nutrition (e.g., phytoplankton for filter-feeders) promotes gamete production, whereas poor nutrition or energy reserves can arrest gametogenesis. Salinity fluctuations and other abiotic stresses further factor into reproductive timing. For example, a moderate drop

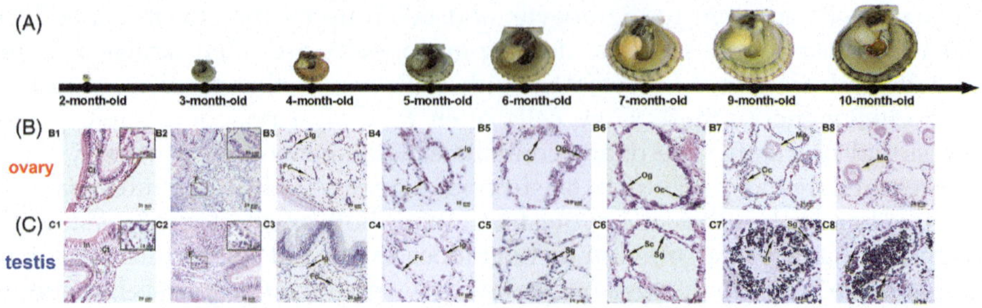

FIGURE 8.5 Morphological observation of juvenile scallops aged 2–10 months old (Wei et al., 2021). (A) Anatomy of the scallop showing the shell height and gonad morphology (the right shell valve was removed). Anterior toward right and posterior toward left. The gonad has distinct male and female portions, with the testis being proximal and the ovary distal. The solid white line indicates the boundary of ovary and testis, and the red and blue dotted lines mark the ovary and testis, respectively. (B) Paraffin sections of the ovaries. (C) Paraffin sections of the testes. (A1, B1, C1) 2-month-old; (A2, B2, C2) 3-month-old; (A3, B3, C3) 4-month-old; (A4, B4, C4) 5-month-old; (A5, B5, C5) 6-month-old; (A6, B6, C6) 7-month-old; (A7, B7, C7) 9-month-old; (A8, B8, C8) 10-month-old. *In*, intestine; *Ct*, connective tissues; *F*, follicle; *Fc*, follicle cell; *Ig*, indistinguishable gonium; *Og*, oogonium; *Oc*, oocyte; *Mo*, mature oocyte; *Sg*, spermatogonium; *Sc*, spermatocyte; *St*, spermatid.

in salinity (simulating rainfall) can induce spawning in certain bivalves, aligning reproduction with favorable conditions (El-Sokkary et al., 2023). In sum, favorable environmental conditions set the stage for molluscan gonad development and spawning, and many aquaculture protocols exploit these cues (temperature shocks, photoperiod manipulation, and supplemental feeding) to induce or synchronize breeding.

Internally, environmental signals feed into endocrine and genetic pathways that govern gonadal development at the cellular level. A suite of evolutionarily conserved germline genes directs the formation and proliferation of gametes in mollusks. Key germ cell markers such as *Vasa* and *Nanos* are expressed in primordial germ cells, oogonia, and spermatogonia, and they are essential for germ cell maintenance and early gametogenesis (Kranz et al., 2010; Liu et al., 2022). These germline determinants (along with Piwi-family proteins) are continuously expressed throughout the gametogenic cycle and serve as molecular hallmarks of the developing eggs and sperm. The DM-domain transcription factor gene (*Dmrt1/Dmrt1L*) promotes testis development, whereas the forkhead transcription factor *FoxL2* promotes ovarian development, functioning as an antagonistic "yin-yang" pair similar to their roles in vertebrates (Li et al., 2018; Wei et al., 2021). Experimental studies show that *FoxL2* and *Dmrt1L* expression diverge well before any histological sex differences become apparent—in *Argopecten* scallops, *FoxL2* is upregulated in undifferentiated ovaries and *Dmrt1L* in testes several weeks in advance of gonadal sex differentiation, indicating they are early drivers of the process (Li et al., 2018; Wei et al., 2021). Moreover, in the hermaphroditic bay scallop (*A. irradians*), *FoxL2* and *Dmrt1L* show opposite correlations with steroid hormones: *FoxL2* levels are positively correlated with ovarian estradiol titers, whereas *Dmrt1L* is associated with higher testicular testosterone (Wei et al., 2021). This suggests *FoxL2* and *Dmrt1* function not only as genetic switches but also interface with endocrine signaling. Beyond these master genes, broad transcriptomic analyses (RNA-seq) have identified gene coexpression networks active during gonad development. In the dwarf surfclam *M. lateralis*, the weighted gene coexpression network analysis (WGCNA) has revealed modules of coregulated genes: some modules turned on during early gonad formation, others during sex differentiation, and others during gametogenesis (Li et al., 2022b). *Sox2, FoxZ, HsfY* (a Y-linked heat shock factor), *FoxL2*, and *Hes1* were highlighted as hub transcription factors in clam gonadal development. The precise regulation of these genes ensures that germ cells proliferate and differentiate into oocytes or sperm appropriate to the organism's sex.

Neuroendocrine pathways provide a critical link between the environment and gene regulation in molluscan gonad development. Field and laboratory studies show that females often have higher gonadal estrogen levels and males higher androgen levels during certain reproductive stages, mirroring the pattern in vertebrates (Avila-Poveda et al., 2015; Gauthier-Clerc et al., 2006; Matsumoto et al., 1997; Osada et al., 2004). In the Yesso scallop (*P. yessoensis*), our studies identified 63 genes encoding neuropeptide precursors, including 50 genes corresponding to known neuropeptide families and 13 previously uncharacterized peptides (Zhang et al., 2018). Furthermore, we identified two gonadotropin receptor-like genes (LGR and LGR5L), four steroidogenesis enzymes (CYP17A1, HSD17B12, HSD3B1, and HSD3B2), and proposed a GnRH signaling pathway (GnRH-GnRHR-GPB5-LGR/LGR5L) in the cerebral and pedal ganglia (CPG) of the scallop., This pathway may participate in regulating sex hormones, resembling the vertebrate "hypothalamic-pituitary-gonadal" axis (Zhang et al., 2020) operating in the CPG, coupled with a steroidogenic pathway in the gonad. In cephalopods, the optic gland in the brain functions as a specialized neuroendocrine organ comparable to the pituitary. Upon sexual maturity, the optic gland enlarges and releases a gonadotropin signals, including oct-GnRH (an octopus gonadotropin-releasing hormone-like peptide), which induces gonadal maturation and spawning behaviors (Minakata et al., 2009). The octopus optic gland hormone stimulates the gonads to synthesize and release sex steroids and also orchestrates ovulatory and egg-laying processes. Likewise, in gastropods, neuroendocrine cells in the cerebral and pleural ganglia secrete hormones (e.g., conopressin, APGWamide, and GnRH-like peptides) that regulate ovotestis function and spawning reflexes, integrating environmental stimuli like photoperiod into the animal's reproductive physiology. Ultimately, molluscan gonadal development emerges from this multitiered regulatory network: external environmental factors activate endocrine responses, which then converge on molecular gene networks within the gonad (e.g. *Vasa, Nanos, Dmrt1, FoxL2*, and others) to control germ cell development, sex differentiation, and gamete maturation in a coordinated fashion. This integrated mechanism ensures that gametogenesis proceeds only under favorable conditions and that the organism's molecular developmental program is tuned to its environment, which is vital for the reproductive success of mollusks in their often-variable habitats.

8.3 Sex chromosomes and sex determination

8.3.1 Sex chromosome systems and genetic sex determination

Molluscan sex determination can be genetic sex determination (GSD), environmental sex determination (ESD), or a combination. Despite the diversity of sexual systems, most molluscan species lack obvious sexual dimorphism in body morphology. Although traditional XX/XY or ZZ/ZW systems have been documented in certain mollusks, they are often characterized by homomorphic sex chromosomes without pronounced Y or W differentiation. These include the XX/XY system in *Littorina saxatilis, Atrina pectinata, M. lateralis*, and *Haliotis discus hannai*, the XO/XX system in *Neotricula aperta*, and ZZ/ZW system in the ivory shell (*Babylonia areolata*) (Guo & Allen, 1994; Luo et al., 2021; Thiriot-Quievreux, 2003; Zhou et al., 2018; Zou et al., 2025).

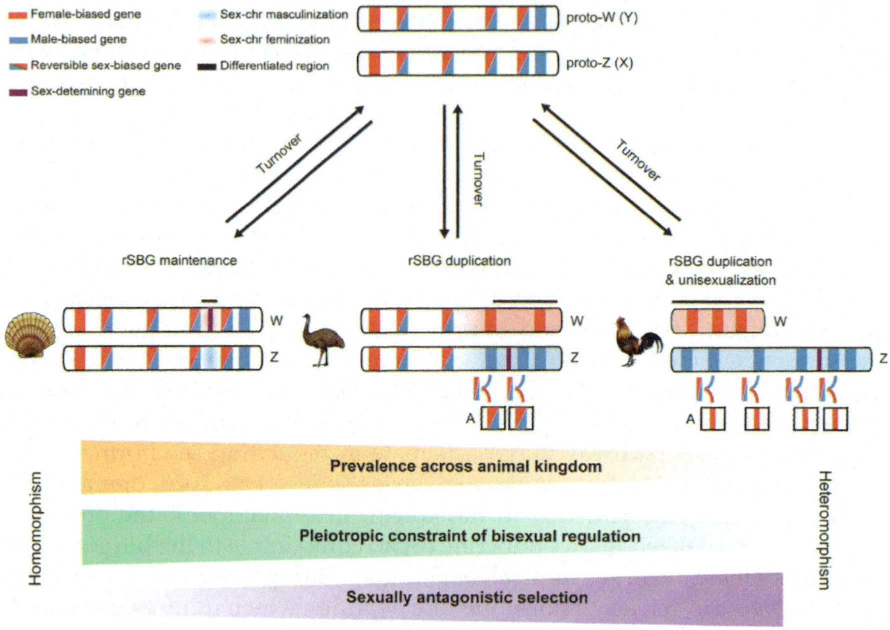

FIGURE 8.6 Evolutionary model for homomorphy maintenance and heteromorphy transition of sex chromosomes. The homomorphy of sex chromosomes would be evolutionarily stable, if an SDG emerges on the rSBGdominant protosex chromosome (in ancient animal groups, such as scallop) where chromosomal divergence would be limited. Driven by sexually antagonistic selection for sexual dimorphism, the pleiotropic constraint of rSBGs could be resolved through gene duplication followed by unisexual subfunctionalization, leading to moderately (for example, paleognathous birds) to highly (for example, chickens and flies) differentiated sex chromosomes. The evolutionary model is drawn for simplicity, with irrelevance of sex chromosome homology among species. *From Han et al. (2022).*

In the two-spot octopus (*Octopus bimaculoides*), females have one fewer copy of a particular chromosome (denoted Z) while males have two, indicating a ZZ/ZO system (female is ZO, male ZZ) (Coffing et al., 2025). Interestingly, our recent research on scallops (family Pectinidae) has revealed that their ZW sex chromosome pair has remained homomorphic for more than 350 million years, which represents the oldest known case of enduring homomorphic sex chromosomes in animals and far exceeds the age of the heteromorphic sex chromosomes of mammals or birds (Han et al., 2022). A key finding from our laboratory proposes that scallops avoid sex-chromosome divergence through a mechanism involving reversible sex-biased genes (rSBGs) and frequent sex-determiner translocation. Essentially, genes that are biased in expression toward one sex can toggle their activity (reversible bias), reducing the selective pressure for permanent chromosome degeneration. A proposed evolutionary model (Fig. 8.6) illustrates how an initially homomorphic protosex chromosome with a sex-determining gene can either remain homomorphic (if pleiotropic sex-biased genes constrain divergence) or transition to heteromorphic (if those constraints are released by gene duplication and subfunctionalization). Many molluscan species have genic sex determination with only small genomic region differences between sexes, rather than whole-chromosome differences. From an aquaculture perspective, this also means that identifying sex-linked markers in mollusks is challenging.

Genetic sex determination (GSD) in mollusks is controlled by inherited factors on sex chromosomes or specific loci. Notably, GSD mechanisms in mollusks span all major classes, with variations in sex-chromosome systems but a broadly conserved molecular toolkit orchestrating gonad fate. Central to this toolkit are antagonistic transcription factors: *Dmrt1L/Dsx* promotes testicular development, whereas *FoxL2* promotes ovarian development. Beyond these two master regulators, additional genes contribute to the molecular sex-regulation network in mollusks. Members of the *Sox* gene family (SRY-related HMG-box) act as supporting factors in the male pathway. In the Manila clam (*Ruditapes philippinarum*), for instance, a *Sox9*-like gene is preferentially expressed in testes. Similarly, a *SoxH* gene in oysters has been proposed to interact with *Dmrt1* and *FoxL2* in the cascade of sex determination. Downstream factors, such as *Fem1*-like (analogous to *C. elegans fem-1*) and components of the Wnt signaling pathway, further refine sex outcomes. In bivalves, ovarian development correlates with high β-catenin (Wnt pathway effector) and Rspo/Wnt4 activity, paralleling vertebrate proovarian pathways. The integration of these factors produces a gene regulatory network wherein whichever side of the *Dmrt1–FoxL2* antagonism gains a slight early advantage will tip the bipotential gonad toward testis or ovary (Nicolini et al., 2023).

8.3.2 Epigenetic mechanisms in sex differentiation

Epigenetic mechanisms, including DNA methylation, histone modifications, and noncoding RNAs, are increasingly recognized as crucial regulators of sex determination and sexual differentiation in mollusks (Guo, 2025; Sun et al., 2022; Valk et al., 2025). In bivalves, shifts in the epigenome are closely tied to changes in sexual phenotype. For example, the European flat oyster (*Ostrea edulis*) can alternate sexes, and this sex reversal is accompanied by extensive DNA methylation reprogramming. Environmental cues (e.g., temperature or nutrition) modulate these epigenetic marks. Energetically stressful conditions, such as starvation, tend to masculinize oysters and correlate with altered DNA methylation of key genes (Sun et al., 2023). In the Pacific oyster (*C. gigas*), males have globally higher gene-body DNA methylation in gonads than females, and genes in conserved signaling pathways (rather than the master sex-determining loci) are preferential targets of methylation differences. For example, the promoter of *Dgkd* gene (diacylglycerol kinase) is hypermethylated in males but demethylated and more actively expressed in females, suggesting that promoter demethylation helps activate female development (Sun et al., 2022). Recent analyses in octopus reveal that a significant fraction of CpGs in the genome are methylated, with 5-methylcytosine enriched in gene bodies of highly expressed genes, and conserved histone marks (H3K4me3, H3K9me3, H3K27me3) are present across tissues (Macchi et al., 2022). The octopus testis, for instance, shows elevated expression of several histone-modifying enzymes (H3K4/H3K9 methyltransferases and demethylases) relative to other tissues, suggesting that chromatin modifications help regulate sex-related genes in cephalopod gonads. In addition to DNA and histone modifications, noncoding RNAs contribute to epigenetic control of molluscan sexual development. In the scallop *C. farreri*, for example, multiple microRNAs are differentially expressed between developing ovaries and testes, and two of these miRNAs directly target pivotal sex-differentiation genes (one represses the female-determining factor *Foxl2*, while another downregulates a male-associated factor *Klf4*), thereby fine-tuning gonadal differentiation (Li et al., 2022a). Collectively, these studies across bivalves, gastropods, and cephalopods illustrate that epigenetic modifications

(DNA methylation patterns, histone marks, and regulatory RNAs), often responsive to environmental factors, form a key layer of control in molluscan sex determination and differentiation.

8.3.3 Environmental influences on sex determination

Molluscan sex determination and differentiation often involve a dynamic interplay between genetic predispositions and environmental cues. Even in species with an otherwise fixed genetic sex, environmental conditions can modulate sexual development or trigger sex reversal, effectively making many mollusks partial environmental sex determiners (Leonard, 2019). Temperature is a particularly well-documented influence. For example, in the Pacific oyster (*C. gigas*), a 1-year conditioning at low temperatures (8°C) induced a male-biased sex ratio (Fabioux et al., 2005). For *C. corteziensis*, higher proportions of hermaphrodite males were associated with a larger annual temperature fluctuation (~18°C), whereas more females were observed under a narrower annual range (~9°C) (Mazón-Suástegui et al. (2011). Such thermal effects are thought to operate via metabolic and endocrine pathways or through epigenetic modifications of sex-related genes (for instance, influencing regulators such as *FoxL2, Dmrt1*, or aromatase) that tilt the developmental balance toward one sex. Beyond temperature, nutritional status, and other environmental stressors also modulate molluscan sex. Limited food availability can bias sex ratios toward males, presumably because producing energetically costly eggs is only feasible for well-nourished individuals. For example, in the mussel *Mytella charruana*, adults starved for one month shifted from a female-biased to a male-biased sex ratio (Stenyakina et al., 2010). In sequentially hermaphroditic gastropods such as the common slipper limpet (*Crepidula fornicata*), sex determination is influenced by the presence or absence of conspecifics—juveniles become male when in contact with large females, but if no female is present, a male can grow and transition into a female (Collin, 2013; Proestou et al., 2008). These diverse cases illustrate that extrinsic conditions can trigger or delay sex differentiation in mollusks. Mechanistically, such environmental influences are often mediated by hormonal shifts or epigenetic changes (e.g., DNA methylation) that modulate the expression of key sex-determining genes. In evolutionary terms, the flexibility to adjust sex in response to external conditions may be advantageous for maximizing reproductive success.

8.4 Fertility and sex control technologies

8.4.1 Traditional approaches

8.4.1.1 Triploidy induction

Sex-control breeding means the sex of cultured animals is manipulated or controlled to enhance desirable traits. In many aquaculture species, controlling sex ratio or reproductive capacity can improve growth rates, meat quality, and survival. Triploidy induction is a widely used method to produce sterile or infertile mollusks by endowing embryos with an extra chromosome set. In practice, this is achieved by inhibiting polar body extrusion shortly after fertilization. Pacific oysters (*C. gigas*) are commonly rendered triploid via a 5-min 37 °C heat shock, yielding ~42% triploid larvae. Chemical agents such as 6-DMAP (6-dimethylaminopurine) are also effective (e.g., 450 μM for 10 min, with >80% triploid larvae). Triploid oysters grew 60%–80% faster than diploids and suffer lower summer mortality (Chen et al., 2024; Gerard et al., 1994;

Guo, 2008; Melo et al., 2022). They conserve energy and maintain flesh quality during the normal spawning season, because triploids do not undergo full gametogenesis. Triploidy has been successfully induced in various mollusks (including oysters, scallops, and even abalone) and is currently the most commercially adopted sex-control method in molluscan aquaculture.

8.4.1.2 Hormone/biochemical interventions

Estrogen compounds can promote feminization, while androgens can induce masculinization. Thus the use of exogenous sex steroids to induce sex reversal in developing mollusks serves as a common strategy in aquaculture production. For example, immersing juvenile bivalves in 17β-estradiol (E_2) at ~10 μg/L for extended periods has yielded female-skewed populations (Huo et al., 2025). Conversely, exposure to the synthetic androgen 17α-methyltestosterone (MT) (e.g., 50 μg/L during the larval stage) has been used to direct genetic females to develop as phenotypic males. A recent study on the Mediterranean mussel (*Mytilus galloprovincialis*) demonstrated 100% female-to-male sex inversion by administering testosterone during the postspawning gonadal regrowth period (Chelyadina et al., 2024). Intriguingly, the same study found that estradiol treatment did not induce any sex change in either male or female mussels, highlighting that steroid responsiveness can be species- and context-specific. Hormonal sex control can thus be a powerful tool, but it carries practical limitations.

Bay scallops (*A. irradians*) are predominantly hermaphroditic, with a single individual capable of simultaneously releasing both oocytes and sperm. A major challenge in their crossbreeding is the avoidance of self-fertilization and inbreeding. To address this issue, our recent work has developed a reversible chemical treatment method. Injection of the protein kinase inhibitor K252a during the gonadal development stage of bay scallops significantly reduced sperm count and motility, while largely preserving oocyte production (Liu et al., 2023). Consequently, the treated scallops functioned primarily as females, a state aimed at preventing self-fertilization. This approach was applied to interspecific hybridization between the bay scallop (*A. irradians*) and the Peruvian scallop (*A. purpuratus*), resulting in a 3.19-fold increase in hybrid yield, with 67% of offspring being hybrids. K252a treatment represents a short-term, non-genetic intervention that can be applied to scallops at the growing stage to facilitate controlled hybridization and holds promise for large-scale application. Importantly, the K252a concentration was optimized to minimize toxicity, achieving a high survival rate of 95% at the effective dose (Liu et al., 2023). This study demonstrates how targeted biochemical intervention can address breeding bottlenecks in hermaphroditic bivalves.

8.4.1.3 Environmental manipulation

Regulating environmental conditions, such as temperature and pH, during critical periods of gonadal development can also influence sex ratios in some mollusks (Santerre et al., 2013). Certain bivalves exhibit temperature-dependent sex determination or plasticity in sex differentiation. For instance, rearing oyster larvae at moderately elevated temperatures favors female development, whereas brief exposure of Pacific oyster juveniles to 28°C can result in a male-biased population, reaching up to 64% males. Low pH induced a female-biased sex ratio (16% more females than males) after a single reproductive cycle (one generation) in the Sydney rock oyster (*Saccostrea glomerata*) (Dang et al., 2025; Parker et al., 2018). The advantages of environmental sex control lie in its simplicity and low cost; however, the resulting sex ratios can be variable. Therefore environmental methods are sometimes combined with other techniques to improve consistency. For example, a hatchery may apply a mild thermal regimen to adjust sex ratios and then further enhance female yield through hormone treatment or triploid induction.

8.4.2 Novel biotechnologies

Gene editing, particularly CRISPR-Cas9, offers a powerful tool for the genetic improvement of aquaculture species; however, its application in sex control of mollusks remains largely confined to basic research. Existing gene editing studies in mollusks have primarily focused on growth- or development-related genes. For instance, Yu et al. (2019) successfully knocked out muscle growth-associated genes *MSTN* and *Twist* by microinjecting Cas9–sgRNA complexes into early-stage embryos. Chan et al. employed electroporation to deliver Cas9–sgRNA into oyster embryos, achieving a larval survival rate of 76%–77% after optimization and an editing efficiency of approximately 10%, though mosaic mutations remained predominant (Chan et al., 2022). We have also effectively knocked out the *Cfap206* gene in the shellfish research model *M. lateralis* through CRISPR/Cas9 technology, showing an obviously decreased number of cilia in trochophore larvae (Wang et al., 2022). To date, gene editing targeting sex-determining genes has not been reported. Current research on sex regulation in mollusks predominantly relies on methods such as RNA interference (RNAi) (Liu et al., 2025). For example, recent studies from our group have shown that knockdown of *FoxL2* in the Yesso scallop resulted in abnormal oocyte development, down-regulation of ovarian-related genes, and concurrent upregulation of testis-associated genes, including *Dmrt1, Sox7,* and *Sox9*. Nevertheless, the direct manipulation of these genes via CRISPR to achieve monosex breeding represents a key future research direction. It is noteworthy that the use of CRISPR to edit genes involved in germ cell development for creating agametic or sterile individuals has already been successfully demonstrated in fish (e.g., knockout of the dead end gene yielding germ cell-free salmon) (Roy et al., 2022). This conceptual approach holds similar potential for mollusks. Future integration with conventional breeding strategies could pave the way for novel approaches combining sex control and reproductive containment. In summary, current challenges in mollusk gene editing—such as low delivery efficiency and high rates of mosaic mutations—coupled with unresolved regulatory frameworks and public acceptance issues, relegate its practical application to the research and development phase. Despite these hurdles, the prospects for this technology remain highly promising.

8.5 Summary and future directions

With the rapid development of aquaculture, sex-control breeding technologies for shellfish are important to enhance production efficiency and optimize germplasm resources. This chapter systematically reviews the diversity of shellfish reproductive strategies, the regulatory mechanisms of gonadal development, and the current status of sex-control breeding applications. At present, sex-control breeding in shellfish has evolved into a multi-tiered technological system. Among traditional methods, triploid induction is the most well-established technique. It has been widely implemented in oyster culture and, by suppressing gametogenesis, significantly improves growth rates and summer survival. Hormone interventions and environmental manipulation are also used in certain species for targeted sex ratio control. In addition, chemical gamete regulation techniques (such as K252a treatment) developed for hermaphroditic species (e.g., the bay scallop) effectively overcome the challenge of self-fertilization and significantly enhance hybrid breeding efficiency. On the front of new technologies, gene editing methods (e.g., CRISPR-Cas9) have made initial progress in editing shellfish growth-related genes. However, research into their application for sex control remains in an exploratory stage, and challenges such as editing efficiency, mosaicism, and regulatory oversight have yet to be resolved.

In the future, breakthroughs in shellfish sex-control breeding will require integrated and precise approaches. This means combining environmental regulation, hormone treatments, polyploid breeding, and molecular marker-assisted selection to establish a coordinated multitechnology sex-control breeding system. It will also be important to promote the practical application of cutting-edge tools such as gene editing. Research should focus on the functional analysis and precise editing of key sex-determining genes, exploring new methods to create sterile or single-sex populations. In addition, basic research in reproductive biology should be strengthened to deeply elucidate the genetic, epigenetic, and endocrine regulatory networks governing sex determination and differentiation, thereby providing a strong theoretical foundation for technological innovation. Finally, ecological safety and industry standards must be emphasized: While pursuing economic benefits, attention must be paid to the ecological risks associated with these technologies and work to establish appropriate standards and regulatory frameworks.

In summary, the continued innovation and integrated application of sex-control breeding techniques in shellfish will strongly propel aquaculture toward greater efficiency, precision, and sustainability. These advances will provide crucial technological support for ensuring a stable supply of high-quality protein and the sustainable utilization of fishery resources.

References

Avila-Poveda, O. H., Montes-Pérez, R. C., Koueta, N., Benítez-Villalobos, F., Ramírez-Pérez, J. S., Jimenez-Gutierrez, L. R., & Rosas, C. (2015). Seasonal changes of progesterone and testosterone concentrations throughout gonad maturation stages of the Mexican octopus, *Octopus maya* (Octopodidae: Octopus). *Molluscan Research, 35*(3), 161–172. doi:10.1080/13235818.2015.1045055.

Chan, J., Zhang, W., Xu, Y., Xue, Y., & Zhang, L. (2022). Electroporation-based CRISPR/Cas9 mosaic mutagenesis of β-tubulin in the cultured oyster. *Frontiers in Marine Science, 9*, 912409. doi:10.3389/fmars.2022.912409.

Chelyadina, N. S., Popov, M. A., & Pospelova, N. V. (2024). The effect of sex hormones on sex inversion in the mussel *Mytilus galloprovincialis*. *Federation Marine Environmental Research, 201*, 106710. doi:10.1016/j.marenvres.2024.106710.

Chen, D., Zheng, Z., Zhou, Z., Song, Y., Chen, Z., Lin, G., & Xue, T. (2024). Application of metabolomics approach to investigate the flavor substance differences between triploid and diploid oysters (*Crassostrea angulata*). *Frontiers in Marine Science, 11*, 1481047. doi:10.3389/fmars.2024.1481047.

Coffing, G. C., Tittes, S., Small, S. T., Songco-Casey, J. O., Piscopo, D. M., Pungor, J. R., … Kern, A. D. (2025). Cephalopod sex determination and its ancient evolutionary origin. *Current Biology, 35*(4), 931–939. doi:10.1016/j.cub.2025.01.005.

Collin, R. (2013). Phylogenetic patterns and phenotypic plasticity of molluscan sexual systems. *Integrative and Comparative Biology, 53*(4), 723–735. doi:10.1093/icb/ict076.

Dang, X., Zhang, Y., Dupont, S., Gaitán-Espitia, J. D, He, Y. Q., Wang, H. H., … Thiyagarajan, V. (2025). Low pH means more female offspring: A multigenerational plasticity in the sex ratio of marine bivalves. *Environmental Science & Technology, 59*(2), 1308–1321. doi:10.1021/acs.est.4c07808.

El-Sokkary, S. G., El-Wakeil, K. F. A., Obuid-Allah, A. H., & Omer, M. Y. (2023). Influences of habitat and seasonal changes on gonadal maturation of *Echinometra mathaei* (Echinodermata: Echinoidea) and *Tridacna squamosa*

(Mollusca: Bivalvia) in the Red Sea, Egypt. *Environmental Monitoring and Assessment, 195*(9), 1088. doi:10.1007/s10661-023-11713-9.

Fabioux, C., Huvet, A., Souchu, P. L., Pennec, M. L., & Pouvreau, S. (2005). Temperature and photoperiod drive *Crassostrea gigas* reproductive internal clock. *Aquaculture, 250*(1–2), 458–470. doi:10.1016/j.aquaculture.2005.02.038.

García-Flores, M., Aguilar-Cruz, C. A., Arellano-Martínez, M., & Ceballos-Vázquez, B. P. (2019). Morphological and histological description of the reproductive system of Octopus hubbsorum (Mollusca: Cephalopoda). *Invertebrate Reproduction & Development, 63*(4), 268–281. doi:10.1080/07924259.2019.1646674.

Gauthier-Clerc, S., Pellerin, J., & Amiard, J. C. (2006). Estradiol-17β and testosterone concentrations in male and female *Mya arenaria* (Mollusca bivalvia) during the reproductive cycle. *General and Comparative Endocrinology, 145*(2), 133–139. doi:10.1016/j.ygcen.2005.08.004.

Gerard, A., Naciri, Y., Peignon, J.-M., Ledu, C., & Phelipot, P. (1994). Optimization of triploid induction by the use of 6-DMAP for the oyster *Crassostrea gigas* (Thunberg). *Aquaculture Research, 25*(7), 709–719. doi:10.1111/j.1365-2109.1994.tb00735.x.

Guo, X. (2008). Building a superior oyster for aquaculture. *Jersey Shoreline, 25*(1), 7–9.

Guo, X. (2025). Genetics and epigenetics in life history and reproduction: Bivalve molluscs. *Encyclopedia of reproduction*. doi:10.1016/b978-0-443-21477-6.00409-0.

Guo, X., & Allen, S. K (1994). Sex determination and polyploid gigantism in the dwarf surfclam (*Mulinia lateralis* Say). *Genetics, 138*(4), 1199–1206. doi:10.1093/genetics/138.4.1199.

Guo, X., Hedgecock, D., Hershberger, W. K., Cooper, K., & Allen, S. K. (1998). Genetic determinants of protandric sex in the Pacific oyster, *Crassostrea gigas* Thunberg. *Evolution, 52*(2), 394–402. doi:10.1111/j.1558-5646.1998.tb01640.x.

Han, W., Liu, L., Wang, J., Wei, H., Li, Y., Zhang, L., Guo, Z., Li, Y., Liu, T., Zeng, Q., Xing, Q., Shu, Y., Wang, T., Yang, Y., Zhang, M., Li, R., Yu, J., Pu, Z., Lv, J., ... Wang, S. (2022). Ancient homomorphy of molluscan sex chromosomes sustained by reversible sex-biased genes and sex determiner translocation. *Nature Ecology & Evolution, 6*(12), 1891–1906. doi:10.1038/s41559-022-01898-6.

Hoving, H. J. T., Roeleveld, M. A. C., Lipinski, M. R., & Videler, J. J. (2006). Nidamental glands in males of the oceanic squid *Ancistrocheirus lesueurii* (Cephalopoda: Ancistrocheiridae) – sex change or intersexuality? *Journal of Zoology, 269*(3), 341–348. doi:10.1111/j.1469-7998.2006.00093.x.

Huo, Z., Xiao, J., Liu, Y., Fang, L., Li, J., Li, D., ... Yan, X. (2025). Long-term exposure to 17β-estradiol of the Manila clam (*Ruditapes philippinarum*): Shifts of gender ratio and changes of gene expression. *Genomics, 117*(3), 111052. doi:10.1016/j.ygeno.2025.111052.

Ibáñez, C. M., Pérez-Álvarez, J., Catalán, J., Carrasco, S. A., Pardo-Gandarillas, M. C., & Rezende, E. L. (2019). Sexual selection and the evolution of male reproductive traits in *Benthic octopuses*. *Frontiers in Physiology, 10*, 1238. doi:10.3389/fphys.2019.01238.

Kim, H. J., Shin, S. R., Oh, H. Y., Kim, J. W., & Lee, J. S. (2021). Sex of mussel *Mytilus coruscus* (Bivalvia: Mytilidae): sequential hermaphroditism. *Development & Reproduction, 25*(1), 55–58. doi:10.12717/DR.2021.25.1.55.

Komaru, A., Konishi, K., Nakayama, I., Kobayashi, T., Sakai, H., & Kawamura, K. (1997). Hermaphroditic freshwater clams in the genus *Corbicula* produce non-reductional spermatozoa with somatic DNA content. *The Biological Bulletin, 193*(3), 320–323. doi:10.2307/1542934.

Kranz, A. M., Tollenaere, A., Norris, B. J., Degnan, B. M., & Degnan, S. M. (2010). Identifying the germline in an equally cleaving mollusc: *Vasa* and *Nanos* expression during embryonic and larval development of the vetigastropod *Haliotis asinina*. *Journal of Experimental Zoology Part B: Molecular and Developmental Evolution, 314*(4), 267–279. doi:10.1002/jez.b.21336Australia.

Leonard, J. (2019). Transitions between sexual systems: Understanding the mechanisms of, and pathways between, dioecy, hermaphroditism and other sexual systems. *Transitions between sexual systems*. Springer.

Li, R., Zhang, L., Li, W., Zhang, Y., Li, Y., Zhang, M., ... Bao, Z. (2018). FOXL2 and DMRT1L are Yin and Yang genes for determining timing of sex differentiation in the bivalve mollusk *Patinopecten yessoensis*. *Frontiers in Physiology, 9*, 1166. doi:10.3389/fphys.2018.01166.

Li, S. L, Fan, S., Huang, X., Zhang, Z., & Qin, Z. (2022a). Identification and characterization of MicroRNAs involving in initial sex differentiation of Chlamys farreri gonads. *Biology, 11*(3), 456. doi:10.3390/biology11030456.

Li, Y., Liu, L., Zhang, L., Wei, H., Wu, S., Liu, T., Shu, Y., Yang, Y., Yang, Z., Wang, S., Bao, Z., & Zhang, L. (2022b). Dynamic transcriptome analysis reveals the gene network of gonadal development from the early history life stages in dwarf surfclam *Mulinia lateralis*. *Biology of Sex Differences, 13*(1), 69. doi:10.1186/s13293-022-00479-3.

Liu, T., Li, R., Liu, L., Wu, S., Zhang, L., Li, Y., ... Bao, Z. (2022). The Effect of temperature on gonadal sex differentiation of Yesso scallop *Patinopecten yessoensis*. *Frontiers in Cell and Developmental Biology*, *9*, 803046. doi:10.3389/fcell.2021.803046.

Liu, T., Wei, H., Zhang, L., Liu, L., Li, Y., Shu, Y., Yang, Y., Zhao, B., Xing, Q., Zeng, Q., Wang, C., Wang, S., Bao, Z., & Zhang, L. (2023). K252a inhibition of sperm viability for efficient crossbreeding of hermaphroditic bivalves. *Aquaculture*, *577*, 739930. doi:10.1016/j.aquaculture.2023.739930.

Liu, X., Yun, H., Xing, Y., Wang, S., Zhou, X., & Zhang, J. (2025). Study of FOXL2 regulation on ovarian function in *Chlamys farreri* through comparative ChIP-seq and transcriptome analysis using RNA interference. *Biology*, *14*(9), 1259. doi:10.3390/biology14091259.

Luo, H., Xiao, J., Jiang, Y., Ke, Y., Ke, C., & Cai, M. (2021). Mapping and marker identification for sex-determining in the Pacific abalone, *Haliotis discus hannai* Ino. *Aquaculture*, *530*, 735810. doi:10.1016/j.aquaculture.2020.735810.

Macchi, F., Edsinger, E., & Sadler, K. C. (2022). Epigenetic machinery is functionally conserved in cephalopods. *BMC Biology*, *20*(1), 202. doi:10.1186/s12915-022-01404-1.

Marian, J. E. A. R., Apostólico, L. H., Chiao, C. C., Hanlon, R. T., Hirohashi, N., Iwata, Y., ... Shaw, P. W. (2019). Male alternative reproductive tactics and associated evolution of anatomical characteristics in loliginid squid. *Frontiers in Physiology*, *10*, 1281. doi:10.3389/fphys.2019.01281.

Matsumoto, T., Osada, M., Osawa, Y., & Mori, K. (1997). Gonadal estrogen profile and immunohistochemical localization of steroidogenic enzymes in the oyster and scallop during sexual maturation. *Comparative Biochemistry and Physiology Part B*, *118*(4), 811–817. doi:10.1016/S0305-0491(97)00233-2.

Mazón-Suástegui, J. M., Ruíz-García, M. C., Chávez-Villalba, J., Rodríguez-Jaramillo, C., & Saucedo, P. E. (2011). Analysis of growth and first reproduction of hatchery-reared juvenile Cortez oyster (*Crassostrea corteziensis*) in northwestern Mexico: proposal of a minimal fishing size. *Aquaculture Research*, *42*, 1558–1568. doi:10.1111/j.1365-2109.2010.02748.x.

McElroy, K. E., Bankers, L., Soper, D., Hehman, G., Boore, J. L., Logsdon, J. M., & Neiman, M. (2022). Patterns of gene expression in ovaries of sexual vs. asexual lineages of a freshwater snail. *Frontiers in Ecology and Evolution*, *10*, 845640. doi:10.3389/fevo.2022.845640.

Melo, E. M. C., Sühnel, S., da Silva, F. C., & de Melo, C. M. R. (2022). Induction to tetraploidy in Pacific oysters (*Crassostrea gigas*). *Acta Scientiarum. Animal Sciences*, *44*, e55337. doi:10.4025/actascianimsci.v44i1.55337.

Minakata, H., Shigeno, S., Kano, N., Haraguchi, S., Osugi, T., & Tsutsui, K. (2009). Octopus gonadotrophin-releasing hormone: A multifunctional peptide in the endocrine and nervous systems of the cephalopod. *Journal of Neuroendocrinology*, *21*(4), 322–326. doi:10.1111/j.1365-2826.2009.01852.x.

Nicolini, F., Ghiselli, F., Luchetti, A., Milani, L., & Zufall, R. (2023). Bivalves as emerging model systems to study the mechanisms and evolution of sex determination: A genomic point of view. *Genome Biology and Evolution*, *15*(10), evad181. doi:10.1093/gbe/evad181.

Osada, M., Tawarayama, H., & Mori, K. (2004). Estrogen synthesis in relation to gonadal development of Japanese scallop, *Patinopecten yessoensis*: gonadal profile and immunolocalization of P450 aromatase and estrogen. *Comparative Biochemistry and Physiology Part B*, *139*(1), 123–128. doi:10.1016/j.cbpc.2004.07.002.

Pandian, T. J. (2018). *Reproduction and development in mollusca*. CRC Press 10.1201/b22125.

Parker, L. M., O'Connor, W. A., Byrne, M., Dove, M., Coleman, R. A., Pörtner, H.-O., ... Ross, P. M. (2018). Ocean acidification but not warming alters sex determination in the Sydney rock oyster, *Saccostrea glomerata*. *Proceedings of the Royal Society B: Biological Sciences*, *285*(1872), 20172869. doi:10.1098/rspb.2017.2869.

Pathak, C. R., Luitel, H., Utaaker, K. S., & Khanal, P. (2024). One-health approach on the future application of snails: A focus on snail-transmitted parasitic diseases. *Parasitology Research*, *123*(1), 28. doi:10.1007/s00436-023-08021-z.

Proestou, D. A., Goldsmith, M. R., & Twombly, S. (2008). Patterns of male reproductive success in *Crepidula fornicata* provide new insight for sex allocation and optimal sex change. *Biological Bulletin*, *214*(2), 194–202. doi:10.2307/25066676.

Rato, A., Joaquim, S., Matias, D., & Hubbard, P. C. (2025). The roles of chemical cues in the life cycle of bivalves: Spawning, settlement, and metamorphosis. *Reviews in Aquaculture*, *17*(1), e13007. doi:10.1111/raq.13007.

Reyes-Tur, B., Allen, J., Cuellar-Araujo, N., Hernández, N., Lodi, M., Méndez-Hernández, A., & Koene, J. (2015). Mating behaviour, dart shape and spermatophore morphology of the Cuban tree snail *Polymita picta* (Born, 1780). *Journal of Molluscan Studies*, *81*(2), 187–195. doi:10.1093/mollus/eyu089.

Roy, S., Kumar, V., Kumar Behera, B., Parhi, J., Mohapatra, S., Chakraborty, T., & Kumar Das, B. (2022). CRISPR/Cas genome editing—Can it become a game changer in future fisheries sector? *Frontiers in Marine Science, 9*, 924475. doi:10.3389/fmars.2022.924475.

Santerre, C., Sourdaine, P., Marc, N., Mingant, C., Robert, R., & Martinez, A.-S. (2013). Oyster sex determination is influenced by temperature — First clues in spat during first gonadic differentiation and gametogenesis. *Comparative Biochemistry and Physiology Part A: Molecular & Integrative Physiology, 165*(1), 61–69. doi:10.1016/j.cbpa.2013.02.007.

Staikou, A., & Koemtzopoulos, E. (2019). Intraspecific morphological variation of the sperm storing organ in two hermaphroditic land snail species. *Journal of Biological Research-Thessaloniki, 26*, 1. doi:10.1186/s40709-019-0093-y.

Stenyakina, A., Walters, L. J., Hoffman, E. A., & Calestani, C. (2010). Food availability and sex reversal in *Mytella charruana*, an introduced bivalve in the southeastern United States. *Molecular Reproduction and Development, 77*(3), 222–230. doi:10.1002/mrd.21132.

Sun, D., Li, Q. i, & Yu, H. (2022). DNA methylation differences between male and female gonads of the oyster reveal the role of epigenetics in sex determination. *Gene, 820*, 146260. doi:10.1016/j.gene.2022.146260.

Sun, D., Yu, H., & Li, Q. (2023). Starvation-induced changes in sex ratio involve alterations in sex-related gene expression and methylation in Pacific oyster *Crassostrea gigas*. *Comparative Biochemistry and Physiology Part B: Biochemistry and Molecular Biology, 267*, 110863. doi:10.1016/j.cbpb.2023.110863.

Thiriot-Quievreux, C. (2003). Advances in chromosomal studies of gastropod molluscs. *Journal of Molluscan Studies, 69*(3), 187–202. doi:10.1093/mollus/69.3.187.

Valk, S., Engelsma, M., Megens, H.-J., Kamermans, P., Murk, A. J., & Nijland, R. (2025). Oysters in transition: Hermaphrodite oysters display unique DNA methylation patterns in gill tissue. *BMC Genomics, 26*(1), 567. doi:10.1186/s12864-025-11736-1.

Wang, Y., Zhu, X., Lian, S., Li, Y., Hu, N., Hu, X., … Shi, W. (2022). Functional characterization of *Cfap206* for bivalve ciliogenesis by RNAi and CRISPR/Cas9 technologies. *Frontiers in Marine Science, 9*, 864037. doi:10.3389/fmars.2022.864037.

Wayne, N. L., & Block, G. D. (1992). Effects of photoperiod and temperature on egg-laying behavior in a marine mollusk, *Aplysia californica*. *The Biological Bulletin, 182*(1), 8–14. doi:10.2307/1542176.

Wei, H., Li, W., Liu, T., Li, Y., Liu, L., Shu, Y., … Bao, Z. (2021). Sexual development of the hermaphroditic scallop *Argopecten irradians* revealed by morphological, endocrine and molecular analysis. *Frontiers in Cell and Developmental Biology, 9*, 646754. doi:10.3389/fcell.2021.646754.

Yu, H., Li, H., Li, Q. i, Xu, R., Yue, C., & Du, S. (2019). Targeted gene disruption in Pacific oyster based on CRISPR/Cas9 ribonucleoprotein complexes. *Marine Biotechnology, 21*(3), 301–309. doi:10.1007/s10126-019-09885-y.

Zhang, M., Wang, Y., Li, Y., Li, W., Li, R., Xie, X., … Bao, Z. (2018). Identification and characterization of neuropeptides by transcriptome and proteome analyses in a bivalve mollusc *Patinopecten yessoensis*. *Frontiers in Genetics, 9*, 197. doi:10.3389/fgene.2018.00197.

Zhang, M., Wei, H., Liu, T., Li, W., Li, Y., Wang, S., … Bao, Z. (2020). Potential GnRH and steroidogenesis pathways in the scallop *Patinopecten yessoensis*. *The Journal of Steroid Biochemistry and Molecular Biology, 204*, 105756. doi:10.1016/j.jsbmb.2020.105756.

Zhou, L., Wang, X., Wu, B., Sun, X., Chen, S., Liu, Z., … Zhang, G. (2018). Chromosome preparation and karyotypes analysis of both male and female *Atrina pectinata*. *Progress in Fishery Science, 39*(5), 66–72.

Zou, Y., Liang, Y., Luo, X., You, W., Shen, M., Fu, J., & Ke, C. (2025). Development of a sex-specific molecular marker reveals the ZW/ZZ sex-determination system in *Babylonia areolata* (link 1807). *Aquaculture, 598*, 742044. doi:10.1016/j.aquaculture.2024.742044.

CHAPTER

9

Genomic selecon

Xiaofei Yu[a,b], *Yangfan Wang*[a], *Qifan Zeng*[a], *Xijiang Yu*[c], *Xiaolin Wu*[d] *and Zhenmin Bao*[a]

[a]College of Marine Life Sciences, Ocean University of China, Qingdao, China [b]School of Life Sciences, Nanchang University, Nanchang, China [c]Breeding, Genetics and Food Production Systems, Norwegian University of Life Sciences, Aas, Norway [d]University of Wisconsin-Madison, Madison, WI, United States

9.1 Introduction

With the continuous growth of the global population and the increasing demand for high-quality protein sources, aquaculture has emerged as a vital contributor to future food security. Over the past few decades, advances in breeding technologies have significantly enhanced global aquaculture production. However, maintaining productivity and developing robust aquatic animal varieties remains a critical challenge. Modern genetic breeding technologies, such as genomic selection (GS) and gene editing, offer powerful tools to accelerate the development of strains with enhanced growth rates, disease resistance, and overall performance. These innovations are essential for meeting the rising global demand for aquatic products while supporting the long-term sustainability and resilience of the aquaculture industry.

In conventional breeding, individuals are selected primarily based on phenotypic traits, which are effective for single-gene traits but less so for complex traits influenced by multiple genes and environmental factors. To improve selection efficiency, the best linear unbiased prediction (BLUP) method was introduced, integrating pedigree and phenotype data to estimate breeding values. The Key Laboratory of Marine Genetics and Breeding (MGB) of the Ministry of Education at Ocean University of China (OUC) pioneered the world's first shellfish genetic evaluation system and molecular marker-assisted breeding technology framework based on BLUP. BLUP has also been successfully applied to enhance economic traits in aquaculture species such as oysters (de Melo et al., 2019) and shrimp (Castillo-Juárez et al., 2007). However, its effectiveness is limited for traits with low heritability or those difficult to measure, such as disease resistance or meat quality.

With advances in sequencing technologies and reduced costs, breeders now use dense molecular markers, such as single nucleotide polymorphisms (SNPs), to improve selection accuracy. Two genomic approaches have emerged: Marker-assisted selection (MAS) and GS. MAS utilizes

Genetic Breeding and Aquaculture of Shellfish. A Collaborative Approach to a Sustainable Future
DOI: https://doi.org/10.1016/B978-0-443-33606-5.00019-9

markers linked to quantitative trait loci (QTLs), allowing early selection of breeding candidates. However, MAS is often constrained by the need for known marker-QTL associations and tends to capture only large-effect QTLs, explaining a limited portion of genetic variance (Manolio et al., 2009). Its success is thus restricted to traits with major genes, such as Infectious Pancreatic Necrosis (IPN) resistance in Atlantic salmon (*Salmo salar*). For polygenic traits, the limited number of identified QTLs reduces MAS efficiency, making its application in aquaculture breeding challenging despite extensive research efforts.

GS overcomes the challenges faced by MAS as it is based on the prediction of genomic estimated breeding values (GEBVs): The effects of numerous chromosomal regions, defined by dense arrays of genetic markers distributed across the genome, are estimated without considering the statistical significance of the effects; instead, these effects are aggregated in predicting the genetic merit of the breeding candidates. Compared to pedigree-based approaches, GS provides a more accurate assessment of realized genetic relationships. For example, although pedigree information assumes a full-sib kinship coefficient of 0.5, genome-wide marker data often reveal deviations from this expectation, thus improving the accuracy of estimated breeding values (EBVs) (Song et al., 2023). GS has been used to improve economically important traits of various aquaculture species, including scallops (Wang et al., 2018b), pacific white shrimp (Wang et al., 2017), tilapia (Lu et al., 2020), and large yellow croaker (Dong et al., 2016; Zhao et al., 2021).

9.2 Principle and process of genomic selection

GS is a breeding method that uses genome-wide genetic markers to predict the breeding values of selection candidates, thereby allowing breeders to make selection decisions without direct phenotypic evaluations on those candidates (Zenger et al., 2019). The underlying principle of GS is the ability to capturing the effects of numerous loci across the genome, including loci with small individual effects, to estimate an individual's total genetic merit GEBVs. This methodology was first proposed by Meuwissen et al. (2001), who demonstrated that using dense markers covering the entire genome could enable the prediction of total breeding values with greater accuracy than traditional methods, especially when traits are controlled by many genes. The primary advantages of GS include: (1) the possibility of selecting for difficult-to-measure traits, as phenotyping is potentially performed only once for the individuals of the reference population, (2) the assessment of the individual genetic merit for the trait of interest, and (3) the exploitation of the within-family component of the genetic variance, making breeding value estimates more accurate than conventional methods. Models incorporating genomic data have demonstrated, across various species and traits, higher prediction ability compared with those relying solely on pedigree information. Evidence from early applications of GS in aquaculture has been encouraging. For instance, in the Pacific oyster, implementing a genomic best linear unbiased prediction (GBLUP) model increased the accuracy of selection for growth traits by approximately 25%–30% compared with pedigree-based BLUP (Gutierrez et al., 2018). Similarly, studies on disease resistance in blue mussel indicated that using genomic data (GBLUP and single-step methods) can capture within-family variation, leading to higher heritability estimates and prediction accuracies than pedigree-only models (Ajithkumar et al., 2025). These results underscore the principle that GS leverages the realized genomic relationships and marker-QTL linkage disequilibrium to achieve more accurate selection.

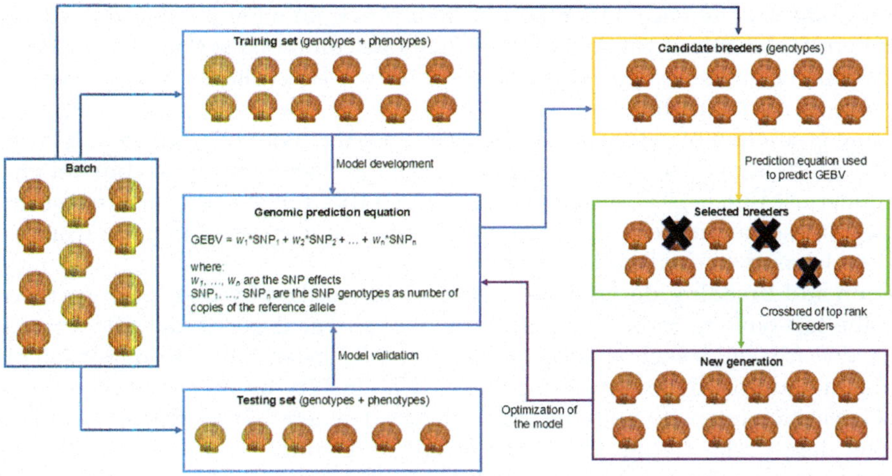

FIGURE 9.1 The general process of genomic selection in aquaculture involves several key steps.

GS treats all the markers as possibly associated with at least one QTL influencing the trait. It relies on genome-wide dense SNP panels for genotyping and phenotyping of a reference population (training set). The genotypes and phenotypes of the reference population are used to estimate marker effects on traits and to build a prediction equation. These estimated marker effects are validated in a testing population (validation set). After validation, the prediction model is routinely used to predict the GEBVs of breeding candidates, which only require to be genotyped. The GS process generally involves several key steps (Fig. 9.1):

(1) Reference population creation: Assemble a population of individuals (often from one or more breeding cohorts or families) that are both genotyped for genome-wide SNP markers and phenotyped for the traits of interest. This population reference forms the training dataset for the GS model. In shellfish, this may involve tagging and rearing large families or communal populations of oysters, mussels, or other bivalves, and recording traits like growth measurements or survival rates.

(2) Genotype coding and data preparation: Genotypic data are coded numerically (e.g., as 0, 1, and 2 for SNP alleles) and possibly quality filtered. Markers should be ensured to span the genome adequately. Often, genotype data might be imputed to fill missing calls, and minor allele frequency filters applied to remove rare variants. This step will be elaborated on in the coding and scaling of genotypes.

(3) Model training (estimation of SNP effects or GEBVs): Using data from a reference population, statistical models are trained to associate genetic markers with phenotypic traits, either by estimating SNP effects or by using genomic relationship matrices to predict breeding values. Model training often involves cross-validation to assess prediction accuracy. Once validated, the model is applied to selection candidates to estimate GEBVs, guiding the choice of breeding individuals. As new data becomes available, models can be retrained to improve prediction accuracy over time.

(4) Cross-validation and validation: It is common to evaluate the model's predictive performance by cross-validation within the reference population or by using a separate validation set. For example, reference data might be split into training and validation subsets. Model predictions for GEBVs of validation individuals are compared to their observed phenotypes or conventional breeding values to calculate accuracy. This validation step ensures the model is robust and not overfitting.

(5) **Genomic prediction of selection candidates:** Once the model is validated, it is applied to a new set of individuals (selection candidates) that have been genotyped but not phenotyped. The model uses their SNP profiles to predict GEBVs for the trait of interest. In shellfish breeding, these candidates could be juvenile oysters or mussels that we want to select as broodstock based on predicted performance.

(6) **Selection and breeding decisions:** Candidates are ranked by their GEBVs. Breeders then select the top-ranking individuals as parents of the next generation. Because GS can be done at an early age, selection decisions can be made faster than waiting for full phenotypic records. In addition, by examining genomic relationships among candidates, breeders can balance selection intensity with the maintenance of genetic diversity (for example, avoiding selecting closely related individuals to manage inbreeding).

(7) **Breeding program integration:** The prediction model can be periodically retrained or updated as new phenotypes and genotypes become available from subsequent generations. As the breeding program progresses, the accuracy of GEBVs may be improved by expanding the reference population and reestimating marker effects, ensuring that the prediction equation remains up-to-date with the population's genetics. Over time, genetic gain from GS accumulates faster relative to traditional selection, as demonstrated in other aquaculture species (e.g., theoretical studies suggested up to ~10% greater gain for body weight with GS vs. traditional selection). Furthermore, GS can reduce rates of inbreeding by up to 81% when compared with traditional selection programs under specific simulation parameters.

9.3 Linear models

9.3.1 Mixed linear model

The foundation of most GS prediction methods is the mixed linear model (MLM), which in this context refers to a linear regression framework relating phenotypes to genotypic predictors. In a simple form, an MLM for a quantitative trait can be written as:

$$y_i = \mu + \sum_{j=1}^{p} x_{ij}\beta_j + \epsilon_i$$

where y_i is the phenotypic value of the individual i (e.g., growth rate or shell length), μ is the overall population mean, x_{ij} is the genotype of individual i at marker j (coded numerically), β_j is the effect of marker j, and ε is a residual term for individual i. This model is a standard multiple linear regression including p marker covariates. When all β_j as random effects to be estimated, the model is commonly referred to as a random-effects whole-genome regression.

The MLM provides a conceptual starting point for genomic prediction, but directly using it to estimate thousands of β_j via ordinary least squares (OLS) is usually not feasible in practice for two reasons: (1) High dimensionality ($p >> n$)—the number of markers p often far exceeds the number of individuals n, leading to an underdetermined system with infinitely many solutions if we try to fit all markers simultaneously via OLS, and (2) multicollinearity—markers in linkage disequilibrium are highly correlated, violating OLS assumptions and causing unstable estimates. In the context of GS, these issues are addressed by introducing regularization or by treating the marker effects as random (leading to mixed models). Regularization imposes penalties or priors on the size of β_j estimates to prevent overfitting. For example, ridge regression (which forms the basis of SNP-BLUP) applies an $L2$ penalty that shrinks some β_j toward zero. The Bayesian methods impose priors that effectively regularize estimates as well (e.g., heavy-tailed priors that allow some large effects and many small ones). Alternatively, by treating β_j as random effects with a common variance, we enter the mixed model domain, where BLUP solutions can be obtained.

9.3.2 Equivalent models: Genomic best linear unbiased prediction

The SNP-BLUP model can be rewritten in an equivalent form known as GBLUP (genomic BLUP) (Meuwissen et al., 2001). Instead of explicitly including each SNP effect in the model, we consider the total genomic breeding value $u_i = \sum_{j=1}^{p} x_{ij} g_j$ for individual i. Under the SNP-BLUP assumptions $g_i \overset{iid}{\sim} \mathrm{N}(0, \sigma_g^2)$, or $\mathbf{g} \sim \mathrm{N}(\mathbf{0}, \sigma_g^2 \times \mathbf{I})$ i.e., g_i independently identically distributed as ($g_j \sim N\left(0, \sigma_g^2\right)$), the vector of genomic breeding values $u = (u_1,\ldots, u_n)$ follows a multivariate normal distribution: $u \sim N\left(0, G\sigma_u^2\right)$, where $\sigma_u^2 = p \cdot \sigma_g^2$ (since u_i is the sum of $p$independent $N\left(0, \sigma_g^2\right)$ variables) and G is the genomic relationship matrix (Meuwissen et al., 2001). In practice, we usually reparameterize σ_u^2 to equal the additive genetic variance explained by markers (often denoted σ_a^2). G can be specially normalized such that its average diagonal is 1. Then $u \sim N\left(0, \sigma_a^2 G\right)$.

Using u_i as a random effect representing the total genomic breeding value of the individual i, the model becomes:

$$y_i = \mu + u_i + \epsilon_i$$

with u_i as a random effect and $u_i \sim N\left(0, \sigma_a^2\right)$ covariances given by $Cov(u_i, u_{i'}) = \sigma_a^2 G_{ii'}$ (Meuwissen et al., 2001). In matrix form: $y = 1\mu + \mathbf{g}$, or $\mathbf{G}$. "?>+ $Z\ u + \varepsilon$, where Z is an incidence matrix mapping individuals to themselves (essentially an identity for genotyped individuals if they have one record each, and zeros for non phenotyped). Solving the mixed model equations in this form directly yields the BLUP of u_i for each individual. This is conceptually simpler since it avoids dealing with each SNP separately—instead, it uses the realized genomic relationships between individuals (which encapsulate the cumulative effects of shared alleles at all loci).

The equivalence of SNP-BLUP and GBLUP means that whether we solve for all SNP effects and then compute each individual's GEBV, or we solve directly for the GEBVs using G, we get the same answers (Zenger et al., 2019). GBLUP was popularized in dairy cattle breeding (VanRaden, 2008) and has become a standard approach in animal breeding software. The genomic relationship matrix G is typically constructed using the formula described earlier, and it measures genetic similarity between each pair of individuals based on their SNP genotypes. If two individuals share a lot of alleles identical-by-state, $G_{ii'}$ will be high, and GBLUP uses that to infer they likely have similar breeding values.

9.3.3 Genomic selection combining genomic best linear unbiased prediction and least absolute shrinkage selection operator

In the framework of linear mixed models, Li et al. (2015) extended GBLUP by including additional QTL information that was extracted from high-throughput SNPs by using the least absolute shrinkage selection operator (LASSO). GBLUP was combined with three LASSO methods—standard LASSO (SLGBLUP), adaptive LASSO (ALGBLUP), and elastic net (ENGBLUP)—that were used for detecting QTLs, and these QTLs were fitted as fixed effects; the remaining SNPs were fitted using a realized genetic relationship matrix. Simulations performed under distinct scenarios revealed that (1) the prediction accuracy of SLGBLUP was the lowest; (2) the prediction accuracies of ALGBLUP and ENGBLUP were equivalent to or higher than that of GBLUP, except under scenarios in which the number of QTLs was large; and (3) the persistence of prediction accuracy over generations was strongest in the case of ENGBLUP. Building on the favorable computational characteristics of GBLUP, ENGBLUP enables robust modeling and efficient computation to be performed for GS.

9.3.4 An efficient unified model for genome-wide association studies and genomic selection

A quantitative trait is controlled both by major variants with large genetic effects and by minor variants with small effects. Genome-wide association studies (GWAS) are an efficient approach to identify QTL, and GS with high-density SNPs can achieve higher accuracy of EBVs than conventional BLUP. GWAS and GS address different aspects of quantitative traits, but, as statistical models, they are quite similar in their description of the genetic mechanisms that underlie quantitative traits. Li et al. (2017) proposed a stepwise linear regression mixed model (StepLMM) to unify GWAS and GS in a single statistical model. First, the variance components of the GBLUP model are estimated. Then, in the SNP selection step, the linear mixed model (LMM) for GWAS is equivalently transformed into a simple linear regression to improve computation speed, and the most significant SNP is selected and included in the evaluation model. In the SNP dropping step, the SNPs in the evaluation model are tested according to the standard errors of their estimated effects. If nonsignificant SNPs are present, the least significant one is dropped from the model, and variance components are reestimated.

9.3.5 Bayesian methods for computing single nucleotide polymorphism effects

Bayesian regression methods for GS extend the mixed model approach by allowing more flexible assumptions about the distribution of SNP effects. Instead of assuming all SNP effects come from a single normal distribution with equal variance (as in rrBLUP), Bayesian methods assign prior distributions to the effects (and sometimes to their variances) that can lead to differential shrinkage—some SNPs undergo substantial shrinkage, whereas others are shrunk to a lesser degree. This family of methods was introduced in the seminal GS paper by Meuwissen et al. (2001), which described BayesA and BayesB, and has since expanded (often dubbed the "Bayesian alphabet" of methods). The Bayesian approach provides a probabilistic framework: The Bayesian framework involves specifying prior distributions for model parameters, followed by posterior inference of SNP effects via Bayes' rule. Posterior estimation is typically carried out using Markov chain Monte Carlo (MCMC) or similar sampling techniques.

BayesA is one of the original Bayesian GS methods (Gianola & Rosa, 2015). In BayesA, each SNP effect was assumed g_j (for $j = 1 \ldots p$) has its own variance σ_j^2. The prior for each g_j is a Student's *t*-distribution (which can be seen as a normal with unknown variance with an inverse chi-square prior). Specifically, Meuwissen et al. (2001) assumed $g_j \mid \sigma_j^2 \sim N\left(0, \sigma_j^2\right)$ and $\sigma_j^2 \sim \mathrm{Inv}-\chi^2(\nu, S)$(with some degrees of freedom ν and scale *S*chosen). This heavy-tailed prior means that most g_j will be small (because their variance could be small), but it allows an occasional SNP to have a much larger variance (thus a large effect) if the data strongly support it. The Student-*t* distribution has a heavier tail than the normal distribution. Compared with the single normal prior used in SNP-BLUP, BayesA can produce a few large-effect estimates alongside many near-zero ones. This behavior aligns with the assumption that the genetic architecture of a trait may include loci with moderate to large effects.

Implementation of BayesA via Gibbs sampling would involve iterating between sampling each g_j from its conditional normal posterior (given its marker-specific variance and the data) and sampling each σ_j^2 from its conditional inverse-chi-square posterior (given the current g_j estimate). The resulting posterior means of g_j often show a wides distribution: Some SNPs might get effectively heavier shrinkage (if the data suggests they contribute little, their σ_j^2 remains small), while others get lighter shrinkage.

BayesA has been applied in various GS studies to evaluate whether modeling heterogeneous marker variances enhances predictive accuracy. In certain aquaculture settings, where a trait is suspected to be influenced by a few genomic regions with large effects—such as a major QTL associated with disease resistance—BayesA may outperform GBLUP by more effectively capturing these signals. On the other hand, if the trait is truly polygenic with no standout loci, BayesA could overfit noise by trying to assign separate variances. In practice, BayesA often performs similarly to GBLUP in prediction accuracy, unless there are major effect loci. It tends to produce a posterior distribution of SNP effects that is useful for posterior QTL discovery (one can look at the SNPs with the highest estimated σ_j^2 as candidates for QTL).

BayesB, also introduced by Meuwissen et al. (2001), adds a feature to allow many SNPs to have an essentially zero effect. In BayesB, for each SNP *j*, with some probability π it is assumed to have no effect ($g_j = 0$), and with probability $(1 - \pi)$, it has an effect drawn from a distribution similar to BayesA's (a normal with its own variance). Effectively, BayesB's prior for g_j is a spike-and-heavy-tail mixture: a point mass at zero (with probability π) and a Student-*t* (or normal-inverse-chi-square) distribution for nonzero effects (with probability $1 - \pi$). Typically, π might be large (like 0.95), reflecting the belief that most markers have no effect, and a small subset has appreciable effects.

In an MCMC implementation, BayesB introduces a latent indicator for each SNP to denote whether it is included or not in the model at a given iteration. Sampling steps involve deciding for each SNP if it should be "on" or "off" (zero) given its contribution to the likelihood, and if "on," then sampling its effect and variance similarly to BayesA. The outcome is that many SNPs' posterior effects shrink exactly to zero, effectively performing a variable selection. Meuwissen's results showed BayesB often slightly outperformed GBLUP and BayesA in simulations where only a fraction of loci were truly affecting the trait (Gianola & Rosa, 2015).

The Bayesian Lasso (BL) (sometimes called Bayes Lasso) is another method of the Bayesian alphabet, which applies a double-exponential (Laplace) prior on SNP effects. This corresponds to an *L* regularization in a Bayesian framework (Gianola & Rosa, 2015). Rather than each SNP having its own variance parameter (like BayesA) or a mixture with zero (like BayesB), the Bayes Lasso assumes the distribution of $g_j$is centered at 0 with a high density around zero but with reasonably heavy tails to allow moderate effects. Typically, one parameter λ controls the rate of the exponential distribution (the equivalent of a shrinkage parameter)—sometimes a prior is also put on λ to make it Bayesian.

9.3.6 Application of mixed linear genomic selection models in shellfish

In Zhikong scallop (*Chlamys farreri*), MGB pioneered the establishment of three core technological systems for shellfish breeding in 2008: quantitative trait genetic evaluation, molecular marker-assisted selection, and genomic selection. Notably, MGB developed the world's first aquaculture strain bred "Penglai Red No. 2" (GS-01-006-2013) using genomic selection in 2013. In Yesso scallop (*Patinopecten yessoensis*), Dou et al. (2016) evaluated the prediction performances of six parametric GS models, including GBLUP, rrBLUP, selection operator (LASSO), and BL. They found that six GS models exhibited variable performance in prediction accuracy depending on the scenarios (e.g., heritability, sample size, and population structure), but Bayes-alphabet and BLUP-based models generally outperformed other models. Finally, they performed the evaluation using an empirical dataset composed of 349 Yesso scallops that were derived from five families. The prediction accuracy for this empirical dataset could reach 0.4 based on optimal GS models. In the triangle sail mussel *Hyriopsis cumingii*, Wang et al. (2023) applied GS to improve inner shell purpleness, a key trait for producing purple pearls. Using genotype-by-sequencing (97,090 SNPs), they analyzed 496 reference and 268 candidate individuals. Four shell color traits showed moderate-to-high genomic heritability (0.39–0.67). Among tested models, BayesB yielded the highest prediction accuracy and was used to develop a low-cost GS approach with 5K GWAS-informative SNPs. The top 30 individuals with the highest GEBVs were selected as broodstock. In field trials, the proportion of deep purple mussels increased markedly in both Wuyi and Chongming sites, demonstrating the effectiveness of GS for enhancing shell color and its potential for broader application in aquaculture.

9.4 Nonlinear models

The linear models discussed so far (GBLUP, Bayesian linear regressions) assume that the trait is an additive combination of marker effects. However, if there are nonadditive genetic architectures (such as dominance or epistasis), or if the relationship between genotype and phenotype is complex, nonlinear models might capture patterns that linear models miss. Nonlinear models can take various forms: Tree-based methods (random forests and gradient boosting), support vector machines, and kernel methods.

9.4.1 Reproducing Kernel Hilbert Space regression

In the context of GS, a prominent nonlinear approach has been the use of kernel regression methods, particularly the Reproducing Kernel Hilbert Space (RKHS) regression introduced by Gianola and colleagues for genomic data (de los Campos et al., 2009). RKHS regression is a method that can capture complex (including epistatic) relationships by mapping the data into a high-dimensional (potentially infinite-dimensional) feature space defined by a kernel. In simpler terms, instead of modeling trait value as a linear function of markers, we model it as $y_i = \mu + f(x_i) + \varepsilon_i$, where $f(x_i)$ is a potentially nonlinear function of the marker genotype vector x_i. Effectively, we assume flies in the space generated by a kernel function $K(x_i, x_j)$ measuring similarity between genotypes of individuals i and j.

A common choice is a Gaussian kernel:

$$K(x_i, x_j) = \exp\left(-\frac{||x_i - x_j||^2}{2\sigma^2}\right)$$

This kernel can capture nonlinear interactions because it compares whole genotype patterns. With an RKHS prior, the BLUP of $f(x_i)$ ends up being of the form $\hat{f}(xi) = \sum_{j=1}^{n} \alpha_j K(x_i, x_j)$, for some weights α_j that are solved by mixed model equations (treating fvalues as random effects). This is analogous to GBLUP, but with G replaced by a kernel matrix K computed by $K(x_i, x_j)$ (de los Campos et al., 2009).

For shellfish, traits such as survivability under stress (e.g., variable salinity or temperature) might have nonlinear genotype interactions (e.g., certain combinations of alleles needed for tolerance). RKHS or other machine learning models could potentially capture these better than a linear model (Wang et al., 2018b). One challenge is that aquaculture datasets often have limited sizes (hundreds of individuals), making very flexible models prone to overfitting. RKHS (with appropriate smoothing through the kernel's bandwidth parameter) offers a controlled way to introduce nonlinearity without overly complex explicit models. It essentially uses similarity between individuals in genotype space to predict similarity in phenotype space, which is a powerful concept also underlying GBLUP (GBLUP is a special case of the kernel method with a linear kernel equal to the genomic relationship matrix).

Beyond RKHS, other nonlinear models have been attempted: for example, neural networks, or tree-based methods, but kernel methods have the advantage of a solid statistical footing in mixed models (one can still do REML to estimate hyperparameters, such as σ_f^2, σ_e^2) and they have been well explored in the GS literature (Wang et al., 2024). Therefore RKHS remains a valuable tool in the GS toolbox. If a shellfish breeder suspects significant epistasis or wants to hedge against the linear model possibly missing predictive signals, they might try an RKHS in parallel with GBLUP. The results in the literature suggest that although RKHS sometimes edges out GBLUP in accuracy, the differences are often modest for complex traits unless the nonlinearity is pronounced. In those cases, combining linear and nonlinear kernels or using other advanced learning methods might further improve predictions.

9.4.2 Application of nonlinear genomic selection models in shellfish

In Zhikong scallop (*Chlamys farreri*), Wang et al. (2018) evaluated the predictability of four GS methods through real dataset analyses of four economical traits (e.g., shell length, shell height, shell width, and whole weight). The analysis revealed that different GS models exhibited variable performance in prediction accuracy depending on genetic and statistical factors, but the nonparametric method, including RKHS regression, generally outperformed the parametric linear method, such as GBLUP and BayesB. Furthermore, they demonstrated that the predictability relied mainly on the heritability regardless of the GS methods. The size of the training population and marker density also had considerable effects on the predictive performance. In the autotetraploid Pacific oyster (*Crassostrea gigas*), Wang et al. (2024) first applied GS methods for genomic prediction of five growth traits (e.g., shell length, shell height, shell width, total weight, and soft tissue weight). Overall, the four linear models (e.g., BayesA, BayesB, BayesC, and GBLUP) and the nonlinear model (e.g., RKHS) might not significantly affect the prediction accuracies of growth traits in autotetraploid *C. gigas*. Although the accuracy of GBLUP was higher than that of the Bayes models and RKHS, Bayes models and RKHS still had high predictability of shell height, shell length, and shell width, indicating that these prediction models, such as the GS method, can be used for genomic prediction of growth traits in autotetraploid *C. gigas*. The performance of four SNP encoding models—additive, diploidized additive, simplex dominant, and duplex dominant—was assessed for estimating the heritability and genomic prediction accuracy of growth traits in autotetraploid *C. gigas*. Among these, the additive model demonstrated the highest reliability. Furthermore, the genomic prediction results indicated that SNP encoding in autotetraploid Pacific oysters is inherently more complex than in diploid organisms, warranting further investigation into the underlying mechanisms in future studies.

9.5 Deep learning

9.5.1 Deep learning in genomic estimated breeding value prediction

Traditional parametric models for GEBV prediction in GS make strong assumptions about the functional form and the statistical distribution of SNPs. For example, GBLUP assumes that all markers affect the feature of interest, and these effects have a common variance in their distribution. BayesB allows for the estimation of SNP effects with differential shrinkage, but posterior inferences, particularly SNP variance, depend heavily on the prior assumptions used in the model.

Broadly speaking, the traditional parametric or nonparametric models often ignore complicated interactions between genes as well as higher-order nonlinearities, which may not be accurate in reality. To meet this challenge, neural networks (NNs) (Fig. 9.2) have been proposed as a new tool for genomic prediction of complex traits in the context of animal breeding. In aquatic animals, Wang et al. (2018) compared NNs with two linear parametric models, GBLUP and BayesB, for genomic prediction in four growth traits of Zhikong scallop (*C. farreri*). This study showed that the nonlinear model NNs outperformed the two linear parametric models GBLUP and BayesB.

More recently, deep learning (DL) models (LeCun et al., 2015), e.g., deep neural networks (DNNs) and convolutional neural networks (CNNs), are emerging as possibly powerful tools for genomic prediction of complex quantitative traits. The hierarchical function of DL systems enables machine learning to process data with a highly nonlinear approach. Among the DL models, CNNs (Fig. 9.3) seem to be more promising for genomic prediction of complex traits than DNNs. As pattern recognition algorithms, CNNs do not require a predefined feature vector and can automatically discover some useful features of the input for predicting the output. CNNs implement high-dimensional regressions by using nonlinear methods to uncover complex patterns that map the inputs (e.g., SNPs) to the outputs (e.g., complex traits).

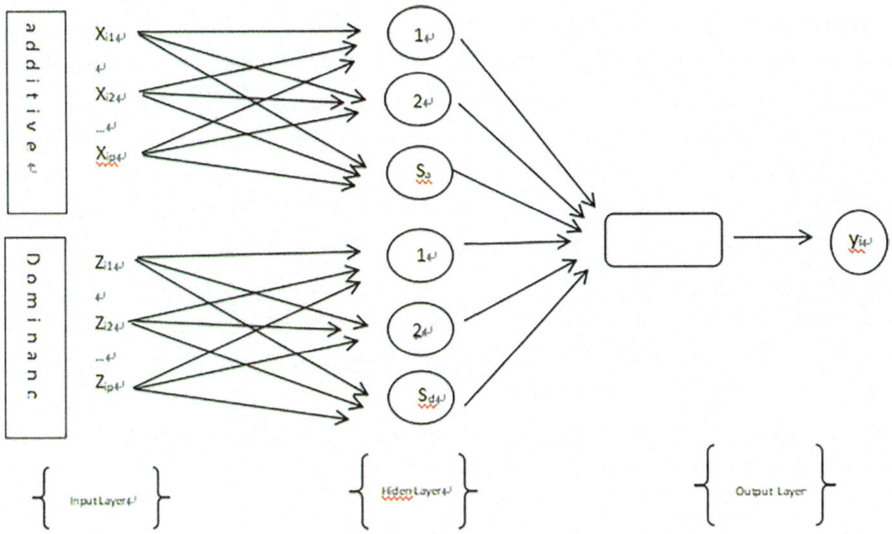

FIGURE 9.2 Graphical representation of the Multilayer Perceptrons (MLPs) for genomic prediction with inputs of the entries of additive and dominance matrices, respectively. *From Wang et al. (2024).*

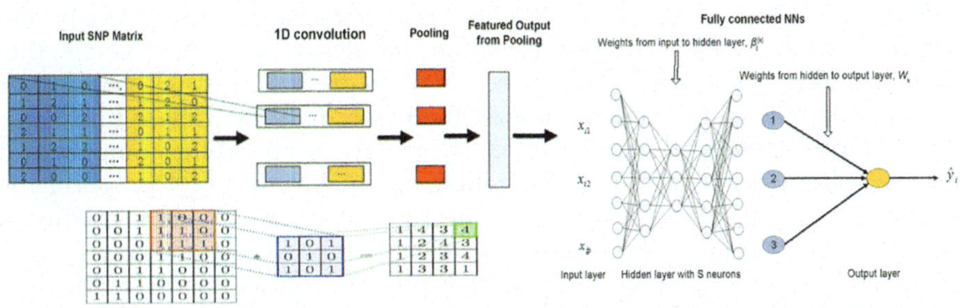

FIGURE 9.3 Schematics of a convolutional neural network design used in genomic selection. *From Zhu et al. (2021).*

9.5.2 Application of deep learning genomic selection models in shellfish

In aquatic animals, Zhu et al. (2021) proposed a CNN-based GS method for genomic prediction of growth traits in a Bay scallop (*Argopecten irradians irradians*) population. The results showed that CNNs outperformed GBLUP, BayesB, and DNNs on shell height, shell width, and total weight, suggesting that CNNs could discover valuable features from input SNP data and extract useful information, such as complicated interactions between genes as well as higher-order nonlinearities, from SNP features for prediction. They also found that DNNs presented the least accurate predictions, because DNNs had far more parameters to estimate than CNNs, GBLUP, and BayesB, and their predictions were challenged by the overfitting problem.

Because of the overfitting problem when using whole-genome SNPs as input, it is difficult to design an efficient model of DL for genomic prediction in GS. To solve the overfitting problem in some DL models, many researchers adopted various dimensionality reduction strategies. For

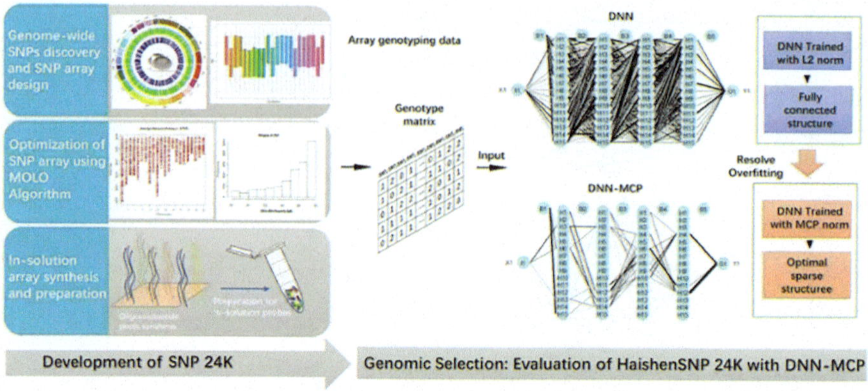

FIGURE 9.4 Schematic overview of the DNNs and DNN-MCP for genomic prediction in genomic selection. *DNN*, Deep neural networks; *MCP*, minmax concave penalty. *From Wang et al. (2024).*

example, genome-wide relationships between individuals (VanRaden, 2008) or singular value decomposition (SVD) of genome-wide SNPs (Guzelbulut et al., 2022) were used as DL inputs. Constructing optimal sparse structures of DL is also considered to be a key method to overcome the overfitting problem. For example, in Zhikong scallop (*C. farreri*), Wang et al. (2018) developed a sparse NN model, called SNNs, to find the optimal sparse structure of NNs by minimizing the squared error to penalize the L1 norm (weights and biases) of the parameters, thus solving the overfitting problem. Zeng et al. (2022) used SNNs to predict GEBV in real data analyses of growth traits of Atlantic salmon and compared the performance with two linear GBLUP and Bayes Lasso. The results showed that SNNs had advantages in handling complex traits among the GS models. Although L1 norm regularization enjoys several good properties for sparse optimization of DL models, it is sensitive to outliers and may cause serious bias in estimation. To overcome this defect, many nonconvex surrogates are proposed and analyzed, including smoothly clipped absolute deviation (SCAD) and minmax concave penalty (MCP) (Ma et al., 2019). Lv et al. (2022) developed a new DNN-based model, called DNN-MCP, which employed MCP regularization for DNNs to overcome the problem of overfitting and applied it for genomic prediction in sea cucumber (*Apostichopus japonicus*). DNN-MCP could find an optimal sparse structure of DNNs (Fig. 9.4) and showed the best prediction accuracy compared with DNNs, GBLUP, and BayesB.

9.6 Future research directions

9.6.1 Limitations and improvements of genomic selection

Although GS seems promising when applied to the aquaculture sector, several issues have recently emerged. Genomic prediction accuracies relying on random k-fold cross-validations performed on relatively small one-generation populations, usually consisting of a few thousand samples (Sukhavachana et al., 2021), might provide overly optimistic results and may not be informative in the long term. In fact, different studies have drawn attention to the progressive decrease in prediction accuracy as the relationships between training and testing sets are less tight (Faggion et al., 2023; Fraslin et al., 2022), which is likely to happen when a selective breeding program is practically implemented due to the disruption of the family-based linkage disequilibrium (de los Campos et al., 2013). Therefore finding solutions to build strong prediction equations that could work across different generations of animals, different populations, or different strains has become one of the major trends in GS, which, in turn, might prevent routine phenotyping, especially for difficult-to-measure traits or those requiring animal sacrifice (disease resistance, carcass yield, efficiency traits).

9.6.2 Incorporating attention mechanisms for weight optimization in deep learning models

To learn complex combinations of predictive features from data, state-of-the-art DL models for GEBV prediction typically consist of tens of millions of free parameters that are learned during training. Although the high capacity of DL models to encode latent feature representations results in state-of-the-art prediction accuracy, it comes with the challenge of their predictions by the overfitting problem. Weight regularization is a general technique to help mitigate the issue of overfitting in training neural networks and often improves both performance and the ability to isolate important features. In our previous study, the DNN-MCP model could effectively solve the overfitting problem of DNNs in genomic prediction of GS by minimizing the square error subject to a penalty on the nonconvex MCP-norm of the parameters (weights and biases). Meanwhile, we can also consider different mechanisms for the weights of DL models to avoid the overfitting problem in the future.

Attention is a complex cognitive function that is indispensable for human beings (Chaudhari et al., 2021). An important feature of attention is that humans do not tend to process information at one time. Attention mechanism can be used as weight regularization for which weights are reserved for input sequences to selectively concentrate on a part of the input as likely to hold relevant information for processing. While learning hidden features, attention weights force the DL models to focus on a limited portion of the input, which can be examined directly to aid in identifying parts of the input that played a key part in the model's internal representation, solving overparameters of DL models for genomic prediction. Therefore making attention mechanisms for weights of DL models to solve the overfitting problem is one of the future research directions.

9.6.3 Making deep learning models explainable: Toward causality and multiomics integration

In GS, DL models have demonstrated promising potential to outperform traditional statistical methods in predicting GEBVs. However, in the context of genome-enabled prediction, they are pretty good predictive machines, but through their so-called black-box behavior, they cannot be used to make any explainable inference of SNPs on phenotypes (Novakovsky et al., 2023). For researchers aiming to uncover the biological underpinnings of complex traits, such explanatory insights are often more valuable than predictions alone. The genotype–phenotype relationship is inherently complex and cannot be fully resolved through genome sequence data alone. Multiomics approaches—including genomics, transcriptomics, proteomics, and metabolomics—offer a more comprehensive framework for uncovering the molecular mechanisms and causal

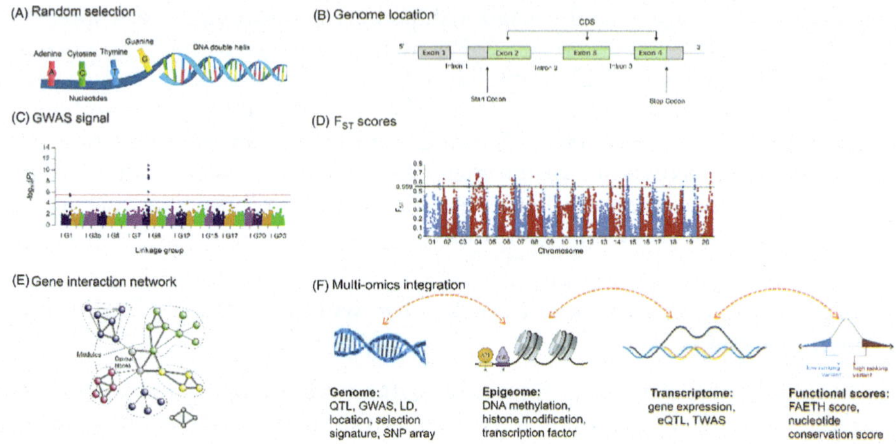

FIGURE 9.5 Overview of the major SNP preselection and prioritization strategies. (A) Random selection of SNPs, (B) location of elements in the genome (exons, introns, 3′UTRs, 5′UTRs, MNase HS regions), (C) GWAS signal, (D) Fst scores, (E) gene interaction network, and (F) multiomics integration. *GWAS*, Genome-wide association studies; *SNPs*, single nucleotide polymorphisms; UTR, Untranslated Region; Fs, Fixation Index. *From Yu et al. (2025).*

relationships underlying phenotypic traits (Subramanian et al., 2020). In aquaculture, this is especially relevant, as traits such as growth, disease resistance, stress tolerance, and hybrid vigor are influenced by intricate genetic and epigenetic regulatory networks (Gan et al., 2023). Although much effort has been devoted to understanding these mechanisms, including nonadditive gene expression, DNA methylation, histone modifications, and microRNAs, epigenetic markers have yet to be incorporated into routine breeding programs in aquaculture.

To address these challenges, future DL models must be designed to integrate diverse multiomics datasets in a biologically meaningful way. This integration can not only improve prediction accuracy but also enhance model interpretability by moving from mere association to biological causality. For instance, transfer learning (TL) strategies can be applied to train models on small subsets of variants or genes and extrapolate predictions across the genome, including rare and low-frequency variants with small or large effects. Furthermore, by prioritizing genetic variants using omics-informed criteria before applying genomic prediction models, researchers can focus on the most biologically relevant features, thereby improving both the efficiency and interpretability of breeding strategies.

Overall, advancing DL model interpretability through causal inference and multiomics integration represents a critical frontier in GS research. This shift not only supports more accurate GEBV prediction but also facilitates a deeper understanding of the biological pathways involved, enabling more precise and sustainable genetic improvement in aquaculture. Omics studies provide an in-depth overview of various biological aspects. We have examined the unique characteristics of various omics datasets and their roles in prioritizing genetic variants for genomic prediction across humans, plants, and livestock, while also highlighting their potential applications in aquaculture (Yu et al., 2025). This approach involves the integration of diverse high-throughput data, including genomics, epigenomics, transcriptomics, and proteomics—as illustrated in Fig. 9.5.

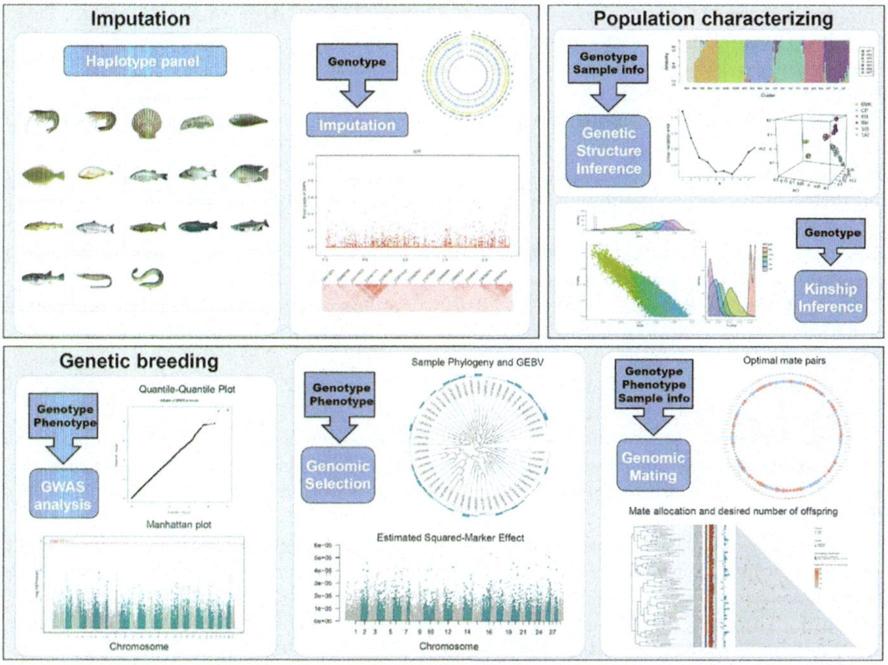

FIGURE 9.6 Functions of aquaculture molecular breeding platform (AMBP), including imputation, population characterizing, and genetic breeding. *From Zeng et al. (2022).*

9.6.4 Developing comprehensive web servers for genomic selection in aquaculture

Most current software of DL for phenotyping, genotyping, GEBV prediction, and other genetic analyses require a specific background knowledge of bioinformatics and quantitative genetics, making it challenging for general geneticists and biologists to perform the analysis. Therefore it is essential to develop a user-friendly web server to fill the gap. In our previous study, we developed the aquaculture molecular breeding platform (AMBP) (Zeng et al., 2022), the first web server for genetic data analysis in aquatic species of farming interest (Fig. 9.6). AMBP integrates the haplotype reference panels of 18 aquaculture species, which greatly improves the accuracy of genotype imputation. It also supports multiple tools to infer genetic structures, dissect the genetic architecture of performance traits, and estimate breeding values by existing state-of-the-art genomic prediction models, including a sparse DNN model. All the tools are coherently linked in a web interface for users to generate interpretable results and evaluate statistical appropriateness. With the development of web server and database technologies, more intelligent DL-based GS systems should be possible in the future.

References

Ajithkumar, M., D'Ambrosio, J., Travers, M.A., Morvezen, R., Degremont, L., 2025. Genomic selection for resistance to one pathogenic strain of *Vibrio splendidus* in blue mussel *Mytilus edulis*. Frontiers in Genetics 15, 1664–8021. doi:10.3389/fgene.2024.1487807.

Castillo-Juárez, H., Casares, J.C.Q., Campos-Montes, G., Villela, C.C., Ortega, A.M., Montaldo, H.H., 2007. Heritability for body weight at harvest size in the Pacific white shrimp, *Penaeus (Litopenaeus) vannamei*, from a multi-environment experiment using univariate and multivariate animal models. Aquaculture 273 (1), 42–49. doi:10.1016/j.aquaculture.2007.09.023.

Chaudhari, S., Mithal, V., Polatkan, G., Ramanath, R., 2021. An attentive survey of attention models. ACM Transactions on Intelligent Systems and Technology 12 (5), 1–32. doi:10.1145/3465055.

de los Campos, G., Hickey, J.M., Pong-Wong, R., Daetwyler, H.D., Calus, M.P.L., 2013. Whole-genome regression and prediction methods applied to plant and animal breeding. Genetics 193 (2), 327–345. doi:10.1534/genetics.112.143313United States.

de los Campos, G., Gianola, D., Rosa, G.J.M., 2009. Reproducing kernel Hilbert spaces regression: A general framework for genetic evaluation. Journal of Animal Science 87 (6), 1883–1887. doi:10.2527/jas.2008-1259.

Dong, L., Xiao, S., Chen, J., Wan, L., Wang, Z., 2016. Genomic selection using extreme phenotypes and pre-selection of SNPs in large yellow croaker (*Larimichthys crocea*). Marine Biotechnology 18 (5), 575–583. doi:10.1007/s10126-016-9718-4.

Dou, J., Li, X., Fu, Q., Jiao, W., Li, Y., Li, T., Wang, Y., Hu, X., Wang, S., Bao, Z., 2016. Evaluation of the 2b-RAD method for genomic selection in scallop breeding. Scientific Reports 6 (1), 2045–2322. doi:10.1038/srep19244.

Faggion, S., Carnier, P., Franch, R., Babbucci, M., Pascoli, F., Dalla Rovere, G., … Bargelloni, L., 2023. Viral nervous necrosis resistance in gilthead sea bream (*Sparus aurata*) at the larval stage: Heritability and accuracy of genomic prediction with different training and testing settings. Genetics Selection Evolution 55 (1), 1297–9686. doi:10.1186/s12711-023-00796-0.

Fraslin, C., Yáñez, J.M., Robledo, D., Houston, R.D., 2022. The impact of genetic relationship between training and validation populations on genomic prediction accuracy in *Atlantic salmon*. Aquaculture Reports 23. doi:10.1016/j.aqrep.2022.101033.

Gan, R.H., Zhou, L., Gui, J.F., 2023. Efficiently editing multiple duplicated homeologs and alleles for recurrent polyploids. Methods in Molecular Biology 2545, 491–512. doi:10.1007/978-1-0716-2561-3_26.

Gianola, D., Rosa, G.J.M., 2015. One hundred years of statistical developments in animal breeding. Annual Review of Animal Biosciences 3, 19–56. doi:10.1146/annurev-animal-022114-110733.

Gutierrez, A.P., Matika, O., Bean, T.P., Houston, R.D., 2018. Genomic selection for growth traits in Pacific Oyster (*Crassostrea gigas*): Potential of low-density marker panels for breeding value prediction. Frontiers in Genetics 9. doi:10.3389/fgene.2018.00391.

Guzelbulut, C., Suzuki, K., Shimono, S., 2022. Singular value decomposition-based gait characterization. Heliyon 8 (12), e12006. doi:10.1016/j.heliyon.2022.e12006.

LeCun, Y., Bengio, Y., Hinton, G., 2015. Deep learning. Nature 521 (7553), 436–444. doi:10.1038/nature14539.

Li, H., Su, G., Jiang, L., Bao, Z., 2017. An efficient unified model for genome-wide association studies and genomic selection. Genetics Selection Evolution 49 (1). doi:10.1186/s12711-017-0338-x.

Li, H., Wang, J., Bao, Z., 2015. A novel genomic selection method combining GBLUP and LASSO. Genetica 143 (3), 299–304. doi:10.1007/s10709-015-9826-5.

Lu, S., Zhu, J., Du, X., Sun, S., Meng, L., Liu, S., … Chen, S., 2020. Genomic selection for resistance to *Streptococcus agalactiae* in GIFT strain of *Oreochromis niloticus* by GBLUP, wGBLUP, and BayesCπ. Aquaculture 523, 735212. doi:10.1016/j.aquaculture.2020.735212.

Lv, J., Wang, Y., Ni, P., Lin, P., Hou, H., Ding, J., … Bao, Z., 2022. Development of a high-throughput SNP array for sea cucumber (*Apostichopus japonicus*) and its application in genomic selection with MCP regularized deep neural networks. Genomics 114 (4), 110426. doi:10.1016/j.ygeno.2022.110426.

Ma, R., Miao, J., Niu, L., Zhang, P., 2019. Transformed ℓ 1 regularization for learning sparse deep neural networks. Neural Networks 119, 286–298. doi:10.1016/j.neunet.2019.08.015.

Manolio, T.A., Collins, F.S., Cox, N.J., Goldstein, D.B., Hindorff, L.A., Hunter, D.J., McCarthy, M.I., Ramos, E.M., Cardon, L.R., Chakravarti, A., Cho, J.H., Guttmacher, A.E., Kong, A., Kruglyak, L., Mardis, E., Rotimi, C.N., Slatkin, M., Valle, D., Whittemore, A.S., … Visscher, P.M., 2009. Finding the missing heritability of complex diseases. Nature 461 (7265), 747–753. doi:10.1038/nature08494.

Rodrigues de Melo, C.M., Divilov, K., Schoolfield, B., Langdon, C., 2019. Selection of group and individual traits of Pacific oysters (*Crassostrea gigas*) on the West Coast, US. Aquaculture 512, 734389. doi:10.1016/j.aquaculture.2019.734389.

Meuwissen, T.H., Hayes, B.J., Goddard, M.E., 2001. Prediction of total genetic value using genome-wide dense marker maps. Genetics 157 (4), 1819–1829. doi:10.1093/genetics/157.4.1819.

Novakovsky, G., Dexter, N., Libbrecht, M.W., Wasserman, W.W., Mostafavi, S., 2023. Obtaining genetics insights from deep learning via explainable artificial intelligence. Nature Reviews Genetics 24 (2), 125–137. doi:10.1038/s41576-022-00532-2.

Song, H., Dong, T., Yan, X., Wang, W., Tian, Z., Sun, A., Dong, Y., Zhu, H., Hu, H., 2023. Genomic selection and its research progress in aquaculture breeding. Reviews in Aquaculture 15 (1), 274–291. doi:10.1111/raq.12716.

Subramanian, I., Verma, S., Kumar, S., Jere, A., Anamika, K., 2020. Multi-omics data integration, interpretation, and its application. Bioinformatics and Biology Insights 14, 117793221989905. doi:10.1177/1177932219899051.

Sukhavachana, S., Senanan, W., Pattarapanyawong, N., Srithong, C., Joerakate, W., Tunkijjanukij, S., Poompuang, S., 2021. Multiple-trait genomic prediction of harvest and fillet traits in *Asian seabass* (Lates calcarifer, Bloch 1790). Aquaculture 544, 737069. doi:10.1016/j.aquaculture.2021.737069.

VanRaden, P.M., 2008. Efficient methods to compute genomic predictions. Journal of Dairy Science 91 (11), 4414–4423. doi:10.3168/jds.2007-0980.

Wang, Q., Yu, Y., Li, F., Zhang, X., Xiang, J., 2017. Predictive ability of genomic selection models for breeding value estimation on growth traits of Pacific white shrimp *Litopenaeus vannamei*. Chinese Journal of Oceanology and Limnology 35 (5), 1221–1229. doi:10.1007/s00343-017-6038-0.

Wang, W., Bao, X., Lin, T., Meng, Y., Li, Z., Sun, G., … Wang, Y., 2024. Genomic prediction of growth traits in autotetraploid Pacific oyster (*Crassostrea gigas*) using whole-genome resequencing. Aquaculture 593, 741259. doi:10.1016/j.aquaculture.2024.741259.

Wang, Y., Mi, X., Rosa, G.J.M., Chen, Z., Lin, P., Wang, S., Bao, Z., 2018a. Technical note: An R package for fitting sparse neural networks with application in animal breeding. Journal of Animal Science 96 (5), 2016–2026. doi:10.1093/jas/sky071.

Wang, Y., Ni, P., Sturrock, M., Zeng, Q., Wang, B., Bao, Z., Hu, J., 2024. Deep learning for genomic selection of aquatic animals. Marine Life Science & Technology 6 (4), 631–650.

Wang, Y., Sun, G., Zeng, Q., Chen, Z., Hu, X., Li, H., Wang, S., Bao, Z., 2018b. Predicting growth traits with genomic selection methods in Zhikong Scallop (Chlamys farreri). Marine Biotechnology 20 (6), 769–779. doi:10.1007/s10126-018-9847-z.

Wang, Z., Hu, H., Sun, T., Lv, G., Lv, X., Wang, H., … Bai, Z., 2023. Genomic selection improves inner shell purpleness in triangle sail mussel Hyriopsis cumingii (Lea, 1852). Aquaculture 575, 739815.

Yu, X., Faggion, S., Liu, Y., Wang, B., Zeng, Q., Lu, C., Hu, J., Bargelloni, L., Fang, L., Bao, Z., 2025. Role of multi-omics in aquaculture genetics and breeding: Current status and future perspective. Science China Life Sciences 68 (9), 2591–2604. doi:10.1007/s11427-024-2828-8.

Zeng, Q., Zhao, B., Wang, H., Wang, M., Teng, M., Hu, J., Bao, Z., Wang, Y., 2022. Aquaculture molecular breeding platform (AMBP): A comprehensive web server for genotype imputation and genetic analysis in aquaculture. Nucleic Acids Research 50 (W1), W66. doi:10.1093/nar/gkac424.

Zenger, K.R., Khatkar, M.S., Jones, D.B., Khalilisamani, N., Jerry, D.R., Raadsma, H.W., 2019. Genomic selection in aquaculture: Application, limitations and opportunities with special reference to marine shrimp and pearl oysters. Frontiers in Genetics 10. doi:10.3389/fgene.2018.00693.

Zhao, J., Bai, H., Ke, Q., Li, B., Zhou, Z., Wang, H., … Xu, P., 2021. Genomic selection for parasitic ciliate *Cryptocaryon irritans* resistance in large yellow croaker. Aquaculture 531, 735786. doi:10.1016/j.aquaculture.2020.735786.

Zhu, X., Ni, P., Xing, Q., Wang, Y., Huang, X., Hu, X., Hu, J., Wu, X., Bao, Z., 2021. Genomic prediction of growth traits in scallops using convolutional neural networks. Aquaculture 545, 737171. doi:10.1016/j.aquaculture.2021.737171.

CHAPTER

10

Germplasm preservation

Yibing Liu[a], Xiaoxu Li[b], Huiping Yang[c] and Serean L Adams[d]

[a]Key Laboratory of Mariculture (Ministry of Education), Fisheries College, Ocean University of China, Qingdao, China [b]Aquatic Sciences Centre, South Australian Research and Development Institute, Adelaide, Australia [c]School of Forest, Fisheries, and Geomatics Sciences, Institute of Food and Agricultural Sciences, University of Florida, Gainesville, FL, USA [d]Cawthron Institute, Nelson, New Zealand

10.1 Introduction

In animals, germplasm broadly refers to live hereditary material that carries transmissible genetic information to the offspring, including germ cells, sperm, oocytes, embryos, or larvae (Morrell & Mayer, 2017; Yang, 2017). The key objective of germplasm preservation is to conserve the genetic diversity of an organism or genetic stock for its use at any time in the future (Huo et al., 2022; Morrell & Mayer, 2017). In general, germplasm preservation can be carried out by two approaches, namely in situ conservation and ex situ conservation. In situ conservation preserves germplasm in its natural environment and has been considered a high-priority strategy, while ex situ conservation preserves germplasm outside its natural habitats, and may encompass genetic material from both farmed and wild sources (Kumar et al., 2024; Morrell & Mayer, 2017).

Mollusks represent the second largest farmed group of aquatic organisms worldwide, accounting for approximately 20% of total world aquaculture production (excluding algae) in 2022 (FAO, 2024). The majority of farmed molluscan species are bivalves, including oysters, clams, scallops, and mussels (FAO, 2024). These bivalves have a higher protein content than many other meat sources and plant crops. They also have high levels of essential omega-3 fatty acids and micronutrients, such as zinc, iron, vitamin A, and vitamin B12, which are derived from freely available suspended detrital material and marine phytoplankton and zooplankton that they feed on (Brown et al., 2024; Zhang et al., 2022; Zhukova, 2019). Therefore bivalve aquaculture represents a key opportunity to fill the gap between supply and demand for high-quality, nutritious protein globally as a sustainable food source. Moreover, marine bivalve aquaculture has a smaller environmental footprint than farming of most other species due to: (1) its use of seawater instead of land or freshwater and (2) lower carbon emissions based on harvested protein (Warmerdam et al., 2021; Zhang et al., 2022). In addition, bivalve aquaculture provides

Genetic Breeding and Aquaculture of Shellfish: A Collaborative Approach to a Sustainable Future
DOI: https://doi.org/10.1016/B978-0-443-33606-5.00021-7

a wide range of ecosystem services, including buffering estuaries and coastal waters against phytoplankton blooms caused by anthropogenic nitrogen loading, increasing water clarity, and acting as a form of carbon capture during shell production (Brown et al., 2024). Therefore from both a nutritional and an environmental standpoint, bivalve aquaculture represents a compelling choice of food production for human consumption.

Hatcheries play a key role in the continued expansion of molluscan aquaculture for spat production and genetic improvement, although some countries or areas still capture spatfalls from wild sources to on grow, such as Pacific oysters (*Crassostrea gigas*) in France, Korea, Japan, and New Zealand, green-lipped mussels (*Perna canaliculus*) in New Zealand, and *Mytilus* mussels in Europe (Mondol et al., 2016; Nascimento-Schulze et al., 2021; Pang et al., 2024). To ensure the long-term sustainability of molluscan aquaculture, germplasm preservation and genetic improvement programs, such as selective breeding, ploidy manipulation, and hybridization, have been initiated in many farmed species (Chang et al., 2024; Huo et al., 2022; Martínez-Páramo et al., 2017; Nascimento-Schulze et al., 2021; Vignier et al., 2024; Yang et al., 2021).

Germplasm cryopreservation has emerged as an effective and reliable technique for conservation of threatened and endangered species and to assist ex situ breeding and management programs (Brereton et al., 2025; FAO 2026). Germplasm samples, such as sperm, oocytes, embryos/larvae, and gonadal tissues, can be cryopreserved in liquid nitrogen (LN) facilities for long-term storage and subsequently used for assisted reproduction techniques (Brereton et al., 2025). Moreover, cryopreservation can address some key bottlenecks in molluscan aquaculture and genetic improvement programs, including synchronizing spawning between males and females within and between species, extending the short window period of natural spawning, and minimizing the risks and costs associated with maintaining large numbers of broodstock from breeding programs (Martínez-Páramo et al., 2017). For example, (1) maintaining rare genetic strains to use should breeding objectives change or a disease outbreak occur, and (2) creating back crosses to enable rapid genetic gain or maintaining a backup of germplasm in the event that living banks are destroyed through harsh weather events, such as hurricanes and cyclones.

In marine molluscan species, gamete and larval cryopreservation techniques have been developed in some farmed species (Martínez-Páramo et al., 2017). However, compared with the livestock and fish species, the application of these techniques to germplasm banking and genetic improvement programs has been limited. The main reasons for this may include: (1) studies are mainly reported from laboratory-scale trials with limited industry long-term postthaw performance evaluation, (2) no standardization for high-throughput operations, and (3) lack of details in logistical factors regarding application of the cryopreservation technique for different purposes (e.g., broodstock numbers required and straw numbers frozen per run with identified cryopreservation system, and postthaw fertilization procedures for different applications). This chapter briefly describes the history and principles of cryopreservation, key aspects and understanding of sperm, oocytes, and larval cryopreservation from the last two decades of research, and then discusses opportunities and directions for future research.

10.2 A brief development history of germplasm cryopreservation

10.2.1 Sperm

In 1949, it was discovered that avian sperm could be preserved using glycerol as a cryoprotectant agent (CPA) at a low temperature of −79°C (Polge et al., 1949). This pivotal discovery is considered a breakthrough milestone that launched the scientific field of cryopreservation and its modern-day routine application in human in vitro fertilization (IVF), medicine, agriculture, and beyond. Since then, the technique has been further improved with the earliest offspring produced from cryopreserved sperm in cows, pigs, and horses, and sheep (see Curry, 2000 for review). In some livestock species, cryopreserved sperm is routinely used for artificial insemination for convenience since the technique matches the rates that are attained with fresh sperm (Arif et al., 2025; Sharafi et al., 2022; Tamargo et al., 2024).

In aquatic organisms, the first success was reported by Blaxter (1953) in finfish herring (*Clupea harengus*) sperm. Since that time, fish sperm cryopreservation has been investigated in more than 200 freshwater and marine fish species, including carp, salmonids, catfish, cichlids, medakas, white-fish, pike, milkfish, grouper, cod, and zebrafish (Magnotti et al., 2018; Martínez-Páramo et al., 2017; Neyrão et al., 2024).

In molluscan species (mainly bivalves), sperm cryopreservation was first reported by Lannan (1971) in Pacific oysters using a method modified from Blaxter (1953). To date, sperm cryopreservation has been reported in approximately 20 farmed species, including oysters, mussels, clams, scallops, and abalone (Hassan et al., 2015; Hossen et al., 2025; Liu et al., 2015c; Martínez-Páramo et al., 2017; Wang et al., 2024).

10.2.2 Oocytes/eggs

Successful oocyte cryopreservation was first reported in mice in 1977 (Whittingham, 1977). To date, research has been published in bovine, ovine, horse, pig, canine, feline, and primate species (Hinrichs, 2020; Ríos et al., 2025; Somfai, 2024; Tharasanit & Thuwanut, 2021). Although oocytes of livestock species are generally more sensitive to cold stress than those of mice, live offspring have been produced in cats, pigs, cows, and horses (Hinrichs, 2020; Ríos et al., 2025; Somfai, 2024; Tharasanit & Thuwanut, 2021). Oocyte cryopreservation is now routinely used in human IVF and overcomes ethical dilemmas associated with embryo cryopreservation, as well as providing a pathway for young ovarian cancer patients to have children (Sciorio et al., 2023).

In finfish, oocyte cryopreservation has been attempted in many species, including farmed freshwater and marine species, and model species (e.g., zebrafish *Danio rerio*) (Guan et al., 2010; Martínez-Páramo et al., 2017). However, no postthaw fertilization success has been achieved, and this has been attributed to the large yolk volume, cold sensitivity, and low permeability of CPAs caused by the chorion layer (Diwan et al., 2020; Franěk & Pšenička, 2020; Martínez-Páramo et al., 2017).

In mollusks, the first successful oocyte cryopreservation was reported in Pacific oysters with postthaw fertilization and D-stage larval survival rates ranging from 0.8 to 74.5% and 0.1 to 30.1%, respectively (Tervit et al., 2005). Although efforts have been made to further improve postthaw oocyte quality in this species, limited progress has been achieved in the postthaw D-stage larval rate (Adams et al., 2013; Diwan et al., 2020; Martínez-Páramo et al., 2017). In mussels, oocyte cryopreservation has been investigated in green-lipped (Adams et al., 2009) and blue mussels (*Mytilus galloprovincialis*; Liu & Li, 2015), with the latter resulting in >12% of postthaw oocytes reaching D-stage larvae. Furthermore, there was no significant difference in survival of the D-stage larvae to spat when compared to the unfrozen control (Liu & Li, 2015).

10.2.3 Embryos/larvae

The first embryos successfully cryopreserved were mouse embryos in 1972 (Whittingham et al., 1972). Over the past half-century, this technique has been adapted and developed for more than 25 mammalian species. In humans, mice, and cattle, embryo cryopreservation has become a routine procedure in assisted reproduction and research (Dobrinsky, 2002; Leibo & Sztein, 2019; Mandawala et al., 2016).

In finfish, embryo cryopreservation is challenging because of their meroblastic discoidal cleavage (cleavage does not penetrate the yolk) they undergo and the large yolk they have (Diwan et al., 2020; Martínez-Páramo et al., 2017). However, development of postthaw embryos to hatched larvae has been reported in flounder (*Paralichthys olivaceus*; Chen & Tian, 2005) and further improvement has been reported in zebrafish as two larvae from cryopreserved embryos have been grown to adulthood and spawned (Khosla et al., 2020), suggesting the potential in these two and other finfish species.

Bivalve embryos tend to be smaller in size than finfish and have a lower yolk content. Their cleavage patterns are also different as embryos undergo holoblastic cleavage (complete divisions). The success of embryo cryopreservation is still limited in bivalves, but has been investigated in oysters, mussels, and clams (Martínez-Páramo et al., 2017; Yang & Huo, 2022). Trochophore and D-stage larvae are the two important developmental stages in the life cycle of bivalves. Trochophore larvae are usually oval in shape and have a row of cilia around the middle with a long apical flagellum for swimming, whereas D-stage larvae have two valves and an organ called velum, which is circular in shape and protrudes between the valves. Sufficient long-term postthaw development and survival have only been reported in two commercial species, Pacific oysters (Labbé et al., 2018; Liu et al., 2020a) and blue mussels (Liu et al., 2020b) when these two larval developmental stages were used for cryopreservation.

10.3 Principles of cryopreservation

Typically, exposing cells to temperatures below 0°C without a CPA is either damaging or lethal. Generally, water is the most abundant molecule in cells, accounting for about 70% of the cell volume. The freezing of water both intracellularly and extracellularly imposes severe stress on cells (Mazur, 2004; Pegg, 2015). Two factors have been considered as the major reasons for cryodamage: (1) ice crystals formed intracellularly during freezing and thawing processes mechanically disrupt cellular membranes and are generally lethal to cells and (2) exposure to high solute concentrations that occur as a result of ice formation causing excessive dehydration and solute toxicity (Figure 10.1; Mazur, 2004; Parihar et al., 2023; Pegg, 2015). Optimized CPAs and cooling and thawing rates can help protect cells from these damaging processes.

10.3.1 Cryoprotective agents

CPAs are chemical compounds that protect cells from cryodamage (Elliott et al., 2017; Pegg, 2015; Ribeiro et al., 2022). CPAs protect cells by depressing the freezing point of the extracellular solution, disrupting hydrogen bonding between water molecules, and reducing the likelihood of intracellular ice formation (Bojic et al., 2021; Elliott et al., 2017; Ribeiro et al., 2022). CPAs also reduce the concentration of solutes at a given subzero temperature, thereby reducing the amount of shrinkage and dehydration that cells undergo during exposure to high solute concentrations at a given subzero temperature and freezing. However, high concentrations of CPAs can be toxic to cells or cause lethal osmotic stress or osmotic shock during addition and removal (Parihar et al., 2023). Although the use of CPAs has become a routine in cryopreservation, biochemical and biophysical modes of action, and the optimal applications of the diverse range of solutes that constitute CPAs still need to be further elucidated (Elliott et al., 2017). Moreover, it is necessary

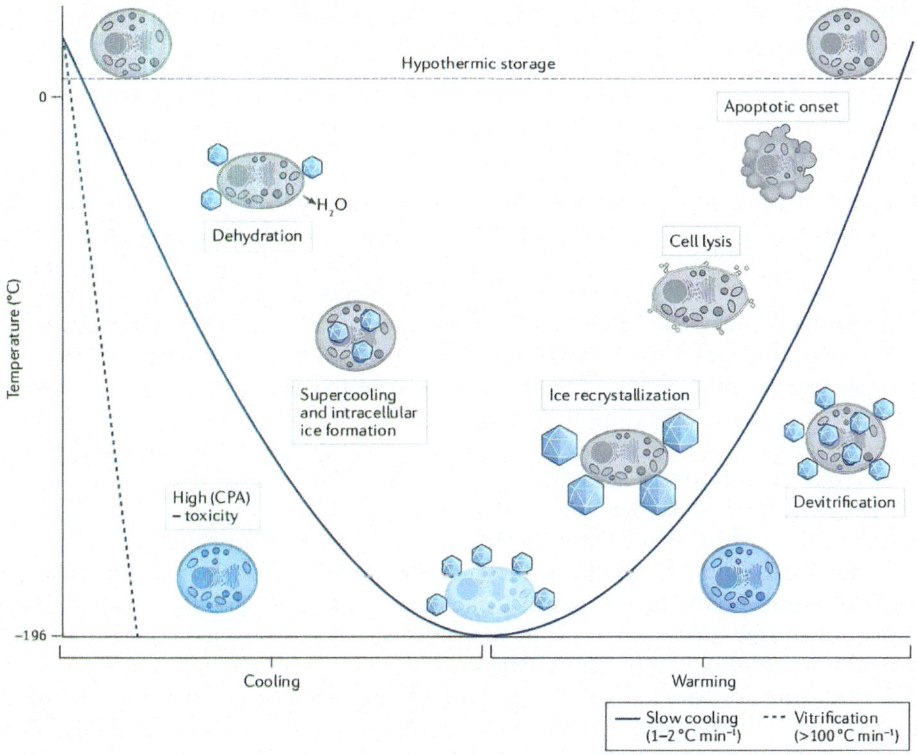

FIGURE 10.1 Potential mechanisms of cellular damage during cryopreservation. Note that these are all extremes, do not necessarily occur simultaneously, and can be partially mitigated by the addition of CPAs. Ice crystals are not drawn to scale for illustrative purposes. *CPAs*, Cryoprotectant agent. *Reproduced with permission from Murray and Gibson (2022), by Springer Nature.*

to highlight that different cell types or even the same cell type from different species require specific and optimal concentrations of CPAs and methods for loading and unloading them to achieve successful cryopreservation (Ribeiro et al., 2022).

In general, CPAs are divided into permeable and nonpermeable categories according to their ability to cross the cellular membrane (Ribeiro et al., 2022). Permeable CPAs are commonly organic compounds with a low molecular weight and high amphiphilic properties, which can permeate passively through the plasma membrane (Ribeiro et al., 2022). The most widely used permeable CPAs are glycerol, ethylene glycol (EG), propylene glycol (PG), dimethyl sulfoxide (DMSO), and methanol. The amount of CPA that has permeated intracellularly is dependent on the CPAs used and the addition procedure. CPA addition (and removal) is a kinetic process of the movement of CPA and water across the cell membrane (Murray & Gibson, 2022). It is determined by the intrinsic membrane permeability and external conditions, such as the loading CPA concentration, duration of addition, and temperature of addition (Elliott et al., 2017; Fahy & Wowk, 2015). Nonpermeable CPAs are high molecular weight compounds that are usually highly hydrophilic (Ribeiro et al., 2022). These CPAs generally protect cells from high osmotic stresses with less toxicity (Murray & Gibson, 2022). The most common nonpermeable CPAs are polymers (e.g., polyvinylpyrrolidone and polyethylene glycol) and sugars (e.g., sucrose and trehalose). Commonly, combinations of permeable and nonpermeable CPAs are used to reduce cryopreservation-induced osmotic damage in cells (Martínez-Páramo et al., 2017).

10.3.2 Cooling

Cooling is a critical step for germplasm cryopreservation. In general, the rate of cooling controls the rate at which water is converted to ice. Consequently, it controls the changing rate and balance of intracellular and extracellular solutes (Murray & Gibson, 2022; Pegg, 2015). Ice first forms outside the cells, causing the buildup of solutes in the extracellular solution. As a result, water moves from inside the cells to the extracellular solution to maintain equilibrium, causing the cells to shrink and dehydrate (Bojic et al., 2021). Therefore the rate of cooling influences the rate at which water is transported out of the cells as a consequence of changes in the osmolality of the surrounding fluid (Pegg, 2015).

When the cooling rate is too fast, the formation of extracellular ice crystals leads to an osmotic imbalance across the cell membrane (Murray & Gibson, 2022). This causes the cell's water to supercool by an amount proportional to the cooling rate and inversely proportional to the permeability of the cell to water (Murray & Gibson, 2022; Parihar et al., 2023). Supercooling cannot be maintained, and eventually water freezes internally, usually resulting in an increased chance of fatal intracellular ice formation (Mazur, 2004; Murray & Gibson, 2022; Pegg, 2015). The fact that a cell can supercool at all in the presence of external ice means that the plasma membrane is at least initially a barrier to the growth of ice from the exterior to the interior, yet intracellular freezing appears to be dependent on the presence of external ice (Mazur, 2004).

Controlled slow cooling utilizes the osmotic regulation of the cell and freeze-induced dehydration to increase the viscosity of the cell's cytoplasm so that it vitrifies when plunged into LN (Parihar et al., 2023; Reed & Uchendu, 2008). Often, programmable rate freezers specifically designed for cryopreservation are used. Cells in the CPA solution are cooled at a slow rate to an intermediate temperature. Below the freezing point of the cryoprotectant solution, ice nucleation is initiated, and ice forms in the supercooled cryoprotectant solution and intercellular spaces (Murray & Gibson, 2022; Parihar et al., 2023). The cytoplasm remains unfrozen because of the solute concentration, especially electrolytes (Mazur, 2004; Pegg, 2015). As the temperature is further decreased to −35°C – −40°C, the extracellular solution becomes increasingly icy, and the intracellular solutes become highly concentrated. Cells lose water to their extracellular surroundings, and the cytoplasm is further concentrated. The intracellular freezable water is safely reduced before cells are plunged into LN. If the cells are optimally dehydrated, the cytoplasm is vitrified on contact with LN. If the cells are underdehydrated, leaving additional water in the cytoplasm, ice will form inside the cells, and this is often lethal (Reed & Uchendu, 2008).

Vitrification is an alternative approach to controlled cooling. Vitrification is a super-rapid cooling process that produces a glass-like solidification of living cells, completely avoiding ice crystal formation during cooling (Fahy & Wowk, 2015; Kuleshova & Lopata, 2002). Subsequently, the thawing of vitrified samples also needs to be super-fast to completely avoid ice crystal formation during warming (Fahy & Wowk, 2015). Vitrification generally requires high concentrations of CPAs or very low volumes to achieve these rapid cooling and thawing rates. The high concentration of CPAs required for vitrification can have detrimental effects from chemical toxicity and/or osmotic damage during loading and unloading

(Murray & Gibson, 2022; Parihar et al., 2023). Multistep CPA addition and removal procedures can be used to reduce these osmotic effects (Fahy & Wowk, 2015; Kuleshova & Lopata, 2002). The most common vitrification protocol consists of two loading steps, placing cells into a pretreatment solution for a specific duration and then into the final vitrification solution for another specific duration before plunging into LN (Liu et al., 2012). In human and domestic animals, one to a few oocytes/embryos are vitrified per run in small volumes using specially designed devices such as microcapillary straws, heat-softened and pulled straws, thin film of a nylon loop, aluminum foil, and electron microscope gold/copper grids (Cuello et al., 2021; Jia et al., 2020; Jin et al., 2014; Kuleshova & Lopata, 2002; Sciorio et al., 2023). In farmed molluscan species, hundreds of thousands to millions of oocytes/embryos are needed to meet the requirements of genetic improvement programs and hatchery production. Furthermore, the survival rate from embryo to adult in farmed molluscan species is normally low. Therefore to practically apply vitrification in farmed molluscan species, the existing devices need to be modified, or new devices and processes need to be invented to align with the requirements for a high number of oocytes/embryos/larvae in the industry.

10.3.3 Thawing

The goal of cryopreservation is to retain the biological properties of cells after thawing. During warming, devitrification and ice recrystallization can occur, where ice crystals grow and cause mechanical damage and cell lysis (Murray & Gibson, 2022). Therefore thawing procedures for cryopreserved samples are equally as important as cooling (Mazur, 2004; Pegg, 2015).

Commonly, a fast-thawing rate is beneficial for rapidly cooled cells. Higher cooling rates form smaller internal ice crystals, and small crystals appear to be less damaging than larger crystals (Parihar et al., 2023). If the warming rate is too slow, small crystals can enlarge to a damaging size through the process of recrystallization. In addition, at a slow cooling rate, if dehydration has been adequate, the remaining intracellular water vitrifies. If the warming rate is too slow, this glassy water may devitrify (i.e., freeze), resulting in ice crystal formation with lethal consequences (Mazur, 2004; Pegg, 2015).

Laser warming is a super-fast thawing approach that can generate rates on the order of millions of °C min^{-1}, thereby exceeding the critical warming rates needed to prevent ice crystallization (Khosla et al., 2020). Laser warming can aid in the rescue of cryoprotectant-sensitive cells and larvae and provide rapid and uniform warming rates to minimize damage caused by ice crystallization (Khosla et al., 2019). This has been successfully used in mouse oocyte (Jin et al., 2014) and coral larvae (Narida et al., 2023) cryopreservation.

10.4 Key parameters of gamete and larval cryopreservation in marine mollusks

10.4.1 Extenders and cryoprotectant agents

Extenders are the extracellular solutions that are mixed with CPA chemicals to make CPA solution. Suitable extenders support postthaw cellular viability and are also sometimes used for short-term storage of cells. In sperm cryopreservation over the last two decades, seawater has been the most common extender used in marine molluscan species, including both natural and artificial seawater. For example, natural seawater was applied in the Pacific oyster (Zhang et al., 2012), flat oyster (*O. angasi*, Hassan et al., 2017a, 2017b and *Ostrea edulis*, Vitiello et al., 2011), pearl oyster (*Pinctada fucata martensii*; Table 10.1), blue mussel (Table 10.1), bay scallop (*Argopecten irradians irradians*; Ma et al., 2018), yesso scallop (*Patinopecten yessoensis*; Wang et al., 2024), greenlip abalone (*Haliotis laevigata*; Table 10.1), blacklip abalone (*H. rubra*; Liu et al., 2015b), pinto abalone (*H. kamtschatkana*; O'Brien, 2019), and Pacific abalone (*H. discus hannai*; Table 10.1), whereas artificial seawater has been used in Portuguese oysters (*C. angulata*; Table 10.1), Pacific oyster (Riesco et al., 2019) and small abalone (*H. diversicolor*; Gwo & Lei, 2023). Hanks' balanced salt solution (HBSS) has been used when sperm were collected by strip spawning in Pacific oyster (Dong et al., 2005a, 2005b; 2006; 2007a, 2007b), eastern oysters (*C. virginica*; Huo et al., 2022; Yang et al., 2021), and scallops (*Chlamys nobilis*; Zheng et al., 2019, and *Mimachlamys nobilis*; Li et al., 2024; 2025). The benefits of using HBSS extender for sperm to maintain motility, fertilizing ability, and larval production may be because of its ability to inhibit sperm activation, resulting in energy conservation in the sperm before freezing (Dong et al., 2005).

DCSB4 is another saline solution that has only been used in great scallops (*Pecten maximus*; Suquet et al., 2016a, 2016b). In addition, high-purity water such as ultrapure water, Milli-Q water, and distilled water has been used in eastern oyster (Paniagua-Chávez et al., 2006), green-lipped mussel (Smith et al., 2012a, 2012b), pearl oysters (*Pinctada margaritifera*; Demoy-Schneider et al., 2018; Hui et al., 2011), and blackfoot abalone *H. iris* (Serean L. Adams, pers.comm).

In contrast, in marine molluscan oocyte cryopreservation over the last two decades, high-purity water (e.g., Milli-Q water) has been mostly recommended as a suitable extender with spat being produced from cryopreserved oocytes in Pacific oysters and blue mussels (Liu & Li, 2015; Tervit et al., 2005; Zhu et al., 2024). For larval cryopreservation over the last two decades, Milli-Q water, distilled water, natural/artificial seawater, and HBSS have all been used as extenders in different species of oysters, mussels, clams, and abalone. However, reasonable percentages of spat have only been reported when Milli-Q water was used as the extender in Pacific oyster and blue mussel larval cryopreservation (Liu et al., 2020a, 2020b; Zhu et al., 2023a).

Among the permeable CPAs used in marine molluscan sperm cryopreservation over the last two decades, DMSO is the most common CPA. It has been used in a range of species, including edible oysters, pearl oysters, scallops, mussels, clams, and abalone, typically with a concentration between 6% and 10 % (Table 10.1). Methanol has been preferentially used in pearl oyster (*P. fucata martensii*) sperm cryopreservation (Arita et al., 2018; Kawamoto et al., 2007; Narita et al., 2008a, 2008b). EG, methanol, and PG have also been used in a few species for sperm cryopreservation, and it appears that the selection of CPA is dependent on the extender, cooling methodology, and thawing temperature used (Table 10.1).

In contrast, only EG has been demonstrated as a suitable permeable CPA in oocyte cryopreservation of marine molluscan species studied to date (Liu & Li, 2015; Tervit et al., 2005; Zhu et al., 2024). Similarly, in larval cryopreservation, EG has been optimized as the preferred permeable CPA in most species, including species of oysters, mussels, and clams (Table 10.2). Although DMSO and PG have also been assessed, lower percentages of postthaw spat have been produced in comparison to embryo and larvae cryopreserved with EG (Table 10.2).

In sperm cryopreservation, nonpermeable CPAs, such as sugars, have often been used in combination with permeable CPAs. For example, trehalose has been used to cryopreserve sperm of edible oysters, pearl oysters, and green-lipped mussels (Table 10.1). Glucose has been used for farmed greenlip and blacklip abalone (Liu et al., 2014a, 2014b; 2015a, 2015b), and yesso scallops (Wang et al., 2024). Sucrose has been used for Pacific abalone (Hossen et al., 2022a, 2022b). Glycine has been used in Pacific oysters (Zhang et al., 2012), blue mussels (Liu et al., 2016), and

TABLE 10.1 Key parameters investigated in marine molluscan sperm cryopreservation in the last two decades (early reviews in Hassan et al., 2015; Liu et al., 2015c).

Species	Extender	CPAs	Cooling protocol	Cryo-container	Thawing temperature	Outcome and/or parameters assessed	Reference
Crassostrea gigas	Ca-free HBSS	8% DMSO	−16°C/min to −140°C (CDBSFM)	0.5 mL straw	40°C (7 s)	96% fertilization (2n); 28% fertilization (4n)	Dong et al. (2005a)
		6% methanol	−5°C/min to −30°C; −45°C/min to −80°C; Held for 5 min			70% motility; 98% fertilization	Dong et al. (2005b)
		6% PEG + 4% PG or 4% DMSO				17% motility (4n); 7% fertilization (4n)	Dong et al. (2006)
		6% methanol; 4% methanol + 2% PEG	CDBSFM			49% fertilization	Dong et al. (2007a)
		10% methanol or 12% DMSO	−5°C/min to −30°C; −45°C/ min −80°C			96% fertilization (2n); 48% fertilization (4n)	Dong et al. (2007b)
	Distilled water	5% DMSO + 0.54M trehalose	Chilled to −75°C with dry ice for 10 min	4.5 mL cryovial	20°C (5–8 min)	90% fertilization	Adams et al. (2008)
	Seawater	6% DMSO + 0.6% glycine	3 cm above LN surface for 20 min	2.0 mL cryovial	60°C	92% fertilization; 35% D-larvae	Zhang et al. (2012)
	Artificial seawater	10% DMSO	−6°C/min from 0°C to −70°C	0.5 mL straw	37°C (10 s)	PMI; DNA; LPO; fertilization and thermal stress-related genes	Riesco et al. (2019)

(continued on next page)

TABLE 10.1 Key parameters investigated in marine molluscan sperm cryopreservation in the last two decades (early reviews in Hassan et al., 2015; Liu et al., 2015c)—cont'd

Species	Extender	CPAs	Cooling protocol	Cryo-container	Thawing temperature	Outcome and/or parameters assessed	Reference
C. virginica	Ultrapure deionized water	5% or 10% PG	−2.5°C/min from 15°C to −30°C; held for 5 min		25 or 70°C (6–7 s)	PMI; functional mitochondria	Paniagua-Chávez et al. (2006)
	Ca-free HBSS	10% DMSO	−25°C/ min from 5°C to −80°C		40°C (8 s)	15% motility (CASA); 95% fertilization	Yang et al. (2012)
						15% motility (CASA); 87% fertilization; PMI	Yang et al. (2013)
			−10–−15°C/min from RM to −80°C		50°C (6 s)	~16% motility; PMI	Yang et al. (2021)
			−10 or −15°C/min from 4°C to −80°C		40°C (8 s); 50°C (6 s); 60°C (5 s);	~25% motility	Huo et al. (2022)
C. angulata	Artificial seawater	10% DMSO	−6°C/min from 0°C to −70°C		37°C (10 s)	> 70% fertilization	Riesco et al. (2017)
						PMI; DNA; LPO; fertilization and thermal stress-related genes	Riesco et al. (2019)
		10% DMSO + 0.45M trehalose				~1% motility (CASA); DNA; PMI; AI; LPO; ROS; SOD; GR; GP; apoptosis	Anjos et al. (2021)
		10% DMSO	−15°C/min from RM to −30°C, held for 3 min	1.5 mL microtube	70°C	~16% motility; 17% D-larvae; PMI; MMP; AI; oxidation level; DNA	Kuo & Gwo (2022)
Ostrea edulis	Seawater	15% EG	−3°C/min from 3°C to −70 °C	2.0 mL cryovial	60°C/min up to 18±2°C in 55±2°C	50% motility	Vitiello et al. (2011)

(*continued on next page*)

		10% DMSO + 0.45M trehalose	−3°C/min from 4°C to −80°C	0.5 mL straw	40°C (8 s)	~44% motility; PMI	Hassan et al. (2017a)
		15% EG + 0.2M trehalose	8 cm above LN surface for 10 min			~62% motility; PMI	Hassan et al. (2017b)
Pinctada margaritifera		5% DMSO+1M trehalose	At −120 °C for 10 min	0.25 mL straw	25°C (30 s)	89% motility (CASA)	Lyons et al. (2005)
		0, 0.64, 1.02, or 1.53 M DMSO + 0.45M trehalose	6 cm above LN surface for 10–20 min			35% motility (CASA)	Acosta-Salmón et al. (2007)
	Distilled water	0.8 M DMSO + 0.7M trehalose	3 cm above LN surface for 10 min	0.5 mL straw	27°C (25 s)	80% fertilization; Eyed larvae (Day 23); PMI	Hui et al. (2011)
		0.8M DMSO + 0.7M trehalose + 10% oyster hemolymph	3 cm above LN surface for 12 min		25°C (30 s)	~54% motility (CASA); O_2 consumption; ATP content; CK; ultrastructure, morphology	Demoy-Schneider et al. (2018)
P. fucata martensii	Seawater + FBS	10% methanol	10 or 12 cm above LN surface to −50°C	0.25 mL straw	18–20°C (15 s)	32%–35% motility	Kawamoto et al. (2007)
			10 cm above LN surface to −50°C			24% motility; ultrastructure	Narita et al. (2008a)
						25% motility; 76% fertilization	Narita et al. (2008b)
			12 cm above LN surface to −50°C		20°C (15 s)	~45% motility	Arita et al. (2018)
	Seawater	5%–15% DMSO, EG, PG				50%–70% motility (CASA)	Zheng et al. (2018)
Mytilus galloprovincialis		7% EG	−0.25 °C/min from 20 °C to 2 °C; −1 °C/min to −15 °C; −6 °C/min to −80 °C	2.0 mL cryovial	60°C/min up to 20±1°C in 25°C	65% motility; 60% fertilization	Matteo et al. (2009)
		8% DMSO + 0.8% glycine	7.8 cm above LN surface for 10 min	0.25 mL straw	60°C	95% fertilization; PMI; AI	Liu et al. (2016)

(continued on next page)

TABLE 10.1 Key parameters investigated in marine molluscan sperm cryopreservation in the last two decades (early reviews in Hassan et al., 2015; Liu et al., 2015c)—cont'd

Species	Extender	CPAs	Cooling protocol	Cryo-container	Thawing temperature	Outcome and/or parameters assessed	Reference
		8% DMSO	−4 °C/min from 2 °C to −30 °C		20°C	91% fertilization; PMI; AI	Liu et al. (2018)
Perna canaliculus	Milli-Q water	12% DMSO + 0.2M trehalose	3 cm above LN surface for 10 min	0.5 mL straw	~18°C (10–20 s)	~85% fertilization	Smith et al. (2012a)
						motility (CASA); ~33% fertilization; DNA; PMI	Smith et al. (2012b)
Nodipecten nodosus		5% PG or 4% or 6% methanol				75%–85% motility; > 70% D larvae	Silvestri et al. (2016)
Pecten maximus	DCSB4	20% PEG	5.5 cm above LN surface for 10 min	0.5 mL straw	25°C (15 s)	~60% motility; 3% D larvae	Suquet et al. (2016a)
						~20% motility (CASA); intracellular ATP; viability; morphology; ultrastructure	Suquet et al. (2016b)
A. irradians irradians	Seawater	13.3% DMSO	10 cm above LN surface for 5 min	1.8 mL cryovial	30°C	86% motility; 21.7% fertilization	Ma et al. (2018)
Chlamys nobilis	Ca-free HBSS	5%–20% DMSO	−10°C/min from 0°C to −80°C, held for 2 min	0.25 mL straw	25°C (15 s)	~30% motility (CASA); 30% fertilization	Zheng et al. (2019)
Mimachlamys nobilis	HBSS	10% DMSO	−5°C/min from 4°C to −25°C, held for 5 min; −10°C/min to −80°C, held for 5 min	2.0 mL cryovial	38°C	~68% motility (CASA); ultrastructure	Li et al. (2024)
						~70% motility (CASA); ultrastructure	Li et al. (2025)

(continued on next page)

Patinopecten yessoensis	Seawater	6% DMSO + 3% glucose	5 cm above LN surface for 10 min	0.25 mL straw	30°C	~44% fertilization; morphology and ultrastructure; PMI; MMP; AI; SOD; catalase; ATPase; CK; SD; DNA	Wang et al. (2024)
Haliotis rufescens		10% glycerol	−16°C/min from 4°C to −140°C, held for 10 min	0.5 mL straw	45°C (8 s)	~10% motility; ~30% fertilization; PMI;	Salinas-Flores et al. (2005)
H. laevigata (wild)	Seawater	6% DMSO	At −60°C for 20 min	2.0 mL cryovial	50°C	95% fertilization	Zhu et al. (2014)
H. laevigata		6% DMSO + 1% glucose + 0.6% glycine	5.2 cm above LN surface for 10 min	0.25 mL straw	60°C (5s)	~35% motility; ~94% fertilization; PMI; MMP; AI	Liu et al. (2014a)
		6% DMSO + 1% glucose				~40% motility; ~85% fertilization; PMI; MMP; AI	Liu et al. (2014b)
						~28% motility; ~80% fertilization; PMI; MMP; AI	Liu et al. (2015a)
		10% DMSO + 0.6% glycine	−5°C/min from 0°C to −30°C		40°C	~20% motility; > 90% fertilization; PMI; MMP; AI	Liu et al. (2016)
H. rubra		6% DMSO + 2% glucose	5.2 cm above LN surface for 10 min		60°C (5s)	~35% motility; ~70% fertilization	Liu et al. (2015b)
H. kamtschatkana(wild)		5% DMSO	−3°C/min to −60°C	0.5 mL straw	40°C (8 s)	~14% motility (CASA); ~12% fertilization;	O'Brien, 2019
H. discus hannai	Seawater	8% DMSO	5 cm above LN surface for 10 min	0.25 mL or 0.5 mL straw	60°C (5 s)	~50% motility; PMI; MMP; AI	Kim et al. (2020)

(*continued on next page*)

TABLE 10.1 Key parameters investigated in marine molluscan sperm cryopreservation in the last two decades (early reviews in Hassan et al., 2015; Liu et al., 2015c)—cont'd

Species	Extender	CPAs	Cooling protocol	Cryo-container	Thawing temperature	Outcome and/or parameters assessed	Reference
				0.25 mL straw		~31% fertilization; ~20% hatching; AI; PMI; MMP; DNA; *PKA-C*, *hsp70* and *hsp90* genes	Hossen et al. (2021b)
				0.5 mL straw		~60% motility; PMI; AI; MMP;	Hossen et al. (2024)
		8% DMSO + 10 μg/mL AFP III			60°C (<4 s)	~61% motility; ~43% fertilization; ~29% hatching; AI; PMI; MMP; DNA; *hsp90* gene	Hossen et al. (2021a)
					60°C (5 s)	~61% motility; ~44% fertilization; ~40% hatching; similar body weight at 18 months and 3 years; AI; MMP; PMI; DNA; growth-, reproduction-, immune-related genes	Hossen et al. (2025)
		8% DMSO + 3% SUC				~60% motility; ~40% fertilization; ~25% hatching; AI; PMI; MMP; DNA; ATP; motility-related genes	Hossen et al. (2022a)

(*continued on next page*)

						~61% motility; ~40% fertilization; ~28% hatching; PMI; MMP; superoxide anion; DNA; catalase; SOD; GSH; LPO; apoptosis, antiapoptosis, antioxidant and HSP genes	Hossen et al. (2022b)
		8% DMSO, 2% glycerol, or 8% EG + AFPIII				*Catalase* gene	Hossen et al. (2023)
H. diversicolor	Artificial seawater	10% DMSO	−15°C/min from 0°C to −30°C, held for 5 min;	1.5 m mL cryovial	70°C (1 min)	~61% motility; ~67% fertilization; PMI; AI; mitochondrial activity; oxidative level; DNA	Gwo & Lei (2023)

AFP III, Antifreeze protein III; *AI*, acrosome integrity; *ATP*, adenosine triphosphate; *CASA*, computer-assisted sperm analysis; *CDBSFM*, commercial dairy bull sperm freezing method; *CK*, creatine kinase; *DMSO*, dimethyl sulfoxide; *EG*, ethylene glycol; *FBS*, fetal bovine serum; *GF*, glutathione peroxidase; *GR*, glutathione reductase; *GSH*, glutathione; *HBSS*, Hanks' balanced salt solution; *HSP*, heat shock protein; *LPO*, lipid peroxidation; *MMP*, mitochondrial membrane potential; *PEG*, polyethylene glycol; *PG*, propylene glycol; *PKA-C*, protein kinase A; *PMI*, plasma membrane integrity; *RM*, room temperature; *ROS*, reactive oxygen species; *SD*, succinate dehydrogenase; *SOD*, superoxide dismutase.

greenlip abalone (Liu et al., 2016) sperm cryopreservation. Furthermore, polyethylene glycol (PEG) has been trialed in Pacific oysters (Dong et al., 2007a) and great scallops (Suquet et al., 2016a, 2016b), while the antifreeze protein III has been used in Pacific abalone (Hossen et al., 2021a; 2023; 2025) sperm cryopreservation.

In marine molluscan oocyte cryopreservation, nonpermeable CPAs are commonly included and may be indispensable, and also species-specific. To date, the disaccharides (e.g., trehalose) and polysaccharides (e.g., Ficoll 70) have produced postthaw spat in Pacific oysters (Tervit et al., 2005) and blue mussels (Liu & Li, 2015; Zhu et al., 2024), respectively, although higher percentages were reported for blue mussels with Ficoll 70. The mechanism of the cryoprotective effects of these saccharides is unclear, although they may stabilize ice crystals and thereby prevent recrystallization (Mao et al., 2018). For example, the highly compact structure of Ficoll 70 could serve as a mechanical barrier to prevent ice growth formation (Mao et al., 2018). In addition, the prevention of recrystallization may also be partially due to the highly hydrophilic surface of the Ficoll 70 molecules, as the spherical surface of this chemical is formed by branched and highly hydrophilic sucrose molecules (Fissell et al., 2007; Mao et al., 2018). Protection may also be related to the ability of these molecules to stabilize proteins (Murray & Gibson, 2022). Such characteristics would enable macromolecules to slow water molecule displacement near their surface (Mao et al., 2018). Recently, the addition of salmon lipid extract in a cryoprotective solution containing Ficoll 70 has further improved the oocyte cryopreservation in blue mussels, doubling the postthaw survival rate to the spat stage (Zhu et al., 2024). The cryoprotective capability of the salmon lipid extract is probably due to the combined effects of lipids and antioxidants (Zhu et al., 2024). The exogenous lipids may support the homeoviscous adaptation of the oocyte cell membrane, providing enhanced cryotolerance (Zhu et al., 2023b), and their antioxidants may mitigate against oxidative stress during cryopreservation (Kowalczyk, 2022).

In marine molluscan larval cryopreservation, both disaccharides (e.g., trehalose and sucrose) and polysaccharides (e.g., Ficoll 70) are the main nonpermeable CPAs that have been investigated. The addition of Ficoll 70 or the replacement of sugar with Ficoll 70 may improve larval cryopreservation in Pacific oysters and blue mussels (Liu et al., 2020a, 2020b). Moreover, its combination with polyvinylpyrrolidone (PVP) has significantly increased the postthaw spat production in these two economically important marine bivalve species. The beneficial role of PVP is unclear, although it may be because of its ability to lower external salt concentrations and alter the physical properties of solutions during cryopreservation (Guha & Devireddy, 2010; Knorpp et al., 1967).

10.4.2 Cooling

Both programmable-controlled freezing and nonprogrammable freezing techniques have been applied in marine molluscan gamete and larval cryopreservation. The former requires special equipment to strictly control cooling rates and can be expensive, while the latter cannot strictly control cooling rates, but the technique is low-cost (commonly using styrofoam boxes with racks) and easy to operate on-site (Martínez-Páramo et al., 2017).

In marine molluscan sperm cryopreservation, both freezing techniques have been used. Nonprogrammable freezing techniques have involved placing samples at a level 3–12 cm above the LN surface for 5–20 min before plunging into LN. This technique has been used in Pacific oysters (Zhang et al., 2012), flat oyster *O. angasi* (Hassan et al., 2017b), pearl oysters (Table 10.1), mussels (Liu et al., 2016; Smith et al., 2012a, 2012b), scallops (Ma et al., 2018; Suquet et al., 2016a, 2016b; Wang et al., 2024), and abalone (Table 10.1). Programmable freezers have been frequently

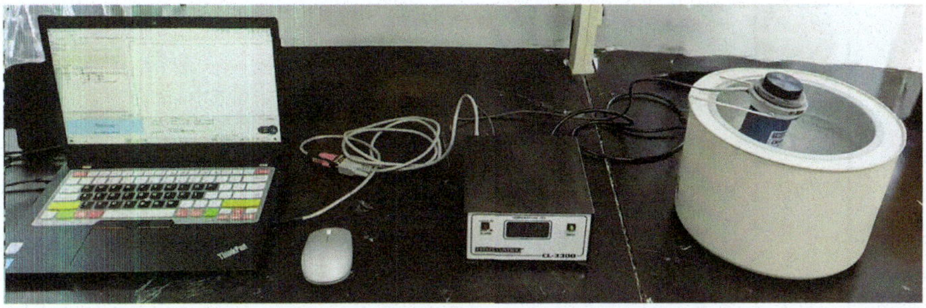

FIGURE 10.2 Oocyte and trochophore larvae cryopreservation uses controlled programmable freezers to cool at slow rates (Cryologic, Australia). The device consists of a temperature controller (middle), a cryobath (right), and a cryochamber (inside the cryobath). The freezing protocol is set and monitored by CryoGenesis software (V5) installed in a computer (left). Photo provider: Dr Yibing Liu.

used for edible oysters and scallops, *Chlamys nobilis* and *Mimachlamys nobilis* (Table 10.1). In addition, sperm vitrification has been developed for the dwarf surfclam (*Mulinia lateralis*), a potential bivalve model species. For this species, the D larval rate derived from vitrified sperm (82%) was comparable to that from fresh sperm (85%; Yibing Liu, pers.comm).

Oocyte and larval cryopreservation requires controlled programmable freezing because much slower cooling rates are needed to ensure cell survival (Fig. 10.2). During cooling, samples are commonly cooled at −1 or −2.5°C/min from 0–4°C to −10 or −12°C (to allow ice to seed either naturally or manually), then cooled more slowly at −0.3 to −1°C/min to −30 or −35°C before being plunged into LN (Tables 10.2 and 10.3). Although oocyte and embryo vitrification have been investigated in Pacific oysters (Chao et al., 1997) and blue mussels (Wang et al., 2014), postthaw survival remains unachievable.

10.4.3 Thawing

The rate of thawing that cells experience is dependent on the thawing temperature that is used, whether the sample is thawed in air or in a water bath, and the volume and shape of the cryo-container (e.g., straws versus cryovials). In marine molluscan sperm cryopreservation, a wide range of thawing temperatures has been successfully used (Table 10.1). For example, for Pacific oyster sperm, water baths over a range of 20°C–60°C for 7 s–8 min have been used. For eastern oyster sperm, a range of 20°C–70°C for 5 s–8 s has been used. For Portuguese oyster sperm, temperatures of 37°C and 70°C have been used. Similarly, for flat oyster *O. angasi*, and pearl oysters *P. margaritifera* and *P. fucata martensii*, thawing temperature and duration were within the range used for sperm of *Crassostrea* species. Likewise, for mussel and scallop sperm, this low and medium thawing temperature was suitable for mussel and scallop sperm cryopreservation, as temperatures between 18°C and 60°C have been applied. For abalone sperm, 40°C thawing temperature was used when the sperm were cryopreserved in a programmable controlled freezing technique (Liu et al., 2016; O'Brien, 2019), whereas 60°C thawing temperature was frequently used when the sperm were cryopreserved in a non-programmable freezing technique in greenlip, blacklip, and Pacific abalone (Table 10.1).

In marine molluscan oocyte cryopreservation, a water bath thawing temperature of 28°C was commonly used in bivalves (Table 10.3). In larval cryopreservation, a range of 20°C–50°C thawing temperatures of 5 s–30 s duration has been used (Table 10.2). However, only thawing

TABLE 10.2 Key parameters investigated in marine molluscan larval cryopreservation in the last two decades (see reviews by Yang and Huo, 2022, for summaries of earlier bivalve studies).

Species	Larval stage	Extender	CPAs	Cooling protocol	Cryo-container	Thawing temperature	Outcome and/or parameters assessed	Reference
Crassostrea gigas (*Magallana gigas*)	10 hpf T; 4 dpf D; 8 dpf U	Artificial seawater	2M EG + 0.2M sucrose	−1°C/min from 0°C to −12°C, seeding and held for 10 min, −1°C/min to −35°C, held for 30 min	0.5 mL straw	20°C (30 s)	Survival: 63.8% in T; 84.1% in D; 56.3% in U	Kim et al. (2013)
	12 dpf U		2M or 2.5M EG + 0.2M sucrose	−1°C/min from 0°C to −12°C, seeding and held for 10 min, −1°C/min to −35°C		25°C (10 s)	99% survival; ultrastructure	Choi and Chang (2014)
	24 hpf D	Milli-Q water	10% EG + 0.2M sucrose+ 1% PVP	−1°C/min from 0°C to −10°C, held for 5 min, −0.3°C/min to −35°C		37°C (10 s)	Survival: 83% (Day 2); 2.2% (Day 7)	Suquet et al. (2012)
	13±2 hpf T; 24±2 and 43±2 hpf D						survival (Day 982): 0.05% (13±2 hpf); 0.15% (24±2 hpf); 0.18% (43±2 hpf)	Suquet et al. (2014)
	24 hpf D			−2.5°C/min from 0 to −10°C, held for 5 min, −0.3°C/min to −20°C, −2.5°C/min to −35°C			~10% survival (Day 12)	Labbé et al. (2018)
	14 hpf T		10% EG	0°C for 5 min, −1°C/min to −10°C, held for 5 min, −0.5°C/min to −35°C, held for 5 min	0.25 mL straw	28°C	60% D larvae	Paredes et al. (2013)

(continued on next page)

	22 hpf D		10% EG + 0.4M sucrose + 1% PVP	0°C for 5 min, −1°C/min to −10°C, held for 5 min, −1°C/min to −35°C			Survival: <5% (Day 11); feeding rate; ultrastructure; organogenesis	Suneja et al. (2014)
	18 hpf T		10% EG + 5% Ficoll + 0.2% PVP	0°C for 5 min, −1°C/min to −10°C, −0.3°C/min to −34°C			~5% spat	Liu et al. (2020a)
							Antioxidant; ROS; autophagy, apoptosis, and antioxidant-related genes	Liu et al. (2022b)
							DNA methylation, epigenetic regulator genes, and histone modification	Liu et al. (2023)
	14 hpf T		10% EG + 5% Ficoll + 0.2% PVP + 0.25 mg mL^{-1} POPC + 1 mg mL^{-1} tocophero				~8.5% spat	Zhu et al. (2023a)
C. angulata	24–30 hpf D	Milli-Q water	10% EG + 0.2M sucrose + 1% PVP	−2.5°C/min from 0°C to −10°C, held for 5 min, −0.3°C/min to −20°C, −2.5°C/min to −35°C	0.5 mL straw	37°C (10 s)	~76% normal D larvae; SOD, GP, GR; movement	Anjos et al. (2022)
			10% DMSO + 0.2M sucrose + 1% PVP				611 differentially expressed genes	Anjos et al. (2024)
Saccostrea glomerate	24 hpf larvae	Seawater	10% DMSO	−7°C/min from 21°C to 0°C, held for 15 min, −2.5°C/min to −36°C, held for 10 min		30°C (10–15 s)	93% survival	Liu and Li (2008)

(continued on next page)

TABLE 10.2 Key parameters investigated in marine molluscan larval cryopreservation in the last two decades (see reviews by Yang and Huo, 2022, for summaries of earlier bivalve studies)—cont'd

Species	Larval stage	Extender	CPAs	Cooling protocol	Cryo-container	Thawing temperature	Outcome and/or parameters assessed	Reference
Ostrea edulis	Larvae	Distilled water	10% DMSO or 10% EG + 0.4M sucrose or 0.4M trehalose	−1°C/min from 0°C to −12°C, held for 2 min, −1°C/min to −35°C; or −2.5°C/min from 0°C to −10°C, held for 5 min, −0.3°C/min to −20°C, −2.5°C/min to −35°C		37°C (12 s)	Survival: ~80% (20 min postthawing); ~60% (24 h postthawing)	Horváth et al. (2024)
Mytilus trossulus	22 hpf T	Seawater	7% DMSO	0°C for 10 min, −6–−8°C/min to −33°C	2.0 mL cryovial	30°C	Caspase; apoptosis; PMI; mitotic cells; ultrastructure	Odintsova et al. (2017)
M. galloprovincialis	30 hpf D	Milli-Q water	5% DMSO	0°C for 15 min, −1.5°C/min to −7°C, held for 10 min, −0.4°C/min to −33°C	0.25 mL straw	26°C (5 s)	1.1% survival (Day 21)	Wang et al. (2011)
	20 hpf T		10% EG + 0.2M trehalose	0°C for 5 min, −1°C/min to −10°C, held for 5 min, −0.5°C/min to −35°C, held for 5 min		28°C	80% D larvae	Paredes et al. (2013)
	25 hpf T		10% EG + 7.5% Ficoll + 0.2% PVP	0°C for 5 min, −1°C/min to −10°C, −0.3°C/min to −34°C			~21% spat	Liu et al. (2020b)
	25 hpf T			0°C for 5 min, −1°C/min to −10°C, −0.3°C/min to −34°C			ROS; antioxidant; apoptotic and antioxidant enzyme-related genes	Liu et al. (2022a)

(*continued on next page*)

	18–20 hpf T	Seawater	10% EG + 0.2 M trehalose	4°C for 2 min, −1°C/min to −12°C, held for 5 min, −1°C/min to −35°C		35°C (6 s)	~50% D larvae	Rodriguez-Riveiro et al. (2019)
	48 and 72 hpf D		10% EG (48 hpf); 10% EG + 0.4M trehalose (72 hpf)	4°C for 2 min, −1°C/min to −12°C, seeding and held for 2 min, −1°C/min to −35°C			Normal larvae: ~7% (48 hpf); ~93% (72 hpf)	Heres et al. (2019)
	72 hpf D		10% EG + 0.4M trehalose	4°C for 2 min, −1°C/min to −12°C, seeding and held for 2 min, −1°C/min to −35°C			~4% and 0.5% spats in cryopreserved F1 and F2, respectively	Heres et al. (2022)
	40 and 45 dpf juveniles		12% EG + 0.4M trehalose	4°C for 2 min, −1°C/min to −12°C, seeding and held for 2 min, −1°C/min to −35°C	0.5 mL straw	35°C (8 s)	Survival: ~60% (40 dpf); ~10% (45 dpf)	Lago et al. (2025)
Perna canaliculus	16–20 hpf T	Milli-Q water	10% EG + 0.4M trehalose	0°C for 5 min, −1°C/min to −10°C, held for 5 min, −0.5°C/min −35°C, held for 5 min	0.25 mL straw	28°C	~3% survival (Day 18)	Paredes et al. (2012)
	16 hpf T; 48 hpf D			0°C for 5 min, −1°C/min to −10°C, seeding and held for 5 min, −0.5°C/min to −35°C			<1% survival (Day 18); shell formation; organo- and neuro-genesis; feeding ability	Rusk et al. (2020)

(continued on next page)

TABLE 10.2 Key parameters investigated in marine molluscan larval cryopreservation in the last two decades (see reviews by Yang and Huo, 2022, for summaries of earlier bivalve studies)—cont'd

Species	Larval stage	Extender	CPAs	Cooling protocol	Cryo-container	Thawing temperature	Outcome and/or parameters assessed	Reference
	48 hpf D		14% EG + 0.6 M trehalose + 1% PVP	0°C for 5 min, −1°C/min to −10°C, seeding and held for 5 min, −0.5°C/min to −35°C, held for 5 min			>50% survival (Day 4)	Heres et al. (2020)
Venerupis corrugata, Ruditapes decussatus R. philippinarum	18-20 hpf T; 48 and 72 hpf D	Seawater	10% EG + 0.4M trehalose	4°C for 2 min, −1°C/min to −12°C, seeding and held for 2 min, −1°C/min to −35°C		35°C (6 s)	Survival: ~50% (18–20 hpf); ~80% (48 and 72 hpf)	Heres et al. (2021)
Cerastoderma edule	72 hpf D						~36% survival (Day 9)	Heres et al. (2023)
Mercenaria mercenaria	15 hpf T	Ca-free HBSS	5% or 10% DMSO or PG	−5°C/min from 4 to −80°C	0.5 mL straw	50°C (6 s)	~30% D larvae	Simon & Yang (2018)
Chamelea gallina	24–30 hpf D	Milli-Q water	10% DMSO + 1% PVP + 0.2M sucrose	−2.5°C/min from 0°C to −10°C, held for 5 min, −0.3°C/min to −20°C, −2.5°C/min to −35°C		37°C (10 s)	~90% normal D larvae; SOD, GP, GR activity; movement	Anjos et al. (2022)
Haliotis midae	5 hpf morula	Seawater	5%–10% PEG				~84% - 94% trochophore	Roux et al. (2008)

D, D larvae; *dpf*, days post fertilization; *GP*, glutathione peroxidase; *GR*, glutathione reductase; *hpf*, hours post fertilization; *PEG*, polyethylene glycol; *PMI*, plasma membrane integrity; *POPC*, 1-palmitoyl-2-oleoyl-sn-glycero-3-phosphocholine; *PVP*, polyvinylpyrrolidone; *ROS*, reactive oxygen species; *SOD*, superoxide dismutase; *T*, trochophore larvae; *U*, umbo larvae.

TABLE 10.3 Key parameters investigated in marine molluscan oocyte cryopreservation in the last two decades.

Species	Extender	CPAs	Cooling protocol	Cryo-container	Thawing temperature	Outcome and/or parameters assessed	Reference
Crassostrea gigas	Milli-Q water	10% EG	0°C for 5 min, −1°C/min to −10°C, held 5 min, −0.3°C/min to −35°C	0.25 mL straw	28°C	0.8%–74.5% fertilization; 0.1 - 30.1% D larvae	Tervit et al. (2005)
					28°C (~10 s)	Diet conditioning increases post-thaw fertilization	Adams et al. (2013)
Perna canaliculus		10% EG + 0.2M trehalose	0°C for 5 min, −1°C/min to −10°C, held for 10 min, −0.5°C/min to −35°C		28°C	~22% fertilization; <1% D larvae	Adams et al. (2009)
					28°C (20–30 s)	~6% fertilization; protein carbonyls; LHP; SOD; catalase; GP; RG; TG	Gale et al. (2014)
Mytilus galloprovincialis		1.5M EG + 0.2M trehalose	0°C for 10 min, −1°C/min to −7°C, held 10 min, −0.4°C/min to −33°C, held for 5 min		26 °C (5 s)	~32% fertilization, <1% D larvae	Wang et al. (2014)

(*continued on next page*)

TABLE 10.3 Key parameters investigated in marine molluscan oocyte cryopreservation in the last two decades—cont'd

Species	Extender	CPAs	Cooling protocol	Cryo-container	Thawing temperature	Outcome and/or parameters assessed	Reference
		10% EG + 7.5% Ficoll	0°C for 5 min, −1°C/min to −10°C, −0.3°C/min to −34°C		28 °C (8 s)	~47% fertilization; 12% D larvae; ~ 5% spat	Liu and Li (2015)
		10% EG + 7.5% Ficoll + 0.5 mg/mL salmon lipid extract				~10% D larvae; 4.5% spat	Zhu et al. (2024)
C. virginica		2M DMSO	−1.5°C/min to −30°C, held for 5 min		35 °C (10 s)	No embryonic development	Paniagua-Chávez et al. (2006)
Haliotis diversicolor	Seawater	2M DMSO	−1.5°C/min from 20°C to −50°C, not in LN	1.5 mL tubes	RM	~16% trochophore; intracellular ice	Yang et al. (2013)
H. rufescens		0.4 M trehalose	−2°C/min from 15°C to −8°C, −0.3°C/min to −30°C, −50°C/min to −150°C	0.5 mL pipette	RM 30s, then 30 °C (40 s); 30 °C (40 s); 50 °C (13 s); 60 °C (10 s); 70 °C (13 s or 9 s)	Viability (fluorescein diacetate)	Torrez et al. (2015)

GP, Glutathione peroxidase; *LHP*, lipid hydroperoxides; *RG*, reduced glutathione; *RT*, room temperature; *SOD*: superoxide dismutase; *TG*, total glutathione.

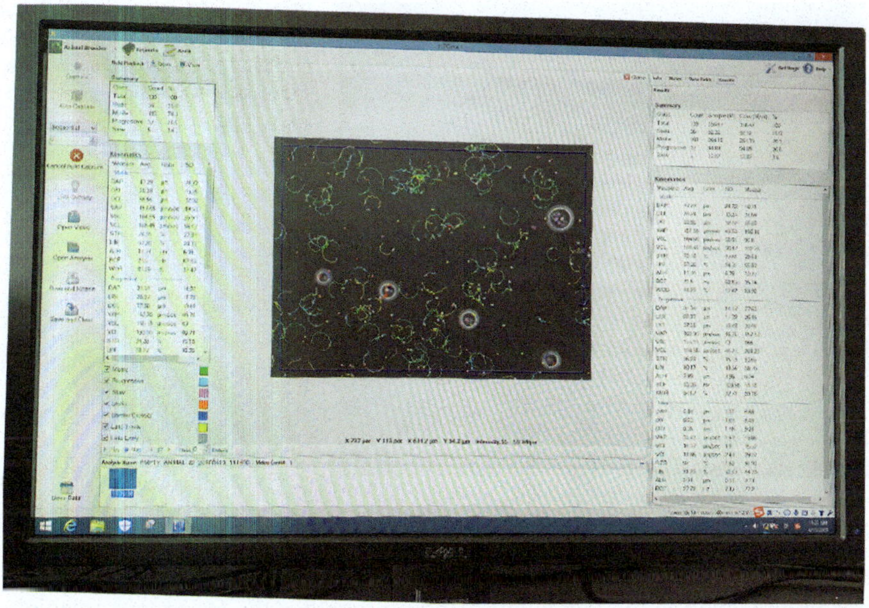

FIGURE 10.3 A computer screen showing the results of yesso scallop fresh sperm analyzed by CASA systems (IVOS II, IMV). Photo provider: Dr Yibing Liu.

at 28 °C produced a reasonable percentage of spat in Pacific oysters and blue mussels (Liu et al., 2020a, 2020b; Zhu et al., 2023a).

10.5 Postthaw gamete and larval quality assessment in marine mollusks

10.5.1 Sperm

Motility is often used as a key metric to assess the quality of postthaw sperm. Motility can be measured either by directly counting the percentage of motile sperm in a subsample or by (CASA) systems (Fig. 10.3), which are able to assess a large number of sperm, distinguish motility patterns, and discriminate morphometric characteristics (Martínez-Páramo et al., 2017). Fertilization measures the sperm's capability to produce viable embryos and is the most important and accurate parameter for assessment of sperm quality in cryopreservation (Martínez-Páramo et al., 2017; Fig. 10.4). Although these two parameters are generally used to measure postthaw sperm quality, they are unable to reveal the mechanism(s) of cryodamage. For this, other measures have been employed, including those that assess the structure, function, and integrity of membranes, acrosome, mitochondria, nuclei, and DNA, as well as physiological and molecular changes.

The sperm plasma membrane is predominantly composed of a lipid bilayer made up of phospholipids, cholesterol, and proteins (Castro et al., 2025). Membrane fluidity is essential for functional performance and may be influenced by temperature (Castro et al., 2025). Published studies indicate that membranes are one of the primary sites of cryodamage in sperm

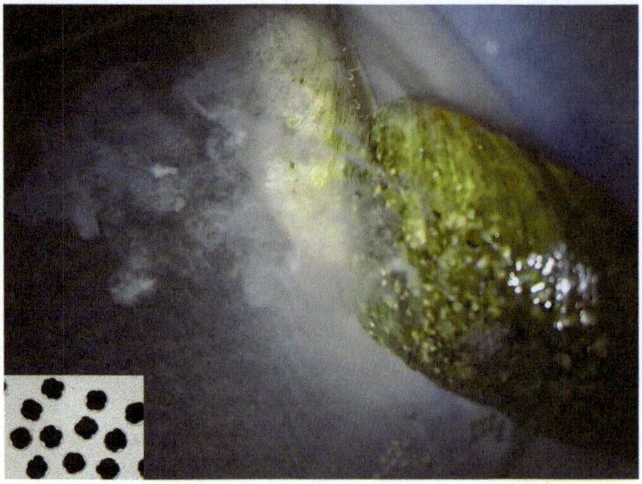

FIGURE 10.4 Embryos (insert) from a natural spawning of greenlip abalone oocytes fertilized with cryopreserved sperm. Photo shows oocytes being released. Photo provider: Dr Yibing Liu.

(Partyka et al., 2012). Lipid peroxidation (LPO) products generated from oxidative stress are thought to be one of the key factors altering membrane integrity (Khan et al., 2021).

The acrosome reaction is the key event bringing together the fusogenic areas of sperm and oocytes during fertilization in marine bivalve species (Carlisle & Swanson, 2021). It involves the fusion of the plasma membrane of both gametes and the release of hydrolytic enzymes from sperm, which penetrate the egg envelope (Carlisle & Swanson, 2021). Therefore maintaining membrane and acrosomal integrity of sperm is vital during cryopreservation.

Mitochondria are the fundamental organelle for cell metabolism and play a crucial role in sperm to fertilize oocytes. Cryopreservation may compromise mitochondrial function, resulting in elevated intracellular oxidative stress that decreases sperm competence and the dynamics of fertilization and embryo development (Gualtieri et al., 2021).

The integrity of sperm DNA is of prime importance for the paternal genetic contribution to normal offspring (Estudillo et al., 2021). Damaged DNA can have a significant negative impact on oocyte fertilization and subsequent development. Morphological evaluations by scanning and/or transmission electron microscopes and specific fluorescence dyes observed by fluorescence microscopy or flow cytometry have been applied to directly demonstrate cryodamage of sperm structures and organelles (Martínez-Páramo et al., 2017).

Cryopreservation can also cause physiological changes to sperm, and oxidative stress is a key aspect of this. Antioxidants are molecules capable of preventing, delaying, or inhibiting the oxidation of other molecules, thus mitigating the oxidative damage (Kowalczyk, 2022). The antioxidant defense of biological systems can be mediated by both enzymatic and nonenzymatic antioxidants that can be assessed using commercial kits (Kowalczyk, 2022). Glutathione (GSH) is a nonenzymatic antioxidant and is the most abundant antioxidant in almost all cell types, participating directly in the neutralization of oxidative damage. Both superoxide dismutase (SOD) and catalase are enzymatic antioxidants that can quickly neutralize reactive molecules through different reactions. Glutathione peroxidase (GP) is an enzyme that catalyzes the reduction of hydrogen peroxide and organic peroxides, including peroxides of phospholipids in sperm (Kowalczyk, 2022). GP protects sperm's DNA against oxidative stress and is also involved in the chromatin condensation process. Creatine kinase (CK) is another enzyme that catalyzes the regeneration of ATP from the chemical shuttle between creatine and creatine phosphate. This particular enzyme is a marker of sperm maturity and correlates with sperm fertilizing capacity (Kowalczyk, 2022). Therefore CK is indispensable for sperm function.

In marine mollusks, the impact of cryopreservation on gene expression has been examined in a handful of studies (Hossen et al., 2021a, 2021b; 2022a, 2022b; 2023; 2025; Liu et al., 2022a, 2022b; Riesco et al., 2019). However, only one study has employed omics analyses coupled with next-generation sequencing (NGS) to explore cryodamage in the sperm of pearl oyster *P. fucata martensii* (Xu et al., 2026).

10.5.2 Oocyte

In marine molluscan oocyte cryopreservation, fertilization rate is a direct measure that could assess the postthaw oocyte quality immediately. However, published results show that it is important to investigate the ability of oocytes to develop further, for example, to D-stage larvae and beyond, to truly demonstrate that a cryopreservation method is successful (Adams et al., 2009; Tervit et al., 2005; Liu & Li, 2015; Zhu et al., 2024). As oocyte cryopreservation is very challenging in mollusks, it would be beneficial if more parameters could be developed to understand cryodamage in these cells.

10.5.3 Larvae

Although both trochophore and D-stage larvae have been successfully cryopreserved in a few bivalve species, reasonable numbers of spat have only been reported in a few studies (Liu et al., 2020a, 2020b; Zhu et al., 2023a). In bivalve larval cryopreservation, the key parameters used to assess postthaw larval performances are their swimming activity, feeding rate, morphology, early-stage survival rates, and postthaw development (Yang & Huo, 2022). One study has applied RNA-seq to reveal cryodamage of larvae at a molecular level in Portuguese oysters (Anjos et al., 2024). For larval cryopreservation to be fully implemented, the performance of resulting progenies over at least one complete life cycle needs to be assessed.

10.6 Importance of applying omics technologies for postthaw cellular quality assessment

It is important to guarantee that postthaw germplasm and the offspring generated from it have no genetic instability. With the advent of NGS, omics (e.g., transcriptomics, proteomics, and metabolomics) technologies have been applied to identify specific biomarkers that ensure the quality of material before cryopreservation, as well as explore postthaw molecular alterations correlated with postthaw fertilizing capability and further development (Çiftçi & Kaya, 2024). These technologies have been widely used in human, livestock, and finfish gamete and embryo cryopreservation to elucidate molecular mechanisms of cryodamage. However, investigation of marine mollusks has been more limited.

10.6.1 Transcriptomics

Transcriptomics examines the mRNA produced by DNA transcription in a cell over a certain period, which can change with environmental factors, such as temperature (Çiftçi & Kaya, 2024). Therefore transcriptomic techniques could pave the way to understand the cellular processes and molecular adjustments that provide tolerance to the multiple stresses that cells are exposed to during cryopreservation.

The mature sperm has usually been considered as a dormant cell that only carries the paternal genome to the oocyte. However, in recent years, the biological complexity of sperm has become more evident due to the discovery of a rich sperm RNA population that has functional roles in fertilization and early embryonic development in livestock species (Khan et al., 2021). These studies have shown a sperm-specific transcriptome with a large population of transcripts that mainly present at low levels and are highly fragmented (Khan et al., 2021). Sperm contain complex populations of RNAs, including messenger RNA (mRNA), ribosome RNAs (rRNA), mitochondrial RNAs (mtRNAs), and microRNAs (miRNAs). Increasing evidence has demonstrated that the transcriptomes, such as mRNA and miRNA, can be influenced by cryopreservation. For example, Card et al. (2017) detected differences in mRNA transcripts of cryopreserved sperm between high-fertility and low-fertility bulls by RNA-seq, with several transcripts being found to be unique to the high-fertility group. Cytochrome oxidase subunit 7C (*COX7C*), which is involved in the terminal step in the electron transport chain for ATP synthesis in the inner mitochondrial membrane, was found to correlate negatively with bull fertility. More enriched transcripts have been identified from cryopreserved sperm, which could be evaluated as a "freezability" biomarker in farm animals (Khan et al., 2021). In addition, Zhang et al. (2017) detected 23 differentially expressed miRNAs in boar sperm following cryopreservation. Functional annotations indicate that these miRNAs and their targets were mainly found to be associated with metabolic and cellular processes. In fish, Yang et al. (2019) identified some differentially expressed genes (DEGs) in sperm of yellow catfish (*Pelteobagrus fulvidraco*) that play a crucial role in the ability of sperm to withstand freeze-thawing procedures. The transcriptomics profiles of cryopreserved sperm have also been investigated to understand the molecular mechanisms of cryodamage in blue catfish (*Ictalurus furcatus*; Wang et al., 2022) and black rockfish (*Sebastes schlegelii*; Niu et al., 2022).

In oocyte cryopreservation, both slow-freezing and vitrification procedures caused transcriptomic alterations in human oocytes with changes being characterized by the downregulation of several genes related to embryo development, energetic pathways, and DNA integrity (Estudillo et al., 2021). In contrast, Gao et al. (2017) found that the influence of vitrification on the transcriptome of mouse oocytes was negligible, as no DEGs were found between fresh and vitrified oocytes.

In embryo cryopreservation, gene expressions related to cell death, proliferation, embryonic development, and DNA repair were compromised in vitrified porcine blastocysts, with genes, such as *TP53INP*, *MGMT*, and *DKK3*, being upregulated in postthaw embryos (Cuello et al., 2021). In mice, Lee et al. (2024) found that DEGs assessed by RNA-seq from vitrified blastocysts were primarily associated with thermogenesis, chemical carcinogenesis–reactive oxygen species, oxidative phosphorylation, immune response, and MAPK-related signaling pathways.

10.6.2 Proteomics

Proteomics is a postgenomic biotechnology that can comprehensively analyze the entire proteins present in protein samples, including information on their abundances, structure, modification, and regulatory networks (Jia et al., 2020). In cryopreservation, this technology enables researchers to monitor protein alterations at specific time points or after certain treatments, such as equilibration with CPA, cooling, and thawing. Therefore application of proteomic techniques has been considered an invaluable tool for exploring the mechanisms of cryodamage in terms of identifying the key proteins affected by the cryopreservation process (Larbi et al., 2024). Moreover, investigations using proteomic analyses could assist in refining cryopreservation techniques and enhancing reproductive outcomes (Larbi et al., 2024).

In sperm cryopreservation, proteins related to cryodamage predominantly encompass those integral to the plasma membrane, capacitation, and acrosome reaction, highlighting their abundance and significance during fertilization. In addition, cryopreservation can also affect other proteins, such as enzymes regulating metabolism, proteins engaged in redox reactions, stress tolerance and apoptosis factors, and carbonylated proteins related to oxidative modifications. This has been demonstrated in sperm of farm animals, such as bull, boar, ram, goat, and stallion (Larbi et al., 2024). The modulations of these proteins are critical as they can profoundly influence sperm structure, biological processes, and functional integrity, resulting in decreased postthaw fertility (Larbi et al., 2024).

In fish, significant alterations in sperm proteins essential for flagellar motility, energy metabolism, nuclear pore organization, and fertilization processes have been revealed in Siberian sturgeon (*Acipenser baerii*), common carp (*Cyprinus carpio*), rainbow trout (*Oncorhynchus mykiss*), and sterlet (*Acipenser ruthenus*) (Kodzik et al., 2024).

In oocyte cryopreservation, differentially expressed proteins (DEPs) have been identified after the vitrification in porcine and mouse oocytes. These DEPs are involved in metabolism, signal transduction, response to stimulus and immune response, mitochondrial function, and cytoskeleton (Jia et al., 2020; Zhuan et al., 2022).

The rapid development of proteomic technologies has facilitated the discovery of molecular pathways governing cryodamage and assisted in the identification of biomarkers critical for assessing gamete quality and fertility (Larbi et al., 2024). Therefore application of this technology to molluscan cryopreservation research would accelerate the refinement of cryopreservation techniques.

10.6.3 Metabolomics

Metabolomics is a key scientific field in the postgenomic era, which investigates small molecules to supplement genomics, transcriptomics, and proteomics (Longobardi et al., 2020; Xu et al., 2022). Moreover, metabolites are end-products or intermediates of the metabolism processes and are closely linked to the phenotype of a biological system. Metabolites govern the modulation of the phenotype function (Qiu et al., 2023). Metabolomics allows for the comprehensive identification of metabolites, such as amino acids, peptides, fatty acids, and carbohydrates in cells (Longobardi et al., 2020; Xu et al., 2022).

In sperm cryopreservation, metabolomics has been applied to analyze cryodamage in farm animals to determine the metabolic biomarkers suitable for postthaw sperm quality assessment. For example, Xu et al. (2022) identified 48 metabolites differentially produced between goat sperm cryopreserved with and without trehalose and found that the addition of trehalose may affect the amino acid synthesis and glycerol metabolism, which were important for maintaining sperm motility and structural integrity. In addition, some metabolites (e.g., L-isoleucine, L-leucine, L-threonine, and dihydroxyacetone) have also been identified as potential biomarkers for postthaw sperm quality assessments (Xu et al., 2022). In bulls, Longobardi et al. (2020) found that cryopreservation affected the sperm metabolome, resulting in variation in metabolites that counteract oxidative stress and regulate sperm capacitation and motility. Therefore identification of sperm quality-associated metabolites can open the way for the development of effective cryopreservation strategies, such as reintegrating key compounds lost during cryopreservation to restore the fertility of cryopreserved sperm.

10.7 Future research perspectives in marine molluscan gamete, embryo, and larval cryopreservation

10.7.1 Application of omics technologies for cryodamage investigation

So far, many studies in mollusks have been conducted to optimize CPA solutions and cooling and thawing methods to improve postthaw gamete and larval quality. However, such improvements have not yet reached the desired level of postthaw success, as gametes and larvae still largely lose their viability after cryopreservation. Despite best endeavors, a significant gap remains in our understanding of the molecular mechanisms of cryodamage in molluscan gametes, embryos, and larvae. This lack of detailed understanding continues to impede advancements in cryopreservation for these species. Integration of omics technologies into molluscan gamete, embryo, and larval cryopreservation would provide new insights to decipher the molecular mechanisms of cryodamage and factors affecting their cryotolerance. This valuable information could provide essential information to identify biomarkers that predict gamete and larval "freezability" as well as for the development of new strategies to enhance cryopreservation outcomes. In addition, the application of omics technologies in mollusks could have a major impact on understanding adverse effects (e.g., mechanism of inbreeding) in germplasm preservation and genetic improvement programs (Andersen et al., 2025).

10.7.2 Protocol standardization

Maintaining postthaw cellular function is the main objective in cryopreservation. However, the lack of standardization of cryopreservation protocols is one of the difficulties in obtaining reproducible results in a species. Published studies have demonstrated that it is necessary and urgent to standardize different aspects, such as extender, CPA solution, type of cryo-container, cell concentration, cooling methodology, thawing systems, and the format for reporting results. This would enable such protocols to be adapted for large-scale application in molluscan aquaculture practices (Asturiano et al., 2017).

10.7.3 Integration of cryopreservation in molluscan aquaculture

Cryopreservation has been widely acknowledged as an effective technique to assist in germplasm preservation and genetic improvement programs in aquaculture. Its application may reduce costs associated with field-based in situ preservation, providing significant cost savings. For the Australian aquaculture industry, Li (2012) has developed a model to estimate the cost of establishing an on-site cryobanking service (both cryopreserving and storage) with backup storage off-site that can be used as a tool to compare and select cryobanking options that would best suit the industry's short- and long-term needs. In the United States, sperm cryopreservation has been applied to an ongoing oyster breeding program to archive the males in the program as a sperm repository for future (Yang et al., 2021).

Inbreeding depression is a risk for genetic improvement programs in aquaculture, which could reduce the fitness because of the lower survival, mating, and reproduction in the progeny of related individuals compared to that of unrelated individuals (Hedrick & Garcia-Dorado, 2016). Moreover, inbreeding depression is usually accompanied by loss of alleles in terms of genetic drift (Hedrick & Garcia-Dorado, 2016). Inbreeding and genetic drift are determined by the effective population size (N_e) therefore maintaining the desired N_e using cryopreservation could prevent adverse effects on productivity and profits. For example, in Atlantic salmon gene banks, the Norwegian Atlantic salmon Gene Bank operates rigorous population management standards to minimize inbreeding and genetic drift during domestication/selection. Improvements in salmonid fish sperm cryopreservation enabled the male biomass to be reduced without reducing the N_e, providing significant benefits by reducing the costs associated with field-based in situ cryopreservation (Asturiano et al., 2017; Bøe et al., 2021).

In Poland, Judycka et al. (2024) used standardized sperm cryopreservation methodologies to establish fish sperm cryobanks. Their cryobanks contain germplasm from the most economically valuable fish, including both wild and farmed sources, and cryopreserved sperm has also been used in hatchery practices (Judycka et al., 2024). Aquaculture cryobanks have also been established in other European countries (Ukraine, UK, Germany, Czech Republic, and France) as well as in the USA, Brazil, Australia, and New Zealand, mainly for fish, but also for oysters (FAO 2026; Martínez-Páramo et al., 2017).

In mollusks, it is feasible to integrate the cryopreservation techniques in molluscan aquaculture for most purposes identified so far (selective breeding, year-round hatchery production, etc.). In practice, as for livestock, factors such as the type of germplasm, sample size, and number of germplasm required, and the number of donor individuals, need to be carefully considered for different purposes in different species.

10.7.4 Application of model species

In general, an animal can serve as a valuable research model if it possesses representative biological characteristics of a larger group of organisms and is easy to culture and access in laboratories. Commonly, a good animal model would have all or most of the following characteristics: (1) structural simplicity with basic cellular processes that more complex organisms have, (2) simple to manipulate, (3) easy to condition in a laboratory, (4) readily available, (5) a relatively small and stable genome, and (6) have a short life cycle (Ribas & Piferrer, 2013).

Although gamete and larval cryopreservation techniques have been published in many economically important bivalve species, especially for sperm (Hassan et al., 2015; Liu et al., 2015c; Martínez-Páramo et al., 2017), their application in commercial production and/or cryobanking has been restricted partially due to the lack of knowledge and understanding of any long-term impacts. These concerns have been further stressed in recent years as negative effects of cryopreservation on progeny fitness have been reported in fish and mammalian species, such as brown trout (*Salmo trutta*; Nusbaumer et al., 2019), Japanese eels (*Anguilla japonica*; Müller et al., 2018), horses (Ortiz-Rodriguez et al., 2019), and mice (Hosseini et al., 2025). Urgent studies to address these knowledge gaps were recommended for Atlantic salmon gene banking (Bøe et al., 2021). However, investigations on the performance of economically important traits and across generations are time-consuming and challenging in most species, including farmed marine bivalves because of (1) their long generation interval ($\geq$ 1 year) and (2) high costs and challenges to manage experiments in open marine environments (Xu et al., 2024). These obstacles could be practically addressed by taking a model species approach.

The dwarf surfclam is a dioecious marine bivalve and has most characteristics of a model species: A short generation interval (approximately 3 months), a well-established laboratory cultivation technique, and accessible genomic information. This species has been used as a model species for other applications, such as the optimization of tetraploid production methods and the elucidation of shell formation mechanisms for bivalves (Guo et al., 2023; Yang & Guo, 2006). In addition, sperm cryopreservation techniques developed for other bivalves are also suitable for the dwarf surfclam (Xu et al., 2024). Therefore use of this species could accelerate the investigation of cryo-impacts on the performance of resultant progeny at late developmental stages and/or on subsequent generations (Xu et al., 2024).

10.8 Conclusion

Molluscan aquaculture is one of the most profitable and environmentally friendly aquaculture sectors in seafood supply. Strategies, such as germplasm preservation and genetic improvement programs, have been integrated to guarantee the industry's long-term sustainable and competitive advantages. Although gamete, embryo, and larval cryopreservation protocols have been reported in some economically important species, oocyte and larval cryopreservation is still challenging, and integration of these techniques (including sperm cryopreservation) into molluscan aquaculture is still limited.

The application of the omics technology would provide new insights to decipher the molecular mechanism of cryodamage and factors affecting the cryotolerance of marine mollusk germplasm. Omics biomarkers could not only predict the "freezeability" of gametes and larvae but also be used to develop new strategies to enhance cryopreservation outcomes. These further improvements could, in turn, assist standardization of cryopreservation protocols in each species, which is paramount for their application in molluscan aquaculture practices. In addition, the application of a model species (e.g., dwarf surfclam) would speed up investigations to understand cryopreservation impacts on economically important traits and cross-generations, both key knowledge gaps hindering the application of cryopreservation in molluscan aquaculture.

References

Acosta-Salmón, H., Jerry, D. R., & Southgate, P. C. (2007). Effects of cryoprotectant agents and freezing protocol on motility of black-lip pearl oyster (*Pinctada margaritifera* L.) spermatozoa. *Cryobiology, 54*, 13–18.

Adams, S. L., Salinas-Flores, L., & Lim, M. H. (2013). Diet conditioning of Pacific oyster, *Crassostrea gigas*, broodstock to improve oocyte cryopreservation success. *Journal of Shellfish Research, 32*(2), 391–399.

Adams, S. L., Smith, J. F., Roberts, R. D., Janke, A. R., King, N. G., Tervit, H. R., & Webb, S. C. (2008). Application of sperm cryopreservation in selective breeding of the Pacific oyster, *Crassostrea gigas* (Thunberg). *Aquaculture Research, 39*, 1434–1442.

Adams, S. L., Tervit, H. R., McGowan, L. T., Smith, J. F., Robers, R. D., Salinas-Flores, L., Gale, S. L., Webb, S. C., Mullen, S. F., & Critser, J. K. (2009). Towards cryopreservation of greenshell mussel (*Perna canaliculus*) oocytes. *Cryobiology, 58*(1), 69–74.

Andersen, L. K., Thompson, N. F., Abernathy, J. W., Ahmed, R. O., Ali, A., Al-Tobasei, R., Beck, B. H., Calla, B., Delomas, T. A., Dunham, R. A., Elsik, C. G., Fuller, S. A., García, J. C., Gavery, M. R., Hollenbeck, C. M., Johnson, K. M., Kunselman, E., Legacki, E. L., Liu, S., … Reading, B. J. (2025). Advancing genetic improvement in the omics ear: Status and priorities for United States aquaculture. *BMC Genomics, 26*, 155.

Anjos, C., Duarte, D., Diogo, P., Matias, D., & Cabrita, E. (2022). Assessment of larval quality of two bivalve species, *Crassostrea angulate* and *Chamelea gallina*, exposed and cryopreserved with different cryoprotectant solutions. *Cryobiology, 106*, 24–31.

Anjos, C., Duarte, D., Fatsini, E., Matias, D., & Cabrita, E. (2024). Comparative transcriptome analysis reveals molecular damage associated with cryopreservation in *Crassostrea angulata* D-larvae rather than to cryoprotectant exposure. *BMC Genomics, 25*, 591.

Anjos, C., Santos, A. L., Duarte, D., Matias, D., & Cabrita, E. (2021). Effect of trehalose and sucrose in post-thaw quality of *Crassostrea angulata* sperm. *Frontiers in Physiology, 12*, 749735.

Arif, A., Zahoor, N., Tang, J., Tang, M., Dong, L., Khan, S. Z., & Dai, G. (2025). Cryopreservation strategies for poultry semen: A comprehensive review of techniques and applications. *Veterinary Sciences, 12*(2), 145.

Arita, K., Takamatsu, S., Isowa, K., Aoki, H., & Ohta, H. (2018). Development of a novel non-programmable cryopreservation method capable of accurate cooling rate manipulation. *Aquaculture, 484*, 145–151.

Asturiano, J. F., Cabrita, E., & Horváth, Á. (2017). Progress, challenges and perspectives on fish gamete cryopreservation: A mini-review. *General and Comparative Endocrinology, 245*, 69–76.

Blaxter, J. H. S. (1953). Sperm storage and cross-fertilization of spring and autumn spawning herring. *Nature, 172*, 1189–1190.

Bøe, K., Bjøru, B., Bårdsen, M. T., Wist, A. N., Wolla, S., & Sivertsen, A. (2021). Opportunities and challenges related to sperm cryopreservation in Atlantic salmon gene banks. *Conservation Science and Practice, 3*, 5552.

Bojic, S., Murray, A., Bentley, B. L., Spindler, R., Pawlik, P., Cordeiro, J. L., Bauer, R., & de Magalhães, J. P. (2021). Winter is coming: The future of cryopreservation. *BMC Biology, 15*, 56.

Brereton, J. E., Spooner, S. L., Walker, S. L., Mooney, A., Wilson, P. B., Mastromonaco, G. F., Hunter, E., & White, S. (2025). When to cryopreserve and when to let it go? A systematic review of priorities in wild animal cryobanking. *Theriogenology Wild, 6*, 100119.

Brown, A. R., Huntington, T., Bickley, L. K., Powell, T., & Tyler, C. R. (2024). Positive tipping points for the sustainable growth of bivalve shellfish mariculture in England and Wales. *Marine Policy, 162*, 106058.

Card, C. J., Krieger, K. E., Kaproth, M., & Sartini, B. L. (2017). Oligo-dT selected spermatozoal transcript profiles differ among higher and lower fertility dairy sires. *Animal Reproduction Science, 177*, 105–123.

Carlisle, J. A., & Swanson, W. J. (2021). Molecular mechanisms and evolution of fertilization proteins. *Journal of Experimental Zoology Part B: Molecular and Developmental Evolution, 336*, 652–665.

Castro, M., Leal, K., Pezo, F., & Contreras, M. J. (2025). Sperm membrane: Molecular implications and strategies for cryopreservation in productive species. *Animals, 15*, 1808.

Chang, Y., Nan, F., & Hsu, T. (2024). Genetic management of small abalone (*Haliotis diversicolor*) by using ISSRseq technology. *Aquaculture Report, 38*, 102345.

Chao, N., Lin, T., Chen, Y., Hsu, H., & Liao, I. (1997). Cryopreservation of late embryos and early larvae in the oyster and hard clam. *Aquaculture, 155*, 31–44.

Chen, S. L., & Tian, Y. S. (2005). Cryopreservation of flounder (*Paralichthys olivaceus*) embryos by vitrification. *Theriogenology, 63*, 1207–1219.

Choi, Y. H., & Chang, Y. J. (2014). Influences of developmental stages, protective additives and concentrations of cryoprotective agents on the cryopreservation of Pacific oyster (*Crassostrea gigas*) larvae. *CryoLetters, 35*, 495–500.

Çiftçi, Y. Ö., & Kaya, E. (2024). Transcriptomics of cryopreserved cells. *CryoLetters, 45*(6), 329–339.

Cuello, C., Martinez, C. A., Cambra, J. M., Parrilla, I., Rodriguez-Martinez, H., Gil, M. A., & Martinez, E. A. (2021). Effects of vitrification on the blastocyst gene expression profiles in a porcine model. *International Journal of Molecular Sciences, 22*(3), 1222.

Curry, M. R. (2000). Cryopreservation of semen from domestic livestock. *Reviews of Reproduction, 5*(1), 46–52.

Demoy-Schneider, M., Schmitt, N., Le Pennec, G., Suquet, M., & Cosson, J. (2018). Quality assessment of cryopreserved black-lip pearl oyster *Pinctada margaritifera* spermatozoa. *Aquaculture, 497*, 278–286.

Diwan, A. D., Harke, S. N., Gopalkrishna, & Panche, A. N. (2020). Cryobanking of fish and shellfish egg, embryos and larvae: An overview. *Frontiers in Marine Science, 7*, 251.

Dobrinsky, J. R. (2002). Advancements in cryopreservation of domestic animal embryos. *Theriogenology, 57*(1), 285–302.

Dong, Q. X., Eudeline, B., Huang, C. J., Allen, S. K., & Tiersch, T. R. (2005a). Commercial-scale sperm cryopreservation of diploid and tetraploid Pacific oysters, *Crassostrea gigas*. *Cryobiology, 50*(1), 1–16.

Dong, Q. X., Huang, C. J., Eudeline, B., Allen, S. K., Jr, & Tiersch, T. R. (2006). Systematic factor optimization for sperm cryopreservation of tetraploid Pacific oysters, *Crassostrea gigas*. *Theriogenology, 66*, 387–403.

Dong, Q. X., Huang, C. J., Eudeline, B., & Tiersch, T. R. (2005b). Systematic factor optimization for cryopreservation of shipped sperm samples of diploid Pacific oysters, *Crassostrea gigas*. *Cryobiology, 51*, 176–197.

Dong, Q. X., Huang, C. J., Eudeline, B., & Tiersch, T. R. (2007a). Cryoprotectant optimization for sperm of diploid Pacific oysters by use of commercial dairy sperm freezing facilities. *Aquaculture, 271*, 537–545.

Dong, Q. X., Huang, C. J., & Tiersch, T. R. (2007b). Control of sperm concentration is necessary for standardization of sperm cryopreservation in aquatic species: Evidence from sperm agglutination in oysters. *Cryobiology, 54*, 87–98.

Elliott, G. D., Wang, H., & Fuller, B. J. (2017). Cryoprotectants: A review of the actions and applications of cryoprotective solutes that modulate cell recovery from ultra-low temperatures. *Cryobiology, 76*, 74–91.

Estudillo, E., Jiménez, A., Bustamante-Nieves, P. E., Palacios-Reyes, C., Velasco, I., & López-Ornelas, A. (2021). Cryopreservation of gametes and embryos and their molecular changes. *International Journal of Molecular Sciences, 22*, 10864.

Fahy, G. M., & Wowk, B. (2015). Principles of cryopreservation by vitrification. In W. Wolkers, & H. Oldenhof (Eds.), *Cryopreservation and freezing-drying protocols* (3rd edition) (pp. 21–82). Springer.

FAO. (2024). *The state of world fisheries and aquaculture 2024: Blue transformation in action*. Food and Agriculture Organization.

FAO. (2026). Aquaculture development – Guidelines for ex situ in vitro gene banking of aquatic genetic resources. FAO Technical Guidelines for Responsible Fisheries, No. 5 Suppl. 13. Rome.

Fissell, W. H., Manley, S., Dubnisheva, A., Glass, J., Magistrelli, J., Eldridge, A. N., Fleischman, A. J., Zydney, A. L., & Roy, S. (2007). Ficoll is not a rigid sphere. *American Journal of Physiology-Renal Physiology, 293*, F1209–F1213.

Franěk, R., & Pšenička, M. (2020). Cryopreservation of germ stem cells in fish. In J. Betsy, & S. Kumar (Eds.), *Cryopreservation of fish gametes* (pp. 285–312). Springer.

Gale, S. L., Burritt, D. J., Tervit, H. R., Adams, S. L., & McGowan, L. T. (2014). An investigation of oxidative stress and antioxidant biomarkers during Greenshell mussel (*Perna canaliculus*) oocyte cryopreservation. *Theriogenology, 82*, 779–789.

Gao, L., Jia, G., Li, A., Ma, H. J., Huang, Z. Y., Zhu, S., Hou, Y., & Fu, X. (2017). RNA-seq transcriptome profiling of mouse oocytes after in vitro maturation and/or vitrification. *Scientific Reports, 7*, 13245.

Gualtieri, R., Kalthur, G., Barbato, V., Nardo, M. D., Adiga, S. K., & Talevi, R. (2021). Mitochondrial dysfunction and oxidative stress caused by cryopreservation in reproductive cells. *Antioxidants, 10*(3), 337.

Guan, M., Rawson, D. M., & Zhang, T. (2010). Cryopreservation of zebrafish (*Danio rerio*) oocytes by vitrification. *CryoLetters, 31*, 230–238.

Guha, A., & Devireddy, R. (2010). Polyvinylpyrrolidone (PVP) mitigates the damaging effects of intracellular ice formation in adult stem cells. *Annals of Biomedical Engineering, 38*(5), 1826–1835.

Guo, X., Li, X., Zhao, F., Liu, D., Yang, Z., Li, M., Li, Y., Wei, H., Wang, H., Qin, Z., Zhang, Z., & Bao, Z. (2023). Full-length transcriptome analysis provides insights into larval shell formation in *Mulinia lateralis*. *Frontiers in Marine Science, 9*, 1111241.

Gwo, J., & Lei, S. (2023). Evaluation of cryodamage in small abalone, *Haliotis diversicolor*, spermatozoa by using a flow cytometer. *Journal of the World Aquaculture Society, 54*, 181–192.

Hassan, M. M., Li, X., & Qin, J. G. (2017a). Sperm cryopreservation in the spermcasting Australian flat oyster *Ostrea angasi* by a programmable freezing method. *Cryobiology, 76*, 119–124.

Hassan, M. M, Li, X., & Qin, J. G. (2017b). Improvement of post-thaw sperm survivals using liquid nitrogen vapor in a spermcasting oyster *Ostrea angasi*. *Cryobiology, 78*, 1–7.

Hassan, M. M., Qin, J. G., & Li, X. (2015). Sperm cryopreservation in oysters: A review of its current status and potentials for future application in aquaculture. *Aquaculture, 438*, 24–32.

Hedrick, P. W., & Garcia-Dorado, A. (2016). Understanding inbreeding depression, purging, and genetic rescue. *Trends in Ecology & Evolution, 31*(12), 940–952.

Heres, P., Rodriguez-Riveiro, R., Troncoso, J., & Paredes, E. (2019). Toxicity tests of cryoprotecting agents for *Mytilus galloprovincialis* (Lamark, 1819) early developmental stages. *Cryobiology, 86*, 40–46.

Heres, P., Troncoso, J., & Paredes, E. (2021). Larval cryopreservation as new management tool for threatened clam fisheries. *Scientific Reports, 11*, 15428.

Heres, P., Troncoso, J., & Paredes, E. (2022). Long-term study on survival and development of successive generations of *Mytilus galloprovincialis* cryopreserved larvae. *Scientific Reports, 12*, 13632.

Heres, P., Troncoso, J., & Paredes, E. (2023). Exploring the feasibility of cryopreserving larvae of the common cockle (*Cerastoderma edule*) for hatchery production. *Cryobiology, 113*, 104582.

Heres, P. G., Vignier, J., Copedo, J., Berry, J., Paredes, E. R., & Adams, S. L. (2020). Development of a method to cryopreserve Greenshell mussel (*Perna canaliculus*) veliger larvae. *Cryobiology, 96*, 37–44.

Hinrichs, K. (2020). Advances in holding and cryopreservation of equine oocytes and embryos. *Journal of Equine Veterinary Science, 89*, 102990.

Horváth, Á, Jug-Dujaković, J., Radočaj, T., Barić, O., Špelić, I., Balogh, R. E., Lefler, K. K., Kitanović, N., Marinović, Z., Urbanyi, B., & Gavrilović, A. (2024). Cryopreservation of European flat oyster (*Ostrea edulis*) larvae using a controlled-rate freezer. *Aquaculture, 587*, 740887.

Hosseini, S., Hosseini, S., Algholi, H., & Salehi, M. (2025). Embryo vitrification impacts learning and spatial memory by altering the imprinting genes expression level in the mouse offspring's hippocampus. *Scientific Reports, 15*, 5419.

Hossen, S., Sharker, R., Cho, Y., Sukhan, Z. P., & Kho, K. H. (2021a). Effects of antifreeze protein III on sperm cryopreservation of Pacific abalone, *Haliotis discus hannai*. *International Journal of Molecular Sciences, 22*, 3917.

Hossen, S., Sukhan, Z. P., Cho, Y., Choi, C. Y., & Kho, K. H. (2022a). Saccarides influence sperm quality and expressions of motility and fertilization-associated genes in cryopreserved sperm of Pacific abalone, *Haliotis discus hannai*. *Frontiers in Cell and Developmental Biology, 10*, 935667.

Hossen, S., Sukhan, Z. P., Cho, Y., Hanif, M., Cho, D. H., & Kho, K. H. (2025). Production of Pacific abalone progeny from cryopreserved sperm: Molecular insights into growth, reproduction, and immunity. *Aquaculture, 601*, 742290.

Hossen, S., Sukhan, Z. P., Cho, Y., & Kho, K. H. (2021b). Effects of cryopreservation on gene expression and post thaw sperm quality of Pacific abalone, *Haliotis discus hannai*. *Frontiers in Marine Science, 8*, 652390.

Hossen, S., Sukhan, Z. P., Cho, Y., & Kho, K. H. (2024). Physiological evaluation of seasonal sperm quality in a biannual spawner, Pacific abalone: Effects on in-vitro fertilization and cryotolerance. *Ecotoxicology and Environmental Safety, 269*, 115809.

Hossen, S., Sukhan, Z. P., Cho, Y., Lee, W. K., & Kho, K. H. (2022b). Antioxidant activity and oxidative stress-oriented apoptosis pathway in saccharides supplemented cryopreserved sperm of Pacific abalone, *Haliotis discus hannai*. *Antioxidants, 11*, 1303.

Hossen, S., Sukhan, Z. P., Kim, S. C., Hanif, A., Kong, I., & Kho, K. H. (2023). Molecular cloning and functional characterization of catalase in stress physiology, innate immunity, testicular development, metamorphosis, and cryopreserved sperm of Pacific abalone. *Antioxidants, 12*, 109.

Hui, B., Vonau, V., Moriceau, J., Tetumu, R., Vanaa, V., Demoy-Schneider, M., Suquet, M., & Moullac, G. L. (2011). Hatchery-scale trials using cryopreserved spermatozoa of black-lip pearl oyster, *Pinctada margaritifera*. *Aquatic Living Resources, 24*, 219–223.

Huo, Y. Z., Yee, J. C., & Yang, H. P. (2022). Development of a cost-effective and portable aeration freezing system for germplasm cryopreservation for eastern oysters and other aquatic species. *Aquaculture Research, 53*, 453–466.

Jia, B., Xiang, D., Fu, X., Shao, Q., Hong, Q., Quan, G., & Wu, G. (2020). Proteomic changes of porcine oocytes after vitrification and subsequent in vitro maturation: A tandem mass tag-based quantitative analysis. *Frontiers in Cell and Developmental Biology, 8*, 614577.

Jin, B., Kleinhans, F. W., & Mazur, P. (2014). Survivals of mouse oocytes approach 100% after vitrification in 3-fold diluted media and ultra-rapid warming by an IR laser pulse. *Cryobiology, 68*, 419–430.

Judycka, S., Dietrich, M. A., Nynca, J., & Ciereszko, A. (2024). Preservation of fish male germplasm in Poland. *Frontiers in Marine Science, 11*, 1407895.

Kawamoto, T., Narita, T., Isowa, K., Aoki, H., Hayashi, M., Komaru, A., & Ohta, H. (2007). Effects of cryopreservation methods on post-thaw motility of spermatozoa from the Japanese pearl oyster, *Pinctada fucata martensii*. *Cryobiology, 54*, 19–26.

Khan, I. M., Cao, Z., Liu, H., Khan, A., Rahman, S., Khan, M. Z., Sathanawongs, A., & Zhang, Y. (2021). Impact of cryopreservation on spermatozoa freeze-thawed traits and relevance omics to assess sperm cryo-tolerance in farm animals. *Frontiers in Veterinary Science, 8*, 609180.

Khosla, K., Kangas, J., Liu, Y., Zhan, L., Daly, J., Hagedorn, M., & Bischof, J. (2020). Cryopreservation and laser nanowarming of zebrafish embryos followed by hatching and spawning. *Advanced Biosystems, 4*, 2000138.

Khosla, K., Zhan, L., Bhati, A., Carley-Clopton, A., Hagedorn, M., & Bischof, J. (2019). Characterization of laser gold nanowarming: A platform for millimeter-scale cryopreservation. *Langmuir, 35*, 7364–7375.

Kim, K. T., Lim, H. K., & Chang, Y. J. (2013). Survival rates with time course of frozen-thawed Pacific oyster larvae in indoor rearing system. *Development & Reproduction, 17*, 337–343.

Kim, S. C., Hossen, S., & Kho, K. H. (2020). Cryopreservation of sperm from farmed Pacific abalone, *Haliotis discus hannai*. *Cryobiology, 94*, 49–56.

Knorpp, C. T., Merchant, W. R., Gikas, P. W., Spencer, H. H., & Thompson, N. W. (1967). Hydroxyethyl starch: Extracellular cryophylactic agent for erythrocytes. *Science, 157*, 1312–1313.

Kodzik, N., Ciereszko, A., Judycka, S., Słwińska, M., Szczaepkowska, B., & Świderska, B. (2024). Cryoprotectant-specific alternations in the proteome of Siberian sturgeon spermatozoa induced by cryopreservation. *Scientific Reports, 14*, 17707.

Kowalczyk, A. (2022). The role of the natural antioxidant mechanism in sperm cells. *Reproductive Sciences, 29*, 1387–1394.

Kuleshova, L., & Lopata, A. (2002). Vitrification can be more favorable than slow cooling. *Fertility and Sterility, 78*(3), 449–454.

Kumar, R., Singh, R. K., Kumar, M. S., & Kumar, S. (2024). Ex situ conservation: Status, strategies and challenges. In U. K. Sarkar, T. T. A. Kumar, N. Sood, R. K. Sing, & L. K. Tyagi (Eds.), *Sustainable management of fish genetic resources* (pp. 91–110). Springer.

Kuo, T., & Gwo, J. (2022). Quality assessment of cryopreserved Portuguese oyster (*Crassostrea angulata*) sperm through ultrastructural and flow cytometry analysis. *Cryobiology, 104*, 79–86.

Labbé, C., Haffray, P., Mingant, C., Quittet, B., Diss, B., Tervit, H. R., Adams, S. L., Rimond, F., & Suquet, M. (2018). Cryopreservation of Pacific oyster (*Crassostrea gigas*) larvae: Revisiting the practical limitations and scaling up the procedures for application to hatchery. *Aquaculture, 488*, 227–234.

Lago, A., Troncoso, J., & Paredes, E. (2025). Cryopreservation of juvenile *Mytilus galloprovincialis* to safeguard mollusk biodiversity and support aquaculture. *Scientific Reports, 15*, 25587.

Lannan, J. E. (1971). Experimental self-fertilization of the Pacific oyster, *Crassostrea gigas*, utilizing cryopreserved sperm. *Genetics, 68*(4), 599–601.

Larbi, A., Li, C., & Quan, G. (2024). An updated review on the application of proteomics to explore sperm cryoinjury mechanisms in livestock animals. *Animal Reproduction Science, 263*, 107441.

Lee, C. Y., Tsai, H. N., Cheng, E. H., Lee, T. H., Lin, P. Y., Lee, M. S., & Lee, C. I. (2024). Transcriptomic analysis of vitrified-warmed vs. fresh mouse blastocysts: Cryo-induced physiological mechanisms and implantation impact. *International Journal of Molecular Sciences, 25*(16), 8658.

Leibo, S. P., & Sztein, J. M. (2019). Cryopreservation of mammalian embryos: Derivation of a method. *Cryobiology, 86*, 1–9.

Li, M., Fu, Z., Yu, G., & Ma, Z. (2024). Cryopreservation of noble scallop *Mimachlamys nobilis* sperm: Optimization study of cooling and thawing temperature. *Aquaculture Reports, 39*, 102492.

Li, M., Fu, Z., Yu, G., & Ma, Z. (2025). Effects of different cryoprotectants on cryopreservation of sperm from noble scallop *Mimachlamys nobilis*. *Journal of Marine Science and Engineering, 13*, 607.

Li, X. (2012). *Scope of options to establish gamete cryobanking services to genetic improvements programs in Australian aquaculture industry*. South Australian Research and Development Institute (Aquatic Sciences), Adelaida. SARDI Publication No. F2012/000027-1. SARDI Research Report Series No. 636. 84pp..

Liu, B. L., Liu, Y., Liu, S. W., Xu, T., Liu, Q., & Li, X. (2016). Cryopreservation of strip spawned sperm using non-programmable freezing technique in the blue mussel *Mytilus galloprovincialis*. *Aquaculture Research, 47*, 3888–3898.

Liu, B. Z., & Li, X. (2008). Preliminary studies on cryopreservation of Sydney rock oyster (*Saccostrea glomerata*) larvae. *Journal of Shellfish Research, 27*, 1125–1128.

Liu, J., Phy, J., & Yeomans, E. (2012). Theoretic considerations regarding slow cooling and vitrification during cryopreservation. *Theriogenology, 78*, 1641–1652.

Liu, Y., Bao, L., Catalano, S. R., Zhu, X. C., & Li, X. (2023). The effects of larval cryopreservation on the epigenetics of the Pacific oyster *Crassostrea gigas*. *International Journal of Molecular Sciences, 24*, 17262.

Liu, Y., Catalano, S. R., Qin, J. G., Han, J. B., Zhan, X., & Li, X. (2022a). Effect of cryopreservation on redox status and gene expression of trochophore larvae in *Mytilus galloprovincialis*. *Journal of the World Aquaculture Society, 53*, 516–526.

Liu, Y., Gluis, M., Miller-Ezzy, P., Han, J., Qin, J., Zhan, X., & Li, X. (2020a). Development of a programmable freezing technique on larval cryopreservation in the Pacific oyster *Crassostrea gigas*. *Aquaculture, 523*, 735199.

Liu, Y., Gluis, M., Miller-Ezzy, P., Qin, J., Han, J., Zhan, X., & Li, X. (2020b). Development of a programmable freezing technique on larval cryopreservation in *Mytilus galloprovincialis*. *Aquaculture, 516*, 734554.

Liu, Y., & Li, X. (2015). Successful oocyte cryopreservation in the blue mussel *Mytilus galloprovincialis*. *Aquaculture, 438*, 55–58.

Liu, Y., Li, X., Robinson, N., & Qin, J. (2015c). Sperm cryopreservation in marine mollusk: A review. *Aquaculture International, 23*, 1505–1524.

Liu, Y., Li, X., Xu, T., Robinson, N., & Qin, J. (2014a). Improvement in non-programmable sperm cryopreservation technique in farmed greenlip abalone *Haliotis laevigata*. *Aquaculture, 434*, 362–366.

Liu, Y., Li, X., Xu, T., Robinson, N., & Qin, J. (2015a). Effects of broodstock age and sperm collection time over a natural spawning period on sperm cryopreservation in farmed greenlip abalone, *Haliotis laevigata*. *Journal of the World Aquaculture Society, 46*, 665–671.

Liu, Y., Li, X., Xu, T., Robinson, N., & Qin, J. (2016). Greenlip abalone (*Haliotis laevigata* Donovan, 1808) sperm cryopreservation using a programmable freezing technique and testing the addition of amino acid and vitamin. *Aquaculture Research, 47*, 1499–1510.

Liu, Y., Liu, S. W., Liu, B. L., Qin, J. G., Xu, T., & Li, X. (2018). Cryopreservation of strip spawned sperm using programmable freezing technique in the blue mussel *Mytilus galloprovincialis*. *Journal of Oceanology Limnology, 36*, 2351–2357.

Liu, Y., Xu, T., Robinson, N., Qin, J., & Li, X. (2014b). Cryopreservation of sperm in farmed Australian greenlip abalone *Haliotis laevigata*. *Cryobiology, 68*, 185–193.

Liu, Y., Xu, T., Robinson, N., Qin, J., & Li, X. (2015b). Cryopreservation of sperm in farmed blacklip abalone (*Haliotis rubra* Leach, 1814). *Aquaculture Research, 46*, 2628–2636.

Liu, Y., Zhan, X., Catalano, S. R., Qin, J. G., Han, J. B., & Li, X. (2022b). Investigation on redox status and gene expression related to larval cryopreservation in the Pacific oyster *Crassostrea gigas*. *Fisheries Science, 88*, 377–386.

Longobardi, V., Kosior, M. A., Pagano, N., Fatone, G., Staropoli, A., Vassetti, A., Vinale, F., Campanile, G., & Gasparrini, B. (2020). Changes in bull semen metabolome in relation to cryopreservation and fertility. *Animals, 10*, 1065.

Lyons, L., Jerry, D. R., & Southgate, P. C. (2005). Cryopreservation of black-lip pearl oyster (*Pinctada margaritifera*, L.) spermatozoa: Effects of cryoprotectants on spermatozoa motility. *Journal of Shellfish Research, 24*, 1187–1190.

Ma, P. Z., Zhang, Z., Yu, R. H., Wang, Z. P., & Li, P. F. (2018). Studies on sperm cryopreservation of bay scallop (*Argopecten irradians irradians*). *Journal of Ocean University of China, 48*, 34–40.

Magnotti, C., Cerqueira, V., Lee-Estevez, M., Farias, J. G., Valdebenito, I., & Figueroa, E. (2018). Cryopreservation and vitrification of fish semen: A review with special emphasis on marine species. *Reviews in Aquaculture, 10*, 15–25.

Mandawala, A. A., Harvey, S. C., Roy, T. K., & Fowler, K. E. (2016). Cryopreservation of animal oocytes and embryos: Current progress and future prospects. *Theriogenology, 86*(7), 1637–1644.

Mao, Y., Han, X., & Zhang, Y. (2018). Cryoprotective mechanism of using Ficoll for cell cryopreservation at non-cryogenic temperatures: A molecular dynamics study. *International Journal of Heat Mass Transfer, 127*, 319–325.

Martínez-Páramo, S., Horváth, Á., Labbé, C., Zhang, T., Robles, V., Herráez, P., Suquet, M., Adams, S., Viveiros, A., Tiersch, T. R. ,, & Cabrita, E. (2017). Cryobanking of aquatic species. *Aquaculture, 472*, 156–177.

Matteo, O. D., Langellotti, A. L., Masullo, P., & Sansone, G. (2009). Cryopreservation of the Mediterranean mussel (*Mytilus galloprovincialis*) spermatozoa. *Cryobiology, 58*, 145–150.

Mazur, P. (2004). Principles of cryobiology. In B. J. Fuller, N. Lane, & E. E. Benson (Eds.), *Life in the frozen state* (pp. 3–65). CRC Press.

Mondol, M. R., Kim, C., Kang, C., Park, S. R., Noseworthy, R. G., & Choi, K. (2016). Growth and reproduction of early grow-out hardened juvenile Pacific oysters, *Crassostrea gigas* in Gamakman bay, off the south coast of Korea. *Aquaculture, 463*, 224–233.

Morrell, J. M., & Mayer, I. (2017). Reproduction biotechnologies in germplasm banking of livestock species: A review. *Zygote, 25*(5), 545–557.

Müller, T., Matsubara, H., Kubara, Y., Horváth, Á., Kolics, B., Taller, J., Stéger, V., Kovács, B., Horváth, L., Asturiano, J. F., Peñaranda, D. S., & Urbányi, B. (2018). Testing cryopreserved European eel sperm for hybridization (*A. japonica* × *A. anguilla*). *Theriogenology, 113*, 153–158.

Murray, K. A., & Gibson, M. I. (2022). Chemical approaches to cryopreservation. *Nature Reviews Chemistry, 6*, 579–593.

Narida, A., Tsai, S., Hsieh, W., Wen, Z., Wang, L., Huang, C., & Lin, C. (2023). First successful production of adult corals derived from cryopreserved larvae. *Frontiers in Marine Science, 10*, 1172102.

Narita, T., Kawamoto, T., Isowa, K., Aoki, H., Hayashi, M., Komaru, A., & Ohta, H. (2008b). Fertility of cryopreserved spermatozoa of the Japanese pearl oyster, *Pinctada fucata martensii*. *Aquaculture, 275*, 178–181.

Narita, T., Kawamoto, T., Isowa, K., Aoki, H., Hayashi, M., Ohta, H., & Komaru, A. (2008a). Effects of cryopreservation on sperm structure in Japanese pearl oyster *Pinctada fucata martensii*. *Fish Science, 74*, 1069–1074.

Nascimento-Schulze, J. C., Bean, T. P., Houston, R. D., Santos, E. M., Sanders, M. B., Lewis, C., & Ellis, R. P. (2021). Optimizing hatchery practices for genetic improvement of marine bivalves. *Reviews in Aquaculture, 13*, 2289–2304.

Neyrão, I. M., Santos, F. B. P., Rodrigues, R. B., Streit, D. P., & Godoy, L. (2024). Use of powdered milk in semen cryopreservation protocols for fish: A systematic review. *Biopreservation and Biobanking, 22*(1), 4–20.

Niu, J., Wang, X., Liu, P., Liu, H., Li, R., Li, Z., He, Y., & Qi, J. (2022). Effects of cryopreservation on sperm with cryodiluent in viviparous black rockfish (*Sebastes schlegelii*). *International Journal of Molecular Sciences, 23*, 3392.

Nusbaumer, D., de Cunha, L. M., & Wedekind, C. (2019). Sperm cryopreservation reduces offspring growth. *Proceedings of the Royal Society B: Biological Sciences, 286*, 20191644.

O'Brien, C. (2019). *Development of cryopreservation methods for sperm of pinto abalone, Haliotis kamtschatkana, for conservation aquaculture.* Western Washington University Master's thesis.

Odintsova, N. A., Boroda, A. V., Maiorova, M. A., & Yakovlev, K. V. (2017). The death pathways in mussel larval cells after a freeze-thaw cycle. *Cryobiology, 77*, 41–49.

Ortiz-Rodriguez, J., Ortega-Ferrusola, C., Gil, M. C., Martín-Cano, F. E., Gaitskell-Phillips, G., Rodríguez-Martínez, H., Hinrichs, K., Álvarez-Barrientos, A., Román, Á., & Peña, F. J. (2019). Transcriptome analysis reveals that fertilization with cryopreserved sperm downregulates genes relevant for early embryo development in the horse. *PLoS One, 14*, e0213420.

Pang, Y., Ono, T., & Tanaka, T. (2024). Environmental effects on growth performance of Pacific oyster *Crassostrea gigas* cultured in the Seto Island Sea, Japan, from 1990 to 2021. *Fisheries Oceanography, 33*, e12686.

Paniagua-Chávez, C. G., Jenkins, J., Segovia, M., & Tiersch, T. R. (2006). Assessment of gamete quality for the eastern oyster (*Crassostrea virginica*) by use of fluorescent dyes. *Cryobiology, 53*, 128–138.

Paredes, E., Adams, S. L., Tervit, H. R., Smith, J. F., McGowan, L. T., Gale, S. L., Morrish, J. R., & Watts, E. (2012). Cryopreservation of greenshell mussel (*Perna canaliculus*) trochophore larvae. *Cryobiology, 65*, 256–262.

Paredes, E., Bellas, J., & Adams, S. L. (2013). Comparative cryopreservation study of trochophore larvae from two species of bivalves: Pacific oyster (*Crassostrea gigas*) and blue mussel (*Mytilus galloprovincialis*). *Cryobiology, 67*, 274–279.

Parihar, A., Kumar, A., Panda, U., Khan, R., Parihar, D. S., & Khan, R. (2023). Cryopreservation: A comprehensive overview, challenges, and future perspectives. *Advanced Biology, 7*, 2200285.

Partyka, A., Łukaszewicz, E., & Niżański, W. (2012). Effect of cryopreservation on sperm parameters, lipid peroxidation and antioxidant enzymes activity in fowl semen. *Theriogenology, 77*, 1497–1504.

Pegg, D. E. (2015). Principles of cryopreservation. In W. F. Wolkers, & H. Oldenhof (Eds.), *Cryopreservation and freezing-drying protocols* (3rd ed) (pp. 3–19). Springer.

Polge, C., Smith, A. U., & Parkes, A. S. (1949). Revival of spermatozoa after vitrification and dehydration at low temperatures. *Nature, 164*, 666.

Qiu, S., Cai, Y., Yao, H., Lin, C., Xie, Y., Tang, S., & Zhang, A. (2023). Small molecule metabolites: Discovery of biomarkers and therapeutic targets. *Signal Transduction and Targeted Therapy, 8*, 132.

Reed, B. M., & Uchendu, E. (2008). Controlled rate cooling. In B. M. Reed (Ed.), *Plant cryopreservation: A practical guide* (pp. 77–92). Springer.

Ribas, L., & Piferrer, F. (2013). The zebrafish (*Danio rerio*) as a model organism, with emphasis on applications for finfish aquaculture research. *Reviews in Aquaculture, 5*, 1–32.

Ribeiro, J. C., Carrageta, D. F., Bernardino, R. L., Alves, M. G., & Oliveira, P. F. (2022). Aquaporins and animal gamete cryopreservation: Advances and future challenges. *Animals, 12*, 359.

Riesco, M. F., Félix, F., Matias, D., Joaquim, S., Suquet, M., & Cabrita, E. (2017). First study in cryopreserved *Crassostrea angulata* sperm. *General and Comparative Endocrinology, 245*, 108–115.

Riesco, M. F., Félix, F., Matias, D., Joaquim, S., Suquet, M., & Cabrita, E. (2019). Comparative study on cellular and molecular responses in oyster sperm revealed different susceptibilities to cryopreservation. *Aquaculture, 498*, 223–229.

Ríos, G. L., García, M. F. S., Manrique, R. J., & Buschiazzo, J. (2025). Optimization of bovine oocyte cryopreservation: Membrane fusion competence and cell death of linoleic acid-in vitro matured oocytes subjected to vitrification. *Frontiers in Veterinary Science, 12*, 1628947.

Rodriguez-Riveiro, R., Heres, P., Tronscoso, J., & Paredes, E. (2019). Long term survival of cryopreserved mussel larvae (*Mytilus galloprovinciallis*). *Aquaculture, 512*, 734326.

Roux, A., & Sandenbergh, L. (2008). Roodt-Wilding R, Preliminary investigation to determine the cytotoxicity of various cryoprotectants on southern African abalone (*Haliotis midae*). *Cryobiology, 57*, 308–311.

Rusk, A. B., Alfaro, A. C., Young, T., Watts, E., & Adams, S. L. (2020). Development stage of cryopreserved mussel (*Perna canaliculus*) larvae influences post-thaw impact on shell formation, organogenesis, neurogenesis, feeding ability and survival. *Cryobiology, 93*, 121–132.

Salinas-Flores, L., Paniagua-Chavez, C. G., Jenkins, J. A., & Tiersch, T. R. (2005). Cryopreservation of sperm of red abalone (*Haliotis rufescens*). *Journal of Shellfish Research, 24*, 415–420.

Sciorio, R., Pluchino, N., & Fuller, B. J. (2023). Review of human oocyte cryopreservation in ART programs: Current challenges and opportunities. *Cryobiology, 113*, 104590.

Sharafi, M., Borghei-Rad, S. M., Hezavehei, M., Shahverdi, A., & Benson, J. D. (2022). Cryopreservation of semen in domestic animals: A review of current challenges, applications, and prospective strategies. *Animals, 12*, 3271.

Silvestri, F., Vitiello, V., Prete, F. D., Fabbrocini, A., Ferreira, J. F., & Sansone, G. (2016). Sperm motility and fertilization performance of *Nodipecten nodosus* (L., 1758) exposed at two different cryoprotectants. *Aquaculture Research, 47*, 624–631.

Simon, N. A., & Yang, H. P. (2018). Cryopreservation of trochophore larvae from the hard clam *Mercenaria mercenaria*: Evaluation of the cryoprotectant toxicity, cooling rate and thawing temperature. *Aquaculture Research, 49*, 2869–2880.

Smith, J. F., Adams, S. L., Gale, S. L., McGowan, L. T., Tervit, H. R., & Robers, R. D. (2012a). Cryopreservation of greenshell mussel (*Perna canaliculus*) sperm. I. establishment of freezing protocol. *Aquaculture, 334–337*, 199–204.

Smith, J. F., Adams, S. L., McDonald, R. M., Gale, S. L., McGowan, L. T., & Tervit, H. R. (2012b). Cryopreservation of greenshell mussel (*Perna canaliculus*) sperm. II. Effect of cryopreservation on fertility, motility, viability and chromatin integrity. *Aquaculture, 364-365*, 322–328.

Somfai, T. (2024). Vitrification of immature oocytes in pigs. *Animal Science Journal, 95*, e13943.

Suneja, S., Alfaro, A. C., Rusk, A. B., Morrish, J. R., Tervit, H. R., McGowan, L. T., & Adams, S. L. (2014). Multi-technique approach to characterise the effects of cryopreservation on larval development of the Pacific oyster (*Crassostrea gigas*). *New Zealand Journal of Marine and Freshwater Research, 48*, 335–349.

Suquet, M., Gourtay, C., Donval, A., Goïc, N., Quere, C., Malo, F., Le Grand, J., Ratiskol, D., Mingant, C., & Fauvel, C. (2016b). The quality of great scallop (*Pecten maximus*) sperm after thawing. *General and Comparative Endocrinology, 229*, 127–131.

Suquet, M., Labbé, C., Puyo, S., Mingant, C., Quittet, B., Boulais, M., Queau, I., Ratiskol, D., Diss, B., & Haffray, P. (2014). Survival, growth and reproduction of cryopreserved larvae from a marine invertebrate, the Pacific oyster (*Crassostrea gigas*). *PLoS One, 9*(4), e93486.

Suquet, M., Le Mercier, A., Rimond, F., Mingant, C., Haffray, P., & Labbe, C. (2012). Setting tools for the early assessment of the quality of thawed Pacific oyster (*Crassostrea gigas*) D-larvae. *Theriogenology, 78*, 462–467.

Suquet, M., Malo, F., Gourtay, C., Ratiskol, D., Riobe, M., Grand, J., Queau, I., & Mingant, C. (2016a). Optimizing sperm cryopreservation in great scallop (*Pecten maximus*). *Aquatic Living Resources, 29*, 206.

Tamargo, C., Garriga, F., Yeste, M., Pinart, E., Muiño, R., Carbajo, M. T., & Hidalgo, C. O. (2024). Predictive indicators of cryotolerance and fertility in bovine sperm: Evaluating fresh semen quality to improve AI outcomes with frozen-thawed sperm. *Reproduction in Domestic Animals, 59*, e14742.

Tervit, H. R., Adams, S. L., Roberts, R. D., McGowan, L. T., Pugh, P. A., Smith, J. F., & Janke, A. R. (2005). Successful cryopreservation of Pacific oyster (*Crassostrea gigas*) oocytes. *Cryobiology, 51*, 142–151.

Tharasanit, T., & Thuwanut, P. (2021). Oocyte cryopreservation in domestic animals and humans: Principles, techniques and updated outcomes. *Animals, 11*(10), 2949.

Torrez, A. R., Chávez, C. P., & López, A. P. (2015). First attempts to cryopreserve red abalone (*Haliotis rufescens*) oocytes. *Revista Bio Ciencias, 3*, 79–87.

Vignier, J., Adams, S., & Lovatelli, A. (2024). FAO Fisheries and Aquaculture Technical Papers. *FAO Fisheries and Aquaculture Technical Papers*: 698. Food and Agriculture Organization.

Vitiello, V., Carlino, P. A., Prete, F. D., Langellotti, A. L., & Sansone, G. (2011). Effects of cooling and freezing on the motility of *Ostrea edulis* (L., 1758) spermatozoa after thawing. *Cryobiology, 63*, 118–124.

Wang, H., Montague, H. R., Hess, H. N., Zhang, Y., Aguilar, G. L., Dunham, R. A., Butts, I. A. E., & Wang, X. (2022). Transcriptome analysis reveals key gene expression changes in blue catfish sperm in response to cryopreservation. *International Journal of Molecular Sciences, 23*, 7618.

Wang, H. R., Li, X., Wang, M. Q., Clarke, S., & Gluis, M. (2014). The development of oocyte cryopreservation techniques in blue mussels *Mytilus galloprovincialis*. *Fisheries Science, 80*, 1257–1267.

Wang, H. R., Li, X., Wang, M. Q., Clarke, S., Gluis, M., & Zhang, Z. Y. (2011). Effects of larval cryopreservation on subsequent development of the blue mussels, *Mytilus galloprovincialis* Lamarck. *Aquaculture Research, 42*, 1816–1823.

Wang, Y. R., Wang, Y. J., Bao, L. S., Sun, C., Huang, S. H., Li, X., Hu, X., & Liu, Y. (2024). Cryopreservation of Yesso scallop (*Patinopecten yessoensis*) sperm. *Aquaculture Reports, 36*, 102104.

Warmerdam, S., Vickers, J., & Palairet, N. (2021). Life cycle assessment of New Zealand mussels and oysters. *Aquaculture New Zealand & Ministry for Primary Industries, Gayathri Gamage*. .

Whittingham, D. G. (1977). Fertilization in vitro and development to term of unfertilized mouse oocytes previously stored at -196°C. *Journal of Reproduction and Fertility, 1977*(49), 89–94.

Whittingham, D. G., Leibo, S. P., & Mazur, P. (1972). Survival of mouse embryos frozen to -196°C and -260°C. *Science, 178*, 411–414.

Xu, B., Wang, Z., Wang, R., Song, G., Zhang, Y., Su, R., Liu, Y., Li, J., & Zhang, J. (2022). Metabolomics analysis of buck semen cryopreserved with trehalose. *Frontiers in Genetics, 13*, 938622.

Xu, Z., Hu, X., Wei, C., Bao, L., Zhan, X., Li, X., Wu, J., Wang, Y., & Liu, Y. (2026). Development of a non-programmable sperm cryopreservation technique in pearl oyster *Pinctada fucata martensii* and investigation of transcriptomic profile of post-thaw sperm by RNA-seq. *Aquaculture Reports, 46*, 103367.

Xu, Z., Yang, Z. J., Bao, L. S., Li, X. X., Huang, X. T., & Liu, Y. B. (2024). Development of a non-programmable sperm cryopreservation technique in dwarf surfclams *Mulinia lateralis*-a potential model species for bivalve research. *Frontiers in Marine Science, 11*, 1357454.

Yang, C., Yeh, Y., Lee, P., & Lin, T. (2013). Effect of cooling rate and cryoprotectant concentration on intracellular ice formation of small abalone (*Haliotis diversicolor*) eggs. *Cryobiology, 67*, 7–16.

Yang, H. P. (2017). Application of germplasm preservation in breeding programs for molluscan shellfish aquaculture and restoration. *The Bulletin of the Japan Fisheries Research and Education Agency, 45*, 15–20.

Yang, H. P., & Guo, X. M. (2006). Polyploid induction by heat shock-induced meiosis and mitosis inhibition in the dwarf surfclam, *Mulinia lateralis* Say. *Aquaculture, 252*, 171–182.

Yang, H. P., Hu, E., Cuevas-Uribe, R., Supan, J., Guo, X., & Tiersch, T. R. (2012). High-throughput sperm cryopreservation of eastern oyster *Crassostrea virginica*. *Aquaculture, 344–349*, 223–230.

Yang, H. P., & Huo, Y. Z. (2022). Review of molluscan larval cryopreservation and application to germplasm cryobanking and commercial seed production. *Aquaculture, 547*, 737491.

Yang, H. P., Huo, Y. Z., Yee, J. C., Rikard, S., Walton, W. C., & Saillant, E. (2021). Sperm repository for a breeding program of the Eastern oyster *Crassostrea virginica*: Sample collection, processing, cryopreservation, and data management plan. *Animals, 11*, 2836.

Yang, H. P., Supan, J., Guo, X., & Tiersch, T. R. (2013). Nonlethal sperm collection and cryopreservation in the eastern oyster *Crassostrea virginica*. *Journal of Shellfish Research, 32*, 429–437.

Yang, Y., Liu, D. Q., Wu, L. N., Huang, W. H., Yang, S., Xia, J. H., Liu, X., & Meng, Z. (2019). Comparative transcriptome analyses reveal changes of gene expression in fresh and cryopreserved yellow catfish (*Pelteobagrus fulvidraco*) sperm and the effects of cryoprotectant Me_2SO. *International Journal of Biological Macromolecules, 133*, 457–465.

Zhang, S., Zhao, X., Feng, K., Hu, Y., Tillotson, M. R., & Yang, L. (2022). Do mariculture products offer better environment and nutritional choices compared to land-based protein products in China? *Journal of Cleaner Production, 372*, 133697.

Zhang, X. M., Li, X., Clarke, S., & Li, X. (2012). The development of Pacific oysters *Crassostrea gigas* produced using cryopreserved sperm. In J. G. Qin (Ed.), *Oysters: Physiology, ecological distribution and mortality* (pp. 1–18). Nova Science Publishers.

Zhang, Y., Dai, D., Chang, Y., Li, Y., Zhang, M., Zhou, G., Peng, Z., & Zeng, C. (2017). Cryopreservation of boar sperm induces differential microRNA expression. *Cryobiology, 76*, 24–33.

Zheng, X., Gu, Z., Huang, Z., Ding, H., Vasquez, H. E., Liu, Y., Shi, Y., & Wang, A. (2018). The effects of cryoprotectants on sperm motility of the Chinese pearl oyster, *Pinctada fucata martensii*. *Cryobiology, 82*, 64–69.

Zheng, X., Li, S., Yang, R., Qin, J. G., Wang, A., Gu, Z., & Ma, Z. (2019). Sperm cryopreservation of the noble scallop *Chlamys nobilis* by a programmable freezing method: Effect of cryoprotectant. *Aquaculture Research, 50*, 1678–1686.

Zhu, W., Li, X., Qu, D., Liu, Y., & Clarke, S. (2014). Cryopreservation of sperm from wild greenlip abalone (*Haliotis laevigata*) in Australia. *Aquaculture, 432*, 60–66.

Zhu, X. C., Miller-Ezzy, P., Gluis, M., Qin, J. G., Tang, Y. H., Liu, Y. B., Zhao, Y., & Li, X. (2024). Examining the supplementation of exogenous lipid and antioxidant to improve the survivability of post-thaw oocytes in the Mediterranean mussels, *Mytilus galloprovincialis*. *Aquaculture, 578*, 740072.

Zhu, X. C., Miller-Ezzy, P., Gluis, M., Zhao, Y. Y., Qin, J. G., Tang, Y. H., Liu, Y., & Li, X. (2023a). Effects of phosphatidylcholine and tocopherol during larval cryopreservation of Pacific oysters (*Magallana gigas*). *Aquaculture, 574*, 739650.

Zhu, X., Miller-Ezzy, P., Zhao, Y., Qin, J., Tang, Y., Liu, Y., & Li, X. (2023b). Lipid modification to improve cryotolerance of gametes, embryos and larvae and its potential application in aquaculture species: A review. *Frontiers in Marine Science, 10*, 1235958.

Zhuan, Q., Du, X., Bai, J., Zhou, D., Luo, Y., Liu, H., Sun, W., Wan, P., Hou, Y., Li, J., & Fu, X. (2022). Proteomic profile of mouse oocytes after vitrification: A quantitative analysis based on 4D label-free technique. *Theriogenology, 187*, 64–73.

Zhukova, N. V. (2019). Fatty acids of marine mollusks: Impact of diet, bacterial symbiosis and biosynthetic potential. *Biomolecules, 9*, 857.

CHAPTER

11

Genetic manipulation and designed breeding

Shanshan Lian, Zhenmin Bao and Shi Wang

College of Marine Life Sciences, Ocean University of China, Qingdao, China

11.1 Introduction

With the continuous expansion of the global population and increasing scarcity of marine resources, improving shellfish varieties through breeding and biotechnological innovation has become a crucial strategy to ensure the supply of aquatic protein. Traditional breeding methods, such as domestication of wild populations, species introduction, and selective breeding, have already led to significant improvements in both yield and quality of key cultivated shellfish species (Tan et al., 2020). Important milestones include the development of the "raft culture" system in the mid-20th century and the breakthrough in oyster triploid breeding technology in the 1970s, which have greatly propelled the growth of the aquaculture industry (Nell, 2002).

However, against the backdrop of deteriorating marine environments and changing market demands, the limitations of conventional breeding approaches have become increasingly apparent. These methods often fall short in rapidly developing new varieties with desired traits such as fast growth, strong stress resistance, and high quality. In response, molecular-designed breeding has emerged as a cutting-edge strategy that integrates traditional breeding knowledge with omics technologies to enable precise regulation from genes to phenotypes (Xu et al., 2017). Through transcriptomic, proteomic, and other omics-based analyses of the molecular mechanisms underlying key economic traits, this approach allows for targeted genetic improvement. Breeders can now select for specific traits with greater accuracy, substantially enhancing both the efficiency and precision of breeding programs.

A major focus of molecular-designed breeding today lies in identifying key genes related to traits such as disease resistance and thermal tolerance, and applying these findings to develop new germplasm with superior characteristics (Bahrami & Najafi, 2019). Advances in shellfish genomics have further accelerated the discovery of gene resources. The core of modern shellfish breeding revolves around using genetic manipulation to precisely control the expression of such critical genes. Still, the effective application of these techniques remains a bottleneck in shellfish breeding programs. From early selective breeding to polyploid breeding and now to

Genetic Breeding and Aquaculture of Shellfish: A Collaborative Approach to a Sustainable Future
DOI: https://doi.org/10.1016/B978-0-443-33606-5.00013-8

molecular-designed breeding, each technological leap has brought substantial gains in industrial productivity. Thus, by building on the biological characteristics of shellfish, molecular-designed breeding demonstrates significant potential for genetic improvement in aquaculture. Moving forward, it will be essential to integrate various breeding technologies to fully exploit the synergies between conventional methods and modern biotechnology. The future development of this technology holds great importance for building a sustainable "blue granary," offering a viable pathway to meet rising demand for aquatic protein in an environmentally sustainable manner.

11.2 The development of the "blue seed industry"

As a pivotal component of blue agriculture, the blue seed industry specifically focuses on the development and application of superior varieties and strains tailored for aquaculture. This strategic sector has emerged as a critical engine driving the sustainable advancement of blue agriculture. Globally, the aquaculture sector has experienced remarkable growth over the past three decades, with fisheries and aquaculture production consistently on an upward trajectory. Recent statistical data reveal that in 2023, global aquaculture production for human consumption reached an impressive 136.16 million tons, underscoring the industry's expanding significance (Tacon, Metian, Parsons, & Shumway, 2025).

Blue seed industry has made significant strides in areas such as germplasm resource introduction, local variety breeding, hybrid breeding, and aquaculture model innovation (Hu et al., 2021; Langridge, 2018; Stonich & Bailey, 2000). By leaving these strategies, numerous new aquatic species have been successfully developed and commercialized, many of which are genetically superior in terms of growth, disease resistance, and environmental adaptability. The integration of advanced technologies, such as biotechnology, genetic engineering, and molecular breeding, has been critical to this success. One of the key innovations has been the introduction of new germplasm resources from around the world. By importing and cultivating non-native species, the aquaculture industry has been able to expand to include a broader range of species with economic potential (Rogers, 2024). Furthermore, a comprehensive system for protecting and conserving aquatic genetic resources has been developed. This system ensures that valuable genetic traits are preserved for future generations of farmed species. China has also been a pioneer in hybrid breeding, which involves crossbreeding different species or varieties to produce offspring with superior traits (Zhou & Gui, 2018). Hybrid breeding allows farmers to combine the best characteristics of two different species, such as faster growth, greater disease resistance, and improved environmental tolerance. This approach has been successfully applied to species such as shrimp, fish, and shellfish, leading to the development of faster-growing and more resilient varieties. These achievements have made substantial contributions to ensuring national food security and enhancing the sustainability of aquaculture practices.

As the global demand for seafood continues to rise, the blue seed industry will play an increasingly important role in ensuring the sustainability and growth of aquaculture. Advances in biotechnology, genetic engineering, and molecular designed breeding will continue to drive the development of new and improved aquatic species (Chen et al., 2023; Dunham, 2023). The integration of these technologies into breeding programs will allow aquaculture to meet the challenges posed by climate change, disease outbreaks, and the need for increased productivity. The future of the blue seed industry will also be shaped by the need to reduce the environmental

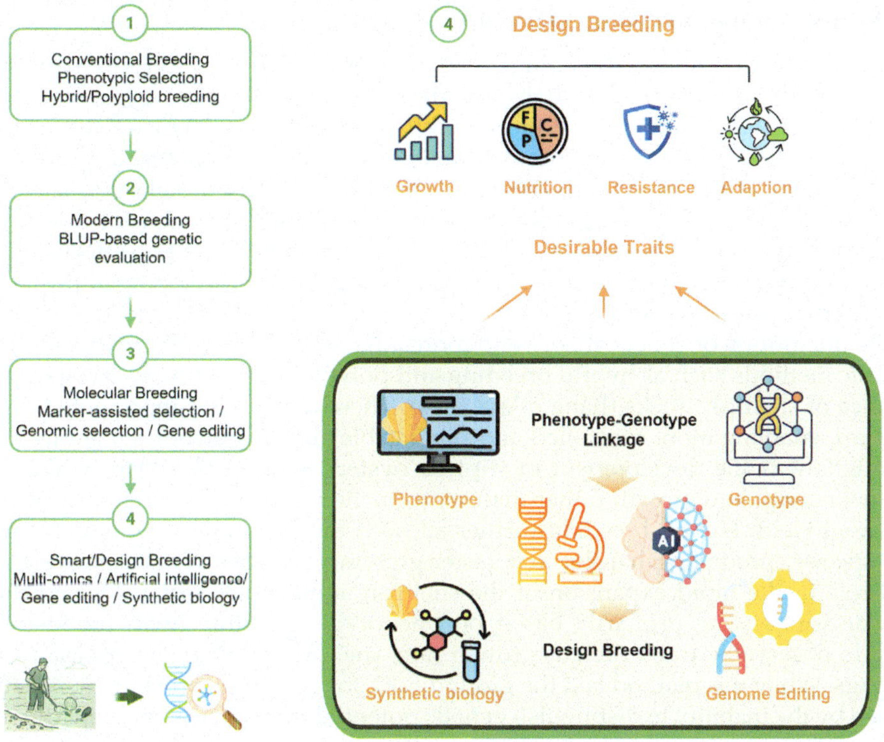

FIGURE 11.1 Main shellfish breeding technologies. Shellfish breeding has evolved through four distinct eras: Conventional breeding (1.0), modern breeding (2.0), molecular breeding (3.0), and smart/design breeding (4.0).

footprint of aquaculture. Innovations in breeding technologies, such as the development of species that are more resilient to environmental stresses, will help make aquaculture more sustainable. Furthermore, the application of closed-loop aquaculture systems, which reduce water consumption and minimize waste, will be essential in promoting environmentally responsible practices. With the integration of cutting-edge technologies such as biotechnology and genetic engineering, the blue seed industry is poised to lead the aquaculture industry toward a future of green, efficient, and high-quality development. This progress will play a vital role in supporting marine economic growth and the construction of an ecological civilization.

11.3 Main breeding technologies and applications

Shellfish breeding is a central aspect of the blue seed industry, as it directly impacts the productivity and sustainability of aquaculture systems. Over the years, shellfish breeding has evolved from traditional empirical methods based on observation and selection to advanced molecular-designed breeding techniques. The transition from conventional methods to molecular breeding represents a paradigm shift in how shellfish are developed and cultivated. This section explores the different stages of shellfish breeding technologies and their applications in the blue seed industry (Fig. 11.1).

11.3.1 Conventional breeding (Breeding 1.0 era)

The earliest phase of shellfish breeding relied on empirical observation and natural selection. Early practitioners identified individuals with desirable traits, such as rapid growth or disease resistance, for use as broodstock. Although this approach was labor-intensive and lacked scientific rigor, it established a foundational framework for subsequent breeding advancements. A key achievement of this era was the domestication of local varieties of oysters, scallops, and other shellfish (Guo, 2024). These populations not only formed the basis of modern shellfish aquaculture but also served as valuable genetic resources for later breeding efforts. The inherent genetic diversity within these local strains has proven crucial for developing improved varieties with enhanced performance and resilience.

With the integration of genetic theory and more advanced techniques, breeders began adopting scientific methods such as hybrid breeding and polyploid induction to enhance the genetic performance of shellfish stocks (Jiang et al., 2024). These approaches represented a significant leap forward, enabling more controlled and predictable genetic outcomes. Among the notable breakthroughs was the development of triploid oysters (Guo et al., 1996), which are sterile and channel energy into growth rather than reproduction, leading to accelerated growth rates and increased yield. Hybrid breeding further allowed the integration of favorable traits from different species, producing hardier and faster-growing varieties. These innovations played a pivotal role in the rapid expansion of the shellfish aquaculture sector. During this period, breeding decisions were primarily based on phenotypic selection, with broodstock chosen according to observable traits such as growth rate, shell morphology, or disease resistance in farm settings. While this method laid the groundwork for genetic improvement, its efficacy was constrained by the inability to distinguish genetic potential from environmental influences, such as variations in water temperature or feed, resulting in slow and uncertain progress, as true breeding values remained largely unquantified.

11.3.2 Modern breeding (Breeding 2.0 era)

Shellfish breeding entered a new era with the adoption of genetic theories and advanced statistical methods, most notably the introduction of best linear unbiased prediction (BLUP). This methodology marked a revolutionary advance by enabling breeders to separate genetic effects from environmental noise through the use of pedigree information. The central objective of BLUP in aquatic breeding is to accurately estimate individuals' breeding values, thereby identifying broodstock with the highest genetic potential (Liu, 2010).

According to the construction standard for farms of aquatic original and improved varieties, breeding traits are classified into three categories. I. Grounded in biological principles, BLUP quantifies genetic potential by constructing a linear model that relates phenotypic data to genetic effects. II. It is particularly effective for economic traits with moderate to high heritability, and performs optimally in populations with well-documented pedigrees and clear family structures. III. From an ecological perspective, the application of BLUP requires data collection under stable environmental conditions to minimize the confounding effects of non-genetic factors on phenotypic variation, including water temperature, feed composition, and stocking density.

The development of a BLUP-based genetic evaluation system for shellfish has led to the establishment of the world's first such platform for genetic assessment of scallops, forming the cornerstone of a modern shellfish breeding technical system (Zeng et al., 2022). By analyzing the performance of relatives across diverse environments, BLUP provides a more accurate estimation of individual genetic merit. This has facilitated more reliable selection of superior broodstock for key traits such as meat yield and survival rate, substantially accelerating genetic progress in major aquaculture species, such as scallops, oysters, and mussels, led to the development of a series of outstanding varieties such as 'Penglai-Red' scallops (Guo, 2009).

11.3.3 Molecular breeding (Breeding 3.0 era)

The third era of shellfish breeding incorporated molecular biology techniques, enabling selection for specific traits at the genetic level. The introduction of marker-assisted selection (MAS) and genomic selection (GS) allowed breeders to more accurately and efficiently identify genes associated with desirable traits, significantly enhancing the precision and speed of breeding programs. A pivotal development in this phase has been the completion of genome sequencing for various shellfish species.

GS, the cornerstone of this era, utilizes dense DNA markers spanning the entire genome to predict breeding values with high accuracy, even for traits that are difficult or costly to measure. This allows selection of juveniles before trait expression, drastically shortening the generation interval (Yue, 2014). By elucidating the genetic makeup of key species, breeders can identify specific genes related to growth, disease resistance, and environmental tolerance, facilitating more informed selection decisions. Consequently, this approach has facilitated the development of superior varieties such as 'Penglai-Red' 2/3/4, 'Haida Golden' scallops and 'Haiyifeng 11/12'. Furthermore, the emergence of gene editing tools such as CRISPR enables precise, targeted genetic improvements, such as introducing disease resistance or enhancing climate resilience, transcending the limitations of existing natural genetic variation (Robinson et al., 2024). Although gene editing-mediated shellfish breeding is still in its early stages, this technology has already proven valuable in elucidating these mechanisms and identifying key target genes. For instance, using both RNAi and CRISPR/Cas9 to investigate the function of *cfap206* in the bivalve *M. lateralis*, researchers revealed the gene's dual role in embryonic cilia formation and spermatogenesis, with reduced cilia numbers and impaired sperm motility resulting from tail assembly defects (Wang et al., 2022b). The integration of gene editing technologies such as CRISPR-Cas9 now allows for targeted modification of specific genes to improve traits such as growth rate, disease resistance, and environmental adaptability.

11.3.4 Smart/design breeding (Breeding 4.0 era)

The fourth era of shellfish breeding marks the transition to intelligent design-based approaches. We are now at the dawn of the 4.0 era—the age of smart breeding—defined by the synergy of "Design" and artificial intelligence (AI). This paradigm shifts the focus from selection to predictive design. AI and machine learning algorithms integrate massive multiomics datasets (genomics, transcriptomics, and proteomics) with environmental and production data to construct digital models of biological systems. This enables the in silico design of ideal shellfish genotypes tailored to specific environments or market demands.

AI-driven predictive models can forecast growth performance, optimize feeding regimes, and predict disease outbreaks, transforming shellfish breeding into a highly precise, efficient, and data-driven discipline. The integration of gene editing, bioinformatics, and AI allows breeders to implement precise genetic modifications. AI plays a critical role by analyzing large datasets and simulating the outcomes of various breeding strategies. By processing genetic and phenotypic information, AI algorithms identify optimal breeding candidates, enabling faster and more accurate decision-making. The combination of gene editing, bioinformatics, and AI represents a monumental leap forward, offering unprecedented precision and efficiency in shellfish breeding.

11.4 Molecular-designed breeding and gene manipulation technology

11.4.1 Molecular-designed breeding

Molecular-designed breeding is an efficient and precise breeding technology developed alongside advances in genetics, molecular biology, genomics, and modern biotechnology. It integrates various theoretical and technical approaches to design, simulate, and optimize breeding processes and strategies based on predetermined breeding objectives. This enables the proposal of optimal varietal genotypes that meet breeding goals, as well as the selection of the best parental genotype combinations, breeding pathways, and strategies to enhance predictability, achieve directional and efficient precision breeding, and significantly improve breeding efficiency. The concept of molecular designed breeding was first proposed by Dutch scientists Peleman and van der Voort in 2003, who also trademarked the term "Breeding by Design" (Peleman & Van Der Voort, 2003). The goal was to control allelic variations of all genes governing traits to develop superior varieties in a short time. Because of its advantages in reducing the blindness of conventional breeding and shortening breeding cycles, molecular-designed breeding quickly became a cutting-edge technology of interest to breeders.

11.4.1.1 Strategies and advances of designed breeding

Significant progress has been made in molecular-designed breeding. For instance, breeders have proposed a three-step molecular designed breeding strategy in China: First, finely map genes and loci related to economic traits and analyze interactions between trait-related genes as well as gene–environment interactions; second, determine varietal genotypes that meet breeding requirements under different environmental conditions based on breeding objectives; third, design breeding plans and optimal implementation pathways based on target genotypes to carry out molecular designed breeding. In crops, after the completion of the fine genomic map of rice, researchers proposed the concept of designed breeding for rice in 2004, aiming to develop new super rice varieties with breakthroughs in yield, quality, and resistance by aggregating favorable genes based on a precise understanding of the regulatory mechanisms of major agronomic traits (Rao et al., 2014). Currently, by elucidating the molecular mechanisms of ideal plant architecture and quality formation in rice, key genes for ideal plant architecture have been discovered, and the fine regulatory network of rice eating quality has been clarified. Using design breeding strategies, quantitative trait loci (QTLs) related to yield, rice appearance quality, cooking and eating quality, and ecological adaptability have been optimally combined, successfully developing varieties that aggregate superior alleles of these traits, achieving efficient, precise, and directional molecular designed breeding. Recent breakthroughs include the use of a key enzyme (β-1,4-glucanase) to enable grafting between distantly related species, allowing one plant to yield eggplants aboveground and potatoes belowground (Feng et al., 2024). Researchers have successfully developed perennial rice, capable of being harvested multiple times

(Zhang et al., 2023), and transformed eggplant seedlings into "eggplant trees" that fruit year-round with a single planting (Xing et al., 2024).

In the field of aquaculture breeding, molecular-designed breeding research targeting important aquaculture species, such as fish and shellfish, was initiated in China between 2006 and 2015. The prerequisite for molecular-designed breeding is a clear understanding of the genetic basis, molecular mechanisms, and regulatory networks controlling target traits. With the rapid development of genome sequencing technologies and declining sequencing costs, the construction of fine genomic maps for major aquaculture species has been achieved, including algae, sea cucumber, shellfish, shrimp, and fish, over the past decade (Li et al., 2017; Li et al., 2018; Wang et al., 2017; Wang et al., 2020; Wang et al., 2022a; Zhang et al., 2019; Zhang et al., 2012). This has not only revealed omics mechanisms related to growth, development, and environmental adaptation but also laid the foundation for the precise analysis of economic traits in aquatic organisms. The development of a series of high-throughput whole-genome variant screening and marker genotyping technologies has also provided efficient and low-cost genetic analysis tools for molecular breeding in aquaculture. For instance, in fish that have already produced gene-edited strains with significant improvements in multiple important commercial traits, such as the "AquAdvantage" transgenic salmon and intermuscular bone-free fish strains (Dong et al., 2023; Gan et al., 2023; Ledford, 2015). In addition, numerous trait genes related to growth, quality, and resistance have been finely mapped in aquaculture species, such as sex-determination genes *Dmrt*, *Amh*, and *Foxl2*, the key gene *BCO-like* regulating carotenoid accumulation, and the growth key gene *E2F3*, with the molecular mechanisms of trait variation revealed (Li et al., 2019; Ning et al., 2019; Sun et al., 2022). Based on the large-scale identification of key trait genes and loci, multiomics analyses have been used to construct gene regulatory networks for traits, enabling the analysis of trait regulatory mechanisms using whole-genome, transcriptome, proteome, and metabolome information.

Smart breeding, powered by big data and AI, will also accelerate the discovery of key trait-related genes and loci (Xu et al., 2022). By analyzing the phenotypic effects of allelic variations across the genome and building deep learning models for molecular phenotypes, we can systematically decipher the molecular mechanisms underlying important traits under varying environmental conditions. This knowledge enables the design of novel superior alleles not found in nature, guiding gene editing strategies and enhancing the predictability and controllability of breeding outcomes. Although whole-genome-scale gene editing has been implemented in crops, such as rice, that significantly boosted breeding efficiency (Shen et al., 2017), similar approaches have not yet been widely achieved in most aquatic species. As the genetic regulatory mechanisms of traits become clearer and efficient gene editing systems are established, molecular designed breeding is expected to become a key driver of progress in the aquaculture seed industry Furthermore, synthetic breeding, integrating biotechnology, molecular and systems biology, computational biology, and genetic engineering, offers new pathways for molecular designed breeding in aquaculture (Bahrami & Najafi, 2019). These integrative approaches promise to reshape not only trait performance but also production systems, contributing to a more sustainable and efficient future for global protein supply.

11.4.1.2 Key approaches of molecular-designed breeding

Molecular-designed breeding operates through three complementary approaches. The first involves aggregating existing favorable genetic variations within germplasm resources. This is achieved by constructing genotype–phenotype linkage models and databases for breeding traits, drawing on detailed information concerning trait-related genes and loci, such as their chromosomal positions, genetic effects, biochemical networks, expression pathways, gene–gene interactions, and interactions with genetic backgrounds and environments. These models enable the simulation and prediction of phenotypic outcomes for diverse genotypes, thereby facilitating the design of target genotypes tailored to specific breeding objectives under varying aquaculture conditions, as well as the subsequent planning of breeding processes.

With the rapid advancement of the internet, big data, and AI technologies, leading breeding companies in Europe and America are leveraging their accumulated phenotypic and genetic data to drive an intelligent revolution in biological breeding. In terms of molecular markers, all whole-genome variations are now being incorporated into model construction, providing a foundation for accurately analyzing the genetic basis of traits and evaluating the phenotypic effects of individual loci. More importantly, the ability to measure phenotypic traits accurately and at high throughput, known as phenomics, has become central to achieving smart breeding. In plant breeding, various sensors, including cameras, LiDAR, hyperspectral imagers, thermal infrared imagers, and chlorophyll fluorescence imaging systems, have been integrated into automated mobile platforms for large-scale phenotypic data acquisition (Li et al., 2014). These systems can automatically collect dozens of traits related to plant architecture, pollination, leaves, panicle number, and pest and disease incidence. The integration and analysis of such massive phenotypic data support the establishment of intelligent phenotypic databases and automated analysis technologies. Since traits result from the interaction between genotype and environment, deep learning and other big data modeling approaches are being used to comprehensively analyze environmental, phenotypic, and genotypic information. This helps construct relational models that simulate potential yield and quality performance of varieties under different environmental conditions. However, the application of smart breeding in aquatic organisms remains in its early stages (Zhang et al., 2023). Despite this, technological foundations are being laid in key areas, such as pan-genome analysis based on low-depth whole-genome resequencing, high-throughput automated phenotyping (e.g., X-ray detection of shellfish muscle traits), and the development of smart breeding model algorithms tailored to aquaculture (Wang et al., 2024; Zhao et al., 2021). Further refinement of these technologies and in-depth data mining will position the aquaculture breeding sector at the forefront of international innovation.

The second approach focuses on creating novel favorable variations based on a clear understanding of trait regulatory mechanisms. This employs technologies such as CRISPR-Cas to precisely edit coding or regulatory sequences of genes controlling target traits. For instance, in tomatoes, European and American researchers performed whole-transcriptome and metabolome analyses across numerous recombinant inbred lines, identifying 959 genes associated with the tomato metabolome (López et al., 2015; Ullah et al., 2025). This work allowed the construction of gene expression regulatory models and interaction networks capable of predicting relationships between gene expression and metabolic phenotypes, ultimately linking these to traits such as growth and quality. Using these models, researchers redesigned genomic configurations and improved key traits by modulating the expression of critical genes. In aquatic organisms, current efforts in molecular-designed breeding mainly center on phenotypic prediction using whole-genome markers. This strategy involves estimating the phenotypic effects of genome-wide variant loci in a reference population, then applying these estimates to predict phenotypic values of offspring from various crossing combinations. Such predictions enable the rational design of parental matches to yield optimal progeny. In recent years, gene editing has also been successfully applied in aquaculture breeding. For example, editing the *Dmrt* gene in half-smooth tongue sole resulted in a significant increase in the growth rate of male fish (Cui et al., 2017), thereby boosting production yields. Similarly, targeting key genes controlling intermuscular bone development in farmed freshwater fish led to the production of crucian carp and blunt snout bream with reduced or absent intermuscular bones, enhancing both consumer acceptance and processing efficiency.

11.4.2 Gene manipulation

The ability/technology to manipulate genetic changes at the molecular level has led to a revolution in the field of biology. It lies at the intersection of techniques in genetics, cell biology, engineering, and synthetic biology, impacting fields from modern medicine (especially in gene therapy, somatic cell cloning, and the establishment of disease models) to the seed industry (including agriculture and aquaculture), by allowing for precise modifications to an organism's genome to introduce new traits or functions. This review will explore scientific advances gained by major genetic manipulation technologies and yield insights into their applications and how they have revolutionized the fundamental biological research, the innovative therapeutic options in human healthcare, as well as the efficiency and creation of germplasm improvement.

11.4.2.1 Transgenic technology

Transgenic technology is a biotechnological approach that involves the introduction of foreign genetic material into an organism's genome, mainly including mechanical delivery techniques for direct transfer, as well as genetic engineering that employs vectors to shuttle genetic material into target organisms.

Several scientific paths converged in the early 1980s to establish mechanical delivery techniques (Fig. 11.2): (1) Gene gun, also known as biolistics, uses high-pressure helium to propel DNA-coated microprojectiles into target cells, which is particularly effective for bypassing plant cells with wall barriers (Iida et al., 1990; Sanford, 1988). (2) Electroporation applies electrical fields to create transient pores in cell membranes, allowing foreign genetic material to enter the cell in both plant and animal cells (Tsong, 1991). (3) Microinjection, which injects foreign genetic material directly into the nucleus/cytoplasm through a fine needle with a microscope. Although labor-intensive, it allows for precise control of gene delivery for transduction-challenged cells. As physical delivery techniques, the advantages include high transduction efficiency and low cytotoxicity. However, the physical damage is inevitable; it is important to consider cell survival, especially to avoid inducing any abnormal development when handling embryos.

Genetic engineering techniques are designed to minimize physical damage to cells while effectively delivering foreign DNA or RNA. This is often accomplished using carriers such as plasmids, viral particles, and nanoparticles (Fig. 11.2). Plasmids are circular DNA molecules that can replicate independently within a host cell, making them ideal for introducing new genetic material without causing harm (Gill et al., 2009; Taniyama et al., 2012). For instance, the transposon vectors utilize the natural ability of transposons, or "jumping genes," to integrate into various locations in a genome, facilitating stable gene insertion (Palazzo & Marsano, 2021). Viral vectors (such as lentiviral vectors and adenoviral vectors), derived from viruses that have been modified to be non-pathogenic, efficiently deliver genetic material into target cells by exploiting the virus's natural ability to infect cells (Li & Samulski, 2020). In addition, nanoparticles, including lipid-based nanoparticles (e.g., liposomes), metallic nanoparticles (e.g., gold nanoparticles), polymeric nanoparticles (e.g., polyethylene glycol), and silica nanoparticles,

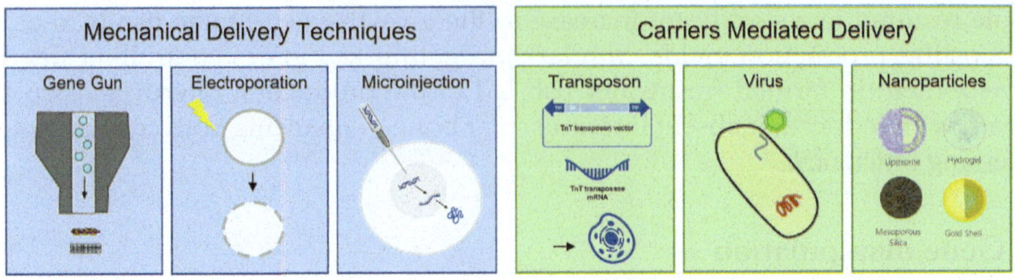

FIGURE 11.2 Delivery approaches used in transgenic technology. Major transgenic delivery approaches, including mechanical delivery techniques and carrier-mediated delivery techniques.

which can encapsulate nucleic acids, provide a versatile platform for gene delivery, offering protection against degradation and allowing for targeted delivery (Duan et al., 2021; Ho et al., 2021).

The pioneering work using transgenic technology on gene targeting in mice through homologous recombination was recognized by the Nobel Prize in Physiology or Medicine in 2007. Their methodologies enabled the creation of genetically modified organisms (GMOs) that serve as invaluable models for studying human diseases, examining gene functions, and developing new therapies (Capecchi, 1989; Evans & Kaufman, 1981). Till now, transgenic techniques have played crucial roles in understanding various biological processes, and are introduced to correct genetic disorders, such as in the treatment of cystic fibrosis and cancer, illustrating the profound impact of this technology on modern medicine and genetics (Arjmand et al., 2020; Hyde et al., 1993; Wadhwa et al., 2021). What is more, transgenic technology also revolutionized the seed industry, offering significant advancements in agriculture and aquaculture. In agriculture, transgenic crops have been engineered to exhibit traits such as pest resistance, herbicide tolerance, and enhanced nutritional profiles, leading to increased yields and reduced reliance on chemical inputs (Brookes & Barfoot, 2012). Similarly, in aquaculture, transgenic fish have been developed to grow faster and resist diseases, contributing to more sustainable seafood production (Shakweer, Krivoruchko, Dessouki, & Khattab, 2023). In addition, a great application potential of somatic cell nuclear transfer (SCNT) based cloning is the production of genetically modified (transgenic) animals, providing valuable models and numerous opportunities for livestock husbandry and aquaculture (Niemann et al., 2011; Niemann & Lucas-Hahn, 2012). Genes that confer resistance to diseases or enhance growth rates can be inserted into donor somatic cells before nuclear transfer. For instance, a tuberculosis-resistant gene, namely NRAMP-1, was inserted into bovine fetal cells, which were later used as donor cells for the creation of SCNT-based TB-resistant cattle embryos (Gao et al., 2017). Besides, SCNT has been widely done mostly in aquaculture fish species since the 1960s (Luo et al., 2011). For instance, researchers pointed out that the application of interspecific SCNT establishes the crucial first steps by adjusting the cloning-methodology in sturgeon's reproduction (Fatira et al., 2018). The application of transgenic technology in mollusks, though still relatively limited compared to other model organisms, has progressed steadily since its introduction in the mid-1990s. Early breakthroughs were achieved in the California red abalone (*H. rufescens*) via electroporation (Powers, Kirby, Cole, & Hereford, 1995) and in the dwarf surf clam (*M. lateralis*) via retroviral vectors (Lu et al., 1996), marking the first successful demonstrations of exogenous gene transfer in molluscan species. These foundational studies paved the way for subsequent research in commercially

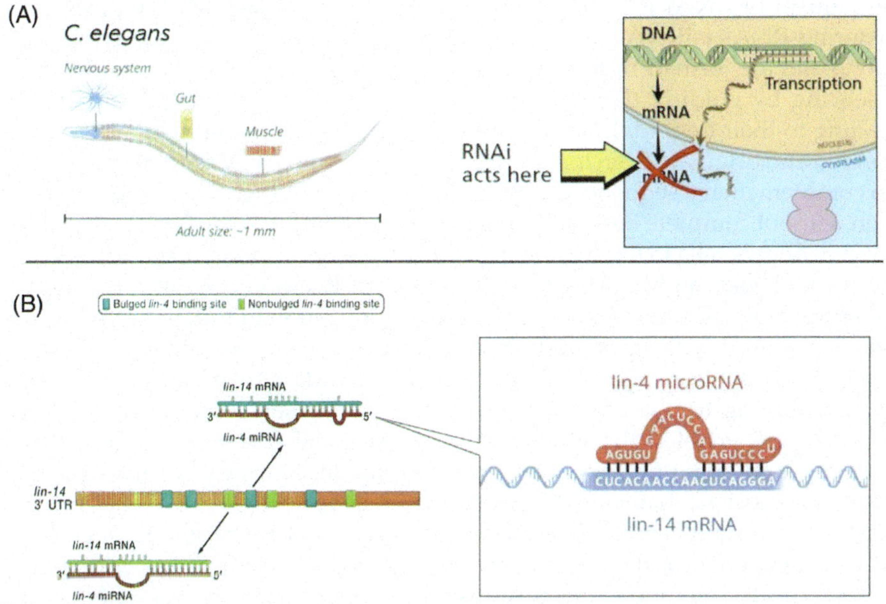

FIGURE 11.3 Nobel prizes in 2006 for RNAi and 2024 for microRNA in *C. elegans*. (A) *C. elegans*, a valuable model for studying cell development, was the subject of the 2006 Nobel Prize-awarded discovery of RNA interference by Andrew Fire and Craig Mello. (B) The discovery by Ambrosand Ruvkun that the lin-4 microRNA regulates gene expression by binding to lin-14 mRNA was awarded the 2024 Nobel Prize in Physiology or Medicine. *Modified from The Nobel Committee for Physiology or Medicine. https://www.nobelprize.org/prizes/medicine/2024/press-release/?utm_source=twitter&utm_medium=social+media+&utm_campaign=nobel+prize+announcements+2024&utm_content=post.*

important species. Notable progress has been reported in the Pacific oyster (*C. gigas*) (Chen et al., 2018), slipper limpet (*Crepidula fornicata*) (Perry & Henry, 2015), and Eastern oyster (*C. virginica*) (Tracy et al., 2020), demonstrating successful integration of transgenes (Fig. 11.5A).

11.4.2.2 Gene silencing

Genetic modification involves introducing foreign genetic material, and in recent years, it has sparked ethical debates. Public acceptance of GM foods has been limited due to these concerns. Under these concerns, gene silencing has emerged as a pivotal technique in molecular biology, enabling researchers to selectively suppress the expression of specific genes without causing permanent changes to the genome. One of the most prominent methods for achieving this is RNA interference (RNAi), which has garnered significant recognition in the scientific community. Together with the discovery of microRNA, the relevant research culminated in the awarding of the Nobel Prize in Physiology or Medicine in 2006 and in 2024 (Fig. 11.3). Their groundbreaking research demonstrated that double-stranded RNA could effectively silence specific genes in *Caenorhabditis elegans*, revolutionizing the understanding of gene regulation (Bernstein et al., 2001; Fire et al., 1998).

The mechanism of RNAi is fundamentally mediated by small RNA molecules, especially small interfering RNAs (siRNAs) and microRNAs (miRNAs). siRNAs are typically 20–25 nucleotides long and are formed from double-stranded RNA precursors. They play a critical role in gene silencing by guiding the RNA-induced silencing complex (RISC) to specific mRNA targets, leading to their degradation (Hammond et al., 2000). In contrast, miRNAs are usually single-stranded and originate from longer primary transcripts. They regulate gene expression by binding to complementary sequences in target mRNAs, resulting in translational repression or degradation without complete destruction of the mRNA (Bartel, 2009; Grimson et al., 2007). This intricate system allows cells to finely tune gene expression in response to various developmental and environmental cues, highlighting the importance of RNAi in cellular regulation.

The versatility of RNAi has led to its widespread application in functional genomics, making it an invaluable tool in both basic and applied sciences (Ding & Voinnet, 2007; Sibley et al., 2010). By selectively silencing genes, researchers can elucidate their roles in various biological processes, contributing to a deeper understanding of cellular mechanisms. For instance, in *C. elegans*, RNAi has enabled the identification of essential pathways involved in longevity, stress responses, and synaptic functions, contributing significantly to our understanding of fundamental biological mechanisms (Kim et al., 2005). In agriculture, RNAi has proven effective in enhancing crop traits such as stress resistance and yield by targeting specific genes that govern these characteristics, thereby facilitating the development of superior plant varieties (Saurabh et al., 2014). The aquaculture industry has also benefited from RNAi applications, where it is employed to enhance disease resistance in fish or shellfish species by silencing genes related to immune responses and pathogen susceptibility. For instance, RNAi techniques have been successfully used to study the microsporidian parasite (*Heterosporis saurida*) of lizardfish (*Saurida undosquamis*) and *Myxobolus cerebralis* (causative agent of whirling disease infection) in the salmonid fishes (Kumar et al., 2022). Besides, RNA-mediated knockdown of p38 mitogen-activated protein kinases leads to higher mortality of shrimp species against pathogenic bacterial pathogens (*Vibrio alginolyticus* and *Staphylococcus aureus*) (Yan et al., 2013). The adoption of RNAi in molluscan research began primarily with key gastropod models such as *Aplysia* (Lee et al., 2001) and *B. glabrata* (Jiang, Loker, & Zhang, 2006). As the technology matured, its application expanded to various species, including abalone, sea slug, golden apple snail, mud snail, scallops, clams, oysters, cuttlefish and octopus (Chen et al., 2024a; Gharbiah, Nakamoto, Johnson, Lambert, & Nagy, 2014; Li et al., 2019; Sun, Yu, & Li, 2022; Wang et al., 2022b; Yao et al., 2024; Zheng, Cao, Qiu, & Chi, 2022; Tu et al., 2025) (Fig. 11.5B). All of these studies will identify key areas within the aquaculture field where gene silencing could be applied.

11.4.2.3 Genome editing

In recent years, consumers have become increasingly focused on the nutritional quality and economic value of aquatic products. As a result, the aquaculture industry is facing pressure to shift from single-trait breeding, such as fast growth, to the multitraits breeding, like with high quality, resistance to stress, as well as other economic traits. This has led to a growing demand for more precise genetic regulation, making genome editing an increasingly valuable tool. Genome editing technology has revolutionized the field of genetics, allowing for precise modifications of DNA in living organisms. The foundational work in recombinant DNA technology during the 1970s and 1980s allowed scientists to manipulate DNA sequences, setting the stage for later developments (Khan et al., 2016). The discovery of zinc finger nucleases (ZFNs) and transcription activator-like effector nucleases (TALENs) further refined the ability to target specific genomic loci, facilitating more precise genetic modifications (Cathomen & Keith Joung, 2008; Sun & Zhao, 2013). However, it was the advent of the CRISPR-Cas9 system that marked a revolutionary leap in genome editing. Originating from a bacterial adaptive immune system, CRISPR-Cas9 harnesses RNA-guided endonucleases to create double-strand breaks at specific locations in the genome, enabling targeted gene editing with remarkable efficiency (Doudna & Charpentier, 2014; Jinek et al., 2012). This groundbreaking technology was awarded the Nobel Prize in Chemistry in 2020, making it a cornerstone of modern molecular biology, fundamentally changing how scientists approach genetic engineering. Over time, the advancements of the CRISPR system have also introduced diverse variants, such as Cas12 and Cas13, expanding the tools available for genetic manipulation and therapeutic development (Zhang, 2019). Cas12, also known as Cpf1, not only cleaves DNA but does so in a way that generates sticky ends, facilitating more efficient DNA integration. Meanwhile, Cas13 targets RNA instead of DNA, making it suitable for applications in RNA editing and antiviral strategies (Abudayyeh et al., 2017). Other variant, such as CasX, was found to be a smaller protein that exhibits distinct structural features compared with other Cas proteins, which allows for more flexible delivery methods and enhanced specificity and efficiency (Kleinstiver et al., 2015; Shmakov et al., 2015). As research on CRISPR-Cas continues to advance, it promises to broaden the landscape of CRISPR-based genome editing technologies.

Notably, CRISPR technology is currently more suitable for well-established model aquatic animals, such as fish or gastropod mollusks, that have large eggs and high survival rates. For bivalves, which make up a significant portion of the aquatic market, there are still challenges related to low transfection efficiency and difficulties in generating positively edited adults. Most related research is still in the stage of studying larval phenotypes (Li et al., 2024; Wang et al., 2022b). Therefore further development of new CRISPR systems is essential. Recently, the innovative Fanzor system was discovered as a eukaryotic programmable RNA-guided endonuclease (reported to be a eukaryotic TnpB-IS200/IS605-like protein), which exists in diverse eukaryotes, including bivalves (such as Venus clam and Zebra mussel) (Jiang et al., 2023; Saito et al., 2023). Specifically, it was designed to enhance the precision and versatility of gene editing through RNA-guided systems termed OMEGA (Obligate Mobile Element-guided Activity), which offers advantages such as endogenous compatibility and a smaller, more compact molecular size, greater specificity and reduced off-target effects, making it a promising tool for future applications in aquaculture breeding (Bao & Jurka, 2013; Kato et al., 2022; Xu et al., 2024; Wang & Han, 2024; Fig. 11.4).

By enabling targeted modifications in the genomes, genome editing technologies have significantly advanced research using model animals. For example, researchers have created knockout mice to investigate the role of specific genes in obesity and metabolic disorders, revealing critical pathways involved in energy homeostasis (Khademi et al., 2024). In zebrafish, genome editing has facilitated studies on developmental biology and regenerative medicine, allowing scientists to explore gene function during organ development and repair (Outtandy et al., 2019). In addition, CRISPR has been utilized to generate models of human diseases, such as Alzheimer's disease, by introducing mutations that mimic pathogenic variants found in humans (De Plano et al., 2022), and researchers could develop transformative tools in the field of gene therapy. For instance, CAR-T (Chimeric Antigen Receptor T-cell) therapies have revolutionized personalized cancer treatments by allowing precise modifications of T cells, and researchers can optimize the design of CARs to improve their efficacy and safety, improving therapeutic outcomes (Liu, 2020; Ren et al., 2017; Rupp et al., 2017). What is more, genome editing technologies

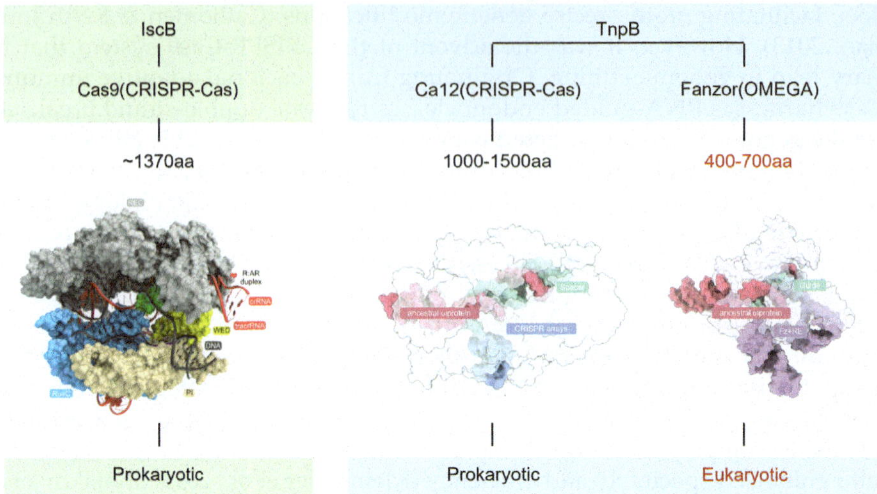

FIGURE 11.4 Cas9, Cas12, and Fanzor. Representative Structure of Cas9, Cas12 and Fanzor (Kato et al, 2022; Saito et al, 2023).

have also transformed the breeding industry by enhancing economic traits in both agriculture and aquaculture, such as disease resistance, yield, and growth rates. In agriculture, scientists have utilized CRISPR to develop crops with improved resistance to pests and diseases, such as wheat with enhanced resistance to powdery mildew and rice with increased tolerance to drought (Zaidi et al., 2018). Similarly, in aquaculture, genome editing has been applied to create fast-growing fish species, such as GM tilapia, which exhibit improved feed efficiency and growth rates (Wu et al., 2023). Genome editing in molluscs remains at an early stage compared with that in traditional model organisms and fish. Nevertheless, CRISPR/Cas9 editing platforms have been successfully established in various molluscs (Abe & Kuroda, 2019; Accorsi et al., 2025; Ahuja et al., 2023; Oonuma, Kuroda, Uchida, & Zhang, 2025; Wang et al., 2022b), enabling efficient and heritable gene knockout, fostering novel biological insights. For example, for the first time in a spiralian lophotrochozoan, CRISPR-based studies in *P. canaliculata* have demonstrated that the *Pax6* gene is essential for compound eye development, thereby establishing a non-vertebrate genetic model capable of complete eye regeneration (Accorsi et al., 2025) (Fig. 11.5C). These advances underscore the utility of genome editing as a fundamental tool for investigating basic mechanisms and evolutionary developmental biology in understudied lineages. Looking forward, continued refinement of these techniques would facilitate future applications in the designed breeding in shellfish. Fig. 11.5.

In summary, the advancement of gene manipulation technology has brought unprecedented opportunities to the aquaculture industry. However, significant challenges remain across different aquaculture species. Mollusks, which constitute a crucial component of China's aquaculture sector (accounting for 70% of total production), present a representative case. To date, genetic manipulation technology has been successfully applied to functional genomics in mollusks, enabling the identification of key economic trait genes and facilitating genetic breeding programs across various molluscan species (Chen et al., 2018; Huan et al., 2021; Li et al., 2021; Yu et al., 2019). Researchers are actively developing functional model systems, such as the dwarf surf clam

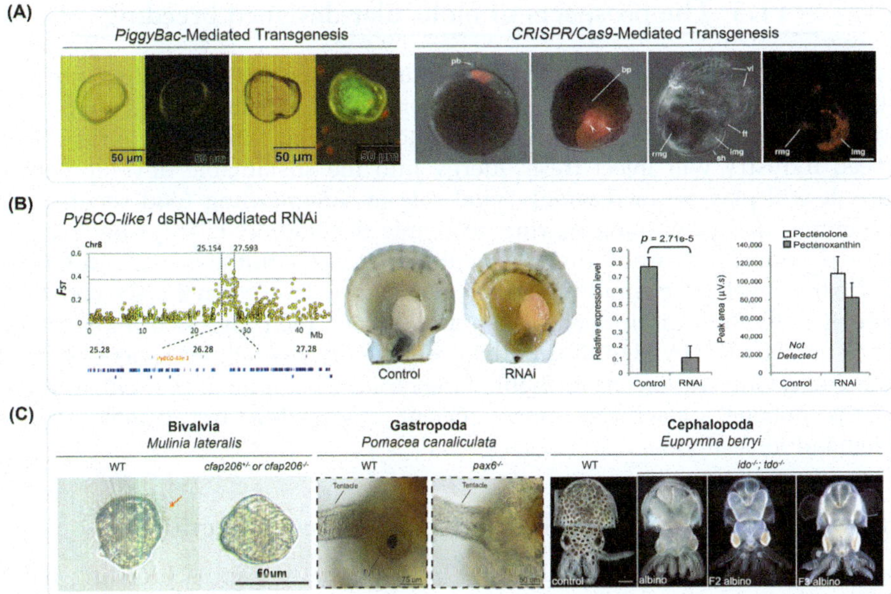

FIGURE 11.5 Representative phenotypes of gene manipulation in shellfish. (A) GWAS analysis identified *PyBCO-like 1* was responsible for carotenoid coloration in Yesso scallop (*Patinopecten yessoensis*), and its function was verified in vivo via RNA interference (RNAi) (Li et al., 2019). (B) Transgenesis mediated by the piggyBac transposon in the Pacific oyster (*Crassostrea gigas*) and CRISPR/Cas9-mediated transgenesis in the slipper snail (*Crepidula fornicata*) (Chen et al., 2018; Perry & Henry, 2015). (C) In the dwarf surf clam (*Mulinia lateralis*), knockout of Cfap206 impairs ciliogenesis during development. In the apple snail (*Pomacea canaliculata*), *Pax6* is essential for camera-type eye development. In squid (*Euprymna berryi*), CRISPR has been used to generate albino lines, which provide clear optical access to the optic lobes for in vivo imaging (Wang et al., 2022b; Ahuja et al., 2023; Accorsi et al., 2025).

as well as the gastropod (Huan & Liu, 2024; Lian et al., 2025 a). However, successful generation of gene-edited F1 strains has been achieved in limited species (Abe & Kuroda, 2019; Accorsi et al., 2025; Ahuja et al., 2023; Oonuma, Kuroda, Uchida, & Zhang, 2025; Wang et al., 2022b). These limitations primarily arise from intrinsic biological constraints characteristic of mollusks, excessively small egg diameter, as well as high mortality rate during metamorphosis, which render conventional techniques such as microinjection prohibitively time-consuming, labor-intensive, and inefficient for most molluscan species. In recent years, multiple research teams have conducted a series of studies to address these issues, developing high-throughput delivery methods as well as efficient promoters for all-in-one gene-editing vector plasmids (Chan et al., 2022; Yoon et al., 2022; Yu et al., 2025; Yue et al., 2020). Compared with commercial fish, which have already produced gene-edited strains with significant improvements in multiple important commercial traits, such as "AquAdvantage" transgenic salmon and intermuscular bone-free fish strains (Dong et al., 2023; Gan et al., 2023; Ledford, 2015), molluscan genetic engineering remains in its nascent stages. However, with the increasing availability of high-quality omics data and continuous advancements in gene editing platforms, this technology holds tremendous potential for transforming molluscan aquaculture and driving innovation across the broader aquaculture seed industry.

11.5 The prospects of molecular-designed breeding

As modern biotechnologies, including gene editing and molecular breeding, are increasingly applied, the blue seed industry is poised to experience significant growth opportunities. By continuously driving innovation in breeding technologies and facilitating industrial upgrades, the blue seed industry will inject fresh energy into the sustainable development of global aquaculture. It will play an increasingly vital role in safeguarding food security, enhancing dietary structures, and promoting marine economic development. This ongoing "blue revolution" not only transforms traditional aquaculture practices but also paves the way for new approaches to the development and utilization of marine resources. In this context, the breeding revolution driven by multidisciplinary integration during the progress of molecular-designed breeding has been fueled by significant advancements in several key areas of biotechnology. These technological innovations have enabled the rapid development of shellfish varieties with superior traits, making aquaculture more efficient, sustainable, and capable of meeting the growing demand for seafood.

11.5.1 Pan-genome insights and big data platform construction

One of the most transformative advances in molecular-designed breeding has been the development of pan-genome sequencing, which provides a comprehensive representation of the genetic diversity within a species (Pantoja et al., 2020). In contrast to conventional genomic approaches that rely on a single reference genome, pan-genome sequencing encompasses the full spectrum of genetic variation present across a population. This approach offers profound insights into the genetic architecture of shellfish populations, facilitating the identification of critical genes governing economically important traits such as growth performance, disease resistance, and environmental adaptability (Gerdol et al., 2020). By serving as an invaluable genomic resource, pan-genome sequencing empowers precision breeding strategies. Through detailed analysis of genetic variations among multiple individuals, breeders can pinpoint trait-associated markers, thereby enhancing the efficiency of parental selection and accelerating the development of improved shellfish varieties tailored to diverse aquaculture environments.

To support such genomic-driven breeding efforts, it is essential to develop high-throughput, multiparameter, and real-time continuous trait monitoring technologies. Concurrent investment in trait detection infrastructure and individual identification systems is critical. Integrated correlation analyses among trait parameters, alongside association studies between phenotypic expressions and molecular markers, will enable the establishment of an in vivo phenotypic data acquisition framework. This system will ensure accurate and systematic trait characterization, providing a reliable phenotypic foundation for molecular-designed breeding. Furthermore, the construction of a high-throughput phenotypic analysis platform is imperative. This involves advancing large-scale image processing algorithms and data visualization techniques, while incorporating emerging technologies such as big data analytics, virtual reality, and AI. These efforts will yield sophisticated analytical tools and support the creation of a unified phenotypic database for aquatic organisms.

Complementing these initiatives, a centralized big data platform for aquatic genetic resources should be established, integrating germplasm and genomic data to consolidate, manage, and analyze the vast and diverse genetic information of key aquaculture species. In parallel, an integrated information service and management system for aquaculture breeding will facilitate end-to-end traceability and intelligent decision-making. Such a platform would encompass the entire breeding chain, from broodstock provenance and induced maturation to spawning protocols, mating design, breeding operations, and commercial deployment through continuous monitoring, large-scale data integration, and predictive analytics.

11.5.2 The role of artificial intelligence in molecular-designed breeding

The application of big data and AI is a defining feature of modern molecular-designed breeding. With the increasing availability of genomic data and phenotypic observations, breeders are now able to analyze vast amounts of information to identify key genetic markers linked to desirable traits. This process is made possible by AI algorithms, which can process large datasets and uncover patterns that might otherwise be overlooked.

Since the initial sequencing of the first shellfish genomes, sequencing technologies have progressed rapidly, generating massive amounts of genomic data. Handling this big data requires high-performance computing platforms and advanced algorithms to process and analyze the information. Automated phenotyping systems are also being developed to collect detailed phenotypic data on important traits such as growth rate, disease resistance, and environmental tolerance. These systems use sensors and imaging technologies to track the development of shellfish throughout their life cycle.

AI and machine learning algorithms can be applied to both genomic and phenotypic data to extract key genetic markers and predict the relationship between genotype and phenotype. By identifying these genetic markers, breeders can make more informed decisions and select individuals with the best genetic potential. Machine learning models can also be used to predict the outcomes of different breeding strategies, helping breeders optimize their selection processes and reduce the time and resources required for breeding new varieties. Deep learning, a subset of machine learning, is being used to build predictive models that can forecast the outcomes of different breeding strategies. These models are trained on large datasets and can accurately predict the relationship between genetic changes and phenotypic traits. By using deep learning, breeders can make more accurate predictions about which breeding strategies will be most successful, leading to faster and more efficient breeding cycles.

11.5.3 Tailored genome editing strategies and the creation of ideal traits

The development of gene editing technologies in shellfish requires overcoming species-specific challenges and promoting the collaborative optimization of multiple technological platforms. In terms of delivery strategies, it is necessary to move beyond traditional microinjection and develop high-throughput techniques such as electroporation, nanocarriers, and ReMOT control that balance efficiency with safety. Simultaneously, the editing tools themselves urgently require upgrades, including the adoption of high-fidelity Cas variants, compact nucleases (e.g., Fanzor), and the application of precise systems such as base editing and prime editing that avoid double-strand breaks to enhance safety and reduce off-target risks. Furthermore, the key to overcoming species barriers lies in mining and constructing universal or tissue-specific endogenous promoters suitable for mollusks to drive the efficient and controllable expression of genetic elements. The coordinated advancement of these technological directions will ultimately provide a powerful and reliable toolbox for gene function analysis and genetic breeding in shellfish.

Moreover, synthetic biology, an interdisciplinary field integrating biological science and engineering principles, offers innovative approaches for designing novel traits in shellfish that transcend natural genetic limitations. This technology enables the construction of de novo genetic pathways and the reprogramming of existing ones to generate shellfish with unprecedented characteristics. For instance, synthetic biology can be applied to develop varieties with enhanced nutritional profiles, such as elevated levels of omega-3 fatty acids, vitamins, or essential minerals. Furthermore, it holds substantial potential for mitigating climate change-related challenges by engineering traits that confer resilience to temperature shifts, ocean acidification, and other environmental stressors, thereby supporting the sustainable future of aquaculture.

To underpin such advanced trait development, it is imperative to establish efficient technologies for the acquisition of multiomics data from aquatic organisms, encompassing genomics, transcriptomics, and proteomics. Concurrently, molecular-designed breeding models and algorithms must be developed to integrate and interpret these multiomics datasets, enabling accurate phenotype simulation and prediction from genomic information. The incorporation of AI, such as deep neural network-based approaches, along with tailored genome editing strategies, can significantly enhance the predictive power and scalability of such models, thereby providing critical guidance for the designed breeding in shellfish.

11.5.4 Development of emerging model species and cell lines

Advancing both forward and reverse genetics is essential to elevate understudied species into well-established models. Emerging shellfish systems for stem cell and aging research provide new insights into developmental and evolutionary homology (Lian et al., 2025 a). Key efforts are underway to develop molluscan models, including *Biomphalaria glabrata*, *Lymnaea stagnalis*, *Lottia peitaihoensis*, *Euprymna berryi*, and *Mulinia lateralis* (dwarf surf clam). *M. lateralis* (also known as the "molluscan fruit fly"), noted for its small size, short generation time, high fecundity, and ancestral karyotype, has enabled transgenic techniques since the 1990s (Lu et al., 1996). Since 2017, a systematic platform has been established featuring laboratory rearing, a haplotype-level genome, and transcriptomic atlases. CRISPR/Cas9, RNAi, and microinjection have been applied for gene editing and cell tracing, underscoring the potential of this model for decoding bivalve development (Wang et al., 2022a; Lian et al., 2025a; Lian, et al., 2025b).

Complementing whole-organism studies, cell lines are also crucial for in vitro analysis of gene function, delivery methods, and diverse applications from virology to ecotoxicology. Although molluscan cell culture remains challenging due to contamination risks, limited proliferation, and suboptimal media, the *B. glabrata* Bge cell line, despite karyotypic variation, represents the established immortalized line (Wheeler et al., 2018; Yoshino et al., 1999). Future work should focus on deriving pluripotent stem cells, identifying immortalization-related genes, and optimizing culture media to support more stable cell lines. Combined with improved genetic tools, such advances will provide essential platforms for functional gene studies in mollusks.

11.6 Conclusion: Revolutionizing shellfish breeding for sustainable aquaculture

The application of molecular-designed breeding in shellfish aquaculture represents a major breakthrough in the field of marine biotechnology. By combining cutting-edge technologies such as gene editing, synthetic biology, and artificial intelligence, breeders can create shellfish with specific traits that meet the needs of a rapidly changing global environment. As these technologies continue to develop, they will enable the aquaculture industry to meet the growing demand for seafood while ensuring sustainability and food security. The integration of these advanced breeding technologies will drive technological innovation and industrial upgrading throughout the entire aquaculture breeding field, contributing to the realization of blue food security. Molecular-designed breeding will not only transform traditional aquaculture practices but also pave the way for new approaches to the development and utilization of marine resources. With significant opportunities for further advancements in technology and breeding strategies, the integration of these technologies will allow breeders to select for multiple desirable traits simultaneously, accelerating the development of superior shellfish varieties to achieve even greater precision and efficiency in their breeding programs.

References

Abe, M., & Kuroda, R. (2019). The development of CRISPR for a mollusc establishes the formin Lsdia1 as the long-sought gene for snail dextral/sinistral coiling. *Development, 146*(9), dev175976. doi:10.1242/dev.175976.

Abudayyeh, O. O., Gootenberg, J. S., Essletzbichler, P., Han, S., Joung, J., Belanto, J. J., Verdine, V., Cox, D. B. T., Kellner, M. J., Regev, A., Lander, E. S., Voytas, D. F., Ting, A. Y., & Zhang, F. (2017). RNA targeting with CRISPR-Cas13. *Nature, 550*(7675), 280–284. doi:10.1038/nature24049.

Accorsi, A., Pardo, B., Ross, E., Corbin, T. J., McClain, M., Weaver, K., ... Sánchez Alvarado, A. (2025). A genetically tractable non-vertebrate system to study complete camera-type eye regeneration. *Nature Communications, 16*(1), 6698. doi:10.1038/s41467-025-61681-6.

Ahuja, N., Hwaun, E., Pungor, J. R., Rafiq, R., Nemes, S., Sakmar, T., ... Rosenthal, J. J. C. (2023). Creation of an albino squid line by CRISPR-Cas9 and its application for in vivo functional imaging of neural activity. *Current Biology, 33*(13), 2774–2783. doi:10.1016/j.cub.2023.05.066.

Arjmand, B., Larijani, B., Sheikh Hosseini, M., Payab, M., Gilany, K., Goodarzi, P., Parhizkar Roudsari, P., Amanollahi Baharvand, M., & Mohammadi, N. H. (2020). The horizon of gene therapy in modern medicine: Advances and challenges. *Advances in Experimental Medicine and Biology, 1247*, 33–64. doi:10.1007/5584_2019_463.

Bahrami, A., & Najafi, A. (2019). Synthetic animal: Trends in animal breeding and genetics. *Insights in Biology and Medicine, 3*(1), 007–025. doi:10.29328/journal.ibm.1001015.

Bao, W., & Jurka, J. (2013). Homologues of bacterial TnpB_IS605 are widespread in diverse eukaryotic transposable elements. *Mobile DNA, 4*(1), 12. doi:10.1186/1759-8753-4-12.

Bartel, D. P. (2009). MicroRNAs: Target recognition and regulatory functions. *Cell, 136*(2), 215–233. doi:10.1016/j.cell.2009.01.002.

Bernstein, E., Caudy, A. A., Hammond, S. M., & Hannon, G. J. (2001). Role for a bidentate ribonuclease in the initiation step of RNA interference. *Nature, 409*(6818), 363–366. doi:10.1038/35053110.

Brookes, G., & Barfoot, P. (2012). Global impact of biotech crops: Environmental effects, 1996-2010. *GM crops & food, 3*(2), 129–137. doi:10.4161/gmcr.20061.

Capecchi, M. R. (1989). The new mouse genetics: Altering the genome by gene targeting. *Trends in Genetics, 5*(C), 70–76. doi:10.1016/0168-9525(89)90029-2.

Cathomen, T., & Joung, J. K (2008). Zinc-finger nucleases: The next generation emerges. *Molecular Therapy, 16*(7), 1200–1207. doi:10.1038/mt.2008.114.

Chan, J., Zhang, W., Xu, Y., Xue, Y., & Zhang, L. (2022). Electroporation-based CRISPR/Cas9 mosaic mutagenesis of β-tubulin in the cultured oyster. *Frontiers in Marine Science, 9*, 912409. doi:10.3389/fmars.2022.912409.

Chen, H., Liu, J., Wan, X., Zhang, M., Luo, L., Qiu, X., & Yang, C. (2024a). Research on the Mechanism of Metaldehyde on *Pomacea canaliculata*. *Journal of Agricultural and Food Chemistry, 72(25)*, 14152–14164. doi:10.1021/acs.jafc.3c06405.

Chen, J., Wu, C., Zhang, B., Cai, Z., Wei, L., Li, Z., ... Wang, X. (2018). PiggyBac transposon-mediated transgenesis in the Pacific oyster (*Crassostrea gigas*)– First time in mollusks. *Frontiers in Physiology, 9*, 811. doi:10.3389/fphys.2018.00811.

Chen, S., Xu, W., Lu, S., Hu, W., Wang, D., Hu, X., Zhou, Q., Liu, Q., Zhao, Z., Qin, Q., Wang, S., Liu, Y., & Cui, Z. (2023). Development strategy for aquatic breeding biotechnology. *Chinese Journal of Engineering Science, 25*(4), 214. doi:10.15302/j-sscae-2023.07.023.

Cui, Z., Liu, Y., Wang, W., Wang, Q., Zhang, N., Lin, F., ... Chen, S. (2017). Genome editing reveals dmrt1 as an essential male sex-determining gene in Chinese tongue sole (*Cynoglossus semilaevis*). *Scientific Reports, 7*(1), 42213. doi:10.1038/srep42213.

De Plano, L. M., Calabrese, G., Conoci, S., Guglielmino, S. P. P., Oddo, S., & Caccamo, A. (2022). Applications of CRISPR-Cas9 in Alzheimer's disease and related disorders. *International Journal of Molecular Sciences, 23*(15), 8714. doi:10.3390/ijms23158714.

Ding, S. W., & Voinnet, O. (2007). Antiviral immunity directed by small RNAs. *Cell, 130*(3), 413–426. doi:10.1016/j.cell.2007.07.039.

Dong, Q., Nie, C.-H., Wu, Y.-M., Zhang, D.-Y., Wang, X.-D., Tu, T., ... Gao, Z.-X. (2023). Generation of blunt snout bream without intermuscular bones by *runx2b* gene mutation. *Aquaculture, 567*, 739263. doi:10.1016/j.aquaculture.2023.739263.

Doudna, J. A., & Charpentier, E. (2014). The new frontier of genome engineering with CRISPR-Cas9. *Science, 346*(6213), 1258096. doi:10.1126/science.1258096.

Duan, L., Ouyang, K., Xu, X., Xu, L., Wen, C., Zhou, X., ... Liang, Y. (2021). Nanoparticle delivery of CRISPR/Cas9 for genome editing. *Frontiers in Genetics, 12*, 673286. doi:10.3389/fgene.2021.673286.

Dunham, R. (2023). Aquaculture and fisheries biotechnology: Genetic approaches. *CABI*. doi:10.1079/9781789243468.0000.

Evans, M. J., & Kaufman, M. H. (1981). Establishment in culture of pluripotential cells from mouse embryos. *Nature, 292*(5819), 154–156. doi:10.1038/292154a0.

Fatira, E., Havelka, M., Labbé, C., Depincé, A., Iegorova, V., Pšenička, M., & Saito, T. (2018). Application of interspecific somatic cell nuclear transfer (iSCNT) in sturgeons and an unexpectedly produced gynogenetic sterlet with homozygous quadruple haploid. *Scientific Reports, 8*(1), 5997. doi:10.1038/s41598-018-24376-1.

Feng, M., Augstein, F., Kareem, A., & Melnyk, C. W. (2024). Plant grafting: Molecular mechanisms and applications. *Molecular Plant, 17*(1), 75–91. doi:10.1016/j.molp.2023.12.006.

Fire, A., Xu, S., Montgomery, M. K., Kostas, S. A., Driver, S. E., & Mello, C. C. (1998). Potent and specific genetic interference by double-stranded RNA in *Caenorhabditis elegans*. *Nature, 391*(6669), 806–811. doi:10.1038/35888.

Gan, R.-H., Li, Z., Wang, Z.-W., Li, X.-Y., Wang, Y., Zhang, X.-J., Tong, J.-F., Wu, Y., Xia, L.-Y., Gao, Z.-X., Zhou, L., & Gui, J.-F. (2023). Creation of intermuscular bone-free mutants in amphitriploid gibel carp by editing two duplicated runx2b homeologs. *Aquaculture, 567*, 739300. doi:10.1016/j.aquaculture.2023.739300.

Gao, Y., Wu, H., Wang, Y., Liu, X., Chen, L., Li, Q., ... Zhang, Y. (2017). Single Cas9 nickase induced generation of NRAMP1 knockin cattle with reduced off-target effects. *Genome Biology, 18*(1), 13. doi:10.1186/s13059-016-1144-4.

Gerdol, M., Moreira, R., Cruz, F., Gómez-Garrido, J., Vlasova, A., Rosani, U., Venier, P., Naranjo-Ortiz, M. A., Murgarella, M., Greco, S., Balseiro, P., Corvelo, A., Frias, L., Gut, M., Gabaldón, T., Pallavicini, A., Canchaya, C., Novoa, B., Alioto, T. S., ... Figueras, A. (2020). Massive gene presence-absence variation shapes an open pan-genome in the Mediterranean mussel. *Genome Biology, 21*(1), 275. doi:10.1186/s13059-020-02180-3.

Gharbiah, M., Nakamoto, A., Johnson, A. B., Lambert, J. D., & Nagy, L. M. (2014). *Ilyanassa* Notch signaling implicated in dynamic signaling between all three germ layers. *The International Journal of Developmental Biology, 58*, 551–562. doi:10.1387/ijdb.140149ln.

Gill, D. R, Pringle, I. A, & Hyde, S. C (2009). Progress and prospects: The design and production of plasmid vectors. *Gene Therapy, 16*(2), 165–171. doi:10.1038/gt.2008.183.

Grimson, A., Farh, K. K. H., Johnston, W. K., Garrett-Engele, P., Lim, L. P., & Bartel, D. P. (2007). MicroRNA targeting specificity in mammals: Determinants beyond seed pairing. *Molecular Cell, 27*(1), 91–105. doi:10.1016/j.molcel.2007.06.017.

Guo, X. (2024). Genetics in shellfish culture molluscan shellfish aquaculture: A practical guide. *CABI International*, 393–413. doi:10.1079/9781800629189.0014.

Guo, X., DeBrosse, G. A., & Allen, S. K. (1996). All-triploid pacific oysters (*Crassostrea gigas* thunberg) produced by mating tetraploids and diploids. *Aquaculture, 142*(3–4), 149–161. doi:10.1016/0044-8486(95)01243-5.

Guo, X. (2009). Use and exchange of genetic resources in molluscan aquaculture. *Reviews in Aquaculture, 1*(3-4), 251-259. doi:10.1111/j.1753-5131.2009.01014.x.

Hammond, S. M., Bernstein, E., Beach, D., & Hannon, G. J. (2000). An RNA-directed nuclease mediates post-transcriptional gene silencing in Drosophila cells. *Nature, 404*(6775), 293–296. doi:10.1038/35005107.

Ho, W., Gao, M., Li, F., Li, Z., Zhang, X.-Q., & Xu, X. (2021). Next-generation vaccines: Nanoparticle-mediated DNA and mRNA delivery. *Advanced Healthcare Materials, 10*(8), e2001812. doi:10.1002/adhm.202001812.

Hu, Z. M., Shan, T. F., Zhang, J., Zhang, Q. S., Critchley, A. T., Choi, H. G., Yotsukura, N., Liu, F. L., & Duan, D. L. (2021). Kelp aquaculture in China: A retrospective and future prospects. *Reviews in Aquaculture, 13*(3), 1324–1351. doi:10.1111/raq.12524.

Huan, P., Cui, M., Wang, Q., & Liu, B. (2021). CRISPR/Cas9-mediated mutagenesis reveals the roles of calaxin in gastropod larval cilia. *Gene, 787*, 145640. doi:10.1016/j.gene.2021.145640.

Huan, P., & Liu, B. (2024). The gastropod Lottia peitaihoensis as a model to study the body patterning of trochophore larvae. *Evolution & Development, 26*(4), e12456. doi:10.1111/ede.12456.

Hyde, S. C., Gill, D. R., Higgins, C. F., Trezise, A. E. O., MacVinish, L. J., Cuthbert, A. W., Ratcliff, R., Evans, M. J., & Colledge, W. H. (1993). Correction of the ion transport defect in cystic fibrosis transgenic mice by gene therapy. *Nature, 362*(6417), 250–255. doi:10.1038/362250a0.

Iida, A., Seki, M., Kamada, M., Yamada, Y., & Morikawa, H. (1990). Gene delivery into cultured plant cells by DNA-coated gold particles accelerated by a pneumatic particle gun. *Theoretical and Applied Genetics, 80*(6), 813–816. doi:10.1007/bf00224198.

Jiang, K., Chen, C., Jiang, G., Chi, Y., Xu, C., Kong, L., Yu, H., Liu, S., & Li, Q. (2024). Genetic improvement of oysters: Current status, challenges, and prospects. *Reviews in Aquaculture, 16*(2), 796–817. doi:10.1111/raq.12868.

Jiang, K., Lim, J., Sgrizzi, S., Trinh, M., Kayabolen, A., Yutin, N., ... Abudayyeh, O. O. (2023). Programmable RNA-guided DNA endonucleases are widespread in eukaryotes and their viruses. *Science Advances, 9*(39), eadk0171. doi:10.1126/sciadv.adk0171.

Jiang, Y., Loker, E. S., & Zhang, S. M. (2006). In vivo and in vitro knockdown of *FREP2* gene expression in the snail *Biomphalaria glabrata* using RNA interference. *Developmental and Comparative Immunology, 30*, 855–866. doi:10.1016/j.dci.2005.12.004.

Jinek, M., Chylinski, K., Fonfara, I., Hauer, M., Doudna, J. A., & Charpentier, E. (2012). A programmable dual-RNA–guided DNA endonuclease in adaptive bacterial immunity. *Science, 337*(6096), 816–821. doi:10.1126/science.1225829.

Kato, K., Okazaki, S., Kannan, S., Altae-Tran, H., Demircioglu, F. E., Isayama, Y., Ishikawa, J., Fukuda, M., Macrae, R. K., Nishizawa, T., Makarova, K. S., Koonin, E. V., Zhang, F., & Nishimasu, H. (2022). Structure of the IscB–ωRNA ribonucleoprotein complex, the likely ancestor of CRISPR-Cas9. *Nature Communications, 13*(1), 2041–2723. doi:10.1038/s41467-022-34378-3.

Khademi, Z., Mahmoudi, Z., Sukhorukov, V. N., Jamialahmadi, T., & Sahebkar, A. (2024). CRISPR/Cas9 technology: A novel approach to obesity research. *Current Pharmaceutical Design, 30*(23), 1791–1803. doi:10.2174/0113816128301465240517065848.

Khan, S., Ullah, M. W., Siddique, R., Nabi, G., Manan, S., Yousaf, M., & Hou, H. (2016). Role of recombinant DNA technology to improve life. *International Journal of Genomics, 2016*, 2405954. doi:10.1155/2016/2405954.

Kim, J. K., Gabel, H. W., Kamath, R. S., Tewari, M., Pasquinelli, A., Rual, J. F., ... Ruvkun, G. (2005). Functional genomic analysis of RNA interference in *C. elegans*. *Science, 308*(5725), 1164–1167. doi:10.1126/science.1109267.

Kleinstiver, B. P., Prew, M. S., Tsai, S. Q., Topkar, V. V., Nguyen, N. T., Zheng, Z., Gonzales, A. P. W., Li, Z., Peterson, R. T., Yeh, J. R. J., Aryee, M. J., & Joung, J. K. (2015). Engineered CRISPR-Cas9 nucleases with altered PAM specificities. *Nature, 523*(7561), 481–485. doi:10.1038/nature14592.

Kumar, V., Roy, S., Behera, B. K., & Das, B. K. (2022). RNA interference and its potential applications in aquatic animal health management biotechnological advances in aquaculture health management. *Springer Nature*, 25–41. doi:10.1007/978-981-16-5195-3_2.

Langridge, P. (2018). Innovation in breeding and biotechnology. In R Serraj, & P. Pingali (Eds.), *Agriculture & food systems to 2050: Global trends, challenges and opportunities* (pp. 245–284). World Scientific Publishing. (pp. 245–284). doi:10.1142/9789813278356_0008.

Ledford, H. (2015). Salmon is first transgenic animal to win US approval for food. *Nature*. doi:10.1038/nature.2015.18838.

Lee, J. A., Kim, H. K., Kim, K. H., Han, J. H., Lee, Y. S., Lim, C. S., ... Kaang, B. K. (2001). Overexpression of and RNA interference with the CCAAT enhancer-binding protein on long-term facilitation of *Aplysia* sensory to motor synapses. *Learning & Memory, 8*(4), 220–226. doi:10.1101/lm.40201.

Li, C., & Samulski, R. J. (2020). Engineering adeno-associated virus vectors for gene therapy. *Nature Research, 21*(4), 255–272. doi:10.1038/s41576-019-0205-4.

Li, H., Yu, H., Du, S., & Li, Q. (2021). CRISPR/Cas9 mediated high efficiency knockout of myosin essential light chain gene in the Pacific oyster (*Crassostrea gigas*). *Marine Biotechnology, 23*(2), 215–224. doi:10.1007/s10126-020-10016-1.

Li, L., Zhang, Q., & Huang, D. (2014). A review of imaging techniques for plant phenotyping. *Sensors, 14*(11), 20078–20111. doi:10.3390/s141120078.

Li, Q., Yu, H., & Li, Q. (2024). Dual sgRNA-directed tyrosinases knockout using CRISPR/Cas9 technology in Pacific oyster (*Crassostrea gigas*) reveals their roles in early shell calcification. *Gene, 927*, 148748. doi:10.1016/j.gene.2024.148748.

Li, X., Wang, S., Xun, X., Zhang, M., Wang, S., Li, H., Zhao, L., Fu, Q., Wang, H., Li, T., Lian, S., Xing, Q., Li, X., Wu, W., Zhang, L., Hu, X., & Bao, Z. (2019). A carotenoid oxygenase is responsible for muscle coloration in scallop. *Biochimica et Biophysica Acta (BBA) - Molecular and Cell Biology of Lipids, 1864*(7), 966–975. doi:10.1016/j.bbalip.2019.03.003.

Li, Y., Sun, X., Hu, X., Xun, X., Zhang, J., Guo, X., Jiao, W., Zhang, L., Liu, W., Wang, J., Li, J., Sun, Y., Miao, Y., Zhang, X., Cheng, T., Xu, G., Fu, X., Wang, Y., Yu, X., … Bao, Z. (2017). Scallop genome reveals molecular adaptations to semi-sessile life and neurotoxins. *Nature Communications, 8*(1), 2041–2723. doi:10.1038/s41467-017-01927-0.

Lian, S., Lv, J., Li, Y., Zhang, L., Bao, Z., & Wang, S. (2025a). Marine evo-devo: New frontiers from Lophotrochozoa. *The Innovation, 6*(6), 100869. doi:10.1016/j.xinn.2025.100869.

Li, Y., Wang, R., Xun, X., Wang, J., Bao, L., Thimmappa, R., … Shi, W. (2018). Sea cucumber genome provides insights into saponin biosynthesis and aestivation regulation. *Cell Discovery, 4*(1), 29. doi:10.1038/s41421-018-0030-5.

Lian, S., Hu, N., Chen, X., Dai, X., Zhu, X., Qiao, R., Liu, S., Lu, Y., Zhang, F., Sun, F., Pu, Z., Yang, Z., Xu, R., Wang, H., Zhao, L., Xing, Q., Li, S., Wang, W., Dong, B., … Shi, W. (2025b). Widespread presence of bone marrow–like hematopoietic stem cell niche in invertebrate skeletons. *Science Advances, 11*(26), eadw0958. doi:10.1126/sciadv.adw0958.

Liu, Q. (2020). World-first phase I clinical trial for CRISPR-Cas9 PD-1-edited T-cells in advanced nonsmall cell lung cancer. *Global Medical Genetics, 07*(03), 073–074. doi:10.1055/s-0040-1721451.

Liu, Z. J. (2010). Comparison of index selection, BLUP, MAS, and whole genome selection. In *Next generation sequencing and whole genome selection in aquaculture* (pp. 185–218). Wiley. doi:10.1002/9780470958964.ch12.

López, M . G., Zanor, M. I., Pratta, G. R., Stegmayer, G., Boggio, S. B., Conte, M., Bermúdez, L., Coluccio Leskow, C., Rodríguez, G. R., Picardi, L. A., Zorzoli, R., Fernie, A. R., Milone, D., Asís, R., Valle, E. M., & Carrari, F. (2015). Metabolic analyses of interspecific tomato recombinant inbred lines for fruit quality improvement. *Metabolomics, 11*(5), 1416–1431. doi:10.1007/s11306-015-0798-3.

Lu, J. K, Chen, T. T, Allen, S. K, Matsubara, T., & Burns, J. C (1996). Production of transgenic dwarf surfclams, Mulinia lateralis, with pantropic retroviral vectors. In *Proceedings of the National Academy of Sciences: 93* (pp. 3482–3486). doi:10.1073/pnas.93.8.3482.

Luo, D. J., Hu, W., Chen, S. P., & Zhu, Z. Y. (2011). Critical developmental stages for the efficiency of somatic cell nuclear transfer in zebrafish. *International Journal of Biological Sciences, 7*(4), 476–486. doi:10.7150/ijbs.7.476.

Nell, J. A. (2002). Farming triploid oysters. *Aquaculture, 210*(1-4), 69–88. doi:10.1016/S0044-8486(01)00861-4.

Niemann, H., Kues, W. A., Lucas-Hahn, A., & Carnwath, J. W. (2011). Somatic cloning and epigenetic reprogramming in mammals. In A. Atala, R. Lanza, J. A. Thomson, & R. Nerem (Eds.), *Principles of regenerative medicine* (pp. 121–124). Elsevier. doi:10.1016/B978-0-12-381422-7.10007-0.

Niemann, H., & Lucas-Hahn, A. (2012). Somatic cell nuclear transfer cloning: Practical applications and current legislation. *Reproduction in Domestic Animals, 47*(s5), 2–10. doi:10.1111/j.1439-0531.2012.02121.x.

Ning, X., Li, X., Wang, J., Zhang, X., Kong, L., Meng, D., … Bao, Z. (2019). Genome-wide association study reveals *E2F3* as the candidate gene for scallop growth. *Aquaculture, 511*(1), 734216. doi:10.1016/j.aquaculture.2019.734216.

Oonuma, K., Kuroda, R., Uchida, T., & Zhang, S. M. (2025). CRISPR/Cas9-germline editing of *Biomphalaria glabrata*, A breakthrough in genetic modification of snails that transmit schistosomiasis. *Science Advances, 11*(41), eadx5889. doi:10.1126/sciadv.adx5889.

Outtandy, P., Russell, C., Kleta, R., & Bockenhauer, D. (2019). Zebrafish as a model for kidney function and disease. *Pediatric Nephrology, 34*(5), 751–762. doi:10.1007/s00467-018-3921-7.

Palazzo, A., & Marsano, R. M. (2021). Transposable elements: A jump toward the future of expression vectors. *Critical Reviews in Biotechnology, 41*(5), 792–808. doi:10.1080/07388551.2021.1888067.

Pantoja, Y., Da Costa Pinheiro, K., Araujo, F., Da Costa Silva, A. L., & Ramos, R. (2020). Bioinformatics approaches applied in pan-genomics and their challenges. In *Pan-genomics: Applications, challenges, and future prospects* (pp. 43–64). Elsevier. doi:10.1016/B978-0-12-817076-2.00002-0.

Peleman, J. D., & Van Der Voort, J. R. (2003). Breeding by design. *Trends in Plant Science, 8*(7), 330–334. doi:10.1016/S1360-1385(03)00134-1.

Perry, K. J., & Henry, J. Q. (2015). CRISPR/Cas9-mediated genome modification in the mollusc, *Crepidula fornicata*. *Genesis, 53*(2), 237–244. doi:10.1002/dvg.22843.

Powers, D. A., Kirby, V. L., Cole, T., & Hereford, L. (1995). Electroporation as an effective means of introducing DNA into abalone (*Haliotis rufescens*) embryos. *Molecular Marine Biology and Biotechnology, 4*(4), 369–375.

Rao, Y., Li, Y., & Qian, Q. (2014). Recent progress on molecular breeding of rice in China. *Plant Cell Reports, 33*(4), 551–564. doi:10.1007/s00299-013-1551-x.

Ren, J., Liu, X., Fang, C., Jiang, S., June, C. H., & Zhao, Y. (2017). Multiplex genome editing to generate universal CAR T cells resistant to PD1 inhibition. *Clinical Cancer Research, 23*(9), 2255–2266. doi:10.1158/1078-0432.ccr-16-1300.

Robinson, N. A., Østbye, T. K. K., Kettunen, A. H., Coates, A., Barrett, L. T., Robledo, D., & Dempster, T. (2024). A guide to assess the use of gene editing in aquaculture. *Reviews in Aquaculture, 16*(2), 775–784. doi:10.1111/raq.12866.

Rogers, A. J. (2024). Aquaculture in the ancient world: Ecosystem engineering, domesticated landscapes, and the first blue revolution. *Journal of Archaeological Research, 32*(3), 427–491. doi:10.1007/s10814-023-09191-1.

Rupp, L. J., Schumann, K., Roybal, K. T., Gate, R. E., Ye, C. J., Lim, W. A., & Marson, A. (2017). CRISPR/Cas9-mediated PD-1 disruption enhances anti-tumor efficacy of human chimeric antigen receptor T cells. *Scientific Reports, 7*(1), 1–10. doi:10.1038/s41598-017-00462-8.

Saito, M., Xu, P., Faure, G., Maguire, S., Kannan, S., Altae-Tran, H., Vo, S., Desimone, A., Macrae, R. K., & Zhang, F. (2023). Fanzor is a eukaryotic programmable RNA-guided endonuclease. *Nature, 620*(7974), 660–668. doi:10.1038/s41586-023-06356-2.

Sanford, J. (1988). The biolistic process. *Trends in Biotechnology, 6*(12), 299–302. doi:10.1016/0167-7799(88)90023-6.

Saurabh, S., Vidyarthi, A. S., & Prasad, D. (2014). RNA interference: Concept to reality in crop improvement. *Planta, 239*(3), 543–564. doi:10.1007/s00425-013-2019-5.

Shakweer, W. M. E., Krivoruchko, A. Y., Dessouki, S. M., & Khattab, A. A. (2023). A review of transgenic animal techniques and their applications. *Journal of Genetic Engineering and Biotechnology, 21*(1), 55. doi:10.1186/s43141-023-00502-z.

Shen, L., Hua, Y., Fu, Y., Li, J., Liu, Q., Jiao, X., Xin, G., Wang, J., Wang, X., Yan, C., & Wang, K. (2017). Rapid generation of genetic diversity by multiplex CRISPR/Cas9 genome editing in rice. *Science China Life Sciences, 60*(5), 506–515. doi:10.1007/s11427-017-9008-8.

Shmakov, S., Abudayyeh, O. O., Makarova, K. S., Wolf, Y. I., Gootenberg, J. S., Semenova, E., Minakhin, L., Joung, J., Konermann, S., Severinov, K., Zhang, F., & Koonin, E. V. (2015). Discovery and functional characterization of diverse class 2 CRISPR-Cas systems. *Federation Molecular Cell, 60*(3), 385–397. doi:10.1016/j.molcel.2015.10.008.

Sibley, C. R., Seow, Y., & Wood, M. J. (2010). Novel RNA-based strategies for therapeutic gene silencing. *Molecular Therapy, 18*(3), 466–476. doi:10.1038/mt.2009.306.

Stonich, S. C., & Bailey, C. (2000). Resisting the blue revolution: Contending coalitions surrounding industrial shrimp farming. *Human Organization. Society for Applied Anthropology, 59*(1), 23–36. doi:10.17730/humo.59.1.862811321884231k.

Sun, D., Yu, H., & Li, Q. (2022). Examination of the roles of *Foxl2* and *Dmrt1* in sex differentiation and gonadal development of oysters by using RNA interference. *Aquaculture, 548*(2), 737732. doi:10.1016/j.aquaculture.2021.737732.

Sun, N., & Zhao, H. (2013). Transcription activator-like effector nucleases (TALENs): A highly efficient and versatile tool for genome editing. *Biotechnology and Bioengineering, 110*(7), 1811–1821. doi:10.1002/bit.24890.

Tacon, A. G., Metian, M., Parsons, G. J., & Shumway, S. E. (2025). A global aquaculture and aquafeed production update: 2010 to 2023. *Reviews in Fisheries Science & Aquaculture*, 1–10. doi:10.1080/23308249.

Tan, K., Zhang, H., & Zheng, H. (2020). Selective breeding of edible bivalves and its implication of global climate change. *Reviews in Aquaculture, 12*(4), 2559–2572. doi:10.1111/raq.12458.

Taniyama, Y., Azuma, J., Kunugiza, Y., Iekushi, K., Rakugi, H., & Morishita, R. (2012). Therapeutic option of plasmid-DNA based gene transfer. *Current Topics in Medicinal Chemistry, 12*(15), 1630–1637. doi:10.2174/156802612803531342.

Tracy, A. N., Yadavalli, R., Reed, K. S., Parnaik, R., Poulton, N. J., Bishop-Bailey, D., & Robledo, J. A. F. (2020). Genome to phenome tools: In vivo and in vitro transfection of *Crassostrea virginica* hemocytes. *Fish & Shellfish Immunology, 103*, 438–441. doi:10.1016/j.fsi.2020.05.022.

Tsong, T. Y. (1991). Electroporation of cell membranes. *Biophysical Journal, 60*(2), 297–306. doi:10.1016/s0006-3495(91)82054-9.

Tu, Z., Tang, L., Shen, H., Hu, M., Chen, F., Wang, X., Abo-Raya, M. H., & Wang, Y. (2025). Molecular insights into the physiological impact of low-frequency noise on sea slug *Onchidium reevesii*: Activation of p53 signaling and oxidative stress response. *Journal of Environmental Management* 373,123481. doi:10.1016/j.jenvman.2024.123481.

Ullah, Z., Ahmad, S., Waqar, R., Mahmood, T., Iqbal, J., Yaseen, T., & Mahmood, T. (2025). Future horizons: Emerging "Omics" technologies and challenges in tomato omics approaches for tomato yield and quality trait improvement. In *Omics approaches for tomato yield and quality trait improvement* (pp. 347–394). Springer. doi:10.1007/978-981-96-3890-1_16.

Wadhwa, N., Ganguli, N., & Majumdar, S. S. (2021). Generating transgenic animal models: Recent technological advancements. *Essentials of laboratory animal science: Principles and practices*, 709–724. doi:10.1007/978-981-16-0987-9_29.

Wang, D., Yu, X., Xu, K., Bi, G., Cao, M., Zelzion, E., … Mao, Y. (2020). *Pyropia yezoensis* genome reveals diverse mechanisms of carbon acquisition in the intertidal environment. *Nature Communications, 11*(1), 2041–2723. doi:10.1038/s41467-020-17689-1.

Wang, L., & Han, H. (2024). Strategies for improving the genome-editing efficiency of class 2 CRISPR/Cas system. *Heliyon, 10*(19), e38588. doi:10.1016/j.heliyon.2024.e38588.

Wang, S., Zhang, J., Jiao, W., Li, J., Xun, X., Sun, Y., … Bao, Z. (2017). Scallop genome provides insights into evolution of bilaterian karyotype and development. *Nature Ecology & Evolution, 1*(5), 120. doi:10.1038/s41559-017-0120.

Wang, Y., Li, X. Y., Xu, W. J., Wang, K., Wu, B., Xu, M., … Gui, J. F. (2022a). Comparative genome anatomy reveals evolutionary insights into a unique amphitriploid fish. *Nature Ecology and Evolution, 6*(9), 1354–1366. doi:10.1038/s41559-022-01813-z.

Wang, Y., Ni, P., Sturrock, M., Zeng, Q., Wang, B., Bao, Z., & Hu, J. (2024). Deep learning for genomic selection of aquatic animals. *Marine Life Science & Technology, 6*(4), 631–650. doi:10.1007/s42995-024-00252-y.

Wang, Y., Zhu, X., Lian, S., Li, Y., Hu, N., Hu, X., … Shi, W. (2022b). Functional characterization of Cfap206 for bivalve ciliogenesis by RNAi and CRISPR/Cas9 technologies. *Frontiers in Marine Science, 9*, 2296–7745. doi:10.3389/fmars.2022.864037.

Wheeler, N. J., Dinguirard, N., Marquez, J., Gonzalez, A., Zamanian, M., Yoshino, T. P., & Castillo, M. G. (2018). Sequence and structural variation in the genome of the *Biomphalaria glabrata* embryonic (Bge) cell line. *Parasites & Vectors, 11*(1). doi:10.1186/s13071-018-3059-2.

Wu, Y., Wu, T., Yang, L., Su, Y., Zhao, C., Li, L., … Zhou, L. (2023). Generation of fast growth Nile tilapia (*Oreochromis niloticus*) by myostatin gene mutation. *Aquaculture, 562*, 738762. doi:10.1016/j.aquaculture.2022.738762.

Xing, Z., Wang, H., Zhang, M., Wang, J., Xiao, Z., Lu, Y., Wang, T., Ouyang, B., Zhang, Y., Zhang, J., Ye, Z., & Ye, J. (2024). Characteristics of the ET, a new species created by interspecific hybridization of two wild eggplants. *Scientia Horticulturae, 331*, 113143. doi:10.1016/j.scienta.2024.113143.

Xu, P., Saito, M., Faure, G., Maguire, S., Vo, S. C-D.-T., Wilkinson, M. E., … Zhang, F. (2024). Structural insights into the diversity and DNA cleavage mechanism of Fanzor. *Cell, 187*(19), 5238–5252. doi:10.1016/j.cell.2024.07.050.

Xu, Y., Li, P., Zou, C., Lu, Y., Xie, C., Zhang, X., Prasanna, B. M., & Olsen, M. S. (2017). Enhancing genetic gain in the era of molecular breeding. *Journal of Experimental Botany, 68*(11), 2641–2666. doi:10.1093/jxb/erx135.

Xu, Y., Zhang, X., Li, H., Zheng, H., Zhang, J., Olsen, M. S., Varshney, R. K., Prasanna, B. M., & Qian, Q. (2022). Smart breeding driven by big data, artificial intelligence, and integrated genomic-enviromic prediction. *Molecular Plant, 15*(11), 1664–1695. doi:10.1016/j.molp.2022.09.001.

Yan, H., Zhang, S., Li, C. Z., Chen, Y. H., Chen, Y. G., Weng, S. P., & He, J. G. (2013). Molecular characterization and function of a p38 MAPK gene from *Litopenaeus vannamei*. *Fish and Shellfish Immunology, 34*(6), 1421–1431. doi:10.1016/j.fsi.2013.02.030.

Yao, T., Tong, Z., Lu, J., Fu, S., Cheng, C., & Ye, L. (2024). A novel C-type lectin, perlucin, from the small abalone, *Haliotis diversicolor* involved in the innate immune defense against *Vibrio harveyi* infection. *Fish & Shellfish Immunology, 155*, 110029. doi:10.1016/j.fsi.2024.110029.

Yoon, J., Gu, W. B., Konuma, M., Kobayashi, M., Yokoi, H., Osada, M., & Nagasawa, K. (2022). Gene delivery available in molluscan cells by strong promoter discovered from bivalve-infectious virus. *Proceedings of the National Academy of Sciences of the United States of America, 119*(45). e2209910119. doi:10.1073/pnas.2209910119.

Yoshino, T. P., Coustau, C., Modat, S., & Castillo, M. G. (1999). The *Biomphalaria glabrata* embryonic (Bge) molluscan cell line: Establishment of an in vitro cellular model for the study of snail host-parasite interactions. *Malacologia, 41*(2), 331–343.

Yu, H., Li, H., Li, Q. i, Xu, R., Yue, C., & Du, S. (2019). Targeted gene disruption in Pacific oyster based on CRISPR/Cas9 ribonucleoprotein complexes. *Marine Biotechnology, 21*(3), 301–309. doi:10.1007/s10126-019-09885-y.

Yu, Y., Li, Q., Yu, H., & Li, Q. i (2025). Comparative analysis of promoter activity in *Crassostrea gigas* embryos: Implications for bivalve gene editing. *Marine Biotechnology, 27*(1), 20. doi:10.1007/s10126-024-10398-6.

Yue, C., Yu, H., Li, H., Du, S., & Li, Q. (2020). The endogenous EF-1α promoter is highly active in driving gene overexpression in developing embryos of the Pacific oyster *Crassostrea gigas*. *Aquaculture, 522*, 735134. doi:10.1016/j.aquaculture.2020.735134.

Yue, G. H. (2014). Recent advances of genome mapping and marker-assisted selection in aquaculture. *Fish and Fisheries, 15*(3), 376–396. doi:10.1111/faf.12020.

Zaidi, S. S . A., Mukhtar, M. S., & Mansoor, S. (2018). Genome editing: Targeting susceptibility genes for plant disease resistance. *Belgium Trends in Biotechnology, 36*(9), 898–906. doi:10.1016/j.tibtech.2018.04.005.

Zeng, Q., Zhao, B., Wang, H., Wang, M., Teng, M., Hu, J., … Wang, Y. (2022). Aquaculture molecular breeding platform (AMBP): A comprehensive web server for genotype imputation and genetic analysis in aquaculture. *Nucleic Acids Research, 50*(W1), W66–W74. doi:10.1093/nar/gkac424.

Zhang, C., Jiang, S., Tian, Y., Dong, X., Xiao, J., Lu, Y., Liang, T., Zhou, H., Xu, D., Zhang, H., Luo, M., & Xia, Z. (2023). Smart breeding driven by advances in sequencing technology. *Modern Agriculture, 1*(1), 43–56. doi:10.1002/moda.8.

Zhang, F. (2019). Development of CRISPR-Cas systems for genome editing and beyond. *Quarterly Reviews of Biophysics, 52*, 0033–5835. doi:10.1017/s0033583519000052.

Zhang, G., Fang, X., Guo, X., Li, L., Luo, R., Xu, F., Yang, P., Zhang, L., Wang, X., Qi, H., Xiong, Z., Que, H., Xie, Y., Holland, P. W. H., Paps, J., Zhu, Y., Wu, F., Chen, Y., Wang, J., … Wang, J. (2012). The oyster genome reveals stress adaptation and complexity of shell formation. *Nature, 490*(7418), 49–54. doi:10.1038/nature11413.

Zhang, S., Huang, G., Zhang, Y., Lv, X., Wan, K., Liang, J., Feng, Y., Dao, J., Wu, S., Zhang, L., Yang, X., Lian, X., Huang, L., Shao, L., Zhang, J., Qin, S., Tao, D., Crews, T. E., Sacks, E. J., … Lyu, J. (2023). Sustained productivity and agronomic potential of perennial rice. *Nature Sustainability, 6*(1), 28–38. doi:10.1038/s41893-022-00997-3.

Zhang, X., Yuan, J., Sun, Y., Li, S., Gao, Y., Yu, Y., … Xiang, J. (2019). Penaeid shrimp genome provides insights into benthic adaptation and frequent molting. *Nature Communications, 10*(1), 356. doi:10.1038/s41467-018-08197-4.

Zhao, L., Lian, S., Ren, Q., Yang, Z., Guo, Z., Lou, J., Kong, X., Li, M., Bao, Z., & Hu, X. (2021). In vivo and rapid assessment of scallop muscle trait. *Aquaculture, 530*, 735817. doi:10.1016/j.aquaculture.2020.735817.

Zheng, L., Cao, H., Qiu, J., & Chi, C. (2022). Inhibitory effect of FMRFamide on NO production during immune defense in *Sepiella japonica*. *Frontiers in Immunology, 13*, 825634. doi:10.3389/fimmu.2022.825634.

Zhou, L., & Gui, J. F. (2018). Applications of genetic breeding biotechnologies in Chinese aquaculture. In *Aquaculture in China: Success stories and modern trends* (pp. 463–496). Wiley. doi:10.1002/9781119120759.ch6_1.

CHAPTER

12

Emerging bivalve model

Zujing Yang[a], *Ximing Guo*[b] *and Zhenmin Bao*[a]

[a]College of Marine Life Sciences, Ocean University of China, Qingdao, China

[b]Department of Marine and Coastal Sciences, Rutgers University, New Brunswick, NJ, United States

12.1 Introduction

Meeting the nutritional needs of the growing human population while preserving biodiversity and environmental integrity represents a paramount challenge of the 21st century. As the fastest-growing farming industry in the world, aquaculture currently supplies over half of the global aquatic food supply, or approximately 87.5 million tons, and plays a crucial role in ensuring food security (Bartley, 2022). Currently, molluscan shellfish farming accounts for approximately 21% of global aquaculture production of animals, with bivalves (oysters, clams, scallops, and mussels) dominating in both yield and economic value (Bartley, 2022; Hollenbeck & Johnston, 2018). Successful shellfish breeding programs have been established worldwide, and the production of major farmed species has greatly benefited from genetically improved germplasm resources, such as improved growth rate and disease resistance (de Melo et al., 2016; Dégremont et al., 2015; Guo, 2021). Despite these successes, shellfish farming production still relies heavily on wild seeds, which are not reliable or sustainable (Nascimento-Schulze et al., 2021). In addition, the complex genomic and life-history features of shellfish further limit the efficiency of shellfish breeding programs (Guo, 2026; Hedgecock & Pudovkin, 2011; Hollenbeck & Johnston, 2018). Therefore there is great potential for genetic improvements to enhance the production efficiency of cultured shellfish worldwide.

Breeding 4.0, as a cutting-edge paradigm in modern agriculture, offers transformative potential for genetic improvement through the deep integration of genetics, big data, and artificial intelligence (Wallace et al., 2018). In this process, model organisms, because of their unique advantages in revealing genetic mechanisms of key traits and accelerating breeding progress, have become an essential bridge between genetic theory and breeding practice, which can significantly improve the accuracy and efficiency of breeding. In shellfish, several model organisms have been developed, such as *Biomphalaria glabrata*, *Lymnaea stagnalis*, and *Lottia peitaihoensis*, which have played an important role in exploring disease prevention, development, biomineralization, and sexual selection (Davison & Neiman, 2021; Huan & Liu, 2024). However, the

Genetic Breeding and Aquaculture of Shellfish: A Collaborative Approach to a Sustainable Future
DOI: https://doi.org/10.1016/B978-0-443-33606-5.00003-5

existing model organisms are mainly gastropods, which cannot fully meet the research needs of cultured shellfish, which are mainly bivalves. Therefore it is essential to establish a bivalve model to advance bivalve genetics and breeding.

The dwarf surf clam, *Mulinia lateralis* (Say, 1822), is a small, infaunal-dwelling bivalve native to the estuarine benthic and intertidal zones along the Atlantic coast, ranging from Canada to the Caribbean (Rhodes et al., 1975). It has been regarded as the "molluscan fruit fly" and proposed as a model bivalve because of its ease of care, fast growth, short life cycle, and transparent shells (Calabrese, 1969a). Since the mid-20th century, several research groups have utilized *M. lateralis* to study development, behavior, toxicology, and genetics, laying foundations for many key experimental approaches, including optimizing care and breeding protocols, developing tools for gene editing, and advancing genetic analyses (Calabrese, 1969b; Cripe, 2006; Guo & Allen, 1994a; Lu et al., 1996; Rhodes et al., 1975; Shumway et al., 1983; Shumway & Newell, 1984; Wang & Guo, 2008). In 2017, *M. lateralis* was introduced from the United States to the MOE Key Laboratory of Marine Genetics and Breeding, Ocean University of China (Qingdao, China). Since then, our team has focused on developing *M. lateralis* into a bivalve model, particularly in genetic manipulation techniques, to accelerate its application in bivalve studies (Guo et al., 2023; Wang et al., 2022; Yang et al., 2021, 2022; Zhang et al., 2024; Zhao et al., 2025). This chapter provides an overview of *M. lateralis* as a model system, covering its biological characteristics, methods for artificial culture and reproduction, and available resources and technical platforms. It explores the application of this species in bivalve genetics and breeding, highlighting the accessibility and adaptability of this system. Finally, the chapter discusses key technical bottlenecks, highlighting future development directions and challenges facing the use of this model organism.

12.2 Advantages of M. *lateralis* as a model organism

Compared with other bivalves, the following characteristics of *M. lateralis* stand out and support its use as a model species:

(1) *M. lateralis* is a small (maximum shell length 21 mm), short-lived bivalve (maximum life span 2 years), requiring less space for artificial cultivation in the laboratory, and the maintenance cost is low.

(2) *M. lateralis* has a short generation cycle (approximately 60 days), mature females release large numbers of eggs at a time ($>10^5$), and under laboratory conditions, gametogenesis occurs year-round.

(3) The development is rapid, and the shells are transparent during early development, making it easy to observe behavior, measure traits, and track organ development.

(4) Because of its natural distribution (estuarine and intertidal zones from the Gulf of St. Lawrence to the Gulf of Mexico), it tolerates a wide range of environmental conditions and accepts a wide range of diets, making it easy to culture and breed in the laboratory.

(5) *M. lateralis* is dioecious, and the two sexes can be separated by their gonad color through their transparent shells, with males being milky white and females being orange-red. It has a haploid chromosome number of 19, which represents the common or plesiomorphic karyotype of bivalve molluscs (Guo & Allen, 1994a; Wang et al., 2017).

12.3 Biological characteristics of M. *lateralis*

12.3.1 Morphological characteristics

The dwarf surf clam (*Mulinia lateralis*), a member of the bivalve family Mactridae, is characterized by several key shell features. These characteristics include: (1) a triangular shell outline, (2) a distinct radial ridge along the posterior end of the valves, (3) an exclusively internal ligament, (4) well-developed accessory lamella, (5) anterior lateral teeth in the right valve varying in size, with the ventral one being longer, and two posterior lateral teeth of similar size, (6) a shell color ranging from white to creamy white, often adorned with a thin, yellowish periostracum that may thicken along the posterior-dorsal margin beyond the posterior ridge, (7) a smooth shell surface by very fine concentric growth lines, (8) a distinctly convex spot on the shell, and (9) a broad cardinal area between the umbo in larger specimens (Craeymeersch et al., 2019). Under natural conditions, the maximum shell length of *M. lateralis* can reach up to 21 mm (Montagna et al., 1993).

12.3.2 Distribution and habitats

Several studies have documented the natural habitats of *M. lateralis*, with most observations indicating their distribution ranges from the Gulf of St. Lawrence to the Gulf of Mexico, along the Atlantic coast of North America (Rhodes et al., 1975). Notably, sudden environmental changes throughout its native range can cause mortalities of *M. lateralis* (Buron et al., 2013), making its populations transient or unstable. In the field, dead shells of *M. lateralis* are often found in great numbers, with no live individuals present. On the other hand, *M. lateralis* is now frequently found outside its native range. For instance, in 2017, *M. lateralis* was recorded for the first time in Dutch coastal waters, the Wadden Sea, and the Westerschelde estuary (Craeymeersch et al., 2019). The increasing observations of *M. lateralis* at multiple sites in the western and central Wadden Sea suggest that the species has established natural populations in Europe (Gismann et al., 2023). *M. lateralis* appears to prefer relatively slow-moving estuarine and intertidal zones with sandy-muddy substrate, though they have been found in a variety of different environments, suggesting broad ecological tolerance (Walker & Tenore, 1984). While laboratory-reared clams are typically maintained at 20–22°C and 22–28 psu, clams have been reported to survive in a wide range of temperatures from 7.5–32.5°C and salinities of 7.5–37.5 psu, demonstrating their adaptability to diverse environments (Calabrese, 1969b; Yang et al., 2022). For example, *M. lateralis* has been observed as one of the numerically dominant species in Tampa Bay following summer defaunations, likely because of its ability to withstand short-term anoxia (Santos & Simon, 1980; Shumway et al., 1983). In the St. Lucie River estuary in Florida, McKeon et al. (2015) found that *M. lateralis* was most abundant in areas with poor water quality, characterized by low oxygen, low salinity, and sediment-rich "muck" conditions. As an infaunal-dwelling species, *M. lateralis* possesses a well-developed foot, which it uses to burrow into the substrate. This burrowing behavior is a key survival strategy, helping the species avoid predators and endure adverse environmental conditions (Cintra-Buenrostro et al., 2005). *M. lateralis* is a filter feeder, drawing water through its inhalant siphon to capture food particles (Chalermwat et al., 1991; Shumway & Newell, 1984). These particles are filtered by the gills and subsequently transported to the stomach for digestion, assisted by the digestive gland. Its natural diet primarily consists of microalgae, bacteria, and small organic particles (Chalermwat et al., 1991; Cripe, 2006; Rhodes et al., 1975). In turn, *M. lateralis* is likely preyed upon by crabs, gastropods, and fish in its environments (Morton & Machado, 2019).

12.3.3 Reproduction

The dwarf surf clam is a dioecious species with an XX/XY sex determination system, where females are XX and males are XY (Guo & Allen, 1994a). As shells of this species are thin and transparent, the sex can be determined by the color of their gonads: the female gonads are red or orange, and the male gonads are white or beige (Calabrese, 1969a). This species has a remarkably short generation time, approximately 60 days from egg to egg, and gametogenesis occurs over much of the year (Calabrese, 1969a; Montagna et al., 1993). While cultured *M. lateralis* breed year-round, wild populations show seasonal breeding patterns associated with temperature changes. For example, significant numbers of mature *M. lateralis* have been observed in May and June in Long Island, with spawning typically occurring from July to September (Calabrese, 1970). In the wild, *M. lateralis* breeds in soft-bottom habitats, whereas laboratory breeding is typically conducted in plastic or glass tanks, where breeding efficiency can be enhanced by increasing water temperature (Cripe, 2006; Walker & Tenore, 1984; Yang et al., 2021). During spawning, male and female individuals interact to coordinate the gamete release to facilitate fertilization. *M. lateralis* exhibits high fecundity, with individuals releasing large numbers of gametes during spawning (Calabrese, 1969a). For instance, Guo et al. reported that sexually mature clams with a shell length of 13 mm could produce up to 3×10^5 eggs and 2.5×10^9 sperm each time (Guo & Allen, 1994a). *M. lateralis* undergoes external fertilization, and the fertilized eggs can sink or be filtered by adults. To prevent this, most laboratory breeding strategies use a 25 μm screen to separate the adults from the eggs.

12.3.4 Growth and development

The embryonic development of *M. lateralis* is rapid, and the embryo's transparency facilitates easy observation and staging. As shown in Fig. 12.1 and Table 12.1, the newly fertilized zygote is spherical with a diameter of ~50 μm (A). At 22–25°C, most fertilized eggs release the first polar body around 12–18 min postfertilization (B) and the second polar body around 35–45 min postfertilization (C). Beginning around 60 min after fertilization, the zygote undergoes the first cell divisions in the form of spiral cleavage, passing through the first cleavage (D), 2-cell stage (E), 4-cell stage (F), 8-cell stage (G), and 16-cell stage (H), eventually developing into a morula (I). The blastula stage is reached around 4 hours postfertilization (J), characterized by cilia on the embryo's surface, whose oscillations drive the embryo to rotate. Between 4 and 8 h after fertilization, cells on the future dorsal side of the embryo begin involuting, and the blastopore undergoes germ layer invagination, with one end elongating and the other forming a disc-shaped structure, termed the gastrula (K). From 8 to 12 h, the disc-shaped position of the embryo develops a ciliary band and a bundle of fine apical cilia, forming a trochophore larva (L). The trochophore larva exhibits active swimming behavior, and as its cilia continue to grow and move faster over time. At approximately 12 h postfertilization, the ciliary band of the trochophore larva gradually forms a velum, marking the transition to the veliger larva stage and the conventional end of the embryonic period. The speed of development is strongly influenced by incubation temperature.

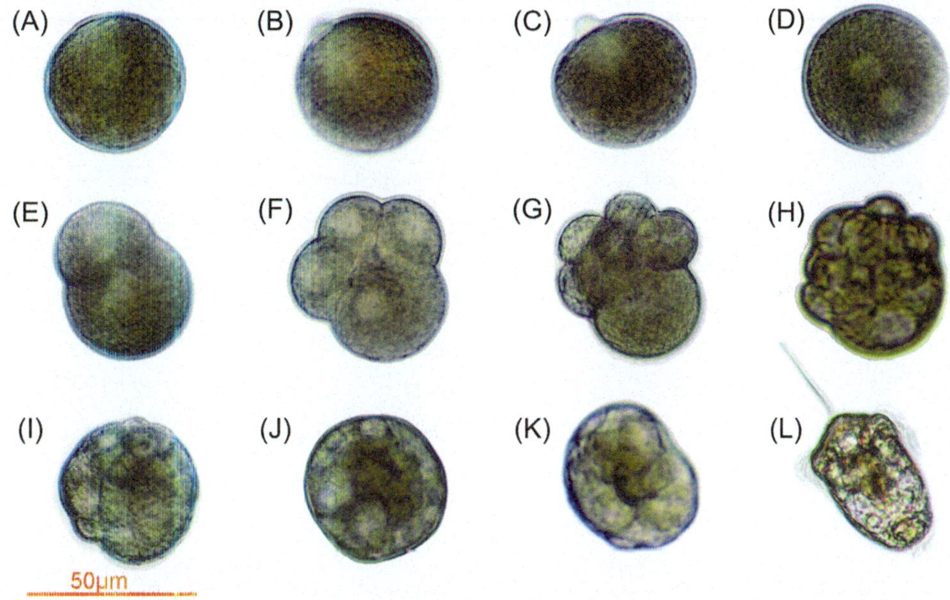

FIGURE 12.1 Embryonic stages of *Mulinia lateralis*. (A) Zygote; (B) Release of the first polar body; (C) Release of the second polar body; (D) The first cleavage; (E) 2-cell stage; (F) 4-cell stage; (G) 8-cell stage; (H) 16-cell stage; (I) Morula stage; (J) Blastula; (K) Gastrula; (L) Trochophore. A scale bar of 50 μm is used for embryonic stages.

After the period of embryonic development, *M. lateralis* spends the next ~4 weeks as juveniles (Fig. 12.2 & Table 12.1). Between 12 and 14 h postfertilization, the side of the veliger larva near the velum becomes arc-shaped, while the opposite side remains straight (A). The straight side later develops into the hinge of the larva, and the shell gland secretes a shell that envelops the larva in a "D" shape, initiating the 3-day D-shaped larva stage (B). At 2 days postfertilization (dpf), D-shaped larvae have an average shell height of 56.35 ± 2.33 μm and a shell length of 69.73 ± 2.60 μm. As the larvae continue to develop, the hinge becomes more pronounced, and the shell takes on a semicircular shape, characteristic of the umbo stage, which includes the early umbo larva (3–5 dpf, C), middle umbo larva stage (5–7 dpf, D), and late umbo larva stage (7–10 dpf, E). During these stages, the larvae become rounded, and the umbo becomes increasingly prominent. After 10 dpf, the larvae's foot develops further and can extend freely outside the shell (F). At this stage, the larvae exhibit functional velum and foot, enabling them to crawl or swim—a hallmark of the pediveliger larva stage, indicating that they are ready for metamorphosis. During metamorphosis, the velum gradually retracts and disappears, and the foot becomes highly active, displaying crawling behavior as the larvae search for a suitable substrate (G). At this time, the inhalant siphon, responsible for excretion and reproduction, begins to form. As the velum atrophies and eventually detaches completely, the foot becomes the only organ of movement, marking the transition to the single-siphon spat stage (H). In addition, spats develop other distinctive features, including a well-developed single siphon, fully functional gills, and an early heart. Between 20 and 25 dpf, the multiple finger-like protrusions at the base of the single siphon in juveniles fuse to form a second siphon, which is responsible for water intake

TABLE 12.1 Developmental stages of embryo, larvae, and juveniles of *Mulinia lateralis*, grown at 22–25 °C.

Stage	Postfertilization time	Stage	Postfertilization time
Zygote	0 min	Trochophore	8–12 h
Polar body	10–50 min	Early veliger larva	12–14 h
First cleavage	50–60 min	D-shaped larva	14 h–3 d
2-cell	60–70 min	Early umbo larva	3–5 d
4-cell	80–90 min	Middle umbo larva	5–7 d
8-cell	60–90 min	Late umbo larva	7–10 d
16-cell	90–120 min	Pediveliger	10–12 d
Morula	2.5–3 h	Spat	12–14 d
Blastula	3–4 h	Single-siphon spat	14–20 d
Gastrula	4–8 h	Double-siphon juvenile	20–25 d

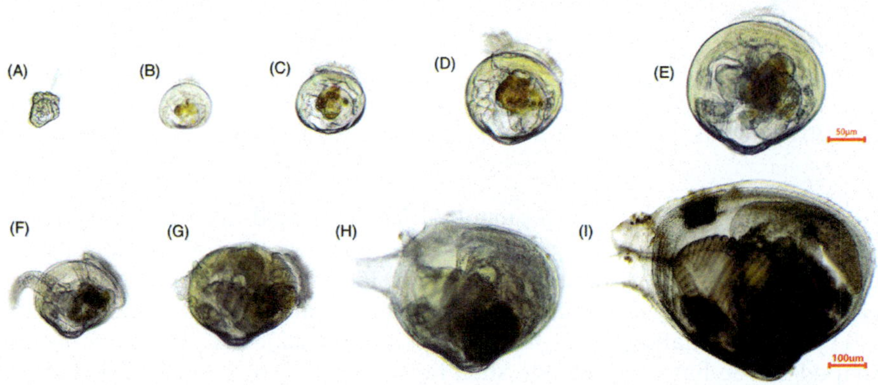

FIGURE 12.2 Larval and juvenile stages of *Mulinia lateralis*. (A) Early veliger larva; (B) D-shaped larva; (C) Early umbo larva; (D) Middle umbo larva; (E) Late umbo larva; (F) Pediveliger; (G) Siphon larvae; (H) Single-siphon juvenile; (I) Double-siphon juvenile. A scale bar of 50 μm is used for larva stages, while a scale bar of 100 μm is used for juvenile stages.

and feeding (I). This marks the full development of the double-siphon system. In addition, the internal organs of the juveniles become fully consistent with those of adults, and their lifestyle gradually shifts from crawling to infaunal-dwelling.

Once *M. lateralis* completes the complex metamorphosis process, it is called a spat. In the early stages of spat, the shell covering the soft tissue is transparent, allowing the internal structures to be directly observed. As the spat grows, the growth lines become visible on the shell surface, and the shells gradually thicken, becoming denser and less transparent. In addition, long-term infaunal-dwelling leads to constant friction between the shell and the substrate, giving the shell a rough and dull appearance. The gonads of the juvenile begin to differentiate, signaling the transition of *M. lateralis* into the adult stage. According to Li et al. (2022), the gonad formation

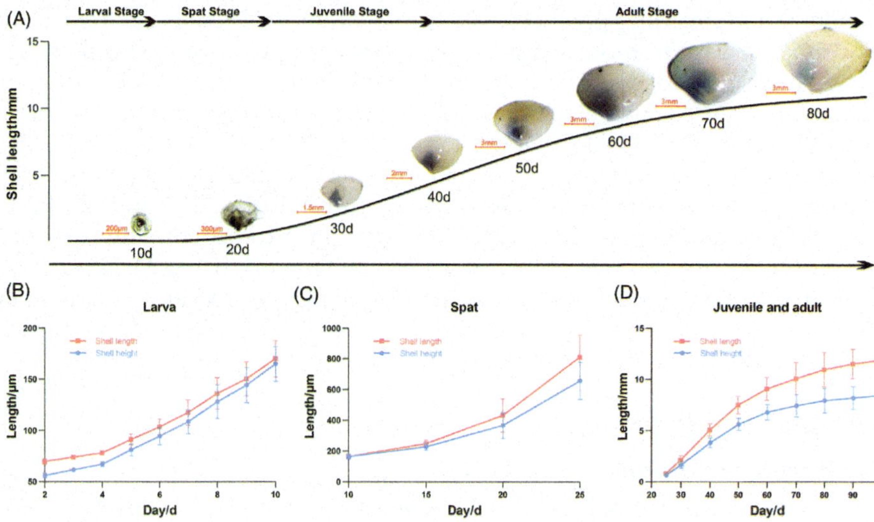

FIGURE 12.3 Growth characteristics of *Mulinia lateralis* during different stages. (A) Growth curve of *M. lateralis* throughout the life span; (B) Larva growth; (C) Spat growth; (D) Juvenile and adult growth. A scale bar of 300 μm is used for larva and spat stages, and a scale bar of 3 mm is used for juvenile and adult stages.

and sex differentiation of *M. lateralis* occur at 35 and 40–45 dpf, respectively, and the gonads then undergo gametogenesis, including a proliferating stage (45–50 dpf), a growing stage (50–55 dpf), and a mature stage (55–60 dpf). At approximately 60 dpf, they are usually sexually mature and ready to spawn. Under wild conditions, *M. lateralis* has a short lifespan and often experiences die-offs due to sudden fluctuations in its habitat (Buron et al., 2013). Its lifespan typically does not exceed two years (Calabrese, 1969a; Lu et al., 1996).

As illustrated in Fig. 12.3, *M. lateralis* exhibits continuous growth up to 15 mm, after which growth slows significantly (A). During the larval stage, the average daily growth in shell height is about 12.38 μm/d, whereas the average daily growth in shell length is 12.82 μm/d (B). After metamorphosis, the growth rate increases significantly, particularly during the double-siphon juvenile stage, where growth in shell length surpasses that of shell height (C). By 30 dpf, the spat reach a shell height of 1.67 ± 0.33 mm and a shell length of 2.15 ± 0.42 mm. At this stage, *M. lateralis* enters a rapid growth phase, which continues until approximately 50 dpf, with an average daily increase in shell length of up to 0.269 mm/d. During this growth process, the asynchronous development of shell height and shell length causes the shells of *M. lateralis* to gradually assume a triangular shape. Subsequently, as gonad development progresses, the growth rate of *M. lateralis* gradually slows but does not cease (D).

12.4 Laboratory culture and breeding of M. *lateralis*

12.4.1 Overview of the procedure

The development, growth, and reproduction of *M. lateralis* are greatly influenced by factors such as temperature, stocking density, and diet, making it challenging to standardize the laboratory culture and breeding practices. This protocol for *M. lateralis* laboratory culture and breeding is primarily based on our team's eight years of experience in breeding for research purposes, combined with relevant findings from published reports (Calabrese, 1969b; Cripe, 2006; Rhodes et al., 1975; Yang et al., 2021, 2022). All procedures and methods outlined herein have consistently demonstrated effectiveness (Yang et al., 2024; Zhao et al., 2025). The protocol emphasizes simplicity and practicality, streamlining common laboratory practices—for example, by reducing the cumbersome temperature cycle control during the conditioning process to just two temperature gradients. We have aimed to make this protocol broadly applicable across laboratories to enhance the reproducibility of experimental work using *M. lateralis* as a model, such as by incorporating commercially available microalgae concentrates as a diet (Yang et al., 2021, 2022).

12.4.2 General conditions

12.4.2.1 Culturing conditions

M. lateralis from the germplasm resources maintained at the Ocean University of China are mainly cultured in glass or plastic aquariums. Culturing facilities significantly affect individual growth rates. For small-scale cultures, such as a limited number of adults or larvae, 15L aquariums (30 cm × 25 cm × 20 cm) without recirculation systems are sufficient. For larger-scale cultures involving numerous individuals or lines, multiple circulation systems, each equipped with several 15L aquariums, are needed. These systems are relatively simple and consist of several key components: Aquariums, water circulation, a filter, air supply, lighting, water quality monitoring, and control. They have been widely used for culturing *M. lateralis*. The aquarium unit includes not only the 15L aquariums but also multilayer grid plates for individual tracking and high-efficiency culture. In addition, an automated feeder can be used for efficiently feeding large numbers of aquariums within the system.

As an infaunal-dwelling bivalve, *M. lateralis* thrives in a 3 cm layer of sand with particle sizes ranging from 0.25 to 0.5 mm, which serves as an ideal laboratory habitat (Yang et al., 2024). The sand used for culturing is sourced from Jinsha Beach (Qingdao, China) and is filtered through a 150 μm screen to remove large particles. It is then disinfected in a 0.02% sodium hypochlorite solution for 6 h and thoroughly rinsed before being used as a substrate.

12.4.2.2 Water quality

In the wild, *M. lateralis* inhabit bodies of water that vary widely in quality, including water chemistry, temperature, salinity, as well as the frequency of fluctuations of these parameters (McKeon et al., 2015; Walker & Tenore, 1984). In captivity, we recommend using water with stable parameters to ensure optimal conditions. In our breeding practice, seawater is collected from the Mai Island shore region (Qingdao, China) and filtered using sand, activated carbon, absorbent cotton (with a 1-μm aperture), and a hollow ultrafiltration membrane to produce filtered seawater (FSW). In circulation systems, water quality is closely monitored to prevent the accumulation of ammonia, nitrates, and nitrites, which can negatively impact the health of *M. lateralis*. In addition, water is filtered to remove metabolic waste using a biological filter containing nitrogen-reducing bacteria. When setting up a new system, it is essential to supplement nitrogen-reducing bacteria from an existing system or a commercial product to initiate nitrogen cycling. Water is replaced weekly to remove waste and replenish losses due to evaporation. For aquariums without a circulation system, water changes are performed every two days with half of the water replaced each time. To enhance the survival and growth of *M. lateralis*, we recommend maintaining a water temperature of 20–22°C and a salinity of 22–25 psu (Yang et al., 2022). It is important to note that metabolic rate is temperature-dependent, with higher food requirements (and faster waste accumulation) at higher ambient temperature (and vice versa).

12.4.2.3 *Feeding*

Feeding has a significant impact on the growth and fecundity of *M. lateralis*. Currently, the diet for *M. lateralis* consists primarily of microalgae, with options including artificially cultivated fresh microalgae or commercially available microalgae concentrates, involving no artificial diet. In our breeding practice, we have tested and confirmed that five microalgae species, *Isochrysis galbana, Nitzschia closterium, Chaetoceros muelleri, Platymonas helgolandica*, and *Chlorella pyrenoidesa*, are suitable as parts of a diet for *M. lateralis* (Yang et al., 2021, 2022). Microalgae concentrates are purchased from Jianyang Biological Technology Ltd. (Dalian, China). The concentrates are stored at 4°C and diluted with FSW before use. The diluted microalgae suspension is filtered through a 40 μm mesh screen to eliminate clumps, and the cell concentration of each suspension is determined using a Countstar (IC1000, Countstar, China).

12.4.3 Broodstock conditioning

Adult clams are cultured to maturity in circulation systems for 10–15 days with gentle aeration. During this period, *M. lateralis* is fed four times daily with a mixture of *C. muelleri, P. helgolandica*, and *I. galbana* in a 3:1:1 ratio at a concentration of 10–20 × 10^4 cells/mL, with a stocking density ranging from 400–600 individuals/m^2. In addition, the feeding is supplemented with *Rhodotorula benthica* once daily at a concentration of 5 × 10^4 cells/mL to promote gonadal development. Seawater flow is halted before feeding and resumed approximately 4 h afterward. When the gonads reach maturity, the aquarium temperature is maintained at 18–20°C to prepare for spawning.

12.4.4 Spawning induction

Sexually mature individuals are selected from the circulation system, with females identified by their orange-red gonads and males by their white or beige gonads. These mature individuals are subjected to thermal stimulation to induce spawning (Yang et al., 2021). Briefly, the clams are rinsed with FSW, placed on ice, and shielded from light for 1h. Following this, female and male clams are transferred to separate presterilized glass dishes, and 26–27°C FSW is added to initiate spawning. After approximately 30 min, the mature individuals begin releasing gametes. The gametes from each individual are then examined under a microscope to assess egg quality and sperm motility. The gametes are subsequently collected and washed to remove debris and fecal matter. Females with a shell length of 8.92 ± 1.52 mm produce an average of 16.30 ± 3.97 × 10^4 eggs per individual, with egg number showing a linear correlation to the size and weight of *M. lateralis*. The eggs are transferred to glass dishes with 300 mL of FSW at 22°C. A sperm suspension is gently added to the egg solution at a 3:1 sperm-to-egg ratio and thoroughly mixed. The gametes remain static for 15 min to facilitate fertilization. Once the first polar body is observed under a microscope, it indicates that fertilization has been completed and the zygotes are about to enter the subsequent embryonic development stage. Generally, the fertilization rate of gametes produced by normal ripening healthy *M. lateralis* is close to 100%.

12.4.5 Embryo incubation

Newly formed zygotes are collected using a 40 μm mesh screen and transferred to 15-L aquariums filled with FSW for embryo incubation. The embryo density is maintained at <50 eggs mL^{-1}, and the temperature is kept at 22°C. Temperature adjustments (±3°C) can be applied to accelerate or decelerate embryonic development. In addition, since the eggs of *M. lateralis* are nonbuoyant (sinking eggs), gentle stirring or adequate aeration is required during incubation to prevent excessive accumulation and compaction at the bottom, which could compromise development. Embryos are monitored through their development, with morphological analysis used to identify all embryonic phases until they reach the D-shaped larva. However, it is uncommon for all zygotes in a batch to develop synchronously; a variable proportion may experience delayed development or fail to develop after other embryos have hatched. As shown in Table 12.1, embryos typically hatch out to D-shaped larvae after 14 h at 22°C, with a hatching rate of approximately 80%.

12.4.6 Raising larvae and juveniles

Freely swimming D-shaped larvae are collected and transferred to clean 15-L aquariums filled with FSW. While embryos rely on their yolk reserves for nutrition, they begin feeding at the D-shaped larval stage and must be provided with food before 2 dpf. During this stage (Fig. 12.2B), *I. galbana* is used as the initial diet at a concentration of $2.4–3.6 \times 10^3$ cells/mL, with a larval stocking density maintained below 9 larvae/mL. As the larvae progress to the umbo stage (Fig. 12.2C-E), a mixed diet of *I. galbana* and *P. helgolandica* in a 1:1 ratio is provided at a concentration of $6.0–12 \times 10^3$ cells/mL, and the stocking density is reduced to less than 7 larvae/mL. Upon reaching the pediveliger stage (Fig. 12.2F-I), the diet transitions to a mixture of *I. galbana, P. helgolandica*, and *N. Closterium* (1:1:1) at a concentration of $3–4 \times 10^4$ cells/mL, with the larvae cultured at a stocking density of less than 5 larvae/mL. For optimal growth, the larvae are fed microalgae three times a day at regular intervals (9:00 a.m., 3:00 p.m., and 9:00 p.m.). The microalgae are distributed evenly across the aquarium, and larvae are considered fully satiated if they stop feeding within approximately 2 h after each feeding. Throughout these stages, adequate aeration is maintained to ensure saturated oxygen levels, and the seawater is fully replaced daily with FSW at a temperature of 21–22°C and a salinity of 25–28 psu. Every 2 days, the aquariums are replaced with clean, presterilized units treated with a 0.2% potassium permanganate solution. Between 20 and 25 dpf, as shown in Fig. 12.2, the larvae undergo multiple developmental stages, and eventually, ~40% of the D-shaped larvae successfully develop into double-siphon juveniles. These juveniles exhibit a body plan and internal organs similar to those of adults, and their lifestyle gradually shifts from crawling to infaunal-dwelling.

12.4.7 Care of juvenile and adult clams

After settlement and metamorphosis, juveniles are transferred to larger aquariums (1 m × 1 m × 0.5 m) or circulation systems with a 3 cm sandy substrate to accommodate their growth and

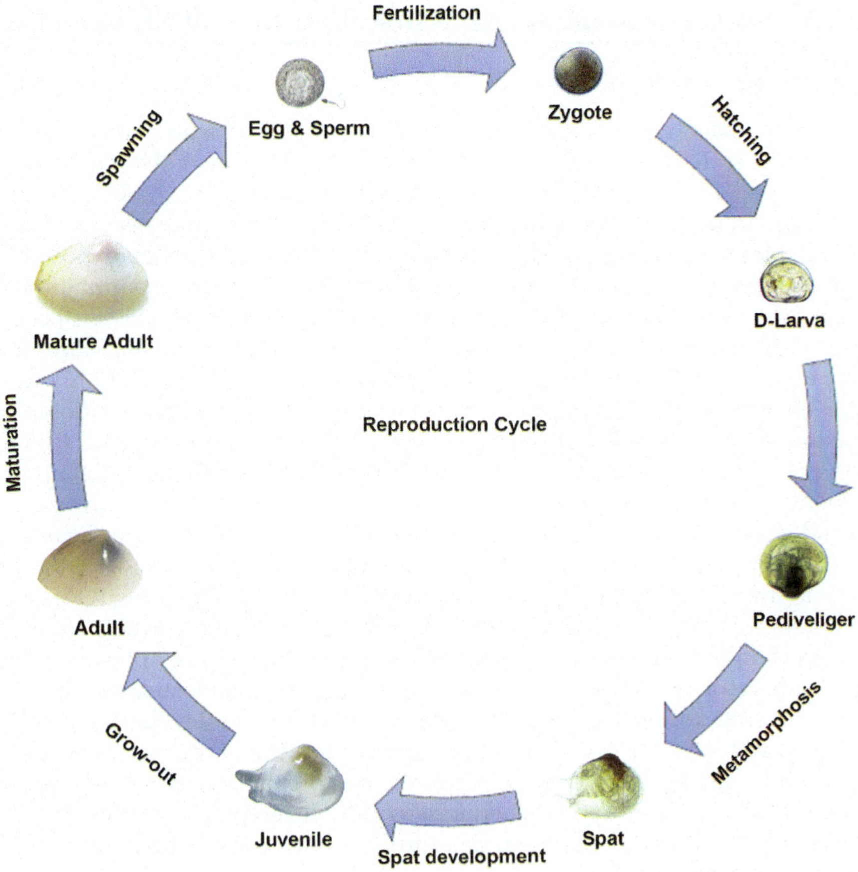

FIGURE 12.4 Artificial culture and breeding of *Mulinia lateralis* in the laboratory. The artificial breeding of *M. lateralis* encompasses several stages: Spawning induction, embryo incubation, larvae and spat culture, grow-out, and adult maturation.

infaunal lifestyle. They are fed a mixture of *C. muelleri* and *C. pyrenoidesa* (1:1) at a concentration of 5–10 × 10^4 cells/mL, with a stocking density of 600–800 individuals/m^2. To ensure optimal growth and survival, the aquariums are maintained at a water temperature of 20–22°C and a salinity of 22–25 psu. As the juveniles continue to develop (45–60 dpf), their diet is switched to a mixture of *C. muelleri, P. helgolandica*, and *I. galbana* in a 3:1:1 ratio at a concentration of 10–20 × 10^4 cells/mL, with a stocking density ranging from 400 to 600 individuals/m^2. When the gonads reach maturity, the aquarium temperature is lowered to 18–20°C to prevent spontaneous gamete release. As previously mentioned, the artificial breeding of *M. lateralis* encompasses several stages: spawning induction, embryo incubation, larvae and spat culture, grow-out, and adult maturation (Fig. 12.4). The entire process takes approximately two months. By this time, *M. lateralis* reaches sexual maturity, with females displaying distinctly orange-red gonads and males exhibiting white or beige gonads, which are ready for the next breeding cycle.

12.5 Available resources and technical platform of M. *lateralis*

12.5.1 Germplasm resources

For a long time, research on *M. lateralis* has primarily relied on wild germplasm (Shumway et al., 1983; Walker & Tenore, 1984; Wang & Guo, 2008). However, *M. lateralis* is a typical opportunistic species. Under favorable environmental conditions, it exhibits high growth and reproduction rates, leading to rapid population expansion and often presents as a dominant species in benthic communities (Santos & Simon, 1980). Conversely, when exposed to adverse conditions such as extreme temperatures or low oxygen levels, *M. lateralis* populations can suffer significant mass mortality (Buron et al., 2013). Furthermore, while *M. lateralis* occurs in high densities in its native range along the North American Coast and in nonnative European coastal waters (Gismann et al., 2023; Rhodes et al., 1975), it is rarely found naturally in other regions. Consequently, the availability of wild germplasm is subject to significant spatiotemporal fluctuations, making it an unreliable source for a stable and continuous supply. To address this limitation, several research groups have attempted to artificially cultivate this clam under laboratory conditions, such as Haskin Shellfish Research Laboratory (Rutgers University, NJ, USA) and Marine Biological Laboratory (Woods Hole, MA, USA), which have significantly improved germplasm availability compared to wild sources (Cripe, 2006; Lu et al., 1996). Following its introduction into the MOE Key Laboratory of Marine Genetics and Breeding (Ocean University of China, Qingdao, China), Yang et al. (2021, 2022) systematically developed protocols for laboratory cultivation for long-term breeding, and cultivated some germplasms with apparent adaptability to laboratory conditions, as well as special germplasms such as inbred lines and bisexual triploid resources, providing germplasm support for the development of *M. lateralis* platform (Yang et al., 2021, 2022, 2024; Zhao et al., 2025). However, for a mature model platform, the existing germplasms are insufficient to meet research demands, particularly because of the lack of lines with unique phenotypes. Therefore to fully leverage the potential of *M. lateralis* as a model organism, further development of additional lines with distinct phenotypic traits is necessary in subsequent research.

12.5.2 Genomics resources

Compared with other model species, the development of genomic resources for *M. lateralis* began relatively late. Genome sequencing of *M. lateralis* was initiated at the Ocean University of China in 2017 and has been completed, with the results soon to be published. The *M. lateralis* genome contains approximately 25,000 genes spanning over 700 million base pairs across 19 pairs of chromosomes. Simultaneously, spatiotemporal transcriptome atlases were constructed for 37 developmental stages and 7 adult organs, identifying multiple regulatory modules through coexpression analysis. In terms of a haplotype genome and spatiotemporal transcriptome atlases, *M. lateralis* possesses a relatively complete repertoire of ancient gene families and conserved gene expression regulatory patterns, demonstrating its value as a model organism for bivalves. With the upcoming publication of the *M. lateralis* genome, all associated details will be released publicly, providing valuable resources for in-depth research on this species as a bivalve model. However, it is important to note that the current genome assembly does not fully cover all chromosomal regions. A telomere-to-telomere (T2T) genome assembly is necessary to facilitate comprehensive genome analysis, encompassing telomeres and centromeres, and allow the exploration of previously uncharted regions of the *M. lateralis* genome. Currently, efforts are underway to assemble the T2T genome of *M. lateralis*, which will be made publicly available in due course.

12.5.3 Technical platform

The recent development of universal and efficient tools for genetic manipulation has greatly expanded the range of model organisms suitable for studying biological evolution and development. However, in bivalves, genetic manipulations remain a significant bottleneck. Although *M. lateralis* has a short generation time, making it an attractive model for reverse genetics, it lagged behind mature model species, such as the mouse and zebrafish. *M. lateralis* was the first bivalve species used for transgenic research. Lu et al. (1996) successfully introduced the β-galactosidase gene into fertilized eggs using a retroviral vector and electroporation, with the transgenic procedure having minimal impact on embryonic survival. Positive transgenes were detected in 13%–33% of the treated embryos. In addition, for hybrid offspring produced by crossing positive individuals with wild-type clams, β-galactosidase activity was observed in 24%–33% of the larvae at the D-stage. This retroviral vector method offers a promising approach for genetic manipulation in bivalves. Although there have been no follow-up reports based on this technology, it highlights the excellent potential of *M. lateralis* as a model for bivalve gene manipulation.

As part of the effort of developing *M. lateralis* into a bivalve model, we developed additional gene manipulation tools for this species. Taking the advantages of microinjection, such as its applicability to a wide range of cell types, the ability to introduce precise and controllable amounts of DNA, and the high efficiency of exogenous DNA integration, along with the soft egg membrane of bivalve eggs, we developed a microinjection-based method for introducing exogenous substances into bivalve eggs. More recently, the advent of CRISPR/Cas9, derived from bacterial defense systems, has facilitated efficient gene editing. CRISPR/Cas9 utilizes a nuclease (Cas9) guided by a short RNA sequence to induce double-stranded breaks in DNA (Komor et al., 2017). Cas9 can be injected into a one-cell embryo as a protein or mRNA, and the guide RNA can either be synthesized in vitro and coinjected or transcribed in vivo from a plasmid or transgene. The reagents needed are inexpensive and widely available. Subsequently, our team employed CRISPR/Cas9 gene knockout and RNA interference (RNAi) to investigate the functional role of the *Cfap206* gene in *M. lateralis* development (Wang et al., 2022). As shown in Fig. 12.5A, primers were designed based on the full-length transcriptome sequence of the *Cfap206* gene, and double-stranded RNA (dsRNA) was synthesized in vitro. The dsRNA was then introduced into fertilized eggs of *M. lateralis* using electroporation (Fig. 12.5B), which caused a significant reduction in the number of cilia in trochophore larvae following the inhibition of *Cfap206* gene expression compared with the control group (Fig. 12.5C). The findings from the RNAi experiments were further confirmed with CRISPR/Cas9 gene editing. Single-guide RNAs (sgRNAs) were designed to target the *Cfap206* gene (Fig. 12.5D), and sgRNAs along with Cas9 protein were microinjected into the fertilized eggs. This resulted in a significant reduction in the number of cilia in gene-edited embryos, which was consistent with those from RNAi experiments (Fig. 12.5E). The development of microinjection technology for small bivalve eggs represents a significant breakthrough, enabling the introduction of exogenous genes and the tracking of cells. The RNAi technology established a single-gene knockdown experimental system, while the CRISPR/Cas9 technology provided a single-gene knockout

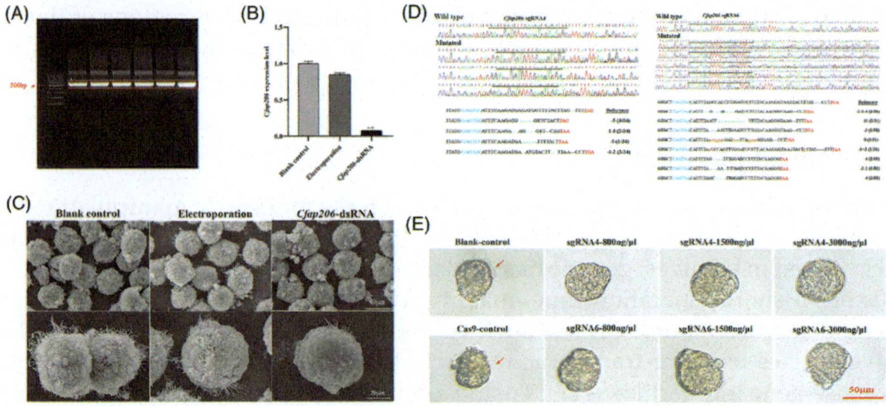

FIGURE 12.5 Functional analysis of the *Cfap206* gene in *Mulinia lateralis* development using CRISPR/Cas9 gene knockout and RNA interference (RNAi) (Wang et al., 2022). (A) Electrophoresis result of *Cfap206* dsRNA; (B) The expression of *Cfap206* after RNAi, $^{**}P < .01$; (C) SEM observation of trochophore larvae after RNAi; (D) Sanger sequencing of PCR products from injected embryos and the identification of mutations; (E) Trochophore larvae after *Cfap206* knockout.

experimental system. These advancements lay the foundation for decoding gene functions in bivalves. Currently, efforts are underway to further refine the *M. lateralis* gene manipulation platform, including the development of high-throughput delivery tools and the screening of efficient editing systems.

12.6 M. *lateralis* as a model species for bivalve development and genetics

M. lateralis serves as a powerful model organism for advancing bivalve aquaculture and genetic improvement because of its unique biological attributes (rapid growth, transparency, well-studied genetics, etc.). Its current applications span several critical research areas.

12.6.1 Investigating regulatory mechanisms of development

Bivalves exhibit unique developmental processes, such as spiral cleavage, mosaic development, and metamorphosis, and deciphering the molecular regulation of developmental traits is essential for developmental biology and provides leverage for germplasm innovation and sustainable aquaculture production. *M. lateralis*, with its rapid external development and transparency, is a key model for such studies. Kidder utilized *M. lateralis* to investigate gene transcription during the early embryonic development of protostomes, analyzing the incorporation of newly synthesized RNA into polyribosomes of embryos (Kidder, 1972). This work challenges speculation concerning possible differences in gene transcription patterns between early mosaic and regulative embryos. Longo et al. (1991) used *M. lateralis* to investigate changes in male and female nuclei after sperm entry during fertilization. They discovered that ultraviolet irradiation of fertilizing sperm did not significantly affect the nuclear transformations leading to the development of male or female pronuclei, but caused chromosome asynchrony during prophase and the nondisjunction of chromosomes at anaphase. Chromatin-based epigenetics has proven to be an effective method for elucidating these complex embryogenetic processes. Shi et al. (2025) integrated chromatin-state profiling and transcriptomic analysis to reveal extensive reprogramming of histone modifications between the morula and gastrula stages, and determined the timing of maternal-to-zygotic transition. These findings provide mechanistic insights into chromatin-mediated regulation of bivalve embryogenesis, paving the way for advanced epigenomic and functional gene research. In bivalves, the formation of larval shells is a critical developmental process essential for larval survival. Guo et al. (2023) through full-length transcriptome sequencing of *M. lateralis* at various developmental stages (blastula, gastrula, trochophore larva, and D-shaped larva), identified 12 genes potentially involved in larval shell formation, including insoluble shell matrix protein-encoding gene (*ISMP*), ring finger protein 41 *(rnf41)*, and solute carrier family 4 member 8 (*SLC4A8*). These findings provide valuable insights into the mechanisms underlying larval shell formation as well as other developmental processes in bivalves.

12.6.2 Exploring gonadal development and understanding gametogenesis

For aquaculture production, gonadal development in the broodstock is critical in enhancing fecundity while maintaining gamete quality and larval viability. In general, the gonadal development of bivalves is regulated by a complex cascade of events. Utilizing the *M. lateralis* model, Li et al. (2022) identified seven gonadal development-related modules through gene coexpression network analysis and pinpointed several hub transcriptional factors, including SRY-box transcription factor 2 (*SOX2*), forkhead box protein Z (*FOXZ*), heat shock transcription factor Y (*HSFY*), and forkhead box protein L2 (*FOXL2*). In addition, the expression patterns of top hub genes from sex-specific modules revealed that molecular sex differentiation (occurring at 35 dpf) precedes morphological sex differentiation (occurring at 40–45 dpf). Gametogenesis, a complex process of cellular differentiation, begins with the division of stem cells and culminates in the production of functional gametes. In spermatogenesis, as shown in Fig. 12.6, single-cell RNA-seq analysis identified four germ cell populations and three somatic populations, with the germ cells displaying distinct cell cycle statuses and an uninterrupted developmental trajectory (Li et al., 2025b). Furthermore, gene expression, translation, ATP metabolic process, and microtubule-based processes are involved in the transition of germ cell types, with extensive bidirectional communication between somatic cells and germline cells. These findings offer profound insights into the molecular basis of gonad development, contributing to a comprehensive understanding of the reproductive regulation network and the transition of germ cell fate during spermatogenesis in bivalves.

12.6.3 Functional verification of growth-associated genes

Improving growth performance is a primary objective in bivalve aquaculture. *M. lateralis* exhibits unique and rapid growth characteristics, making it an ideal model organism for studying bivalve growth. Its experimental accessibility, combined with its fast growth, has facilitated the exploration of growth-related genes, which not only advances our understanding of the mechanisms underlying growth regulation in bivalves but also identifies potential genomic targets for breeding fast-growing bivalve varieties. Several genes involved in growth regulation, such as somatostatin receptor (*SSTR*), E2F transcription factor 3 (*E2F3*), and insulin-related

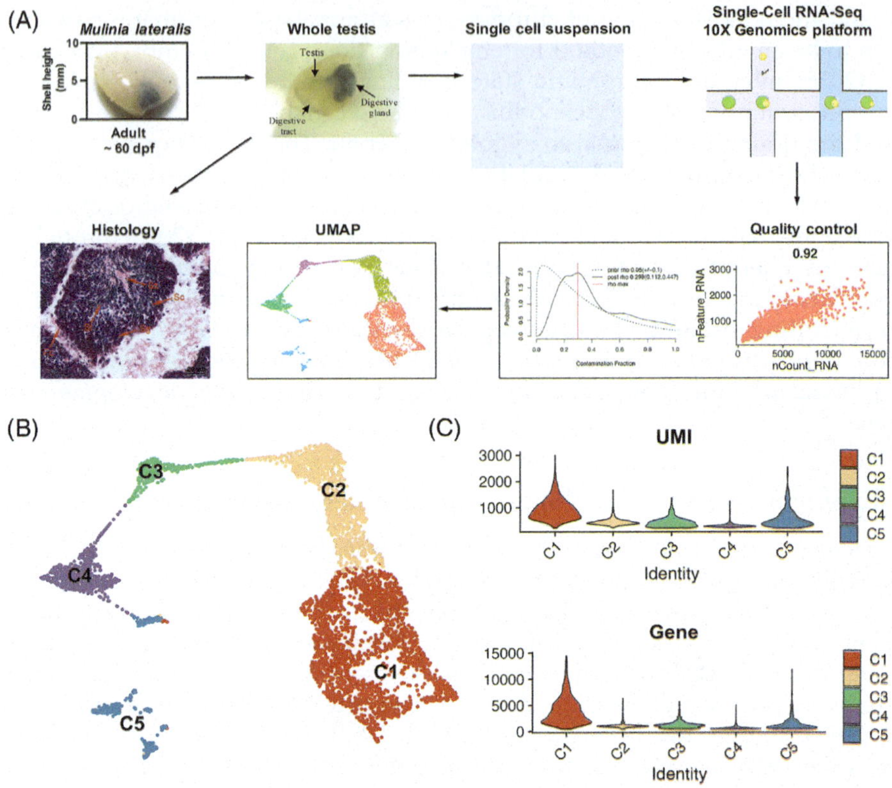

FIGURE 12.6 Overview of the major cell types identified through the analysis of single-cell RNA-Seq data from the testis of *Mulinia lateralis*. (A) Schematic overview of the data collection and analysis process; (B) UMAP plot of scRNA-seq data comprising 4,647 cells. The cells are colored according to broad cell types; (C) The distribution of UMIs and the number of genes detected in each cell across the five cell clusters. *From Li et al. (2025a).*

protein (*ILP*), have been evaluated using the *M. lateralis* model system (Wang et al., 2023; Yao et al., 2024; Zhang et al., 2024). For instance, *SSTR* belongs to the G protein-coupled receptor (GPCR) family and plays versatile roles in inhibiting the secretion of multiple hormones, including growth hormone and thyroid-stimulating hormone (Zhao et al., 2022). Inhibiting the expression of *SSTR-1* in *M. lateralis* resulted in a significant increase in shell length, shell width, shell height, soft tissue weight, and muscle weight by 20%, 22%, 20%, 79%, and 92%, respectively. Transcriptome analysis revealed that the up-regulated genes following *SSTR-1* inhibition were significantly enriched in the fat digestion and absorption pathway and the insulin pathway. These results indicate that *SSTR-1* is a negative regulator of growth in *M. lateralis*, making it a candidate target for gene editing to promote growth in bivalves, as well as for the breeding of fast-growing bivalve varieties.

12.6.4 Verifying the functional role of skeleton-embedding cells in bivalves

A classical model of shell formation posits that mollusk shells are lifeless biominerals secreted by the soft body (Clark, 2020). However, recent findings Lian et al. (2025) have challenged

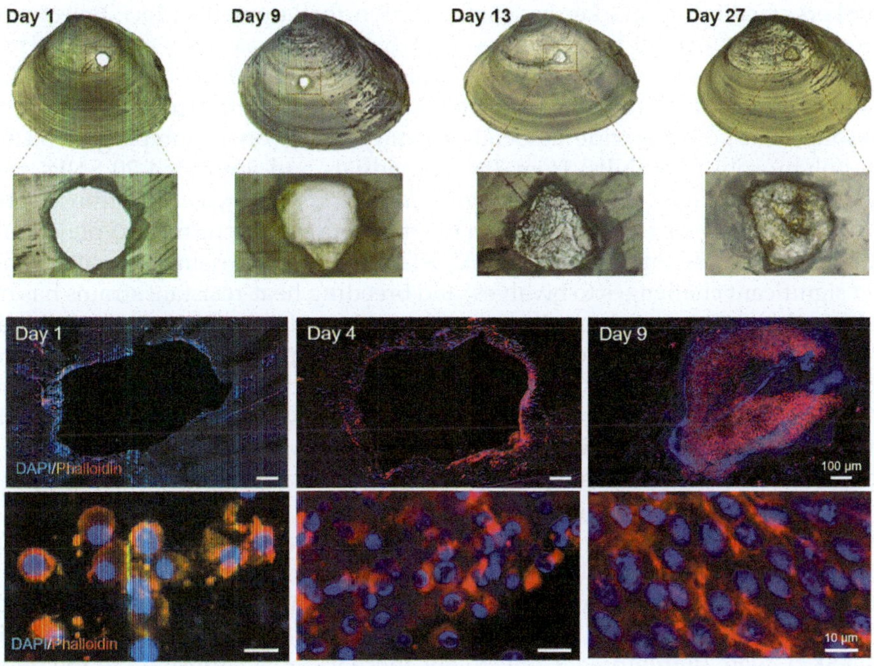

FIGURE 12.7 Light microscopic and confocal observation during the shell regeneration of *Mulinia lateralis*. The upper shows the light observation, and the lower shows the confocal microscopy observation. *From Lian et al. (2025).*

this view, with the discovery of a hematopoietic stem cell (HSC) niche in mollusk shells. They further revealed its developmental origin, molecular and functional resemblance to bone marrow (BM) cells, and its potentially widespread presence in diverse invertebrate groups. To verify biomineralization-related hematopoiesis by skeleton-embedding cells (SECs), *M. lateralis* was employed in a shell injury assay. In this assay, a ~1 mm hole was created in the middle of the left valve, located a distance from the shell edge. As illustrated in Fig. 12.7, the injured region of the shell was completely regenerated within 27 days. Subsequent comprehensive experiments using this injury model further substantiated the hematopoietic function of SECs in the blood supply to the soft body, with functional similarity to the hematopoietic system of vertebrate BM. This study reveals the widespread and potentially deep evolutionary origins of invertebrate skeletal HSC niches, thereby boosting new paradigms for hematopoiesis and stem cell research in invertebrates and providing a redefined understanding of vertebrate BM evolution and water-to-land transition.

12.6.5 Uncovering functional polymorphism for bivalve breeding

The advent of high-throughput sequencing has accelerated the integration of genomics into aquaculture breeding programs. This has enabled genetic breeding to take a step forward, especially for traits that are difficult or impossible to measure directly in aquatic species. In bivalve, quantitative trait locus (QTL) mapping enables identification of trait-associated genetic markers, facilitating marker-assisted selection for large-effect loci. Selection for traits with large-effect QTLs can be improved by applying marker-assisted selection. The shells of bivalves exhibit rich and complex morphological variations, which are of great significance for fulfilling key functions such as protection, locomotion, and structural support. Li et al. (2025a) conducted a genome-wide association study based on genotyping of 32,318 single-nucleotide polymorphisms (SNPs) from 180 individuals and identified 20 SNPs and 16 candidate genes related to shell morphology in *M. lateralis*. Notably, *Lox4* (a transcription factor involved in anteroposterior (A-P) body plan determination) was proven to play an important role in influencing the shell morphology of *M. lateralis*. Recently, rising seawater temperatures have posed significant challenges to bivalves, and breeding heat-resistant strains has become an urgent need for bivalve aquaculture. Wang et al. conducted a genome-wide association study based on a thermal stress experiment involving 600 clams, revealing the regulatory role of the ETHR/EHF signaling pathway in the bivalve's response to thermal stress, which contributes to comprehension of the genetic basis of thermal tolerance in bivalves (Wang et al., 2024). However, long-term artificial breeding risks increased inbreeding and eroded genetic diversity. Genomic selection strategies counteract this by capturing favorable variation to enhance performance while controlling inbreeding. Yang et al. (2024) leveraged 22,196 SNPs genotyped from 310 individuals across five generations to identify selection signatures in *M. lateralis* associated with its rapid adaptation to artificial culture, underscoring the potential of this species for studying adaptive evolution in bivalves. Notably, genes linked to male production, such as galactose-3-O-sulfotransferase 1 (*GAL3ST1*), intraflagellar transport 88 (*IFT88*), and testis-specific serine kinase 2 (*TSSK2*), were highlighted. Collectively, these insights contribute to the development of genomic breeding technologies and program optimization.

12.6.6 Polyploid induction and reproductive mechanisms in bivalves

Polyploids are widely used to accelerate genetic improvements in organisms because of their fast growth and enhanced environmental adaptability, making bivalve polyploidy a significant field of research and application. However, successful polyploid induction and production have been largely confined to oysters (Guo, 2021). Oysters have a small genome with 10 chromosomes, while most bivalves have a larger genome and more chromosomes. It is unclear whether large genome sizes affect the viability of polyploids, particularly tetraploids, which are crucial for the commercial production of triploids for aquaculture. Like many clams and scallops, *M. lateralis* possesses 19 pairs of chromosomes, as confirmed by karyotype analysis (Wang & Guo, 2008), establishing it as a representative model for polyploidy studies. Chromosome set manipulation has been studied in this species. For instance, Guo and Allen (1994a) successfully induced triploids by inhibiting meiosis II and gynogenesis in *M. lateralis* by ultraviolet inactivation of sperm; triploids were significantly larger than diploids, leading to the polyploid gigantism hypothesis, and all offspring were female, which implies that the sex determination in *M. lateralis* follows an XX/XY system. Yang and Guo (2004, 2006) produced tetraploids using cytochalasin B and triploids with heat shock in *M. lateralis*. While high percentages of triploids (86.3%–98.5%) were produced with heat shock, only a few tetraploid *M. lateralis* survived beyond metamorphosis. These experiments laid the groundwork for future studies on chromosome set manipulation in bivalves. Building on this, Zhao et al. (2025) developed an optimized triploid induction protocol for *M. lateralis*, demonstrating significant growth and survival advantages and providing a valuable model for studying advantageous trait formation of triploid bivalves.

While triploid sterility is common, rare fertile triploids can produce viable offspring, although most of the offspring are triploids (Guo & Allen, 1994b). Like in other bivalves, artificially induced triploid *M. lateralis* produce viable gametes with full developmental potential (Zhao et al., 2025). This unique characteristic makes *M. lateralis* a good model for investigating fertility regulation mechanisms in triploid bivalves, with implications for conserving elite germplasm and mitigating ecological risks.

12.6.7 Assessment of bivalve farming environments

Bivalves in aquaculture, encompassing both wild populations and farmed stocks, are exposed to significant environmental variability and pollutants within coastal and estuarine systems. To ensure their ability to thrive in these challenging conditions and achieve sustainable production, it is essential to understand the interactions between the cultured stocks and their environments. In fact, some of the earliest studies on *M. lateralis* primarily focused on using them to evaluate the effects of environmental pollutants on bivalves. For example, Shumway et al. (1983) discovered that *M. lateralis* can maintain high levels of feeding, shell, and locomotion activities during short-term exposure to anoxia and hydrogen sulfide conditions, indicating a degree of tolerance to these conditions. Ho and Zubkoff (1982) utilized calcium-45 uptake by *M. lateralis* larvae as an indicator to evaluate the environmental impact of Cu, Hg, and Zn, and found that the effects of these heavy metals on larval survival and calcium absorption followed the order of Cu > Hg > Zn. Huber et al. (1997) developed a method to assess seawater ammonia content using the survival and growth of *M. lateralis* as indicators. The advantages of *M. lateralis*, including its well-studied genetics, rapid and transparent development, and potential for encoding transgenic biosensors, establish it as a powerful tool for environmental assessment for bivalve aquaculture and general toxicology.

12.7 Future prospects

Significant progress has been made in the development of the *M. lateralis* model platform. However, as an emerging model species compared to established model organisms, further improvements are needed before *M. lateralis* can be effectively utilized in general biological research. Advances in the following areas are particularly important.

(1) Establishing foundational genetic tools: Overcoming key technical bottlenecks is paramount for functional studies. This requires developing accessible forward and reverse genetics tools, including gene editing, cell culture systems, and mutagenesis protocols, and expanding their application. These advancements are crucial for rapidly elevating *M. lateralis* to "mature model" status.

(2) Integrating multiomics for trait analysis: Focusing on shellfish breeding needs and using *M. lateralis* as a model, multiomics approaches should be employed to elucidate the genetic basis and regulatory mechanisms underlying critical traits such as growth, reproduction, disease resistance, stress tolerance, and quality. This will provide precise targets for genetic improvement.

(3) Promoting cross-species application: Research findings and tools developed in *M. lateralis* must be transferred to other shellfish species to enhance industry-wide breeding efficiency. This includes leveraging functional genes and trait-associated genetic variations, implementing AI-driven breeding strategies (such as computational gene editing using AI algorithms and high-quality omics data), and constructing a comprehensive precision breeding system for shellfish.

Collectively, developing *M. lateralis* as an emerging model offers a crucial bridge between understanding fundamental genetic mechanisms and implementing practical breeding strategies in shellfish. This integration promises to significantly improve breeding accuracy and efficiency, better meeting the demands of aquaculture and addressing future challenges.

References

Bartley, D.M. (2022). World Aquaculture 2020 – A brief overview. FAO Fisheries and Aquaculture Circular (No. 1233). Italy, FAO. 10.4060/cb7669en.

Buron, I. D., Roth, P. B, Bergquist, D. C, & Knott, D. M. (2013). *Mulinia lateralis* (Mollusca: Bivalvia) die-off in South Carolina: Discovery of a vector for two elasmobranch cestode species. *Journal of Parasitology, 99*(1), 51–55. doi:10.1645/GE-3092.1.

Calabrese, A. (1969a). Proceedings of the national shellfish association *Mulinia lateralis*, molluscan fruit fly. *Proceedings of the National Shellfish Association, 59*, 65–66.

Calabrese, A. (1969b). Individual and combined effects of salinity and temperature on embryos and larvae of the coot clam, *Mulinia lateralis* (Say). *The Biological Bulletin, 137*(3), 417–428. doi:10.2307/1540164.

Calabrese, A. (1970). Reproductive cycle of the coot clam, *Mulinia lateralis* (Say), in long island sound. *Veliger, 12*(3), 265–269.

Chalermwat, K., Jacobsen, T. R, & Lutz, R. A (1991). Assimilation of bacteria by the dwarf surf clam *Mulinia lateralis* (Bivalvia: Mactridae). *Marine Ecology Progress Series, 71*(1), 27–35. doi:10.3354/meps071027.

Cintra-Buenrostro, C. E., Flessa, K. W., & Guillermo, A.-S. (2005). Who cares about a vanishing clam? Trophic importance of *Mulinia coloradoensis* inferred from predatory damage. *PALAIOS, 20*(3), 296–302. doi:10.2110/palo.2004.p04-21.

Clark, M. S (2020). Molecular mechanisms of biomineralization in marine invertebrates. *Journal of Experimental Biology, 223*(11), jeb206961. doi:10.1242/jeb.206961.

Craeymeersch, J. A., Faasse, M. A., Gheerardyn, H., Troost, K., Nijland, R., Engelberts, A., ... van Zwol, J. (2019). First records of the dwarf surf clam *Mulinia lateralis* (Say, 1822) in Europe. *Marine Biodiversity Records, 12*(1), 1–11. doi:10.1186/s41200-019-0164-7.

Cripe, G. M. (2006). Contaminated sediment testing with the bivalve *Mulinia lateralis*: Culture refinement for organism availability. *Environmental Toxicology and Chemistry, 25*(5), 1332–1336. doi:10.1897/05-271r.1.

Davison, A., & Neiman, M. (2021). Mobilizing molluscan models and genomes in biology. *Philosophical Transactions of the Royal Society B: Biological Sciences, 376*(1825), 20200163. doi:10.1098/rstb.2020.0163.

Dégremont, L., Nourry, M., & Maurouard, E. (2015). Mass selection for survival and resistance to OsHV-1 infection in *Crassostrea gigas* spat in field conditions: Response to selection after four generations. *Aquaculture, 446*, 111–121. doi:10.1016/j.aquaculture.2015.04.029.

de Melo, C. M. R., Durland, E., & Langdon, C. (2016). Improvements in desirable traits of the Pacific oyster, *Crassostrea gigas*, as a result of five generations of selection on the West Coast, USA. *Aquaculture, 460*, 105–115. doi:10.1016/j.aquaculture.2016.04.017.

Gismann, L., Wenke, L.-K., Uhlir, C., Arbizu, P. M, & Wehrmann, A. (2023). Status and occurrence of the non-indigenous dwarf surf clam *Mulinia lateralis* (Say, 1822) in the central Wadden Sea (southern North Sea)—a systematic survey. *Marine Biodiversity, 53*(6), 83. doi:10.1007/s12526-023-01381-w.

Guo, X. (2021). Genetics in shellfish culture. *Molluscan shellfish aquaculture: A practical guide* (pp. 393–413). 5m Books Ltd. doi:10.1079/9781800629189.0014.

Guo, X., & Allen, S. K (1994a). Sex determination and polyploid gigantism in the dwarf surf clam (*Mulinia lateralis* Say). *Genetics, 138*(4), 1199–1206. doi:10.1093/genetics/138.4.1199.

Guo, X., & Allen, S. K (1994b). Reproductive potential and genetics of triploid Pacific oysters, *Crassostrea gigas* (Thunberg). *The Biological Bulletin, 187*(3), 309–318. doi:10.2307/1542288.

Guo, X. (2026). Genetics and Epigenetics in Life History and Reproduction: Bivalve Molluscs. In Michael K. Skinner (Ed.), *Encyclopedia of Reproduction* (3rd) (6, pp. 964–972). US: Elsevier.

Guo, X., Li, X., Zhao, F., Liu, D., Yang, Z., Li, M., … Bao, Z. (2023). Full-length transcriptome analysis provides insights into larval shell formation in *Mulinia lateralis*. *Frontiers in Marine Science, 9*, 1111241. doi:10.3389/fmars.2022.1111241.

Hedgecock, D., & Pudovkin, A. I (2011). Sweepstakes reproductive success in highly fecund marine fish and shellfish: A review and commentary. *Bulletin of Marine Science, 87*(4), 971–1002. doi:10.5343/bms.2010.1051.

Ho, M.-S., & Zubkoff, P. L (1982). The effects of mercury, copper, and zinc on calcium uptake by larvae of the clam, *Mulinia lateralis*. *Water, Air, and Soil Pollution, 17*(4), 409–414. doi:10.1007/BF00460107.

Hollenbeck, C. M, & Johnston, I. A (2018). Genomic tools and selective breeding in molluscs. *Frontiers in Genetics, 9*, 253. doi:10.3389/fgene.2018.00253.

Huan, P., & Liu, B. (2024). The gastropod *Lottia peitaihoensis* as a model to study the body patterning of trochophore larvae. *Evolution & Development, 26*(4), e12456. doi:10.1111/ede.12456.

Huber, M., Pelletier, M. C., Charles, J. B., & Burgess, R. M. (1997). Ammonia tolerance of the bivalve *Mulinia lateralis* sublethal sediment toxicity test. *Bulletin of Environmental Contamination and Toxicology, 59*(2), 292–297. doi:10.1007/s001289900478.

Kidder, G. M (1972). Gene transcription in mosaic embryos. II. Polyribosomes and messenger RNA in early development of the coot clam, *Mulinia lateralis*. *Journal of Experimental Zoology, 180*(1), 75–84. doi:10.1002/jez.1401800106.

Komor, A. C, Badran, A. H, & Liu, D. R (2017). CRISPR-based technologies for the manipulation of eukaryotic genomes. *Cell, 168*(1), 20–36. doi:10.1016/j.cell.2016.10.044.

Li, C., Wang, H., Zhao, A., Bao, L., Wang, H., Zeng, Q., … Bao, Z. (2025a). Genome-wide association study reveals the role of *Lox4* in shell morphology of *Mulinia lateralis*. *Water Biology and Security, 5*(3), 100502. doi:10.1016/j.watbs.2025.100502.

Li, Y., Wei, H., Dai, X., Zhang, L., Liu, L., Chen, X., … Zhang, L. (2025b). Insights from the single-cell level: Lineage trajectory and somatic-germline interactions during spermatogenesis in dwarf surfclam *Mulinia lateralis*. *BMC Genomics, 26*(1), 69. doi:10.1186/s12864-025-11266-w.

Li, Y., Liu, L., Zhang, L., Wei, H., Wu, S., Liu, T., … Zhang, L. (2022). Dynamic transcriptome analysis reveals the gene network of gonadal development from the early history life stages in dwarf surfclam *Mulinia lateralis*. *Biology of Sex Differences, 13*(1), 1–16. doi:10.1186/s13293-022-00479-3.

Lian, S., Hu, N., Chen, X., Dai, X., Zhu, X., Qiao, R., … Shi, W. (2025). Widespread presence of bone marrow–like hematopoietic stem cell niche in invertebrate skeletons. *Science Advances, 11*(26), eadw0958. doi:10.1126/sciadv.adw0958.

Longo, F. J, & Scarpa, J. (1991). Expansion of the sperm nucleus and association of the maternal and paternal genomes in fertilized *Mulinia lateralis* eggs. *The Biological Bulletin, 180*(1), 56–64. doi:10.2307/1542428.

Lu, J.-K., Chen, T. T, Allen, S. K, Matsubara, T., & Burns, J. C (1996). Production of transgenic dwarf surfclams, *Mulinia lateralis*, with pantropic retroviral vectors. *Proceedings of the National Academy of Sciences, 93*(8), 3482–3486. doi:10.1073/pnas.93.8.3482.

McKeon, C. S., Tunberg, B. G., Johnston, C. A, & Barshis, D. J (2015). Ecological drivers and habitat associations of estuarine bivalves. *PeerJ, 3*(11), e1348. doi:10.7717/peerj.1348.

Montagna, P. A., Stockwell, D. A., & Kalke, R. D. (1993). Dwarf surf clam *Mulinia lateralis* (Say, 1822) populations and feeding during the Texas brown tide event. *Journal of Shellfish Research, 12*(2), 433–442.

Morton, B., & Machado, F. M (2019). Predatory marine bivalves: A review. *Advances in Marine Biology, 84*, 1–98. doi:10.1016/bs.amb.2019.10.001.

Nascimento-Schulze, J. C, Bean, T. P, Houston, R. D, Santos, E. M, Sanders, M. B, Lewis, C., & Ellis, R. P (2021). Optimizing hatchery practices for genetic improvement of marine bivalves. *Reviews in Aquaculture, 13*(4), 2289–2304. doi:10.1111/raq.12568.

Rhodes, E. W., Calabrese, A., Cable, W. D., & Landers, W. S. (1975). *The development of methods for rearing the Coot Clam,* Mulinia Lateralis, *and three species of coastal bivalves in the laboratory culture of marine invertebrate animals* (pp. 273–282). Springer Science and Business Media LLC.

Santos, S. L, & Simon, J. L (1980). Response of soft-bottom benthos to annual catastrophic disturbance in a South Florida estuary. *Marine Ecology Progress Series, 3*(4), 347–355. doi:10.3354/meps003347.

Shi, J.-L., Liu, J.-X., Wei, Z.-C., Yin, M.-M., Yang, Z. u-J., Hu, J.-J., … Bao, Z.-M. (2025). Epigenetic regulation and maternal-to-zygotic transition in dwarf surf clam (*Mulinia lateralis*): Insights from chromatin state profiling and transcriptomics. *Zoological Research, 46*(5), 1009–1028. doi:10.24272/j.issn.2095-8137.2025.093.

Shumway, S. E., & Newell, R. C. (1984). Energy resource allocation in *Mulinia lateralis* (Say), an opportunistic bivalve from shallow water sediments. *Ophelia, 23*(2), 101–118. doi:10.1080/00785326.1984.10426607.

Shumway, S. E, Scott, T. M, & Malcolm Shick, J. (1983). The effects of anoxia and hydrogen sulphide on survival, activity and metabolic rate in the coot clam, *Mulinia lateralis* (Say). *Journal of Experimental Marine Biology and Ecology, 71*(2), 135–146. doi:10.1016/0022-0981(93)90069-Z.

Walker, R. L, & Tenore, K. R (1984). Growth and production of the dwarf surf clam *Mulinia lateralis* (Say 1822) in a Georgia estuary. *Gulf Research Reports, 7*(4), 357–363. doi:10.18785/grr.0704.07.

Wallace, J. G, Rodgers-Melnick, E., & Buckler, E. S (2018). On the road to breeding 4.0: Unraveling the good, the bad, and the boring of crop quantitative genomics. *Annual Review of Genetics, 52*(1), 421–444. doi:10.1146/annurev-genet-120116-024846.

Wang, C., Kong, L., Lian, S., Yang, Z., Meng, D., Li, M., … Hu, X. (2023). Verify the function of a potential growth-regulating gene in marine bivalve using a candidate model organism *Mulinia lateralis*. *Journal of Ocean University of China, 22*(4), 1012–1022. doi:10.1007/s11802-023-5382-2.

Wang, H., Yang, Z., Wang, S., Zhao, A., Wang, H., Liu, Z., … Huang, X. (2024). Genome-wide association analysis reveals the genetic basis of thermal tolerance in dwarf surf clam *Mulinia lateralis*. *Genomics, 116*(5), 110904. doi:10.1016/j.ygeno.2024.110904.

Wang, S., Zhang, J., Jiao, W., Li, J., Xun, X., Sun, Y., … Bao, Z. (2017). Scallop genome provides insights into evolution of bilaterian karyotype and development. *Nature Ecology & Evolution, 1*(5), 0120. doi:10.1038/s41559-017-0120.

Wang, Y., & Guo, X. (2008). Chromosomal mapping of the major ribosomal RNA genes in the dwarf surf clam (*Mulinia lateralis* Say). *Journal of Shellfish Research, 27*(2), 307–311. doi:10.2983/0730-8000(2008)27[307:cmotmr]2.0.co;2.

Wang, Y., Zhu, X., Lian, S., Li, Y., Hu, N., Hu, X., … Wang, S. (2022). Functional characterization of *Cfap206* for bivalve Ciliogenesis by RNAi and CRISPR/Cas9 technologies. *Frontiers in Marine Science, 9*, 864037. doi:10.3389/fmars.2022.864037.

Yang, H., & Guo, X. (2004). Tetraploid induction by meiosis inhibition in the dwarf surfclam *Mulinia lateralis* (Say 1822): Effects of cytochalasin B duration. *Aquaculture Research, 35*(13), 1187–1194. doi:10.1111/j.1365-2109.2004.01118.x.

Yang, H., & Guo, X. (2006). Polyploid induction by heat shock-induced meiosis and mitosis inhibition in the dwarf surfclam, *Mulinia lateralis* Say. *Aquaculture, 252*(2–4), 171–182. doi:10.1016/j.aquaculture.2005.07.017.

Yang, Z., Huang, X., Wang, H., Pan, H., Wang, X., Teng, M., … Bao, Z. (2021). Effects of microalgae diets and stocking density on larval growth, survival and metamorphosis of dwarf surfclam, *Mulinia lateralis*. *Aquaculture, 536*, 736440. doi:10.1016/j.aquaculture.2021.736440.

Yang, Z., Wang, H., Li, M., Teng, M., Wang, X., Zhao, A., … Bao, Z. (2022). Optimizing Microalgae Diet, Temperature, and Salinity for Dwarf Surf Clam, *Mulinia lateralis*, Spat Culture. *Frontiers in Marine Science, 8*, 823112. doi:10.3389/fmars.2021.823112.

Yang, Z., Zhao, A., Teng, M., Li, M., Wang, H., Wang, X., … Huang, X. (2024). Signatures of selection in *Mulinia lateralis* underpinning its rapid adaptation to laboratory conditions. *Evolutionary Applications, 17*(2), e13657. doi:10.1111/eva.13657.

Yao, R., Wang, C., Kong, L., Wang, Y., Bao, Z., & Hu, X. (2024). Investigating the genetic bases of growth regulation by *E2F3* in dwarf surf clams *Mulinia lateralis*. *Journal of Ocean University of China, 23*(5), 1347–1358. doi:10.1007/s11802-024-5935-z.

Zhang, X., Niu, Y., Gao, C., Kong, L., Yang, Z., Chang, L., Kong, X., Bao, Z., & Hu, X. (2024). Somatostatin receptor gene functions in growth regulation in bivalve scallop and clam. *International Journal of Molecular Sciences, 25*(9), 4813. doi:10.3390/ijms25094813.

Zhao, A., Yang, Z., Wang, H., Wang, H., Zhong, S., Li, C., … Huang, X. (2025). Establishment and characterization of bisexually fertile triploid dwarf surf clam *Mulinia lateralis*. *Marine Biotechnology, 27*(1), 1–18. doi:10.1007/s10126-024-10406-9.

Zhao, W., Han, S., Qiu, N., Feng, W., Lu, M., Zhang, W., … Zhong, W. (2022). Structural insights into ligand recognition and selectivity of somatostatin receptors. *Cell Research, 32*(8), 761–772. doi:10.1038/s41422-022-00679-x.

SECTION III

Shellfish Aquaculture

CHAPTER

13

Oyster aquaculture

Qi Li[a], *Gaowei Jiang*[a], *Ximing Guo*[b] *and Vengatesen Thiyagarajan*[c]

[a]College of Fisheries, Ocean University of China, Qingdao, China [b]School of Environmental and Biological Sciences, Rutgers University, New Brunswick, NJ, United States [c]School of Biological Sciences, The University of Hong Kong, Hong Kong SAR, China

13.1 The state of the oyster farming industry

Oysters are key components of coastal ecosystems and an important food source, and as such form the resource base of many coastal communities globally. Since 1950, global oyster production has risen dramatically, reaching 7 million tons in 2022, making oysters the leading aquaculture species by whole weight. Oysters are farmed in nearly every coastal country, with notable production in China, the United States, Japan, South Korea, France, Mexico, New Zealand, and Australia. Currently, Asia dominates global oyster farming, producing 6.4 million tons in 2022, which accounts for 95% of worldwide production (FAO, 2024). China alone contributes 5.8 million tons, representing 88% of global oyster production (Jiang et al., 2024). As a result, aquaculture production of oysters has become a multibillion-dollar farming industry in China.

About 100 extant species of oysters have been identified worldwide among many possible synonyms (Guo et al., 2018). The primary cultivated species include the Pacific oyster (*Crassostrea gigas*), the Eastern oyster (*C. virginica*), the Hong Kong oyster (*C. hongkongensis*), the Fujian oyster (*C. angulata*), the European flat oyster (*Ostrea edulis*), and the Sydney rock oyster (*Saccostrea glomerata*) (Botta et al., 2020). *C. gigas*, native to Northeast Asia, has been introduced to more than 60 countries, making it the most extensively cultivated oyster globally. China, South Korea, Japan, France, and the United States are the primary producers of *C. gigas*. China dominates global production, with its total output reaching approximately 2.1 million tons in 2023 (BOF, 2024). Although native to southern China and Vietnam, *C. angulata* is primarily cultured in coastal regions of Fujian and Guangdong Provinces in China, as an introduced species along the coasts of Portugal, Spain, and France. *C. virginica* is predominantly farmed along the east coast of North America, especially in the United States and Canada. In 2022, the U.S. produced approximately 128.8 thousand tons, while Mexico's output was around 6.7 thousand tons (FAO, 2024). The

Genetic Breeding and Aquaculture of Shellfish: A Collaborative Approach to a Sustainable Future
DOI: https://doi.org/10.1016/B978-0-443-33606-5.00023-0

global production of *O. edulis* reached approximately 2.4 thousand tons in 2022, with France contributing 1.4 thousand tons—representing 57.0% of the total output (FAO, 2024). *S. glomerata* is chiefly cultivated in Australia. The cultivation of these oyster species meets local demands and supports a significant global oyster market.

13.2 Biology of oysters

13.2.1 Classification and morphology

Oysters belong to the phylum Mollusca, class Bivalvia, order Ostreida, and family Ostreidae. They possess two asymmetrical shells that partially or completely enclose the soft body inside. These shells are primarily composed of calcium carbonate and consist predominantly of four layers: Three inner layers (the main structural layers of the shell), arranged from outer to inner as prismatic, foliated and chalky layers, and an outer periostracum layer, a brown, horny layer that protects against abrasion and environmental erosion (Bai et al., 2023; Chandra Rajan et al., 2021). The left shell tends to be larger and more convex, while the right shell is smaller and flatter. Shell shapes vary greatly, often taking on elongated, round, or triangular forms, with their morphology and color influenced by species and environmental factors. Oysters build mechanically strong $CaCO_3$ shells using a biologically controlled biomineralization process (Zhang et al., 2012), which protects them from predation and provides substrate and shelter for others. Interestingly, each oyster species has its own way of assembling these crystals to achieve superior mechanical strength for minimal energy expenditure and weight gain.

Inside the shells, the soft body is enveloped by the mantle, which detects environmental changes and secretes the shell. Oysters possess an adductor muscle, located typically at the posterior end of the shell, which contracts to quickly close the shell. Their large, leaf-like gills are used for respiration and filter-feeding. The esophagus leads to the stomach, which is surrounded by digestive glands. One part of the stomach connects to the intestine, which extends to the rectum and anus. Another part connects to a sac-like structure containing the crystalline style, a rotating rod that releases digestive enzymes to aid in food breakdown and digestion. The oyster's heart, a transparent sac located in the pericardial cavity near the adductor muscle, consists of two atria and a ventricle.

13.2.2 Habits

Oysters are filter-feeding bivalves that primarily consume planktonic microalgae, organic detritus, and bacteria from seawater, with their diet varying depending on the environment. After the planktonic larval stage, juvenile and adult oysters lead a sessile lifestyle by attaching themselves to substrates with their left shell, while respiration and feeding are controlled by the opening and closing of the right shell. Oysters exhibit gregarious behavior, clustering together in groups of different ages. New individuals often attach and grow on the shells of older oysters. In natural habitats, oyster shells and living individuals assemble on the seafloor over the years, eventually forming extensive reefs.

Oysters exhibit distinct adaptability to environmental conditions, particularly temperature and salinity, which, along with other factors, shape their geographical distribution. For example, *C. gigas*, distributed in northern China, has an optimal temperature range of 20–26°C, whereas

C. hongkongensis, endemic to southern China, thrives at 24–31°C during its growth season. Generally, oysters living in the intertidal zone are exposed to temperatures near tolerance limits during winter and summer, whereas those in the subtidal zone experience more stable water temperatures with smaller fluctuations. In addition, oysters exhibit species-specific preferences for salinity. Some species, such as those in the genus *Hyotissa* and the *O. edulis*, are adapted to high-salinity marine environments. In contrast, species such as the *C. virginica, C. ariakensis*, and *C. hongkongensis* are more adapted to low-salinity environments or have a broad salinity tolerance range. Although oyster species can withstand high or low salinity for short periods, prolonged exposure to extreme salinity conditions can weaken their physiological functions, reduce their disease resistance, and ultimately affect their survival. Therefore the combined effects of temperature and salinity play a crucial role in shaping the distribution patterns and suitable habitats of oysters.

13.2.3 Gonadal development

Oysters are generally dioecious, meaning that individuals are typically either male or female (Doinsing & Ransangan, 2022; Hedrick & Hedgecock, 2010). However, hermaphroditism, in which the egg and sperm are present at the same time, can also occur (Yang et al., 2015). The sex of oysters is not fixed; they can undergo sex reversal under specific conditions (Broquard et al., 2020). Most oysters reach sexual maturity at the age of one, and the breeding season varies depending on the species and region. Typically, the breeding season lasts from April to September, coinciding with warmer water temperatures.

Oysters exhibit diverse modes of sexual reproduction, including dioecy, protandric, and simultaneous hermaphroditism (Guo et al., 1998). Sex determination, the developmental process oriented to differentiate males and females, mainly consists of two types: Genetic sex determination (GSD) and environmental sex determination (ESD). Oysters show both GSD and ESD concurrently. Oysters do not have sex chromosomes. However, expression of male or female-specific sex gene networks and signaling pathways is recognized as an important molecular triggering mechanism for the establishment and maintenance of gonadal fate and thus sex ratio within a given population (Sun et al., 2022; Zhang et al., 2014). Environmental factors are also known to influence sex determination in oysters, like in many marine organisms. Temperature, food availability, and pH are known factors of sex determination in oysters (Dang et al., 2025; Sun et al., 2023). Under optimal temperatures and abundant food availability, the proportion of females increases. In contrast, males tend to predominate under nutrient-limited conditions (Santerre et al., 2013). Environmental factors such as pollution and exogenous steroids (i.e., 17β-estradiol) are also reported to influence the sex ratio in oysters (Islam et al., 2022).

The *Crassostrea* oysters are oviparous and release gametes into the surrounding environment, where external fertilization occurs. The *Ostrea* oysters are viviparous and retain eggs in the mantle cavity where fertilization occurs. Early larval development occurs within the female's mantle cavity. Once the larvae reach the veliger stage, they are released into the water to live freely.

The maturity of oysters can be assessed through histological examination or anatomical features of gonadal tissue (Fig. 13.1). The development of oyster gonads is typically divided into five stages:

Stage I: Resting phase. Oysters have expelled nearly all their reproductive cells, resulting in a transparent soft body surface that exposes the color of the digestive gland (Fig. 13.1A).

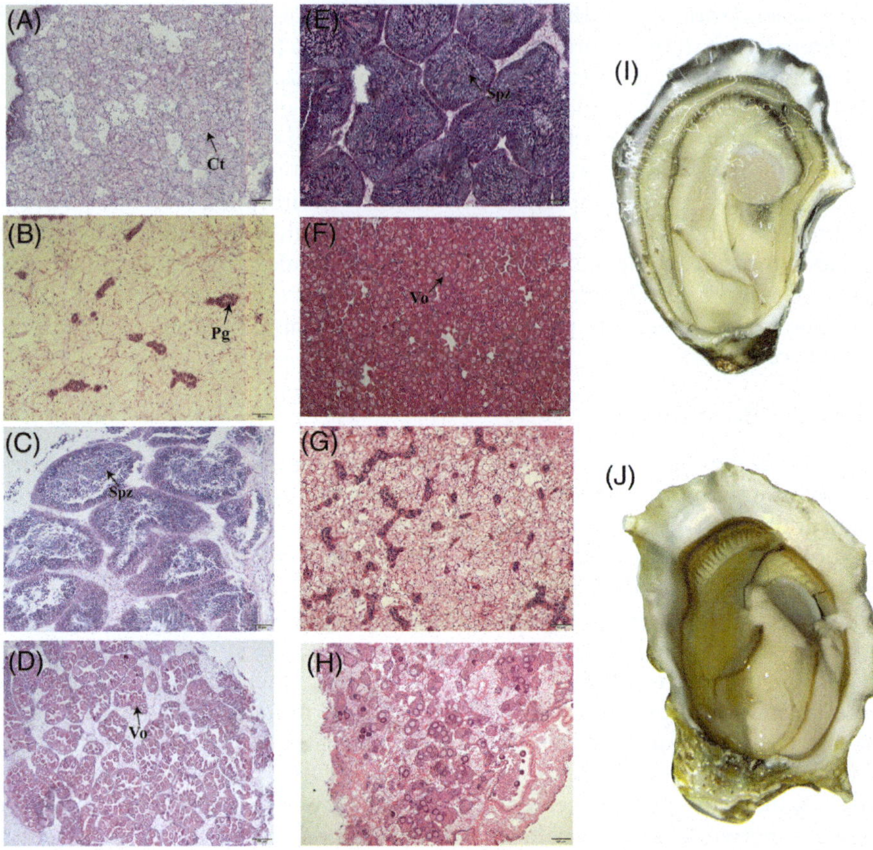

FIGURE 13.1 Microscopic photos and anatomical features of gonadal development at different stages in *Crassostrea gigas*. (A) Stage I: Resting Phase; (B) Stage II: Developing Phase; (C) Stage III: Proliferating Phase-male; (D) Stage III: Proliferating Phase-female; (E) Stage IV: Maturing Phase-male; (F) Stage IV: Maturing Phase-female; (G) Stage V: Spawning Phase-male; (H) Stage V: Spawning Phase-female; (I) External appearance of Stage II gonad; (J) External appearance of Stage IV gonad. *Ct*: Connective tissue; *Pg*: primordial germ cell; *Spz*: spermatozoa; *Vo*: vitellogenic oocytes.

Stage II: Developing phase. The soft body surface begins to show a slight white coloration, though it remains thin and sparse, allowing the digestive organs to still be visible (Fig. 13.1B).
Stage III: Proliferating phase. The milky white gonads become predominant, covering most of the visceral mass. During this stage, oogonia and spermatogonia within the gonad ducts differentiate into oocytes and sperm (Fig. 13.1C, D).
Stage IV: Maturing phase. The gonads undergo rapid development, completely covering the visceral mass, and the soft body becomes plump. The gonad ducts are distinct, with the ovaries nearly filled with oocytes and the testes packed with sperm (Fig. 13.1E, F).
Stage V: Spawning phase. The gonads gradually become thin from the front of the soft body, revealing the brownish visceral mass once more. The gonad ducts become transparent, occasionally containing vacuoles, while the reproductive cells become increasingly sparse (Fig. 13.1G, H).

13.2.4 Life cycle

After fertilization, the eggs contract into a spherical shape and undergo meiosis, releasing two polar bodies, indicating successful fertilization. Oyster embryos usually undergo the morula, blastula, and gastrula stages within 8–10 h. During this period, the embryos exhibit spiral movements driven by ciliary action.

The trochophore emerges approximately 10–12 h after fertilization, swimming with the aid of a strong band of prototroch cilia around the apical plate and a weak band of telotroch cilia around the anus. The veliger larvae typically appear around 24 h post-fertilization, shaped as "D", known as the D-larvae. The D-larvae possess two shells, a complete digestive system, and a specialized organ called the velum. This circular organ, fringed with cilia along its edge, can extend between the shells, enabling the larvae to swim. The D-larvae develop an umbo over time. In the late umbo stage, the larvae develop an eye spot and are called eyed larvae. By the pediveliger stage, a foot develops, and the larvae start crawling around to search for substrates to settle. When attached to suitable substrates, pediveligers go through metamorphosis within 2–3 days. During metamorphosis, the pediveliger loses its velum and foot and develops gills. The speed of larval development is species-specific and influenced by temperature and nutrition.

During the juvenile and adult stages, oysters use well-developed gills for feeding and respiration. These gills are lined with cilia that move rhythmically to create a steady flow of water. Food particles are captured by the ciliary motion and directed along specialized grooves on the gills to the labial palps, which help transport the food into the mouth.

The growth rates vary greatly among different oyster species. For instance, *C. gigas*, *C. hongkongensis*, and *O. denselamellosa* can continue growing for several years, while *C. angulata* typically ceases growth after one year. Temperature, nutrition, and salinity, as well as genetics, influence oyster growth. In estuaries, heavy rainfall washes nutrients into the ocean, promoting phytoplankton blooms that in turn stimulate oyster growth. In temperate regions, oysters grow rapidly during the spring and summer when food is abundant, and seawater temperatures are warmer. The lifespan of oyster species varies widely, ranging from a few years to several decades.

Oyster aquaculture in China and most parts of Asia heavily depends on natural or wild-caught seeds. The high demand for oyster seeds by rapidly growing oyster aquaculture industries cannot be met solely by collecting wild seeds. Over the years, largely due to unpredictable changes in seasonality, natural seed collection has become increasingly difficult. Due to this unstable supply of natural oyster seeds, particularly in China, the cost of getting wild seeds becomes more expensive than the oyster growers can afford.

Over the past 20 years, the rapid development of the coastal areas around the world has had a direct impact on the water quality of the oyster farming sites. Global climate change, the change of land use, reclamation, and eutrophication have deteriorated the oyster industry in Hong Kong and other parts of southern China. The water quality and environmental conditions inhibited the natural spawning of oysters. Although small-scale and scattered spawning may occur, natural oyster seeds are sensitive to minor environmental changes, so that >90% of the fragile oyster larvae cannot survive to the seed stage. Also, the growth rate of natural seed is highly variable, which makes farming operations even more complicated for oyster growers. Moreover, the direct discharge of sewage and livestock waste leads to bacterial and heavy metal contamination of adult oysters and seeds and poses serious food safety concerns (Chan & Ho, 1993). Thus oyster farming based on natural seed collection is not sustainable enough to meet the industry's needs. A stable and sustainable seed source through hatchery production is needed.

13.3 Hatchery production

13.3.1 Broodstock selection

Hatchery production typically requires the collection of adult oysters from the wild, farms, or a breeding program, which are then selected for use as broodstock. The selected broodstock should have a regular shell shape, signs of new/recent growth, and a substantial weight. They should have enlarged milky white gonads. The broodstock should be from a population that is well-adapted to the local environment and has excellent characteristics, such as fast growth, disease resistance, and strong environmental adaptability.

13.3.2 Broodstock conditioning

Broodstock conditioning is a key step in obtaining high-quality gametes for the production of artificial oyster seed. Hatcheries can produce seed earlier through this process, thereby ensuring that the seed has a long growth period before overwintering. The methods for broodstock conditioning or gonadal maturation induction are similar across oyster species. In late winter and early spring, oysters usually require 4–8 weeks of conditioning before reaching full maturation. The conditioning period tends to decrease as the natural breeding season approaches and varies depending on the species, initial rearing conditions, the stage of gamete development, and hatchery conditions. The most important factors are temperature, feed type, and feed quantity.

For conditioning, the broodstock is scrubbed to remove dirt and fouling organisms and cultured in indoor tanks using suspended or floating cages. The conditions for broodstock conditioning should meet the needs of gonad development as much as possible, such as higher water temperature, sufficient feed, and lower stocking density. The suitable salinity for gonad development and gamete maturation of most oysters should be higher than 25, and the temperature should be between 16 and 24°C; however, each species has its specific optimal rearing conditions. The optimum water temperature for *C. gigas* is 22–24°C, and the optimum salinity is 25–34. The *C. angulata* exhibits optimal growth at 24–28°C with a salinity preference of 25–30. In contrast, *C. virginica* and *C. hongkongensis* are better adapted to low-salinity environments, exhibiting optimal growth at salinities of 15–25 and 15–20, respectively. The rearing density is generally 60–80 individuals/m^3, which can be adjusted according to the individual size. During the conditioning period, broodstock can be fed with *Nitzschia closterium*, *Isochrysis galbana*, or *Platymonas* sp, or artificial substitute feeds such as spirulina powder and yeast to promote gonad maturation.

The broodstock conditioning process is divided into two stages. In the first stage, the daily dry weight feed ration is usually 4%–6% of the dry weight of the broodstock. Feeding more than 6% is not conducive to gonadal maturation (Lovatelli, 2004). The water temperature is increased by 0.5–1°C per day until it reaches a level slightly below the optimal temperature for induction of gonadal maturation. Large feeding amounts and high temperature during maturation induction will promote the somatic growth of broodstock instead of gonadal development. In the second stage, the water temperature is gradually increased at a rate of 0.5–1°C/day and finally reaches the optimal temperature. At the same time, the amount of food is slightly reduced from 4%–6% to 2%–3%. During conditioning, the water is changed 2–3 times a day, and the amount of water changed each time is 1/3–1/2, and the feces at the bottom of the tank are removed in time. The tank is emptied and cleaned every two days, except when it is close to maturation, to avoid accidental spawning.

13.3.3 Spawning and fertilization

The gametes of fully mature oviparous oysters can be extracted from the gonads by dissection or squeezing. A small number of gametes is placed in a drop of seawater on a glass slide, and the gamete type is observed under a microscope to determine its sex. During the stripping process, the digestive gland should not be punctured to avoid contamination of the gametes by gastrointestinal tissue, bacteria, and microorganisms. The seawater needs to be filtered and sterilized by ultraviolet light, and the temperature should be maintained at 21–24°C (taking *C. gigas* as an example). Initially, a 100-μm sieve is used to filter out large tissue and impurities, and then a 30-μm sieve is used to catch and rinse the eggs that are ~50 μm. Sperm can be collected following the same procedure and passed through a 30 μm screen to remove tissue debris. Finally, artificial fertilization is performed to obtain fertilized eggs. The optimal ratio is 5 to 8 sperm per egg as observed under a microscope. If there is excessive sperm, decant the upper layer of water after a 30-min gravitational sedimentation of oocytes.

When oysters are mature, sperm and eggs can also be obtained through induced spawning by combining techniques such as air-drying, water flow stimulation, temperature increase, or reduced salinity. Generally, oysters are dried in the shade for 5–8 h, placed in running seawater for 1–1.5 h, and then in warm seawater with a 3–5°C increase in temperature to induce spawning. Alternatively, after air-drying, oysters can be stimulated with flowing seawater at a temperature increase of 3 to 4°C for 1 to 1.5 h. They can also be air-dried overnight for 10–12 h before being placed in seawater. Excessive sperm in the tank can result in polyspermy and negatively affect embryonic development. To avoid this problem, male oysters should be picked out immediately after sperm release. If the spawning is intense, oysters should be moved to a new pool to allow them to continue to release gametes.

Viviparous oysters such as *O. edulis* are somewhat unique. The female draws sperm into the mantle cavity, where fertilization occurs. The fertilized eggs develop internally until they reach the veliger larval stage before being released into the surrounding water.

To increase genetic diversity and avoid inbreeding depression as much as possible, the number of breeders should be kept at >50 at a 1:1 sex ratio (Jiang et al., 2023). For incubation, the density of fertilized eggs is 80–100 eggs/mL. To prevent sedimentation of fertilized eggs affecting embryonic development, the suspension is gently stirred every 30 min or aerated. Depending on the species and water temperature, development to D-larvae takes 20–48 h. Generally, it takes about 22 h for *C. gigas* to develop from fertilized eggs to D-larvae at a temperature of 22–23°C.

13.3.4 Larval rearing

When the larvae develop to the D-larvae stage, the healthy, vigorous, and large D-larvae are collected with a 48-μm sieve and then restocked into a new tank with fresh seawater. To keep the larval culture clean, large debris or zooplankton are filtered out with a 150-μm sieve. Daily management during larval cultivation is as follows:

13.3.4.1 Culture density

The recommended rearing density for D-larvae is generally 8–12 individuals/mL for batch culture. The density should be appropriately reduced to 1–3/mL with the growth of larvae during the rearing process.

13.3.4.2 Feeding

Suitable algae for oyster larvae include *I. galbana, Platymonas* sp, *Chlorella vulgaris*, and *Chaetoceros* sp. D-larvae should be fed after they are transferred from the hatching tank to the rearing tank. The diet is primarily composed of *I. galbana* during the early stage, while *Platymonas* sp and *C. vulgaris* become an excellent food source after the umbo stage. Once the larvae reach a shell length of 130–150 μm, they can feed on diatoms such as *Chaetoceros* sp, which helps increase their growth rate. Usually, farmers feed larvae a mixed diet of various algae, such as *I. galbana, Platymonas* sp, *C. vulgaris*, and *Chaetoceros* sp, to achieve better larval growth.

13.3.4.3 Water change

The size of newly hatched D-larvae of *Crassostrea* oysters is about 50–70 μm. A 300-mesh sieve is used to catch the larvae during water change, which is performed twice a day and for 1/3–1/2 of the water each time. As the larvae grow, the volume of water change should be gradually increased. During the larval rearing process, complete water exchange (100%) can also be implemented to ensure water quality. In many hatcheries, 100% water change is performed every other day except over the weekend to maintain water quality.

Some hatcheries use high-density flow-through culture systems to reduce labor costs and time consumption. Unlike static water exchange methods that require changing water 3–4 times a week, this method allows for continuous, albeit slower, seawater exchange. To prevent larvae from being washed away, a screen with an appropriate mesh size is used for water discharge. The system should provide sufficient residence time for the larvae to ingest and adequate water exchange to prevent the accumulation of metabolic waste and debris. In addition, since flow-through culture does not grade larvae by size like static methods, there is a significant size variation among larvae, extending the rearing time. When the larvae reach the pediveliger stage, they must be transferred to a settlement tank.

13.3.4.4 Oxygen supply

Aeration can increase the oxygen content of the seawater, distribute the larvae and algae evenly in the tank, and facilitate the normal metabolic process of the larvae. Continuous microaeration is employed during the whole rearing process by utilizing air stones positioned close to the bottom to meet the oxygen requirements of the larvae. One air stone should be placed per square meter, with the aeration rate reaching 1%–1.5% of the total water volume per minute.

13.3.4.5 Selection of healthy larvae

The growth rate of larvae varies greatly during the rearing process. Healthy and vigorous larvae with uniform size and activity can be selected for subsequent cultivation. Oyster larvae exhibit an upward-floating behavior, and vigorous larvae are usually distributed in the middle and upper layers of the seawater. Therefore these larvae are usually selected using a siphon and transferred to another tank for further cultivation, and the larvae on the bottom are discarded.

13.3.4.6 Daily observations

The daily increase in shell length of 8–15 μm is normal during the larval period before the umbo-larvae stage. Insufficient feeding, low water temperature, deteriorating water quality, and bacterial or viral infections can lead to slower growth of larvae. In addition, the activity of larvae, feeding, and whether they are sinking to the bottom should be observed every day. Water temperature, dissolved oxygen, pH, ammonia nitrogen, biological oxygen demand, and other environmental parameters should be measured regularly. Problems should be dealt with as soon as possible. The physicochemical conditions required for larval cultivation vary significantly depending on the oyster species. Generally, the optimal water temperature for larval rearing ranges from 20°C to 26°C, and the daily temperature difference does not exceed 2°C–3°C. The salinity is 28–35 for *C. gigas*.

13.3.5 Recirculating systems

Recirculating aquaculture systems (RAS) aim to optimize production by supporting high larval densities while minimizing water usage, offering greater control over culture conditions, and more efficient waste management. With RAS, oyster larvae can be cultured at densities as high as 300–600 larvae/mL in the ultrahigh-density culture systems, leading to greater larval culture efficiency compared to conventional methods. These higher densities are achievable because of the continuous treatment and renewal of seawater, which maintains water quality, removes waste, and ensures a constant supply of food. As oyster larval culture evolves, the above-mentioned factors are crucial for optimizing RAS systems and lowering production costs (e.g., feed, labor and equipment management, energy use, seawater management, and water quality control). Currently, RAS is being experimented with in several oysters, such as *C. gigas*, *C. virginica*, and *C. angulata*, and has the potential for further expansion in the future (Ramos et al., 2021).

13.3.6 Spat collection

13.3.6.1 Collectors

For seed production in the hatchery, the spat collectors are mostly made of oyster and scallop shell strings, which are hung in the tank for larval settlement. Oyster shell pieces with a shell height of more than 8 cm or scallop shell pieces with a shell height of 6–8 cm are strung together with polyethylene lines, with 100 pieces per string. Prior to settlement, the collectors must be thoroughly cleaned. Subsequently, the collectors are soaked in a 0.5‰–1‰ sodium hydroxide solution or a 2‰ bleach powder solution for 24 h, followed by a thorough rinsing with sand-filtered seawater.

13.3.6.2 Deployment time of collectors

At a water temperature of 20–23°C, the shell length of the oyster larvae can reach 330–350 μm after about 20 days of rearing. When about 60% of the larvae develop into the pediveliger stage, which is characterized by an eyespot, a foot, and a crawling or tasting behavior, the collectors can be placed. Larvae are often size-selected, with the pediveligers being used for setting, and small veligers are cultured further. A total of 3000–6000 shells should be used per cubic meter of seawater, depending on the larval density (Fig. 13.2).

13.3.6.3 Larval density

The recommended larval density for settlement is 2–3 larvae /cm^2. When using shellfish shells as seed collectors, it is generally adequate to have 15–20 spat per shell (Fig. 13.3). To

FIGURE 13.2 Seed collectors made from scallop and oyster shells are deployed to collect spat in a hatchery.

FIGURE 13.3 Fifteen to twenty spat per shell are collected to avoid overcrowding.

prevent overcrowding of settled spat, the larvae can be divided into multiple tanks for setting, or the collectors can be deployed in batches and removed promptly after setting.

13.3.6.4 Remote setting

Remote setting is a method of transporting competent oyster larvae to other locations for setting. Pediveligers, commonly called eyed larvae, are filtered out, wrapped in sieve cloth, and placed in damp paper towels to maintain moisture. They are then packed in foam boxes with ice packs for transport. Insulated boxes should be used to ensure that the larvae are transported under low-temperature, high-humidity conditions.

After arriving at the remote location, the larvae are placed in 20–25°C seawater to acclimate for 15–30 min, after which the larvae are poured into the settlement tank. Typically, larvae do not experience mortality after ~12 h of transportation under low temperatures and high humidity. This method allows remote setting at sites without hatcheries. It is convenient, easy to transport, and cost-effective, making it a promising approach for enhancing seed production.

13.3.7 Production of cultchless oysters

Oysters exhibit a gregarious lifestyle, leading to irregular shell shapes due to space constraints, which significantly affects their aesthetic appeal. The clustering also causes food competition among individuals, affecting their growth rate. Cultchless oysters are individual oysters not attached to any substrate. Because cultchless oysters are not constrained by space, their shells are regular, and sizes are uniform, making them a better product, easy to stock and harvest (Fig. 13.4). This method is primarily used for oyster farming in Europe and the United States. The following three methods are commonly used for the production of cultchless oysters.

13.3.7.1 *Epinephrine and norepinephrine treatment*

The treatment with epinephrine (EPI) or norepinephrine (NE) is an effective method for inducing non-settlement metamorphosis and production of cultchless oysters. The optimal concentrations for inducing non-settlement metamorphosis are 20 mg/L for EPI and 18 mg/L for NE, with a treatment duration of 3 hours. Under these conditions, the induction rates of nonattachment metamorphosis can reach 70.0% and 60.0%, respectively (Li et al., 2020). This pharmacological treatment does not negatively affect the growth of oysters.

13.3.7.2 *Microcultch*

Microcultch is a tiny granular substrate for the settlement and metamorphosis of oyster larvae. Oyster larvae grow more rapidly on the substrate than without it. Common materials used as microcultch include shellfish shell powder and quartz sand, ranging in size from 0.30 to 0.40 mm. Notably, particles of ~0.35 mm in size yield the highest rates of individual oysters. This particle size is equivalent to the length of the larval shell and is the minimum specification of the substrates. Particles smaller than 0.25 mm do not facilitate larval settlement, while those larger than 0.50 mm may increase settlement rates but reduce cultchless oyster rates.

13.3.7.3 *Detachment method*

Oyster larvae are first set on various substrates in settlement tanks. Once they reach a certain size, they can be artificially detached from the substrates to become cultchless oysters. Hard substrates, such as shellfish shells and tiles, yield good settlement results. However, detaching spat from the hard substrates can be challenging, because the spat can be easily broken. Generally, soft polyethylene corrugated panels are used as seed collectors. When the spat reach 2–3 cm, the panels can be bent, allowing the oysters to fall off without sustaining any mechanical damage.

FIGURE 13.4 Spat (A), seed (B), and adult individuals (C) of cultchless oysters.

13.3.8 Production of triploid oysters

Interest in farming triploid oysters began in the early 1980s, with significant progress in commercial production achieved after the development of tetraploids (Cheng et al., 2024; Guo & Allen, 1994). Theoretically, triploids are sterile, allowing them to redirect energy meant for gonadal development toward growth (Liang et al., 2024; Zhou et al., 2026). In reality, triploid oysters are larger than diploids before gametogenesis, suggesting that increased heterozygosity and/or polyploid gigantism play more important roles (Guo & Allen, 1994). Regardless, triploid

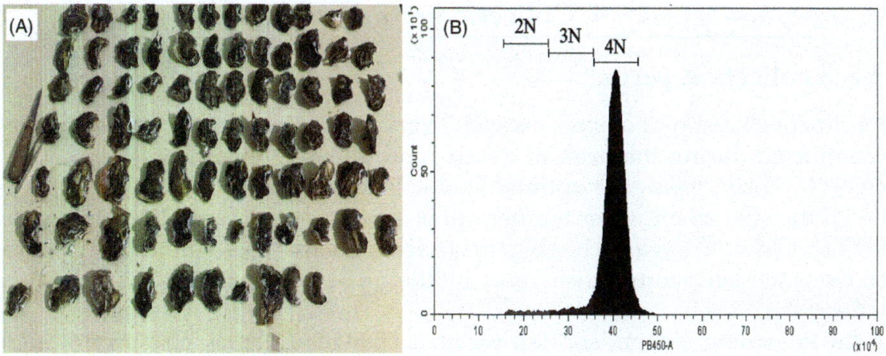

FIGURE 13.5 Tetraploids are selected as broodstock to produce triploid seeds. (A) Selected tetraploid oysters; (B) flow cytometry graph of tetraploid oysters.

oysters have a growth advantage over diploids. Triploids can be obtained by inhibiting the release of the second polar body using physical methods (such as cold shock, heat shock, and hydrostatic pressure) and chemical methods (CB or 6DMAP) (Guo & Allen, 1994). In commercial production, triploid oysters are produced by mating tetraploids with diploids (Guo et al., 1996). This method is straightforward, easy to use, 100% effective, and does not involve toxic side effects from inducing agents. Currently, most hatcheries around the world produce triploid oysters by mating diploid females and tetraploid males.

13.3.8.1 Ploidy confirmation of tetraploid broodstocks

Oysters are anesthetized with a 5% magnesium sulfate ($MgSO_4$) solution, which makes the oysters open their shells. A small piece of gill is taken with tweezers and placed in a 1.5-mL centrifuge tube. Alternatively, a rotary drill can be employed to create a small opening in the shell, and a syringe is used to extract a small amount of hemolymph from the adductor muscle. A DNA staining solution is added to the sample, followed by agitation to prepare a suspension of nuclei. Ploidy analysis is performed using a flow cytometer (Jiang et al., 2024). Confirmed tetraploids are picked out and conditioned as broodstock for triploid production (Fig. 13.5).

13.3.8.2 Sperm and egg collection

In production, triploid embryos are obtained by fertilizing the eggs of diploid oysters with the sperm of tetraploid oysters. Both eggs and sperm are collected through strip-spawning. If tetraploid oysters are scarce, the maturation level, sex, and gamete viability of the gonads are assessed before dissection. Individuals with active sperm are selected as male broodstock under a microscope. The ratio of female to male broodstock is maintained at 25~30:1 if tetraploids are limited. When hatcheries have sufficient numbers of tetraploids, more tetraploid males (20–50 per spawn) should be used to maintain genetic diversity. Fertilization, larval rearing, and spat collection are carried out using conventional methods.

13.4 Collection of natural seed

13.4.1 Seed collection period

The reproductive season of oysters extends over a long period. In practice, spat collection is typically conducted during the peak of oyster reproduction, when water temperatures range from 20 to 30°C. For *C. gigas*, the optimal spat collection period is from mid-July to August, depending on the site, when water temperatures are between 20 and 26°C. In coastal areas of Fujian province, China, *C. angulata* has two major reproductive peaks annually: One from May to June when the water temperature rises, and another in September when the water temperature gradually decreases.

During the spawning season, sudden weather changes, such as heavy rain, high tides, or strong winds, can trigger the spawning of ripe oysters. By monitoring gonadal development, the time of larval settlement and metamorphosis can be predicted, allowing for the timely deployment of spat collectors. During the spat collection in natural marine areas, it is advantageous to deploy spat collectors when the number of pediveligers reaches 25 individuals/m^3 and constitutes a dominant proportion of the larval population. Moreover, the time of spat collection can be determined by calculating the cumulative daily average water temperature during the development of fertilized eggs to the settlement and metamorphosis. For instance, *C. gigas* typically requires a cumulative water temperature of around 280°C from fertilization to the onset of larval settlement. It is essential to determine the time for deploying collectors to avoid extensive barnacle attachment, as barnacles typically breed earlier than oysters.

13.4.2 Seed collection method

The seed collector should have several characteristics, such as being easy to obtain, low cost, durability, a rough surface, and a large attachment area. Depending on the tidal zone, seed collection methods (Wang & Zheng, 2024) can be generally classified into the following types (Fig. 13.6):

13.4.2.1 Subtidal zone seed collection

The seed collection in the subtidal zone typically involves deploying collectors on floating rafts in areas extending from below the low tide line to a water depth of about 10 m. Spat collectors, such as strings of shells, are suspended from the floating rafts, with a spacing of 15–20 cm between strings (Fig. 13.6A). This method provides effective use of space, improves the efficiency of seed collection, and has been widely used to support large-scale oyster aquaculture.

In inner bays with relatively calm waters near the low tide line (1–2 m water depth at low tide), racks can be constructed by erecting wooden/bamboo stakes or concrete pillars to form a grid-like structure (Fig. 13.6B). Strings of seed collectors are hung vertically or horizontally on the rack to collect spat. The length of each string of seed collectors is generally 1–2 m. The collectors should not touch the bottom to avoid affecting the seed collection efficiency. In summer, reducing the hanging depth and increasing exposure to air can minimize the growth of barnacles and tube worms on collectors or oyster shells. This method is typically used in Guangdong and Guangxi Provinces, China, for collecting *C. hongkongensis* seed.

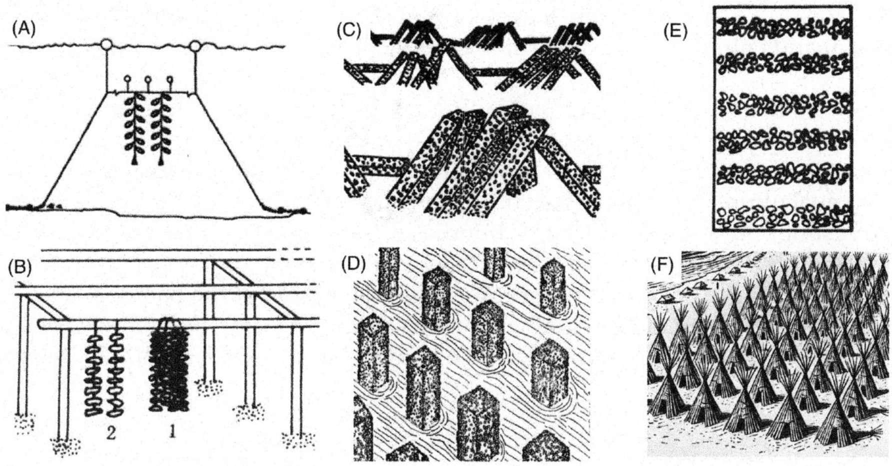

FIGURE 13.6 Schematic diagram of different materials used for seed collection (adapted from Wang & Zheng (2024)) (A) Floating rafts; (B) Wooden/bamboo or concrete racks; (C) Stone slabs or cement pieces; (D) Stone or cement pillars; (E) Stone blocks; (F) Bamboo stakes.

13.4.2.2 Mid-intertidal zone seed collection

In the mid-tidal zone with hard tidal flats, stone slabs or cement pieces with a size of 1.2 m × 0.2 m × 0.05 m are densely stacked at a 60° angle with the beach surface to form a herringbone shape (Fig. 13.6C). Each row consists of a dozen to several dozen slabs. Between rows, two or three stone blocks or concrete rods are aligned in a single column, parallel to the water flow. To avoid overcrowding and enhance the growth rate, the number of settled spat should be appropriately reduced, and the orientation of the stone slabs should be changed as the spat grow. This method is commonly used for *C. angulata* spat collection in Fujian province, China.

Alternatively, vertical stone or cement pillars (1.2 m × 0.2 m × 0.2 m) can be planted at a density of 1–1.5 pillars/m^2, with 30–40 cm buried in the sediment to ensure stability (Fig. 13.6D). This seed collection method is commonly used for the small-scale collection of *C. angulata* and *C. hongkongensis*. Once spat have settled, they are cultivated in place until they reach harvest size. In addition, the settled seed should be inspected to distinguish whether they are oyster seed or barnacle seed. If the attached individuals are slightly rounded, dark, flat, and smooth, they are oyster spat; if they are oval, milky white, raised, and rough, they are barnacle spat. When a large number of barnacle seeds are settled, the collectors should be cleaned for the next batch of seeds.

13.4.2.3 Upper intertidal seed collection

In the upper intertidal zone, stone blocks (weighing approximately 2–4 kg) are usually arranged in rows with a distance of 70–100 cm between rows (Fig. 13.6E). A ridge-shaped seed collection base with a high center and low edges is built. The ridge height should be 30–40 cm, and the two ridges should be separated by a ditch. After spat collection, the collectors should be frequently moved to prevent them from sinking and being buried in sediment. To prevent the stone blocks from sinking or sediment buildup, they must be repositioned 2 to 3 times a year, typically during April to May, July to August, and September to October. This method, commonly used in Guangdong and Fujian provinces of China, is a traditional technique for collecting spat and cultivating juveniles of species such as *C. ariakensis*, *C. hongkongensis*, and *C. angulata*.

The bamboo-stake method is another traditional practice for collecting and cultivating spat of *C. angulata* in the upper intertidal zone along the southern coast of China (Fig. 13.6F). This method is suitable for estuaries or inner bays with mild waves and slow currents, particularly for small-scale aquaculture. The bamboo spat collector is easy to move, effective, and low in labor cost for harvesting. During spat collection, bundles of 3–5 treated bamboo sticks are inserted into the tidal flats, with 50–80 bundles in a row. The length of each row ranges from 4 to 5 m, with an approximate distance of 1 m between rows. The orientation of the bamboo sticks should be periodically adjusted based on the number of spat settled, ensuring that the seeds settle evenly. In production, it is generally required that each bamboo stick is settled with 70–100 spat.

13.5 Intermediate nursery culture

The intermediate nursery culture of oyster seed serves as a critical stage between seed rearing and grow-out farming. If small spat are kept in fine-mesh nets, the mesh can easily become clogged with planktonic algae, silt, and fouling organisms, thus hindering their growth. Transferring spat directly to natural waters may result in significant mortality because of their inability to adapt to the harsh environment. In addition, there is a risk of contamination of hatchery spat by wild populations during the spawning season. Therefore the intermediate culture of oyster seed can reduce costs, improve survival rates, and mitigate contamination by wild oysters. The intermediate nursery culture mainly includes land-based and sea-based culture systems.

13.5.1 Land-based culture

Land-based nurseries are common in China, Europe, and North America. Land-based nurseries are usually located in low-lying coastal areas. Most land-based nurseries pump seawater through upwellers. Some hatcheries allow seawater to flow into seaward-facing ponds through intake gates during high tide, with stop valves installed on the pipelines to retain water. Water is drained by gravity during low tide.

The biomass of juveniles raised in land-based nurseries depends on the productivity of the ponds, which is affected by factors such as temperature, salinity, and nutrient levels. To maintain food abundance and remove waste and metabolic byproducts, fresh seawater is regularly or continuously introduced directly from the sea. Usually, the primary productivity of ponds can be enhanced by applying natural or artificial fertilizers to promote the growth of algae. Although the primary productivity of the pond can be enhanced to levels significantly higher than those of normal marine areas, there is no guarantee that the size, digestibility, and nutritional value of the algae grown in the ponds are suitable for the cultivation of oysters. Nurseries may need to cultivate algae for supplementary feeding. Therefore its management is easier than that of sea-based nursery systems because the quantity and quality of the food for the spat can be controlled to some extent in both quantity and quality. In addition, the water temperature is often significantly higher than that of the nearby seawater, which benefits the growth of warm-water oyster species. However, the rapid increase in algae can occur suddenly, posing a risk that excessive algal blooms will deplete oxygen in the water, leading to unexpected mass mortality.

As a standard practice, excessive algal blooms can be prevented by installing oxygen detection alarms or through careful management. In North America, land-based nurseries typically use upweller silos to culture spat from 1 to 5 mm with flow-through of raw seawater without artificial feeding or the use of ponds.

13.5.2 Sea-based nurseries

Sea-based nurseries serve as an alternative method for intermediate culture, where spat are transferred to marine environments characterized by stable hydrological conditions. This approach facilitates the gradual acclimatization of oysters to natural coastal ecosystems, enhancing survival rates prior to final deployment in open grow-out zones. Compared to land-based nurseries, sea-based nurseries offer greater stability and are less prone to eutrophication. The method also incurs lower operational costs by eliminating the need for feeding, water circulation systems, and energy-intensive infrastructure. Furthermore, exposure to natural phytoplankton populations in open waters can accelerate oyster growth, potentially shortening production cycles. In North America, the floating upweller system (FLUPSY) is commonly used in sea-based nurseries for oyster seed. By keeping upwellers floating at the surface, the FLUPSY provides an efficient system for pumping seawater through at a low energy cost.

13.6 Grow-out

Oyster grow-out refers to the process of cultivating spat into market-sized oysters. Various methods of oyster grow-out are practiced in different areas worldwide. For settled spat, cultivation methods include bottom culture, off-bottom culture using various substrates such as stone, bamboo, and cement, as well as suspended culture on longlines or rafts. Cultchless single oysters can also be cultured on-bottom, off-bottom, or suspended in various bags or cages.

13.6.1 Bottom culture

Bottom culture involves directly broadcasting settled spat onto mudflats or shallow seabed, allowing them to grow naturally without the need for settlement substrates. This method can make full use of tidal flat resources for large-scale aquaculture, with the advantages of low cost, simple operation, and high yield per unit area. Bottom culture is best carried out in sheltered bays with small waves and steady tidal currents. The tidal flat should be level and free of standing water after low tide, with a substrate of mud or muddy sand being most suitable. The farming area should be in the lower part of the middle tidal zone or near the low tide zone. Bottom culture is widespread along China's coasts and is also practiced in other countries. In North America, spat-on-shells are planted on intertidal or subtidal beds, which are harvested by hand or dredging, respectively.

13.6.2 Off-bottom culture

Off-bottom culture is a traditional method of mariculture, where oysters are farmed by utilizing substrates such as stones or bamboo poles in muddy or sandy areas within the intertidal zone or below the low tide line. The selection of substrates varies depending on local environmental conditions (e.g., sediment type, water flow, salinity, and temperature).

Bamboo is primarily used in intertidal zones with muddy or sandy bottoms and relatively calm waters. Seed obtained through the bamboo-stake method can typically be further cultured directly on the bamboo substrates. This method is particularly popular in coastal areas of China, especially in Xiapu, Fujian province.

Hard materials such as stones and cement are suitable for areas with a hard bottom and moderate water flow. Stones or cement blocks placed in the culture area provide a substrate for the oyster larvae to settle and grow. Depending on the depth of water, mudflat firmness, water flow speed, and type of substrate, the substrates with oyster seed are evenly scattered or placed in piles and rows. This method can simulate the natural ecological environment and improve the survival rate and growth rate of cultured organisms. At the same time, hard materials such as stones/cements can also provide shelter for benthic organisms and increase biodiversity. Once the oysters reach market size, they can be harvested and transferred to cages, which are then suspended in nutrient-rich deep-water areas for further fattening over a period of 40–60 days. Stone-block culture is a common oyster farming method in coastal countries in East Asia.

Stone-slab substrates are commonly used for cultivating *C. angulata* along the coast of Fujian province, China. The stone slabs or pillars are arranged like a bridge or erected in the tidal zone. The specifications and arrangements of the stone slabs mirror those used in the bridge seed collection; however, it is essential to space the slabs adequately during the grow-out phase. Regular inspections are necessary to reposition any fallen slabs and adjust spat density as needed. If there is an excessive attachment of spat, manual thinning should be carried out to reduce their density. In cases where too many barnacles or other organisms have settled, the stone pillars should be cleaned before reusing them for spat collection.

In North America and France, cultchless single oysters are often cultured in off-bottom bag and rack systems. The racks are made of metal bars securely pushed into the flat. Oysters in mesh bags are placed on the racks secured with rubber straps at a height of 20–50 cm off the bottom. Different mesh-size bags are used for oysters of different sizes. Oyster bags are worked over regularly to clean, sort by size, and thin oysters down to appropriate densities. Oysters are often tumbled to trim new shell growth, producing oysters with nice shapes and deep cups.

13.6.3 Suspended culture

13.6.3.1 Longline culture

Longline culture is an effective method for cultivating oysters in suspension. The method is used to culture spat settled on shells, which are inserted into nylon ropes (with a diameter of 0.5–1 cm) approximately 10 cm apart. The upper end of the polyethylene rope is secured to the floating rafts, while a stone is affixed to the lower end. Often, the scallop shells with attached seeds are broken into fragments of about 4 cm^2 before being inserted into the rope. Rope culture is particularly suited for marine areas with abundant food sources, calm waters, and depths exceeding 4 m during low tide. This method is a traditional method of oyster culture and is widely employed in coastal provinces of China, such as Fujian, Guangdong, and Shandong, as well as in areas like Hiroshima and Miyagi in Japan and along the southern coast of Korea.

13.6.3.2 *Raft culture*

Raft culture uses a platform constructed with cement-steel bars or wooden poles for oyster grow-out. This method can be implemented in place with freshwater inflow, fertile water, low wind and waves, smooth tidal movements, and a water depth of 2–3 m. The rafts typically consist of piles, vertical poles, and horizontal poles, with their specifications varying across different regions. For instance, a common raft specification used in Guangdong Province, China, is as follows: The piles measure 4.0 m × 0.09 m × 0.09 m, the vertical poles are 3.5 m × 0.12 m × 0.07 m, and the horizontal beams are 2.8 m × 0.11 m × 0.07 m. There are holes for subsequent installation on each pole. A well-designed raft system not only protects oysters from predators but also shields them from adverse environmental conditions. Raft culture offers several advantages, including ease of adult oyster collection, low cost, faster growth, shorter culture cycles, and higher yield and quality.

13.6.3.3 *Cages/bags/baskets culture*

For single spat, common approaches include bag-on-rack systems and suspended cultivation using cages, bags, or baskets. This culture involves placing cultchless oysters or substrates with attached oysters into cages/bags, which are suspended on floating rafts for grow-out. The cages/bags feature appropriately sized mesh openings to allow water to flow in and out freely, ensuring that oysters can carry out their life processes effectively. At the same time, the cages can also provide protection against predators. Cage/bag culture is particularly advantageous in farming scenarios requiring concentrated management and safeguarding of oysters, especially when frequent handling is necessary during cultivation.

The hanging basket is usually made of durable and corrosion-resistant materials, such as stainless-steel mesh or plastic, to ensure long-term use in seawater without damage. These baskets are designed with multiple levels or compartments to facilitate the suspension and management of oyster seed. Weights are attached to the bottom of the baskets to maintain stability in the water, while the top is secured to the floating rafts using ropes or chains. Different sizes of compartments can be set inside the hanging basket to accommodate oysters at different developmental stages. Basket culture is commonly employed in oyster aquaculture industries in countries such as New Zealand, France, Ireland, the Netherlands, and Chile, ensuring efficient management and cultivation of oysters. In North America, a popular grow-out system consists of metal cages with floats that carry multiple oyster bags connected by longlines. The floats are adjustable tanks and can be sunk when necessary. The floating cage can be flipped to control biofouling. The system is amenable to mechanized workover and harvest.

13.6.4 Polyculture

13.6.4.1 *Polyculture of oysters and shrimp*

The polyculture of oysters and shrimp in ponds can significantly improve water quality and enhance overall productivity. The pond should have a water depth of over 1.2 m, a daily water exchange rate of at least 20%, and a salinity above 20, with a substrate primarily consisting of sandy mud. Before placing the seed, the pond needs to be cleaned and disinfected, and flat ridges need to be built as seedbeds. Natural or hatchery seed can be used for planting in early to mid-April. Stringent water quality management is required during the cultivation process. After the shrimp are harvested in the autumn, the focus can shift to the growth of the oysters, which are harvested by the end of November. This polyculture method is commonly practiced in China and several Southeast Asian countries. Utilizing large earth ponds for the polyculture of oysters and shrimp is an effective farming model.

13.6.4.2 Polyculture of oysters and algae

In shallow coastal waters, oysters can be cultured together with kelp, with each species benefiting from the growth of the other. Metabolic byproducts from the oysters provide nutrients for the growth of kelp, while oxygen produced through the photosynthesis of kelp aids the respiration of oysters. Polyculture methods include interval culture (raft-cultured oysters and kelp are arranged alternately), intercropping (oysters and kelp are cocultured on the same raft), and vertical culture (oysters are cultured vertically, and kelp is cultured horizontally between two rafts). Among these, interval culture and vertical culture are the most common. Although the scale of polyculture is still relatively small, it has the potential to optimize the farming environment and enhance economic benefits.

In summary, oyster aquaculture is a major industry worldwide and has been expanding rapidly because of advances in hatchery, nursery, and grow-out technologies. Genetic improvement of oyster stocks exemplified by the development of tetraploid and triploid oyster has contributed significantly to oyster aquaculture. Oyster farming is environment-friendly as it does not require artificial feeding. Filter-feeding oysters help to clean the water, while culture systems create habitat for other organisms. The rapid expansion of oyster aquaculture, especially in China, may have exceeded the carrying capacity of the local ecosystems, causing slow growth and poor meat quality of farmed oysters and sometimes heavy summer mortalities. The future of oyster aquaculture depends on continued innovation in production technologies, including genetic improvement, as well as science-based management under the one health concept.

References

Bai, Y., Liu, S., Hu, Y., Yu, H., Kong, L., Xu, C., & Li, Q. (2023). Multi-omic insights into the formation and evolution of a novel shell microstructure in oysters. *BMC Biology, 21*, 204. doi:10.1186/s12915-023-01706-y.

BOF. (2024). *China fisheries statistic yearbook 2024*. Agriculture Press (in Chinese).

Botta, R., Asche, F., Borsum, J. S., & Camp, E. V. (2020). A review of global oyster aquaculture production and consumption. *Marine Policy, 117*, 103952. doi:10.1016/j.marpol.2020.103952.

Broquard, C., Martinez, A. S., Maurouard, E., Lamy, J. B., & Dégremont, L. (2020). Sex determination in the oyster *Crassostrea gigas* - A large longitudinal study of population sex ratios and individual sex changes. *Aquaculture, 515*, 734555. doi:10.1016/j.aquaculture.2019.734555.

Chan, S. S. (1993). Bacterial contamination of the pacific oysters (*Crassostrea gigas*) cultured in deep bay, Hong Kong. *Environmental Technology., 14*(9), 861–867. doi:10.1080/09593339309385358.

Chandra Rajan, K., Meng, Y., Yu, Z., Roberts, S. B., & Vengatesen, T. (2021). Oyster biomineralization under ocean acidification: From genes to shell. *Global Change Biology, 27*(16), 3779–3797. doi:10.1111/gcb.15675.

Cheng, G., Liang, Y., Zhang, H., Xu, C., & Li, Q. (2024). A comparative study of the survival, growth, and gonad development of the diploid and triploid Pacific oyster, "Haida No.2" strain. *Aquaculture International, 32*(6), 8035–8052. doi:10.1007/s10499-024-01554-9.

Dang, X., Zhang, Y., Dupont, S., Gaitán-Espitia, J. D., He, Y., Wang, H., … Thiyagarajan, V. (2025). Low pH means more female offspring: A multigenerational plasticity in the sex ratio of marine bivalves. *Environmental Science & Technology, 59*(2), 1308–1321. doi:10.1021/acs.est.4c07808.

Doinsing, J. W., & Ransangan, J. (2022). An overview of the life cycle, reproduction and factors influencing recruitment success of tropical oysters. *Egyptian Journal of Aquatic Biology and Fisheries, 26*(3), 1–29. doi:10.21608/ejabf.2022.234926.

FAO, Global aquaculture production quantity (1950–2022), Updated 2024. https://www.fao.org/fishery/statistics-query/en/aquaculture/aquaculture_quantity. (2024),

Guo, X., Hedgecock, D., Hershberger, W. K., Cooper, K., & Allen, S. K. (1998). Genetic determinants of protandric sex in the Pacific oyster, *Crassostrea gigas* Thunberg. *Evolution, 52*(2), 394–402. doi:10.1111/j.1558-5646.1998.tb01640.x.

Guo, X., & Allen, S. K. (1994). Viable tetraploids in the Pacific oyster (*Crassostrea gigas* Thunberg) produced by inhibiting polar body 1 in eggs from triploids. *Molecular Marine Biology and Biotechnology, 3*(1), 42–50.

Guo, X., & Allen, S. K. (1994). Sex determination and polyploid gigantism in the dwarf surfclam (*Mulinia lateralis* Say). *Genetics, 138*(4), 1199–1206.

Guo, X., DeBrosse, G. A., & Allen, S. K. (1996). All-triploid pacific oysters (*Crassostrea gigas* thunberg) produced by mating tetraploids and diploids. *Aquaculture, 142*(3–4), 149–161. doi:10.1016/0044-8486(95)01243-5.

Guo, X., Li, C., Wang, H., & Xu, Z. (2018). Diversity and evolution of living oysters. *Journal of Shellfish Research, 37*(4), 755–771. doi:10.2983/035.037.0407.

Hedrick, P. W., & Hedgecock, D. (2010). Sex determination: Genetic models for oysters. *Journal of Heredity, 101*(5), 602–611. doi:10.1093/jhered/esq065.

Islam, R., Melvin, S. D., Kit Yu, R. M., O'Connor, W. A., Anh Tran, T. K., Andrew-Priestley, M., … MacFarlane, G. R. (2022). Estrogenic mixtures induce alterations in lipidomic profiles in the gonads of female oysters. *Chemosphere, 291*, 132997. doi:10.1016/j.chemosphere.2021.132997.

Jiang, G., Zhang, Y., Du, L., Chen, Y., Liang, Y., Yu, Y., … Li, Q. (2023). Genetic diversity and structure in a selected strain of hybrid oysters between *Crassostrea gigas* and *C. angulata* evaluated from microsatellites and mitochondrial COI sequences. *Aquaculture, 574*, 739716. doi:10.1016/j.aquaculture.2023.739716.

Jiang, G., Xu, C., & Li, Q. (2024). Establishment of four types of allotetraploids derived from *Crassostrea gigas* and *C. angulata* and their breeding potential. *Aquaculture International, 32*(4), 4971–4989. doi:10.1007/s10499-024-01411-9.

Jiang, K., Chen, C., Jiang, G., Chi, Y., Xu, C., Kong, L., Yu, H., Liu, S., & Li, Q. (2024). Genetic improvement of oysters: Current status, challenges, and prospects. *Reviews in Aquaculture, 16*(2), 796–817. doi:10.1111/raq.12868.

Li, Q., Xu, C., & Kong, L. (2020). The aquaculture of *Crassostrea gigas* "Haida No.3". *China Fisheries, 4*, 98–105.

Liang, Y., Hu, H., Cheng, G., Xu, C., & Li, Q. (2024). Growth, survival and thermotolerance of diploids, triploids and tetraploids of the Fujian oyster *Crassostrea angulata* with normal, golden and black shell colors. *Aquaculture, 591*, 741131. doi:10.1016/j.aquaculture.2024.741131.

Lovatelli, A. (2004). Hatchery culture of bivalves: a practical manual. In MM Helm, & N Bourne (Eds.), *FAO fisheries technical paper* (pp. 145–149). Food and Agriculture Organization of the United Nations. https://www.fao.org/4/y5720e/y5720e00.htm.

Ramos, C. O., da Silva, F. C., de Miranda Gomes, C. H. A., Langdon, C., Takano, P., Gray, M. W., & de Melo, C. M. R. (2021). Effect of larval density on growth and survival of the Pacific oyster *Crassostrea gigas* in a recirculation aquaculture system. *Aquaculture, 540*, 736667. doi:10.1016/j.aquaculture.2021.736667.

Santerre, C., Sourdaine, P., Marc, N., Mingant, C., Robert, R., & Martinez, A. S. (2013). Oyster sex determination is influenced by temperature - First clues in spat during first gonadic differentiation and gametogenesis. *Comparative Biochemistry and Physiology - A Molecular and Integrative Physiology, 165*(1), 61–69. doi:10.1016/j.cbpa.2013.02.007.

Sun, D., Yu, H., & Li, Q. (2022). Examination of the roles of *Foxl2* and *Dmrt1* in sex differentiation and gonadal development of oysters by using RNA interference. *Aquaculture, 548*, 737732. doi:10.1016/j.aquaculture.2021.737732.

Sun, D., Yu, H., & Li, Q. (2023). Starvation-induced changes in sex ratio involve alterations in sex-related gene expression and methylation in Pacific oyster *Crassostrea gigas*. *Comparative Biochemistry and Physiology Part B: Biochemistry and Molecular Biology, 267*, 110863. doi:10.1016/j.cbpb.2023.110863.

Wang, Z., & Zheng, X. (2024). *Science of shellfish culture* (pp. 42–44). Science Press.

Yang, H., Wang, Y., Guo, X., & Tiersch, T. R. (2015). Production of inbred larvae through self-fertilization using oocytes and cryopreserved sperm from the same individuals after sex reversal in eastern oyster *Crassostrea virginica*. *Aquaculture Research, 46*(9), 2153–2165. doi:10.1111/are.12371.

Zhang, N., Xu, F., & Guo, X. (2014). Genomic analysis of the Pacific Oyster (*Crassostrea gigas*) reveals possible conservation of vertebrate sex determination in a mollusc. *G3 Genes|Genomes|Genetics, 4*(11), 2207–2217. doi:10.1534/g3.114.013904.

Zhang, G., Fang, X., Guo, X., Li, L., Luo, R., Xu, F., Yang, P., Zhang, L., Wang, X., Qi, H., Xiong, Z., Que, H., Xie, Y., Holland, P. W. H., Paps, J., Zhu, Y., Wu, F., Chen, Y., Wang, J., … Wang, J. (2012). The oyster genome reveals stress adaptation and complexity of shell formation. *Nature, 490*(7418), 49–54. doi:10.1038/nature11413.

Zhou, J., Xu, C., Liu, H., & Li, Q. (2026). Comparative aquaculture performance of genetically improved triploid oysters of 'Haida No. 3' line of the Pacific oyster *Crassostrea gigas*. *Aquaculture, 610*, 742905. doi:10.1016/j.aquaculture.2025.742905.

CHAPTER

14

Scallop aquaculture

Qiang Xing[a], *Ximing Guo*[b], *Shi Wang*[a] *and Zhenmin Bao*[a]

[a]College of Marine Life Sciences, Ocean University of China, Qingdao, China
[b]Haskin Shellfish Research Laboratory, Department of Marine and Coastal Sciences, Rutgers University, Port Norris, New Jersey, United States

14.1 Introduction

Scallop farming has become an important sector of the global marine aquaculture industry, driven by the rising demand for high-quality and sustainable seafood. Scallops, with their delicate texture and rich flavors, are a highly valued commodity in culinary markets worldwide. Scallop aquaculture involves diverse techniques, ranging from suspended to bottom culture, each tailored to optimize growth and maximize yield. This chapter provides an overview of scallop aquaculture, covering historical development, economic significance, biological and ecological considerations (FAO, 2024), and technological advancements that have propelled the industry.

14.1.1 Global importance of scallop aquaculture

Scallop aquaculture represents a significant and growing segment within the global seafood industry, playing a crucial role in meeting the increasing demand for high-quality marine products (Gephart et al., 2021; Stentiford et al., 2020; Willer & Aldridge, 2020). This sector's expansion is driven by the rising consumption of seafood, recognized for its nutritional benefits, including high protein content, and a good source of omega-3 fatty acids and essential minerals. Scallops, in particular, are valued for their delicate texture and flavor, making them a premium product in many markets with an annual yield of two million tons (FAO, 2024). The global importance of scallop aquaculture extends from economic value to ecological and social dimensions (Willer et al., 2021). Sustainable scallop farming practices contribute to marine conservation by providing an alternative to wild capture, which often leads to overfishing and habitat destruction. Furthermore, scallop aquaculture supports coastal communities through job creation and economic diversification, promoting social well-being and resilience in regions heavily dependent on marine resources.

Genetic Breeding and Aquaculture of Shellfish: A Collaborative Approach to a Sustainable Future
DOI: https://doi.org/10.1016/B978-0-443-33606-5.00018-7

TABLE 14.1 The world's top 10 scallop-producing countries: Production and cultured species (Year of 2024).

Rank	Country	Production (ton/year)	Cultured species
1	China	1,850,000	*A. irradians* (Lamarck, 1819), *Chlamys farreri* (Jones & Preston, 1904), *P. yessoensis* (Jay, 1857)
2	Japan	330,000	*P. yessoensis* (Jay, 1857), *C. farreri* (Jones & Preston, 1904)
3	United States	95,000*	*A. irradians* (Lamarck, 1819), *Placopecten magellanicus* (Gmelin, 1971)
4	Peru	52,000	*A. purpuratus* (Lamarck, 1819)
5	Canada	46,000*	*P. yessoensis* (Jay, 1857), *Patinopecten caurinus (Gould, 1850)* and hybrids, *P. magellanicus* (Gmelin, 1791)
6	France	35,000*	*Pecten maximus* (Linnaeus, 1758), *Aequipecten opercularis* (Linnaeus, 1758)
7	Chile	26,000	*A. purpuratus* (Lamarck, 1819)
8	South Korea	12,000	*P. yessoensis* (Jay, 1857), *C. farreri* (Jones & Preston, 1904)
9	Norway	5,000*	*P. maximus* (Linnaeus, 1758)
10	Spain	2,000*	*Pecten jacobaeus* (Linnaeus, 1758)

* *Note:* Production mostly from wild fishery landings.

14.1.2 Brief history and current trends in scallop farming

The history of scallop farming is a testament to human adaptability in harnessing natural resources for sustenance and commerce. Scallop cultivation can be traced back several centuries, with rudimentary techniques employed in various parts of Asia and Europe. Modern scallop aquaculture began to take shape in the mid-20th century, driven by advancements in marine biology and aquaculture technology (Kosaka, 2016). Countries such as China, Japan, and the United States were pioneers in developing and refining methods for scallop breeding and farming. China's scallop industry has experienced remarkable growth, driven by advancements in hatchery techniques and large-scale farming, resulting in the world's highest annual production of ~1.85 million tons (Wang & Gao, 2024). Japan has a long tradition of scallop farming, particularly the culture of the Yesso scallop (*Patinopecten yessoensis*, Jay, 1857), which has been cultivated extensively since the 1950s. In America, the United States and Peru have led the way in scallop aquaculture, particularly with the bay scallop (*Argopecten irradians*, Lamarck, 1819) and Peruvian scallop (*Argopecten purpuratus* Lamarck, 1819), respectively. Nowadays, scallop farming is a global enterprise, with significant production in Asia, America, Europe, as well as expanding operations in Oceania and other regions (Table 14.1) (Guo & Luo, 2016; Kosaka, 2016; Mendo et al., 2016; Parsons et al., 2016; Robinson et al., 2016; Strand et al., 2016; von Brand et al., 2016).

Current trends in scallop farming are characterized by an increased focus on sustainability, efficiency, and product quality (Gui et al., 2016). Advances in biotechnology and enhanced farming techniques are driving productivity gains and environmental stewardship.

14.1.3 Innovation in breeding and culture techniques

Innovation in breeding and culture techniques is pivotal to the continued growth and sustainability of scallop aquaculture. The development of artificial breeding programs has led to the production of scallops with desirable traits (Azra et al., 2022; Zeng et al., 2022). Genetic and molecular tools are increasingly being utilized to understand and manipulate the genetic makeup of scallops, paving the way for the creation of superior strains (Dou et al., 2016). The introduction and cultivation of nonnative species such as bay scallop and Yesso scallop have contributed significantly to global scallop aquaculture (Guo, 2009). In addition, innovations in culture techniques are transforming scallop farming practices. Traditional methods such as suspension culture (e.g., pearl nets and ear-hanging) and bottom culture are being optimized through the use of advanced materials and technologies. For example, the introduction of lantern nets has improved the efficiency and yield of suspended scallop farming (Maeda-Martínez et al., 1997). Furthermore, the adoption of integrated multi-trophic aquaculture (IMTA) systems, where scallops are farmed alongside other species such as seaweeds and fish, has shown promise in enhancing productivity and environmental sustainability (Tang et al., 2024), reducing waste, and improving overall ecosystem health.

Moreover, technological advancements in monitoring and automation are revolutionizing scallop reproduction and larval rearing operations. Real-time monitoring of water quality parameters, coupled with an automated feeding system in artificial scallop seedling rearing, is enhancing the precision and efficiency of hatchery practices (Essamlali et al., 2024). These innovations not only improve production outcomes but also reduce labor costs and mitigate environmental impacts. Another significant area of innovation is the development of disease management strategies. Research into pathogen identification and biosecurity measures is critical to safeguarding scallop populations and ensuring the long-term viability of farming operations (Zheng et al., 2023). The application of probiotics and other biocontrol agents is also gaining traction as a means of enhancing scallop health and resilience (Hossain et al., 2017).

Scallop aquaculture is a vital and dynamic component of the global seafood industry, with significant economic, ecological, and social implications. The history and current trends in scallop farming underscore the importance of innovation in breeding and culture techniques to meet the challenges of growing demand, environmental sustainability, and disease management. Continued advancements in genetics, farming technology, and sustainable practices will be essential to the future success and expansion of scallop aquaculture.

14.2 Scallop species in aquaculture

The adoption of various scallop species for aquaculture has been a strategic and multifaceted process, driven by the biological characteristics, environmental preferences, and market demands. This section delves into the major scallop species cultivated around the world, introducing their unique biological traits and the specific environmental conditions that favor their growth and productivity. The focus will provide insights into the diversity and complexity of scallop farming.

FIGURE 14.1 Major scallop species cultivated around the world.

14.2.1 Major cultivated scallop species worldwide

Several scallop species are prominently cultivated in different regions of the globe, each contributing uniquely to the aquaculture industry. Among the most significant are the Zhikong scallop (*C. farreri*), the Yesso scallop (*P. yessoensis*), the bay scallop (*A. irradians irradians*), the noble scallop (*Chlamys nobilis*, Reeve, 1853), the king scallop (*P. maximus*), the Atlantic sea scallop (*P. magellanicus*), the Peruvian scallop (*A. purpuratus*), and the Asian moon scallop (*Amusium pleuronectes*, Linnaeus, 1758) (Fig. 14.1). Each of these species has been selected for aquaculture based on a combination of their commercial value, growth rates, and adaptability to farming conditions.

The Zhikong scallop, also known as the Chinese scallop, is a species of bivalve mollusk that has significant roles in the aquaculture industry due to its high economic value and wide environmental adaptability (Guo & Luo, 2016). Native to the coastal waters of northern China, Korea, and Japan, the Zhikong scallop has been successfully cultivated in various parts of East Asia, forming a large-scale scallop farming industry. The species is highly valued for its sizable adductor muscle. In the 1970s, success in hatchery production of seed enabled the large-scale cultivation of Zhikong scallops in China, which in turn created abundant supplies of natural seed (Guo & Luo, 2016). Annual production of cultured Zhikong scallops reached more than 0.75 million tons in the mid-1990s, mostly from natural seed (Guo et al., 1999). This success, along with the successful introduction and cultivation of bay scallops, marked a significant milestone, driving China's third major expansion in mariculture development (Wang & Gao, 2024).

The Yesso scallop, native to the northern Pacific, particularly around Japan and Russia, is one of the most widely cultivated scallop species with an annual yield of ~0.34 million tons (Wang & Gao, 2024). Known for its large size and high market value, this species has been a cornerstone of the scallop aquaculture industry in Japan. Its cultivation has also spread to other coastal countries in the North Pacific, including China, South Korea in Asia, and Russia in Europe.

The bay scallop, native to the East Coast of the United States, is another important species in the aquaculture industry. Known for its rapid growth, the bay scallop is cultivated along the Atlantic coast of North America on a limited scale. Since being introduced to China in the 1980s, the bay scallops have been the highest-yielding scallop species in China, with an annual production of ~0.80 million tons (Wang & Gao, 2024). The bay scallop's rapid growth and short cultivation cycle made it a popular choice for scallop farmers in China (Guo & Luo, 2016). Coupled with the industry's swift recovery after the mass mortality of Zhikong scallops since 1996 (Xiao et al., 2005), this species significantly advanced scallop aquaculture in China (Guo, 2009).

The noble scallop is highly prized for its distinctive golden shell and high meat yield, making it a prominent cultivated species in Japan and Southern China, particularly in provinces such as Guangdong and Fujian. Renowned for its rich flavor and nutritional benefits, it supports a substantial aquaculture sector, with an annual production reaching ~0.10 million tons (Tan et al., 2020a; 2020b). Selective breeding efforts have further enhanced its commercial appeal, yielding varieties such as "Nan'ao Golden" noble scallops, which exhibit desirable traits including a striking golden shell and a carotenoid-rich orange red adductor muscle, significantly elevating its market value (Karsoon et al., 2021; Tan et al., 2024).

The king scallop is primarily found in European waters, particularly in the North Atlantic (Strand et al., 2016). This species is highly valued in European markets for its large adductor muscle, which is considered a delicacy. The king scallop's adaptability to different farming methods, including bottom and suspended culture, has made it a popular choice among European aquaculturists.

The Atlantic sea scallop is predominantly harvested in the wild, but there has been increasing interest in its aquaculture potential, particularly in Canada and the northeastern United States. This species is prized for its large size and high-quality meat, which fetches premium prices in the market (Robinson et al., 2016). Efforts to farm the Atlantic sea scallop are focused on supplementing wild stocks and ensuring sustainable supplies.

The Peruvian scallop, found along the coasts of Peru and Chile, is a major species in South American aquaculture. Known for its adaptability to local environmental conditions, the Peruvian scallop is cultivated in both coastal and offshore systems (Mendo et al., 2016; von Brand et al., 2016). The Peruvian scallop has a defined tolerance range for temperature (6–26°C) and salinity (18‰–35‰) (Mendo et al., 2016; Navarro & Gonzalez, 1998; Wang et al., 2011). However, hybrids between the Peruvian scallop and bay scallop (*A. irradians*) displayed heterosis in both growth and tolerance, holding significant potential for further large-scale cultivation and commercial promotion in China (Wang et al., 2011).

The Asian moon scallop is a tropical species prized for its delicate, sweet meat and its characteristic flat, circular, and beautifully patterned shells. Unlike many other scallops, it is primarily harvested from wild fisheries across the Indo-Pacific region, though aquaculture techniques for this species are under active development to meet market demand (Soffa et al., 2024). It is an important commercial species in Southeast Asian countries such as Thailand, Malaysia, and the Philippines. The scallop is particularly known for its swimming ability, a trait rare among bivalves. Its high value and popularity in both domestic and international markets make it a crucial species for the fishing industries in these regions (Morton, 1980).

14.2.2 Biological traits and environmental preferences

Understanding the biological characteristics and environmental preferences of these scallop species is essential for optimizing their cultivation and ensuring sustainable aquaculture. Each species exhibits unique traits that influence its growth rates, reproductive cycles, and overall health in aquaculture settings.

The Zhikong scallop is renowned for its rapid growth, adaptability, and economic value. This cool-water scallop species (which prefers temperatures ranging from 15 to 23°C) exhibits several distinctive biological characteristics that make it particularly well-suited for aquaculture. Morphologically, Zhikong scallops have a fan-shaped shell with pronounced ribs and often display vibrant coloration ranging from shades of orange and yellow to purplish-red, which can be an indicator of their habitat and diet (Guo & Luo, 2016). The adductor muscle is large and prized for its flavor and texture, contributing to the high market demand for this species. The adaptability of Zhikong scallops to cool-water environmental conditions has facilitated their cultivation in northern China, South Korea, and Japan (Li et al., 2017).

The Yesso scallop thrives in cold, nutrient-rich waters typically found in northern latitudes (Costa et al., 2020; Dou et al., 2016). This species prefers temperatures ranging from 8 to 15°C and is highly sensitive to water quality, particularly the presence of phytoplankton, its primary food source. The Yesso scallop has a relatively long-life cycle on cold and stable deep-ocean bottoms, with a growth period that can extend from two to four years before reaching market size (Kosaka, 2016). Its reproductive strategy involves broadcast spawning, where millions of eggs are released into the water column, ensuring high genetic diversity and resilience.

The bay scallop is adapted to warmer waters along the eastern seaboard of North America. It prefers temperatures between 18 and 28°C and is highly tolerant of variations in salinity, making it suitable for estuarine environments. The bay scallop has a short life cycle, typically reaching market size within 1 year. This rapid growth, combined with its high reproductive output, allows for efficient and sustainable farming practices (Parsons et al., 2016). The bay scallop has excellent adaptability to mainstream shallow-sea farming facilities such as rafts and net cages, enhancing the monitoring efficiency of growth conditions and reducing the farming cost.

The king scallop also favors cold waters, typically found in the North Atlantic, with optimal growth occurring at temperatures between 5 and 15 °C. This species is well-adapted to both sandy and rocky substrates, making it versatile for different aquaculture systems. The king scallop has a distinctive growth pattern, with a rapid initial growth phase followed by a slower, more sustained increase in size. It reproduces through external fertilization, with a high fecundity rate, producing large quantities of larvae that settle on suitable substrates (Duncan et al., 2016; Strand et al., 2016).

The Atlantic sea scallop is found in cold, deep waters of the North Atlantic, with an optimal temperature of about 10°C (Duncan et al., 2016; Strand et al., 2016). This species prefers sandy or gravelly substrates at depths mostly ranging from 18 to 110 m. The Atlantic sea scallop has a slower growth rate compared to other species, taking up to 5 years to reach market size. Its reproductive cycle involves synchronized spawning events, which are crucial for maintaining healthy population levels. Efforts to cultivate this species focus on enhancing larval survival and optimizing grow-out conditions to mimic its natural habitat.

The Peruvian scallop possesses specific adaptability to the coastal environmental conditions of Peru and Chile, preferring temperatures ranging from 12 to 20 °C and salinity ranging from 27‰ to 30‰, and thrives in both coastal and offshore waters (Mendo et al., 2016;

Navarro & Gonzalez, 1998; von Brand et al., 2016). The Peruvian scallop has a relatively fast growth rate, reaching market size within 18–24 months (Dall, 1909). Its reproductive strategy includes multiple spawning events throughout the year, ensuring continuous recruitment and population stability.

In addition to these major species, several other scallop species show promising aquaculture potential beyond major cultivated varieties. The Australian scallop (*Pecten fumatus*, Reeve, 1852), thriving in 10–20°C sandy/muddy substrates, reaches market size in 2–3 years through synchronized spawning. The New Zealand scallop (*Pecten novaezelandiae*, Reeve, 1852) prefers 10–15°C sheltered bays, exhibiting slower growth (4 years to maturity) but prized for its unique flavor profile. The Iceland scallop (*Chlamys islandica*, Carpenter, 1864) demonstrates remarkable cold-water adaptation (0–5°C), inhabiting rocky North Atlantic depths (10–100 m) with an extended 5-year growth cycle. While these species currently have limited commercial production, their distinct environmental adaptations and reproductive strategies present valuable diversification opportunities for global scallop aquaculture (Dredge et al., 2016).

Successful scallop aquaculture relies on suitable habitats, efficient farming techniques, and environmental impact management. Critical advancements encompass hatchery systems enabling large-scale seed production, artificial breeding programs enhancing growth performance, disease resistance, and product quality, alongside comprehensive disease management through pathogen diagnostics and preventive biosecurity protocols. Environmentally responsible approaches incorporate strategic farm siting, precision nutrition and waste control systems, and rigorous water quality surveillance to balance ecological sustainability with production efficiency. These strategies collectively enhance scallop farming's viability and sustainability.

14.3 Hatchery management

The aquaculture of scallops begins with the critical phase of hatchery production, a sophisticated process that ensures the production of healthy and viable juvenile scallops for subsequent nursery and grow-out operations. Hatchery management encompasses the comprehensive setup and operation of facilities specifically designed for scallop breeding, spawning, and larval rearing. This section provides an overview of the various components and practices involved in hatchery management, highlighting the technologies and methods used for spawning and larval rearing, and discussing the common challenges encountered in these controlled environments.

14.3.1 Hatchery setup and operational framework

Setting up a scallop hatchery requires meticulous planning and execution to create an environment conducive to scallops' delicate early life stages. A typical hatchery includes broodstock conditioning tanks, spawning and fertilization units, larval rearing tanks, and nursery systems. Each area is designed to meet specific needs at different developmental stages (Fig. 14.2).

The broodstock conditioning area maintains adult scallops under controlled conditions to enhance their reproductive capacity. This involves regulating temperature, salinity, and nutrition to stimulate gonadal development. While most hatcheries can condition broodstock using larval rearing tanks, separating broodstock and larvae is necessary to prevent the transmission of diseases (Utting & Millican, 1997). Broodstock conditioning tanks typically range from 15 to 25 m^3 in volume and 1.5 to 2.0 m in depth, constructed from brick masonry or reinforced concrete, with a sloped floor for drainage. Tanks require multiple layers of waterproof cement coating

FIGURE 14.2 Diagram of scallop hatchery.

and over a month of seawater soaking to leach alkaline substances. The clean seawater for broodstock is processed through sedimentation, sand filtration, and activated carbon absorption (refer to section 14.5.2 for specific water supply configurations). The aeration system uses soft plastic or hard pipes connected to air valves and porous aerators to aerate the seawater. The common broodstock conditioning method is gradually raising the temperature to promote gonadal maturation, which requires heating facilities (Utting & Millican, 1997). Depending on the hatchery situation, different heating methods are adopted, including efficient heat-exchangers such as heat pumps, as well as traditional approaches like laying hot water pipes and using a hot air system, which are widely adopted in numerous hatcheries across China. During conditioning, broodstock scallops are fed with microalgae at specific rations to ensure nutritional needs for gametogenesis are met, and pumps with timed switches are often used to optimizing feeding. The PVC pipes with resistance to seawater corrosion and microbial attachment are used for seawater supply and feed delivery. The conditioning process can take several weeks, during which the scallops' gonads mature and become ready for spawning. After selecting broodstock based on appropriate criteria (Zhang et al., 2024), broodstock scallops are cleaned and inspected for general health. For dioecious species, the broodstock is separated by gender and placed in floating net cages at an appropriate density for conditioning. During conditioning, perform regular water changes and use effective microorganisms (EM) as needed to regulate water quality.

Upon completion of conditioning, broodstock scallops are transferred to spawning units, where environmental cues such as temperature changes or chemical addition are used to induce spawning (Christophersen et al., 2006). Similar to broodstock conditioning tanks, spawning units are also typically large tanks. Unlike the water used for broodstock conditioning, water used for spawning strictly adheres to the standards for larval rearing, undergoing coarse filtration and treatment with protein skimmers to remove protozoans, and with reactive oxygen species and chemical agents to disinfect (refer to section 14.5.2 for specific configurations). Heating and aeration systems are also essential. Once both male and female gametes are released and successfully fertilized, the fertilized eggs are collected and transferred to larval rearing tanks.

Larval rearing is a critical stage in hatchery operations, requiring precise control over water quality, temperature, and feeding. The setting of larval rearing tanks is similar to spawning units, including strict water quality filtration, adequate oxygen supply, and temperature control. Maintaining an appropriate temperature is crucial for rapid larval development and cost savings, which are generally achieved through laying hot water pipes combined with hot air systems (3–5°C above the water temperature) to maintain water temperature. Scallop larvae are fed microalgae, with feeding species and level adjusted to meet nutritional needs at different developmental stages (Cheng et al., 2020). Early larvae are fed with *Isochrysis* as the main feed, supplemented with *Platycoidea* or *Diatoms* (10%–30% of that of *Isochrysis*) once larvae develop to the umbo larvae (Yu & Li, 2016). Over 1–3 weeks, depending on species and temperature, larvae undergo significant morphological changes, culminating in the development of a foot and the ability to settle onto a substrate.

14.3.2 Nursery system setting and management

The nursery stage follows larval rearing and involves transferring the settled juveniles or spat to a nursery for growth until they are suitable for transfer to grow-out facilities. After juveniles fully settle in the hatchery and the water temperature matches that of the nursery, they can be transported to the nursery. Transport should be carried out under favorable weather to prevent spat from wind exposure. Nursery systems can include mesh bags, trays, or floating cages, and must be carefully managed to ensure adequate water flow, food supply, and protection from predators.

The nursery can typically be conducted in offshore coastal areas or in nursery ponds (an area ranging from 1 to 8 hm^2, with a depth exceeding 1.5 m). Compared to offshore nurseries, nursery ponds offer greater control over environmental conditions. The construction of nursery ponds requires consideration of water supply, electricity, and transportation systems. The ponds near natural water sources aid water supply and drainage, and the seawater should meet the conditions required for the survival and development of scallops. Before each breeding season, the ponds should be drained, silt removed, and sun-dried. Hydrated lime (750–1,500 kg/hm^2) is applied for disinfection (Bosma & Verdegem, 2011), followed by flushing daily to dissipate the lime's disinfectant properties. Then, ponds can be filled and fertilized with sodium nitrate (0.2 kg/hm^2), urea (0.2 kg/hm^2), and/or other compound fertilizers to promote microalgae growth. Generally, fertilizing the pond about half a month in advance can achieve ideal results, aiming for a water clarity of 40 cm. In the ponds, ropes are set at intervals to suspend mesh bags and net cages. The mesh bags can be tied in strings of 10, with a weighted bottom to sink. Alternatively, the attachment substrates with stone weights can be placed into the water net cages below, ensuring that the netting is fully submerged below the water surface. Net cages (1.5 m × 3 m, with 30 mesh size) are suspended between ropes and hung every 2.5 m across the entire pond. Some nurseries use trays for spat rearing. During the nursery, daily

FIGURE 14.3 Nursery pond setup.

monitoring is necessary, including changing water, feeding food, and cleaning the environment (Fig. 14.3).

The water transparency should be maintained at 40–80 cm during the nursery stage in ponds; fertilizing in small amounts and frequently, and partial water changing can be performed to regulate the transparency. Change the water every three to four days with one-quarter each time, to maintain water cleanliness; move mesh bags up and down every 3–4 days for simple rinsing. Regularly observe scallop size and change the mesh bags as needed, potentially using bags with a mesh size 20 units smaller or directly transferring them to net cages.

14.3.3 Challenges and strategic solutions in hatcheries

Despite advancements in hatchery technology and methods, several challenges persist in scallop hatchery production. Disease management, genetic diversity, and environmental fluctuations are among the most pressing issues that require ongoing attention and innovative solutions.

Disease management: Bacterial, viral, and protozoan pathogens pose a constant threat to scallop larvae and juveniles, which can lead to high mortality rates and significant economic losses (Sanderson et al., 2023). Effective disease management strategies include sterilizing equipment and facilities, controlling water quality, and isolating infected individuals to prevent the spread of disease. Probiotic treatments are emerging as potential methods to enhance survival, but further research and development are needed.

Genetic diversity: Maintaining genetic diversity is crucial for sustainable and resilient scallop aquaculture (Nascimento-Schulze et al., 2021). Inbreeding and genetic bottlenecks can lead to reduced fitness, increased susceptibility to disease, and lower growth rates (Guo, 2021). To address this issue, hatcheries implement breeding programs that incorporate wild-caught broodstock to introduce new genetic material into the population (Nascimento-Schulze et al., 2021). In addition, genetic monitoring and selective breeding programs are used to enhance desirable traits while maintaining genetic diversity. Advances in genetic technologies, such as marker-assisted and genomic selection, offer promising tools for improving breeding programs and ensuring the health and productivity of scallop stocks.

Environmental fluctuations: Fluctuations in environmental parameters, such as temperature, salinity, and water quality, can lead to stress, reduced growth rates, and increased susceptibility to disease. Climate change and ocean acidification pose additional challenges, potentially altering the suitability of traditional hatchery sites and affecting the overall health of scallop populations. To mitigate these impacts, hatcheries employ adaptive management strategies, such as adjusting rearing protocols, diversifying site locations, and developing resilient strains of scallops through selective breeding (Yu, 2009; Yu & Li, 2016).

Hatchery management is a foundational aspect of scallop aquaculture, encompassing the setup and operation of facilities designed to support the early life stages of scallops. Advances in technology and methods boost the efficiency and success rates of hatcheries, although challenges such as disease management, genetic diversity, and environmental fluctuations continue to require innovative solutions. Many hatcheries in northern China are having difficulties in stably producing Zhikong scallop seed for reasons that are not fully understood. Addressing these challenges can contribute to enable sustainable production of high-quality scallops and support the growth and resilience of the aquaculture industry.

14.4 Broodstock management and reproduction

Broodstock management and reproduction are critical for successful cultivation in scallop aquaculture. This section provides an introduction to the strategies for selecting and maintaining broodstock, explores the genetic considerations vital for robust breeding programs, and outlines the techniques for induced spawning and gamete handling, all of which are integral to the sustainable and efficient production of scallops.

14.4.1 Broodstock selection and maintenance strategies

The selection and maintenance of broodstock are critical in scallop aquaculture, directly impacting the quality and health of future generations. The selection often begins with assessing wild populations or existing cultured stocks to identify potential candidates with desirable traits. For a qualitative trait controlled by a single/several genes, such as color, the selection criteria for parent scallops are relatively clear, and phenotypic selection can be accomplished within two to three generations. In contrast, the selection of quantitative traits is a long-term process. Another major goal of broodstock selection is to ensure genetic diversity. Maintaining a diverse genetic pool is essential to prevent inbreeding and enhance the population's adaptability and resilience to environmental changes and diseases. To achieve this, hatcheries often implement selective breeding programs that carefully manage the genetic lineage of broodstock, preserving and utilizing diverse genetic material. Key factors to consider during selection include genetic diversity, phenotypic traits, and overall health status. Scallop selecting programs employ both single-trait (mass/family selection) and multitrait (tandem/independent culling) approaches to enhance growth, stress tolerance, and other economic traits. Mass selection based on phenotypes is widely adopted for its simplicity, while family-based methods maintain genetic diversity (Gjedrem, 2005). Maintaining effective population sizes (hundreds to thousands) and controlling crossing pedigree are crucial for preventing inbreeding (Guo, 2021). These strategies have proven effective in commercial scallop genetic improvement.

Once selected, maintaining broodstocks in optimal conditions is essential for their reproductive health, longevity, and gonadal development. The conditioning process lasts several weeks, maturing the gonads for spawning. Water quality control—via aeration and regular water change—is fundamental for gonadal maturation, with change frequency and aeration adjusted based on gonadal development. Artificially inducing gonadal maturation in hatcheries requires modifying seawater parameters, such as temperature and nutrient ratio. Currently, most hatcheries employ artificial temperature elevation, starting from the farming area's natural seawater temperature and gradually increasing the temperature to direct energy to gonadal development, paired with increased feeding to meet nutritional needs. Nutrition is crucial for broodstock conditioning, with diets often supplemented with high-quality microalgae (*Nitzschia closterium, Chaetoceros muelleri*) to enhance reproductive performance (O'Connor et al., 2000). While ensuring an adequate supply of microalgae, grinding liquid of macroalgae (e.g., *Sargassum* and *Spirulina*) is also added to support nutrient accumulation and promote gonad development (Magnesen & Christophersen, 2008). In addition, egg yolk, yeast, and algal paste/concentrate—serving as animal- and plant-based protein sources—are also used as substitute feeds for broodstock conditioning (Table 14.2) (Utting & Millican, 1997).

Regular health monitoring is essential in broodstock management, including routine health checks to detect disease or stress. In addition, biosecurity measures, such as quarantine protocols and sterilization of equipment, prevent the introduction and spread of pathogens. Maintaining healthy, genetically diverse broodstock ensures robust offspring with desired traits, forming the foundation for a successful aquaculture operation.

14.4.2 Genetic approaches in breeding programs

Genetic considerations are paramount in breeding programs to enhance scallop traits such as growth, survival, and resistance. The primary goal is to develop genetically superior broodstock that transmit favorable traits to their offspring, improving population performance.

Breeding programs introduce wild-caught individuals into the broodstock population for new genetic material and increase genetic variability, which is key for long-term viability and adaptability. Additionally, genetic analysis tools, such as DNA markers and genomic sequencing, are employed to assess genetic diversity and guide selection decisions (Gjedrem, 2005; Theissinger et al., 2023). Genomic technologies enhance breeding precision (Chi et al., 2022), with marker-assisted selection (MAS) identifying superior traits using genetic markers (Wang et al., 2025), and genomic selection (GS) leveraging genome-wide data to predict breeding values more accurately (Wang et al., 2018, 2025). These methods have successfully produced high-performance varieties such as "Penglaihong 2," "Penglaihong 4" (Zhikong scallop) and "Haiyifeng 12," "Haiyifeng 11" (bay scallop), demonstrating improved yield and thermotolerance. In practice, GS involves: (1) Tissue sampling (e.g., gill filaments) from potential broodstock for resequencing; (2) Breeding value estimation of the breeding population based on genetic variants and phenotypic data from a training population; and (3) Selection of top 5% individuals as core broodstock, complemented by a 0.2% randomly sampled group to maintain diversity. This approach significantly reduces generation intervals compared to traditional methods. To ensure long-term genetic health, programs integrate: (1) Strategic mating plans to minimize inbreeding; (2) Cryopreservation of gametes for genetic resource banking; and (3) Regular genetic monitoring. Such genomic-driven strategies are transforming scallop aquaculture by accelerating genetic gain while preserving diversity, directly contributing to more sustainable and profitable production systems.

Another important genetic management strategy is designing breeding protocols to minimize inbreeding, including carefully planning mating pairs to avoid close relatives and implementing rotational breeding schemes (Guo, 2021). Cryopreservation of gametes and embryos in Yesso scallop, king scallop, and noble scallop is also used to preserve genetic material and maintain

TABLE 14.2 Accumulated temperature and ripening conditions of common scallops (Yu, 2009; Yu & Li, 2016).

	Density (ind/m^3)	Temperature increase rate (°C /day)	Expectant temperature (°C)	Accumulated temperature (°C/day)	Microalgae	Substituted feeds
A. irradians	80–100	1.0	20	180–200	*N. closterium, C. muelleri*	*Sargassum, Spirulina,*
P. yessoensis	20–40	0.2–0.5	8–10	50–60	*C. muelleri*	egg yolk, yeast,
C. farreri	80–120	1.0	15–16	190–200	*N.closterium C. muelleri*	algal paste or concentrates

TABLE 14.3 Temperature ranges for culture and spawning induction in scallop species (Yu, 2009; Yu & Li, 2016).

	Culture temperature (°C)	Induction temperature (°C)
A. irradians	18–20	22–24
P. yessoensis	10–12	15–16
C. farreri	15–16	18–20

genetic diversity over time (Suquet et al., 2014; Wang et al., 2024; Xu et al., 2024; Zheng et al., 2019). Incorporating genetic considerations into breeding programs not only enhances the performance and resilience of scallop populations but also contributes to aquaculture sustainability and profitability. By prioritizing genetic health and diversity, hatcheries can produce high-quality broodstock and offspring that thrive in aquaculture environments and meet market demands.

14.4.3 Spawning induction, gamete handling, and larval rearing

Induced spawning and efficient gamete handling are essential in scallop aquaculture, enabling controlled reproduction and ensuring the availability of fertilized eggs for larval rearing. These techniques provide greater control over spawning timing and success to produce large numbers of high-quality larvae (O'Connor & Heasman, 1995). Induced spawning involves stimulating broodstock to release gametes under controlled conditions, with common methods including thermal shock, where sudden temperature changes induce spawning (Table 14.3), and chemical stimulants such as serotonin or hydrogen peroxide, which trigger spawning by influencing the hormones (Vélez et al., 1990).

After spawning is induced, careful gamete handling can ensure high fertilization rates and healthy larval development. Gametes are typically collected by filtering the water from spawning tanks, then microscopically assessed for viability and quality. Fertilization is usually carried out by mixing eggs and sperm in controlled conditions, monitoring their ratio to optimize success while minimizing polyspermy (abnormal development from multiple sperm fertilization).

After fertilization, the eggs are rinsed and transferred to incubation tanks with controlled temperature, salinity, and oxygen levels for larval development. Various microalgae are selected for culture according to the scallop's size, digestibility, and nutritional needs during the whole production cycle (Table 14.4). Larval rearing tanks have filtration systems to maintain clean seawater and prevent harmful substance buildup. Larvae are fed microalgae in appropriate quantities for growth and development.

Settlement and metamorphosis are critical for scallop larvae. When the scallop umbo larvae reach the eye point stage, attachment substrates are introduced, and the water temperature is adjusted to facilitate the larvae's attachment. Treatment of attachment substrate, such as seawater preconditioning (Galley et al., 2020), sodium hydroxide soaking, and adding microbial membrane (Wang et al., 2017), can improve attachment. Once settled, the juvenile scallops are transferred to nursery systems where they continue to grow under controlled conditions until they are ready for the grow-out phase.

TABLE 14.4 Larval cultivation conditions of common scallops (Yu, 2009; Yu & Li, 2016).

	Culture temperature (°C)	Salinity (‰)	pH	Feeding
A. irradians	22–24	25–32	8.0–8.2	*Isochrysis galbana*
P. yessoensis	15–16	28–34	–	*I. galbana, C. muelleri*
C. farreri	19–22	27–33	7.5–8.5	*I. galbana*

The management of broodstocks and reproduction is important for scallop aquaculture, influencing the genetic quality, health, and productivity. By strategic selection, advanced genetic considerations, precise induced spawning and gamete handling techniques, hatcheries can achieve sustainable and efficient production of scallops, contributing to the industry's success and growth, meeting global market demands.

14.5 Land-based hatchery and nursery systems

Scallop aquaculture has evolved significantly, with innovations in land-based hatchery and nursery systems playing a pivotal role in the early stages of scallop rearing. These systems provide a relatively controlled environment that offers numerous advantages, particularly in the critical early stages of scallop development. This section explores the benefits of using land-based systems, outlines the various types of systems employed, and delves into the intricate management of water quality, feeding, and health within these controlled environments. By leveraging land-based systems, aquaculturists can enhance survival rates, improve growth performance, and mitigate risks associated with environmental variability and disease, ultimately contributing to the sustainability and efficiency of scallop production.

14.5.1 Benefits of land-based systems in early-stages rearing

The adoption of land-based hatcheries and nurseries for the early stages of scallop aquaculture offers several compelling advantages. One of the primary benefits is the ability to create a relatively stable and controlled environment that mitigates the unpredictable variables present in open-water systems, such as ocean acidification and warming, the key stressors for many marine organisms. This stability is crucial for the delicate early life stages of scallops, including larval development and early juvenile growth, where fluctuations in temperature, salinity, and water quality critically impact survival and growth. Land-based systems enable precise regulation of these parameters through the equipped water quality analysis, providing an optimal environment for scallop development (Magnesen & Jacobsen, 2012).

Furthermore, land-based systems offer relatively enhanced biosecurity, allowing the management of disease outbreaks and parasite infestations (von Brand et al., 2016). The controlled setting allows for the implementation of stringent biosecurity measures, such as the filters, biofilters, screens, and monitoring panel in the hatchery, which can ensure the filtration and disinfection of incoming water (Yu, 2009; Yu & Li, 2016), quarantine protocols for new broodstocks, and regular health monitoring. Moreover, the primary focus in land-based systems has

TABLE 14.5 Water quality standards for land-based early-stage scallop aquaculture.

Items	Standard
pH	7.0–8.5
DO	> 5 mg/L
Floatage and suspended substances	No obvious oil film or foam, and artificially increased suspended solids < 10 mg/L
Nitrogen compounds	Ammonia nitrogen < 0.5 mg/L and nitrite nitrogen < 0.1 mg/L
Inorganics and heavy metals	Refrain from exceeding safe thresholds to avoid scallop toxicity

traditionally been on prophylaxis, with regular administration of antibiotics (oxytetracycline, streptomycin, penicillin, florfenicol, etc.). These practices minimize the introduction and spread of pathogens, ensuring healthier and more robust scallop stocks.

Another significant advantage is the ability to implement more efficient and targeted feeding regimes. In land-based systems, aquaculturists can precisely adjust the type and amount of microalgae fed according to the specific nutritional needs of scallops at different developmental stages. This precision enhances growth rates while reducing feed waste and environmental impact, promoting more sustainable aquaculture practices (Christophersen et al., 2006). In addition, land-based nurseries play the role of a bridge between hatchery production and grow-out in the sea, facilitating close observation and intervention. This level of oversight is challenging to achieve in open-water systems and contributes to higher survival and growth rates (Magnesen & Christophersen, 2007).

Overall, the use of land-based systems in the early stages of scallop aquaculture provides a controlled, secure, and efficient environment that supports the health and development of scallops, laying a strong foundation for successful aquaculture operations.

14.5.2 Water quality, feeding, and health management

Effective management of water quality, feeding, and health is critical to the success of land-based hatchery systems. Maintaining optimal water quality is fundamental, as scallops are highly sensitive to changes in their aquatic environment. Key water quality parameters include temperature, salinity, dissolved oxygen (DO), pH, and nutrient levels, which are shown in Table 14.5. In land-based hatchery systems, seawater that has been settled and sand-filtered is further refined using advanced treatment techniques, such as biofilters, UV sterilizers, and protein skimmers, which are employed to remove waste products and maintain water quality. Regular monitoring and adjustments are necessary to ensure these parameters remain within the optimal range for scallop growth and health. Moreover, scallop larvae are also exposed to variable changes in water quality parameters these each water renewal, pathogen level, and risk of contaminant introduction (Merino et al., 2009).

Feeding practices in land-based systems are also closely managed to meet the specific nutritional needs of scallops. Scallops are filter feeders that rely on microalgae and other particulate matter in the water column. In controlled environments, microalgae cultures are often used to provide a consistent and high-quality food source. The quantity and composition of feed are adjusted based on the developmental stage and health status of the scallops (Martínez et al., 1995). In short, during the temporary culture stage of broodstocks, *N. closterium* was the main diet fed to scallops, and the appropriate increase of *I. galbana*, *Platymonas subcordiformis*, and the substitute feeds (egg yolk, yeast, etc.) was beneficial to the maturation of scallop gonads during conditioning. *I. galbana* was the main feed throughout the larval breeding period, which was also the superior opening feed for D-type scallop larvae (e.g., *C. farreri*, *P. yessoensis*, and *A. irradians*), and the amount of feed increased with the growth of the larvae. After placing the attachment substrates, the proportion of *P. subcordiformis* should be increased gradually. And the amount of baiting increased gradually from 10,000 to 100,000 cells/mL, with the optimal density of larvae at 5–8 larvae/mL.

Health management in land-based hatcheries and nurseries involves regular monitoring and preventive measures to minimize the risk of disease outbreaks. Routine health checks, including visual inspections and diagnostic tests, help detect early signs of disease or stress. Quarantine protocols for new broodstock and regular disinfection of equipment and facilities are essential biosecurity measures (Ben-Asher & Lahav, 2016). In the event of disease, prompt intervention with appropriate treatments, such as antibiotics or probiotics, is crucial to prevent the spread and ensure the well-being of the scallop population. To prevent disease, penicillin is regularly applied across the whole larval culture stage, which was applied at a concentration of 3 ppm every 3–4 days during the larval stage, and at 4–6 ppm each day after substrate placement.

In conclusion, land-based hatchery and nursery systems represent a transformative approach in early-stage scallop aquaculture, offering numerous advantages in terms of control, efficiency, and sustainability. By leveraging these systems for the early stages of scallop rearing, aquaculturists can create optimal conditions that enhance survival rates, improve growth performance, and reduce risks associated with environmental variability and disease. Crucially, the success of these operations hinges on the holistic management of water quality, feeding, and health within the controlled environment. This integrated approach, leveraging advanced technology, precise monitoring, and proactive intervention, ensures the production of healthy, high-quality scallops to meet market demands.

14.6 Grow-out systems

The grow-out phase in scallop aquaculture is a critical period that transitions juvenile scallops into market-ready adults. This stage involves various methodologies and technologies aimed at optimizing growth rates, survival, and product quality. The choice of grow-out systems can significantly influence the success and sustainability of scallop farming operations. This section provides an overview of traditional and innovative grow-out practices. It also includes case studies highlighting successful grow-out operations from different regions, illustrating practical applications and outcomes. By understanding the intricacies of grow-out systems, aquaculturists can make informed decisions to enhance productivity and sustainability in scallop aquaculture.

14.6.1 Traditional and innovative grow-out practices

Traditional grow-out practices in scallop aquaculture predominantly involve sea-based systems, where spats are transferred to ocean environments for further growth. These practices include suspended culture, bottom culture, and off-bottom culture (Fig. 14.4). Suspended

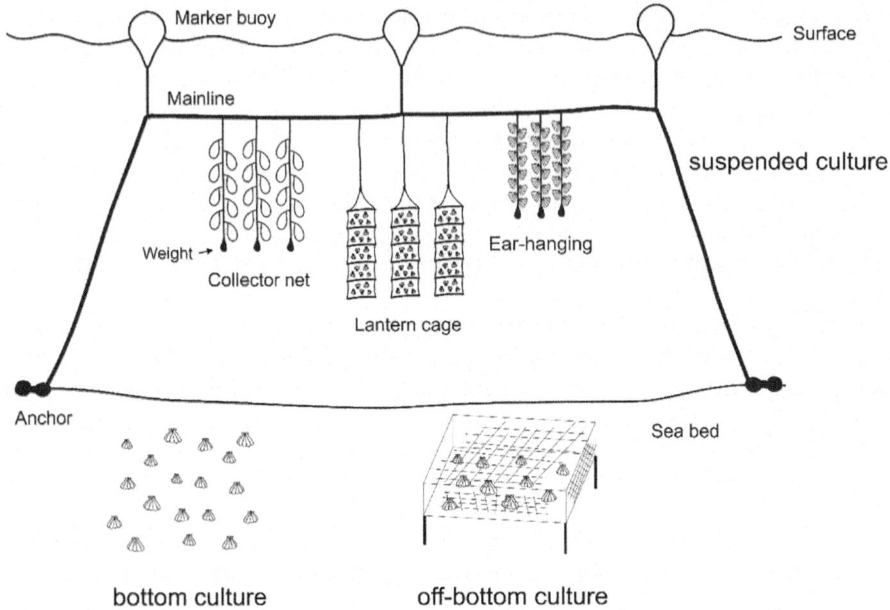

FIGURE 14.4 Grow-out device employed for bottom culture, off-bottom culture, and suspended culture.

culture, often utilizing longlines or rafts with hanging baskets or nets, offers greater control over scallop growth conditions by elevating them into more stable water columns (about 1–6 m deep according to the temperature of the seawater) with consistent food supply and reduced sedimentation issues. With better growth and survival, grow-out in suspension is often the preferred technique to maximize cultured scallop production (Mendoza et al., 2003). Generally, each acre can hold about 2,000 buoyancy lantern cages, and according to the scallop species, 15–30 scallops were placed in each layer. However, the fouling of the cage or net creates food restriction for filter-feeding scallops because of flow obstruction, further resulting in lower growth, which requires more labor costs to replace and clean up the cages (Kosaka, 2016). In addition, the ear-hanging aquaculture system is another important form. When the scallop shell reaches about three centimeters in height, drill a 2 mm hole in the front ear, string the scallops with nylon thread or polyethylene thread, and hang them on the main rope. Bottom culture entails placing scallops directly on the seabed, relying on natural conditions and food availability. This method is cost-effective but exposes scallops to predators. Moreover, this method needs a high seed input, but tends to have a low recapture rate, which may also affect the seabed ecology (Maeda-Martınéz et al., 2000). The aquaculture scale of Zhangzi Island (Dalian, China), the largest bottom culture base of Yesso scallop in China, has reached about 80,000 hm^2 in recent years. Off-bottom culture involves raising scallops above the seabed on structures such as racks, tables, or rigid cages, where scallops can be cultivated away from the seabed, reducing some predation risks and allowing better management of stock density (Maeda-Martínez et al., 1997). In Bahia Magdalena, Mexico, Catarina scallops (*A. ventricosus*) are grown in polyethylene netting placed on the sea floor (Maeda-Martınéz et al., 2000).

In scallop aquaculture practices, the need to boost efficiency while mitigating environmental impacts has driven innovations in grow-out models. Among these, IMTA is an innovative practice gaining traction (Stabili et al., 2023). IMTA involves cultivating scallops alongside other species, such as seaweeds and fish, to create a balanced ecosystem where the by-products of one species support the growth of another (Reid et al., 2010). While IMTA has not been widely used for scallop aquaculture, it can potentially promote sustainability by reducing waste and enhancing resource efficiency. In addition, advancements in automation and monitoring technologies, such as remote sensing and artificial intelligence, are being integrated into grow-out systems to optimize monitoring health and manage environmental conditions more effectively.

14.6.2 Case studies on successful grow-out operations

Examining case studies of successful grow-out operations provides valuable insights into practical applications and outcomes of various grow-out methods. These case studies illustrate how different strategies can be tailored to specific environmental conditions, resources, and market demands to achieve optimal results in scallop aquaculture.

One notable example is the Yesso scallop aquaculture industry, which has effectively utilized suspended culture methods in the cold waters of Hokkaido. By deploying longlines and lantern nets in nutrient-rich coastal areas, Yesso scallop farmers have achieved high growth rates and excellent product quality, and the hanging culture production in Hokkaido in 2010 represented 50% of the total harvest in Japan (Kosaka, 2016). This approach has not only boosted scallop production but also contributed to the sustainability of local marine ecosystems. Another innovative case study comes from the southeastern coast of Brazil, where an IMTA system has been successfully employed to grow scallops alongside mussels and seaweed (da Silva et al., 2022). This integrated approach has created a synergistic environment where the waste from bivalve mollusks provides nutrients for seaweed, and the seaweed, in turn, helps maintain water quality for the scallops and mitigates the environmental effects of ocean acidification on scallops, which could contribute to a major socioeconomic advance in the region. This multitrophic system has improved overall resource efficiency and reduced environmental impact, demonstrating the potential for sustainable and profitable aquaculture practices. These case studies highlight the diversity of successful grow-out strategies and the importance of adapting methods to local conditions and resources. They underscore the potential benefits of integrating traditional practices with innovative technologies and approaches to enhance the efficiency and sustainability of scallop aquaculture.

In summary, the grow-out phase in scallop aquaculture encompasses a range of practices and systems designed to optimize the growth and health of scallops as they mature to market size. Bay scallop aquaculture in China represents a highly successful grow-out practice, characterized by the establishment of an efficient seedling rearing system and the adoption of innovative suspended lantern net culture techniques. This well-developed industry achieves an annual production exceeding 800,000 tons (Wang & Gao, 2024). Meanwhile, regions in America, Europe, and Oceania have also pioneered advanced cultivation methods, further accelerating the global development of the scallop aquaculture sector. By examining successful case studies from different regions with various methods, aquaculturists can gain valuable insights into effective strategies and best practices for scallop grow-out, ultimately contributing to the growth and sustainability of the industry.

FIGURE 14.5 The "Haida Golden scallop" Yesso scallop, the "Penglaihong 2" Zhikong scallop, the "Haiyifeng 11" bay scallop, and "Kafeibei" hybrid scallops.

14.7 Practical breeding cases

Scallop aquaculture has experienced remarkable advancements over the past few decades, driven by the introduction of non-native species, the development of new breeding techniques, and the creation of improved scallop varieties (Huang et al., 2025; Lv et al., 2018). Several improved scallop varieties have been successfully developed, such as "Haida Golden" Yesso scallop, the "Penglaihong" Zhikong scallops, "Haiyifeng," "Zhongke Red" bay scallops, "Nan'ao Golden" noble scallops, as well as hybrids between Yesso and Weathervane scallops and between bay and Peruvian scallops (Bao, 2011; Guo, 2009; Tan et al., 2020a; Wang et al., 2011). This chapter focuses on four notable examples—the "Haida Golden scallop," "Penglaihong 2," "Haiyifeng 11," and "Kafeibei" hybrid scallops—demonstrating the success of modern breeding strategies in advancing the industry (Fig. 14.5).

14.7.1 Genomic selection: "Penglaihong 2" and the "Haiyifeng 11"

The "Penglaihong 2" (*C. farreri*) and "Haiyifeng 11" (*A. irradians*) scallop varieties represent groundbreaking achievements in GS-based breeding in molluscan aquaculture. "Penglaihong 2" (Breed Registration No. GS-01-006-2013) is the world's first aquatic variety developed via whole-genome selection in 2013, pioneering a new era for aquatic molecular breeding. "Penglaihong 2" inherits the excellent traits of high yield and stress resistance from the original "Penglaihong" variety (Breed Registration No. GS-02-001-2005), and demonstrates 53.46% higher yield and 27.36% improved survival compared to controls, with expanded adaptability from northern to subtropical waters (Fig. 14.6). Its improved survival rate and faster growth make it highly valuable for commercial farming. Similarly, "Haiyifeng 11" (Breed Registration No. GS-01-012-2022) achieves 46.62% increase in yield with enhanced thermal tolerance (Arrhenius breaking temperature 33.11°C, +1.00°C) and 16.57% higher survival (Xing et al., 2016). These traits make it a robust variety for sustainable aquaculture (Fig. 14.7). Both varieties highlight the power of GS in accelerating genetic gains, enabling the development of high-performance scallops with desirable traits. Their success underscores the potential of genomic selection to enhance productivity and resilience in marine bivalves, benefiting both farmers and the aquaculture industry.

14.7.2 Marker-Assisted Selection: "Haida Golden scallop"

The "Haida Golden scallop" (*P. yessoensis*) represents a significant achievement in enhancing carotenoids in adductors by MAS, via identifying the SNP downregulating *BCO-like 1* which is

FIGURE 14.6 Major farming areas and production (2023–2024) of "Penglaihong" varieties. *Map lines delineate study areas and do not necessarily depict accepted national boundaries.*

responsible for muscle coloration (Li et al., 2019). This novel variety features a distinctive orange-red adductor muscle rich in carotenoids, offering superior nutritional value and market appeal. Compared to conventional white-muscled strains, the "Haida Golden scallop" (Breed Registration No. GS-01-002-2009) demonstrates remarkable advantages, including more than 20% higher yield and 25% greater survival rate (Li et al., 2022). Its premium meat quality commands higher market prices, benefiting both farmers and the aquaculture industry (Fig. 14.8). The success of this variety highlights how targeted breeding can enhance both the economic value and production efficiency of scallop aquaculture, providing consumers with a more nutritious and visually appealing product while increasing the profitability of scallop farming operations.

14.7.3 Hybridization: The "Kafeibei" strain

The hybridization between the bay scallop (*A. irradians irradians*) and the Peruvian scallop (*A. purpuratus*) has yielded a remarkable new hybrid strain, "Kafeibei," showing the success of

FIGURE 14.7 Major farming areas and production (2022–2024) of "Haiyifeng 11" variety. *Map lines delineate study areas and do not necessarily depict accepted national boundaries.*

interspecific hybridization in aquaculture. The hybrid offspring exhibited significant heterosis, with growth rates in shell height and whole-body weight surpassing both parent species by 46.7%–141.0%. Notably, the adductor muscle weight increased by 145.4%–156.2% compared with the bay scallop, enhancing the main production trait substantially (Chang et al., 2026; Mao et al., 2020; Wang et al., 2011). The hybrids also demonstrated superior survival rates and adaptability to varying temperatures, making them ideal for large-scale cultivation in China. Their unique shell coloration and rib patterns further distinguished them as a novel, high-performance strain. This achievement highlights the potential of interspecific hybridization to develop resilient and productive aquaculture species.

These case studies illustrate how advanced breeding techniques—MAS, GS, and hybridization—can drive innovation in scallop aquaculture, delivering improved yields, resilience, and economic value.

14.8 Conclusion

Scallop aquaculture has emerged as a vital component of the global aquaculture industry, offering numerous advantages while also presenting several challenges. The various aquaculture models used in scallop farming, including suspended culture, bottom culture, and IMTA, each come with their own set of benefits and hurdles. Suspended culture (cages/longlines)

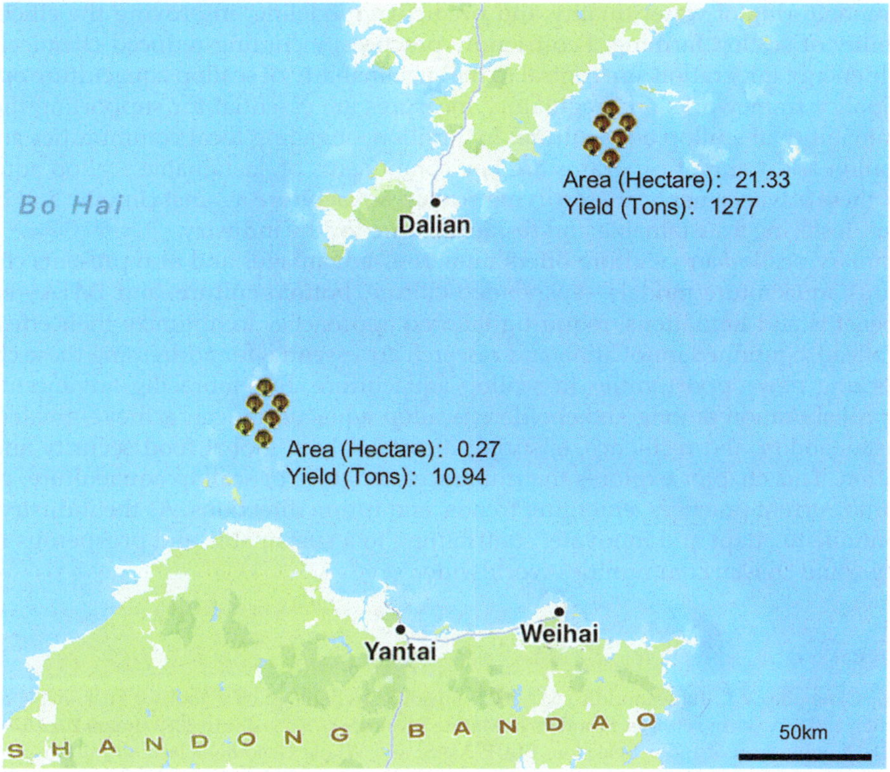

FIGURE 14.8 Culture areas and yield (2019–2022) of "Haida Golden scallop" variety. *Map lines delineate study areas and do not necessarily depict accepted national boundaries.*

maximizes space efficiency and growth through enhanced water flow, but faces storm vulnerability, biofouling management costs, and high initial investment, mitigated by durable materials and automated systems. Bottom culture leverages natural seabed conditions with lower costs, yet struggles with predation, monitoring difficulties, and sedimentation—addressed via predator exclusion devices and optimized siting. IMTA integrates scallops with complementary species (e.g., fish/seaweed) to recycle nutrients, reducing environmental impact while diversifying outputs, and its complexity in species coordination and setup costs is being refined through ongoing research. Understanding these aspects is crucial for optimizing production, enhancing sustainability, and ensuring the long-term viability of scallop aquaculture. This conclusion synthesizes the advantages and challenges of these models, highlights recent advances, and discusses potential directions for future development.

The future of scallop aquaculture is being shaped by a combination of genetic innovation, disease management, and sustainable practices, supported by advanced technologies and policy frameworks. Genomic-driven breeding is enhancing the development of resilient scallop strains with improved growth, disease resistance, and adaptability to changing environments, while next-generation diagnostics and biosecurity measures are mitigating disease risks. Simultaneously, technological advancements such as remote sensing and machine learning are enabling real-time monitoring of water quality and predictive modeling, improving the efficiency and sustainability of scallop farming. Eco-friendly practices, including reduced chemical use and renewable energy integration, are central to the sustainability of scallop aquaculture operations, whereas policy frameworks and regulatory measures are essential for supporting the growth and sustainability of scallop aquaculture. In addition, engaging local communities and fostering education and training programs are key components of sustainable scallop aquaculture. Together, these advancements are fortifying scallop aquaculture against climatic and biological challenges, fostering a sustainable and resilient future for the industry.

In summary, scallop aquaculture offers numerous advantages and also presents challenges. The various aquaculture models—suspended culture, bottom culture, and IMTA—each have unique benefits and limitations, requiring tailored approaches to optimize their effectiveness. Looking ahead, continued innovation and research are essential for addressing these challenges and unlocking new opportunities in scallop aquaculture. By embracing advancements and fostering collaboration among stakeholders, scallop aquaculture can achieve greater productivity, sustainability, and resilience, ensuring its vital role in global food security and marine conservation. This chapter explores the multifaceted nature of scallop aquaculture, providing insights into current practices, emerging trends, and future directions. As the industry evolves, it will continue to adapt and innovate, contributing to a sustainable and prosperous future for aquaculture and coastal communities worldwide.

References

Azra, M.N., Okomoda, V.T., Ikhwanuddin, M., 2022. Breeding technology as a tool for sustainable aquaculture production and ecosystem services. Frontiers in Marine Science 9, 679529. doi:10.3389/fmars.2022.679529.

Bao, Z., 2011. Comparison of Index Selection, BLUP, MAS, and Whole Genome Selection. In: Liu, Z. (Ed.), Next Generation Sequencing and Whole Genome Selection in Aquaculture. Wiley Online Books, pp. 185–218.

Ben-Asher, R., Lahav, O., 2016. Electrooxidation for simultaneous ammonia control and disinfection in seawater recirculating aquaculture systems. Aquacultural Engineering 72-73, 77–87. doi:10.1016/j.aquaeng.2016.05.002.

Bosma, R.H., Verdegem, M.C.J., 2011. Sustainable aquaculture in ponds: Principles, practices and limits. Livestock Science 139 (1-2), 58–68. doi:10.1016/j.livsci.2011.03.017.

Chang, J., Zhang, J., Chu, L., Liu, A., Zhao, X., Xu, Z., … Bao, Z., 2026. Metabolomic and transcriptomic insights into PUFA enrichment in hybrid scallops: Implications for nutritional enhancement and seafood quality. Auqaculture 613 (2), 743465. doi:10.1016/j.aquaculture.2025.743465.

Cheng, P., Zhou, C., Chu, R., Chang, T., Xu, J., Ruan, R., Chen, P., Yan, X., 2020. Effect of microalgae diet and culture system on the rearing of bivalve mollusks: Nutritional properties and potential cost improvements. Algal Research 51, 102076. doi:10.1016/j.algal.2020.102076.

Chi, Y., Jiang, G., Liang, Y., Xu, C., Li, Q., 2022. Selective breeding for summer survival in Pacific oyster (*Crassostrea gigas*): Genetic parameters and response to selection. Aquaculture 556, 738271. doi:10.1016/j.aquaculture.2022.738271.

Christophersen, G., Torkildsen, L., van der Meeren, T., 2006. Effect of increased water recirculation rate on algal supply and post-larval performance of scallop (*Pecten maximus*) reared in a partial open and continuous feeding system. Aquacultural Engineering 35 (3), 271–282. doi:10.1016/j.aquaeng.2006.03.005.

Costa, F., Cerviño-Otero, A., Iglesias, Ó., Cruz, A., Guévélou, E., 2020. Hatchery culture of European clam species (family Veneridae). Aquaculture International 28 (4), 1675–1708. doi:10.1007/s10499-020-00552-x.

Dall, W.H., 1909. Report on a collection of shells from Peru, with a summary of the littoral marine Mollusca of the Peruvian zoological province. Proceedings of the United States National Museum 37 (1704), 147–294. doi:10.5479/si.00963801.37-1704.147.

da Silva, E.G., Castilho-Barros, L., Henriques, M.B., 2022. Economic feasibility of integrated multi-trophic aquaculture (mussel *Perna perna*, scallop *Nodipecten nodosus* and seaweed *Kappaphycus alvarezii*) in Southeast Brazil: A small-scale aquaculture farm model. Aquaculture 552, 738031. doi:10.1016/j.aquaculture.2022.738031.

Dou, J., Li, X., Fu, Q., Jiao, W., Li, Y., Li, T., … Bao, Z., 2016. Evaluation of the 2b-RAD method for genomic selection in scallop breeding. Scientific Reports 6 (1), 19244. doi:10.1038/srep19244.

Dredge, M.C.L., Marsden, I.D., Williams, J.R., 2016. Scallop fisheries, mariculture, and enhancement in Australasia. Developments in Aquaculture and Fisheries Science 40, 1127–1170. doi:10.1016/B978-0-444-62710-0.00030-4.

Duncan, P.F., Brand, A.R., Strand, Ø., Foucher, E., 2016. The European scallop fisheries for *Pecten maximus, Aequipecten opercularis, Chlamys islandica*, and *Mimachlamys varia*. Developments in Aquaculture and Fisheries Science 40, 781–858. doi:10.1016/B978-0-444-62710-0.00019-5.

Essamlali, I., Nhaila, H., Khaili, M.E, 2024. Advances in machine learning and IoT for water quality monitoring: A comprehensive review. Heliyon 10 (6), e27920. doi:10.1016/j.heliyon.2024.e27920.

Food and Agriculture Organization of the United NationsThe state of world fisheries and aquaculture 2024: Blue transformation in action. 2024 FAO 26.

Galley, T.H., Beaumont, A.R., Vay, L.L, King, J.W., 2020. Factors affecting byssus attachment in juvenile scallops, *Pecten maximus* (L.). Aquaculture 528, 735504. doi:10.1016/j.aquaculture.2020.735504.

Gephart, J.A., Henriksson, P.J.G., Parker, R.W.R., Shepon, A., Gorospe, K.D., Bergman, K., Eshel, G., Golden, C.D., Halpern, B.S., Hornborg, S., Jonell, M., Metian, M., Mifflin, K., Newton, R., Tyedmers, P., Zhang, W., Ziegler, F., Troell, M., 2021. Environmental performance of blue foods. Nature 597 (7876), 360–365. doi:10.1038/s41586-021-03889-2.

Gjedrem, T., 2005. Selection and breeding programs in aquaculture. Springer, pp. 1–364. doi:10.1007/1-4020-3342-7.

Gui, J., Bao, Z., Zhang, X., 2016. Development Strategy for Aquaculture Genetic Breeding and Seed Industry. China Engineering 18 (3), 8–14.

Guo, X., 2009. Use and exchange of genetic resources in molluscan aquaculture. Reviews in Aquaculture 1 (3-4), 251–259. doi:10.1111/j.1753-5131.2009.01014.x.

Guo, X., 2021. Genetics in shellfish culture. Molluscan shellfish aquaculture: A practical guide 552.

Guo, X., Ford, S.E., Zhang, F., 1999. Molluscan aquaculture in China. Journal of Shellfish Research 18 (1), 19–31.

Guo, X., Luo, Y., 2016. Scallops and scallop aquaculture in China. Developments in Aquaculture and Fisheries Science 40, 937–952. doi:10.1016/B978-0-444-62710-0.00022-5.

Hossain, M.I., Sadekuzzaman, M., Ha, S.D., 2017. Probiotics as potential alternative biocontrol agents in the agriculture and food industries: A review. Food Research International 100, 63–73. doi:10.1016/j.foodres.2017.07.077.

Huang, X., Duan, X., Ma, S., Sun, X., Geng, Q., Sun, X., … Lv, J., 2025. Optimizing genotype imputation strategies for low coverage whole genome and 2b-RAD sequencing in bivalve molluscs. Auqaculture 608, 742713. doi:10.1016/j.aquaculture.2025.742713.

Karsoon, T., Helu, L., Hongkuan, Z., Shengkang, L., Hongyu, M., Huaiping, Z., 2021. Carotenoids content and lipid nutritional quality variation in tissues of male and female polymorphic (golden and brown) noble scallops *Chlamys nobilis*. Aquaculture 536, 736483. doi:10.1016/j.aquaculture.2021.736483.

Kosaka, Y., 2016. Scallop fisheries and aquaculture in Japan. Developments in Aquaculture and Fisheries Science 40, 891–936. doi:10.1016/B978-0-444-62710-0.00021-3.

Li, X., Li, N., Zhao, L., Shi, J., Wang, S., Ning, X., … Hu, X., 2022. Tissue distribution and seasonal accumulation of carotenoids in Yesso scallop (*Mizuhopecten yessoensis*) with orange adductor muscle. Food Chemistry 367, 130701. doi:10.1016/j.foodchem.2021.130701.

Li, X., Wang, S., Xun, X., Zhang, M., Wang, S., Li, H., … Bao, Z., 2019. A carotenoid oxygenase is responsible for muscle coloration in scallop. BBA- Molecular and Cell Biology of Lipids 1864 (7), 966–975. doi:10.1016/j.bbalip.2019.03.003.

Maeda-Martıńez, A.N., Ormart, P., Mendez, L., Acosta, B., Sicard, M.T., 2000. Scallop growout using a new bottom-culture system. Aquaculture 189 (1-2), 73–84. doi:10.1016/s0044-8486(00)00362-8.

Li, Y., Sun, X., Hu, X., Xun, X., Zhang, J., Guo, X., … Bao, Z., 2017. Scallop genome reveals molecular adaptations to semi-sessile life and neurotoxins. Nature Communications 8 (1), 1721. doi:10.1038/s41467-017-01927-0.

Lv, J., Jiao, W., Guo, H., Liu, P., Wang, R., Zhang, L., … Wang, S., 2018. HD-Marker: a highly multiplexed and flexible approach for targeted genotyping of more than 10,000 genes in a single-tube assay. Genome Research 28 (12), 1919–1930. doi:10.1101/gr.235820.118.

Maeda-Martínez, A.N., Reynoso-Granados, T., Monsalvo-Spencer, P., Sicard, M.T., Mazón-Suástegui, J.M., Hernández, O., … Morales, R., 1997. Suspension culture of catarina scallop *Argopecten ventricosus* (= circularis), (Sowerby II, 1842), in Bahia Magdalena, Mexico, at different densities. Aquaculture 158 (3–4), 235–246. doi:10.1016/S0044-8486(97)00197-X.

Magnesen, T., Christophersen, G., 2007. Large-scale raceway nursery for improved scallop (*Pecten maximus*) spat production. Aquacultural Engineering 36 (2), 149–158. doi:10.1016/j.aquaeng.2006.10.002.

Magnesen, T., Christophersen, G., 2008. Reproductive cycle and conditioning of translocated scallops (*Pecten maximus*) from five broodstock populations in Norway. Aquaculture 285 (1–4), 109–116. doi:10.1016/j.aquaculture.2008.08.024.

Magnesen, T., Jacobsen, A., 2012. Effect of water recirculation on seawater quality and production of scallop (*Pecten maximus*) larvae. Aquacultural Engineering 47, 1–6. doi:10.1016/j.aquaeng.2011.12.005.

Mao, J., Zeng, Q., Yang, Z., Pan, H., Yao, L., Bao, Z., ... Wang, S., 2020. High-resolution linkage and quantitative trait locus mapping using an interspecific cross between *Argopecten irradians irradians* (♀) and *A. purpuratus* (♂). Marine Life Science & Technology 2, 123–134. doi:10.1007/s42995-020-00029-z.

Martínez, G., Caceres, L.A., Uribe, E., Diaz, M.A., 1995. Effects of different feeding regimens on larval growth and the energy budget of juvenile Chilean scallops, *Argopecten purpuratus* Lamarck. Aquaculture 132 (3–4), 313–323. doi:10.1016/0044-8486(94)00359-v.

Mendo, J., Wolff, M., Mendo, T., Ysla, L., 2016. Scallop fishery and culture in Peru. In: Developments in aquaculture and fisheries science, 40. Elsevier, pp. 1089–1109. doi:10.1016/B978-0-444-62710-0.00028-6.

Mendoza, Y., Freites, L., Lodeiros, C.J., López, J.A., Himmelman, J.H., 2003. Evaluation of biological and economical aspects of the culture of the scallop *Lyropecten (Nodipecten) nodosus* in suspended and bottom culture. Aquaculture 221 (1–4), 207–219. doi:10.1016/s0044-8486(02)00646-4.

Merino, G., Uribe, E., Soria, G., von Brand, E., 2009. A comparison of larval production of the northern scallop, *Argopecten purpuratus*, in closed and recirculating culture systems. Aquacultural Engineering 40 (2), 95–103. doi:10.1016/j.aquaeng.2008.11.002.

Morton, B., 1980. Swimming in *Amusium pleuronectes* (Bivahia: Pectinidae). Journal of Zoology 190 (3), 375–404. doi:10.1111/j.1469-7998.1980.tb01434.x.

Nascimento-Schulze, J.C., Bean, T.P., Houston, R.D., Santos, E.M., Sanders, M.B., Lewis, C., Ellis, R.P., 2021. Optimizing hatchery practices for genetic improvement of marine bivalves. Reviews in Aquaculture 13 (4), 2289–2304. doi:10.1111/raq.12568.

Navarro, J.M., Gonzalez, C.M., 1998. Physiological responses of the Chilean scallop *Argopecten purpuratus* to decreasing salinities. Aquaculture 167 (3–4), 315–327. doi:10.1016/s0044-8486(98)00310-x.

O'Connor, W.A., Heasman, M.P., 1995. Spawning induction and fertilisation in the doughboy scallop *Chlamys (Mimachlamys) asperrima*. Aquaculture 136 (1–2), 117–129. doi:10.1016/0044-8486(95)01040-8.

O'Connor, W.A., Heasman, M.P., O'Connor, S.J., 2000. Algal diets for broodstock maintenance of the doughboy scallop *Mimachlamys asperrima* (Lamarck). Aquaculture Research 31 (8–9), 627–635. doi:10.1046/j.1365-2109.2000.00484.x.

Parsons, G.J., Lauzier, R.B., Bourne, N.F., 2016. Scallops of the West Coast of North America. Developments in Aquaculture and Fisheries Science 40, 697–717. doi:10.1016/B978-0-444-62710-0.00015-8.

Reid, G.K., Liutkus, M., Bennett, A., Robinson, S.M.C., MacDonald, B., Page, F., 2010. Absorption efficiency of blue mussels (*Mytilus edulis* and *M. trossulus*) feeding on Atlantic salmon (Salmo salar) feed and fecal particulates: Implications for integrated multi-trophic aquaculture. Aquaculture 299 (1–4), 165–169. doi:10.1016/j.aquaculture.2009.12.002.

Robinson, S.M.C., Parsons, G.J., Davidson, L.A., Shumway, S.E., Blake, N.J., 2016. Scallop aquaculture and fisheries in Eastern North America. Developments in Aquaculture and Fisheries Science 40, 737–779. doi:10.1016/B978-0-444-62710-0.00017-1.

Sanderson, M.P., Hudson, K.L., Gregg, L.S., Chesler-Poole, A.B., Small, J.M., Reece, K.S., Carnegie, R.B., Smith, J.L., 2023. Detection of toxins and harmful algal bloom cells in shellfish hatcheries and efforts toward removal. Aquaculture 562, 738714. doi:10.1016/j.aquaculture.2022.738714.

Soffa, F.B., Pratama, I.S., Dharmawati, V., Rahayu, D.L., Gultom, V.D.N., Supii, A.I., ... Handayani, K.S., 2024. Asian moon scallop (*Amusium pleuronectes*) for Indonesia: An overview from a wild population and farming system. Fisheries and Aquatic Sciences 27 (11), 709–727. doi:10.47853/FAS.2024.e67.

Stabili, L., Giangrande, A., Arduini, D., Borghese, J., Petrocelli, A., Alabiso, G., Ricci, P., Cavallo, R.A, Acquaviva, M.I., Narracci, M., Pierri, C., Trani, R., Longo, C., 2023. Environmental quality improvement of a mariculture plant after its conversion into a multi-trophic system. Science of the Total Environment 884, 163846. doi:10.1016/j.scitotenv.2023.163846.

Stentiford, G.D., Bateman, I.J., Hinchliffe, S.J., Bass, D., Hartnell, R., Santos, E.M., Devlin, M.J., Feist, S.W., Taylor, N.G.H., Verner-Jeffreys, D.W., van Aerle, R., Peeler, E.J., Higman, W.A., Smith, L., Baines, R., Behringer, D.C.,

Katsiadaki, I., Froehlich, H.E., Tyler, C.R., 2020. Sustainable aquaculture through the One Health lens. Nature Food 1 (8), 468–474. doi:10.1038/s43016-020-0127-5.

Strand, Ø., Louro, A., Duncan, P.F., 2016. European Aquaculture. Developments in Aquaculture and Fisheries Science 40, 859–890. doi:10.1016/B978-0-444-62710-0.00020-1.

Suquet, M., Arsenault-Pernet, M.E.J., Ratiskol, D., Mingant, C., 2014. Cryopreservation of great scallop (*Pecten maximus*) sperm: Effect of extender, cryoprotectant and cooling rate on sperm survival. Aquatic Living Resources 27 (1), 35–40. doi:10.1051/alr/2014002.

Tan, K., Zhang, H., Lim, L., Zheng, H., 2020a. Selection breeding program of Nan'ao golden scallop Chlamys nobilis with higher nutritional values and less susceptible to stress. Aquaculture 517, 734769. doi:10.1016/j.aquaculture.2019.734769.

Tan, K., Zhang, H., Zheng, H., 2020b. Selective breeding of edible bivalves and its implication of global climate change. Reviews in Aquaculture 12 (4), 2559–2572. doi:10.1111/raq.12458.

Tan, K., Zhang, H., Zheng, H., 2024. Carotenoid content and composition: A special focus on commercially important fish and shellfish. Critical Reviews in Food Science and Nutrition 64 (2), 544–561. doi:10.1080/10408398.2022.2106937.

Tang, Y., Ju, C., Mei, R., Zhao, L., Liu, J., Yang, Y., … Liu, Q., 2024. Exploring the optimal integrated multi-trophic aquaculture (IMTA) patterns benefiting culture animals and natural water environment. Aquaculture 589, 741011. doi:10.1016/j.aquaculture.2024.741011.

Theissinger, K., Fernandes, C., Formenti, G., Bista, I., Berg, P.R., Bleidorn, C., Bombarely, A., Crottini, A., Gallo, G.R., Godoy, J.A., Jentoft, S., Malukiewicz, J., Mouton, A., Oomen, R.A., Paez, S., Palsbøll, P.J., Pampoulie, C., Ruiz-López, M.J., Secomandi, S., … Zammit, G., 2023. How genomics can help biodiversity conservation. Trends in Genetics 39 (7), 545–559. doi:10.1016/j.tig.2023.01.005.

Utting, S.D., Millican, P.F., 1997. Techniques for the hatchery conditioning of bivalve broodstocks and the subsequent effect on egg quality and larval viability. Aquaculture 155 (1-4), 45–54. doi:10.1016/s0044-8486(97)00108-7.

Vélez, A., Alifa, E., Azuaje, O., 1990. Induction of spawning by temperature and serotonin in the hermaphroditic tropical scallop, *Pecten ziczac*. Aquaculture 84 (3–4), 307–313. doi:10.1016/0044-8486(90)90095-5.

von Brand, E., Abarca, A., Merino, G.E., Stotz, W., 2016. Scallop fishery and aquaculture in Chile: A history of developments and declines. Developments in Aquaculture and Fisheries Science 40, 1047–1072. doi:10.1016/B978-0-444-62710-0.00026-2.

Wang, C., Liu, B., Li, J., Liu, S., Li, J., Hu, L., Fan, X., Du, H., Fang, H., 2011. Introduction of the Peruvian scallop and its hybridization with the bay scallop in China. Aquaculture 310 (3-4), 380–387. doi:10.1016/j.aquaculture.2010.11.014.

Wang, D., Gao, H., 2024. China fisheries statistical yearbook. China Agriculture Press, China, Beijing.

Wang, F.u-K., Zhu, P.-C., Zhang, X.-T., Yu, K., Wang, C.-D.e, Liu, B., 2025. QTL mapping and candidate gene identification for lower temperature tolerance and growth traits using whole genome re-sequencing in Argopecten scallops. Aquaculture 595, 741513. doi:10.1016/j.aquaculture.2024.741513.

Wang, Q., Li, J., Liang, F., Xie, S., Du, X., Deng, Y., 2017. Effects of different substrates on settlement and growth of pearl oyster (*Pinctada maxima*) larvae in hatcheries. Aquacultural Engineering 77, 15–19. doi:10.1016/j.aquaeng.2017.02.001.

Wang, X., Xu, Y., Hu, Z., Xu, C., 2018. Genomic selection methods for crop improvement: Current status and prospects. The Crop Journal 6 (4), 330–340. doi:10.1016/j.cj.2018.03.001.

Wang, Y., Wang, Y., Bao, L., Sun, C., Huang, S., Li, X., Hu, X., Liu, Y., 2024. Cryopreservation of Yesso scallop (*Patinopecten yessoensis*) sperm. Aquaculture Reports 36, 102104. doi:10.1016/j.aqrep.2024.102104.

Wang, Z., Casas, S., Peyre, J.L., Rikard, S., Williams, M.L., Tarnecki, A., Bushek, D., Guo, X., 2025. Genomic selection for dermo resistance in the Eastern Oyster Crassostrea virginica: Production and laboratory testing of F1 generation. Journal of Shellfish Research 44 (1), 75–87. doi:10.2983/035.044.0108.

Willer, D.F., Aldridge, D.C., 2020. Sustainable bivalve farming can deliver food security in the tropics. Nature Food 1, 384–388. doi:10.1038/s43016-020-0116-8.

Willer, D.F., Nicholls, R.J., Aldridge, D.C., 2021. Opportunities and challenges for upscaled global bivalve seafood production. Nature Food 2 (12), 935–943. doi:10.1038/s43016-021-00423-5.

Xiao, J., Ford, S.E., Yang, H., Zhang, G., Zhang, F., Guo, X., 2005. Studies on mass summer mortality of cultured zhikong scallops (*Chlamys farreri* Jones et Preston) in China. Aquaculture 250 (3–4), 602–615. doi:10.1016/j.aquaculture.2005.05.002.

Xing, Q., Li, Y., Guo, H., Yu, Q., Huang, X., Wang, S., … Bao, Z., 2016. Cardiac performance: A thermal tolerance indicator in scallops. Marine Biology 163 (12), 244. doi:10.1007/s00227-016-3021-9.

Xu, Z., Yang, Z., Bao, L., Lu, B., Li, X., Zhan, X., … Liu, Y., 2024. Coenzyme Q10 improves the post-thaw sperm quality in dwarf Surfclam *Mulinia lateralis*. Antioxidants 13 (9), 1085. doi:10.3390/antiox13091085.

Yu, R., 2009. Experimental and practical techniques for shellfish farming and breeding. Ocean University of China Press, China, Qingdao.

Yu, R., Li, Q., 2016. Seed production technology for marine molluscs in China. Ocean University of China Press, China, Qingdao.

Zeng, Q., Zhao, B., Wang, H., Wang, M., Teng, M., Hu, J., Bao, Z., Wang, Y., 2022. Aquaculture molecular breeding platform (AMBP): A comprehensive web server for genotype imputation and genetic analysis in aquaculture. Nucleic Acids Research 50 (W1), W66. doi:10.1093/nar/gkac424.

Zhang, C., He, C., Sun, Y., Zhang, H., Ye, T., Liu, F., … Zheng, H., 2024. Selective breeding in the noble scallop (*Chlamys nobilis*) for low-temperature resistance to reduce overwintering losses. Aquaculture 586, 740737. doi:10.1016/j.aquaculture.2024.740737.

Zheng, X., Lin, S., Yang, R., Qin, J., Wang, A., Gu, Z., Ma, Z., 2019. Sperm cryopreservation of the noble Scallop Chlamys nobilis by a programmable freezing method: Effect of cryoprotectant. Aquaculture Research 50 (6), 1678–1686. doi:10.1111/are.14050.

Zheng, Y.-D., Huang, B.-W., Zhang, X., Liu, C.-F., Xin, L.-S., Wang, C.-M., Bai, C.-M., 2023. The probiotic Bacillus hwajinpoensis colonizes the digestive system of *Crassostrea gigas* larvae and protects them from Vibrio alginolyticus infection. Microorganisms 11 (12), 2918. doi:10.3390/microorganisms11122918.

CHAPTER

15

Clam aquaculture

Zhihua Lin[a,c], *Xiwu Yan*[b], *Yinghui Dong*[a], *Zhongming Huo*[b], *Yongbo Bao*[d], *Lin He*[a] and *Xiafei Zheng*[c]

[a]College of Advanced Agricultural Science, Zhejiang Wanli University, Ningbo, China [b]College of Fisheries and Life Science, Dalian Ocean University, Dalian, China [c]Ninghai Institute of Mariculture Breeding and Seed Industry, Zhejiang Wanli University, Ningbo, China [d]College of Biological & Environmental Sciences, Zhejiang Wanli University, Ningbo, China

15.1 Introduction

Clam aquaculture constitutes a foundational pillar of the global shellfish industry, offering significant economic, nutritional, and social contributions to the "Blue Economy". The transition from artisanal wild harvesting to industrial-scale precision farming has been driven by continuous technological innovation and a focus on sustainability. Global production trends highlight the dominance of aquaculture in meeting demand, while the recognition of clams as nutrient-dense "Blue Food" underscores their role in food security. Current advancements emphasize the integration of ecological principles and sophisticated management strategies to ensure long-term viability.

15.1.1 Global importance of clam aquaculture: Economic, nutritional, and social dimensions

The cultivation of marine bivalves, particularly clams, constitutes a foundational pillar of the contemporary global molluscan shellfish industry. As a critical component of the "Blue Economy", clam aquaculture represents a sophisticated paradigm of food production that harmonizes economic viability with ecological stewardship. Tidal flats, the primary interface for these activities, are among the most biologically productive ecosystems on Earth. While Murray et al. (2019) established the baseline global extent of tidal flats at approximately

Genetic Breeding and Aquaculture of Shellfish: A Collaborative Approach to a Sustainable Future
DOI: https://doi.org/10.1016/B978-0-443-33606-5.00002-3

TABLE 15.1 Global production volume and main producing regions of major farmed clam species (FAO, 2024).

Common name	Scientific name	Main producing countries/regions	Approx. annual production (Tons)
Manila Clam	*Ruditapes philippinarum*	China, Italy, USA, France, South Korea	4,400,000+
Razor Clam	*Sinonovacula constricta*	China, Southeast Asia	900,000+
Blood Clam	*Tegillarca granosa*	China, Thailand, Malaysia, Indonesia	500,000+
Asian Hard Clam	*Meretrix meretrix*	China, Vietnam, Taiwan (China)	400,000+
Hard-shell Clam	*Mercenaria mercenaria*	USA, China (imported), Canada	40,000–60,000
Chinese Cyclina	*Cyclina sinensis*	China, Japan, Korean Peninsula	25,000+

127,921 km^2, recent remote sensing analyses continue to highlight the critical role of these habitats in Asia, which hosts over 44% of the global total. This geographic concentration provides countries such as Indonesia, China, and Australia with a distinct natural advantage, underpinning the Asia-Pacific region's dominance in shellfish production (Murray et al., 2019; Murray et al., 2022).

In the context of global food security, the transition from artisanal foraging to industrial-scale farming has firmly established clams as a premier aquatic protein source. According to the latest *State of World Fisheries and Aquaculture* report (FAO, 2024), global production of clams, cockles, and ark shells has continued its robust upward trajectory, exceeding 6.5 million metric tons in 2022 (Table 15.1). This sector is overwhelmingly dominated by aquaculture, which now accounts for more than 90% of total production for major species. China remains the global powerhouse, contributing over 80% of the volume, yet the industry's footprint is expanding in established and emerging centers across Europe, North America, and Southeast Asia (FAO, 2024).

In Europe, the industry faces a complex dichotomy of success and emerging ecological challenges. Italy remains the preeminent European producer of the Manila clam (*Ruditapes philippinarum*), a species introduced in the 1980s that successfully revitalized the Po River Delta and Venice Lagoon economies. However, recent years have tested the sector's resilience. While the industry historically generated hundreds of millions of euros annually, the proliferation of the invasive blue crab (*Callinectes sapidus*) since 2023 has caused significant predation-related losses, prompting a shift in management strategies (Gavioli et al., 2025).

Conversely, in North America, the hard clam (*Mercenaria mercenaria*) industry demonstrates sustained economic vitality. In the United States, this sector serves as a critical economic engine for coastal communities. In Virginia, the industry has matured into a model of hatchery-based efficiency. According to the Virginia Institute of Marine Science (VIMS), farm-gate values for hard clams reached historical highs in 2024, solidifying their status as the state's most valuable marine food sector (Hudson & White, 2025). Similarly, Florida's shellfish aquaculture sector continues to act as a buffer for coastal livelihoods, maintaining a direct economic impact exceeding $15 million annually despite environmental stressors (USDA, 2023).

Beyond economic metrics, farmed clams are increasingly prioritized in scientific discourse as a pivotal "Blue Food" —a sustainable source of bioavailable nutrition. The landmark *Blue Food*

Assessment highlights bivalves as having one of the lowest environmental footprints across all animal protein sectors (Gephart et al., 2021). Nutritionally, they are nutrient-dense, providing essential omega-3 fatty acids, iron, zinc, and vitamin B12 at levels often surpassing terrestrial livestock (Azra et al., 2021; Golden et al., 2021). Furthermore, recent life cycle assessments (LCAs) reaffirm that bivalve aquaculture is a "fed by nature" system requiring no arable land, freshwater, or exogenous feed (Jones et al., 2022).

15.1.2 Brief history and current trends: The evolution from wild harvest to precision farming

The anthropogenic utilization of clams is deeply rooted in human history, with shell middens dating back thousands of years testifying to the importance of these mollusks in the diet of coastal populations globally. However, the trajectory from opportunistic harvesting to controlled, intensive farming has been a gradual evolution marked by distinct technological leaps. Historically, clam "farming" was limited to the passive management of natural grounds. In ancient China and Rome, early aquaculturists practiced "relaying", a method where seed clams collected from areas of high density were transplanted to private leases characterized by better growth conditions and lower predation pressure. This extensive form of ranching persisted for centuries until the mid-20th century, when the collapse of wild stocks and the degradation of coastal habitats necessitated a more interventionist approach (Gosling, 2003).

The modern era of clam aquaculture began in earnest in the 1960s and 1970s with the scientific closure of the life cycle for key species. In China, a breakthrough was achieved with the artificial breeding of the razor clam (*Sinonovacula constricta*) and the blood clam (*Tegillarca granosa*). These advancements liberated the industry from its reliance on unpredictable wild spatfall, allowing for the intensification of production and the expansion of farming into areas previously devoid of natural recruitment (Lin et al., 2023). A pivotal moment in the globalization of the industry was the anthropogenic dissemination of the Manila clam (*Ruditapes philippinarum*). Native to the Indo-Pacific, this species was accidentally introduced to the Pacific coast of North America in the 1930s alongside oyster shipments, where it rapidly naturalized. Recognizing its superior growth rate and hardiness compared to native species, European scientists intentionally introduced it to France and Italy in the 1970s. The Manila clam proved so successful that it became the dominant species in the European market within two decades (Gosling, 2003).

Current trends in the sector reflect a paradigm shift toward "precision aquaculture" and ecological integration. The industry is aggressively moving away from extensive, low-input methods toward intensive systems that prioritize control, efficiency, and biosecurity. This is exemplified by the widespread adoption of the "three-phase culture method" (hatchery, nursery, grow-out) in northern China. By incorporating a distinct nursery phase, farmers can ensure that only robust, predator-resistant seed—typically grown to a specific "refuge size"—is planted in open mudflats. This practice has significantly reduced the mortality rates that plagued traditional bottom sowing, where microscopic spat were often decimated by predation and environmental stress (Zhang & Yan, 2006). Furthermore, as coastal space becomes increasingly contested by tourism, shipping, and conservation interests, there is a growing trend toward maximizing yield per unit area through genetic improvement and vertical integration. The sector is also grappling with the escalating impacts of climate change, specifically ocean acidification and rising sea temperatures. This has prompted a shift toward "climate-smart" strategies, such as breeding for thermal tolerance (Kong et al., 2025; Ye et al., 2025; Zhang et al., 2024)and relocating operations to deeper waters or land-based systems to escape surface heatwaves (Wang et al., 2025). In addition, consumer demand for food safety has driven the implementation of rigorous traceability systems and the expansion of depuration (purification) facilities to ensure that clams meet international hygiene standards, particularly regarding bacterial loads and phycotoxins (Richards, 1988).

15.1.3 Technological innovation and sustainability

Innovation serves as the primary engine driving the continued expansion and modernization of clam aquaculture. Recent decades have witnessed the integration of advanced biotechnologies, engineering solutions, and ecological principles into traditional farming practices. In the realm of breeding, the application of molecular genetics has been transformative. The utilization of microsatellites and single-nucleotide polymorphisms (SNPs) for parentage assignment and population analysis has enabled the development of sophisticated selective breeding programs. These programs have produced new varieties with commercially desirable traits, such as the "Wanlihong" hard clam, selected for its distinctive purplish-red shell color and rapid growth, and the "Shenzhe No. 1" razor clam, selected for large size and thermal resilience. Genomic selection (GS) is now emerging as a cutting-edge tool, allowing breeders to select for complex polygenic traits such as disease resistance and environmental stress tolerance without the need for lethal pathogen challenges, thereby accelerating genetic gain (Lv et al., 2023; Wang et al., 2025).

Technological innovations are also addressing the physical challenges of farming in dynamic coastal environments. The development of land-based flow-through systems, often referred to as Floating Upweller Systems (FLUPSYs) in North America and horizontal flow-polyculture pond aquaculture model in China, has revolutionized the nursery phase. These systems pump nutrient-rich water through silos containing high densities of seed, providing a predator-free environment that accelerates growth by orders of magnitude compared to wild conditions. Mechanization is another key trend, with the adoption of hydraulic dredges and automated sorting machines reducing the heavy labor burden associated with harvesting and processing, although the environmental impact of such dredging remains a subject of ongoing optimization (NMAMR, 2023).

Environmentally, the industry is increasingly embracing the concept of integrated multitrophic aquaculture (IMTA). In this holistic model, clams are cocultured with fed species (such as finfish or shrimp) and inorganic extractors (such as macroalgae). The clams serve as biological filters, consuming organic particulate waste derived from uneaten feed and fish feces, thereby improving water quality and recycling nutrients into harvestable biomass. This ecological engineering approach optimizes resource utilization, reduces nutrient discharge into the environment, and diversifies economic returns, positioning clam aquaculture as a model for sustainable marine food production (Rusco et al., 2024; Tang, 2023; Tang et al., 2024).

FIGURE 15.1 Major cultivated clam species worldwide.

15.2 Clam species in aquaculture

A diverse portfolio of clam species is cultivated worldwide, each selected for specific economic values and biological adaptations. Commercial production is dominated by species such as the Manila clam, razor clam, and blood clam, which exhibit distinct traits such as burrowing depth, feeding physiology, and environmental tolerances to salinity, temperature, and hypoxia. Understanding these biological nuances—ranging from the deep-burrowing habits of *Sinonovacula constricta* to the hypoxic resistance of *Tegillarca granosa*—is essential for matching species to appropriate culture sites and optimizing farming efficiency.

15.2.1 Major cultivated clam species worldwide: Distribution and economic significance

The global clam aquaculture portfolio is biologically diverse, comprising dozens of species adapted to specific regional environments and market preferences (Fig. 15.1). However, commercial production is heavily skewed toward a few dominant taxa that exhibit favorable aquaculture traits such as rapid growth rates, physiological hardiness, and high market demand. Understanding the specific biology and distribution of these species is essential for comprehending the global industry structure.

Manila clam (*Ruditapes philippinarum*): Undoubtedly the most commercially significant species globally, the Manila clam is a testament to the success of aquaculture introductions. Native to the subtropical and temperate waters of the Indo-Pacific (China, Japan, Philippines), it has established a cosmopolitan distribution through anthropogenic spread. It accounts for the vast majority of clam production in China, where it is farmed across a wide latitudinal range, from the freezing waters of Liaoning in the north to the subtropical coast of Fujian in the south. In Europe, it serves as the backbone of the Italian shellfish industry and is widely cultured in France and Spain. On the Pacific coast of North America, from British Columbia to California, it supports a lucrative intertidal farming sector. Its commercial success is attributed to its rapid growth—reaching market size in 12–18 months—and its ability to tolerate high stocking densities and a wide range of sediment substrates (Willer & Aldridge, 2020; Zhang & Yan, 2006).

Razor clam (*Sinonovacula constricta*): A distinct species characterized by a fragile, elongated shell, the razor clam is a staple of aquaculture in China and Southeast Asia. Unlike the sturdy Manila clam, the razor clam is a fast-burrowing species evolutionarily adapted to soft, muddy substrates. It is uniquely capable of thriving in estuarine conditions with salinities as low as 5 ppt, a physiological trait that restricts the range of many marine pathogens and predators. This adaptation allows it to be farmed in river mouths and reclaimed agricultural lands where other marine bivalves would perish. Recent genetic studies have revealed high levels of genetic diversity in wild populations, indicating strong potential for selective breeding and hybridization to further enhance their aquaculture traits (Dong et al., 2011; Niu et al., 2010).

Blood clam (*Tegillarca granosa*): Also known as the granular ark, this species holds profound cultural significance in East and Southeast Asia, particularly in China, Thailand, and Malaysia. It derives its name from the presence of hemoglobin in its hemolymph, a rare physiological adaptation in bivalves that allows it to maintain aerobic respiration in hypoxic conditions. This trait enables the blood clam to survive in the reducing, organic-rich anoxic sediments of eutrophic mudflats where other species would suffocate. It is considered a delicacy and a "tonic food" in traditional Chinese medicine, commanding a high market price. Its robust shell and tolerance to desiccation make it well-suited for live transport to distant markets (Bao et al., 2021; Dong, Yao, Lin, Xiao, & Chai, 2012).

Hard clam (*Meretrix meretrix*): Commercially valuable in China and Southeast Asia, the Asiatic hard clam is prized for its thick, porcelain-like shell, attractive patterns, and savory meat. It inhabits sandy substrates in the lower intertidal and subtidal zones. Because of historical overfishing of wild stocks, aquaculture production has surged, supported by advanced hatchery technologies. Genetic research has identified significant divergence between northern and southern populations in China, suggesting local adaptation that must be considered in breeding programs to avoid maladaptation (Gu et al., 2015; Jing et al., 2015; Lin et al., 2008; Qi et al., 2013).

Hard-shell clam (*Mercenaria mercenaria*): This species forms the backbone of the clam industry on the Atlantic coast of North America. It is a long-lived, thick-shelled species sold at various market sizes, from "littlenecks" (smallest and most valuable) to "cherrystones" and "chowders" (largest). It is particularly noted for its physiological robustness during transport and long shelf life, allowing it to be shipped live across the continent. The industry is highly developed in states such as Florida, Virginia, and New Jersey, relying almost entirely on hatchery-produced seed, particularly the "*Notata*" strain, which possesses distinctive shell markings (Hadley et al., 1997; USDA, 2023).

Chinese Cyclina (*Cyclina sinensis*): A brackish water species farmed extensively in the high intertidal zones of China. It is often cultured in reclaimed ponds or even rice-paddy conversions due to its extreme salinity tolerance and ability to survive in environments with fluctuating water levels and temperatures (Fang et al., 2020; Sun et al., 2012).

15.2.2 Biological traits and ecological preferences: Adaptations for a benthic life

The successful cultivation of clams requires a profound understanding of their biological traits, particularly their burrowing behavior, feeding physiology, and environmental tolerances. Unlike scallops or oysters that live on the surface or attached to structures, most farmed clams are infaunal burrowers, living buried within the sediment. This fundamental difference dictates every aspect of their farming, from site selection to harvesting methods.

Burrowing mechanics and substrate preference: The morphology of the foot and siphon dictates the substrate preference and burrowing depth of each species. *Sinonovacula constricta* possesses a powerful, muscular foot and long, fused siphons. These adaptations allow it to burrow deeply (up to 50–70 cm) into soft, unstable muds. The long siphons enable it to access the water column for respiration and feeding while remaining safely sequestered deep in the sediment, offering protection from surface predators and desiccation. This deep-burrowing habit necessitates farming in soft mud ponds and makes manual harvest labor-intensive. In contrast, *Ruditapes philippinarum* and *Mercenaria mercenaria* have shorter, separate or partially fused siphons and inhabit shallower depths (typically the top 5–15 cm) in sandy or mixed sand-mud substrates. Their relatively shallow burrowing depth makes them more accessible for harvesting but also more vulnerable to surface predation and temperature fluctuations. Substrate composition is critical; *Meretrix meretrix* prefers fine sand to support its heavy shell, while *Tegillarca granosa* requires soft mud with high organic content to filter benthic detritus (Gosling, 2003).

Feeding physiology: Clams are active filter feeders, a trait that makes them efficient biological water purifiers. They draw water through the inhalant siphon, pass it over the ctenidia (gills) where particles are trapped by mucous and cilia, and expel waste water through the exhalant siphon. They are selective feeders, capable of sorting particles based on size, shape, and nutritional value on the labial palps. Nutritious particles are ingested, while inorganic silt or unpalatable algae are rejected as pseudofeces. This sorting ability is crucial for survival in turbid estuarine waters where suspended sediment loads can be high. For example, *S. constricta* thrives in waters with high suspended particulate matter that would choke other bivalves, utilizing its efficient gill sorting mechanism to extract organic matter from the silt. The filtration rate is influenced by temperature and particle concentration, with optimal filtration typically occurring at temperatures between 20°C and 25°C for most warm-temperate species (Bricelj & Malouf, 1984).

Environmental tolerance and physiology: Commercially important clam species are generally euryhaline and eurythermal, adaptations that allow them to inhabit fluctuating estuarine environments. *R. philippinarum* exhibits remarkable thermal plasticity, capable of surviving temperatures ranging from near freezing to more than 30°C, although optimal growth occurs between 18°C and 25°C. This broad thermal tolerance has facilitated its global spread. *S. constricta* is an estuarine specialist, tolerating wide salinity fluctuations (5–30 ppt) and rapid changes in osmotic pressure, making it ideal for river mouth aquaculture. *T. granosa* is notable for its resistance to hypoxia. Its red blood contains hemoglobin with a high oxygen affinity, allowing it to maintain aerobic respiration at lower oxygen tensions than other bivalves and to survive in the reducing, anoxic sediments of eutrophic mudflats (Bao et al., 2021). Understanding these specific tolerances allows farmers to select the appropriate species for specific sites, such as using

C. sinensis for high intertidal ponds that experience extreme salinity shifts because of evaporation or rainfall (Fang et al., 2020).

15.3 Clam hatchery management

Hatchery management represents a technologically demanding process designed to transform microscopic larvae into benthic juveniles through precise environmental control. Modern facilities rely on sophisticated water treatment to ensure biosecurity, specialized algal culture for optimal larval nutrition, and rigorous broodstock conditioning to synchronize spawning. Critical operational protocols include spawning induction, larval rearing, and the provision of specific settlement cues. Strategic solutions, such as the use of probiotics and genetic management, address persistent challenges like bacterial infections and larval mass mortality events.

15.3.1 Hatchery setup and operational framework

The hatchery phase represents the most technically demanding aspect of clam aquaculture, requiring precise control over environmental variables to transform microscopic larvae into benthic juveniles. A modern commercial clam hatchery is a sophisticated facility designed to optimize survival and growth. Key infrastructure comprises distinct functional zones: Water treatment, algal culture, broodstock conditioning, spawning, and larval rearing.

The water treatment system is the heart of the facility. Incoming seawater must be pristine to prevent the introduction of pathogens and competitors. The treatment train typically involves sedimentation tanks to remove heavy silt, followed by rapid sand filters and a series of cartridge filters (filtering down to 1 μm or even 0.2 μm). Final sterilization is achieved using ultraviolet (UV) light irradiation or ozone treatment to eliminate bacteria and protozoa. This rigorous treatment is critical for preventing infections by *Vibrio* spp., a ubiquitous genus of bacteria that includes highly pathogenic strains capable of causing "bacillary necrosis," a disease that can cause rapid mass mortality in larvae within hours (Sanderson et al., 2023).

The algal culture unit accounts for a significant portion of the hatchery's operational effort and cost. Clam larvae require live microalgae of high nutritional value, specifically those rich in polyunsaturated fatty acids (PUFAs) such as docosahexaenoic acid (DHA) and eicosapentaenoic acid (EPA). Hatcheries maintain stock cultures of flagellates such as *Isochrysis galbana* (clone T-Iso) and *Pavlova lutheri* for early-stage larvae (D-stage), as their small cell size (3–5 μm) and digestibility are ideal. As larvae grow into the umbo and pediveliger stages, larger diatoms, such as *Chaetoceros muelleri*, *Chaetoceros calcitrans*, and *Thalassiosira weissflogii*, are introduced. The operational framework follows a strict cycle, with algal production scaled and synchronized to meet the escalating metabolic demands of developing larval batches (Helm et al., 2004; Zhang & Yan, 2006).

15.3.2 Broodstock management and reproduction

Reliable seed production depends on the quality of the broodstock. Broodstock management involves the selection, conditioning, and spawning of adult clams. Adults are typically collected

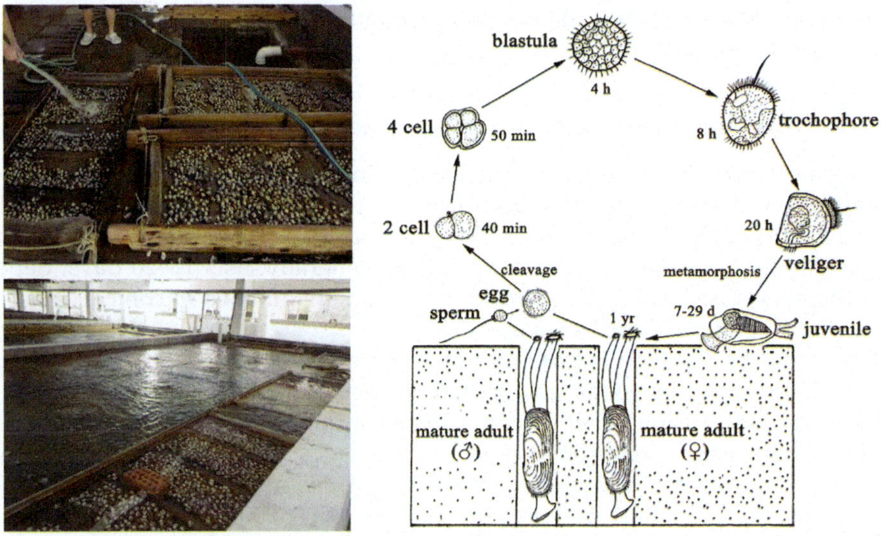

FIGURE 15.2 **The industrialized hatchery process of tidal flat clams (Dong et al., 2020).** Left: Indoor broodstock conditioning tanks; Right: Microscopic development stages from larvae to adult clam.

from wild populations with desirable traits or selected from the fastest-growing cohorts of farm stocks. Upon arrival at the hatchery, they are cleaned of fouling organisms and placed in conditioning tanks (Fig. 15.2).

Conditioning is the process of accelerating gonadal maturation by manipulating temperature and diet, effectively "tricking" the clams into spawning outside their natural season. Water temperature is gradually raised (typically 1°C per day) from ambient levels to the species-specific biological zero and then to the optimal spawning temperature (e.g., 18°C–22°C for *M. meretrix* and 20°C–23°C for *R. philippinarum*). This thermal stimulation increases metabolic rate and gametogenesis. Simultaneously, broodstock are fed *ad libitum* with a high-quality mixed algal diet. Studies have shown that supplementing the diet with lipid emulsions or concentrated algal pastes can significantly boost the lipid reserves in the eggs, which correlates directly with larval survival and metamorphosis success (Utting & Millican, 1997). The duration of conditioning is often calculated using "degree-days" (the sum of daily temperatures above biological zero), allowing hatchery managers to predict spawning dates with high accuracy.

Once the gonads are mature, spawning induction is performed. The method of induction varies by species and evolutionary adaptation. For the Manila clam (*R. philippinarum*) and hard clam (*M. mercenaria*), thermal shock is the industry standard. This involves cycling the clams between cold (e.g., 18°C) and warm (e.g., 28°C) seawater baths. The rapid temperature fluctuation stresses the animals, triggering the release of gametes. For burrowing mud-dwellers such as the razor clam (*S. constricta*) and blood clam (*T. granosa*), a combination of shade, desiccation, and flowing water is highly effective. Broodstock are exposed to air in a cool, dark place for 6–12 h, mimicking a low tide, followed by immersion in flowing, UV-irradiated seawater. The reimmersion and water movement stimulate synchronized mass spawning. In specific research contexts or for recalcitrant individuals, chemical induction using serotonin injection into the adductor muscle may be employed, though this is less common in commercial mass production (Hadley et al., 1997).

15.3.3 Spawning induction, larval rearing, and settlement

Following induction, eggs and sperm are released into the water column. Fertilization is carefully managed to prevent polyspermy (fertilization of an egg by multiple sperm), which leads to developmental failure. Fertilized eggs develop into trochophores and then D-hinge larvae (veligers) within 20–24 h. The larval rearing phase generally lasts 10–20 days, depending on species and temperature. During this period, larvae are maintained in conical or flat-bottomed tanks with gently aerated, filtered seawater.

Larval husbandry: Hygiene is paramount; tanks are drained, cleaned, and refilled every 24–48 h to reduce bacterial loads and metabolic wastes. Larvae are retained on sieves of appropriate mesh size during water changes. While antibiotics were historically used, modern hatcheries increasingly rely on probiotics (e.g., *Bacillus* spp.) and strict sanitation to maintain a healthy microbial balance. Feeding regimes are precise and escalating: Early veligers (D-larvae) generally exhibit a daily microalgal intake of 3000–5000 cells$\bullet$ml$^{-1}\bullet$day^{-1}. In contrast, post-metamorphic umboned larvae (shell-top stage) require significantly higher consumption levels, reaching 20,000–50,000 cells$\bullet$ml$^{-1}\bullet$day^{-1} to support rapid calcification and tissue growth (Braley et al., 2018; Yan et al., 2006).

Metamorphosis and settlement: The transition from the planktonic larval phase to the benthic juvenile phase is known as settlement or metamorphosis. This is a critical bottleneck where high mortality often occurs. Unlike oysters that cement themselves to hard surfaces, clam larvae are pediveligers that develop a functional foot to crawl and seek a suitable sediment substrate. Providing the correct settlement cue is essential. For *S. constricta*, sterilized, fine soft mud is often introduced into the rearing tanks to trigger settlement. For *R. philippinarum* and *M. mercenaria*, hatcheries use "setting trays" or downweller systems with mesh screens. Chemical cues, such as neuroactive compounds (e.g., epinephrine) or the presence of specific bacterial biofilms, can enhance settlement rates by signaling a suitable habitat. Once settled, the post-larvae (spat) undergo significant morphological changes, losing the velum and developing gills for filter feeding, marking the beginning of the nursery phase (Helm et al., 2004).

15.3.4 Challenges and strategic solutions in hatcheries

Despite technological advancements, hatchery production faces significant challenges. Mass mortality events, often termed "crashes," remain a persistent threat. These are frequently linked to *Vibrio* infections, specifically "bacillary necrosis," where bacteria attack the larval gut and hinge, causing death within hours. Strategic solutions include the use of specific pathogen-free (SPF) broodstock, the implementation of "flow-through" larval rearing systems to lower bacterial density, and the application of bacteriophages or probiotics as biocontrol agents. Another challenge is the "slow growth" syndrome, where larvae fail to metamorphose within the expected timeframe. This is often attributed to nutritional deficiencies (e.g., low lipid content in algae) or poor genetic quality (inbreeding depression). To address this, hatcheries are adopting rigorous quality control for algal cultures and implementing genetic management plans to maintain genetic diversity within broodstock populations, ensuring the resilience of the seed produced (Niu et al., 2010; Sanderson et al., 2023).

15.4 Nursery and intermediate culture techniques

Nursery and intermediate culture serve as a vital bridge between the hatchery and grow-out phases, mitigating the high mortality risks faced by vulnerable spat in open environments. Culturing spat to a "refuge size" significantly enhances defense against predation and physical stress. Techniques range from traditional tidal flat water storage and enclosed earthen ponds to intensive land-based flow-through systems, such as raceways and FLUPSYs. These intensive systems provide predator-free environments with optimized food delivery, accelerating growth and ensuring robust seed quality for final planting.

15.4.1 The biological necessity of intermediate culture: Bridging the gap between hatchery and farm

In the commercial production of marine bivalves, the transition from a controlled hatchery environment to the stochastic and often hostile conditions of grow-out sites represents the most precarious phase in the life cycle. Upon metamorphosis, clam spat (post-larvae) typically measure between 0.5 and 2 mm in shell length. At this microscopic size, they possess extremely thin, uncalcified or weakly calcified shells and limited burrowing depth, rendering them functionally defenseless against a myriad of biotic and abiotic threats (Beal et al., 1995). In natural ecosystems, mortality rates for settling spat can exceed 99%, driven by predation from benthic crustaceans (e.g., *Carcinus maenas* and *Callinectes sapidus*), gastropods, and juvenile fish, as well as physical smothering by shifting sediments or suffocation because of organic siltation.

Consequently, the direct planting of hatchery-produced spat onto open mudflats or seabeds is economically unviable. To mitigate these risks, an intermediate culture phase, universally known as the nursery stage, is mandatory in modern commercial operations. The primary biological objective of this phase is to grow the spat to a species-specific "refuge size"—typically 10–15 mm for Manila clams (*Ruditapes philippinarum*) and 5–10 mm for razor clams (*Sinonovacula constricta*). At this size, the shell thickens sufficiently to resist crushing by small predators, and the foot develops the muscular strength required for rapid burrowing, allowing the animal to seek refuge deep within the sediment.

This phase effectively bridges the gap between the hatchery and the farm, acting as a critical buffer that ensures high survival rates and maximizes the return on investment for seed production. The success of the "three-phase culture method" (hatchery, nursery, and grow-out), which has revolutionized clam farming in China and standardized global practices, hinges entirely on the efficiency of this intermediate stage. Operational data indicate that robust nursery management can elevate survival rates from negligible levels to more than 70%–90% prior to final grow-out planting (Lin et al., 2023; Rupp & Parsons, 2016; Zhang & Yan, 2006).

15.4.2 Traditional nursery systems: Tidal flat and pond cultivation

Traditional nursery methods harness the natural primary productivity of intertidal zones, offering a low-capital alternative to energy-intensive land-based systems, though they remain subject to higher environmental risks.

Flat-beach water storage cultivation:

This method is prevalent in developing regions and for extensive farming operations. It involves the engineering of the high intertidal zone, where low earthen dikes or embankments (approximately 20–30 cm in height) are constructed to enclose nursery plots ranging from 67 to 334 m^2. These structures are hydraulically designed to retain a shallow pool of seawater during low tide. This "water storage" serves a dual purpose: It buffers the spat against desiccation and extreme thermal stress during emersion, and it extends the filtration window beyond the natural tidal cycle, allowing for prolonged feeding.

The preparation of these plots is rigorous. The substrate is mechanically leveled to prevent pooling and anoxic spots, debris is removed, and predators are systematically eradicated. Prior to stocking, the sediment is often treated with biodegradable botanical piscicides, such as tea seed cake (containing saponins at 10–15 ppm) or rotenone, which asphyxiate predatory fish and crustaceans without harming the bivalves. Spat are stocked at moderate densities (e.g., 10–20 million ind./ha for *S. constricta*), relying primarily on natural benthic microalgae and tidal influxes for nutrition. While cost-effective, this method is highly vulnerable to environmental stochasticity; severe storms can breach dikes, and summer heatwaves can overheat the shallow retained water, leading to hypoxia and mass mortality events (NMAMR, 2023).

Enclosed pond nursery culture:

An evolution of the open flat method, enclosed pond systems are particularly widespread in East Asia for species adapted to estuarine conditions, such as the razor clam (*Sinonovacula constricta*) and blood clam (*Tegillarca granosa*). These systems utilize supratidal earthen ponds where water management is active rather than passive. Unlike the intertidal plots, pond nursery culture allows for the manipulation of primary productivity. Farmers routinely apply inorganic fertilizers (nitrogen and phosphorus ratios of roughly 10:1) or fermented organic manures to stimulate blooms of beneficial microalgae such as *Chaetoceros*, *Platymonas*, and *Nitzschia*. This eutrophic environment ensures a superabundance of food, supporting rapid growth.

Furthermore, biosecurity is significantly enhanced compared to open flats. Water intake is filtered through a fine mesh (60–80 mesh) to exclude predator larvae. The layout typically includes peripheral drainage ditches (30 cm width, 10 cm depth) to facilitate water circulation and waste removal. Despite these advantages, pond systems require meticulous management of water quality. The high stocking biomass combined with fertilization can lead to rapid depletion of dissolved oxygen, particularly at night, and the accumulation of toxic ammonia and nitrites. Farmers must employ paddlewheel aerators and conduct regular water exchanges to mitigate these risks (Wang, 2004).

15.4.3 Intensive land-based flow-through systems: The raceway and Floating Upweller System revolution

The state-of-the-art in clam nursery technology is represented by intensive land-based flow-through systems and FLUPSYs, which provide the highest degree of environmental control and production efficiency. These systems mark the transition from extensive farming to precision aquaculture.

In China, the land-based flow-through system has become the standard for high-value species such as *S. constricta* and *Meretrix meretrix*. These systems consist of long, shallow fiberglass or concrete troughs (typically 4–10 m long, 0.5–1 m wide, and 0.25 m deep). Recognizing the

FIGURE 15.3 Land-based flow-through systems of intermediate culture techniques for tidal flat clams.

TABLE 15.2 Stocking density and seeding time of shellfish seedlings in land-based flow-through systems

Species	Seed specification ($\times 10^4$ ind./kg)	Stocking density ($\times 10^4$ ind./m^3)	Seeding period
Sinonovacula constricta	150–250	10–20	Sep–Nov
Tegillarca granosa	10–20	4–6	May–Aug
Cyclina sinensis	10–40	6–10	Jul–Aug
Ruditapes philippinarum	150–200	15–20	Year-round
Meretrix meretrix	20–50	4–6	Jul–Sep

burrowing nature of these species, a standardized layer of fine sand or mud (5–10 cm thick) is spread on the raceway floor. This substrate is critical not only for animal welfare but also for stabilizing the clams against the current (Fig. 15.3).

Nutrient-rich seawater, often pumped from a productive adjacent estuary or enriched shrimp pond, flows continuously through the raceway in a unidirectional manner. This flow delivers a constant supply of phytoplankton and dissolved oxygen to the high-density clam beds while simultaneously flushing away feces, pseudofeces, and metabolic wastes (ammonia/ammonium). The hydrodynamics are carefully regulated; flow velocity must be sufficient to prevent the resettlement of waste particles but gentle enough to avoid sediment scour or the dislodgement of juvenile clams (Lin et al., 2023; Liu et al., 2024).

Operational data suggest that for *S. constricta*, stocking densities can be calibrated based on seed size (see Tables 15.2 and 15.3 in source data). For instance, newly settled spat (>250,000 ind./kg) can be stocked at densities exceeding 300,000 ind./m^2. As they grow, periodic size-selective thinning is performed to reduce density to 100,000–150,000 ind./m^2, preventing growth stunting because of spatial competition. This system significantly accelerates the production cycle; the nursery period for *S. constricta* can be shortened to 45–60 days compared with several months in the wild, with yields reaching 10–15 kg/m^2—approximately 10 times that of flat-beach cultivation (NMAMR, 2023).

TABLE 15.3 Stocking density of graded shellfish seedlings in land-based flow-through systems.

Species	Specification ($\times 10^4$ ind./kg)	Stocking density ($\times 10^4$ ind./m^3)
Sinonovacula constricta	>250	>30
	100–250	20–30
	10–100	15–20
	1–10	10–15
	<1	<10
Tegillarca granosa	>20	>6
	10–20	4–6
	5–10	3–4
	1–5	2–3
	<1	<2
Cyclina sinensis	>40	>10
	10–40	1.5–10
	4–10	1–1.5
	1–4	0.6–1
	<1	<3
Ruditapes philippinarum	>200	>20
	100–200	10–20
	10–100	5–10
	1–10	2–5
	<0.5	<2
Meretrix meretrix	>50	>6
	10–50	4–6
	2–10	2–4
	0.3–2	1–2
	<0.3	<1

In North America and Europe, the FLUPSY is the dominant nursery technology for the hard-shell clam (*Mercenaria mercenaria*) and Manila clam (*Ruditapes philippinarum*). Unlike raceways, which rely on horizontal flow, FLUPSYs utilize active vertical upwelling. Clam seed is placed in silos with mesh bottoms, which are suspended in a floating raft or dock. A central paddlewheel or high-volume low-head pump draws water out of a central channel, creating a head pressure difference. These differential forces nutrient-rich water from the surrounding estuary *up* through the mesh bottom of the silos, passing directly through the bed of clams.

This "force-feeding" mechanism ensures that every individual in the silo receives an adequate supply of food and oxygen, preventing the "starvation effect" often seen in the center of static dense aggregations. FLUPSYs offer multiple advantages: (1) Predator exclusion: Suspended silos completely exclude benthic predators such as crabs and starfish. (2) High density: They support exceptionally high stocking densities, often up to 200,000–300,000 individuals per bin for small spat. (3) Growth acceleration: The constant flow of food maximizes metabolic intake, accelerating growth rates by 2–3 times compared to field nurseries.

However, FLUPSYs require careful maintenance. The mesh screens are prone to biofouling by tunicates, bryozoans, and algae, which can impede flow and cause suffocation. Regular cleaning and grading are essential operational protocols. Although they require higher initial capital and energy costs, the dramatic improvement in survival (>80%–90%) and seed quality makes them the preferred choice for modern commercial hatcheries (Hadley et al., 1997; Rupp & Parsons, 2016).

15.5 Grow-out systems

Grow-out methodologies encompass a spectrum of practices tailored to species biology and local environmental conditions. Intertidal bottom sowing remains a global standard, requiring careful site selection, substrate preparation, and predator control. In East Asia, dike-enclosed ponds and polyculture models—integrating clams with shrimp or crabs—leverage ecological complementarity to enhance productivity and water quality. Furthermore, IMTA offers a sustainable paradigm by coculturing clams with fed species and macroalgae, effectively recycling nutrients and reducing the environmental footprint of mariculture.

15.5.1 Intertidal bottom-sowing aquaculture: The global standard

Intertidal bottom sowing remains the most extensive and widely practiced grow-out method globally, accounting for the bulk of production for species such as the Manila clam (*Ruditapes philippinarum*) and hard clam (*Mercenaria mercenaria*). This method essentially enhances the natural benthic habitat of the clams through active husbandry (Fig. 15.4).

Site selection and substrate preparation: The process begins with rigorous site selection. Ideal sites are typically located in the mid-to-low intertidal zones of sheltered bays and estuaries. Hydrodynamically, these areas must be characterized by moderate tidal currents (typically 20–50 cm/s) to ensure sufficient flux of phytoplankton and removal of biodeposits, without causing sediment scour. Substrate composition is species-specific; *M. meretrix* generally requires fine sand, while *R. philippinarum* thrives in mixed sandy-mud. Prior to planting, the substrate is prepared by mechanical tilling or harrowing. This practice loosens the sediment to facilitate rapid burrowing by the seed, oxidizes the sediment to remove toxic hydrogen sulfide accumulated in anoxic layers, and helps to dislodge predators (Serdar et al., 2007).

FIGURE 15.4 The different models of tidal flat clam aquaculture.

Seeding strategies and density management: Hatchery-reared seed (typically 10–20 mm) is broadcast over the prepared plots. Sowing is usually performed during neap tides or slack water to minimize seed dispersal by currents. In large-scale operations, mechanical spreaders mounted on shallow-draft boats are used to ensure uniform distribution. Optimal stocking density is a critical economic and ecological variable; overcrowding leads to intense intraspecific competition for food and space, resulting in slow growth, shell deformities, and increased susceptibility to disease. Densities typically range from 200 to 500 clams/m^2, depending on the site's primary productivity (carrying capacity). Farmers must continuously monitor growth and, if necessary, thin the stocks to maintain optimal spacing (USDA, 2023; Yan et al., 2006).

Predator control: Predation is the single most critical management challenge in bottom culture. In the Venice Lagoon (Italy) and the Atlantic coast of the USA (e.g., Florida and Virginia), the deployment of predator exclusion nets is standard practice. Heavy-duty plastic mesh netting (aperture size adapted to seed size) is laid over the planted beds and buried at the edges to form a protective blanket. This netting effectively excludes voracious predators such as the green crab (*Carcinus maenas*), blue crab (*Callinectes sapidus*), and ray species (*Myliobatis* spp.), which can decimate a crop within days. In China, while netting is used, farmers also employ manual removal and the maintenance of ridges and furrows to facilitate drainage and inspection. The maintenance of these nets—cleaning off biofouling (algae, barnacles) to ensure water flow—is a major labor component of the grow-out phase.

Harvesting: Harvesting methods vary by scale and regulation. Small-scale farmers use hand rakes or "bull rakes" to dig clams from the sediment. Industrial operations, particularly in the United States and Europe, utilize hydraulic escalator dredges or suction dredges. These machines inject high-pressure water jets into the sediment to liquefy the substrate, allowing the clams to be lifted onto a conveyor belt. While highly efficient, hydraulic dredging causes significant benthic disturbance, leading to regulations restricting its use in sensitive habitats, such as seagrass beds.

15.5.2 Dike-enclosed and pond aquaculture: The Asian polyculture model

In East and Southeast Asia, particularly for deep-burrowing species such as the razor clam (*S. constricta*) and blood clam (*T. granosa*), grow-out often occurs in enclosed ponds or dike-enclosed tidal flats. These systems represent a modification of the natural environment to provide greater control over hydrodynamics and biological interactions.

Hydrological management: Dike-enclosed systems utilize low embankments to maintain a layer of water over the clam beds even during low tide. This "permanently submerged" state effectively transforms the intertidal zone into a subtidal environment, extending the feeding time of the clams from just the high-tide window to 24 h a day. This significantly accelerates growth rates. In addition, the water layer acts as a thermal buffer, protecting the animals against extreme air temperatures in summer (heat stress) and winter (freezing) (Lin & You, 2005).

Ecological principles of polyculture: A defining feature of Chinese clam ponds is the prevalence of polyculture. Clams are rarely grown in isolation; instead, they are integrated into a multispecies system, most commonly with swimming crabs (*Portunus trituberculatus*) or shrimp (*Penaeus vannamei*). This system is based on sound ecological principles of complementarity and benthic-pelagic coupling. (1) Biofiltration: Clams act as the system's biological filter; they consume phytoplankton and suspended organic particles, effectively "purifying" the water and preventing the eutrophication and algal blooms that often plague intensive shrimp monocultures. (2) Bioturbation: Conversely, the foraging and movement of the shrimp and crabs bioturbate the sediment surface. This physical disturbance breaks up the benthic boundary layer, enhancing oxygen penetration into the sediment and accelerating the oxidation of toxic sulfides and ammonium that could harm the benthic clams. (3) Pathogen control: Shrimp feed on benthic worms and small fouling organisms that might compete with or irritate the clams.

This symbiotic relationship not only improves the survival and growth of all species but also maximizes the economic output per hectare, providing farmers with a diversified income stream and insurance against the failure of any single crop (Tang, 2023; Wang, 2004).

15.5.3 Integrated multitrophic aquaculture: The future of sustainability

As the global aquaculture industry seeks to reduce its environmental footprint and optimize resource use, IMTA has emerged as a leading sustainable model. In large-scale mariculture zones such as Sanggou Bay in Northern China, clams are integrated into complex, three-dimensional farming systems. These systems typically combine: (1) Fed species: Finfish in cages (e.g., seabass and pufferfish); (2) Inorganic extractors: Macroalgae on longlines (e.g., Kelp *Saccharina japonica* or *Gracilaria*); (3) Organic extractors: Filter-feeding bivalves (scallops, oysters, and clams).

In these IMTA configurations, clams may be cultured in suspended lantern nets (for species such as *R. philippinarum* that can adapt to suspension) or in bottom trays positioned beneath the fish cages and seaweed lines. The ecological rationale is trophic cascading: Clams utilize the "waste" organic particulates—specifically fish feces and uneaten feed settling from the cages, and detritus shedding from the seaweed—as a food source. This effectively recycles nutrients that would otherwise contribute to benthic organic loading and oxygen depletion.

Quantitative studies have demonstrated that clams grown in IMTA systems often exhibit faster growth rates compared with monoculture because of the enhanced food availability (higher organic flux). Moreover, the seaweed component absorbs dissolved inorganic nutrients (nitrogen and phosphorus) excreted by the clams and fish, further balancing the ecosystem's chemical budget (Han et al., 2017; Jiang et al., 2014). This integrated approach transforms waste into harvestable biomass, significantly reducing the environmental impact of mariculture while increasing total system productivity (Rusco et al., 2024; Tang et al., 2024; Zhang et al., 2020).

15.6 Genetic improvement and new variety development

Genetic improvement is a primary driver of industry modernization, transitioning from phenotypic selection to molecular-assisted breeding. The application of genomic tools, including high-density linkage maps and genome-wide association studies (GWAS), enables the identification of markers linked to commercially valuable traits such as growth rate, shell color, and disease resistance. Advanced strategies, such as GS, are increasingly employed to improve complex polygenic traits. These efforts have culminated in the certification of distinct new varieties, offering farmers stock with verified advantages in yield and environmental resilience.

15.6.1 Genetic diversity and population structure analysis

Genetic variability is the raw material for evolution and the fundamental prerequisite for any successful breeding program. Understanding the genetic structure of wild and cultured clam populations is the first step toward effective management and improvement. Recent advances in molecular biology have provided powerful tools to assess this diversity, moving from early markers such as random amplified polymorphic DNA (RAPD) and amplified fragment length polymorphism (AFLP) to more robust microsatellites (SSRs) and SNPs.

Research utilizing these markers has revealed significant genetic differentiation among wild populations of commercially important clams, often driven by geographical barriers and local environmental adaptation. For instance, extensive studies on the hard clam *Meretrix meretrix* in China have identified distinct northern and southern lineages, separated by the Yangtze River outflow, which acts as a formidable biogeographic barrier preventing gene flow (Jing et al., 2015; Qi et al., 2013). Phenotypically, the northern populations are generally more cold-tolerant, whereas southern populations exhibit faster growth in warmer waters (Gu et al., 2015; Lin et al., 2008).

Similarly, the blood clam *Tegillarca granosa* exhibits high genetic diversity but is structured into distinct evolutionary significant units (ESUs) (Li, Song, & Su, 2003; Zhou et al., 2013). These units were likely separated by historical sea-level changes during the Pleistocene, which isolated populations in different marginal seas. Understanding these structures is crucial for breeders to avoid "outbreeding depression" (mixing genetically incompatible stocks, which can lead to reduced fitness in offspring) while also identifying unique genetic resources that may harbor valuable traits for adaptation to specific environments (Li, Song, & Su, 2003; Ni et al., 2012). Furthermore, continuous monitoring of the genetic diversity of hatchery stocks is vital to prevent the loss of alleles and the accumulation of inbreeding, which is a common cause of "hatchery crashes" and reduced larval vigor (Niu et al., 2010). Similar genetic diversity assessments have been conducted for *Ruditapes philippinarum* (Tan et al., 2020; Wei et al., 2023) and *Cyclina sinensis* (Zhao et al., 2007).

15.6.2 Molecular breeding and genomic selection

The field of clam breeding has transitioned from traditional phenotypic mass selection—where the largest individuals are simply chosen as parents—to molecular-assisted precision breeding. This shift is underpinned by the construction of high-density genetic linkage maps and the sequencing of whole genomes (Bai et al., 2019; Dong et al., 2020; Ran et al., 2019; Song et al., 2021; Wei et al., 2020; Yan et al., 2019).

The construction of genetic linkage maps for major species, such as *M. mercenaria, R. philippinarum*, and *S. constricta*, has been a major milestone. These maps serve as genomic roadmaps, allowing researchers to locate quantitative trait loci (QTL)—regions of the genome statistically associated with specific phenotypic traits. For the razor clam *S. constricta*, a high-density map revealed 16 QTLs associated with growth traits, providing targets for selection (Niu et al., 2017). In the Manila clam *R. philippinarum*, a map with more than 9600 markers pinpointed 10 QTLs associated with shell color and growth (Nie et al., 2017). For the hard clam *M. meretrix*, markers linked to shell length and weight have been identified, explaining more than 50% of the phenotypic variation (Lu et al., 2013). This knowledge allows for marker-assisted selection (MAS), where breeders can select parents based on their genotype for a specific marker (e.g., a specific DNA band or SNP) that is linked to a desirable trait, even before the trait is physically expressed. This is particularly useful for traits that are difficult to measure directly, such as disease resistance.

More recently, GWAS has been employed to scan the entire genome for SNPs associated with traits of interest, providing a higher resolution than traditional QTL mapping. In *S. constricta*, a GWAS identified 56 SNPs associated with ammonia tolerance. Notably, significant markers were found adjacent to the Na^+/K^+-ATPase (*nka*) gene, a key enzyme in ion regulation and ammonia excretion. This provides a direct molecular marker for breeding strains capable of surviving in eutrophic ponds (Lv et al., 2023). In *T. granosa*, GWAS on 300 individuals identified 39 growth-associated SNPs linked to 27 candidate genes, including *Vha26, Ptprt*, and *Stat5b*, which regulate transmembrane transport and immune signaling (He et al., 2024). Molecular analysis has also highlighted the role of heat shock proteins (HSPs) and immune genes, such as toll-like receptors (*TLR4*), in conferring resistance to thermal stress and *Vibrio* infections, offering targets for breeding climate-resilient strains (Qiu et al., 2025).

The industry is now moving toward GS, a transformative approach originally developed in livestock breeding. Unlike MAS, which relies on a few specific markers, GS uses genome-wide markers (often thousands of SNPs) to predict the "genomic estimated breeding value" (GEBV) of an individual. This method is particularly effective for improving complex, polygenic traits that are controlled by many genes with small effects, such as general disease resistance or environmental tolerance.

15.6.3 Certified new varieties and their traits

Decades of systematic breeding efforts, integrating both traditional selection and modern molecular tools, have culminated in the release of certified new varieties that are reshaping the economics of the industry. These varieties are distinct, stable, and uniform populations that offer specific, verified advantages to farmers. Table 15.4 and Fig. 15.5 provide a summary of key certified varieties.

"Wanlihong" (*M. meretrix*): Developed by Zhejiang Wanli University, this variety is characterized by a distinctive purplish-red shell color, in contrast to the mottled gray-brown of wild types. Research indicates that this color trait is controlled by a single locus but is genetically linked to faster metabolic rates. Consequently, the "Wanlihong" clam grows approximately 20% faster than common stocks. Its unique appearance also allows it to command a market premium as a branded product, aiding in anti-counterfeiting (Lin et al., 2023).

TABLE 15.4 Certified new varieties of commercially important clams.

Variety name	Approval time	Register number	Target trait	First contributing breeding unit
M. meretrix "Kezhe No. 1"	2013	GS-01-007-2013	Shell color, growth	Institute of Oceanology, Chinese Academy of Sciences
M. meretrix "Wanlihong"	2014	GS-01-007-2014	Shell color, growth	Zhejiang Wanli University
M. meretrix "Wanli No. 2"	2017	GS-01-012-2017	Shell color, growth	Zhejiang Wanli University
M. meretrix "Kezhe No. 2"	2021	GS-02-001-2021	Antivibrio, growth	Institute of Oceanology, Chinese Academy of Sciences
M. meretrix "Suhaihong No. 1"	2024	GS-01-012-2024	Shell color, growth	Marine Fisheries Research Institute of Jiangsu Province
S. constricta "Shenzhe No. 1"	2017	GS-01-013-2017	Growth	Shanghai Ocean University
S. constricta "Yongle No. 1"	2020	GS-01-004-2020	Growth, low-salt tolerance	Zhejiang Wanli University
S. constricta "Yongle No. 2"	2024	GS-01-014-2024	Growth, ammonia tolerance	Zhejiang Wanli University
T. granosa "Yueqingwan No. 1"	2014	GS-01-006-2014	Growth	Zhejiang Mariculture Research Institute
R. philippinarum "Zebra Clam No. 2"	2021	GS-01-007-2021	Shell color, growth	Dalian Ocean University
R. philippinarum "White Zebra Clam"	2016	GS-01-009-2016	Shell color, growth	Dalian Ocean University
R. philippinarum "Zebra Clam"	2015	GS-01-005-2014	Shell color, growth	Dalian Ocean University

"Shenzhe No. 1" (*S. constricta*): Selected by Shanghai Ocean University through six generations of mass selection, this variety was bred specifically for superior growth performance. Field trials demonstrate that it reaches market size significantly faster than wild stocks, allowing farmers to harvest earlier. This shortened culture cycle reduces the risk of overwintering mortality and exposure to spring algal blooms.

Stress-resistant varieties: Recent breeding targets have shifted toward resilience. "Yongle No. 1" (*S. constricta*): Selected for low-salt tolerance, enabling aquaculture in river mouths and areas subject to heavy freshwater runoff. "Kezhe No. 2" (*M. meretrix*): Bred for resistance to *Vibrio* bacteria, addressing the critical issue of summer mass mortality events. "Zebra Clam No. 2" (*R.*

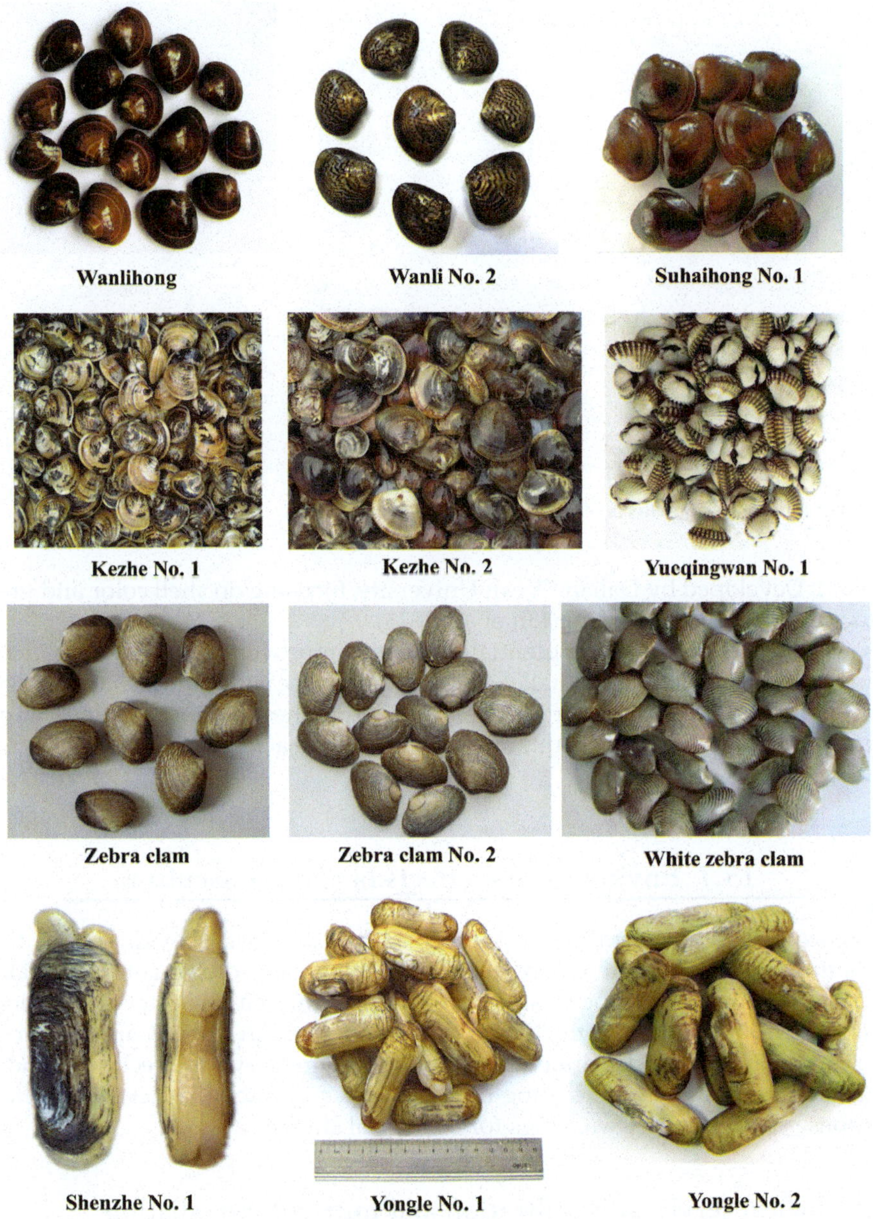

FIGURE 15.5 Certified new varieties of commercially important clams.

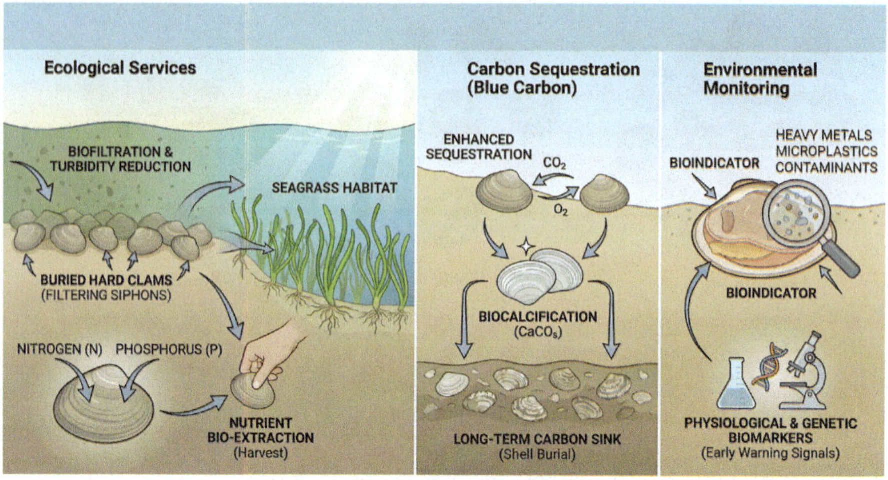

FIGURE 15.6 Environmental effects of clam aquaculture.

philippinarum): Developed by Dalian Ocean University, focusing on shell color and growth rates adapted to northern Chinese waters (Lin et al., 2023).

These varieties demonstrate the potential of genetic improvement to not only increase yield but also to solve specific industrial challenges such as disease susceptibility and climate change adaptation. Future efforts will likely focus on "stacking" these traits—creating "super-clams" that are fast-growing, disease-resistant, and tolerant to the acidifying, warming oceans of the future.

15.7 Environmental effects of clam aquaculture

Clam aquaculture interacts intimately with coastal ecosystems, providing essential services such as biofiltration and nutrient remediation. By filtering vast volumes of water, clams reduce turbidity and extract excess nitrogen and phosphorus, mitigating coastal eutrophication. The industry also contributes to climate change mitigation through "Blue Carbon" processes, sequestering carbon in shells and sediments. In addition, clams serve as effective bioindicators for environmental monitoring, accumulating heavy metals and microplastics, while their physiological responses provide early warnings for ecological stressors.

15.7.1 Ecological services: Biofiltration and nutrient remediation

As shown in Fig. 15.6, Clam aquaculture interacts intimately with the coastal environment, often providing net positive ecosystem services (van der Schatte Olivier et al., 2020). As filter feeders, clams exert a strong "top-down" control on phytoplankton populations. By filtering large volumes of water—a single adult hard clam can filter over 15L per day—they reduce water turbidity. This clarification allows sunlight to penetrate deeper into the water column, promoting the recovery of benthic macrophytes such as seagrasses (*Zostera marina*), which are critical nursery habitats for fish. Furthermore, clams play a pivotal role in nutrient remediation. They assimilate nitrogen (N) and phosphorus (P) from the water into their tissues and shells. When clams are harvested, these nutrients are physically removed from the marine ecosystem (Welsh et al., 2015). This "bio-extraction" is increasingly recognized as a tool to mitigate coastal eutrophication. Studies in the Chesapeake Bay (USA) and various Chinese estuaries have quantified this service, valuing the nitrogen removal provided by shellfish farming as a cost-effective alternative to upgrading wastewater treatment plants (Bricker et al., 2020; Dvarskas et al., 2020; Guyondet et al., 2022; Liu et al., 2024).

15.7.2 Carbon sequestration: The "Blue Carbon" potential

Clam aquaculture is frequently discussed in the context of climate change mitigation and "Blue Carbon". Clams sequester carbon primarily through the biocalcification process, forming shells composed of 95% calcium carbonate ($CaCO_3$). While the chemistry of calcification releases CO_2 in the short term, the long-term storage of carbon in harvested shells (which are often used in construction or agriculture) and the burial of dead shells in sediments constitute a carbon sink (Morris & Humphreys, 2019). LCA studies indicate that clam farming has one of the lowest carbon footprints of any animal protein production system (Martini et al., 2022). For instance, the Manila clam is estimated to sequester approximately 233 g of CO_2 per kg of harvest. When integrated with seaweed in IMTA systems, the net carbon sequestration potential is further enhanced, as the seaweed absorbs the respiratory CO_2 released by the clams (Feng et al., 2023; Ray et al., 2018; Tamburini et al., 2022).

15.7.3 Environmental monitoring and indicators

Because of their sedentary lifestyle and filter-feeding habit, clams act as sentinels for the health of the coastal environment. They bioaccumulate contaminants from the water and sediment, making them excellent bioindicators (Burger, 2007; Niemi & McDonald, 2004). They are widely used in monitoring programs to detect heavy metals (e.g., lead, cadmium, and mercury) and persistent organic pollutants. Recent research has also highlighted their utility in monitoring emerging threats, such as microplastics; the abundance of microplastics in clam tissues often correlates with levels in the surrounding sediment (Ding et al., 2021; Tang et al., 2022). In addition, the physiological and genetic responses of clams (e.g., the expression of stress biomarkers such as HSP or immune genes such as TLR4) provide early warning signals of environmental stressors such as marine heatwaves or pathogen blooms (Qiu et al., 2025; Sun et al., 2015; Zhao et al., 2017).

15.8 Conclusion

Clam aquaculture has transformed from a traditional subsistence activity into a highly technical, globally interconnected industry that is vital to the "Blue Economy". The sector's success is built upon the closure of life cycles for major species such as the Manila clam and razor clam, the standardization of intensive hatchery and nursery technologies such as the raceway system, and the adaptation of diverse grow-out models from tidal flats to IMTA systems. Scientific advancements in genetics have provided the industry with resilient, high-yielding varieties capable of withstanding the challenges of a changing climate. Moreover, the recognition of clams as providers of essential ecosystem services—nutrient remediation, water clarification, and carbon sequestration—positions the industry as a leader in sustainable food production. Looking forward, the continued expansion of clam aquaculture will rely on balancing intensification with ecological stewardship, embracing "climate-smart" breeding targets, and further integrating farming systems with the broader coastal ecosystem management.

References

Azra, M. N., Okomoda, V. T., Tabatabaei, M., Hassan, M., & Ikhwanuddin, M. (2021). The contributions of shellfish aquaculture to global food security: Assessing its characteristics from a future food perspective. *Frontiers in Marine Science, 8,* 654897. doi:10.3389/fmars.2021.654897.

Bai, C. M., Xin, L. S., Rosani, U., Wu, B., Wang, Q. C., Duan, X. K., ... Wang, C. M. (2019). Chromosomal-level assembly of the blood clam, *Scapharca (Anadara) broughtonii*, using long sequence reads and Hi-C. *GigaScience, 8*(7), giz067. doi:10.1093/gigascience/giz067.

Bao, Y., Zeng, Q., Wang, J., Zhang, Z., Zhang, Y., Wang, S., ... Lin, Z. (2021). Genomic insights into the origin and evolution of molluscan red-bloodedness in the blood clam *Tegillarca granosa*. *Molecular Biology and Evolution, 38*(6), 2351–2365. doi:10.1093/molbev/msab030.

Beal, B. F., Lithgow, C. D., Shaw, D. P., Renshaw, S., & Ouellette, D. (1995). Overwintering hatchery-reared individuals of the soft-shell clam, *Mya arenaria* L.: A field test of site, clam size, and intraspecific density. *Aquaculture, 130*(2–3), 145–158. doi:10.1016/0044-8486(94)00221-9.

Braley, R. D., Militz, T. A., & Southgate, P. C. (2018). Comparison of three hatchery culture methods for the giant clam *Tridacna noae*. *Aquaculture, 495,* 881–887. doi:10.1016/j.aquaculture.2018.05.044.

Bricelj, V. M., & Malouf, R. E. (1984). Influence of algal and suspended sediment concentrations on the feeding physiology of the hard clam *Mercenaria mercenaria*. *Marine Biology, 84*(2), 155–165. doi:10.1007/bf00393000.

Bricker, S. B., Grizzle, R. E., Trowbridge, P., Rose, J. M., Ferreira, J. G., Wellman, K., Zhu, C., Galimany, E., Wikfors, G. H., Saurel, C., Miller, R. L., Wands, J., Rheault, R., Steinberg, J., Jacob, A. P., Davenport, E. D., Ayvazian, S., Chintala, M., & Tedesco, M. A. (2020). Bioextractive removal of nitrogen by oysters in Great Bay Piscataqua River Estuary, New Hampshire, USA. *Estuaries and Coasts, 43*(1), 23–38. doi:10.1007/s12237-019-00661-8.

Burger, J. (2007). Bioindicators: Types, development, and use in ecological assessment and research. *Environmental Bioindicators., 1*(1), 22–39. doi:10.1080/15555270590966483.

Ding, J., Sun, C., He, C., Li, J., Ju, P., & Li, F. (2021). Microplastics in four bivalve species and basis for using bivalves as bioindicators of microplastic pollution. *Science of the Total Environment, 782,* 146830. doi:10.1016/j.scitotenv.2021.146830.

Dong, Y., Zeng, Q., Ren, J., Yao, H., Lv, L., He, L., ... Lin, Z. (2020). The chromosome-level genome assembly and comprehensive transcriptomes of the Razor Clam (*Sinonovacula constricta*). *Frontiers in Genetics, 11,* 664. doi:10.3389/fgene.2020.00664.

Dong, Y., Zheng, C., Yao, H., Yao, D., & Lin, Z. (2011). Genetic variation of 5 wild populations of *Sinonovacula constricta* by AFLP analysis. *Genomics and Applied Biology, 30*(6), 1121–1128.

Dong, Y. H., Yao, H. H., Lin, Z. H., Xiao, G. Q., & Chai, X. L. (2012). Screening of AFLP markers related to growth traits in *Tegillarca granosa*. *Journal of Fisheries of China, 36*(6), 825. doi:10.3724/SP.J.1231.2012.27792.

Dvarskas, A., Bricker, S. B., Wikfors, G. H., Bohorquez, J. J., Dixon, M. S., & Rose, J. M. (2020). Quantification and valuation of nitrogen removal services provided by commercial shellfish aquaculture at the subwatershed scale. *Environmental Science and Technology, 54*(24), 16156–16165. doi:10.1021/acs.est.0c03066.

Fang, J., Shen, Y., Zhang, L., Li, T., & Shao, Y. (2020). Development and verification of SSR markers in *Cyclina sinensis*. *Journal of Applied Oceanography, 39*(2), 211–218.

FAO. (2024). *The State of World Fisheries and Aquaculture 2024: Blue transformation in action*. Rome, Italy: FAO https://doi.org/10.4060/cd0683en.

Feng, J. C., Sun, L., & Yan, J. (2023). Carbon sequestration via shellfish farming: A potential negative emissions technology. *Renewable and Sustainable Energy Reviews, 171,* 113018. doi:10.1016/j.rser.2022.113018.

Gavioli, A., Castaldelli, G., Eggleston, D. B., & Christian, R. R. (2025). From ecological to anthropogenic factors: Unraveling the drivers of blue crab *Callinectes sapidus* occurrence along the Mediterranean coasts. *Frontiers in Marine Science, 11,* 1515098. doi:10.3389/fmars.2024.1515098.

Gephart, J. A., Henriksson, P. J. G., Parker, R. W. R., Shepon, A., Gorospe, K. D., Bergman, K., Eshel, G., Golden, C. D., Halpern, B. S., Hornborg, S., Jonell, M., Metian, M., Mifflin, K., Newton, R., Tyedmers, P., Zhang, W., Ziegler, F., & Troell, M. (2021). Environmental performance of blue foods. *Nature, 597*(7876), 360–365. doi:10.1038/s41586-021-03889-2.

Golden, C. D., Koehn, J. Z., Shepon, A., Passarelli, S., Free, C. M., Viana, D. F., Matthey, H., Eurich, J. G., Gephart, J. A., Fluet-Chouinard, E., Nyboer, E. A., Lynch, A. J., Kjellevold, M., Bromage, S., Charlebois, P., Barange, M., Vannuccini, S., Cao, L., Kleisner, K. M., & Thilsted, S. H. (2021). Aquatic foods to nourish nations. *Nature, 598*(7880), 315–320. doi:10.1038/s41586-021-03917-1.

Gosling, E. (2003). *Bivalve molluscs: Biology, ecology and culture*. Blackwell Publishing Ltd.

Gu, X. F., Dong, Y. H., Yao, H. H., Zhou, X. L., Qi, X. Y., & Lin, Z. H. (2015). Microsatellite marker analysis reveals the distinction between the north and south groups of hard clam (*Meretrix meretrix*) in China. *Genetics and Molecular Research, 14*(1), 1210–1219. doi:10.4238/2015.february.6.23.

Guyondet, T., Filgueira, R., Pearce, C. M., Tremblay, R., & Comeau, L. A. (2022). Nutrient-loading mitigation by shellfish aquaculture in semi-enclosed estuaries. *Frontiers in Marine Science, 9*, 909926. doi:10.3389/fmars.2022.909926.

Hadley, N. H., Manzi, J. J., Eversole, A. G., Dillon, R. T., Battey, C. E., & Peacock, N. M. (1997). *A manual for the culture of the hard clam. Sea Grant Consortium, Charleston*. Sea Grant Consortium.

Han, T., Shi, R., Qi, Z., Huang, H., Liang, Q., & Liu, H. (2017). Interactive effects of oyster and seaweed on seawater dissolved inorganic carbon systems: Implications for integrated multi-trophic aquaculture. *Aquaculture Environment Interactions, 9*, 469–478. doi:10.3354/aei00246.

He, X., Liao, Y., Yu, G., Wang, S., & Bao, Y. (2024). Genome-wide association study reveals the underlying regulatory mechanisms of red blood traits in Anadara granosa. *BMC Genomics, 25*(1), 931.

Helm, M.-M., Bourne, N., & Lovatelli, A. (2004). *The hatchery culture of bivalves: A practical manual*. FAO.

Hudson, K., & White, S. (2025). *Virginia shellfish aquaculture situation and outlook report: Based on the results of the 2024. VIMS Marine Resource Report No. 2025-5*. Virginia Institute of Marine Science. doi:10.25773/dapk-yc40.

Jiang, Z. J., Fang, J. G., Han, T. T., Mao, Y. Z., Li, J. Q., & Du, M. R. (2014). The role of *Gracilaria lemaneiformis* in eliminating the dissolved inorganic carbon released from calcification and respiration process of *Chlamys farreri*. *Journal of Applied Phycology, 26*(1), 545–550. doi:10.1007/s10811-013-0110-8.

Jing, Y., Dong, Y., Yao, H., He, L., & Lin, Z. (2015). Characterization of 87 EST-SNP markers in hard clam *Meretrix meretrix* using high-resolution melting analysis. *Conservation Genetics Resources, 7*(4), 811–817. doi:10.1007/s12686-015-0490-y.

Jones, A. R., Alleway, H. K., McAfee, D., Reis-Santos, P., Theuerkauf, S. J., & Jones, R. C. (2022). Climate-friendly seafood: The potential for emissions reduction and carbon capture in marine aquaculture. *BioScience, 17*(2), 123–143. doi:10.1093/biosci/biab126.

Kong, X., Ren, J., Dai, W., Yao, H., Lin, Z., & Dong, Y. (2025). Integrated physiological, transcriptomics and metabolomics analyses provide insights into thermal stress of razor clam, *Sinonovacula constricta*. *Aquaculture, 596*, 741909. doi:10.1016/j.aquaculture.2024.741909.

Lin, Z., & You, Z. (2005). Intensive mudflat bivalves culture in Zhejiang. *Marine Sciences, 29*(08), 95–99.

Li, T. W., Song, L. S., & Su, X. R. (2003). RAPD variation within and among five populations of *Tegillarca granosa*. *Biodiversity Science, 11*(2), 118–124.

Lin, Z., Dong, Y., Li, N., Lu, R. M, Xiao, G., Chai, X., … Hu, J. (2008). The genetic structure and diversity analysis of different geographical populations of *Meretrix meretrix* using morphological parameters and AFLP markers. *Oceanologia Et Limnologia Sinica, 39*(3), 242–247.

Lin, Z., He, L., & Dong, Y. (2023). Development and prospect for scientific and technological innovation of mudflat mollusk breeding industry in Zhejiang. *Journal of Fisheries of China, 47*(1), 019608. doi:10.11964/jfc.20221013754.

Liu, M., Yan, Z., Huang, C., Lin, Z., Peng, Z., Zhao, C., & Zheng, X. (2024). Temporal dynamics of protist communities and environmental factors in the horizontal flow-polyculture pond aquaculture model of *Sinonovacula constricta*. *Aquacultural Engineering, 107*, 102477. doi:10.1016/j.aquaeng.2024.102477.

Lu, X., Wang, H., Liu, B., & Xiang, J. (2013). Three EST-SSR markers associated with QTL for the growth of the clam *Meretrix meretrix* revealed by selective genotyping. *Marine Biotechnology, 15*(1), 16–25. doi:10.1007/s10126-012-9453-4.

Lv, L., Hu, C., Xu, H., Ren, J., Wu, B., Dong, Y., & Lin, Z. (2023). Insight into the genetic basis of ammonia tolerance in razor clam *Sinonovacula constricta* by genome-wide association study. *Aquaculture, 569*, 739351. doi:10.1016/j.aquaculture.2023.739351.

Martini, A., Calì, M., Capoccioni, F., Martinoli, M., Pulcini, D., Buttazzoni, L., Moranduzzo, T., & Pirlo, G. (2022). Environmental performance and shell formation-related carbon flows for mussel farming systems. *Science of the Total Environment, 831*, 154891. doi:10.1016/j.scitotenv.2022.154891.

Morris, J. P., & Humphreys, M. P. (2019). Modelling seawater carbonate chemistry in shellfish aquaculture regions: Insights into CO_2 release associated with shell formation and growth. *Aquaculture, 501*, 338–344. doi:10.1016/j.aquaculture.2018.11.028.

Murray, N. J., Phinn, S. R., DeWitt, M., Ferrari, R., Johnston, R., Lyons, M. B., Clinton, N., Thau, D., & Fuller, R. A. (2019). The global distribution and trajectory of tidal flats. *Nature, 565*(7738), 222–225. doi:10.1038/s41586-018-0805-8.

Murray, N. J., Phinn, S. P., Fuller, R. A., DeWitt, M., Ferrari, R., Johnston, R., Clinton, N., & Lyons, M. B. (2022). High-resolution global maps of tidal flat ecosystems from 1984 to 2019. *Scientific Data, 9*(1), 542. doi:10.1038/s41597-022-01635-5.

Ni, G., Li, Q., Kong, L., & Zheng, X. (2012). Phylogeography of the bivalve *Tegillarca granosa* in coastal China: Implications for management and conservation. *Marine Ecology Progress Series, 452*, 119–130. doi:10.3354/meps09624.

Nie, H., Yan, X., Huo, Z., Jiang, L., Chen, P., Liu, H., … Yang, F. (2017). Construction of a high-density genetic map and quantitative trait locus mapping in the Manila clam *Ruditapes philippinarum*. *Scientific Reports, 7*(1), 11953. doi:10.1038/s41598-017-00246-0.

Niemi, G. J., & McDonald, M. E. (2004). Application of ecological indicators. *Annual Review of Ecology, Evolution, and Systematics, 35*, 89–111. doi:10.1146/annurev.ecolsys.35.112202.130132.

Niu, D., Du, Y., Wang, Z., Xie, S., Nguyen, H., Dong, Z., … Li, J. (2017). Construction of the first high-density genetic linkage map and analysis of quantitative trait loci for growth-related traits in *Sinonovacula constricta*. *Marine Biotechnology, 19*(5), 488–496. doi:10.1007/s10126-017-9768-2.

Niu, D., Chen, H., Wang, S., Lin, G., & Li, J. (2010). Population genetic structure of *Sinonovacula constricta* along the coast of China. *Chinese Journal of Zoology, 45*(2), 11–18.

NMAMR. (2023). *Ningbo City local standard of Zhejiang Province*. Ningbo Municipal Administration for Market Regulation (NMAMR).

Qi, X., Dong, Y., Yao, H., Zhou, X., & Lin, Z. (2013). Identification of 30 microsatellite markers in *Meretrix meretrix* and their transferability in *Meretrix lamarckii* and *Meretrix lyrata*. *Journal of Fisheries of China, 37*(8), 1147. doi:10.3724/SP.J.1231.2013.38484.

Qiu, X., Ding, J., Wang, Y., Fang, L., Li, D., & Huo, Z. (2025). Identification and function analysis of toll–like receptor 4 (TLR4) from Manila clam (*Ruditapes philippinarum*). *International Journal of Biological Macromolecules, 290*, 139000. doi:10.1016/j.ijbiomac.2024.139000.

Ran, Z., Li, Z., Yan, X., Liao, K., Kong, F., Zhang, L., … Xu, J. (2019). Chromosome-level genome assembly of the razor clam *Sinonovacula constricta* (Lamarck, 1818). *Molecular Ecology Resources, 19*(6), 1647–1658. doi:10.1111/1755-0998.13086.

Ray, N. E., O'Meara, T., Wiliamson, T., Izursa, J. L., & Kangas, P. C. (2018). Consideration of carbon dioxide release during shell production in LCA of bivalves. *International Journal of Life Cycle Assessment, 23*(5), 1042–1048. doi:10.1007/s11367-017-1394-8.

Richards, G. P (1988). Microbial purification of shellfish: A review of depuration and relaying. *Journal of Food Protection, 51*(3), 218–251. doi:10.4315/0362-028x-51.3.218.

Rupp, G. S., & Parsons, G. J. (2016). Aquaculture of the Scallop *Nodipecten nodosus* in Brazil. *Developments in Aquaculture and Fisheries Science, 40*, 999–1017. doi:10.1016/B978-0-444-62710-0.00024-9.

Rusco, G., Roncarati, A., Di Iorio, M., Cariglia, M., Longo, C., & Iaffaldano, N. (2024). Can IMTA system improve the productivity and quality traits of aquatic organisms produced at different trophic levels? The benefits of IMTA—Not only for the ecosystem. *Biology, 13*(11), 946. doi:10.3390/biology13110946.

Sanderson, M. P., Hudson, K. L., Gregg, L. S., Chesler-Poole, A. B., Small, J. M., Reece, K. S., Carnegie, R. B., & Smith, J. L. (2023). Detection of toxins and harmful algal bloom cells in shellfish hatcheries and efforts toward removal. *Aquaculture, 562*, 738714. doi:10.1016/j.aquaculture.2022.738714.

Serdar, S., Lök, A., Köse, A., Yildiz, H., Acarli, S., & Goulletquer, P. (2007). Growth and survival rates of carpet shell clam (*Tapes decussatus* Linnaeus, 1758) using various culture methods in Sufa (Homa) Lagoon, Izmir. *Aquacultural Engineering., 37*(2), 89–99. doi:10.1016/j.aquaeng.2007.02.004.

Song, H., Guo, X., Sun, L., Wang, Q., Han, F., Wang, H., … Zhang, T. (2021). The hard clam genome reveals massive expansion and diversification of inhibitors of apoptosis in Bivalvia. *BMC Biology, 19*(1), 15. doi:10.1186/s12915-020-00943-9.

Sun, Y. M., Ma, Z. Y., & Huang, H. P. (2015). Contamination and assessment of heavy metals in shellfish in the intertidal zone of sea islands in China. *Zhongguo Huanjing Kexue (China Environmental Science), 35*(2), 574–578.

Sun, C., Wang, T., Dong, Y., Yao, H., & Lin, Z. (2012). AFLP analysis of molecular markers related to growth traits in *Cyclina sinensis*. *Oceanologia Et Limnologia Sinica, 43*(1), 106–112.

Tamburini, E., Turolla, E., Lanzoni, M., Moore, D., & Castaldelli, G. (2022). Manila clam and Mediterranean mussel aquaculture is sustainable and a net carbon sink. *Science of the Total Environment, 848,* 157508. doi:10.1016/j.scitotenv.2022.157508.

Tan, Y., Fang, L., Qiu, M., Huo, Z., & Yan, X. (2020). Population genetics of the Manila clam (*Ruditapes philippinarum*) in East Asia. *Scientific Reports., 10*(1), 21890. doi:10.1038/s41598-020-78923-w.

Tang, R., Zhang, T., Song, K., Sun, Y., Chen, Y., Huang, W., & Feng, Z. (2022). Microplastics in commercial clams from the intertidal zone of the South Yellow Sea, China. *Frontiers in Marine Science, 9,* 905923. doi:10.3389/fmars.2022.905923.

Tang, Y., Ju, C., Mei, R., Zhao, L., Liu, J., Yang, Y., … Liu, Q. (2024). Exploring the optimal integrated multi-trophic aquaculture (IMTA) patterns benefiting culture animals and natural water environment. *Aquaculture, 589,* 741011. doi:10.1016/j.aquaculture.2024.741011.

Tang, Q. (2023). The past and future of sustainable development of Chinese modern fisheries. *Progress in Fishery Sciences, 44*(6), 1–6.

USDA. (2023). *Aquaculture data*. U.S. Department of Agriculture, Economic Research Service.

Utting, S. D., & Millican, P. F. (1997). Techniques for the hatchery conditioning of bivalve broodstocks and the subsequent effect on egg quality and larval viability. *Aquaculture, 155*(1-4), 45–54. doi:10.1016/s0044-8486(97)00108-7.

van der Schatte Olivier, A., Jones, L., Vay, L. L., Christie, M., Wilson, J., & Malham, S. K. (2020). A global review of the ecosystem services provided by bivalve aquaculture. *Reviews in Aquaculture, 12*(1), 3–25. doi:10.1111/raq.12301.

Wang, F. K., Zhu, P. C., Zhang, X. T., Yu, K., Wang, C. D., & Liu, B. (2025). QTL mapping and candidate gene identification for lower temperature tolerance and growth traits using whole genome re-sequencing in Argopecten scallops. *Aquaculture, 595,* 741513. doi:10.1016/j.aquaculture.2024.741513.

Wang, Y. (2004). Optimization of culture model in seawater pond: Concepts, principles and methods. *Journal of Fisheries of China, 28*(05), 563–568.

Wei, D., Zheng, S., Wang, S., Yan, J., Liu, Z., Zhou, L., … Sun, X. (2023). Genetic and haplotype diversity of Manila clam *Ruditapes philippinarum* in different regions of China based on three molecular markers. *Animals, 13*(18), 2886. doi:10.3390/ani13182886.

Wei, M., Ge, H., Shao, C., Yan, X., Nie, H., Duan, H., Liao, X., Zhang, M., Chen, Y., Zhang, D., & Dong, Z. (2020). Chromosome-level clam genome helps elucidate the molecular basis of adaptation to a buried lifestyle. *iScience, 23*(6), 101148. doi:10.1016/j.isci.2020.101148.

Welsh, D. T., Nizzoli, D., Fano, E. A., & Viaroli, P. (2015). Direct contribution of clams (*Ruditapes philippinarum*) to benthic fluxes, nitrification, denitrification and nitrous oxide emission in a farmed sediment. *Estuarine, Coastal and Shelf Science, 154,* 84–93. doi:10.1016/j.ecss.2014.12.021.

Willer, D. F., & Aldridge, D. C. (2020). Sustainable bivalve farming can deliver food security in the tropics. *Nature Food, 1,* 384–388. doi:10.1038/s43016-020-0116-8.

Yan, X., Nie, H., Huo, Z., Ding, J., Li, Z., Yan, L., Jiang, L., Mu, Z., Wang, H., Meng, X., Chen, P., Zhou, M., Rbbani, M. G., Liu, G., & Li, D. (2019). Clam genome sequence clarifies the molecular basis of its Benthic adaptation and extraordinary shell color diversity. *iScience, 19,* 1225–1237. doi:10.1016/j.isci.2019.08.049.

Yan, X., Zhang, G., & Yang, F. (2006). Effects of diet, stocking density, and environmental factors on growth, survival, and metamorphosis of Manila clam *Ruditapes philippinarum* larvae. *Aquaculture, 253*(1–4), 350–358. doi:10.1016/j.aquaculture.2005.07.030.

Ye, K., Liu, R., Zhan, T., Chen, S., Yu, J., & Niu, D. (2025). The heat-resistant capability analysis of two populations of *Sinonovacula constricta* under thermal stress. *Journal of Thermal Biology, 130,* 104164. doi:10.1016/j.jtherbio.2025.104164.

Zhang, D., Tian, X., Dong, S., Chen, Y., Feng, J., He, R. P, & Zhang, K. (2020). Carbon budgets of two typical polyculture pond systems in coastal China and their potential roles in the global carbon cycle. *Aquaculture Environment Interactions, 12,* 105–115. doi:10.3354/aei00349.

Zhang, G., & Yan, X. (2006). A new three-phase culture method for Manila clam, *Ruditapes philippinarum,* farming in northern China. *Aquaculture, 258*(1–4), 452–461. doi:10.1016/j.aquaculture.2006.04.046.

Zhang, L., Zhao, Z. W., Ma, L. X., & Dong, Y. W. (2024). Genome-wide sequencing reveals geographical variations in the thermal adaptation of an aquaculture species with frequent seedling introductions. *Science of the Total Environment, 927*, 172010. doi:10.1016/j.scitotenv.2024.172010.

Zhao, Y., Li, Q., Kong, L., Bao, Z., & Wang, R. (2007). Genetic diversity and divergence among clam *Cyclina sinensis* populations assessed using amplified fragment length polymorphism. *Fisheries Science, 73*(6), 1338–1343. doi:10.1111/j.1444-2906.2007.01473.x.

Zhao, X., Shi, W., Han, Y., Liu, S., Guo, C., Fu, W., ... Liu, G. (2017). Ocean acidification adversely influences metabolism, extracellular pH and calcification of an economically important marine bivalve, *Tegillarca granosa*. *Marine Environmental Research, 125*, 82–89. doi:10.1016/j.marenvres.2017.01.007.

Zhou, X., Zhu, J., Dong, Y., Lin, Z., Fu, K., Yao, H., & Qi, X. (2013). Development and comparative study of genomic-SSR and EST-SSR in *Tegillarca granosa*. *Oceanologia et Limnologia Sinica, 44*(2), 431–438.

CHAPTER

16

Mussel aquaculture

Jinlong Yang[a], Deborah M. Power[b], Sergey Dobretsov[c] and Xiao Liang[a]

[a]Key Laboratory of Exploration and Utilization of Aquatic Genetic Resources, Ministry of Education, Shanghai Ocean University, Shanghai, China [b]Centro de Ciências do Mar, University of the Algarve, Faro, Portugal [c]Center of Excellence in Marine Biotechnology, Department of Marine Science and Fisheries, College of Agricultural and Marine Sciences, Sultan Qaboos University, Muscat, Oman

16.1 Introduction

As the global human population is projected to reach 10 billion by 2050, the demand for protein is also expected to increase significantly. Although animal-derived protein from terrestrial sources was the primary source of protein for a long time, its production relies on nonrenewable resources (i.e., arable land) (Costa-Pierce, 2016; Flachowsky et al., 2017). In contrast, fisheries and aquaculture systems, which are derived from the vast water bodies covering 70% of the Earth's surface, represent an underutilized resource for food production that can help meet the growing demand for protein. Mussels provide high-quality protein, various vitamins, fatty acids, and excellent levels of minerals, making them an excellent source of protein for dietary needs (Gosling, 2003).

Mussels are an aquaculture species with high economic value that provide important ecosystem services, have a long history of cultivation, and a global distribution. The State of World Fisheries and Aquaculture 2024 shows that the annual global sea mussel production has remained stable at around 2 million tons. Notably, nearly 94% of mussel production is sourced from aquaculture (Avdelas et al., 2021). China, which has a long history of mussel aquaculture, is the largest producer of mussels in the world and contributes 771,200 tons. As early as the 1970s, Chinese aquaculture scientist Prof. Fusui Zhang conducted research on the experimental ecology and mariculture of the mussel *Mytilus galloprovincialis* (Zhang, 1984). Through systematic investigations into the reproductive biology and growth dynamics of mussel populations, he established innovative methodologies for artificial hatchery production and natural spat collection. These technological breakthroughs became the cornerstone for the industrial-scale mussel farming practices in China's temperate coastal zones. By the late 1980s, China's marine mussel aquaculture output had achieved global primacy, marking a pivotal

Genetic Breeding and Aquaculture of Shellfish: A Collaborative Approach to a Sustainable Future
DOI: https://doi.org/10.1016/B978-0-443-33606-5.00008-4

milestone in the nation's blue economy development. In 2022, the leading countries in mussel aquaculture included China (40%), Chile (22%), Spain (10%), with New Zealand, Italy, France, Korea, the Netherlands, Denmark, and Thailand trailing behind, followed by Canada, Greece, and the United Kingdom.

The life cycle of mussels is complicated and is still a challenge for mussel aquaculture since the mussels have planktonic larval stages and a benthic adult phase (Yang et al., 2021). Competent larvae must settle and undergo metamorphosis, which is an irreversible developmental process in the life history of mussels. The processes of larval settlement and metamorphosis are regulated by both endogenous and exogenous biotic factors, as well as by exogenous abiotic factors. The plantigrades, juveniles, and adults may detach from their primary settlement sites to settle in new habitats. Shellfish genetic breeding offers considerable advantages for the sustainability of aquaculture. By employing genetic technologies, key traits such as growth rate, stress resistance, and meat quality can be selectively enhanced (Hollenbeck & Johnston, 2018). Genetically improved traits can improve the efficiency of aquaculture and also increase its profitability. However, genetic breeding of mussels remains in an exploratory phase.

The primary function of mussel aquaculture is to provide food, but it also offers significant environmental benefits through carbon fixation and enhances the fixation of atmospheric CO_2 from human activities. Mussel farming can effectively remove nutrients from water bodies and mitigate the negative impacts of eutrophication in coastal waters. Mussels are efficient water purifiers and significantly decrease organic matter, nutrients, silt, bacteria, and viruses in the environment. Mussel rope in the aquaculture area attracts other marine species to settle and live, and it can promote local biodiversity (Sheehan, 2019).

In addition to their use as a food source, mussels also provide other high-quality products. Bioactive compounds from marine mussels exhibit benefits for human health, for example, antioxidant properties (Wang et al., 2024), antimicrobial effects (Oh et al., 2016), antiinflammatory actions (Li, Fu, Zheng, & Li, 2014), antifungal capabilities (Chikalovets et al., 2016), adhesive applications in surgery (Dinakarkumar et al., 2023), and inhibiting ability of cancer cell proliferation (Kim et al., 2011). In addition, calcium compounds derived from the shells can be a valuable resource for the pharmaceutical industry (Mititelu et al., 2022).

16.2 Aquaculture species and model

Commonly cultured mussel species and mussel species with cultivation potential are shown in Table 16.1.

16.3 Breeding

Mussel aquaculture is of significant economic and ecological value globally and is becoming increasingly important. Robust breeding systems are crucial to ensure a reliable supply of aquaculture mussels and the sustainable development of the industry.

The life history of mussels is complex, encompassing multiple stages from embryonic development to larval growth and the adult stage. Typically, mussel reproduction starts with the reproductive activities of sexually mature individuals. During the breeding season, male mussels release sperm and female mussels expel eggs, and the mixing of sperm and eggs in

TABLE 16.1 Aquaculture mussel species and their main geographic distribution.

Aquaculture Mussel Species	Distribution
Mytilus coruscus	Temperate areas along the coastal waters of China, Japan, Korea, and the Far East of Russia, from the East China Sea to the Sea of Japan. Shengsi County, Zhejiang, is the main aquaculture region for *M. coruscus*
Mytilus edulis	The North Atlantic, North Pacific, and some coastal areas in the Southern Hemisphere (Hilbish et al., 2000)
Mytilus galloprovincialis	The Mediterranean Sea, the Black Sea, and the Adriatic Sea, and many temperate marine areas and large seaports all over the world (Hilbish et al., 2000)
Perna viridis	Tropical coastal areas of East Asia, Southeast Asia, and south of the Philippines, and southern coastal provinces in China, such as Guangdong and Fujian (Wang et al., 1993)
Potential cultivable Mussel varieties	–
Perna canaliculus	Southwest Pacific: Endemic to New Zealand (Hickman & Illingworth, 1980)
Mytilus trossulus	The cold waters of the North Pacific Ocean, including Alaska, the west coast of Canada to central California, and the northeast coast of Asia (Väinölä & Strelkov, 2011)

seawater leads to egg fertilization. As the fertilized eggs develop, they undergo cell division and enter the planktonic larval stage of trochophore larvae (Fig. 16.1). The trochophore larvae disperse in seawater under the influence of water currents and their weak locomotive ability and gradually develop into D-larvae (Fig. 16.1). The D-larvae have well-developed velums, swim and feed on planktonic organisms, and when they enter the metamorphosis stage, as "pediveliger" (Fig. 16.1), they settle on substrates such as rocks, shells, or artificial structures in aquaculture (Seed, 1969). After successful settlement, the larvae metamorphose into juvenile mussels (Fig. 16.1), which continue to grow and develop until they acquire the morphological and physiological characteristics of adult mussels.

16.3.1 Natural spatfall

In the early stage of aquaculture, natural spawning was the main way to obtain mussel seed. Fishermen used their knowledge about mussel breeding season and reproductive habits to collect substrate with settled mussel larvae in natural sea areas. Mussels have two spawning seasons per year in China, and spat are produced in spring and autumn. The selection of a spat collection site is important, and factors of importance are salinity, temperature, an abundance of parent shellfish resources, and a substantial quantity of mussel larvae. The optimal time for collecting mussel spat is approximately 15–20 days following spawning. Spat collected in the spring are generally more abundant and exhibit faster growth rates compared with those harvested in the autumn. It is essential for spat collectors to be deployed at the appropriate time; premature deployment may lead to encrustation by other organisms, whereas delayed

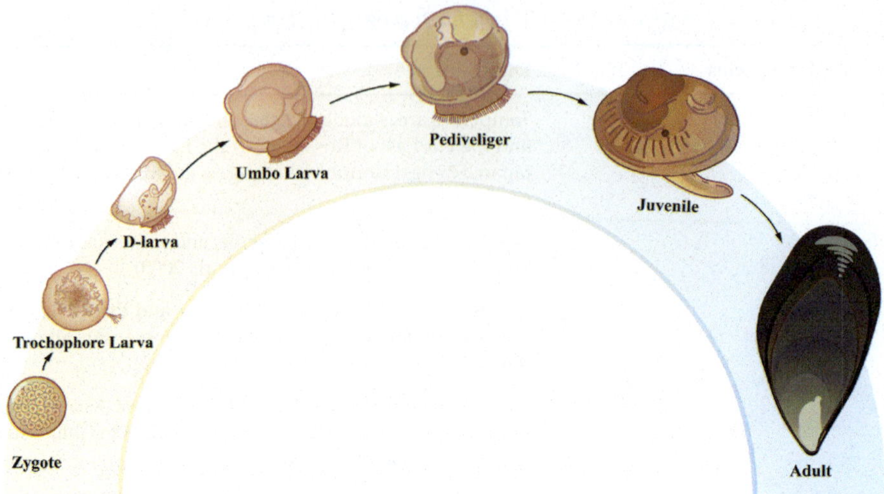

FIGURE 16.1 The different developmental stages of *Mytilus* genus mussels. Trochophore larva: Ciliary bundles; D-larva: Direct hinge line; Umbo larva: The hinge line bulging dorsally; Pediveliger: Eyespot formation; Juvenile and adult: Foot crawling and byssus thread secretion.

placement can result in reduced mussel spat yields. Effective collectors should feature a rough, textured surface and a substantial area to optimize mussel settlement. Collected substrates were then transferred to the aquaculture area, where the larvae settled naturally and grew into juvenile mussels. This method met the small-scale mussel aquaculture seed needs at that time. Since the collected seeds were from natural reproduction, they had relatively strong adaptability. However, this method was unpredictable and difficult to manage since it was strongly affected by the environment. For example, climate change, marine pollution, and overfishing can cause unstable mussel recruitment in natural sea areas, directly affecting the supply of seed. Furthermore, the collected seed had variable quality, which made aquaculture management difficult. Although natural spawning was once the main source of mussel seed, the need for a predictable supply of good-quality seed drove the development of hatcheries and knowledge-based breeding technologies.

16.3.2 Mussel hatcheries

During the production process in a hatchery, seedlings are cultivated in a controlled environment. The main steps include the selection and cultivation of broodstock (to accelerate sexual maturity), as well as spawning, gamete collection, fertilization, larval culture, induction of larval settlement and metamorphosis, and postmetamorphosis rearing. The water used in the hatchery is usually treated by sedimentation, filtration, sterilization, or disinfection to ensure the provision of a diverse supply of microalgae so that the nutritional requirements of the mussel larvae are met (Shumway, 2021). The advantage of hatcheries lies in their ability to improve the quantity and quality of seed production and the production of seed outside the natural spawning season, and to provide a platform for seed propagation and genetic manipulation. Improved control of mussel seed production through manipulation of reproduction makes

FIGURE 16.2 An illustration of a typical mussel nursery within a hatchery.

mussel seed production more reliable, which improves the sustainability of the aquaculture industry (Fig. 16.2).

16.3.2.1 Selection

In the traditional selection of broodstock in mussel hatcheries, phenotypic characteristics serve as important criteria. Generally, individuals with larger body size, rapid growth rates, intact and undamaged shells, and well-developed gonads are selected as broodstock. For example, in the cultivation of *M. edulis*, individuals with larger shell length and width, plump flesh, and bright color are chosen, as these phenotypic characteristics are often associated with better reproductive performance and growth potential of the offspring. Through long-term practical observations, farmers and the scientific community have accumulated abundant experience in selecting broodstock based on phenotypical traits, and this method is still widely used in some small hatcheries or in traditional aquaculture practice. Phenotype selection has certain limitations since it may be affected by environmental factors, and the same genotype may exhibit different phenotypes under different environmental conditions. Moreover, some important economic traits are controlled by multiple genes, and it is difficult to accurately assess genetic potential based on phenotypes.

With the development of molecular biology techniques, genotype selection based on gene sequences has gradually been applied for the selection of mussel broodstock (Hollenbeck & Johnston, 2018). Through the study of the mussel genomes, some genetic markers related to important economic traits such as growth rate, disease resistance, and stress tolerance have been identified (Hollenbeck & Johnston, 2018). For example, genes were identified that contribute to the survival ability of mussels in a low-temperature environment (Yao & Somero, 2012). By identifying genetic markers in broodstock, the researchers can select those more suitable for cultivation in cold sea areas through molecular screening (Mathiesen et al., 2017).

Currently, advanced technologies such as gene chip technology and genome-wide association study (GWAS) have started to be used for research on mussel broodstock selection. These technologies can simultaneously detect a large number of gene loci and evaluate the genetic quality of broodstock more comprehensively and accurately. However, genotype selection is still rarely utilized by industry. This is explained by the relatively high cost of the technologies and the need for sophisticated equipment and trained technical personnel. In addition, for some complex economic traits, the relationship between genes and traits has not yet been fully elucidated, requiring further research and verification. Although the application of genetics in mussel breeding remains limited, its potential for broodstock and trait selection is considerable (Nguyen et al., 2014).

16.3.2.2 *Broodstock enhancement*

Mature thick-shelled mussels (*M. coruscus*), aged 2–4 years and measuring 8–15 cm in shell length, are selected from wild or farmed sources and moved to indoor tanks with naturally filtered seawater for breeding. If the broodstock is already mature or close to maturity when collected, it is essential to stabilize the water temperature and avoid any stimuli that will trigger premature spawning in the broodstock tanks. If the broodstock is still not mature, the water temperature can be gradually increased from the ambient temperature by approximately 0.5°C per day until 20°C for *M. coruscus*. If during the warming broodstock releases sperm or eggs, they should be immediately maintained at a stable temperature to favor further maturation. The captured mussels are fed microalgae at a concentration of 1×10^8–5×10^8 cells mL^{-1} three times daily, and a complete water exchange is conducted every 2 days.

In general, the preferred salinity of mussels ranges between 25‰ and 30‰, and the dissolved oxygen content should be kept above 5 mg/L, and the seawater pH should be between 7.5 and 8.5. Providing sufficient and appropriate food is of vital importance for broodstock cultivation. Broodstock mainly feed on planktonic algae, and during the cultivation of broodstock, quantities have to be adjusted for the species of mussel and the tank density. For example, for *M. coruscus* broodstock, mixed algae such as diatoms and green algae are used. The quality and quantity of algae fed are important since overfeeding can cause deterioration of water quality, while underfeeding can negatively affect broodstock maturation. Enrichment of algae with vitamins and minerals has been used to promote shell development and improve the reproductive performance of broodstock.

16.3.2.3 *Spawning*

The temperature required for mussel spawning is higher than that needed for mussel cultivation. Prior to fertilization, broodstock mussels are initially placed out of water in clean containers for 6 h before being returned to seawater at 20°C. Sperm is expelled through the exhalant opening in a consistent, narrow stream, while eggs are released irregularly in cloudy masses. Typically, male *M. edulis* release sperm first; the presence of sperm in the water subsequently stimulates females to spawn. Once one mussel initiates spawning, other individuals tend to follow suit. After males have released their sperm, they should be promptly removed from the environment to prevent excessive sperm concentration from contaminating the water. An egg density of approximately 20 eggs/mL is deemed sufficient for successful fertilization. It is advisable to change the water every 2 days and maintain aeration throughout this process. The sperm can detect chemical signals emitted by eggs, allowing them to efficiently select eggs from conspecific but different female individuals.

16.3.2.4 *D-stage larval rearing*

After approximately 12–18 h postfertilization, the organism develops into the trochophore larval stage, during which it does not require external feeding and instead derives its nourishment from the yolk of the egg. By around 24–48 h following fertilization, the fertilized egg of mussel *M. coruscus* transforms into a D-shaped larva. The stocking density of larvae varies at different developmental stages and gradually decreases over time. For *M. coruscus*, the recommended breeding densities are as follows: for D-larvae, 10–15 larvae/mL; for umbo larvae, 10–12 larvae/mL; for pediveliger larvae, 4–6 larvae/mL. Control of the water temperature is of crucial importance in larval cultivation, as different developmental stages have different thermal requirements. For example, during the trochophore larval stage of mussels, the water temperature is generally controlled between 19°C and 21°C. As the larvae develop into the veliger larval stage, the water temperature can be raised to 20°C–22°C for *M. coruscus*.

The main feed for larvae is unicellular microalgae. In the cultivation of *M. coruscus* in Shengsi, China, microalgae such as *Isochrysis zhanjiangensis* are fed during the D-larva to pediveliger. Feeding is conducted three times daily (morning, midday, and evening) with microscopic phytoplankton species at a concentration of 1×10^8–5×10^8 cells mL^{-1} of tank water. The quality of the algae is crucial and old or contaminated algae can check the growth and reduce the general quality of the larvae. Meanwhile, stable water quality should be maintained, with the dissolved oxygen content controlled above 5 mg/L and the salinity within an appropriate range. The rearing tank water should be regularly replaced and treated with water purification equipment to eliminate harmful substances. It is recommended to perform complete water replacement every 2 days.

16.3.2.5 Settlement and metamorphosis

When larvae develop into pediveligers (shell length: 220–309 μm) with eyespots, they undergo metamorphosis, which involves a reorganization of the body structure and a transition to a settled lifestyle. During this process, the velum, which serves as a feeding organ, is lost and gills begin to develop. Pediveligers swim using their velum; however, postlarvae, known as plantigrades, crawl using their foot. After metamorphosis, an adult dissoconch shell is secreted that differs from the larval shell. When suitable substrates and inductive cues are present, swimming pediveligers will undergo metamorphosis and initiate their settled lifestyle. In the absence of such inductive cues, larvae continue to swim for several weeks. During the settlement phase, it is essential that larvae are provided with a mixed microalgal diet. Seawater should be replaced every 2 days to maintain water quality. Once the settled juveniles reach an approximate shell length of 1 cm, they are transferred to coastal nursery systems for intermediate cultivation.

To promote the settlement and metamorphosis of larvae during aquaculture substrates such as polythene corrugated plates, polyethylene ropes, shell fragments, and netting are provided. Artificial reefs also serve as excellent settlement substrates, and attract other organisms to settle on their surfaces that provide an enhanced settlement environment for mussel larvae. Some environmental factors have an important impact on the settlement and metamorphosis of mussel larvae. Moreover, natural chemical substances derived from other organisms significantly impact mussel growth and development. The biofilm formed by the aggregation of marine bacteria and the extracellular polymeric substances they secrete provide excellent settlement cues for mussel larvae. Recent studies indicate that some of the substances secreted by bacteria, including colanic acid, flagellin, and the secondary messenger c-di-GMP effectively promote the settlement and metamorphosis of *M. coruscus* larvae (Hu et al., 2024; Liang et al., 2020; Peng et al., 2020). Other non-bacterial compounds at appropriate concentrations such as neuroactive compounds, and ions also induce larval settlement or metamorphosis of the mussels (Yang et al., 2013).

16.3.2.6 Potential genetic breeding

The advancement in commercial mussel seedling production enables full-lifecycle aquaculture, ensuring independence from wild stocks and facilitating continuous broodstock improvement through selective breeding (Hollenbeck & Johnston, 2018).

Hybrid selection breeding technology is commonly used for animals and is an effective method for selecting new varieties (Bao et al., 2002). The hybrid selection breeding of mussels is still at the laboratory stage and has not been widely applied in aquaculture species. For example, the Atlantic coastal waters of Europe host both pure and hybridized populations of *M. edulis* L. and *M. galloprovincialis* L., formed through natural interbreeding (Hilbish et al., 2002). In the laboratory, *M. edulis* and *M. galloprovincialis* have been hybridized in a series of reciprocal crosses, but the hybrid larvae grew slower than larvae of either pure species (Beaumont et al., 2004). Although *M. galloprovincialis* and *M. coruscus* distribution in coastal areas of China overlap, no hybrid offspring in the natural environment have been reported. In the lab, the offspring derived from crosses between *M. galloprovincialis*♀ × *M. coruscus*♂ exhibited a remarkable advantage in growth, and they were significantly larger than the purebred groups. In laboratory settings, certain advantages may arise from the maternal effect. Hybrid breeding is a lengthy process that necessitates the timely identification of desirable traits. It is important to note that no new genes are generated during this process; instead, hybrid offspring exhibit trait segregation, which contributes to a slow and complex breeding trajectory.

Chromosome manipulation, including polyploid and gynogenesis breeding strategies, has been used for genetic enhancement and sex control in farmed aquatic species. Triploidy induction was conducted in mussels (*M. edulis*) by both heat and cold shock in 1988 (Yamamoto & Sugawara, 1988). The induction of triploidy reached 97.4% when treatment was conducted at 32°C for 10 min, starting 20 min postinsemination (Yamamoto & Sugawara, 1988). In contrast, a triploidy induction rate of 85.3% was achieved by treating fertilized eggs at 1°C for 10 min, beginning 35 min postinsemination (Yamamoto & Sugawara, 1988). Polyploid individuals have a faster growth rate, a larger individual size, and are of better quality compared with diploids, but in aquaculture, polyploid mussels are still not used. Further research and enhancement of both theoretical foundations and practical technologies are required before the application of chromosome manipulation becomes routine for mussel breeding in aquaculture.

Marker-assisted selection (MAS) uses molecular markers linked to trait genes to screen for target traits. In aquaculture, molecular markers, including restriction fragment length polymorphism (RFLP), random-amplified polymorphic DNA (RAPD), amplified fragment length polymorphism (AFLP), simple sequence repeat (SSR), and single nucleotide polymorphisms (SNPs), have been studied in the laboratory (Bao et al., 2002). A 16S rRNA RFLP assay exists for the rapid identification of Northern and Southern Hemisphere *M. galloprovincialis* (Westfall et al., 2010). SNP markers are extensively distributed throughout the genome and exhibit a high density. In comparison to other molecular markers, SNPs provide a greater amount of genetic information, which enhances their applicability in precise quantitative trait loci (QTL) mapping, construction of high-density genetic linkage maps, GWAS, and genomic selection (GS) (You et al., 2020). In the breeding of *M. galloprovincialis*, a panel of SNP markers was developed (Nguyen et al., 2014). Among them, 179 SNPs were used for family identification, and the Bayesian method based on these SNPs significantly improved the matching rate of offspring to parents, contributing to the evaluation of genetic parameters and the achievement of genetic gains (Nguyen et al., 2014). A Blue Mussel 60 K array has recently been developed for the *M. edulis* complex (*M. edulis, M. galloprovincialis,* and *M. trossulus*), enabling applications such as parentage assignment, inbreeding assessment, and species/provenance identification in breeding programs (Nascimento-Schulze et al., 2023).

The rapid development of sequencing technology has provided a wealth of genomic resources for key mussel species and GS techniques. The precise identification of genes closely linked to rapid growth, like those involved in nutrient absorption and assimilation, and metabolic pathways, will facilitate the breeding of mussels with outstanding growth potential. The sequencing and assembly of the genomes of several commercially exploited mussels, *M. coruscus* (Yang et al., 2021), *M. edulis* (Corrochano-Fraile et al., 2022), and *M. galloprovincialis* (Murgarella et al., 2016) will facilitate the identification of genes associated with traits of interest. This comprehensive genomic information enhances MAS, allowing breeders to efficiently select individuals exhibiting desirable traits. Assembled genomes offer a detailed genetic map crucial for understanding the complex mechanisms behind interesting traits. Such insights enable the accurate localization of beneficial genes, thereby establishing a critical foundation for molecular breeding and promoting the effective advancement of mussel aquaculture.

16.3.3 Seeding

Suspended culture mode: In the suspended culture mode, the seed is allowed to settle on culture ropes for open-sea aquaculture. This is achieved by allowing juveniles that have settled and metamorphosed to attach to preprepared polyethylene culture ropes. Generally, the specific density depends on factors such as the environmental conditions of the sea area, the aquaculture goals, and the management. When deploying ropes in the high seas, they are typically positioned at the spring tide low water line, which remains permanently submerged by seawater under relatively stable temperature and salinity conditions. Other environmental parameters of importance are exposure to sufficient sunlight, food, and good-quality water.

Off-bottom culture mode: For seed cultured in off-bottom systems, knowledge about the seabed topography, substrate, and ecological environment is required. After selecting a suitable bottom sowing area, the hatchery-raised juveniles are evenly scattered on the seabed (Shumway, 2021). Conservation of the seabed ecological environment is a priority during the release procedure since it affects mussel aquaculture productivity. Consequently, specialized release equipment, such as seed sowing ships, have been developed and are used.

Raft culture mode: After 2 months of growth on the nursery rope or spat collector, juvenile mussel reaches a size of approximately 0.7–1.0 cm. Spat collectors are then retrieved, and the mussel seed is stripped from them. Soaking is typically performed manually onshore by wrapping the seed mussels around the grow-out rope using a piece of mesh. The grow-out rope measures 2.6–2.8 m in length and is constructed from polyethylene with a diameter of about 5 cm, or it may consist of a rope hinged with four palm ropes, each having a diameter of 1.2 cm. The density of juvenile mussels is maintained between 1500 and 2000 individuals per meter of rope. Each main line is 90 m in length with a row spacing of approximately 4.0 m. Each line contains approximately 100 mussel culture ropes, with individual mussel strings measuring 2.6 m in length. The spacing between mussel culture ropes ranges from 0.5 to 0.7 m.

The deployment at sea of juvenile mussels in the raft culture system is similar to that in the suspended culture mode, but the aquaculture facilities are slightly different. The seed is attached to the culture ropes or nets attached to a raft, and then they are fixed to the raft frames (Fig. 16.3). During deployment at sea the position and depth of the raft frames are important to ensure a suitable environment is provided with stable water currents, an appropriate water temperature,

FIGURE 16.3 Raft culture of *Mytilus coruscus* in Gouqi Island, China.

abundant phytoplankton, suitable salinity, and sufficient sunlight. Settlement and growth are regularly checked, and debris and harmful organisms are removed from the rafts.

16.3.4 Breeding management

Feed management. Mussels have different feed requirements at different growth stages. During the larval stage, they mainly feed on unicellular algae, such as *Isochrysis galbana* and *Phaeodactylum tricornutum* mentioned earlier. As mussels grow and develop, adult mussels can utilize organic debris, bacteria, etc., as food sources in addition to feeding on planktonic algae. During the aquaculture of mussels, appropriate feed should be provided according to the growth stage and the aquaculture environment.

Different species of mussels exhibit variations in their developmental rate, duration of each life history stage, and dependence on environmental conditions. The reproductive season is usually affected by local environmental factors such as the water temperature and food resources, further enhancing the adaptability to habitats and the diversity of reproductive strategies. These differences are not only the outcomes of ecological adaptation but also provide valuable references for mussel aquaculture and resource management. The quantity of feed is dependent on factors such as the density of mussels, growth rate, water temperature, and water quality.

Water quality management. Water quality management is a key aspect of mussel broodstock management. Conventional water quality parameters include water temperature, salinity, dissolved oxygen, pH value, ammonia, nitrogen, and nitrate. As mentioned earlier, the water temperature should be controlled within the range suitable for mussel growth, with different requirements for different species and growth stages. The salinity is generally maintained between 25‰ and 30‰, and the stability of the salinity is controlled by monitoring and adjusting the inflow and outflow of seawater. The dissolved oxygen content should be maintained above 5 mg/L, and the dissolved oxygen in the water can be increased by using aeration equipment such as aerators and aeration devices. The pH value is maintained between 7.5 and 8.5. Ammonia, nitrogen, and nitrite are harmful substances in the aquaculture water bodies, and their concentration should be maintained at a low level through methods such as changing the tank water and biological filtration.

New developments and technology applications. In recent years, some new water quality management technologies and equipment have emerged. Biofloc technology has begun to be applied to water quality management of shellfish aquaculture. By regulating the carbon–nitrogen ratio in the water body where aquaculture takes place, the growth of beneficial microorganisms is promoted, and bioflocs are formed. Bioflocs can absorb harmful substances such as ammonia nitrogen and nitrite in the water, and at the same time, they may also serve as a food supplement for mussel aquaculture (Emerenciano et al., 2017). There are few reports about bioflocs application in the mussel hatchery. In addition, intelligent water quality monitoring systems have been widely applied (Huan et al., 2020). This system monitors water quality parameters in real time using sensors for temperature, pH value, dissolved oxygen, etc., and transmits the data to a central control system. Aquaculture farmers can view water quality data at any time using devices such as mobile phones or computers. When the water quality parameters exceed the set thresholds, the system will automatically trigger an alarm, reminding the farmers to take measures (Huan et al., 2020).

16.4 Mussel aquaculture

Mussel aquaculture is a significant and growing sector within the global aquaculture industry. The success of mussel aquaculture operations worldwide hinges on several key factors, including proper site selection, effective husbandry practices, mechanization, and an appropriate scale of operations. These elements are not only crucial for the economic viability of individual farms but also for the sustainable growth of the entire mussel aquaculture industry.

16.4.1 Site selection

Site selection constitutes a foundational and critical step in mussel aquaculture, as it directly determines the overall productivity and yield of farming operations. Mussel growth is a complex biological process influenced by environmental factors such as chlorophyll-a (Chl-a) and organic matter concentrations. Advanced modeling tools, such as the generalized "ShellSIM" framework are needed, and this framework integrates parameters including Chl-a, particulate organic matter (POM), and particulate organic carbon (POC) to optimize site suitability assessments for mussel growth (Hawkins et al., 2013). After identifying the water areas with sufficient nutrients, it is also necessary to ensure that there is an adequate water flow passing through the aquaculture farm. Only in this way,the depletion of food within mussel rafts can be prevented and, at the same time, the optimal flow velocity for mussels to feed can be provided. Hydrodynamic flow models are therefore indispensable for identifying coastal zones with sufficient current dynamics. For instance, a 6-year wave dataset from North American mussel farming regions informed modeling efforts, revealing a mean significant wave height range of 0.6–1.5 m for sustainable aquaculture (Panchang et al., 2008). Complementary research in Maine demonstrated that raft-based *M. edulis* systems filtered approximately 8×10^6 L/h and consumed 40 g chlorophyll a (chl a)/h under favorable conditions, with raft structures reducing ambient flow velocity by 75%–80% because of drag from culture ropes and predator exclusion nets. Notably, chlorophyll consumption exhibited positive correlations with both internal flow velocity and food concentration. Modeling further highlighted that hydrodynamic performance depends on raft orientation, dimensions, aspect ratio, predator net configuration, adjacent raft spacing, rope density, and diameter. Optimizing these design parameters enhances seston flux and feeding efficiency, thereby improving mussel growth rates (Newell et al., 2001).

16.4.2 Mussel aquaculture models and facilities

Historically, mussel farming worldwide has been continuously evolving through a variety of farming models and technological innovations. In China, early farming practices mainly adopted traditional methods such as bottom-seeding culture and raft culture, and later, the floating raft suspension culture method, which boasts advantages such as relatively fast growth, high yield, low cost, and convenient management, became widely used (Zhang, 1984). As early as the 13th century, France pioneered the "bouchot" farming model. Arranging wooden stakes in rows to form a vertical substrate for mussels to attach to, it significantly increased the growth rate and yield of mussels. In the late 20th century, Spain revolutionized the industry with raft culture, deploying large floating platforms in sheltered bays, which became the global standard due to its scalability and high productivity (Pérez-Camacho et al., 1991).

Modern advancements have introduced offshore and semisubmerged systems to address environmental and spatial constraints. For instance, semisubmerged longline systems and tube-based units are now used in deep waters, reducing exposure to surface waves and enhancing mussel survival (van den Burg et al., 2017). Multitrophic integrated aquaculture has gained traction, combining mussels with seaweed or fish to improve ecosystem balance and economic returns.

Mussel aquaculture facilities exhibit significant differences that depend on the region and the specific culture method employed (Table 16.2).

16.4.3 Water quality

Good water quality is of utmost importance for the health and growth of mussels in aquaculture. Mussels are filter feeders, and any deterioration in water quality can have significant consequences, including reduced growth rates, increased mortality, and poor product quality. Several key water quality parameters need to be carefully monitored to ensure optimal conditions for mussel growth. These parameters include temperature, salinity, dissolved oxygen, pH, and the presence of pollutants and harmful algal blooms. To effectively monitor water quality, farmers employ a combination of in situ sensors and laboratory analysis. In situ sensors are valuable tools as they can provide real-time data on parameters such as temperature, salinity, and dissolved oxygen. This real-time information allows farmers to quickly respond to any changes in water quality, enabling them to take immediate corrective actions if necessary. Laboratory analysis, on the other hand, is essential for detecting the presence of pollutants, harmful algal blooms, and pathogens. In areas where harmful algal blooms are a concern, farmers need to implement specific strategies to prevent mussels from accumulating toxins. Regular monitoring of phytoplankton populations is crucial in such areas, and farmers may need to implement harvesting restrictions when the risk of toxin accumulation is high. In addition, water treatment methods may be employed to reduce the levels of pollutants and maintain suitable water quality for mussel growth.

16.4.4 Disease and pest control

Diseases and pests pose significant challenges to mussel aquaculture, which varies with geographic location and affects both the survival and quality of the harvest (Table 16.3).

TABLE 16.2 Facilities for mussel culture.

Facility	Facility composition	Region	References
Longline systems	Horizontal headrope, plastic floats or polyfoam buoys, polyethylene anchor ropes, concrete or stake anchors	China	Feng et al., 2021
Raft culture systems	Modular raft, short droppers	Spain	Pérez-Camacho et al. (2013)
Submersible mussel raft	Pipe pontoons, adjustable buoyancy	China	Wang et al. (2015)
"Set and Forget" system	Fully submersible double backbone, intermediate moorings, release, and clamp system	New Zealand	Goseberg et al. (2017)
Bottom culture	Intertidal or subtidal culture plots	Europe	Ysebaert et al. (2009)
Bouchot culture	Poles, horizontal ropes	Europe	Aubin, Fontaine, Callier, & Roque d'Orbcastel, 2018
Suspended culture	Submerged ropes, socks	Europe	Filgueira et al., 2018

TABLE 16.3 Diseases and pest control reported for mussel aquaculture.

Species	Region	Impact	Management measures	References
Starfish, sea ducks, crabs	Europe	Predation affecting survival and quality	Predator nets, fens	(Kamermans et al., 2009)
Sea otters, sea star, sea snails	US West Coast	Predation threat	Predator nets similar to salmon farms	(VanBlaricom, 1988; Shumway, 2021)
Barnacles, sea squirts	German Bight, North Sea; Prince Edward Island	Biofouling leading to mussel losses	Power washing systems in Prince Edward Island (PEI), Canada	(Buschbaum & Saier, 2001; Shumway, 2021)
Mediterranean blue mussel	New Zealand	Predation, fouling, oversettlement	Video technology for blue mussel selection, ongoing research for mitigation strategies	(Atalah et al., 2017; Forrest & Atalah, 2017)
Starfish, crabs, shrimps, eider ducks, wrasses, parasites	Europe	Predation, fouling , disease, and production losses	Stringent biosecurity protocols, regular monitoring	(Avdelas et al., 2021; Meister et al., 2023)

16.5 Harvest and products of mussel

Harvesting process. In most cases, mussels are settled on culture ropes or other cultivation structures. Workers or machines need to detach the mussels from these structures. Once detached, the mussels are collected and transported. In countries with advanced mechanization, this can be done efficiently using automated collection systems. There are devices designed for mussel harvest in China, for example, the drum winch and hoisting system extract mussel from ropes onto the harvesting chute, where integrated shelling and cleaning units process the bivalves, with spent ropes automatically rewound for reuse (Xu et al., 2006). After cleaning, the mussels are sorted and size graded since this influences the price.

Mussel products. Mussels are a popular and nutritious shellfish. However, proper preservation and storage methods are essential to maintain freshness, quality, and safety. One study suggested that to maintain optimal quality and ensure health benefits, fresh and stored mussels should be consumed within 5 days when refrigerated at 4°C or below. To maintain freshness and safety, cooked mussels should be rapidly cooled and stored airtight in a refrigerator or freezer, depending on the storage time. For frozen-cooked mussels, maintaining temperatures around −30°C proves optimal for ensuring long-term storage stability. In addition, mussels can be processed into various valuable products such as canned mussels, dried mussels, mussel jerky, and extracts. Advances in technology have made it possible to extract valuable components from mussels, including adhesive proteins and polysaccharides, which can be utilized in health product development and functional foods. By combining the numerous nutrients found in mussels, such as proteins, unsaturated fatty acids, vitamins, and minerals, with other beneficial components, products with specific health benefits can be created.

16.6 Sustainable development of mussel aquaculture

At present, some progress has been made in the sustainable development of global mussel farming, mainly reflected in the following aspects:

Aquaculture management and regulations: Some countries and regions have formulated relevant policies and regulations to standardize mussel farming practices and ensure the sustainable development of the industry. For instance, Zhejiang Province in China has standardized mussel farming techniques and defines restricted farming areas, controls farming density, and stipulates the farming models, etc. Comprehensive technical guidance is provided for *M. coruscus*, and includes artificial reproduction, seedling cultivation, growth, and harvesting. The aim is to improve mussel aquaculture sustainability and minimize the impact on the marine environment. Aquaculture planning takes into account the carrying capacity of the marine ecosystem, rationally allocating the aquaculture sea areas, and avoiding overconcentration of farming areas and resource competition. The carrying capacity of shellfish aquaculture depends on the interaction between the cultured species and the ecosystem, especially the food availability for suspension feeders. Aquaculture capacity assessments have also been conducted in other regions of the globe. For example, (Filgueira & Grant, 2009) constructed a multibox dynamic ecosystem model to study the carrying capacity of mussel (*M. edulis*) aquaculture in Tracadie Bay, Prince Edward Island, Canada, providing a scientific basis for sustainable aquaculture.

Environmental protection and resource utilization: Emphasis is placed on the treatment of waste generated during mussel farming, such as discarded aquaculture facilities and seafood waste. Through methods like recycling and reuse, the impact of aquaculture waste on the marine environment is reduced, and in the case of mussel shells they are processed into fertilizers. Some regions have carried out marine ecological restoration, such as artificial reefs and planting seagrasses, which has been a strategy used to provide better ecological conditions for mussel farming. In addition, mussel farming has significant ecological impact on the surrounding mariculture bacterial communities. Compared with nonfarmed areas, mussel farming significantly shapes the seawater bacterial communities and increases diversity, biomass, and relative abundance. For example, a higher abundance of genera related to aerobic anoxygenic photoheterotrophic bacteria occurs in the seawater of mussel farming areas (He et al., 2023).

Technological innovation and application: The stable supply of high-quality seedlings is the foundation of mussel aquaculture. However, this is still challenging since husbandry methods are still immature, and this leads to unstable survival rates and growth rates of the seed. There is still limited use of genetic improvement, and the stress resistance and growth performance of mussel seed need to be improved (Nascimento-Schulze et al., 2021). Inappropriate aquaculture density and farm layout can negatively affect the growth and quality of mussels and the environment. If the aquaculture density is too high, it will intensify the competition among mussels, resulting in insufficient food resources and slow growth. Continuous exploration of new aquaculture techniques and models has enhanced the efficiency and quality of aquaculture. For example, fish-attracting habitats in raft aquaculture in marine ranches have been created, and technologies such as biofilms have been integrated to induce efficient metamorphosis of *M. coruscus* larvae, directed settlement of juvenile mussels, and optimization of the raft structure and layout for *M. coruscus*. This has effectively increased the metamorphosis rate of mussel larvae, the survival rate of juvenile mussels, and the economic benefits of aquaculture. It also provides a new technological approach to solving problems such as insufficient bait organism attraction, essential for the function of artificial reefs in marine ranches.

16.7 Conclusion

The mussel industry is progressing through the advancement of technology across the global mussel aquaculture industry. In the future, it is expected that artificial intelligence (AI) will revolutionize mussel aquaculture, enhancing every stage from selective breeding and larval rearing to open-sea farming. Selection of appropriate aquaculture models and facilities for local environmental conditions, such as climate and water quality, is contributing to productive and sustainable mussel aquaculture. In terms of postharvest processing, the mussel industry is evolving toward greater diversification and higher added value of its products. The development and implementation of cold chain logistics technology has provided robust support for long-distance transportation while maintaining the quality and safety of mussel products. However, numerous pressing challenges remain within the mussel industry, particularly concerning breeding approaches and technologies. Genetic selection is still in its infancy, and there is a notable deficiency in adaptable, fast-growing, high-quality new varieties of mussel. Consequently, accelerating research and development efforts focused on polyploid mussel breeding and genetic selection technology is essential for the future advancement of the mussel industry.

References

Atalah, J., Rabel, H., & Forrest, B. M. (2017). Modelling long-term recruitment patterns of blue mussels *Mytilus galloprovincialis*: a biofouling pest of green-lipped mussel aquaculture in New Zealand. *Aquaculture Environment Interaction, 9*, 103–114.

Aubin, J., Fontaine, C., Callier, M., & Roque d'Orbcastel, E. (2018). Blue mussel (*Mytilus edulis*) bouchot culture in Mont-St Michel Bay: Potential mitigation effects on climate change and eutrophication. *The International Journal of Life Cycle Assessment, 23*(5), 1030–1041. doi:10.1007/s11367-017-1403-y.

Avdelas, L., Avdic-Mravlje, E., Borges Marques, A. C., Cano, S., Capelle, J. J., Carvalho, N., Cozzolino, M., Dennis, J., Ellis, T., Fernández Polanco, J. M., Guillen, J., Lasner, T., Le Bihan, V., Llorente, I., Mol, A., Nicheva, S., Nielsen, R., van Oostenbrugge, H., Villasante, S., & Asche, F. (2021). The decline of mussel aquaculture in the European Union: Causes, economic impacts and opportunities. *Reviews in Aquaculture, 13*(1), 91–118. doi:10.1111/raq.12465.

Bao, Z. M., Wan, J. F., Wang, J. Y., Wang, X. L., & Wang, R. C. (2002). Advance in the studies on marine economic shellfish breeding. *Journal of Ocean University of China, 32*, 567–573.

Beaumont, A. R., Turner, G., Wood, A. R., & Skibinski, D. O. F. (2004). Hybridisations between *Mytilus edulis* and *Mytilus galloprovincialis* and performance of pure species and hybrid veliger larvae at different temperatures. *Journal of Experimental Marine Biology and Ecology, 302*(2), 177–188. doi:10.1016/j.jembe.2003.10.009.

Buschbaum, C., & Saier, B. (2001). Growth of the mussel *Mytilus edulis* L. in the Wadden Sea affected by tidal emergence and barnacle epibionts. *Journal of Sea Research, 45*(1), 27–36. doi:10.1016/s1385-1101(00)00061-7.

Chikalovets, I. V., Kovalchuk, S. N., Litovchenko, A. P., Molchanova, V. I., Pivkin, M. V., & Chernikov, O. V. (2016). A new Gal/GalNAc-specific lectin from the mussel *Mytilus trossulus*: Structure, tissue specificity, antimicrobial and antifungal activity. *Fish and Shellfish Immunology, 50*, 27–33. doi:10.1016/j.fsi.2016.01.020.

Corrochano-Fraile, A., Davie, A., Carboni, S., & Bekaert, M. (2022). Evidence of multiple genome duplication events in *Mytilus* evolution. *BMC Genomics, 23*(1), 340. doi:10.1186/s12864-022-08575-9.

Costa-Pierce, B. A. (2016). Ocean foods ecosystems for planetary survival in the Anthropocene. In E. M. Binder (Ed.), World nutrition forum: Driving the protein economy (pp. 301–320). Getzersdorf: Erber, AG.

Dinakarkumar, Y., Arravind, A., & Mohan, N. M. (2023). Medical adhesives from extracted mussel adhesive proteins. In K. L. Mittal, & S. Neogi (Eds.), Adhesives in biomedical applications (pp. 213–251). Hoboken: Wiley.

Emerenciano, M. G. C., Martínez-Córdova, L. R., Martínez-Porchas, M., & Miranda-Baeza, A. (2017). Biofloc technology (BFT): A tool for water quality management in aquaculture. London: InTech.

Feng, D. J., Meng, A., Wang, P., Yao, Y. B., & Gui, F. K. (2021). Effect of design configuration on structural response of longline aquaculture in waves. *Applied Ocean Research, 107*, 102489. doi:10.1016/j.apor.2020.102489.

Filgueira, R., & Grant, J. (2009). A box model for ecosystem-level management of mussel culture carrying capacity in a coastal bay. *Ecosystems, 12*(7), 1222–1233. doi:10.1007/s10021-009-9289-6.

Filgueira, R., Grant, J., & Petersen, J. K. (2018). Identifying the optimal depth for mussel suspended culture in shallow and turbid environments. *Journal of Sea Research, 132*, 15–23. doi:10.1016/j.seares.2017.11.006.

Flachowsky, G., Meyer, U., & Südekum, K. H. (2017). Land use for edible protein of animal origin—A review. *Animals, 7*(3), 25. doi:10.3390/ani7030025.

Forrest, B. M, & Atalah, J. (2017). Significant impact from blue mussel *Mytilus galloprovincialis* biofouling on aquaculture production of green-lipped mussels in New Zealand. *Aquaculture Environment Interactions, 9*(1), 115–126. doi:10.3354/aei00220.

Goseberg, N., Chambers, M. D., Heasman, K., Fredriksson, D., Fredheim, A., & Schlurmann, T. (2017). Technological approaches to longline- and cage-based aquaculture in open ocean environments. In B. Buck, & R. Langan (Eds.), Aquaculture perspective of multi-use sites in the open ocean: The untapped potential for marine resources in the anthropocene (pp. 71–95). Berlin: Springer.

Gosling, E. M. (2003). Bivalve molluscs: biology, ecology and culture. Oxford: Blackwell.

Hawkins, A. J. S., Pascoe, P. L., Parry, H., Brinsley, M., Black, K. D., McGonigle, C., Moore, H., Newell, C. R., O'Boyle, N., Ocarroll, T., O'Loan, B., Service, M., Smaal, A. C., Zhang, X. L., & Zhu, M. Y. (2013). Shellsim: A generic model of growth and environmental effects validated across contrasting habitats in bivalve shellfish. *Journal of Shellfish Research, 32*(2), 237–253. doi:10.2983/035.032.0201.

He, J. Y., Jia, M. X., Wang, J. X., Wu, Z. Q., Shao, S., He, Y. T., … Zhang, X. L., et al. (2023). *Mytilus* farming drives higher local bacterial diversity and facilitates the accumulation of aerobic anoxygenic photoheterotrophic related genera. *Science of the Total Environment, 856*(Pt 1), 158861. doi:10.1016/j.scitotenv.2022.158861.

Hickman, R. W., & Illingworth, J. (1980). Condition cycle of the green-lipped mussel *Perna canaliculus* in New Zealand. *Marine Biology, 60*(1), 27–38. doi:10.1007/bf00395603.

Hilbish, T. J., Carson, E. W., Plante, J. R., Weaver, L. A., & Gilg, M. R. (2002). Distribution of *Mytilus edulis, M. galloprovincialis,* and their hybrids in open-coast populations of mussels in southwestern England. *Marine Biology, 140*(1), 137–142. doi:10.1007/s002270100631.

Hilbish, T. J., Mullinax, A., Dolven, S. I., Meyer, A., Koehn, R. K., & Rawson, P. D. (2000). Origin of the antitropical distribution pattern in marine mussels (*Mytilus* spp.): Routes and timing of transequatorial migration. *Marine Biology, 136*(1), 69–77. doi:10.1007/s002270050010.

Hollenbeck, C. M., & Johnston, I. A. (2018). Genomic tools and selective breeding in molluscs. *Frontiers in Genetics, 9*, 253. doi:10.3389/fgene.2018.00253.

Hu, X. M., Peng, L. H., Wang, Y. Y., Ma, F., Tao, Y., Liang, X., & Yang, J. L. (2024). Bacterial c-di-GMP triggers metamorphosis of mussel larvae through a STING receptor. *npj Biofilms and Microbiomes, 10*, 51. doi:10.1038/s41522-024-00523-7.

Huan, J., Li, H., Wu, F., & Cao, W. J. (2020). Design of water quality monitoring system for aquaculture ponds based on NB-IoT. *Aquacultural Engineering, 90*, 102088. doi:10.1016/j.aquaeng.2020.102088.

Kamermans, P., Blankendaal, M., & Perdon, J. (2009). Predation of shore crabs (*Carcinus maenas* (L.)) and starfish (*Asterias rubens* L.) on blue mussel (*Mytilus edulis* L.) seed from wild sources and spat collectors. *Aquaculture, 290*(3–4), 256–262. doi:10.1016/j.aquaculture.2009.02.031.

Kim, E. K., Kim, Y. S., Lee, S. J., Jeon, Y. J., Ahn, C. B., Kim, Y. T., … Kang, H., et al. (2011). Effect of partially purified lipid from the mussel *Mytilus coruscus* on apoptosis in cancer cells. *Journal of Applied Biological Chemistry, 54*(1), 59–65. doi:10.3839/jksabc.2011.008.

Li, G. P., Fu, Y. Q., Zheng, J. S., & Li, D. (2014). Anti-inflammatory activity and mechanism of a lipid extract from hard-shelled mussel (*Mytilus coruscus*) on chronic arthritis in rats. *Marine Drugs, 12*(2), 568–588. doi:10.3390/md12020568.

Liang, X., Zhang, X. K., Peng, L. H., Zhu, Y. T., Yoshida, A., Osatomi, K., & Yang, J. L. (2020). The flagellar gene regulates biofilm formation and mussel larval settlement and metamorphosis. *International Journal of Molecular Sciences, 21*, 710. doi:10.3390/ijms21030710.

Mathiesen, S. S., Thyrring, J., Hemmer-Hansen, J., Berge, J., Sukhotin, A., Leopold, P., … Bekaert, M., et al. (2017). Genetic diversity and connectivity within *Mytilus* spp. in the subarctic and Arctic. *Evolutionary Applications, 10*(1), 39–55. doi:10.1111/eva.12415.

Meister, N., Langbehn, T. J., Varpe, Ø., & Jørgensen, C. (2023). Blue mussels in western Norway have vanished where in reach of crawling predators. *Marine Ecology Progress Series, 721*, 85–101. doi:10.3354/meps14416.

Mititelu, M., Stanciu, G., Drăgănescu, D., Ioniță, A. C., Neacșu, S. M., Dinu, M., … Stefan-van Staden, R. I., et al. (2022). Mussel shells, a valuable calcium resource for the pharmaceutical industry. *Marine Drugs, 20*(1), 25. doi:10.3390/md20010025.

Murgarella, M., Puiu, D., Novoa, B., Figueras, A., Posada, D., & Canchaya, C. (2016). A first insight into the genome of the filter-feeder mussel *Mytilus galloprovincialis*. *PLOS One, 11*(3), e0151561. doi:10.1371/journal.pone.0151561.

Nascimento-Schulze, J. C., Bean, T. P., Houston, R. D., Santos, E. M., Sanders, M. B., Lewis, C., & Ellis, R. P. (2021). Optimizing hatchery practices for genetic improvement of marine bivalves. *Reviews in Aquaculture, 13*(4), 2289–2304. doi:10.1111/raq.12568.

Nascimento-Schulze, J. C., Bean, T. P., Peñaloza, C., Paris, J. R., Whiting, J. R., Simon, A., Fraser, B. A., Houston, R. D., Bierne, N., & Ellis, R. P. (2023). SNP Discovery and genetic structure in blue mussel species using low coverage sequencing and a medium density 60 K SNP-array. *Evolutionary Applications, 16*(5), 1044–1060. doi:10.1111/eva.13552.

Newell, C. R., Wildish, D. J., & MacDonald, B. A. (2001). The effects of velocity and seston concentration on the exhalant siphon area, valve gape and filtration rate of the mussel *Mytilus edulis*. *Journal of Experimental Marine Biology and Ecology, 262*(1), 91–111. doi:10.1016/S0022-0981(01)00285-4.

Nguyen, T. T. T., Hayes, B. J., & Ingram, B. A. (2014). Genetic parameters and response to selection in blue mussel (*Mytilus galloprovincialis*) using a SNP-based pedigree. *Aquaculture, 420-421*, 295–301. doi:10.1016/j.aquaculture.2013.11.021.

Oh, R., Lee, M. J., Kim, Y. O., Nam, B. H., Kong, H. J., Kim, J. W., ... An, C. M., et al. (2016). Isolation and purification of antimicrobial peptide from hard-shelled mussel, *Mytilus coruscus*. *Journal of Life Science, 26*(11), 1259–1268. doi:10.5352/JLS.2016.26.11.1259.

Panchang, V. G., Jeong, C., & Li, D. (2008). Wave climatology in coastal maine for aquaculture and other applications. *Estuaries and Coasts, 31*(2), 289–299. doi:10.1007/s12237-007-9016-5.

Pérez-Camacho, A., Labarta, U., Vinseiro, V., & Fernández-Reiriz, M. J. (2013). Mussel production management: Raft culture without thinning-out. *Aquaculture, 406-407*, 172–179. doi:10.1016/j.aquaculture.2013.05.019.

Peng, L. H., Liang, X., Chang, R. H., Mu, J. Y., Chen, H. E., Yoshida, A., ... Osatomi, K., et al. (2020). A bacterial polysaccharide biosynthesis-related gene inversely regulates larval settlement and metamorphosis of *Mytilus coruscus*. *Biofouling, 36*, 753–765. doi:10.1080/08927014.2020.1807520.

Pérez-Camacho, A., Gonzales, R., & Fuentes, J. (1991). Mussel culture in Galicia (N.W. Spain). *Aquaculture, 94*, 263–278.

Shumway, S. E. (2021). Molluscan shellfish aquaculture: A practical guide. London: 5m Books.

Seed, R. (1969). The ecology of *Mytilus edulis* L. (Lamellibranchiata) on exposed rocky shores: I. Breeding and settlement. *Oecologia, 3*(3–4), 277–316. doi:10.1007/BF00390380.

Sheehan, E. V., Bridger, D., Mascorda Cabre, L., Cartwright, A., Cox, D., Rees, S., Holmes, L., & Pittman, S. (2019). Bivalves boost biodiversity. *Food Science and Technology, 33*, 18–21.

Väinölä, R., & Strelkov, P. (2011). *Mytilus trossulus* in Northern Europe. *Marine Biology, 158*(4), 817–833. doi:10.1007/s00227-010-1609-z.

van den Burg, S. W. K., Kamermans, P., Blanch, M., Pletsas, D., Poelman, M., Soma, K., & Dalton, G. (2017). Business case for mussel aquaculture in offshore wind farms in the North Sea. *Marine Policy, 85*, 1–7. doi:10.1016/j.marpol.2017.08.007.

VanBlaricom, G. R. (1988). Effects of foraging by sea otters on mussel-dominated intertidal communities. In G. R. Van-Blaricom, & J. A. Estes (Eds.), The community ecology of sea otters (pp. 48–91). Berlin: Springer.

Wang, R. C., Wang, Z. P., & Zhang, J. Z. (1993). Marine shellfish aquaculture. Qingdao: Qingdao Ocean University Press.

Wang, X. E., Liao, Z., Yang, Q. M., Ye, Y. Y., Shen, W., Liu, H. H., ... Yan, X. J., et al. (2024). Characterization of a novel antioxidant byssal protein from *Mytilus coruscus* foot. *International Journal of Biological Macromolecules, 273*, 133095. doi:10.1016/j.ijbiomac.2024.133095.

Wang, X. X., Swift, M. R., Dewhurst, T., Tsukrov, I., Celikkol, B., & Newell, C. (2015). Dynamics of submersible mussel rafts in waves and current. *China Ocean Engineering, 29*(3), 431–444. doi:10.1007/s13344-015-0030-2.

Westfall, K. M., Wimberger, P. H., & Gardner, J. P. A. (2010). An RFLP assay to determine if *Mytilus galloprovincialis* Lmk. (Mytilidae; Bivalvia) is of Northern or Southern hemisphere origin. *Molecular Ecology Resources, 10*(3), 573–575. doi:10.1111/j.1755-0998.2009.02779.x.

Xu, J., Dong, Y., Zhang, Y. M., & Luo, L. Z. (2006). On the equipment and technology of mechanized workbench used in harvest and process of mussel. *Journal of Ship and Ocean Engineering, 2*, 98–100.

Yamamoto, S., & Sugawara, Y. (1988). Induced triploidy in the mussel, *Mytilus edulis*, by temperature shock. *Aquaculture, 72*(1–2), 21–29. doi:10.1016/0044-8486(88)90143-3.

Yang, J. L., Feng, D. D., Liu, J., Xu, J. K., Chen, K., Li, Y. F., ... Zhu, Y. T., et al. (2021). Chromosome-level genome assembly of the hard-shelled mussel *Mytilus coruscus*, a widely distributed species from the temperate areas of East Asia. *GigaScience, 10*(4), giab024.

Yang, J. L., Li, S. H., Li, Y. F., Liu, Z. W., Liang, X., Bao, W. Y., & Li, J. L. (2013). Effects of neuroactive compounds, ions and organic solvents on larval metamorphosis of the mussel *Mytilus coruscus*. *Aquaculture, 396–399*, 106–112. doi:10.1016/j.aquaculture.2013.02.039.

Yao, C. L., & Somero, G. N. (2012). The impact of acute temperature stress on hemocytes of invasive and native mussels (*Mytilus galloprovincialis* and *Mytilus californianus*): DNA damage, membrane integrity, apoptosis and signalling pathways. *Journal of Experimental Biology, 215*(Pt 24), 4267–4277. doi:10.1242/jeb.073577.

You, X., Shan, X., & Shi, Q. (2020). Research advances in the genomics and applications for molecular breeding of aquaculture animals. *Aquaculture, 526*, 735357. doi:10.1016/j.aquaculture.2020.735357.

Ysebaert, T., Hart, M., & Herman, P. M. J. (2009). Impacts of bottom and suspended cultures of mussels *Mytilus* spp. on the surrounding sedimentary environment and macrobenthic biodiversity. *Helgoland Marine Research, 63*(1), 59–74. doi:10.1007/s10152-008-0136-5.

Zhang, F. (1984). Mussel culture in China. *Aquaculture, 39*(1-4), 1–10. doi:10.1016/0044-8486(84)90255-2.

CHAPTER

17

Pearl oyster aquaculture

Jiale Li[a], *Dean Jerry*[b], *Kyall Zenger*[b], *Zhiyi Bai*[a], *Qigen Liu*[a], *David Jones*[c], *Cecile Massault*[b] *and Zhiyan Wang*[a]

[a]Key Laboratory of Freshwater Aquatic Genetic Resources, Ministry of Agriculture and Rural Affairs, Shanghai Ocean University, Shanghai, China [b]Tropical Futures Institute, James Cook University, Singapore [c]Centre for Sustainable Tropical Fisheries and Aquaculture, College of Science and Engineering, James Cook University, Townsville, Australia

17.1 Introduction

17.1.1 Overview of the global importance of pearl oyster aquaculture

Pearls are among the oldest known gemstones, with evidence of their collection dating back at least 8,500 years (Ainis et al., 2019). Pearls are highly valued for their beauty and rarity, and have played a significant historical role in various civilizations worldwide. In many cultures, pearls hold symbolic significance, representing wealth, beauty, and purity (Fig. 17.1). The value of pearls has long been associated with the spiritual realm, granting them significance beyond mere monetary worth. The widely traveled 11th-century Persian explorer Al-Biruni once stated, "the desire of pearls is universal across all nations," while Pliny the Great reported that to demonstrate the affluence of Ancient Egypt, the Egyptian queen Cleopatra dissolved a pearl in vinegar and drank the solution in front of the Roman emperor Mark Antony.

Pearls are formed within the soft tissues of certain mollusks, including oysters, mussels, and snails. Ancient civilizations, including those in China, India, and the Persian Gulf region, harvested pearls from natural oyster beds in rivers and oceans. In the ancient world, diving was the primary method for harvesting natural pearls. Natural pearls form accidentally in wild oysters or mussels, making them extremely rare (approximately 1 individual per 10,000 oysters). However, artificial cultivation has made pearls more widely available. In response to the unprecedented demand for natural pearls and the depletion of wild oyster supplies, people developed methods to cultivate pearls through artificial propagation. These pearls, known as cultured pearls, are now the most common type on the market.

The pearl farming industry offers economic opportunities to coastal communities in developing countries (Zhu et al., 2019). Nowadays, major cultured pearl-producing countries and

Genetic Breeding and Aquaculture of Shellfish: A Collaborative Approach to a Sustainable Future
DOI: https://doi.org/10.1016/B978-0-443-33606-5.00010-2

FIGURE 17.1 Pearls from freshwater and marine species (left to right). (A) Freshwater pearls (*Hyriopsis cumingii*). (B) South sea pearls (*Pinctada maxima*). (C) Black pearls (*Pinctada margaritifera*).

regions include China, Japan, Australia, French Polynesia (Tahiti), the Philippines, Indonesia, Myanmar, Vietnam, the United States, and Mexico (Saucedo et al., 2023; Zhu et al., 2019). Pearl farming symbolizes innovation and sustainability in modern aquaculture. Integrating gem production with aquaculture, it creates a unique and valuable product. The global pearl jewelry market is projected to reach USD 20 billion by 2024.

17.1.2 Brief history and current status of pearl farming

Pearl cultivation can be traced back to China's Song Dynasty (960–1279 AD), where blister pearls were grown on freshwater mussel shells (Li et al., 2019). In the 20th century, modern pearl farming methods emerged—pioneered by Kokichi Mikimoto in Japan, marking the advent of large-scale cultured pearl production (Southgate, 2007). Today, pearl cultivation primarily involves two groups of mollusks: Marine pearl oysters and freshwater mussels. Saltwater pearls, such as Akoya, South Sea, and Tahitian pearls, are primarily produced in Japan, Australia, French Polynesia, and Indonesia using species such as *Pinctada fucata, Pinctada maxima*, and *Pinctada margaritifera*. Although freshwater pearl cultivation began in Japan in 1935, it is now dominated by China, where industrial-scale cultivation started in the 1980s. Currently, China produces more than 95% of the world's freshwater cultured pearls, using several species of freshwater mussels—including *Hyriopsis cumingii, Hyriopsis schlegelii*, and *Cristaria plicata* (Li et al., 2019).

China has emerged as a global leader in freshwater pearl cultivation. Freshwater pearl cultivation using natural pearl mussels was first demonstrated in 1958, and the first batches of cultured freshwater pearls in China were harvested in Jiangsu and Zhejiang in 1967 and 1968, respectively (Bai et al., 2014). By 1978, the breakthrough in artificial reproduction technology of freshwater pearl mussel addressed the demand for mass production of seedlings gradually. From 1979 to 1991, China's pearl industry entered its second stage of development, during which the successful artificial propagation of freshwater mussels drove annual pearl production to megaton levels. From 1992 to 2012, China's pearl industry entered its third stage of development, with advancements in production technology increasing annual pearl output to kiloton levels. Since 2013, China's freshwater pearl industry has shifted from high-yield practices to a more focused, high-quality pearl production. In 2023, China's freshwater pearl production was 755 tons (Bureau of Fisheries and Management, 2024). Besides, China also produces marine pearls, primarily from Akoya oysters. After artificial breeding was achieved in 1965, production peaked at approximately 20 tons annually in the 1990s but has since declined to around 2 tons by 2023 (Bureau of Fisheries and Management, 2024).

Japan remains one of the key producers of marine pearls, supported by mature and systematic cultivation technologies. Large-scale production began in 1945 and peaked at 127 tons in 1966 (Alagarswami, 1987). Japan developed a comprehensive pearl cultivation system in the 20th century, with marine pearls once generating an average annual production value of USD 127 million—accounting for 51.6% of the global market (Zhu et al., 2019). Although production has declined—dropping below 15 tons by 2023—demand for high-quality Japanese pearls continues to grow. In 2022, auction pearl prices rose by 70% year-on-year, with exports over JPY 20 billion. Despite rising competition from China, Australia, and Indonesia, Japan retains a strong position in the global market.

Australian pearls are renowned for their high purity, large size, silk-like luster, and diverse shapes, particularly the "Australian South Sea Pearl," a type of South Sea pearl distinguished by its exceptional quality and luster. The industry began in the late 19th century, and modern pearl farming was established in 1956 at Kuri Bay, Western Australia. Australia dominated the global South Sea pearl market until 1974. Today, pearl farming remains a key part of Australia's fisheries, with annual production of around 3 tons and export revenues nearing AUD 200 million, making it the second-largest fishery in Western Australia after rock lobster.

French Polynesia is the world's leading producer of Tahitian black pearls, with a distinct history of cultivation. In 1961, French veterinarian Jean-Marie Domard began research on artificial pearl culture, successfully harvesting 1,000 pearls by 1963. A 1965 collaboration between the local fisheries department and a Japanese company further advanced the industry, leading to the start of exports in 1972. By the late 20th century, French Polynesia had become a major global supplier. Pearl farming is a key economic sector, with production centered in the Society, Gambier, and Tuamotu archipelagos. As of 2021, the region had 10 million pearl oysters and produced 12.9 metric tons of cultured pearls annually, accounting for nearly half of its export revenue.

Several countries with relatively short histories in pearl cultivation now play important roles in the global market. Indonesia has surpassed Australia to become the world's largest producer of South Sea pearls, accounting for 43% of global supply. Other smaller producers include Myanmar, Malaysia, and Papua New Guinea, which have also developed successful *P. maxima* pearl farming (Zhu et al., 2019). Since the mid-1970s, several other Pacific Island nations, such as the Cook Islands, have produced "black" South Sea pearls from the black-lip pearl oyster, *Pinctada margaritifera*. The development of pearl oyster culture offers economic and livelihood opportunities for Pacific Island nations. Recognizing the global trade potential of cultured freshwater pearls, Bangladesh, Korea, the Philippines, Thailand, Vietnam, Brazil, and Mexico have recently initiated both research and industrial-scale projects (Saucedo et al., 2021).

In summary, pearl culture enables the production of pearls with specific characteristics, such as size, shape, and color, and has become a significant global industry. Pearl farming is an attractive business venture because of the high value of the final product, and the area for pearl cultivation continues to expand.

17.2 The main pearl oyster species

17.2.1 Freshwater pearl mussel

Freshwater mussels (Bivalvia: Unionidae) are a highly diverse family, with approximately 700 species being widely distributed across the Northern Hemisphere and Africa. Large

FIGURE 17.2 The visual appearance of *Hyriopsis cumingii* and pearls it produces.

freshwater mussels with a shell length exceeding 10 cm are generally considered capable of producing high-quality pearls. *Hyriopsis cumingii*, endemic to China, is the world's most important freshwater pearl mussel. *H. cumingii* possesses a large, flat, triangular shell, with a dorsal margin that extends upward to form a sail-like wing. Its outer shell is dark brown to yellowish-brown, while the inner nacreous layer is bright. *H. cumingii* is considered a highly suitable species for freshwater pearl cultivation (Fig. 17.2). Pearls produced from this species are prized for their high yield, smooth and dense luster, bright and colorful (including white, gold, pink, and the rare purple). The refined nucleus insertion technique, along with the species' natural pearl-producing advantages, has cemented its dominant position in the pearl industry, accounting for more than 70% of the global annual pearl output (including both freshwater and saltwater pearls) (Li et al., 2019). *Hyriopsis schlegeli*, native to Lake Biwa in Japan, is another key species for freshwater pearl cultivation. After its introduction to China, it has become the second most widely cultured species, after *H. cumingii*, in regions such as Jiangxi, Hunan, and Hubei provinces. The large freshwater mussels *Cristaria plicata* and *Anodonta woodiana*, commonly found in the rivers and lakes of East and Southeast Asia, are also important species for pearl cultivation. Among these, *Cristaria plicata* has developed a notable industrial scale in China, especially in the production of image and large nucleated pearls.

The large freshwater mussels *Margaritifera margaritifera, Margaritifera auricularia, Anodonta cygnea*, and *Anodonta anatina*, naturally distributed in rivers and streams across Europe, Northern Asia, and North America, have adult shell lengths exceeding 10 cm, with some individuals reaching up to 20 cm. These mussels have thick nacre in their shells. Limited studies suggest that these species hold potential for pearl cultivation, with *M. margaritifera* showing the greatest potential. However, due to small populations, slow growth, and underdeveloped cultivation methods, their commercial cultivation remains largely unexplored. In addition, in the Tabasco region of Mexico, a series of studies have been conducted on freshwater pearl mussels to evaluate the pearl production potential of several local species, including *Psoronaias crocodilorum, Nephronaias mexicanus, Rangia cuneata*, and *Potamilus alata*. Among these, *Psoronaias crocodilorum* was selected for further development, and initial-scale freshwater pearl production has already been established (Saucedo et al., 2021).

FIGURE 17.3 The silver-lipped pearl oyster (*Pinctada maxima*) showing its silver and gold nacre variations.

17.2.2 Marine pearl oysters

Marine pearl oysters (Bivalvia: Pteriidae) form the backbone of the global cultured pearl industry, with three species—*Pinctada maxima*, *Pinctada margaritifera*, and *Pinctada fucata*—dominating commercial saltwater pearl production. These oysters, found throughout the Indo-Pacific, are each associated with a distinct type of cultured pearl, varying in size, color, and value.

Pinctada maxima, known as the silver-lip and gold-lip pearl oyster, is the largest of all pearl-producing species (Fig. 17.3). Native to the tropical waters of northern Australia, Southeast Asia, and parts of the Indian Ocean, adult oysters can exceed dinner plate sizes of 25 cm in shell diameter. This large size allows for the cultivation of South Sea pearls, prized globally for their impressive size (typically 10–16 mm, occasionally over 20 mm), high nacre quality, and colors ranging from brilliant white and silver to deep golden hues. Australia leads the world in South Sea pearl production, underpinned by a carefully managed wildstock fishery, and more recently, sophisticated hatchery-rearing systems and implementation of selective breeding for pearl quality (Jerry et al., 2012). The industry's emphasis on quality over quantity, coupled with a relatively long grow-out period (2–6 years), has positioned *P. maxima* pearls at the premium end of the market (Taylor & Strack, 2008).

Pinctada margaritifera, the black-lip pearl oyster, is native to the Indo-Pacific and produces the renowned Tahitian black pearls (Fig. 17.4). These pearls exhibit diverse overtones—green, peacock, aubergine, and blue—derived from the oyster's dark nacre. *P. margaritifera* typically reaches 15–20 cm in shell size and is cultured extensively in lagoon environments across French Polynesia, the Cook Islands, and Fiji. The French Polynesian industry, in particular, has refined its grafting and grow-out methods to produce consistent, high-quality pearls. Selective breeding programs are ongoing, aiming to improve growth rates, pearl shape, and color predictability (Teaniniuraitemoana et al., 2016). Despite these advances, black pearl production remains sensitive to environmental changes and lagoon management practices, which are key to sustaining pearl quality and oyster health.

Pinctada fucata, often referred to as the Akoya pearl oyster and South China pearl oyster, plays a central role in the production of the classic white and cream-colored pearls (Fig. 17.5).

FIGURE 17.4 The black-lipped pearl oyster shows an example of the many natural colors that its nacre comes in.

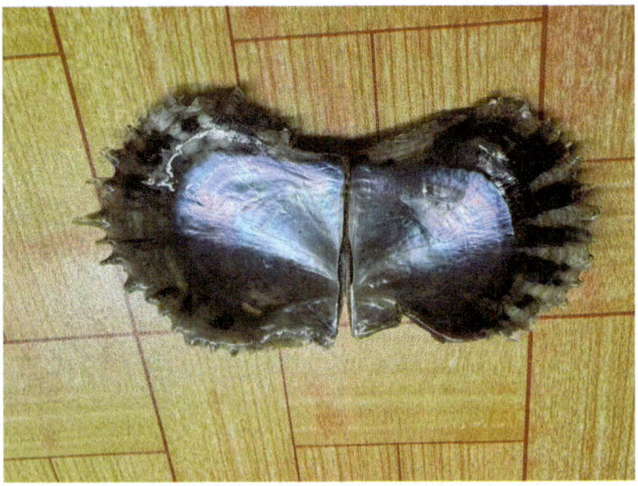

FIGURE 17.5 Akoya pearl oyster and South China pearl oyster (*Pinctada fucata*).

Native to the cooler waters of Japan, China, and Vietnam, *P. fucata* is a smaller species (typically 6–10 cm in shell diameter), which results in smaller pearls (commonly 6–8 mm), but with a highly desirable high-gloss luster. Akoya pearls are known for their perfectly round shape and mirror-like finish, achieved through precise grafting techniques and relatively short cultivation times (usually less than 2 years). Japan pioneered modern pearl culture with this species, thanks to the pioneering work of Kokichi Mikimoto in the early 20th century. Although China has emerged as a significant producer, Japan maintains a reputation for superior quality through meticulous farming practices, post-harvest treatments, and quality grading systems.

Other species, such as *Pinctada radiata* and *Pinctada imbricata*, have been used historically in pearl culture across the Middle East and South Asia, but their role today is minor compared to the global dominance of *P. maxima*, *P. margaritifera*, and *P. fucata*.

17.3 Genetics and genetic improvement

17.3.1 Genetic basis of shell and pearl quality traits

17.3.1.1 Freshwater pearl oyster

Modern pearl cultivation has developed an integrated system of donor and host pearl mussels for optimized pearl production. Researchers have systematically examined the genetic structure of key pearl quality traits, including size, color, and luster, within this specialized system. For the representative freshwater pearl mussel *H. cumingii*, pearl yield and size are significantly correlated with the shell width and body weight of host mussels (Li et al., 2015). Pearl color and luster are significantly correlated with the inner shell nacre color and luster of the donor mussel (Sun et al., 2022).

Genetic structure analysis revealed that growth traits in *H. cumingii* are typical quantitative traits (Wang et al., 2025). Inner shell color trait is particularly complex; it is governed by oligogenic inheritance, involving a few large-effect loci as well as large numbers of small-effect loci (Wang et al., 2025). Moreover, evaluations of trait heritability, based on pedigree and genome-wide SNP markers, reveal that growth traits—including shell length, shell height, shell width, shell weight, anal angle radial rib length, and body weight—exhibit moderate to high heritability estimates, ranging from 0.14 to 0.51 (Bai et al., 2017; Hu et al., 2021; Wang et al., 2022). The heritability of color traits, including lightness (L), redness (a), yellowness (b), and color difference (dE), ranges from 0.09 to 0.67 (; Hu et al., 2021; Wang et al., 2023), whereas luster heritability varies from 0.14 to 0.37 (Sun et al., 2022). Further studies on genotype-by-environment interactions indicate that the growth and inner shell color traits of H. cumingii vary significantly with farming location, water depth, substrate type, and light condition (Bai et al., 2017; Hu et al., 2021; Lv et al., 2024).

These findings suggest that precise genetic analysis, multienvironment breeding trials, and environmental adaptability assessments can improve selection efficiency, optimize trait performance, and advance the sustainable development of the freshwater pearl industry. Actually, many of these research findings have already been applied to guide genetic improvement in the freshwater pearl industry, significantly advancing its development. Researchers are now systematically evaluating the potential of different traits for joint selection, aiming to enable intelligent molecular breeding and produce high-quality freshwater pearls across various dimensions.

17.3.1.2 Marine pearl oysters

Selective breeding programs for *P. maxima*, *P. margaritifera*, and *P. fucata* have primarily targeted improvements in shell growth and pearl quality traits. Across these species, growth traits such as shell height, shell length, and shell weight typically exhibit moderate heritability, indicating favorable potential for selection (Kvingedal et al., 2010; Jones et al., 2014).

In *P. maxima*, pearl traits—including size, weight, luster, and color—exhibited moderate donor-based heritability ($h^2 = 0.13–0.25$), whereas shape demonstrated low heritability, implying a greater environmental influence (Jerry et al., 2012). Genetic correlations revealed trade-offs; for instance, silver inner shell color was negatively associated with growth, and faster pearl growth often compromised luster. Genotype-by-environment (G × E) interactions were evident for pearl traits such as size and luster, leading to genotype reranking across different culture environments (Kvingedal et al., 2010). In *P. margaritifera*, donor effects were significant for nacre thickness, color, and weight, with moderate heritability estimates ($h^2 = 0.26–0.37$), whereas shape traits exhibited low genetic control ($h^2 < 0.05$) (Blay et al., 2018). Strong genetic correlations were observed among deposition traits (e.g., nacre thickness and weight, $rg \approx 0.99$), but negative correlations with luster indicate quality trade-offs. G × E interactions were significant only for luster. Gene expression analyses further confirmed the critical role of donor mantle tissue in biomineralization. In *P. fucata*, selective breeding has achieved notable improvements in shell color and size. Heritability estimates for shell traits ranged from 0.2 to 0.7, although some values may be overestimated (Wang et al., 2010; Wada & Komaru, 1996). Selection for white shell color significantly enhanced the production of white pearls. Although direct heritability estimates for pearl quality in this species are lacking, host sex significantly influenced pearl size and quality, with males producing superior results.

Collectively, these findings confirm that shell and pearl traits exhibit varying degrees of heritability across species, and that donor genetics play a pivotal role in determining pearl quality. However, trait trade-offs and genotype–environment interactions must be carefully considered to optimize breeding strategies.

17.3.2 Population structure and germplasm of main farmed pearl oyster

17.3.2.1 *Freshwater pearl oysters*

H. cumingii, H. schlegelii, and *C. plicata* are the most important species of freshwater pearl mussels. Research on the demography, genetic diversity, and pearl production traits of *H. cumingii*, along with the identification and evaluation of genetic germplasm from various geographical populations, selectively bred varieties (also referred to as elite varieties), and strains, has been systematic and comprehensive (Bai et al., 2014; Wang et al., 2014). China's freshwater pearl mussels possess abundant genetic resources. Systematic evaluations of the growth performance, pearl production traits, population structure, and genetic diversity of wild *H. cumingii* populations from Poyang, Dongting, Taihu, Chaohu, and Hongze Lakes have been conducted. Notably, the wild populations from Poyang, Dongting, and Taihu Lakes demonstrate high genetic diversity and superior aquaculture performance (Li et al., 2009; Qian et al., 2003; Wang et al., 2008). Further research indicates that the hybrid combination, with the Poyang Lake population as the maternal parent and the Dongting Lake population as the paternal parent, exhibits optimal overall performance (Dong et al., 2008). Based on this finding, the Freshwater Aquaculture Germplasm Innovation Team at Shanghai Ocean University has successfully bred two elite varieties: "Shenzi No. 1" and "Shenzhe No. 3," through directed selection. In addition, a high-quality assembly of the *H. cumingii* genome has been completed (Bai et al., 2024).

Other species of freshwater pearl mussels are distributed across the United States, Vietnam, Thailand, India, Brazil, Mexico, and several countries and regions in Latin America (Saucedo et al., 2021). Research efforts in these countries and regions have primarily focused on surveying freshwater mollusk resources and developing pearl cultivation techniques, while largely remaining at a preliminary stage in establishing a pearl production system.

17.3.2.2 *Marine pearl oysters*

Pearl oysters (*P. maxima, P. margaritifera,* and *P. fucata*) are widely distributed across the Indo-Pacific and exhibit high genetic diversity in wild populations (Lal et al., 2018; Lind et al., 2007, 2009). Despite their long pelagic larval durations promoting extensive gene flow, population genetic studies reveal varying degrees of structure and strong signals of local adaptation. These patterns are shaped by environmental gradients, oceanographic barriers, and historical demography (Nayfa & Zenger, 2016).

Wild populations typically show high heterozygosity (e.g., $He = 0.6–0.7$ in *P. maxima*) and rich allelic diversity, especially in central Indo-Pacific regions such as Indonesia and French Polynesia (Lind et al., 2007). However, some isolated populations (e.g., the Marquesas Archipelago in *P. margaritifera*) display reduced diversity, likely because of founder effects (Reisser et al., 2019). Genome-wide SNP studies further reveal that while neutral markers indicate high connectivity (global $F_{ST} < .05$) (Takeuchi et al., 2020), outlier loci under selection can show much stronger differentiation (e.g., F_{ST} up to 0.348 in *P. maxima*) (Nayfa & Zenger, 2016). In contrast, hatchery or farmed stocks often suffer from reduced genetic variation due to practices like mass spawning and limited broodstock use. For instance, *P. maxima* cultured populations showed only 30%–44% of wild allelic richness and very low effective population sizes ($Ne < 10$) (Lind et al., 2009). While *P. fucata* hatcheries in China appear to maintain diversity through the use of mixed wild broodstock, homogenization because of long-term translocation has been observed in *P. margaritifera* (Lal et al., 2018).

Going forward, germplasm conservation and breeding programs must emphasize genetic management. This includes preserving diverse wild populations (especially from core and isolated regions), maximizing broodstock diversity, and monitoring inbreeding. Conservation strategies should be region-specific and informed by both neutral and adaptive genetic markers. Genomic tools now allow precise tracking of diversity and adaptation, which is critical for ensuring long-term resilience and productivity in pearl oyster aquaculture.

17.3.3 The progress of genetic improvement

17.3.3.1 Freshwater pearl oyster

Innovations in the seed industry are crucial drivers of industrial development. Globally, genetic improvement of freshwater pearl mussels remains limited; most countries still rely on wild collection or have only recently begun artificial propagation. China has emerged as a global leader in this field. Since the 1990s, Chinese researchers have launched systematic selective breeding programs for freshwater pearl mussels, driving innovation in the seed industry and advancing the development of the pearl sector.

The "Kangle" mussel (Certified in 2007), the first elite variety of Chinese freshwater pearl mussels, was successfully developed, hybridized between the *H. schlegelii* population as the maternal parent and the *H. cumingii* population from Poyang Lake as the paternal parent. The "Kangle" mussel exhibited significant heterosis (i.e., hybrid vigor) and strong disease resistance. Compared to wild *H. cumingii*, "Kangle" mussels increased by 31.96% in pearl yield, and increased by 23.32% the size of pearls, and the ratio of pearls with ≥8 mm in diameter increased by 3.72 times.

The new variety "Shenzi No. 1" (Certified in 2018) was developed through a combination of mass selection and family selection, targeting deep purple nacre and large size, achieved over five generations (Bai et al., 2017). It is characterized by deep purple inner shell nacre, with 95.6% of individuals displaying this characteristic. This marks the first successful targeted cultivation of purple pearls, and the proportion of harvested pearls with such coloration reaches nearly 50% after 18 months (Fig. 17.6A).

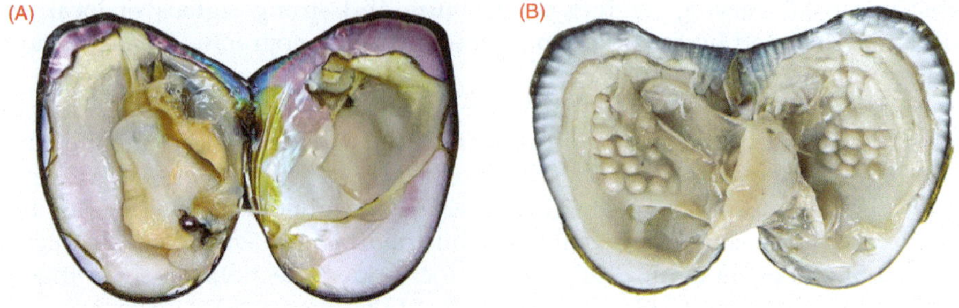

FIGURE 17.6 Visual appearance of "Shenzi No. 1" and "Shenzhe No. 3" and their produced pearls. (A) The visual appearance of "Shenzi No. 1" and the pearls it produces. (B) The visual appearance of "Shenzhe No. 3" and the pearls it produces.

The new variety "Shenzhe No. 3" (Certified in 2021) was developed through family selection, targeting traits such as body weight and shell width. After four generations of selective breeding, it showed a 16.2% increase in body weight and a 10.3% increase in shell width compared to the wild population. The proportion of large nonnucleated pearls larger than 8 mm rose by 20.3%, and the proportion of large nucleated pearls larger than 10 mm increased by 22.9% (Fig. 17.6B).

Two other elite varieties, "Zhebai No. 1" (Certified in 2020) and "Pozhu No. 1" (Certified in 2020), also demonstrated superior culture performance. "Zhebai No. 1" primarily produces white freshwater pearls, whereas "Pozhu No. 1," a new variety of *H. schlegelii*, is distinguished by its strong capacity for producing larger-sized pearls.

Overall, significant progress has been made in the genetic improvement of freshwater pearl mussels, with improved varieties being adopted at a rate exceeding 90%, fostering the growth of China's freshwater pearl industry. However, the emphasis on traits such as growth and color in new varieties remains too narrow to meet the industry's evolving demands. Future efforts will focus on breeding high-quality mussels that produce pearls with larger sizes, purer colors, and stronger luster to better cater to the industry's growing requirements.

17.3.3.2 Marine pearl oysters

Selective breeding in pearl oysters has progressed from concept to application, supported by genomic advances across *P. maxima*, *P. margaritifera*, and *P. fucata*. Early reliance on wild broodstock has shifted to hatchery-based seed production, enabling controlled breeding and trait selection. In major seawater pearl oyster-producing regions such as Australia, Japan, and French Polynesia, selective breeding efforts primarily focus on adaptability and aquaculture performance, emphasizing the development and application of commercially superior strains. In China, significant achievements have been made in the selective breeding of *P. fucata*, with several nationally approved aquatic genetic improvement varieties already released and widely promoted.

The new variety "Haiyou No. 1" (Certified in 2011) was developed through hybridization between the Indian culture population as the maternal parent and the Sanya wild population as the paternal parent. This variety exhibits a fast growth rate and a high proportion of top-quality pearls. Compared with the locally cultured *P. fucata* in Hainan Province, the pearl formation rate increased by 15.9%, and the proportion of high-quality pearls improved by 17.7%.

The new variety "Haixuan No. 1" (Certified in 2014) was developed through a combination of mass selection and family selection, targeting traits such as shell width and shell length. Compared with the control group, the shell width and length of the variety have increased by more than 16%, and the pearl retention rate, pearl layer thickness, and yield have increased by more than 17% as well.

Two other elite varieties, "Nanzhen No. 1" (Certified in 2015) and "Nanke No. 1" (Certified in 2015), also demonstrated superior culture performance. "Nanzhen No. 1" showed increases of over 10.0% in shell height and 25.0% in weight, and the rate of high-quality pearls was over 60.0%. While "Nanke No. 1" showed increases by 14.2% in shell width, the rate of high-quality pearls reached over 49.0%.

Across species, challenges include long generation times, environmental influences on traits, and infrastructure limitations. However, with the continued development of low-cost genotyping, trait-linked markers, and integrated breeding platforms, genetic improvement of pearl oysters is becoming increasingly feasible and essential for sustainable aquaculture.

17.4 Artificial reproduction

17.4.1 Artificial reproduction of freshwater pearl oyster

Freshwater pearl mussels possess a distinctive reproductive strategy involving three life history stages. The first stage, male mussels release sperm into the water, which females capture through filter feeding and water currents, facilitating internal fertilization. Fertilized eggs develop within the female's outer gill chamber, advancing through stages of cleavage, blastula, gastrula, intracapsular glochidia, and mature glochidia (parasitic larval stages of freshwater mussels that require fish hosts) (Hu et al., 2021). The second stage, the female mussel releases mature glochidia larvae from its gill chamber into the water. These larvae drift with the currents, using hook-like structures to attach to fish gills, skin, or fins, where they undergo metamorphosis. In the third stage, the glochidia larvae undergo metamorphosis into juvenile mussels. Upon completing their transformation, juveniles detach from their host fish and adopt a benthic lifestyle. The age at which they reach sexual maturity varies among species, generally requiring 2–4 years. For example, the artificial propagation of *H. cumingii* includes four key steps: (1) broodstock selection and conditioning, (2) pairing and fertilization, (3) yellow catfish (*Pelteobagrus fulvidraco*) hosting, and (4) juvenile cultivation (Fig. 17.7).

Broodstock selection begins in mid-November. Ideal brood mussels should be healthy, exhibit powerful siphon water jetting, have clean shells free of biofouling, and possess large body weight and wide shell width. In early March of the following year, male and female mussels are identified by the density of their outer gill filaments. If more precise identification is required, a bit of germ cells collected from the gonads is examined under a microscope. They are separated into algal-rich ponds for conditioning to prevent premature fertilization in the females.

Pairing begins in early April when water temperatures stabilize at around 24°C. To enhance fertilization success, females are placed on the downwind side of the pond and males on the upwind side, maintaining a 2:1 female-to-male ratio. After one week, the fertilized eggs are checked daily at 6 a.m. Developmental progress is monitored by gently sampling the outer gill filaments with a bamboo stick for microscopic inspection. When a substantial number of glochidia larvae are observed to have broken free from their membranes, the females are

FIGURE 17.7 Paired greenhouse for the breeding of *Hyriopsis cumingii* (left) and microflowing water greenhouse for the breeding of young mussels.

removed from the pond, dried in the shade stimulation for 4 h, and placed in a tank with yellow catfish to facilitate glochidia parasitism.

Small yellow catfish (approximately 50 g) are considered optimal hosts. The parasitism process lasts for 10–20 min, during which white specks (attached glochidia) become visible on the gills of the fish. Studies indicate that maintaining a hosting density of 1,500–2,000 glochidia per fish ensures moderate stress levels and a high metamorphosis rate of the glochidia. The yellow catfish are then transferred to a greenhouse microflow juvenile rearing pond for temporary conditioning. Each independent 1 m^2 rearing pond has a plastic membrane lining the bottom, covered with a 1 cm thick layer of nutrient-rich mud. The water depth is controlled at approximately 15 cm, with 20 yellow catfish placed in each pond. A black mesh net is placed above the pond to prevent the fish from leaping out. The greenhouse water temperature is typically maintained at around 28 $\pm$ 2°C. The glochidia larvae develop in about 7–10 days, metamorphosing into juvenile mussels, which then detach. During the temporary rearing period, the ponds are inspected three times daily—morning, noon, and evening—to remove dead fish and maintain water clarity. After one week, the gills of the yellow catfish are checked, and once the white specks (glochidia) are almost gone, the fish are removed from the pond.

The mussels enter the juvenile mussel cultivation stage. The microflow aquaculture method is maintained, with the water exchange rate gradually increasing and water fertility enhancing as the mussel juveniles grow. The rearing ponds are inspected daily, with harmful organisms such as apple snails (*Pomacea canaliculata*) and other pests (*Bellamya quadrata*) promptly removed. The juvenile mussel stage in the greenhouse typically lasts around 40 days. Once the juvenile mussels reach a shell length of over 1 cm, they are harvested and transferred to outdoor earthen ponds for suspended culture at a water depth of 40–60 cm. Initially, the bottoms of the mesh cages are covered with nutrient-rich sediment. After 2 months of cultivation, the stocking density needs to be adjusted, and the mussels are then transferred to mesh belts for continued suspended culture.

17.4.2 Artificial reproduction of seawater pearl oyster

The three species of marine pearl oysters, *P. maxima*, *P. margaritifera*, and *P. fucata*, are all broadcast spawners. In a partially synchronized manner, pearl oysters release millions of eggs and sperm in the water column, resulting in external fertilization. In the wild, fertilization rates are extremely low. Once pearl oyster eggs are fertilized through external spawning (zygote),

FIGURE 17.8 Examples of mass spawning and single pairing mating tables for *Pinctada maxima*.

they begin a delicate developmental process with extremely high mortality. The fertilized eggs develop into trochophore larvae within a day, then progress through several planktonic stages—D-stage, veliger, and finally pediveliger—over a span of about 2–3 weeks. During this time, the larvae drift with ocean currents, feeding on plankton, but are highly vulnerable to predation, starvation, and environmental stressors such as temperature fluctuations and pollution. Only a tiny fraction—less than 1%—survives long enough to find a suitable hard substrate, settle, and metamorphose into juvenile oysters (spat).

In terms of artificial reproduction, for the silver-lipped pearl oyster, broodstock are brought into the hatchery and cleaned from biofouling prior to gonad examination and sex determination during the spawning seasons (February to April, and September to November). Typically, because of the protandrous nature of the pearl oyster, females are larger than males. Once assessed, oysters with good gonad conditions (produced a large number of gametes) are put either in mass-spawning tanks with ~ 30 other oysters for a mass-spawn, or in a spawning table with one or two other oysters to produce a smaller number of families (Fig. 17.8). In a commercial hatchery, broodstock are selected according to their potential (such as size, color), but as well extra care is undertaken to ensure that mating between related individuals is avoided. This stage is guaranteed by genotyping the broodstock, determining relatedness among individuals, and establishing a mating plan (Massault et al., 2021).

To induce spawning, one common established method is to vary the water temperature (Southgate, 2008). Water level can also be used to induce spawning by dropping and refilling several times to mimic the tide. In addition, a male oyster can be sacrificed, with the sperm extracted from the gonad and added into the spawning tank, either fresh or frozen, to prevent its activity. Interestingly, Utami et al. (2024) found that natural spawning between wild male and hatchery females had the best spawning rate, while natural spawning between wild males and females had the best seed performance, whereas artificial induction (sperm and heat shock) did not result in as good a spawning rate or performance. Another method that can be used in artificial reproduction is stripped spawning, where gametes from both males and females are extracted. Gametes are treated with ammonium to ensure viability. Once eggs are fertilized in a pearl oyster hatchery, they undergo a series of developmental stages under carefully controlled conditions. Within a few hours, the fertilized eggs begin to divide and develop into free-swimming larvae known as trochophores. These larvae soon transform into veliger larvae, which have a tiny shell and use cilia for movement. Over the next several days, they feed on a cocktail of microalgae provided in the larval rearing tanks to support their growth.

Akoya, black-lipped, and silver-lipped pearl oysters all settle and change into spat at approximately the same age of 20 to 23 days and a size of roughly 250 μm (Southgate, 2008). In the hatchery, collectors are then used to collect the spat. Collectors are typically made of materials such as fine mesh, plastic strips, or fibers that provide an ideal surface for the larvae to attach to. The texture and structure of the collector mimic natural settlement surfaces found in the ocean, encouraging the larvae to settle and transform into spat. After settlement, the spat remains attached to the collectors as they grow, making it easier to monitor and handle them. Once they reach a suitable size, the juvenile oysters can be removed from the collectors and transferred to nursery systems for further rearing. In the nursery stage, the spat is kept on the collector while being reared for 3–7 weeks for black and silver-lip pearl oyster and until 3 μm for Akoya oyster (Southgate, 2008), at which stage the spat is transferred to the ocean for grow-out.

17.5 Pearl implantation

17.5.1 Nonnucleated pearl implantation

Nonnucleated pearls are primarily freshwater pearls. An allograft (2–4 mm^2) was prepared from the pallial zone of a mussel mantle (donor), from which the inner epithelium had been removed, and was implanted into the middle mantle of other mussels (host) through the incision (Xiong et al., 1980). In freshwater pearl aquaculture with *H. cumingii*, 1-year-old juvenile mussels (shell length: 6–10 cm) are typically used as donor mussels. Each can produce 60–120 allografts, sufficient for seeding three host mussels. Each half-shell of a host mussel can be implanted with 10–20 allografts. Pearls produced using this method exhibit vibrant color, thick nacre, and high luster. However, there are significant drawbacks, such as irregular shapes, with most pearls being elliptical and perfectly round pearls being rare. In addition, the growth cycle is relatively long, typically taking 3–5 years.

17.5.2 Nucleated pearl implantation

Nucleated pearls originated from seawater pearl cultivation in the last decade. Freshwater nucleated pearls have also become mainstream gradually. In pearl aquaculture with the pearl oyster *Pinctada fucata martensii*, a tissue fragment from the pallial zone of one pearl oyster's mantle (donor) is implanted into the gonad of other oysters (host). A small inorganic bead made from the shell is also inserted and put in contact with the allograft. After the operation, the outer epithelial cells undergo proliferation and healing, eventually forming enclosed pearl sacs surrounding the bead, which secrete nacre and layer it around the nucleus to form a pearl. Beads for pearl cultivation vary in size (e.g., 2 mm, 6 mm, and 10 mm), with implantation sites determining pearl type. The mantle cavity, ideal for small, round pearls, can host up to 30 bead nuclei per side in *H. cumingii*. In contrast, the gonad accommodates larger beads, typically 1–2 per mussel or pearl oyster. Overall, nucleated pearls exhibit shorter growth cycles and higher economic value, making their cultivation the most widely applied technique in the pearl industry. However, the introduction of a bead nucleus as a foreign material complicates

FIGURE 17.9 Shell-attached image pearls cultivated from *Hyriopsis cumingii*, engraved with the Shanghai Ocean University logo to commemorate its 110th anniversary.

implantation surgery and heightens the risk of severe immune rejection, which leads to nucleus expulsion. Future advancements should focus on oyster conditioning, surgical sterilization, and postoperative care to improve nucleus retention and pearl quality.

17.5.3 Image pearl implantation

Image pearls are a specialized type of nucleated pearl, emphasizing artistic form and deliberate shape control. In contrast, traditional nucleated pearls typically exhibit more standardized shapes because of the use of bead nuclei. Image pearls were produced by inserting intricately carved molds into the shell, where nacre gradually coated the molds, forming pearls with distinct patterns or images (Fig. 17.9). Today, advancements in 3D printing technology offer promising prospects for pearl cultivation. This innovative approach enables the production of highly intricate and finely detailed pearl nuclei, surpassing the limits of traditional manufacturing methods. In addition, 3D printing facilitates personalized customization, enabling the creation of unique pearl products tailored to consumer preferences.

17.6 Pearl oyster culture

17.6.1 Freshwater pearl mussel culture

Following implantation, appropriate aquaculture conditions are crucial to ensure optimal pearl development. Freshwater pearl mussels are filter feeders, sustaining themselves by consuming microscopic algae and organic detritus filtered from the water. Traditional freshwater pearl farming typically involves fertilizing water to promote plankton and algae proliferation, thereby enhancing mussel growth. However, this practice has caused severe water eutrophication and pollution, rendering it incompatible with modern sustainable development goals. Fundamentally, freshwater mussels act as natural water purifiers, improving water transparency and quality while controlling excessive algal growth. As ecosystem engineers, they are vital for preserving the balance and health of aquatic ecosystems. To advance eco-friendly and sustainable aquaculture, two primary farming models have emerged: fish-mussel integrated farming in ponds and nonfeeding ecological aquaculture in open water bodies.

The fish–mussel integrated farming in ponds integrates fish and mussel cultivation, capitalizing on mutualistic ecological interactions. The water depth of the fish pond is preferably 2–3 m, with the area of a single pond being at least 3 ha. For 1-year-old pearl mussels, the stocking density should range from 18,000 to 22,500 individuals per hectare; for 2-year-old mussels, it should range from 12,000 to 15,000 individuals; and for 3-year-old mussels and older, the density should not exceed 10,000 individuals per hectare. Mixed fish species may include omnivorous fish such as grass carp, carp, and crucian carp, as well as filter-feeding fish such as silver carp and bighead carp. Annual fish yield should be controlled within 7,500 to 11,250 kg per hectare. Pearl mussels filter organic matter from the water, thereby reducing nutrient levels, improving water quality, mitigating algal blooms, and stabilizing aquatic ecosystems. This process supports the healthy growth of both fish and mussels, yielding higher economic returns than traditional monoculture systems.

The second main culture model is nonfeeding-based ecological aquaculture in open water bodies, such as lakes or reservoirs. This model focuses on the carrying capacity of the farming environment, referring to the maximum stocking density or farming capacity that a water body can support while ensuring water quality, mussel health, and ecological sustainability. By monitoring factors such as water quality, ecosystem composition, biological community dynamics, and stocking density, the carrying capacity of the water area is assessed to ensure the healthy growth and sustainability of the pearl mussels. This evaluation requires long-term monitoring data, ecological models, and practical experience to continuously refine and optimize the farming plan. By integrating water quality enhancement with pearl farming, this model provides a sustainable solution for both environmental conservation and economic growth.

In addition, excitingly, the industrial aquaculture model for freshwater pearl mussels has recently emerged. The aquaculture system primarily consists of four core modules: The pearl mussel farming workshop, algae cultivation system, water circulation system, and digital monitoring system. This aquaculture system offers technical advantages for culturing top-quality pearl products by controlling the optical environmental conditions, as well as water efficiency.

17.6.2 Marine pearl oyster culture

Pearl culture, or pearling, is a well-established aquaculture practice that involves the cultivation of marine pearl oysters to produce high-value pearls. Among the most prominent species used globally are *P. maxima* and *P. margaritifera*. These species underpin the cultured pearl industries of Australia, French Polynesia, Indonesia, and several other Indo-Pacific nations. While both share fundamental aspects of biology and husbandry, their culture practices are shaped by their respective environmental preferences, pearl qualities, and regional traditions.

The culture process begins with the collection or breeding of broodstock. Traditionally, in Australia and French Polynesia, wild-caught broodstock were caught and spawned in hatcheries. However, due to increasing pressure on wild stocks and the need for greater genetic control, hatchery production is now more commonly based on the spawning of hatchery-produced oysters. Hatchery spawning is usually induced by manipulating environmental parameters, such as thermal shocks, drying in the sun, or the addition of frozen sperm into the spawning tanks. The oysters are generally mass-spawned, although again, for greater genetic control, many hatcheries use one-on-one spawning practices in partitioned raceways. Eggs, once fertilized, develop into free-swimming trochophore and then veliger larvae, which are reared in tanks with regular feeding using cultured microalgae such as *Isochrysis and*/or *Chaetoceros*. After about 3 weeks, competent pediveliger larvae settle onto prepared collectors—typically bundles of fine mesh, ropes, or netting (Fig. 17.10A). Postsettlement juveniles, or spat, are maintained

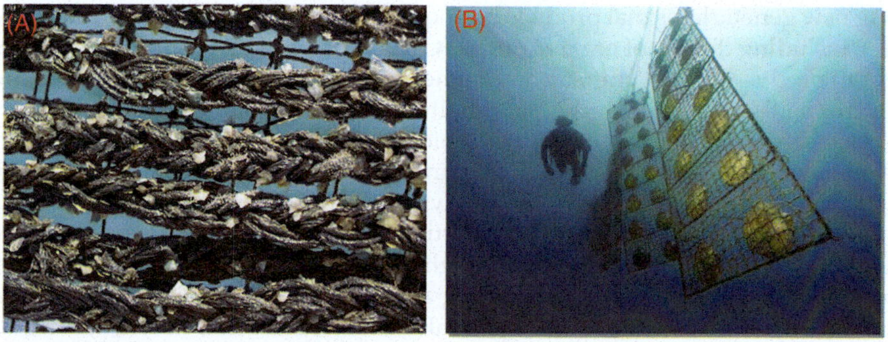

FIGURE 17.10 Oyster culture practices. (A) Settling spat on nylon ropes. (B) Adult oysters in panel-nets suspended on long-lines.

in nursery systems, usually in land-based tanks until they reach a size suitable for transfer to ocean nursery grow-out systems (Fig. 17.10B).

Juvenile oysters are transferred to ocean-based systems—either floating rafts, longlines, or lantern nets—depending on local conditions and infrastructure. The grow-out phase can last 1–2 years, during which oysters are regularly cleaned to remove biofouling and graded to ensure optimal health and growth. Fouling organisms, such as barnacles and algae, must be removed periodically to prevent smothering and reduce competition for food. For both species, site selection is critical. *Pinctada maxima* prefers the relatively clear, nutrient-poor waters of northern Australia and eastern Indonesia, often at depths of 5–10 m. In contrast, *P. margaritifera* thrives in more nutrient-rich, lagoonal environments typical of French Polynesia and the Cook Islands.

Once oysters reach a suitable size, typically 10–15 cm in dorsal length in *P. maxima*, they are ready for grafting. This is the delicate surgical procedure that defines pearl culture. A nucleus (a spherical bead made from freshwater mussel shell) is inserted into the gonad of the oyster, along with a small piece of mantle tissue from a donor oyster. The donor tissue contains epithelial cells that form the pearl sac, which secretes nacre around the nucleus to form the pearl. The grafting process requires highly skilled technicians. Success rates, pearl quality, and eventual nacre thickness are all influenced by the precision of this operation. Post-graft mortality and rejection are significant risks, so oysters are given a recovery period in low-stress conditions to maximize retention and survival. They also, in *P. maxima* culture, are x-rayed after 6 months post-implantation to see if they have retained the nucleus and will grow a pearl, and if the nucleus has been rejected, they are often reseeded. Typically, in *P. maxima*, it takes 18–24 months to grow a pearl, while in *P. margaritifera*, pearls are usually harvested after 18 months, though longer periods can yield thicker nacre and higher-value pearls. Both species can be re-grafted multiple times if the oyster remains healthy, potentially producing several pearls over its lifetime.

17.7 Future prospects

The pearl industry continues to prioritize aesthetic quality in response to consumer preferences. Improving pearl quality remains an enduring goal through continuous culture and selective breeding, as well as technical advancements. In the future, efforts should focus on developing new genetic lines that produce high-quality pearls, as well as inventing new grafting techniques, including the use of organic nuclei (Cartier & Krzemnicki, 2013) or saibo cell culture techniques to produce a more uniform and higher-quality product. The understanding of pearl formation mechanisms will push the pearl culture technology innovation. Nowadays, major progress has been made in pearl oyster genomics, including *P. martensii, P. maxima, P. margaritifera, H. cumingii*, and *M. margaritifera*, enabling cultured pearl research to advance rapidly worldwide. This signals a new era in the study of pearl aquaculture, particularly in the field of genetic improvement of pearl oysters. Of particular importance will be the improved quantitative phenotyping of pearl quality traits through computer vision methodologies (Lapico et al., 2019; Zenger et al., 2019) for selective breeding applications. The ability to obtain quantitative measurements beyond that of current categorical classified pearl trait data (i.e., color and luster) will improve genetic merit prediction and advance genetic improvement programs within the industry (Zenger et al., 2019). Furthermore, the development of new genomic prediction models considering both host and donor oyster genetic interactions will further improve genetic merit evaluations. Although currently unrealistic, genetic engineering or even pearls grown in vitro could not be ruled out in the future (Jayasankar et al., 2018).

With regard to farm husbandry, the pearl farming industry also faces significant challenges, including declining output and quality, due to the sensitivity of pearl oysters to environmental changes, pollution, and natural disasters. According to research conducted by the South China Sea Fisheries Research Institute (SCSFRI) and the University of the Sunshine Coast (USC) in Australia, global pearl production decreased by 60%, and output value fell by 39% between 2009 and 2019. Pearl oyster metabolism is dependent on water temperature, which is believed to be a determining factor in producing high-quality pearls. Lower water temperatures are thought to reduce pearl oyster metabolism and affect the nacreous structure in the final layers of the pearl, resulting in thinner nacre tablets that enhance luster. However, global warming and the associated rise in water temperature can lead to increased oyster mortality, diseases, and lower-quality nacre (Latchere et al., 2018). Ocean acidification, driven by higher levels of carbon dioxide in the atmosphere (and thus in the oceans), causes malformation of shell nacre, a significant decline in shell strength, and may also affect pearl nacre. According to the CIBJO Pearl Commission report, "greater fluctuations in water temperature, ocean acidification, and changes in the plankton profile" due to climate change may limit pearl production in the Philippines. Research and development that targets predictive analytics capability via machine learning of complex environmental data is an urgent area of need. The ability to forecast biological processes or farm outcomes based on highly complex environmental and ecological data will aid in future farm management practices. Furthermore, any improvement in production systems knowledge may ultimately lead to factory-based pearl culture technologies, which is not unforeseeable in the future.

References

Alagarswami, K., 1987. Perspectives in pearl culture contributions in marine sciences, felicitation volume, 37–49

Ainis, A., Fujita, H., Vellanoweth, R., 2019. The antiquity of pearling in the Americas: Pearl modification beginning at least 8,500 years ago in Baja California Sur, México. Latin American Antiquity 30 (3), 637–643. doi:10.1017/laq.2019.49.

Bai, Z., Lu, Y., Hu, H., Yuan, Y., Li, Y., Liu, X., ... Li, J., 2024. The first high-quality genome assembly of freshwater pearl mussel *Sinohyriopsis cumingii*: New insights into pearl biomineralization. International Journal of Molecular Sciences 25 (6), 3146. doi:10.3390/ijms25063146.

Bai, Z., Wang, G., Liu, X., Li, J., 2014. The status and development trend of freshwater pearl seed industry in China. Journal of Shanghai Ocean University 23, 874–881.

Bai, Z., Li, Q., Han, X., Li, J., 2017. Estimates of genetic parameters and genotype by environment interactions for shell nacre color and growth traits in the purple freshwater pearl mussel *Hyriopsis cumingii*. Aquaculture International 25 (6), 2079–2090. doi:10.1007/s10499-017-0170-x.

Blay, C., Planes, S., Ky, C., 2018. Crossing phenotype heritability and candidate gene expression in grafted black-lipped pearl oyster *Pinctada margaritifera*, an animal chimera. Journal of Heredity 109 (5), 510–519.

Bureau of Fisheries and Management, 2024. China Fishery Statistical Yearbook in 2023. China Agriculture Press, Beijing.

Cartier, L., Krzemnicki, M., 2013. New developments in cultured pearl production: Use of organic and baroque shell nuclei. Australian Gemmologist 25, 6–13.

Dong, Z., Li, J., Zheng, H., 2008. Morphological variability in original parents and their reciprocal hybrids F_1 of freshwater mussel *Hyriopsis cumingii* from three geographical population in China. Journal of Dalian Fisheries University 23 (2), 92–97.

Hu, H., Bai, Z., Guo, S., Feng, S., Sun, T., Chen, W., 2021. Continuous observation of early embryonic development of *Hyriopsis cumingii* under in vitro culture condition and analysis of biological zero point and effective accumulated temperature. Journal of Fisheries of China 45 (11), 1835–1842. doi:10.11964/jfc.20201212524.

Jayasankar, V., Vasudevan, S., Poulose, S., Divipala, I., 2018. Nacre formation by epithelial cell cultures from mantle of the black-lip pearl oyster, *Pinctada margaritifera*. In Vitro Cellular and Developmental Biology - Animal 54 (7), 477–485. doi:10.1007/s11626-018-0269-z.

Jerry, D., Kvingedal, R., Lind, C., Evans, B., Taylor, J., Safari, A., 2012. Donor-oyster derived heritability estimates and the effect of genotype×environment interaction on the production of pearl quality traits in the silver-lip pearl oyster, *Pinctada maxima*. Aquaculture 338-341, 66–71. doi:10.1016/j.aquaculture.2012.02.001.

Jones, D., Jerry, D., Khatkar, M., Moser, G., Raadsma, H., Taylor, J., Zenger, K., 2014. Determining genetic contributions to host oyster shell growth: Quantitative trait loci and genetic association analysis for the silver-lipped pearl oyster, *Pinctada maxima*. Aquaculture 434, 367–375. doi:10.1016/j.aquaculture.2014.08.040.

Kvingedal, R., Evans, B., Lind, C., Taylor, J., Dupont-Nivet, M., Jerry, D., 2010. Population and family growth response to different rearing location, heritability estimates and genotype × environment interaction in the silver-lip pearl oyster (*Pinctada maxima*). Aquaculture 304 (1–4), 1–6. doi:10.1016/j.aquaculture.2010.02.035.

Lal, M., Southgate, P., Jerry, D., Zenger, K., 2018. Genome-wide comparisons reveal evidence for a species complex in the black-lip pearl oyster *Pinctada margaritifera* (Bivalvia: Pteriidae). Scientific Reports 8 (1). doi:10.1038/s41598-017-18602-5.

Lapico, A., Sankupellay, M., Cianciullo, L., Myers, T., Konovalov, D.A., Jerry, D.R., Toole, P., Jones, D.B., Zenger, K.R., 2019. Digital image computing: Techniques and applications, DICTA 2019. Using image processing to automatically measure pearl oyster size for selective breeding. Institute of Electrical and Electronics Engineers Inc doi:10.1109/DICTA47822.2019. http://ieeexplore.ieee.org/xpl/mostRecentIssue.jsp?punumber=8943071.

Latchere, O., Mehn, V., Gaertner-Mazouni, N., Moullac, G.L, Fievet, J., Belliard, C., … Schubert, M., 2018. Influence of water temperature and food on the last stages of cultured pearl mineralization from the black-lip pearl oyster *Pinctada margaritifera*. PLOS ONE 13 (3), e0193863. doi:10.1371/journal.pone.0193863.

Li, J., Wang, G., Bai, Z., 2009. Genetic variability in four wild and two farmed stocks of the Chinese freshwater pearl mussel (*Hyriopsis cumingii*) estimated by microsatellite DNA markers. Aquaculture 287 (3–4), 286–291.

Li, J., Wang, D., Bai, Z., Guan, Y., Wu, C., Chen, L., 2019. Report on the development of freshwater pearl culture industry in China. China Aquaculture 23, 23–29.

Li, Q., Bai, Z., Liu, X., Han, X., Luo, H., Dong, S., Li, J., 2015. Correlation analysis of non-nucleated pearl quality parameters with growth traits and inner shell color of *Hyriopsis cumingii*. Journal of Fishery Sciences of China. 39, 1631–1639.

Lind, C.E., Evans, B.S., Knauer, J., Taylor, J.J.U., Jerry, D.R., 2009. Decreased genetic diversity and a reduced effective population size in cultured silver-lipped pearl oysters (*Pinctada maxima*). Aquaculture 286 (1–2), 12–19. doi:10.1016/j.aquaculture.2008.09.009.

Lind, C.E., Evans, B.S., Taylor, J.J.U., Jerry, D.R., 2007. Population genetics of a marine bivalve, *Pinctada maxima*, throughout the Indo-Australian Archipelago shows differentiation and decreased diversity at range limits. Molecular Ecology 16 (24), 5193–5203. doi:10.1111/j.1365-294X.2007.03598.x.

Lv, X., Bao, Q., Wang, Z., Hu, H., Zhang, Y., Wang, H., … Bai, Z., 2024. Genotype by environment interactions for growth and inner shell color disparity in *Hyriopsis cumingii* under different light conditions. Aquaculture 592, 741095. doi:10.1016/j.aquaculture.2024.741095.

Massault, C., Jones, D.B., Zenger, K.R., Strugnell, J.M., Barnard, R., Jerry, D.R., 2021. A SNP parentage panel for the silver lipped pearl oyster (*Pinctada maxima*). Aquaculture Reports 20, 100687.

Nayfa, M.G., Zenger, K.R., 2016. Unravelling the effects of gene flow and selection in highly connected populations of the silver-lip pearl oyster (*Pinctada maxima*). Marine Genomics 28, 99–106. doi:10.1016/j.margen.2016.02.005.

Qian, R., Li, J., Dong, Z., Zheng, H., Li, Y., Yuan, W., 2003. Morphological variations analysis among populations of *Hyriopsis cumingii* in five large lakes of China. Oceanologia et Limnologia Sinica 34, 436–443.

Reisser, C., Lo, C., Schikorski, D., Sham Koua, M., Planes, S., Ky, C.L., 2019. Strong genetic isolation of the black-lipped pearl oyster (*Pinctada margaritifera*) in the Marquesas archipelago (French Polynesia). Scientific Reports 9 (1). doi:10.1038/s41598-019-47729-w.

Saucedo, P.E., Acosta-Salmón, H., McLaurin-Moreno, D., Castillo-Domínguez, A., Melgar-Valdés, C.E., Mazón-Suástegui, J.M., 2021. Freshwater pearl culture in Mexico: Historic context, present status and future perspectives. Reviews in Aquaculture 13 (3), 1379–1396. doi:10.1111/raq.12527.

Saucedo, P.E., McLaurin, D., Lodeiros, C., Freites, L., León, L., Cáceres-Puig, J.I., Albuquerque, M.C.P., Southgate, P.C., Acosta-Salmon, H., 2023. Progress towards reestablishing Latin America as a major pearl producing region: A review. Reviews in Aquaculture 15 (1), 242–260. doi:10.1111/raq.12714.

Southgate, P.C., 2007. Overview of the cultured marine pearl industry. Pearl oyster health management: A manual 503, 7–17.

Southgate, P.C., 2008. Pearl oyster culture: The pearl oyster. Elsevier, pp. 231–272. doi:10.1016/B978-0-444-52976-3.00007-3.

Sun, T., Wang, H., Hu, H., Li, J., Bai, Z., 2022. Estimation of non-nucleated pearl quality traits from donor and host mussel-derived genetic parameters in the golden strain of *Hyriopsis cumingii*. Aquaculture 560, 738460. doi:10.1016/j.aquaculture.2022.738460.

Takeuchi, T., Masaoka, T., Aoki, H., Koyanagi, R., Fujie, M., Satoh, N., 2020. Divergent northern and southern populations and demographic history of the pearl oyster in the western Pacific revealed with genomic SNPs. Evolutionary Applications 13 (4), 837–853. doi:10.1111/eva.12905.

Taylor, J., Strack, E., 2008. Pearl production: The pearl oyster. Elsevier, pp. 273–302. doi:10.1016/B978-0-444-52976-3.00008-5.

Teaniniuraitemoana, V., Leprêtre, M., Levy, P., Vanaa, V., Parrad, S., Gaertner-Mazouni, N., … Le Moullac, G., 2016. Effect of temperature, food availability, and estradiol injection on gametogenesis and gender in the pearl oyster *Pinctada margaritifera*. Journal of Experimental Zoology Part A: Ecological Genetics and Physiology 325 (1), 13–24. doi:10.1002/jez.1992.

Utami, D.A.S., Nisa, P.I.H., Nisa, A.C., Ilham Abrori, M., Sudiarsa, I.N., Kusmiatun, A., Kiswanto, A., 2024. BIO Web of Conferences. In: Evaluation of pearl oyster (*Pinctada maxima*) spawning using natural and artificial induction methods to increase the seed production. EDP Sciences Indonesia.

Wada, K.T., Komaru, A., 1996. Color and weight of pearls produced by grafting the mantle tissue from a selected population for white shell color of the Japanese pearl oyster *Pinctada fucata martensii* (Dunker). Aquaculture 142 (1–2), 25–32. doi:10.1016/0044-8486(95)01242-7.

Wang, H.H., Wang, H., Yan, L., Zhang, Y., Wang, H., Lv, X., … Bai, Z., 2025. Genome-wide association analysis reveals genetic architecture of growth and inner shell color traits in freshwater pearl mussel *Hyriopsis cumingii* (Lea 1852). Aquaculture 595, 741658. doi:10.1016/j.aquaculture.2024.741658.

Wang, G., Bai, Z., Liu, X., Li, J., 2014. Research progress on germplasm resources of *Hyriopsis cumingii*. Journal of Fisheries of China 38, 1618–1627.

Wang, H., Du, X., Lü, W., Liu, Z., 2010. Estimating the heritability for growth-related traits in the pearl oyster, *Pinctada fucata martensii* (Dunker). Aquaculture Research. 42 (1), 57–64. doi:10.1111/j.1365-2109.2010.02552.x.

Wang, J., Li, J., Wang, G., Bai, Z., 2008. Sequence variation of ribosomal DNA internal transcribed spacer 1 of *Hyriopsis cumingii* from the five freshwater lakes of China. Journal of Lake Science 20, 208–214.

Wang, Z., Hu, H., Sun, T., Li, X., Lv, G., Bai, Z., Li, J., 2022. Genomic selection for improvement of growth traits in triangle sail mussel (*Hyriopsis cumingii*). Aquaculture 561, 738692. doi:10.1016/j.aquaculture.2022.738692.

Wang, Z., Hu, H., Sun, T., Lv, G., Lv, X., Wang, H., … Bai, Z., 2023. Genomic selection improves inner shell purpleness in triangle sail mussel *Hyriopsis cumingii* (Lea, 1852). Aquaculture 575, 739815. doi:10.1016/j.aquaculture.2023.739815.

Xiong, D., Wu, J., He, X., 1980. A preliminary study on the formation of nucle-free pearls in freshwater mussels. Journal of Zhanjiang Fisheries University 3, 3–9.

Zenger, K.R., Khatkar, M.S., Jones, D.B., Khalilisamani, N., Jerry, D.R., Raadsma, H.W., 2019. Genomic selection in aquaculture: Application, limitations and opportunities with special reference to marine shrimp and pearl oysters. Frontiers in Genetics 10. doi:10.3389/fgene.2018.00693.

Zhu, C., Southgate, P.C., Li, T., 2019. Production of pearls. Springer Nature, pp. 73–93. doi:10.1007/978-3-319-96776-9_5.

CHAPTER

18

Abalone aquaculture

Weiwei You[a], *Peter Cook*[b], *Caihuan Ke*[a] *and Xuan Luo*[a]

[a]College of Ocean and Earth Sciences, Xiamen University, Xiamen, China
[b]University of Western Australia, Albany, Australia

18.1 Introduction

18.1.1 Major abalone species around the world

Abalone (*Haliotis* spp.) are marine gastropods belonging to the family *Haliotidae*, and are widely recognized as one of the most valuable fishery products globally (Peng et al., 2021). These mollusks inhabit tropical and temperate waters around the world, contributing to the rich biodiversity of marine ecosystems (Hanif et al., 2022). Until recently, many authors suggested that there were about 75 species of abalone worldwide. The distribution of abalone is highest along the Pacific coast, followed by the Indian Ocean, and lowest in the Atlantic Ocean (Wu & Zhang, 2016). So far, no abalone have been found along the Arctic coast. Many species are found in temperate and subtropical waters, with fewer in tropical waters, and the least in cold regions (Geiger, 1999). However, not all species are of commercial importance, with fewer than one-third being exploited commercially, either because of their small size or rarity (Gordon & Cook, 2013). Taxonomic descriptions of the family *Haliotidae* were first published by Linnaeus in 1758, who described seven species (Cook, 2023). Several subsequent taxonomic works followed, culminating in one by Geiger and Poppe, who wrote a chapter on *Haliotidae* in the volume Conchological Iconography, and then by the most recent work of Geiger and Owen (Cook, 2023). Much of the early work on abalone taxonomy was based on museum collections of shells, with features such as shell size, weight, shape, spire height, respiratory holes, and shell coloration being useful taxonomic characteristics. Even today, however, the detailed taxonomy is still somewhat uncertain.

The best-known abalone species are those that have been exploited commercially. The distribution of economic species of abalone in different sea areas is as follows:

1) Pacific Northwest Coast

The main economic species along the coast of Japan, Korea, and China include *Haliotis discus hannai, H. discus discus, H. sieboldii, H. madaka*, and *H. diversicolor*. With the exception

Genetic Breeding and Aquaculture of Shellfish: A Collaborative Approach to a Sustainable Future
DOI: https://doi.org/10.1016/B978-0-443-33606-5.00014-X

of the subtropical and tropical species *H. diversicolor*, the others are all temperate species and are of medium size.

2) North American Pacific Coast

There are six main economic species distributed along the coast from Alaska to Mexico, namely the red abalone (*H. rufescens*), green abalone (*H. fulgens*), pink abalone (*H. corrugata*), white abalone (*H. sorenseni*), black abalone (*H. cracherodii*), and Pinto abalone (*H. kamtschatkana*). Among these, red abalone and pinto abalone are cold-water species, mainly distributed along the coasts of the United States and Canada. The shell length of red abalone can reach up to 30 cm. Green abalone, pink abalone, and other warm-water species are mainly distributed along the west coast of the United States and the western coast of Mexico (Leighton, 2000).

3) Oceania

There are six main economic species distributed in the waters around Australia and New Zealand in the southwestern Pacific Ocean. These include the blacklip abalone (*H. rubra*), greenlip abalone (*H. laevigata*), Roe's abalone (*H. roei*), and Paua (*H. iris*). The blacklip abalone, greenlip abalone, and Roe's abalone are mainly distributed in the southern part of Australia and along the coast of Tasmania. The main distribution around New Zealand's islands and reefs is the yellow-foot abalone (*H. australis*) and the white-foot abalone (*H. virginea*).

4) Southern Africa

In Africa, the main economic species is the South African abalone (*H. midae*), which is found mainly along the south-west and south coasts, whilst most of the abalone farming operations are located along the coast near Cape Town.

5) Europe

The main economic species along the European coast is *H. tuberculata*, which is mainly distributed along the coasts of France and Spain, and has a maximum shell length of up to 12 cm.

6) Middle East

The main economic species along the Arab coast of the Middle East is the Omani abalone (*H. mariae*), mainly distributed along the coast of the Sultanate of Oman.

18.1.2 Brief history and current trends in abalone fishery and aquaculture

According to the definition of the global abalone industry by the International Abalone Society, it is divided into two parts: Abalone fishery and aquaculture (Gordon & Cook, 2013). Among them, abalone fishery refers to the annual fishing quota allowed for commercial activities (set by each country), usually measured in terms of wet in-shell weight. This category includes the proliferation and release of abalone seedlings in areas without artificial attachment bases (rocks, etc.). The definition does not, however, include legal recreational fishing or illegal abalone fishing that occurs on a global scale (Cook, 2025). Abalone aquaculture refers to intensive farming using artificial ponds, net cages, or other facilities on land or in the sea, or extensive farming through the spreading of abalone seedlings into the sea (including onto artificially modified attachment substrates or facilities), with or without food. Currently, nearly 20 countries worldwide are engaged in abalone fishery and aquaculture activities. The world abalone fishery industry has a history of nearly a century, and the current trend of total quantity remains relatively stable, mainly in countries such as Australia, New Zealand, Japan, and Mexico (Cook, 2025).

Due to overfishing and poaching, however, wild resources are no longer able to meet market demand, and artificial breeding of abalone has been developed to fill this market gap. Currently, 95% of global abalone production comes from farming, with China, South Korea, and South Africa ranking as the top three in terms of aquaculture production (Cook, 2025). Aquaculture production of the main countries also varies greatly. China is an important participant in the global abalone industry and has formed a complete industrial chain, including seedling cultivation, breeding, processing, and distribution. In 2023, China's abalone production reached 245,000 tons, accounting for about 90% of the world's total aquaculture production (China Fisheries Statistics Yearbooks, 2024).

The global abalone farming industry has undergone several significant changes over the past 50 years (Cook, 2014). The initial development stage was from 1970 to 2002, during which time significant fluctuations in production volumes occurred. In 1977, global abalone production exceeded 100 tons for the first time, and in 1987, it exceeded 1000 tons for the first time. After 1991, global abalone production maintained a relatively stable growth trend, with an average annual growth rate of nearly 20%. The second stage is a sustainable growth period from 2003 to 2023. In 2003, global abalone production exceeded 10,000 tons for the first time, reaching 12,700 tons, and then steadily increased to 117,000 tons in 2013. Afterward, it continued to grow to 245,800 tons in 2021, an increase of almost 20 times compared with 2003 (Cook, 2025).

18.1.3 The rise of China's abalone industry

In China, abalone has been regarded as the jewel of seafood for thousands of years. The *Compendium of Materia Medica*, a famous guide to traditional Chinese medicine, includes the earliest written reference to abalone, stating that the abalone shell had a medicinal value and was beneficial to the eyes. In the 1700s, abalone was closely associated with royalty and nobility, as evidenced by a record of an all-abalone feast as part of a court ceremony (Ke, 2013). Today, abalone is a popular item in celebratory banquet menus for family gatherings and festivals in Chinese culture (You, 2023).

Research into abalone biology and aquaculture began in the late 1960s in China. The artificial seed production technologies of two local species, *H. diversicolor* in subtropical southern China, and *H. discus hannai* in temperate northern China, were successfully fulfilled in the 1970s (Chen et al., 1977; Du & Zhang, 2004). By the late 1980s, artificial breeding of Pacific abalone *H. discus hannai* had begun to take shape along the coasts of Liaoning and Shandong provinces, with an annual output exceeding 500 tons. But in 1993, large-scale outbreaks of mortality occurred, resulting in a sharp decline in production, and severely affecting the enthusiasm for abalone farming. Since the mid-1990s, however, the southern coastal areas have vigorously developed artificial seedling cultivation and breeding of the small abalone *H. diversicolor*, resulting in a significant increase in production and an annual output of over 2000 tons (Du & Zhang, 2004). In the early 2000s, the introduction of the Pacific abalone into Fujian waters for aquaculture was successful, and subsequently, the aquaculture area and yield showed an explosive growth trend. The Pacific abalone quickly became the dominant aquaculture species in Fujian waters. From 2003 to 2013, the aquaculture production of abalone in China rapidly increased from 11,200 tons to 110,400 tons. By 2023, the production reached 245,000 tons, with an annual output of 12.8 billion seedlings and an annual output value of more than 20 billion RBM (Gao et al., 2023). According to statistics, there are currently over 200,000 employees in the abalone industry nationwide (You, 2023). After 30 years of development, the abalone industry

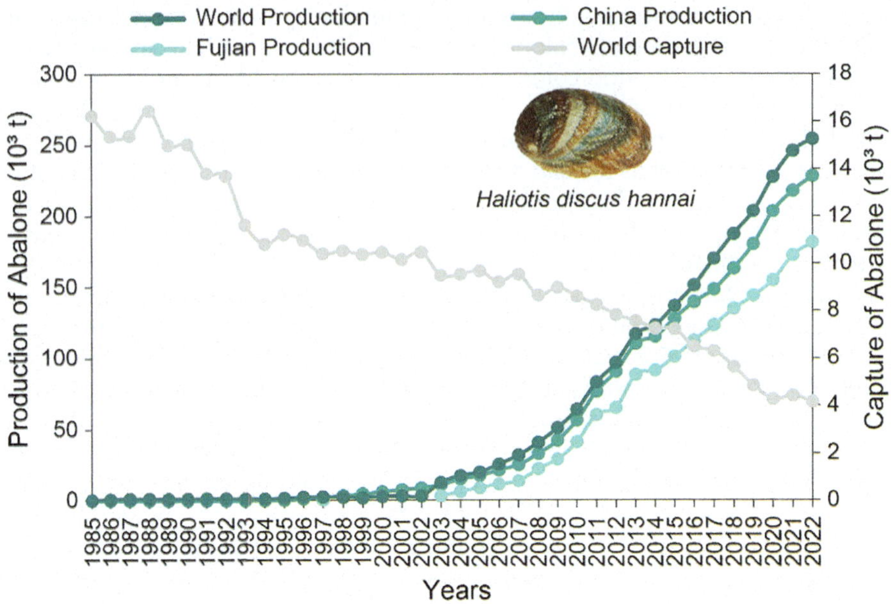

FIGURE 18.1 The world and Chinese capture and cultured abalone production.

has become one of the pillar industries of marine aquaculture in China, with a fast growth rate and important economic benefits. It has also driven the development of supporting industries such as seaweed farming, compound feed production, processing, farming equipment, and consumption, creating huge economic, social, and ecological benefits, and making the abalone industry an important part of China's marine aquaculture industry (Fig. 18.1).

Abalone aquaculture has developed rapidly in Fujian Province for several reasons. First, sea-based farming is inexpensive and simple for smallholder farmers to implement, and can also be applied successfully at large scales. Indeed, sea-based cage farming has been the primary form of aquaculture since 2010, totally replacing traditional land-based intensive farming methods (Ke, 2013; Wu & Zhang, 2016). Fujian Province has many inshore gulf and bay areas, which are protected from typhoons and have suitable seawater conditions (e.g., temperature). Second, seaweed is the main feed for abalone, and Fujian is the primary area of kelp and *Gracilaria* production in China. The Fujian province supplies millions of tons of seaweed for abalone each year. An affordable, sustainable seaweed supply is another important reason underlying the rapid development of abalone aquaculture in Fujian Province over the last decade. Third, *H. discus hannai*, which commands a better price and a larger market size than the local *H. diversicolor*, was successfully introduced from northern China (seawater temperature averaging 18°C) into southern China (seawater temperature averaging 26°C) in Fujian Province (Lin et al., 2017). The rapid, large-scale, long-distance transplantation of *H. discus hannai*, spanning 15 degrees in latitude, changed the global abalone industrial structure. Right now, around 70% of all farmed abalone worldwide is produced in the Fujian Province (Fig. 18.2).

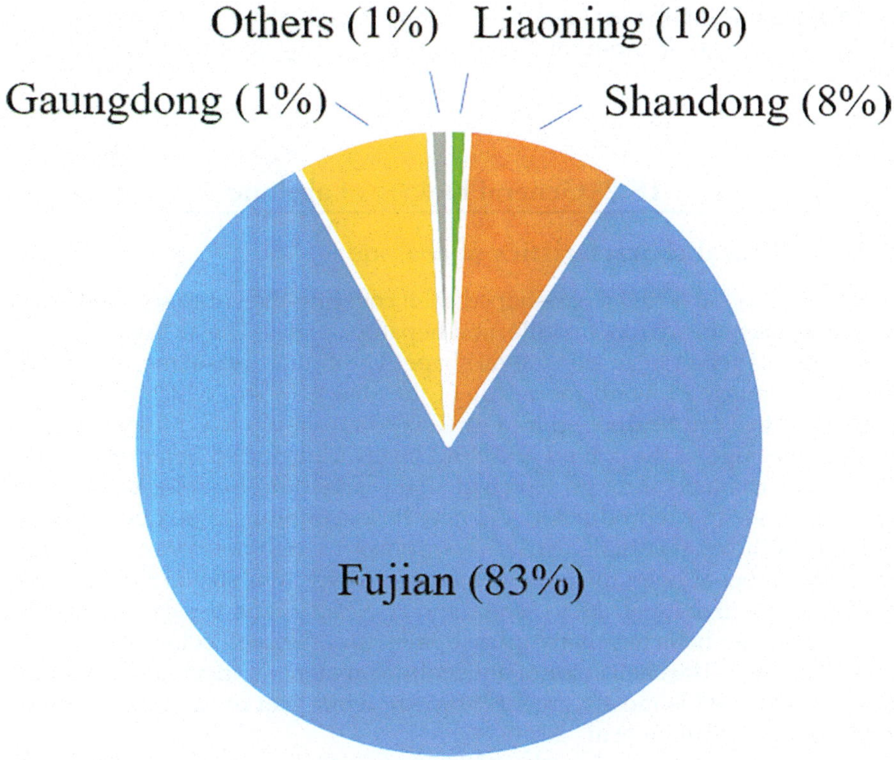

FIGURE 18.2 Abalone production by the primary abalone farming provinces in China.

18.1.4 International abalone market

The international abalone market is characterized by a complex interplay of supply, demand, and trade dynamics. As a highly valued delicacy, abalone commands premium prices in both domestic and export markets. Abalone is widely recognized by the market and the public for its rich nutritional value and unique taste. At present, the abalone in the market is mainly live abalone. However, long-distance transportation and long-term storage have limited the development of the live abalone market (Li et al., 2025). Processed products, such as frozen abalone, dried abalone, and canned abalone, can effectively solve this problem, expanding the consumption area of abalone and helping to improve its taste and flavor (You, 2023). The East Asian region, represented by China, is the main area for abalone consumption, which is closely related to the nutritional value and dietary culture of abalone products in the region.

Abalone is one of the most internationalized seafood products. Specifically, abalone exports occur in both developing and developed countries and play a significant role in the economic development of developing countries. In terms of imports, the main countries or regions are all developed economies except for China, and all East Asian countries or regions except for the United States. In terms of export, China is the main export country of abalone, accounting for more than 55% of the world's exports, followed by South Korea, Australia, and South Africa, with a total export of 19,400 tons, accounting for 34.43% of the global export volume (Cook, 2025).

In terms of abalone import, the top regions are Hong Kong, Malaysia, China, Japan, Singapore, and South Korea, with a total import of 17,500 tons, accounting for more than 90% of the world's imports (FAO, 2024).

18.2 General biology of abalone

18.2.1 Morphological characteristics of abalone

Abalone is a gastropod with an ear-shaped shell covering the entire soft body, which can be divided into four parts: head, foot, mantle, and hepatopancreas. The abalone shell is quite hard and thick, slightly flattened in an ear or oval shape. The cuticle, prismatic, and pearl layers are arranged from the outside toward the inside of the abalone shell, which is mainly composed of limestone. The shell can be divided into three layers from top to bottom, of which there are several holes at the edge of the second and third layers, called shell holes or respiratory holes. Shell holes are the channels through which abalone carry out physiological activities such as respiration, excretion, and reproduction. As new holes continue to be developed at the front of the shell, the old holes eventually close. The number of shell holes varies depending on the species, but the number of holes for each abalone species remains stable: Seven to nine holes for *H. diversicolor*, three to five for *H. discus hannai* and *H. gigantea*, and three to four for *H. rufescens*. The surface of abalone shells has many growth lines, and the shell color is mostly dark brown or dark reddish-brown. The shell color of aquaculture abalone is affected by its food. When fed on brown algae, the shell is mostly greenish-brown, while that of abalone fed on red algae is mostly reddish-brown (Bullon et al., 2023).

The foot muscle, which is flat, broad, and located on the ventral surface of the soft body, is crucial for abalone adhesion and movement. It consists of two parts: The epipodium, which has large and small sensory epipodial tentacles, and the soft and hard foot (pedal sole), which is disc-shaped in the center. A cylindrical adductor muscle connects the foot muscle to the shell. The mantle is a thin film of tissue that envelops the back of the body, with its inner edge connected to the right shell muscle, and with a free outer edge. The mantle covers the entire back of the visceral sac to form a mantle cavity. At the entrance of the mantle cavity, there is an olfactory detector, which is a yellow ridge-shaped protrusion. The cavity has two feather-shaped gills that function in respiration, hence being known as the respiratory cavity. There are two mucus glands, which are used to secrete mucus to lubricate and clean the respiratory cavity, anus, kidneys, and gills. The hepatopancreas is located on the back of the soft body, mainly surrounding the posterior edge of the adductor muscle. Its end is conical and includes organs such as the heart, kidney, stomach, intestine, angular digestive gland, and reproductive gland. It is often dark reddish-brown, but varies in gonadal color depending on sexual maturity, with male gonads being light yellow and female gonads being dark green or tan.

The digestive system of abalone includes the mouth, esophagus, crop, stomach, gastric blind sac, intestine, anus, and angular digestive gland. The mouth is located below the cephalic lobe, and there are some small mastoid-shaped protrusions, called labia minora. The oral cavity is called the oral ball, which is composed of the oral cavity, jaw plate (radula), and tongue. The tongue is shaped like a belt, and the ingested food is first chopped by the radula, then finely ground by the tongue and mixed with saliva. It is then passed into the esophagus through the pharyngeal valve. The angular digestive gland, also known as the liver or hepatopancreas, is a well-developed organ in abalone. It is a large gland that covers almost all organs, such as the crop, stomach, and gastric blind sac, occupying a considerable part of the entire internal organs.

The main respiratory organs consist of a pair of feather-shaped gills located in the front of the heart, within the mantle cavity. The left gill is larger than the right gill. One end of the gills is attached to the central axis, and the other end is free and parallel on both sides. There are many transverse folds in the center of the gill surface to increase the surface area. Its outer membrane can also provide some respiratory function.

The circulatory system is mainly composed of the heart, anterior and posterior major arteries, renal veins, extrarenal veins, gill vessels, and mantle vessels. The excretory organs include the left and right kidneys, which have two small openings. The left kidney is smaller and located in the left anterior part of the pericardial cavity. The reproductive system of abalone is relatively simple and is mainly composed of numerous branching glands, which cover the outside of the digestive gland. During the breeding season, the appearance and color of the reproductive glands of mature male and female abalone are different. The reproductive glands of male abalone are mostly a milky yellow or milky white color, while those of the female are emerald green or grayish-brown.

18.2.2 Life history

Abalone are broadcast spawners, that is, they release their gametes (eggs and sperm) into the open water, where they meet and fertilization occurs. Mature females can release up to 1 million eggs per spawning, although fertilization and survival rates are very low. Currents disperse the fertilized eggs until they hatch into motile larvae known as veligers. The veligers undergo a number of other developmental stages for 4–6 days until they settle. Once settled, the juvenile abalone are called "spat," and they graze on microalgae, which are present on the substrate. Spat can actively select a suitable substrate depending on the level of food availability. Once the spat have reached approximately 5 mm in size, they make their way into rock crevices for protection, and they then come out at night to feed on drifting seaweed.

18.2.3 Ecological habits and food

In natural sea areas, abalone prefer to live on rocky or gravelly sea beds, where there are reef crevices, larger rocks, or caves for them to attach to, and where there are nearby algae for them to eat. Different abalone species have different requirements regarding the temperature and depth of the water in their habitat. Abalone usually hide in the darker areas of the habitat during the day, while their feeding, breeding, and other activities are generally carried out at night.

The diet of abalone varies depending on its growth stage, body size, and season. Prior to the veliger larvae stage, abalone do not feed and rely solely on yolk nutrition to survive. During the stages of veliger larvae and initial attachment, abalone begin to feed on single-cell algae or organic debris. The type and quality of single-cell algae have a significant impact on the metamorphosis rate, growth rate, and survival rate of abalone larvae. As individuals develop, juvenile abalone can feed on benthic diatoms and seedlings of large algae. As juvenile abalone continue to grow and the first breathing hole appears, their food intake also increases. When the abalone shell reaches a length of over 10 mm, the types of food consumed are basically the same as those of adult abalone. Adult abalone can feed on brown algae, green algae, red algae, and benthic diatoms.

18.3 Hatchery and seed culture technology

Artificial abalone seedling production is a globally practiced approach to meet the growing demand for this prized seafood. Hatcheries worldwide employ diverse technologies and methodologies tailored to local environmental conditions and species characteristics. In the United States, for instance, land-based hatcheries with sophisticated water filtration and temperature control systems are used to culture red abalone (*H. rufescens*) (Ault, 1985; Muñoz et al., 2012). In Mexico, hatcheries often employ flow-through systems to maintain optimal water quality for culturing green abalone (*H. fulgens*) and pink abalone (*H. corrugata*) (Searcy-Bernal & Anguiano-Beltrán, 1998; Vélez-Arellano et al., 2020). Asian countries, particularly China and South Korea, are leading producers in abalone aquaculture. Chinese and Korean hatcheries often incorporate large-scale, mechanized systems to culture the Pacific abalone (*H. discus hannai*) and its hybrids (Wu & Zhang, 2016). Australia and New Zealand utilize hatcheries that integrate natural sunlight and seawater flow for culturing blacklip abalone (*H. rubra*) and greenlip abalone (*H. laevigata*) (Grubert & Ritar, 2005). In South Africa, hatcheries are primarily land-based, with advanced biosecurity protocols to safeguard against alien species introductions while culturing the indigenous *H. midae* (Sales & Britz, 2001). Chile has emerged as a significant producer, leveraging its coastal protected areas for the cultivation of red abalone, emphasizing sustainable practices with minimal environmental impact (Flores-Aguilar et al., 2007).

18.3.1 The selection and design of culture farms

Here, we will choose the Chinese abalone farm as a case study. Selecting the site of a culture farm is very important. Access to adequate, good-quality seawater, sufficient electric power, traffic access, easy transportation, and the absence of local rivers flowing into the sea should be considered. The salinity should be at least 28 psu, the pH should be from 7.4 to 8.4, and dissolved oxygen should be more than 5 mg/L. The recommended size for the seed culture pond is 3.0 m × 7.0 m × 1.0 m, with a shade net over the pond. A separate room should be built as a hatchery for abalone spawning and fertilization (Figs. 18.3 and 18.4).

18.3.2 Broodstock conditioning

The first step in culturing abalone is obtaining broodstock (sexually mature adult abalone). Broodstock may be gathered from the wild by those who hold a permit to collect wild abalone or through a licensed quota holder. The other alternative is to obtain broodstock from previously cultured stock. The condition of the gonad indicates the abalone's readiness to reproduce. Fully matured animals with a complete shell, pleopoda of strong vitality, and plump edible parts are collected, and male and female abalone are placed in separate ponds. The sex of an abalone is determined by the color of its gonad; sexually mature males have a creamy yellow gonad, while sexually mature females have a dark-green or brown gonad.

18.3.3 Collection of benthic diatoms

Generally, from the creeping larva stage, abalone start to feed on benthic diatoms, such as Cocconeis, Navicula, Nitzschia, Auriculaceae, Diploneis, and Amphora. Algae can be collected from

FIGURE 18.3 Seed culture ponds and abalone seeds in southern China.

FIGURE 18.4 Abalone seeds in southern China3.2 Broodstock conditioning.

cultivated membranes in the pool or from natural waters. Briefly, stones, or other sheet substrates in shallow seashores are selected and settled algae washed off their surfaces. Subsequently, algae separation is carried out twice by filtering through bolting silk with a net mesh of 500 μm. Large individuals are eliminated while small species are retained in the sieve and sprinkled evenly into the pond as algae seeds. A transparent polyethylene membrane measuring 100 cm × 100 cm is generally used as the settling structure for the collected diatoms. A stone sinker is usually tied in the middle of the membrane, and benthic diatoms are inoculated. Polyethylene corrugated plates, measuring 40 cm × 35 cm, can also be used in seed culture. These plates are grouped into bundles of 10–20 sheets arranged at 3–5 cm intervals. Light is essential for indoor diatom culture, which requires enough light to activate photosynthesis. Black shade nets are used to prevent direct sunlight and to adjust light intensity. Propagation is fast at relatively strong light levels (1500–3500 lux). Artificial illumination (about 1500 lux) is particularly important in the

FIGURE 18.5 Types of substrates for diatoms and spat settlement. (A) and (B) represent plastic corrugated panels placed in plastic frames in northern China, (C) and (D) represent plastic films in Southern China. *From (Gao et al., 2023).*

rainy season when natural sunlight is dim. The illuminating photoperiod time is usually 10–12 h of light a day (Fig. 18.5).

18.3.4 Induced spawning, incubation, and larvae culture

Broodstock abalone need to be induced to produce eggs and sperm. Common methods of induction include exposure to air, thermal stimulation, and application of hydrogen peroxide. However, the most commonly used method is ultraviolet radiation of seawater. Seawater is exposed to 800 mWh/L UV radiation to generate free oxygen radicals, which stimulate prostaglandin cyclooxygenase in gonad tissue.

Eggs and sperm are released after a period of induction. The sperm of abalone are active for 24 h. However, fertilization of eggs should be completed within 1 h of spawning. Eggs are collected and transported into a larval rearing tank to be artificially fertilized by adding sperm into the tank. Polyspermy is caused by an overabundance of spermatozoa, whilst a lack of sufficient spermatozoa leads to a low fertilization rate. After fertilization, zygotes sink to the bottom, and the upper water layer of the tank is siphoned off to remove unfertilized eggs. Zygote washing is repeated at intervals of 1–2 h. It is important to maintain a constant temperature throughout the washing process.

After 18–24 h, fertilized eggs hatch into trochophore larvae that start floating and swimming. The hatch time is closely related to water temperature. After most of the fertilized eggs are hatched into trochophores, they can be transferred to nursing ponds. These ponds are supplied with substrates supporting fully grown benthic diatoms for trochophores to attach to. At this stage, it is important to provide optimum conditions for larval growth. The planktonic larvae search for a suitable substrate for their creeping lifestyle. The foot begins to grow, and the velum disappears within 4 days. When a suitable substratum is found, they change shape several times and soon start to feed actively. At this stage, it is most important to provide the larvae with a biologically, chemically, and physically suitable substrate as well as other environmental requirements. When larvae enter the creeping stage, boards should be placed into the rearing tank. The color of the boards should not allow light penetration, and their surfaces should be toxic-free, as well as suitable for creeping.

Normally, the boards are corrugated, made of a plastic material, and measure 30–60 cm × 30–60 cm. From 100,000 to 1,000,000 larvae per ton of rearing water are placed into the tank, and 80 setting boards are fixed on the water surface. Beforehand, microorganisms such as bacteria, yeast, protozoa, measuring 5 mm, and *Navicula* sp. are cultured on the setting boards. This diatomaceous layer is cultured on the surface of the board and is fed upon by the juvenile abalone until they reach 5–6 mm in shell length. When the juveniles reach 5–6 mm, after about 80 days, they are transferred from the setting boards to intermediate rearing tanks. Removing abalone from the setting boards is not done by hand but with the use of anaesthetizing reagents such as carbon dioxide, ether, ethanol, and para-aminobenzoic acid. After anaesthetizing the juveniles, they are graded and selected according to size.

18.3.5 Intermediate rearing of seedlings

The abalone larvae are cultured in nursery ponds after they are detached. The bottom of the pond is laid with bricks, which are 30 cm × 30 cm in size and 3–5 cm in height. The bricks are laid flat or overlapped with each other. The stocking density for newly detached juvenile abalone is 4000–5000 abalone/m^2, while the stocking density for 5–10 mm juveniles is 3000–4000 abalones/m^2. When they grow to 10 mm shell length, a stocking density of 2000–3000 abalones/m^2 is recommended.

In the intermediate stage, abalone are mainly fed with artificial feed. When abalones are 3–5 mm long, and the stocking density is 4000–5000 abalones/m^2, the initial feeding rate is 4–6 g/m^2/day. When they reach 5–10 mm, the feeding rate can be adjusted to about 4%–5% of the total weight of abalone in the pond. For abalones of 10–20 mm, the feeding rate is controlled at 1.5%–3% of the total abalone weight.

For juvenile abalones with shells longer than 5 mm, natural food such as red algae, Ulva flake, and Gracilaria shreds can be used as feed. The feeding rate for artificial feed is about 3%–5% of the abalone weight.

The pond is washed every 1–2 days. Before washing, the pond is drained, and then accumulated materials and feed residues are washed off the bottom. Fresh seawater is added immediately after washing. At the same time, the feed intake of juvenile abalone and the amount of remaining feed in the pond should be observed, and the amount of artificial feed should be adjusted when there is excessive or not enough feed in the pond. The growth of juveniles should be observed every day, and a log should be kept of the mortality rate and other data. After

FIGURE 18.6 Skilled women laborers in southern China removing larger juveniles by hand to ensure optimal stocking density. (A) The scene of skilled women laborers working. (B) The juvenile Pacific abalone for stocking density optimization.

intermediate culture, abalone of 3 cm may be reared in land-based facilities, or released into the sea until they reach marketable size (Fig. 18.6).

18.4 Abalone farming techniques

Overall, the aquaculture of abalone can be divided into closed water aquaculture on land and open water aquaculture in the sea.

18.4.1 Land-based intensive culture

Internationally, abalone farming in most countries is carried out using land-based industrial conditions, which can be roughly divided into two types: The American model and the Japanese model (Gao et al., 2023). In the North American red abalone farming industry, juvenile abalone are first hatched and produced indoors, and later raised outdoors in cement tanks. When cultivating seedlings indoors, plastic corrugated plates are usually used as attachment bases, and a cylindrical water tank is used as a cultivation pool. When the juvenile abalone grow to 8–10 mm, they are transferred to separate outdoor cement ponds. In the cement pool, 3 cm juvenile abalone are raised in cylindrical, single-layer baskets to increase space, while abalone larger than 3 cm are raised using a free-range process. This model, or similar improved models, is common in countries such as New Zealand, Australia, Ireland, and France (Morales-Bojórquez et al., 2008). Another widely used land-based farming model is the Japanese model, which is a high-density intensive abalone farming model continuously improved and developed in southern China. The production and cultivation of seedlings in this mode are carried out in indoor facilities. The advantages of land-based industrial aquaculture include controllable water quality, small environmental footprint, large culture capacity, and lower susceptibility to natural environmental conditions (Liu, 2011; Stotz et al., 2006). In China, land-based industrial aquaculture takes three forms: flow-through industrial aquaculture, semi-closed industrial recirculating aquaculture, and closed industrial recirculating aquaculture (Gao et al., 2023). The controllable quality of aquaculture water enables land-based industrial aquaculture to vertically extend abalone culture capacity. Compared with the coastal raft culture, featuring a wide area of water, land-based industrial aquaculture achieves a larger culture capacity per unit area. In China, industrial abalone aquaculture currently relies on a flow-through system, where water is pumped into a settling tank at high tide, filtered, and then pumped into each culturing tank. Unfortunately, this cultural pattern may result in a large waste of resources and energy. At high tide, water can be pumped into the culturing tank without interruption, and this process can continue as long as the seawater can be pumped. Consequently, large quantities of seawater will overflow and be discharged back to the natural sea area, leading to a waste of water and high-energy consumption to keep the pump in operation (Gao et al., 2023). In addition, aquaculture wastewater is directly discharged into the sea without proper treatment, resulting in intense pressure on the environmental conditions in the coastal sea area.

18.4.2 Sea-based floating raft culture system in shallow seas

Compared with the land-based, intensive culture method, inshore raft or longline, and sea-based cage farming have several advantages, including fast abalone growth, low cost, and ease of operation (Bi & Zhao, 2021). Floating raft culture is the most important culture method in Southern China, and there are two types: raft culture and longline culture. In floating raft culture, abalone are placed directly in a net cage, which is suspended from a floating raft. The raft floats on the water due to the presence of foam balls or polyethylene plastic buoys, and a breeding frame made of wooden boards or plastic boards is attached to the raft. Most cages are rigid plastic cylinders or multilayer cylindrical net cages. Layers of plastic trays (4–10 layers) are connected by iron rods and plastic sleeves, with a 20–30 cm space between each layer, wrapped in a fishing net with a zipper. The whole raft frame structure is fixed in place by piles on the seabed. Workers can stand on the frame to feed abalone easily.

In longline culture, the unit is composed of piles, pile rope, floating rope, and floating balls. Unlike raft frame culture, longline culture has no fixed frame. Instead, ropes float on the water and are tied to floating balls, with both ends of the rope fixed to piles on the seabed. The net cage hangs directly from the rope floating on the water. Each rope is less than 100 m long, and ropes are laid every 2–3 m in the horizontal direction. Workers are required to stand on the raft for feeding. In comparison with raft culture, longline culture is complicated to operate and laborious. However, it has a low initial cost, and resists wind and storms well, and so it has been widely adopted in shallow sea aquaculture (Fig. 18.7) (Gao et al., 2023).

18.4.3 South-North relay culture

Abalone production in southern regions is affected strongly by summer mortality (Wu & Zhang, 2016), with a typical summer survival rate between 50%–70% (Ke, 2013). Because of improvements in transportation technology and cost reductions, the "North-South relay" farming mode is expanding to improve the summer survival rate. In this method, abalone are transported from southern China to northern China (more than 2000 km) using trucks and sea vessels in May and returned to southern China in November. This aquaculture mode, unique to China, takes full advantage of the temperature difference between the northern and southern regions, which reduces the required culture duration and expands the areas suitable for farming. This method also avoids the high temperatures, typhoons, and other adverse factors common in southern China during the summer, dramatically increasing abalone survival rates

FIGURE 18.7 Sea-based longline cage farming.

FIGURE 18.8 The vessel designed for transportation on "North-South relay" mode.

(Zeng et al., 2020). However, if ships transporting living abalone enter the confluence of fresh water and seawater, a small navigation error may cause the death of abalone, resulting in a substantial loss. Therefore attention should be paid to salinity changes in the sea area. Water exchange between the ship and the surrounding water should not take place in low salinity areas, and an aerator should be used to oxygenate the water on the ship. "South-North relay" culture represents a significant innovation in China's abalone culture industry. The movement of abalone from south to north mimics "migratory bird" south-north migration and is believed to boost the development of the abalone culture industry (Figs. 18.8 and 18.9) (Huang, 2019).

18.4.4 Integrated multitrophic culture

Integrated multitrophic culture is a sustainable mariculture concept that has been developed over the past few years. An integrated aquaculture system comprises different trophic organisms (e.g., feed organisms, suspended filter feeders, nutrient extractive organisms, and benthic detritivores). The wastes produced by some of the organisms in this system provide a source of nutrients for other biological functional groups, which creates an eco-friendly aquaculture model that allows for the efficient recycling of nutrients in the aquaculture system,

FIGURE 18.9 The live-fish carrier is being loaded with abalone cages for the "North-South relay" aquaculture transportation.

FIGURE 18.10 The abalone and kelp culture area in Shandong Province, China.

and it controls the eutrophication of aquaculture waters (Fang et al., 2018). Over the years, as *H. discus hannai* was treated as the main culture species, integrated multitrophic culture systems such as shellfish-algae integrated culture, fish-shellfish-algae, and fish-shellfish-algae-sea cucumber have been developed and constructed in Sanggou Bay, Shandong Peninsula. This sort of integrated multi-trophic culture system has been promoted from harbors to deep waters because of economic and ecological benefits. To date, China has only cultured macroalgae in about 0.3% of its territorial waters, which means there is enormous potential for this type of culture (Gao & McKinley, 1994). As China's abalone culture increases year on year, macroalgae culture is indirectly boosted. The development of integrated multitrophic culture is expected to attain the sustainable development goal of "high efficiency, high quality, ecology, health, and safety" in the aquaculture industry, which also provides a vital path of responding to and addressing the problems of marine ecosystems under multiple stresses (Fig. 18.10).

18.4.5 Off-the-coast culture

In recent years, with the continuous improvement of the manufacturing capacity and quality of offshore engineering equipment, China has proven its ability to manufacture deep-sea

FIGURE 18.11 Cage system inside the pilot offshore automatic abalone farming platform.

aquaculture platforms, so that offshore and deep-sea aquaculture has become a new area of growth. Limited by shrinking nearshore-farming space, offshore aquaculture may represent a major future direction for the abalone industry (Lovatelli et al., 2013). Despite the large size of the offshore regions, these areas have many challenges, including strong winds and waves, and typhoon damage. Several large enterprises have invested in pilot projects, researching and developing new anti-wind-and-wave automatic abalone aquaculture platforms. Such developments may radically affect the abalone aquaculture industry in China and worldwide.

"Zhenbao No. 1", the world's premier storm and wave-proof automated abalone culturing platform developed by Fujian China-Singapore Evervest Co., Ltd. and Shanghai Zhenhua Heavy Industries Co., Ltd., is believed to be the world's first offshore deep-water abalone culturing platform. "Zhenbao No. 1" was in operation by the end of 2018, paving a new path for the development of China's abalone culture industry. A single platform can accommodate approximately 5000 abalone culture cages, mainly comprising floating structures, culture cages, upper frames, underwater frames, and mechanical hoisting devices, with a projected annual output of about 12–20 tons per unit. In 2019, "Fubao No. 1", built by Fujian Funing Shipbuilding Industry Co., Ltd., was put into operation. This platform covers an area of up to 1228.4 m^2 and can accommodate 15,000 cages—three times that of "Zhenbao No. 1." The platform is powered by wind and solar power and constructed as a "buoyancy plus frame", which can be suitably used in sea areas at a distance less than 10 nautical miles from the shore, and with a water depth greater than 17 m. It can withstand typhoon force 12, and has a projected annual output of 40 tons. However, after three years of culturing, some problems have become apparent.

FIGURE 18.12 Pilot offshore automatic abalone farming platform.

First, there are no specific culturing operation techniques and specifications available for the new culturing platform; for example, how to determine a reasonable culturing density and feeding regime. Second, optimization of the cage layout and reasonable stocking and culling specifications are urgently required. Third, in the summer, fouling organisms adhere to the inside and outside of cages on a large scale, which affects the exchange of water, meaning that new eco-friendly antifouling coatings need to be developed. Finally, efforts are required to create cooperation and collaboration between enterprises and farmers, so that both parties will earn profits. Traditional production methods, such as raft aquaculture, are subject to the stress of sustainable development, resulting in an increase in resource and environmental costs. In this sense, off-the-coast aquaculture could have greater potential (Figs. 18.11 and 18.12).

18.5 Genetic and breeding of abalone

The huge growth of the abalone farming industry has brought new challenges to genetic improvement. Compared to other aquatic animals, abalone have a slow growth rate and usually require years of cultivation to reach market size. Second, long-term self-propagation can easily lead to genetic degradation and result in problems such as a low survival rate. Therefore improving production performance, genetic breeding and improvement of existing abalone germplasm, and cultivating high-yield, stress-resistant, high-quality new abalone varieties, are some of the key issues that urgently need to be addressed in the current abalone industry.

18.5.1 Selective breeding

Individual selection, family selection, and within-family selection are common methods of selective breeding in aquatic animals. Population genetic background analysis and trait genetic parameter evaluation, based on pedigree records or molecular markers, can help to evaluate the relationship between selection responses, facilitate the adjustment of breeding strategies, and accelerate breeding progress. Therefore establishing a genetic breeding program suitable for the target population through data analysis and combining it with molecular breeding methods for germplasm improvement will be one of the necessary and effective strategies for abalone breeding. South Africa has carried out large-scale breeding of *H. midae*. The study collected 800 wild abalone from different areas as the base population, and then established 300 families at five experimental sites. The offspring of the families were marked and mixed, and the optimal family was selected for the next generation breeding based on relevant production data (Franchini et al., 2011). In Australia, research ideas have been proposed for MBLUP breeding of abalone. A total of 141 black-lipped abalone and 148 green-lipped abalone were collected from five areas as the base population. MBLUP breeding was studied through methods such as family establishment, marker breeding, breeding value evaluation, and molecular marker-assisted selection. At present, the research team has estimated the heritability values of production traits such as total weight, meat weight, and foot muscle color of two abalone populations, and the values show that both populations have good breeding potential (Hayes et al., 2007).

18.5.2 Hybridization

Hybrid breeding is currently an extremely effective means of cultivating new varieties of abalone. Through interspecific hybridization, different genotypes of different species can be quickly aggregated in the next generation, resulting in a rich combination of genetic variations (Sui et al., 2025). For quantitative traits controlled by multiple genes, it is easy to exhibit certain hybrid advantages (Xiao et al., 2021). There are 10 species of abalone distributed along the coast of Japan, and research reported that the growth rate of the hybrid offspring of all combinations was faster than their respective self-breeding combinations. Among these, *H. madaka* ♀ × *H. gigantea* ♂ were superior to the parent combination, showing a certain hybrid advantage (Ahmed et al., 2008). In North America, since Owen (1961) discovered the existence of natural hybridization between major abalone species along the Pacific coast of North America, many scholars have conducted hybridization experiments between several major abalone species. Leighton and Lewis studied the hybridization between green abalone (*H. fulgens*), red abalone (*H. rufescens*), pink abalone (*H. corrugata*), and white abalone (*H. sorenseni*) along the California coast (Cruz et al., 2005; Díaz et al., 2006). The hybrid offspring of red abalone and green abalone showed strong growth advantages (Tripp-Valdez et al., 2021). The research team of CSIRO in Australia conducted interspecific hybridization research using the greenlip abalone (*H. laevigata*) and blacklip abalone (*H. rubra*), and successfully cultivated a hybrid species called "Jade tiger abalone" for commercial production. This hybrid abalone has the characteristics of a fast growth rate and excellent quality, and is now in commercial production (Mateos et al., 2010, 2013). In China, extensive studies on abalone hybridization have been conducted, and hybridization technologies have played an important role in increasing abalone production over the past two decades. For example, Japanese stocks of *H. discus hannai* and *H. diversicolor* were introduced into China, and intraspecific hybrids between the Chinese and Japanese stocks were developed. The hybrids exhibited significantly improved survival and growth as compared to the wild types, supporting the development of the Chinese abalone industry (Du & Zhang, 2004; Guo, 2009). A strong heterotic effect has also been observed in hybrids of *H. gigantea* and *H. discus hannai*: both reciprocal hybrids had significantly faster growth rates after 100 days of culture. The survival rates of the hybrids were similar to those of the parental species after 180 days of grow-out (Luo, Ke, You, Wang, & Chen, 2010a, 2010b). The North American green abalone *H. fulgens* was also introduced into China, and was successfully hybridized with the Pacific abalone. The hybrids *H. discus hannai* ♀ × *H. fulgens*♂ exhibited hybrid vigor with respect to growth and survival, and had an increased tolerance of high temperatures (You et al., 2015). Over the last two decades, four hybrids, *H. discus hannai* "Dalian-I", *H. diversicolor* "Dongyou-I", *H. sieboldii* ♀ × *H. discus*

FIGURE 18.13 The new national aquatic varieties. "Dongyou No. 1 (*H. diversicolor*)" (A) *H. gigantea* ♀ × *H. discus hannai* ♂ hybrids (B) and *H. discus hannai* ♀ × *H. fulgens* ♂ hybrids (C) have been extensively promoted across China, due to their heat resistance, fast growth rates, and high survival rates. *From Guo et al. (2022)*

hannai ♂, and *H. discus hannai* ♀ × *H. fulgens* ♂, have received "new variety certificates" from the Chinese Ministry of Agriculture, which have been commercially produced to improve abalone farming efficiency (Fig. 18.13).

18.5.3 Triploid

Polyploidy refers to organisms that contain three or more complete sets of chromosomes. According to the origin of chromosomes, polyploids found in nature are mainly divided into two categories: Homologous polyploids and heterologous polyploids. Polyploid breeding has become an important means of genetic improvement and is widely used in plants and aquatic animals. For abalone, triploid induction studies on various species were conducted, including *H. discus hannai, H. midae, H. asinine*, and *H. laevigata* (Dunstan et al., 2007; Liu et al., 2004; Norris & Preston, 2003; Stepto & Cook, 1998). The growth of triploid Pacific abalone, at the juvenile stage, increased by 34% compared to diploid abalone (Zhang et al., 1998). Another study reported that the shell length of triploid Pacific abalone increased by 10.2% and the body weight increased by 20.1% compared to the diploid control group, both showing the growth advantage of triploidy (Wang et al., 2022, 2026). However, the induction rate of triploidy is generally low (20%–80%), which limits its application in commercial production (Zhang et al., 2024). Recently, scientists have optimized the induction method for triploid Pacific abalone and attempted triploid breeding of green abalone for the first time. By using the induction method, it is possible to obtain batches of homologous triploids and heterologous triploids seedling of the Pacific abalone with a triploid rate of 100% (or close to 100%), solving the problem of the poor triploidy rate in large-scale seed production technology for this species (Wang et al., 2021).

18.6 Future directions

To maintain sustainable abalone aquaculture in the future, holistic solutions are needed—incorporating the government, industry leaders, and academic institutions. It is recommended that the government provide strong policy support and effective measures to optimize farming, based on full investigations and aquaculture capacity assessments. In addition, red tide monitoring and forecasting during the high-incidence season should also be increased through cooperation between the government and research institutions. Such measures might reduce the economic losses of thousands of abalone farmers. Limited by shrinking nearshore-farming space, offshore aquaculture may represent a major future direction for the abalone industry. In addition, compound feeds, which might replace fresh seaweed and improve the feed conversion rate, should be investigated. Biofouling is another serious problem affecting nearshore cage farmers. Thus the development and application of environmentally friendly antifouling coatings should be improved. The mismanagement of genetic resources for this highly fecund species, resulting in the loss of genetic diversity and leading to inbreeding depression, may be one of the most important issues affecting the sustainable development of the abalone industry. Finally, market factors may drive an imbalance of industry operations and bring many challenges to the current landscape of the abalone industry. Therefore the development of disciplined breeding programs should be a management priority.

References

Ahmed, F., Koike, Y., Strüssmann, C. A., Yamasaki, I., Yokota, M., & Watanabe, S. (2008). Genetic characterization and gonad development of artificially produced interspecific hybrids of the abalones, Haliotis discus discus Reeve, Haliotis gigantea Gmelin and Haliotis madaka Habe. *Aquaculture Research, 39*(5), 532–541. doi:10.1111/j.1365-2109.2008.01911.x.

Ault, J. S. (1985). Some quantitative aspects of reproduction and growth of the red abalone, Haliotis rufescens swainson. *Journal of the World Mariculture Society, 16*(1-4), 398–425. doi:10.1111/j.1749-7345.1985.tb00221.x.

Bi, C., & Zhao, Y. (2021). *Traditional aquaculture structures* (pp. 1–12). Springer Science and Business Media LLC. doi:10.1007/978-981-10-6963-5_327-1.

Bullon, N., Alfaro, A. C., Hamid, N., Dezfooli, S. M., Seyfoddin, A., & Dawood, M. (2023). Effect of dietary insect meal and grape marc inclusion on flavor volatile compounds and shell color of juvenile abalone Haliotis iris. *Aquaculture Nutrition, 2023,* 1–11. doi:10.1155/2023/6628232.

Chen, M., Haokui, L., Shijie, C., Kaien, X., & Bingneng, C. (1977). A preliminary study on the artificial rearing of the larvae and juveniles of Haliotis discus hannai [J]. *Acta Zoologica Sinica, 23*(1), 35–46.

Cook, P. A. (2014). The worldwide abalone industry. *Modern Economy, 05*(13), 1181–1186. doi:10.4236/me.2014.513110.

Cook, P. A. (2023). Introduction, taxonomy, and general biology of abalone. *Developments in Aquaculture and Fisheries Science, 42,* 1–8. doi:10.1016/B978-0-12-814938-6.00001-4.

Cook, P. A. (2025). Worldwide abalone production: An update. *New Zealand Journal of Marine and Freshwater Research, 59*(1), 4–10. doi:10.1080/00288330.2023.2261869.

Cruz, P., Ibarra, A. M., Fiore-Amaral, G., Galindo-Sánchez, C. E., & Mendoza-Carrión, G. (2005). Isolation of microsatellite loci in green abalone (*Haliotis fulgens*) and cross-species amplification in two other North American red (*Haliotis rufescens*) and pink (*Haliotis corrugata*) abalones. *Molecular Ecology Notes, 5*(4), 857–859. doi:10.1111/j.1471-8286.2005.01088.x.

Díaz, F., Re, A. D., Medina, Z., Re, G., Valdez, G., & Valenzuela, F. (2006). Thermal preference and tolerance of green abalone Haliotis fulgens (Philippi, 1845) and pink abalone Haliotis corrugata (Gray, 1828). *Aquaculture Research, 37*(9), 877–884. doi:10.1111/j.1365-2109.2006.01506.x.

Du, B. X., & Zhang, X. J. (2004). Discharge energy and dielectric breakdown of gamma-ray irradiated polybutylene naphthalate. In *Proceedings of the 2004 IEEE International Conference on Solid Dielectrics ICSD 2004* (pp. 947–950). IEEE. doi:10.1109/ICSD.2004.1350588.

Dunstan, G. A., Elliott, N. G., Appleyard, S. A., Holmes, B. H., Conod, N., Grubert, M. A., & Cozens, M. A. (2007). Culture of triploid greenlip abalone (Haliotis laevigata Donovan) to market size: Commercial implications. *Aquaculture, 271*(1-4), 130–141. doi:10.1016/j.aquaculture.2007.06.023.

Fang, J., Zhang, P., Fang, J., Jiang, Z., Gao, Y., & Du, M. (2018). The growth and carbon allocation of abalone (Haliotis discus hannai) of different sizes at different temperatures based on the abalone-kelp integrated multitrophic aquaculture model. *Aquaculture Research, 49*(8), 2676–2683. doi:10.1111/are.13769.

FAO. (2024). *Global aquaculture production: Fisheries and aquaculture statistics.* FishStatJ.

Flores-Aguilar, R. A., Gutiérrez, A., Ellwanger, A., & Searcy-Bernal, R. (2007). Development and current status of abalone aquaculture in Chile. *Journal of Shellfish Research, 26*(3), 705–711. doi:10.2983/0730-8000(2007)26[705:DACSOA]2.0.CO;2.

Franchini, P., van der Merwe, M., & Roodt-Wilding, R. (2011). Transcriptome characterization of the South African abalone Haliotis midae using sequencing-by-synthesis. *BMC Research Notes*, *4*(1), 59. doi:10.1186/1756-0500-4-59.

Gao, K., & McKinley, K. R. (1994). Use of macroalgae for marine biomass production and CO_2 remediation: A review. *Journal of Applied Phycology*, *6*(1), 45–60. doi:10.1007/BF02185904.

Gao, X., Zhang, M., Luo, X., You, W., & Ke, C. (2023). Transitions, challenges and trends in China's abalone culture industry. *Reviews in Aquaculture*, *15*(4), 1274–1293. doi:10.1111/raq.12769.

Geiger, D. L. (1999). Distribution and biogeography of the recent Haliotidae (gastropoda: vetigastropoda) world-wide-all databases. *Bollettino Malacologico*, *35*(5-12), 57–120.

Gordon, H. R., & Cook, P. A. (2013). World abalone supply, markets, and pricing: 2011 update. *Journal of Shellfish Research*, *32*(1), 5–7.

Grubert, M. A., & Ritar, A. J. (2005). The effect of temperature and conditioning interval on the spawning success of wild-caught blacklip (Haliotis rubra, Leach 1814) and greenlip (H. laevigata, Donovan 1808) abalone. *Aquaculture Research*, *36*(7), 654–665. doi:10.1111/j.1365-2109.2005.01269.x.

Guo, X. (2009). Use and exchange of genetic resources in molluscan aquaculture. *Reviews in Aquaculture*, *1*(3-4), 251–259. doi:10.1111/j.1753-5131.2009.01014.x.

Hanif, M. A., Hossen, S., Cho, Y., Sukhan, Z. P., Choi, C. Y., & Kho, K. H. (2022). Characterization and expression analysis of mollusk-like growth factor: A secreted protein involved in Pacific abalone embryonic and larval development. *Biology*, *11*(10), 1445. doi:10.3390/biology11101445.

Hayes, B., Baranski, M., Goddard, M. E., & Robinson, N. (2007). Optimisation of marker assisted selection for abalone breeding programs. *Aquaculture*, *265*(1-4), 61–69. doi:10.1016/j.aquaculture.2007.02.016.

Huang (2019). Analysis on aquacultural modes of abalone in Fujian Province (in Chinese, with English abstract). *Journal of Fisheries Research, 41(4)*, 346–352.

Ke, C. (2013). Current status and prospect of abalone aquaculture in China. *China Fisheries*, *1*, 27–30.

Leighton, D. L. (2000). In *The biology and culture of the California abalones: 1* (p. 256). Dorrance Publ. Co.

Li, J., Li, J., Zhang, Y., Jiao, M., Li, A., Liu, Z., … Mao, Y. (2025). Increasing alkalinity to mitigate biogenic acidification and its negative effects during the North–south-relay transportation of abalone. *Aquaculture International*, *33*(6), 474. doi:10.1007/s10499-025-02159-6.

Lin, S., Wu, F., & Zhang, G. (2017). Analysis of monthly variation of biological conditions, metabolic parameters and antioxidative capacities in sea-based farmed Pacific abalone during summer days. *Aquaculture International*, *25*(3), 1345–1359. doi:10.1007/s10499-017-0118-1.

Liu, W., Heasman, M., & Simpson, R. (2004). Induction and evaluation of triploidy in the Australian blacklip abalone, Haliotis rubra: A preliminary study. *Aquaculture*, *233*(1–4), 79–92. doi:10.1016/j.aquaculture.2003.09.006.

Liu, Y. (2011). Research progress on marine industrial recirculating aquaculture technology [J]. *Journal of Agricultural Science and Technology*, *13*(5), 50.

Lovatelli, A., Aguilar-Manjarrez, J., & Soto, D. (2013). *Expanding mariculture farther offshore: technical, environmental, spatial and governance challenges [Z]*: 2013. FAO.

Luo, X., Ke, C., You, W., Wang, D., & Chen, F. (2010a). China Aflp analysis of populations of haliotis discus hannai, haliotis gigantea, and their hybrids. *Journal of Shellfish Research*, *29*(3), 731–734. doi:10.2983/035.029.0326.

Luo, X., Ke, C., You, W., Wang, D., & Chen, F. (2010b). Molecular identification of interspecific hybrids between Haliotis discus hannai Ino and Haliotis gigantea Gmelin using amplified fragment-length polymorphism and microsatellite markers. *Aquaculture Research*, *41*(12), 1827–1834. doi:10.1111/j.1365-2109.2010.02568.x.

Mateos, H. T., Lewandowski, P. A., & Su, X. Q. (2010). Seasonal variations of total lipid and fatty acid contents in muscle, gonad and digestive glands of farmed Jade Tiger hybrid abalone in Australia. *Food Chemistry*, *123*(2), 436–441. doi:10.1016/j.foodchem.2010.04.062.

Mateos, H. T., Lewandowski, P. A., & Su, X. Q. (2013). The growth performance of Jade Tiger cultured abalone fed diets supplemented with fish oil and vegetable oils. *Journal of the Science of Food and Agriculture*, *93*(6), 1389–1396. doi:10.1002/jsfa.5905.

Morales-Bojórquez, E., Muciño-Díaz, M. O., & Vélez-Barajas, J. A. (2008). Analysis of the decline of the abalone fishery (Haliotis fulgens and H. corrugata) along the westcentral coast of the Baja California Peninsula, Mexico. *Journal of Shellfish Research*, *27*(4), 865–870. doi:10.2983/0730-8000(2008)27[865:AOTDOT]2.0.CO;2.

Muñoz, P., Ambler, R., & Bulboa, C. (2012). Settlement, survival, and post-larval growth of Red abalone, Haliotis rufescens, on polycarbonate plates treated with germlings of Ulva sp. *Journal of the World Aquaculture Society*, *43*(6), 890–895. doi:10.1111/j.1749-7345.2012.00615.x.

Norris, B. J., & Preston, N. P. (2003). Triploid induction in the tropical abalone, Haliotis asinina (Linne), with 6-dimethylaminopurine. *Aquaculture Research, 34*(3), 261–264. doi:10.1046/j.1365-2109.2003.00810.x.

Peng, W., Yu, F., Wu, Y., Zhang, Y., Lu, C., Wang, Y., Huang, Z., Lu, Y., Chen, N., Luo, X., You, W., & Ke, C. (2021). Identification of growth-related SNPs and genes in the genome of the Pacific abalone (Haliotis discus hannai) using GWAS. *Aquaculture, 541*, 736820. doi:10.1016/j.aquaculture.2021.736820.

Sales, J., & Britz, P. J. (2001). Research on abalone (Haliotis midae L.) cultivation in South Africa. *Aquaculture Research, 32*(11), 863–874. doi:10.1046/j.1365-2109.2001.00629.x.

Searcy-Bernal, R., & Anguiano-Beltrán, C. (1998). Optimizing the concentration of gamma-aminobutyric acid (GABA) for inducing larval metamorphosis in the Red abalone Haliotis rufescens (Mollusca: Gastropoda). *Journal of the World Aquaculture Society, 29*(4), 463–470. doi:10.1111/j.1749-7345.1998.tb00670.x.

Stepto, N. K., & Cook, P. A. (1998). Induction of triploidy in the South African abalone using cytochalasin B. *Aquaculture International, 6*(2), 161–169. doi:10.1023/a:1009242424678.

Stotz, W. B., Caillaux, L., & Aburto, J. (2006). Interactions between the Japanese abalone Haliotis discus hannai (Ino 1953) and Chilean species: Consumption, competition, and predation. *Aquaculture, 255*(1-4), 447–455. doi:10.1016/j.aquaculture.2006.01.006.

Sui, L., Lin, X., Zhang, J., Cai, M., Wang, Y., & Zhang, Z. (2025). Mechanism of formation of hybrid abalone (Haliotis discus hannai and H. fulgens) heterosis on growth trait based on allele-specific expression analysis. *Aquaculture, 596*, 741867. doi:10.1016/j.aquaculture.2024.741867.

Tripp-Valdez, M. A., Cicala, F., Galindo-Sánchez, C. E., Chacón-Ponce, K. D., López-Landavery, E., Díaz, F., Re-Araujo, D., & Lafarga-De la Cruz, F. (2021). Growth performance and transcriptomic response of warm-acclimated hybrid abalone Haliotis rufescens (♀) × H. corrugata (♂). *Marine Biotechnology, 23*(1), 62–76. doi:10.1007/s10126-020-10002-7.

Vélez-Arellano, N., Valenzuela-Quiñonez, F., García-Domínguez, F. A., Lluch-Cota, D. B., Gutiérrez-González, J. L., & Martínez-Rincón, R. O. (2020). Long-term analysis on the spawning activity of green (Haliotis fulgens) and pink (Haliotis corrugata) abalone along the central west coast of Baja California. *Fisheries Research, 228*, 105588. doi:10.1016/j.fishres.2020.105588.

Wang, Y., Gan, Y., Zhang, J., Xiao, Q., Shen, Y., Chen, Y., … Ke, C. (2022). Performance of triploid Haliotis discus hannai cultured in a subtropical area using sea-based suspended systems. *Aquaculture, 548*, 737722. doi:10.1016/j.aquaculture.2021.737722.

Wang, Y., Wang, X., Gan, Y., Huang, M., You, W., Luo, X., & Ke, C. (2021). Autotriploid and allotriploid abalone larvae (Haliotis discus hannai and H. discus hannai ♀ × Haliotis fulgens ♂) produced by two chemical methods. *Aquaculture Reports, 20*, 100769. doi:10.1016/j.aqrep.2021.100769.

Wang, Y., Zhang, J., Gan, Y., Chen, Y., You, W., Luo, X., & Ke, C. (2026). Disrupted gonadal development and associated alterations in meat yield and physiological response in triploid Pacific abalone, Haliotis discus hannai. *Aquaculture, 611*, 743007. doi:10.1016/j.aquaculture.2025.743007.

Wu, F., & Zhang, G. (2016). Pacific abalone farming in China: Recent innovations and challenges. *Journal of Shellfish Research, 35*(3), 703–710. doi:10.2983/035.035.0317.

Xiao, Q., Huang, Z., Shen, Y., Gan, Y., Wang, Y., Gong, S., … Ke, C. (2021). Transcriptome analysis reveals the molecular mechanisms of heterosis on thermal resistance in hybrid abalone. *BMC Genomics, 22*(1), 650. doi:10.1186/s12864-021-07954-y.

You, W., Guo, Q., Fan, F., Ren, P., Luo, X., & Ke, C. (2015). Experimental hybridization and genetic identification of Pacific abalone Haliotis discus hannai and green abalone H. fulgens. *Aquaculture, 448*, 243–249. doi:10.1016/j.aquaculture.2015.05.043.

You, W. W. (2023). History and current status of abalone aquaculture in China. *Developments in Aquaculture and Fisheries Science, 42*, 363–371. doi:10.1016/B978-0-12-814938-6.00012-9.

Zeng, L., Zhong, S., You, W., & Ke, C. (2020). Effects of south-north relay mode on the nutritive compositions of Haliotis discus hannai. *Journal of Fisheries of China, 44*(1), 111–117. doi:10.11964/jfc.20181011475.

Zhang, G., Wang, Z., Chang, Y., Song, J., Ding, J., Wang, Y., & Wang, R. (1998). Triploid induction in Pacific abalone Haliotis discus hannai Ino by 6-dimethylaminopurine and the performance of triploid juveniles. *Journal of Shellfish Research, 17*(3), 783–788.

Zhang, J., Wang, Y., Lu, Y., You, W., Luo, X., & Ke, C. (2024). Comparative cytogenetic analysis of diploid and triploid Pacific Abalone, *Haliotis discus hannai. Cytogenetic and Genome Research, 163*(5-6), 327–333. doi:10.1159/000535045.

. (2024). *China fisheries statistics yearbooks.* China Agriculture Press.

CHAPTER

19

Octopus aquaculture

Xiaodong Zheng[a], Jaruwat Nabhitabhata[b,c], Ian G. Gleadall[d] and Jian Zheng[a]

[a] Fisheries College, Ocean University of China, Qingdao, China [b] Centre for Marine and Coastal Studies, University of Sains Malaysia, Penang, Malaysia [c] Centre for Biodiversity of Peninsular Thailand, Prince of Songkla University, Hat Yai, Thailand [d] AiCeph LLC, Sendai, Japan

19.1 Introduction

The decline in many finfish stocks has led to increased attention on other groups such as cephalopods, the increasing economic importance of which is obvious from the rapid rise in their global landings over the last few decades (Arkhipkin et al., 2021; Sauer et al., 2021). Cephalopods are a class of mollusks, including squids, cuttlefish, and octopuses. With more than 800 extant species, they are widely distributed throughout the World Ocean (Arkhipkin et al., 2021; Pita et al., 2021) and are prized for their edible yield, tender meat, and high nutritional value. They also have potential as a source of pharmacological products.

In recent years, regions such as the Southwest Atlantic and Southeast Pacific, previously unfished and abundant in unexploited species, have become important cephalopod fishery areas, introducing these species to the international market (Gleadall et al., 2024). The expansion of cephalopod fisheries requires expanded monitoring, assessment, and management, taking into account the differences in life history between cephalopods and fish (Arkhipkin et al., 2021; Pita et al., 2021). Unfortunately, reliance on wild catches is failing to meet market demands, creating a supply–demand imbalance. This increasing demand for cephalopods on the international market has stimulated research on aquaculture as an alternative to reliance on fisheries.

Wild fisheries are currently still the primary source of octopus harvest, but the escalating market demand and ecological pressures on octopus populations (such as overfishing and global warming) are unrelenting. Encouragingly, most cephalopods are annual species with a short life cycle and rapid growth rate, and this renders them potentially suitable for aquaculture (Song et al., 2019). Currently, there are a number of efforts worldwide to develop octopus aquaculture, with much of the effort concentrated in Asia, but stable, reliable commercial viability remains elusive (Fig. 19.1).

Genetic Breeding and Aquaculture of Shellfish: A Collaborative Approach to a Sustainable Future
DOI: https://doi.org/10.1016/B978-0-443-33606-5.00016-3

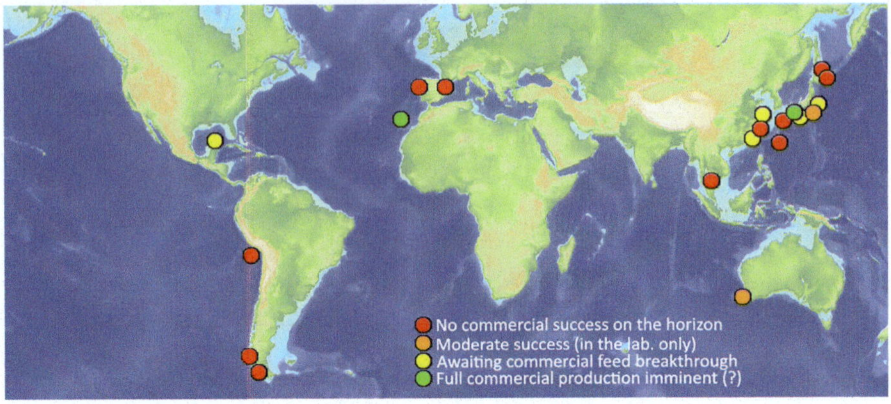

FIGURE 19.1 Approximate status of research on octopus aquaculture.

Octopuses (Order: Octopodida), comprising more than 300 species, are among the most ecologically diverse cephalopods. Their global distribution spans tropical to polar marine ecosystems (Fig. 19.1), with coastal benthic species dominating continental shelves and slopes. They have one of two basic types of cephalopod life cycle: Holobenthic, where the young hatch into small, adult-like young, which immediately begin a benthic stage; or merobenthic, where the hatchlings (known as "paralarvae") have a pelagic life in the plankton for several weeks or months, before settling into the adult benthic life.

The majority of commercial octopus species targeted for aquaculture are merobenthic, which is counterintuitive since challenges persist in paralarval rearing because of the delicate planktonic phase and in managing the planktonic-to-benthic settlement phase, resulting in low survival rates under captive conditions (Vidal et al., 2014). Temperature sensitivity further complicates production, with optimal ranges confined to 13–28°C, depending on the stage of life in the life cycle, necessitating region-specific broodstock selection.

19.2 Octopus species cultivated in Asian waters

Among the merobenthic species cultivated in Asia are *Amphioctopus aegina*, the sandbird octopus, and *Octopus sinensis* d'Orbigny, 1841 (closely related to the Mediterranean and East Atlantic species *O. vulgaris* Cuvier, 1797). With its moderate size, *A. aegina* is amenable to culture on a small scale throughout its life cycle. The life cycle of *A. aegina* has been completed in captivity to develop it for aquaculture (Nabhitabhata, 2014) but this has yet to be established for *O. sinensis*. However, a stable culture of both species requires resolving the bottleneck problems of feeding associated with the rearing of planktonic young and the transition to settling juveniles (Nabhitabhata, 2014; Vidal et al., 2014).

Examples of holobenthic species cultivated in East Asia include *Amphioctopus fangsiao* d'Orbigny, 1841, and *Octopus minor*. The large benthic eggs of these species (around 8 mm long in *A. fangsiao* and 13 mm in *O. minor*) produce hatchlings that are directly benthic, avoiding a planktonic phase and thereby simplifying hatchery protocols. As a cephalopod of economic importance, the culture of *O. minor* has been a major focus of octopus aquaculture attempted in recent years in China (Zheng et al., 2023a). After capture, broodstock acclimation to captivity conditions and successful mating, eggs are spawned on an artificial substratum where the females then brood them until hatching.

19.3 Biological characteristics of Asian cultivated species

19.3.1 *Octopus sinensis*, the East Asian common octopus

O. sinensis d'Orbigny, 1841, is a robust species, with arms 3–4 times the length of the body. Adults mature at around 70 mm (males) or 110 mm (females), reaching a maximum size of around 170 mm (Fig. 19.2A). As with many other species in the *O. vulgaris* complex, this species was included as part of a cosmopolitan population of *O. vulgaris* until recently: the true *O. vulgaris* is now known to be limited to localities in the Mediterranean and eastern Atlantic. *Octopus sinensis* occurs from northern Honshu, Japan, along the East Asian continental shelf to the East China Sea, although the southern extent of its range is uncertain (Gleadall, 2016; Lu et al., 2012). This species has fewer suckers on the arms (around 200–280 per arm, versus 320–350 in *O. vulgaris*) (Li et al., 2022), and mature males have enlarged suckers on arms 2 and 3 located between sucker numbers 12 and 15 (15–19 in *O. vulgaris*) (Gleadall, 2016).

The life cycle of *O. sinensis* features distinct developmental phases. Females lay festoons of small (2 mm) eggs in rocky dens, with incubation time (23 days–5 months) inversely temperature-dependent. Hatchlings (3 mm in Mantal length) initially rely on yolk reserves before transitioning to zooplankton predation (e.g., crustacean larvae) during a 45–55-day planktonic phase (at 21°C). Postsettlement juveniles (8.5 mm mantle length, ML) exhibit exponential growth (6%–12% daily weight gain).

Octopus sinensis favors rocky substrates from subtidal zones to the continental shelf (down to 200 m depth), modifying dens with rocks for protection from predators. Distribution is determined by temperature, amount of upwelling, productivity, and salinity (>29 psu), with low-salinity events causing mass mortality. Paralarvae feed mainly on crustaceans (e.g., copepods, crab zoeae [Gleadall & Nishitani, unpublished data]), whereas adults are opportunistic predators of crustaceans (64 + species), mollusks (17 bivalves), and fish, and are occasionally cannibalistic. Predators include perciform fish (e.g., groupers), dolphins, sharks, and seals. As host to 30 + parasite species, *O. sinensis* may accumulate algal toxins (e.g., PSTs) in food chains.

19.3.2 *Octopus minor*, long-armed octopus

Males of the benthic *O. minor* possess a distinctive spoon-shaped ligula on the shortened third right arm or hectocotylus, which is modified for mating (Fig. 19.3A). The elongated, fusiform body of this species has faint warts, a broad mantle opening, and sparse suckers in two alternating rows. The third right arm in males bears 21–24 sucker pairs and a prominent copulatory groove. Gill structure (15–17 lamellae) and a VV-shaped funnel organ further distinguish it (Lu et al., 2012). Phylogenetic studies link it to the *Octopus macropus* group, and genetic divergence within Chinese populations suggests the presence of cryptic species. This temperate species burrows into substrates and tolerates low-oxygen environments (Zheng et al., 2014).

O. minor inhabits temperate coastal waters of China, the Korean Peninsula, southern Sakhalin Island, and Japan's main islands (excluding Okinawa and the Ryukyu Archipelago). It thrives

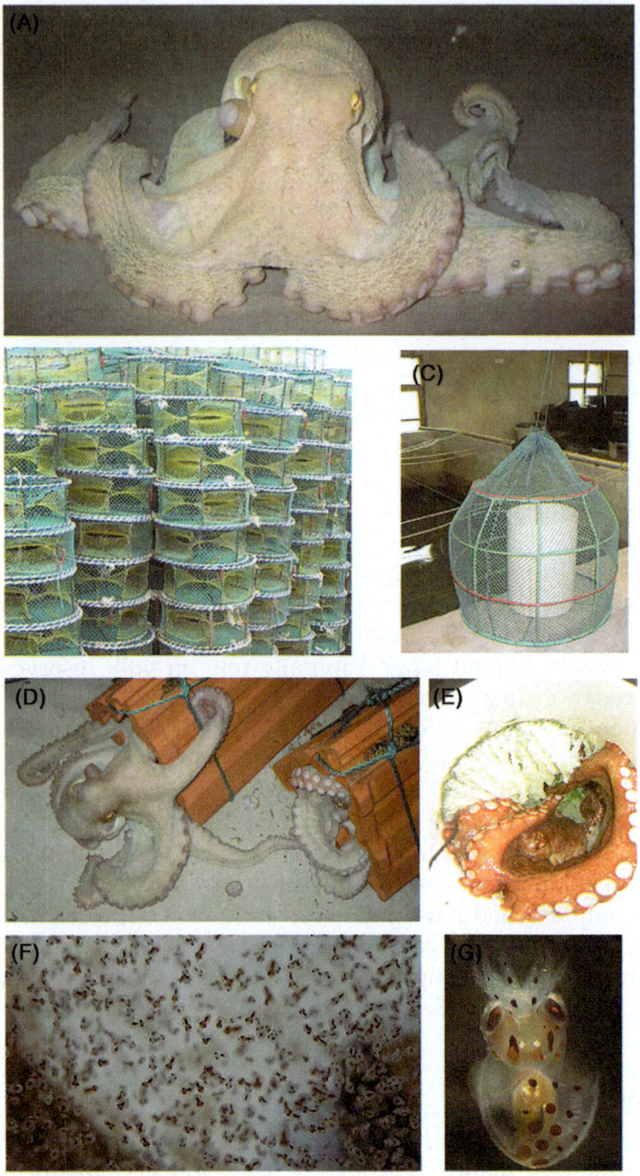

FIGURE 19.2 Morphological characteristics and artificial breeding process of *Octopus sinensis*. (A) A mature *O. sinensis*. (B–C) Cages for temporary rearing and ripening of *O. sinensis* broodstock. (D) Mating of *O. sinensis* broodstock animals in rearing tank (the male, on the left, is extending his reproductive arm, hectocotylus, into the mantle cavity of a female, to the right). (E) A female *O. sinensis* brooding her eggs. (F–G) Hatchlings of *O. sinensis*.

in shallow (<3 m), muddy substrates with bivalves and seagrass, notably in China's Moon Lake (Shandong Province), Japan's Seto Inland Sea, and Korea's Jeollanam-Do coast. Seasonal migrations occur: spawning peaks in summer (June–July in China; August–September in Japan), with movements to deeper waters in winter (Zheng et al., 2014). In Korea, it dominates coastal trap fisheries (28% of the 2006 total catch for octopus), whereas China's Moon Lake yields ~6226 kg annually (Bo et al., 2016; Zheng et al., 2024).

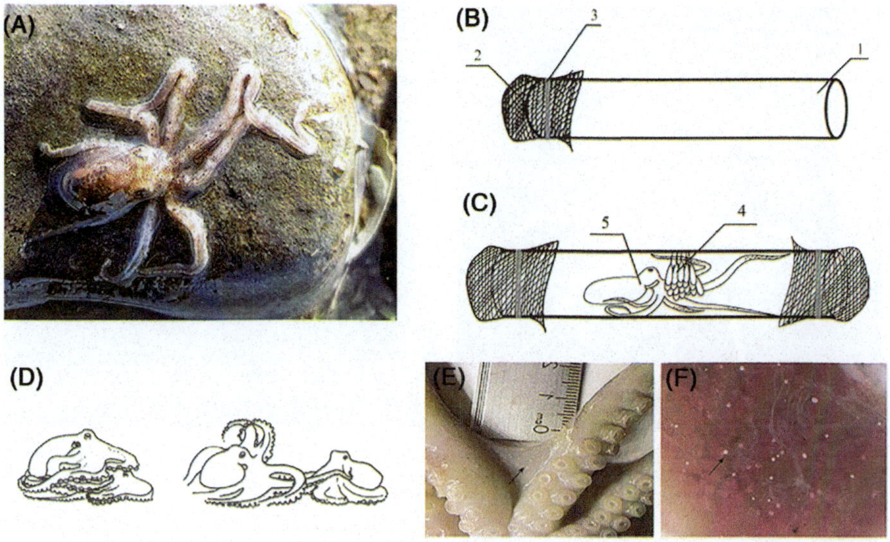

FIGURE 19.3 Morphology, reproductive behavior, and parasitic infection of *Octopus minor*. (A) The morphology of *O. minor*. (B–C) Artificial egg collectors. 1 PVC tube. 2 nylon nets. 3 rubber bands. 4 egg. 5 female parent. (D) schematic diagram of octopus mating. (E–F) Parasite of *O. minor*.

O. minor undergoes direct holobenthic development (without a larval stage). Embryos hatch after 72–89 days at 21–25°C, and are characterized by 20 developmental stages, including two major embryo inversions (days 21 and 65). Hatchlings measure 8.5–11.5 mm ML and 25–31 mm in total length, immediately adopting benthic behavior. Growth is rapid: in culture, males reach ~123 g and females ~197 g by 250 days (Qian et al., 2013). Reproduction is semelparous, with spawning peaks in summer (July in China; August–September in Japan). Males die shortly after mating, whereas females spawn up to 346 days posthatching, suggesting a 1-year lifespan under controlled conditions. However, field studies in Japan's Fukuoka Bay indicate slower growth, with juveniles reaching 7 mm ML by November, 20 mm by June, and 40 mm by September of their first year. Sexual maturity (≥90 mm ML) occurs in the second year, followed by spawning and death by October, implying a 2-year lifespan in the wild (Nakaoka & Sugino, 2014). This discrepancy between lab and field data highlights environmental influences on growth and longevity.

The mud-dwelling *O. minor* is a generalist predator in temperate coastal ecosystems, feeding on crustaceans (e.g., *Charybdis japonica*), mollusks (e.g., *Ruditapes philippinarum*), polychaetes (e.g., *Diopatra* spp.), and fish (notably Gobiidae, accounting for 46% of the diet by occurrence). Cannibalism is significant (23% of stomach contents), especially in high-density areas such as Moon Lake, China. Females exhibit higher feeding activity than males, with activity peaking during ovarian maturation. Prey selection aligns with optimal foraging theory: preferred bivalves (*R. philippinarum*) offer higher energy (5.94 kJ) and shorter handling time, despite tougher shells. Hunting strategies vary by prey type: ambush for fish (57%), stalking for crabs (37%), and opportunistic foraging for sessile prey (Song et al., 2019). The species constructs complex lairs (20 cm deep tunnels with breathing holes) in muddy substrates, aiding in predator avoidance. Key parasites include *Aggregata sinensis* (Ren & Zheng, 2022), which forms white oocysts in digestive

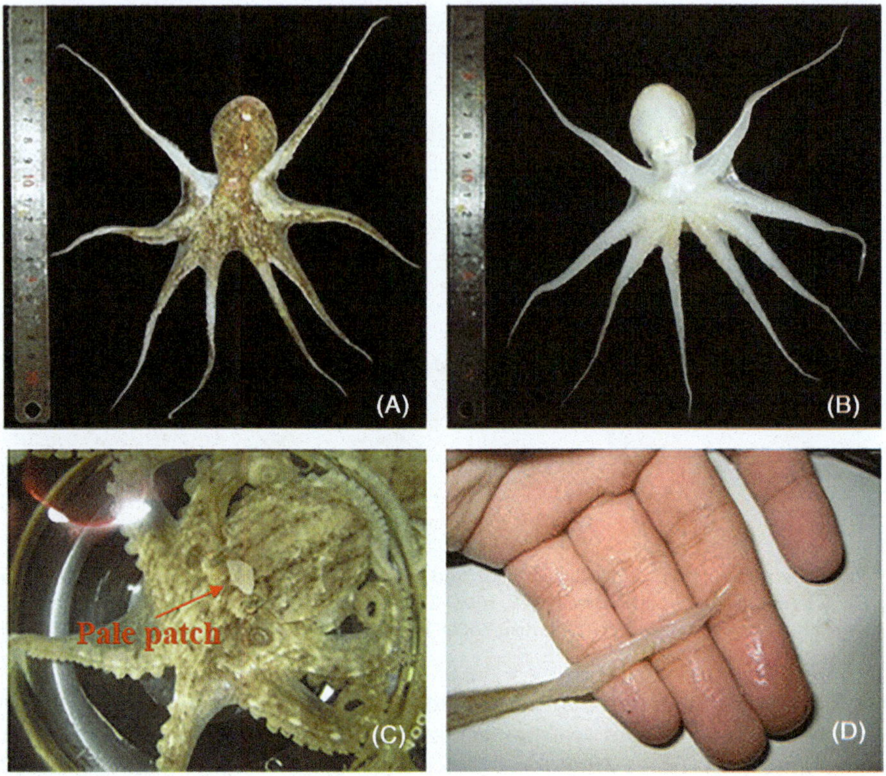

FIGURE 19.4 External morphology of *Amphioctopus fangsiao*. (A) Dorsal view. (B) Ventral view. (C) Pale head patch (showed in red arrow). (D) Hectocotylus.

tracts and is transmitted via crustacean hosts. While conspecifics are primary predators, natural predators remain undocumented. Reproductive behavior involves multiple paternity (60% of broods), with postcopulatory sperm competition. Seasonal migrations occur: individuals move to shallow subtidal zones in summer, deeper waters in winter.

19.3.3 *Amphioctopus fangsiao*, boiled-rice octopus

A. fangsiao (d'Orbigny, 1841) is commonly referred to in East Asia as the "boiled-rice octopus" ("iidako" in Japanese) due to its distinctive eggs about the size of rice grains. In recent decades, it has also been identified as *Octopus ocellatus* (a junior synonym). It is a small-to-medium-sized species endemic to East Asian coastal waters. It has a pair of distinctive golden ocelli on the dorsal surface of the skin between arms II and III. The ocelli are used as a component of a "startle" or "deimatic" display in response to threats from predators (Gleadall & Naggs, 1991; Segawa & Gleadall, 2024). *A. fangsiao* also has a characteristic thick, pale transverse bar on the head between the eyes (Fig. 19.4). The adults are compact in size, maturing at an ML of 20–30 mm, reaching a maximum of around 70 mm, typically much less. This small size facilitates laboratory maintenance, with demonstrated tolerance for group housing in sandy substrate aquaria supplemented with artificial shelters. Commercially, gravid females bearing mature egg clutches command premium market prices, particularly during seasonal spawning periods, with consumer demand driven by the culinary evaluation of the ripe ovary full of large eggs, each about the size of a grain of rice. This economic dynamic results in pronounced price fluctuations aligned with the reproductive cycle, underscoring the dual significance of this species as both a model organism for laboratory experimentation and a high-value fisheries resource (Segawa & Gleadall, 2024).

A. fangsiao inhabits coastal benthic zones on sandy mud or gravel substrates across East Asia, ranging from the tropical South China Sea (Hainan, China) to temperate southern Hokkaido (Japan), including the Bohai, Yellow, Japan, and Korean coastal seas (Segawa & Gleadall, 2024). Despite its small size, it sustains significant populations and is separately recorded in Japanese and Chinese fisheries statistics because of its commercial abundance (Sauer et al., 2021).

Female fecundity is low (< 400 eggs per female) and development is holobenthic, so dispersal is limited (Villanueva et al., 2016). Chinese Yellow Sea populations may undertake seasonal inshore-offshore movements: Spawning nearshore in spring, juvenile growth in summer–autumn, and overwintering offshore under cold coastal currents. Genetic analyses reveal two mitochondrial clades and nine subpopulations along China's coast, suggesting potential future subspecies delineation (Zheng et al., 2023b).

The life history of *A. fangsiao* is characterized by low fecundity, holobenthic development, and temperature-dependent dynamics (Dong et al., 2013). Embryogenesis encompasses cleavage, blastulation, gastrulation, organogenesis, and somatic growth, with hatching duration varying by temperature: 51 days at 25°C (Tokyo Bay) versus 24 days at 24°C (yielding smaller hatchlings of 4.6 mm ML in Lianyungang, China) (Jiang et al., 2020a; Segawa & Gleadall, 2024). Females lay large eggs (7.0–8.1 mm), producing benthic hatchlings (~4.5 mm ML) with >20 suckers, bypassing a planktonic phase—contrasting with the small-egged species *O. sinensis* (Segawa & Gleadall, 2024). Fecundity ranges from 191 to 1217 eggs per female, averaging 300–400 eggs, with egg size inversely correlated to incubation temperature (Jiang et al., 2020b). Sexual maturation in males is attained at approximately 20 mm ML (Segawa & Gleadall, 2024). As a benthic cephalopod inhabiting sandy mud or gravel substrates in East Asian coastal waters (intertidal to <50 m depth), *A. fangsiao* occupies a niche as both predator and prey. Adults primarily prey on benthic invertebrates, including bivalves (e.g., *R. philippinarum*), crabs, and gastropods, employing dual foraging tactics—pulling shells apart or drilling via combined mechanical and chemical actions (posterior salivary gland secretions). Juveniles, despite lacking a planktonic phase, exhibit vertical swimming for foraging under suboptimal den conditions (Segawa & Gleadall, 2024).

19.3.4 *Amphioctopus aegina*, sandbird octopus

A. aegina (Gray, 1849) (the marbled octopus or sandbird octopus) is merobenthic with an ML of 40–100 mm. Its diagnostic characters are a cream-colored longitudinal stripe along the dorsal midline and a chevron-like cream transverse bar on the dorsal head between the eyes (Fig. 19.5A). The "marble" or mesh pattern on the dorsal and dorsolateral mantle is formed by a pattern of dark grooves that encircle pale round patches. The arm length is 2–3 times ML. Ocelli are absent (Gleadall, 1997).

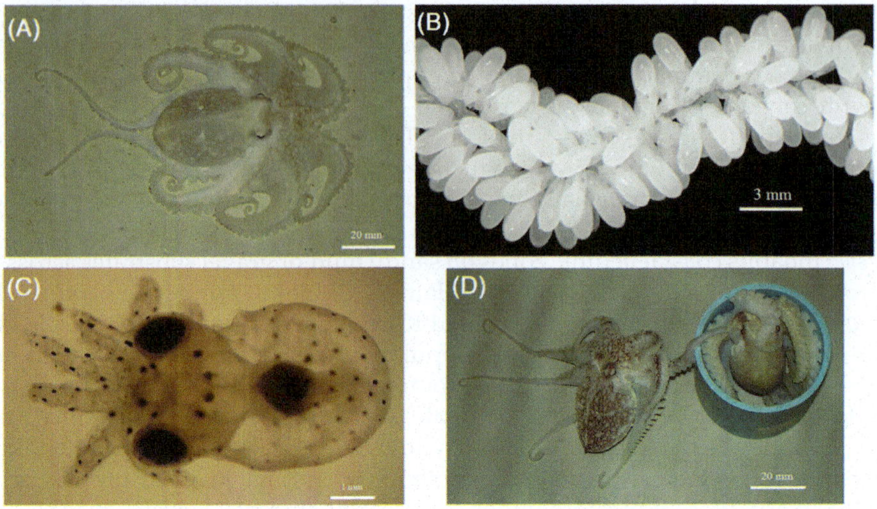

FIGURE 19.5 Artificial breeding and early development observations of *Amphioctopus aegina*. (A) *Amphioctopus aegina* in rearing tank. (B) An egg festoon of *Amphioctopus aegina*. (C) Hatchling of *Amphioctopus aegina*. (D) Mating of *Amphioctopus aegina* broodstock animals in rearing tank (female at right in a PVC shelter). *Modified from (B) Promwuthiporn (2019).*

This species is common in Southeast Asia (Central Indo-West Pacific) (Norman 1998). The distribution of *A. aegina* ranges from 30° to 170°E and 40°N to 40°S. In the Indian Ocean, the distribution extends to the west to Madagascar and into the Red Sea. In the Pacific Ocean, the distribution extends to eastern China and the Solomon Islands. Occurrence in the East Mediterranean Sea has been reported as a consequence of Lessepsian migration through the Suez Canal (Bakir et al., 2012; Mienis, 2004; Osman et al., 2014; Öztürk et al., 2014; Salman et al., 1999).

Similar to many other species of octopus, *A. aegina* is solitary, crepuscular, and shows homing behavior, living in holes, defending its home territories (Promboon et al., 2011). It lives in waters of 1.5–20 m depth, inhabiting sand, muddy sand, and sandy mud substrates. Reproduction occurs throughout the year with 2–3 nonprominent peaks. Female fecundity is 1000–12,000 oocytes in the wild (Ignatius et al., 2011), and 5500–13,000 egg capsules are laid in captivity (Ignatius et al., 2011; Nabhitabhata, 2014; Nabhitabhata et al., 2024).

The egg capsule is 3.2 × 1.2 mm with a weight of 0.2 mg (Nabhitabhata et al., 2010; Nabhitabhata, 2014; Nabhitabhata et al., 2024). The egg capsules are drawn out into stalks, which the female intertwines into festoons (Fig. 19.5). The egg is telolecithal. The embryonic phase lasts 15–22 days (mean 18.3 days) at 27–30°C, consisting of 20 stages of development as described by Nabhitabhata et al. (2010) and Nabhitabhata et al. (2024). At first, the animal pole is at the opposite end of the capsule stalk. A first inversion is observed from 137 h (day 5, Stage IX) and the second one from 335 h (Day 13, Stage XIX). The eggs are brooded by the female, cradling egg festoons in her arm web. The brooding female ceases feeding until hatching, which is followed by her senescence and death. The hatchlings (Fig. 19.5C) are planktonic and then gradually settle to the benthic lifestyle from 20 to 25 days after hatching (ML 3.2–4.2 mm).

19.4 Overview of aquaculture methodology of Asian cultivated species

19.4.1 *Octopus sinensis*, East Asian common octopus

This species is widely distributed along the southeast coast of China and is an important economic cephalopod, with an ever-increasing market demand (Song et al., 2019). In recent years, the short growth cycle, fast growth rate, high food conversion rate, large numbers of spawned young, and delicious taste have made it a highly promising aquaculture species. At present, although the aquaculture of *O. sinensis* is in progress, it still faces many challenges, such as the high mortality rate of the paralarval and settlement stages.

The broodstock should be selected as individuals with strong vitality and robust physiques. The suitable broodstock weight for breeding ranges from 562 to 1677 g. When cultivating the broodstock, the conditions required include water temperature 23.2–25.6°C, salinity 30–32 psu, and a daily water exchange rate of 100%–200%. An appropriate amount of high-quality feed (such as *R. philippinarum*) is fed daily to meet the nutritional requirements of the broodstock for gonad growth promotion.

Optimal incubation of the fertilized eggs is a key requirement. In a research project in China to investigate the incubation of *O. sinensis*, the spawning broodstock, egg masses, and shelters were placed in a cement pond with flowing-water incubation (Fig. 19.6).

In Chinese aquaculture of *O. sinensis*, there are two common farming systems: Indoor cement pond culture and cage culture. When choosing which system to use, factors such as site conditions and economic costs must be comprehensively considered. Water quality is a key factor in the culture. During culture, water quality is maintained and monitored regularly for temperature (22–25°C), salinity (31–34 psu), pH (7.6–8.2), and dissolved oxygen (>5 mg/L). *Octopus sinensis* requires a lot of feed, which remains an issue that requires further research. Appropriate feed can be selected and provided according to the growth stage. The quantity of feed is also important since uneaten feed is a source of water quality deterioration. In addition to bivalves, crustaceans and fish can also be used to meet the nutritional requirements.

19.4.2 *Octopus minor*, long-armed octopus

Over the past decade, the culture of *O. minor* has been developed in China, since the artificial release of juveniles has emerged as a crucial strategy for conservation and population enhancement of this resource, particularly in the coastal waters of Shandong Province (Zheng et al., 2023a).

The main method to capture mature individuals of *O. minor* is by bottom trawl in the northeastern coastal waters of China during May and June. In addition to indoor artificial breeding and juvenile rearing, artificial egg collectors can be deployed on the muddy bottoms of coastal waters to collect embryos (Bo et al., 2023) (Fig. 19.3B–C).

Wild individuals must be carefully selected, and those with no wounds are selected as broodstock and transported to indoor tanks for adaptation to captivity. The recommended light intensity is less than 600 Lx, with seawater temperatures of 20–24°C, salinity of 28–31 psu, and dissolved oxygen >5 mg L^{-1}. Sterilized polyvinyl chloride (PVC) pipes or eel culture cages are provided as shelters (Bo et al., 2023). The broodstock density is maintained at around 5–7 individuals m^{-2}. Small live crabs are fed as the main food. In the morning of the following day, leftover feed is removed, and the seawater is replaced (Song et al., 2019).

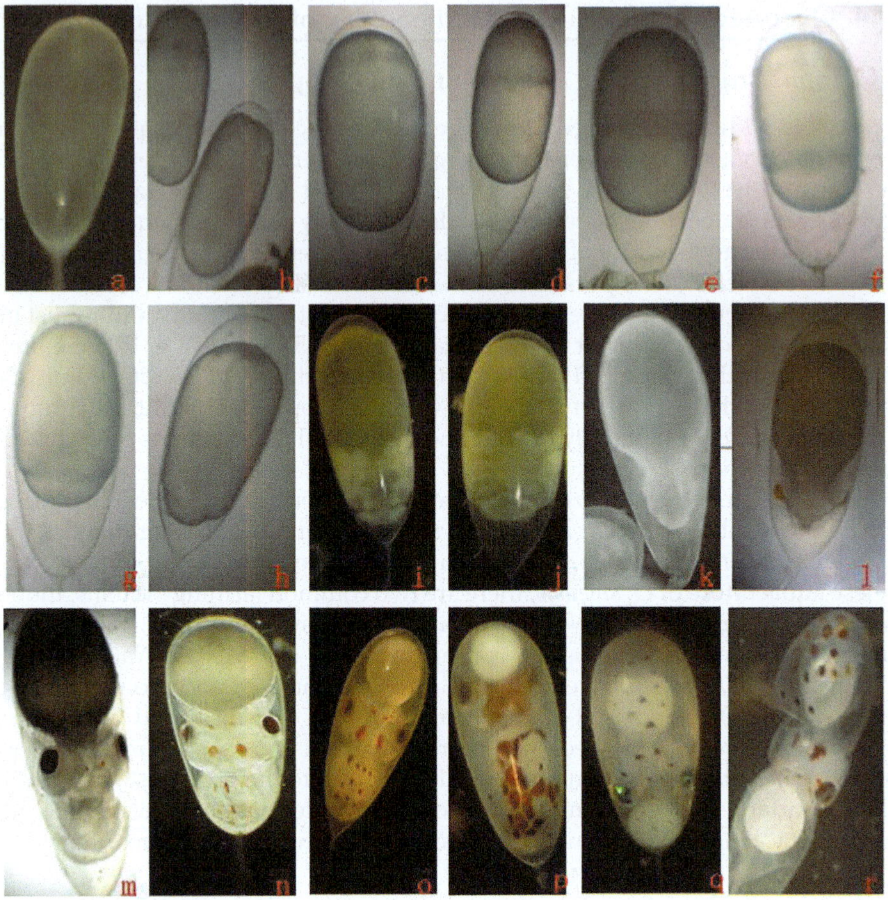

FIGURE 19.6 Embryonic stages (from I to XX) of *Octopus sinensis*.

After 1 month of acclimation, mating is typically observed at night or in the morning. The male mounts the female and inserts the hectocotylized arm into the female mantle cavity, transferring spermatophores to the oviduct. Copulation usually lasts approximately 20 min (Fig. 19.3D). The females can store spermatophores for at least 2 months before spawning.

After mating, the female will lay eggs after 2–3 weeks. The shelter pipes are firmly sealed using a diameter of 5–6 mm nylon mesh. These pipes are thoroughly washed, disinfected, and then immersed in a saline solution for 24 h. Once spawning occurs, the other end of the shelter pipe is also sealed with nylon mesh. The shelter pipes are oriented in the same direction as the seawater flow, enabling the eggs to be exposed to a constant flow of fresh, well-oxygenated seawater (Qian et al., 2013).

Food intake of the broodstock sharply decreases after egg-laying. Crabs without claws are provided to the shelter pipes every 5–7 days. Any leftover feed is removed the following morning. Most of the eggs are attached one by one to the inner walls of the shelter. Each egg has an elongated shape, resembling a small eggplant, with a length of 13–20 mm, a width of 4–6 mm, and a wet weight of 0.15–0.26 g. The number of eggs spawned per female ranges from 9 to 125, depending on the weight of each female. The incubation temperature is maintained at around 21–25°C and salinity at 28–31. Under these conditions, the process of embryonic development is divided into 20 stages (Qian et al., 2013) (Fig. 19.7). When the octopus hatchlings reach the second embryonic reversal, the original sealing mesh is substituted with polyethylene mesh featuring holes of 1.9–2.0 mm in size.

To provide extra shelter and reduce stress, the breeding pools contain some green macroalgae (e.g., *Ulva pertusa*) or seagrass. In addition to seaweed, ceramic pots of 8–12 mm inner diameter and gray–green PVC pipes of 10 mm diameter are also provided on the pool bottom as shelters for hatchlings. The number of shelters must be 1.5–2 times the total number of hatchlings, to allow them to move around among the shelters (Zheng et al., 2014). After 2 months of rearing, larger pieces of piping are provided.

After 1 month of hatchling rearing, *O. minor* juveniles have reached the ongrowing stage. During this period, the seawater temperature is maintained at 15–20°C, while other physical and chemical parameters should remain consistent with those used at hatching. Juvenile octopuses have many characteristics similar to adults, including aggressiveness, territoriality, and cannibalism. Therefore juveniles should be kept together in groups of similar size. Some small live bivalves, such as species of the genus *Potamocorbula*, are provided as feed. In addition, an adequate amount of seaweed is placed in the culture pools to provide extra cover for the juveniles.

At the juvenile stage, both fresh and frozen feeds become acceptable substitutes for live ones. Small pieces of fresh muscle of *R. philippinarum* are well accepted as an inert diet. Generally, juveniles should be fed at night, and any leftover food must be removed the next morning to maintain water quality. Juvenile density gradually decreases to 30–40 individuals m^{-3} after 2 months of growth.

Most juveniles reach a weight of 100 g in 6–7 months under indoor culture conditions, which is the commercial, marketable size ready for harvest. Alternatively, these individuals can be transferred to outdoor ponds for further growth or released into the wild as a contribution to stock enhancement. Some individuals of *O. minor* captured from natural waters harbor a parasite on their skin surface and internal body, identified as *A. sinensis* (Ren & Zheng, 2022) (Fig. 19.3E–F).

19.4.3 *Amphioctopus fangsiao*, boiled-rice octopus

Currently, the artificial cultivation of *A. fangsiao* is still in the development stage, but it has promising future potential. In China, *A. fangsiao* cultivation is mainly concentrated in northern coastal areas such as Shandong and Liaoning provinces. However, there are still many challenges in *A. fangsiao* cultivation, such as unstable seedling supply, imperfect diet selection, and feeding techniques.

The selection and cultivation of broodstock are crucial for *A. fangsiao* cultivation. Broodstock should be selected as individuals with a strong constitution, no disease or disability, and mature gonads. Clean water, appropriate temperature, stable salinity, and suitable shelter (clay pots or PVC pipes) should be provided in the broodstock cultivation pond. For feeding, live foods, such as small crabs and Manila clams, are mainly used to ensure sufficient nutrition for the broodstock and promote gonadal development.

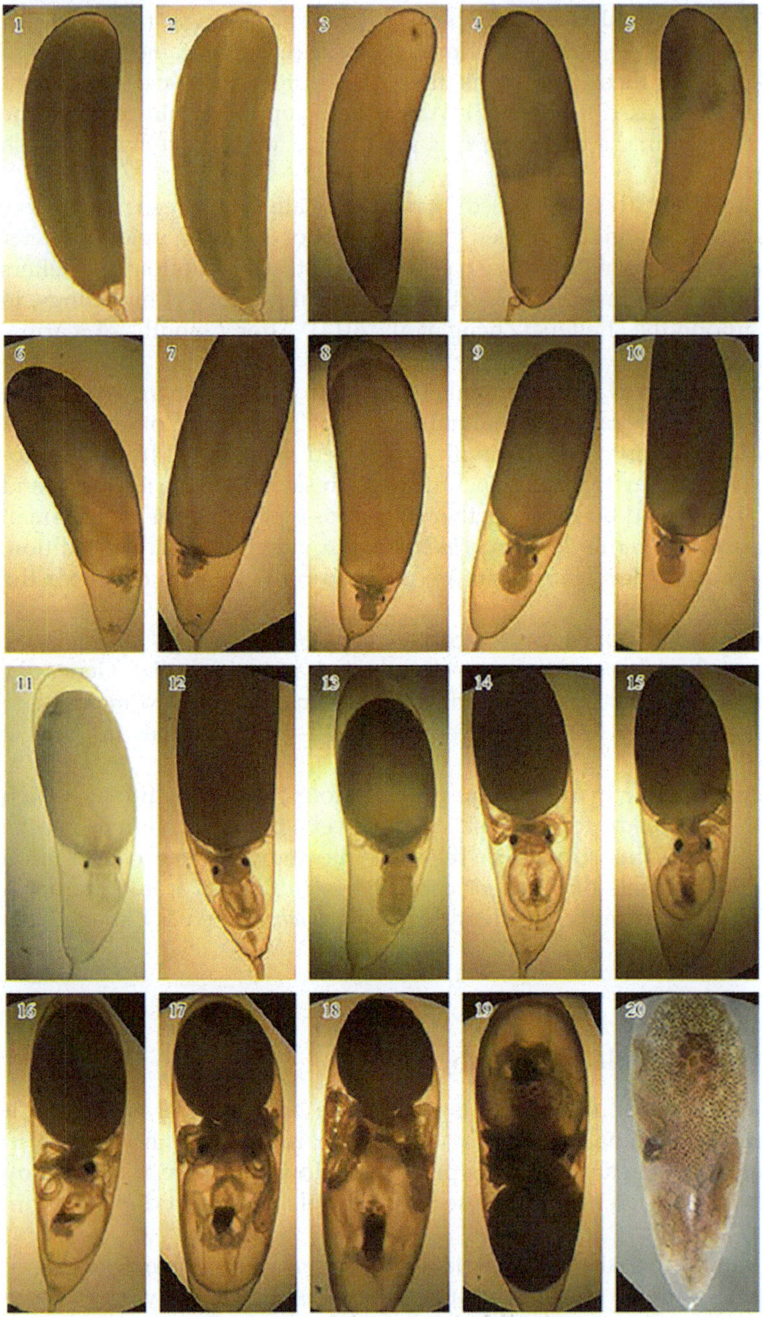

FIGURE 19.7 Embryonic stages (from I to XX) of *Octopus minor.*

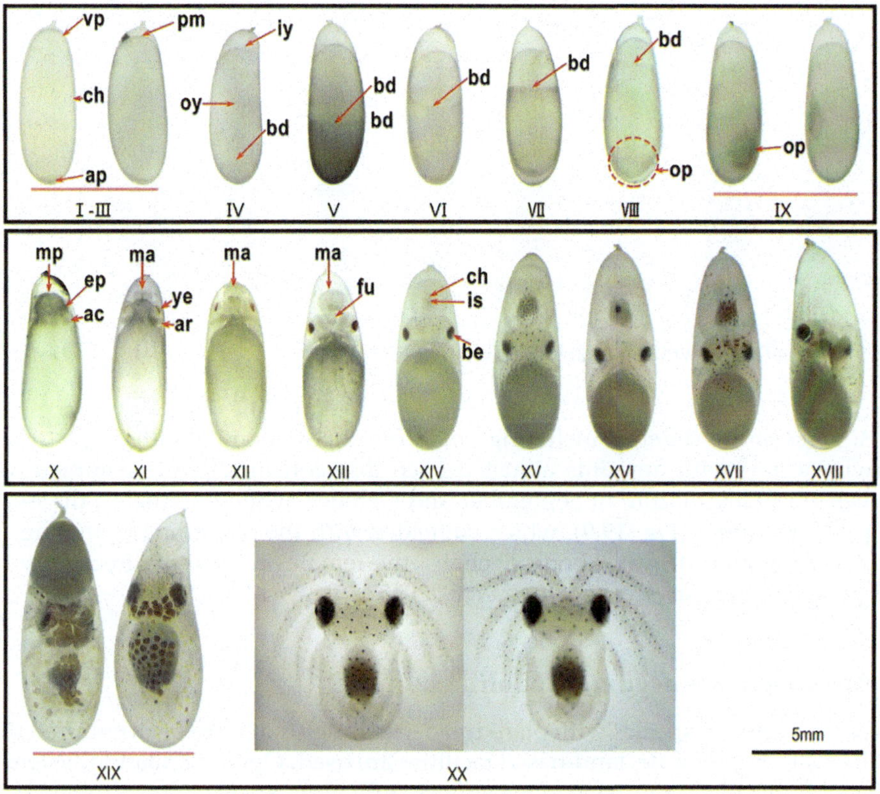

FIGURE 19.8 Embryonic development of *Amphioctopus fangsiao. FromJiang et al. (2020a).*

Mating of *A. fangsiao* usually occurs when the water temperature is suitable in spring or autumn. After mating, females lay eggs in suitable shelters (such as clay pots or the walls of the culture pond). The hatching of fertilized eggs is significantly affected by environmental factors such as temperature and salinity (Jiang et al., 2020b). The biological zero temperature for the embryonic development of *A. fangsiao* fertilized eggs is 5.15°C, the effective accumulated temperature is 589.45°C• d, and the optimal hatching temperature range is 19–24°C (Dong et al., 2013). Under suitable temperatures, fertilized eggs can hatch into juveniles in about 33 days (Jiang et al., 2020a) (Fig. 19.8).

Hatchling growth to juveniles is the most vulnerable stage in *A. fangsiao* cultivation. Hatchlings are particularly susceptible to environmental changes. Appropriate environmental conditions and high-quality feed are required. The suitable water temperature is 18–25°C (21°C optimum); salinity should be maintained at 28–30 psu. In terms of feed, nauplii of *Artemia* are the key organisms for hatchlings (0.03 g), and mixed feeding can improve the survival rate and growth rate (Wang et al., 2017).

Adult *A. fangsiao* can be cultured in indoor cement ponds with a complete water quality (temperature and salinity) control system to ensure clean water quality and sufficient dissolved oxygen. Sufficient shelters (such as bamboo tubes) are used as dens. The feed animals provided

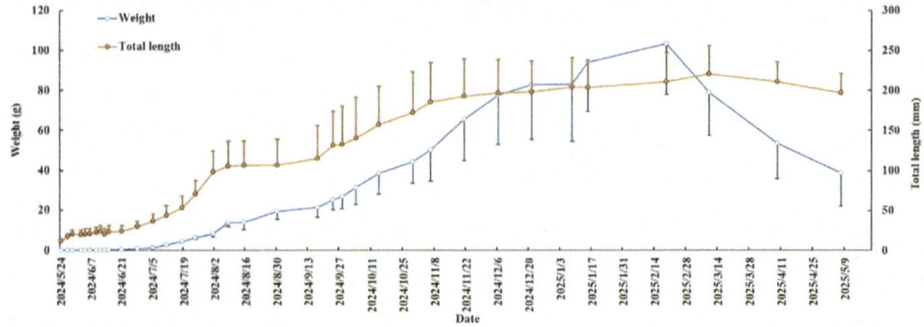

FIGURE 19.9 Monthly growth variation in *Amphioctopus fangsiao.*

to adult *A. fangsiao* are diverse, including small crustaceans and shellfish. Feed is provided once or twice a day, with amounts adjusted such that consumption is complete within 1–2 h. The males gradually mature in September and October, while in females, the ovary begins to develop in November (Fig. 19.9), which coincides with the beginning of mating. When the temperature of the natural water body declines, the females reduce the frequency of feeding and enter prespawning quiescence (Dong et al., 2013).

19.4.4 *Amphioctopus aegina*, sandbird octopus

A. aegina is an octopus species with planktonic hatchlings that have been cultured throughout its entire life cycle. Small-scale, land-based facilities are used. Open-circulation systems are used for broodstock and, later, for nursing of the young. Difficulties are resolved through a continuous supply of several types of live feed for hatchlings through to the settled juveniles.

The broodstock for *A. aegina* is collected from the landing of the octopus trap fisheries and maintained in 50-L capacity cylindrical fiberglass tanks containing 30 L of fresh seawater with aeration for transport to the hatchery (Nabhitabhata, 2014).

In the hatchery, the octopuses are separated by sex and maintained for a 7-day acclimation period in circular concrete tanks of 1.8 m diameter with an open flow-through water system. The tanks are equipped with cylindrical PVC pipe pieces (50 mm diameter) as dens or shelters, aeration devices, and supplied with running filtered seawater that flows through a central drainage. After 7 days, the octopuses are transferred to new tanks in similar conditions with a male and a female in each tank. The octopuses will mate and spawn spontaneously in the tanks within 20–40 days (Fig. 19.5D).

After spawning, the male is removed from the tank, while the brooding female is left undisturbed in the tank. Near hatching, after the second inversion at stage XIX of the embryos (around day 13 at 28°C), the brooding female is transferred to a 50-L capacity cylindrical PVC tank with dark-blue walls, containing 30 L of fresh seawater with aeration (Fig. 19.5D). Since the eggs are brooded by the female, no specific management is required for egg nursing. Routine management focuses on tank cleaning and adjusting water quality if any changes are detected. As far as possible, visual stimuli to, and physical disturbance of, the brooding females is avoided to minimize stress (Nabhitabhata, 2014; Promboon et al., 2011). Disturbed and stressed females may reject their egg mass. Rejected egg masses soon die from insufficient aeration and

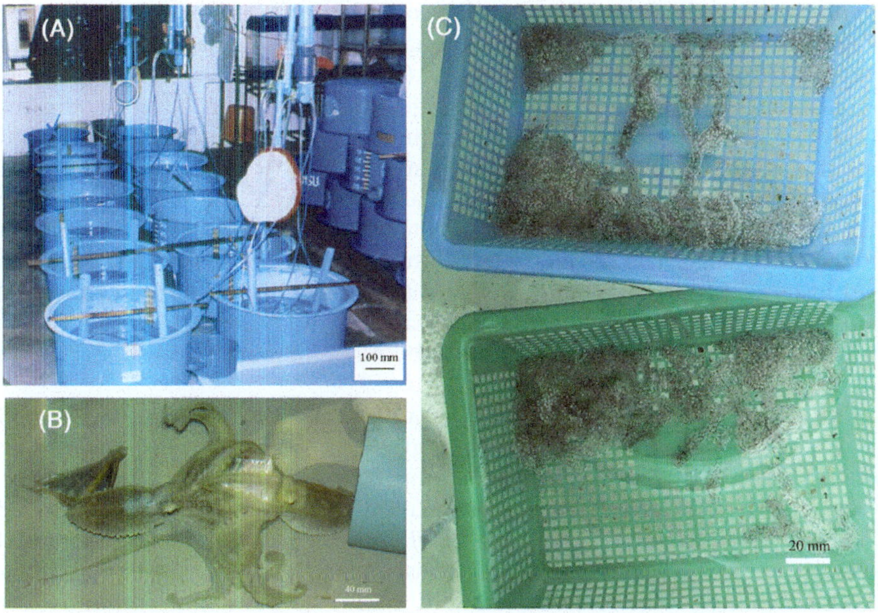

FIGURE 19.10 Nursing system, feeding behavior, and discarded egg festoons of *Amphioctopus aegina* (A) nursing tanks. (B) Two *A. aegina* feeding on the same piece of fish meat. (C) Festoons of dead or infected eggs discarded by the females in the nursing tanks.

(without maternal cleaning) microbial infection, although, with proper management, the rejected egg mass can be maintained to survive to hatching. After the eggs have hatched, the female is removed from the tank.

Hatchlings are retained in brooding tanks for further nursing. The nursing tank is filled with filtered seawater to a water depth of 150 mm (Fig. 19.10A). An artificial current is generated by directing the air-bubble stream with an air stone attached to a piece of longitudinal-cut PVC pipe (25 mm diameter, 400 mm length). Tanks are cleaned by siphoning, subsequently enabling about 80% of the volume to be replaced daily. Any temperature change is minimized by means of a running water bath outside the tank bottom (Nabhitabhata, 2014; Promboon et al., 2011). Hatchlings are fed *ad libitum* with live prey (zoea of *Macrobrachium sintangense*) from the first day to their early postsettlement (day 25–50), to generate a similar growth rate and avoid cannibalism. The initial stocking density for hatchlings is 1 individual L^{-1}. The initial water depth of 150 mm is gradually increased by about 50 mm d^{-1} to attain a final depth of 300 mm at 30 days, corresponding to the octopus growth.

Grow-out makes use of open-circulation flow-through systems. The settling young (30–50 days) are transferred to 1.8 m diameter circular concrete tanks and maintained at a density of 1 individual L^{-1}. The initial water depth is 150 mm, but is gradually increased (50 mm daily) to 400 mm. PVC pipe pieces (50 mm diameter, 150 mm length) are supplied as shelters. The young are trained to feed on dead prey (chopped fish meat, *Selaroides leptolepis*) (Fig. 19.10B). The system, facilities, and method are similar to those for the previous larval rearing or nursing phases (Nabhitabhata, 2014; Promboon et al., 2011).

The growth patterns of *A. aegina* through the entire life cycle can be separated into reflection points of rapid change of growth rate into 2–3 phases, depending on the parameters observed. The relationship between the ML (mm) and age (days after hatching) consists of two phases, with one inflection point at the settlement stage (ML 6 mm). In contrast, the relationship between body weight (g) and age consists of three phases, with the first inflection point at the settlement stage (0.07 g) and the second at maturation (6 g). Overall, the growth rate is exponentially high during the planktonic stage after hatching. The feeding and growth rates peak during the transitional change from planktonic-to-benthic lifestyle, and then gradually decrease to the adult stage. The development of the reproductive system, especially in females, is expressed in terms of rapid change of weight (the second inflection) and the peak of food conversion efficiency, but is not detected by measuring body length. Details of growth patterns have been described as statistical models by Nabhitabhata (2014) and Promboon et al. (2011).

The nature of the artisanal octopus trap as a semipassive fishing gear is considered an advantage in view of natural stock conservation, since, at least, only full-grown octopuses are exploited. However, the disadvantage of trap fishing is that the female enters the trap for spawning and brooding her eggs. The solution to this problem was initiated in 1990 by the Thailand Department of Fisheries (DOF) by using aquaculture methodology and management to endorse egg nursing, and consequently, efficient hatching of those eggs. Thousands of hatchlings were released annually back into the fishing grounds to maximize restocking efficiency during the early years of this century. The objective of such activities was to raise the awareness of fishermen of the importance of natural resource conservation, since the brooding females were supplied by the fishermen themselves. The biological effectiveness of this restocking strategy has not been assessed. Further research is necessary.

The sociological effectiveness of this project is high. At present, restocking activity in Thailand is conducted by several communities of octopus-trapping fishermen who are aware of the need to preserve their livelihood through the conservation of natural resources. These activities in the Gulf of Thailand, the so-called "Octopus Seed Bank," are in three provinces in southern Thailand and one province in eastern Thailand. These activities are partially funded by the government sector (i.e., DOF, universities).

The egg-nursing method is performed in either of two ways: (1) Natural egg incubation by the brooding female, or (2) artificially without the female. The hatchlings are collected for release.

The second method is to nurse the eggs continuously, without the brooding female. Deceased females (in rearing tanks) and stressed females (wild specimens, traumatized during capture) leave egg strands unattended, so they are collected and placed in plastic baskets with a 5 mm mesh size, floating in aerated seawater in 200-L tanks (Fig. 19.10C). The hatching rate decreases from more than 90% (Nabhitabhata, 2014) with brooding to less than 50% without a female present. However, this is obviously better than discarding them. This method is also used for eggs acquired from other fishing methods, such as egg strands found attached to the bars of clam dredging gear.

In aquaculture conditions at 28°C, *A. aegina* spends 30 days after hatching from the planktonic hatchlings to settling as juveniles. Reproduction lasts 93 days, beginning from 70 days after hatching to spawning at 145 days, and the brooding period lasts more than 18 days. The mean life span of *A. aegina* is approximately 200 days after hatching (Nabhitabhata, 2014; Nabhitabhata et al., 2024; Promboon et al., 2011).

19.5 Further research

Octopuses constitute a globally significant economic marine resource, yet escalating market demand and overexploitation have precipitated severe declines in wild populations, rendering large-scale aquaculture and stock enhancement indispensable for sustainable resource utilization. Significant progress has been made in octopus farming. However, some technical challenges remain unresolved, including high mortality during planktonic paralarval and settlement phases for merobenthic species, pervasive cannibalism in juveniles, overreliance on live feeds, and outbreaks of parasitic disease infections. To advance octopus aquaculture, future research should prioritize: Elucidating species-specific reproductive mechanisms and environmental adaptation strategies to optimize broodstock management; establishing stage-specific nutritional systems, developing cost-effective compound feeds; and implementing standardized health monitoring frameworks.

References

Arkhipkin, A. I., Hendrickson, L. C., Payá, I., Pierce, G. J., Roa-Ureta, R. H., Robin, J. P., & Winter, A. (2021). Stock assessment and management of cephalopods: Advances and challenges for short-lived fishery resources. *ICES Journal of Marine Science, 78*(2), 714–730. https://doi.org/10.1093/icesjms/fsaa038.

Bakır, B. B., Öztürk, B., Doğan, A., & Önen, M. (2012). Mollusc fauna of Iskenderun Bay with a checklist of the region. *Turkish Journal of Fisheries and Aquatic Sciences, 12*, 171–184. https://doi.org/10.4194/1303-2712-v12_1_20.

Bo, Q. K., Wang, J. H., Xing, D., Qian, Y. S., Song, M. P., & Zheng, X. D. (2023). Novel, complex burrow structure and burrowing behavior of the mud-dwelling octopus *Octopus minor* (Sasaki, 1920). *Aquatic Biology, 32*, 43–54. https://doi.org/10.3354/ab00760.

Bo, Q. K., Zheng, X. D., Gao, X. L., & Li, Q. (2016). Multiple paternity in the common long-armed octopus *Octopus minor* (Sasaki, 1920) (Cephalopoda: Octopoda) as revealed by microsatellite DNA analysis. *Marine Ecology, 37*(5), 1073–1078. https://doi.org/10.1111/maec.12364.

Dong, G., Yang, J. M., Wang, W. J., Zheng, X. D., Song, X. J., Feng, Y. W., Wei, X. M., & Sun, G. H. (2013). Studies on biological zero and effective accumulated temperature for embryonic development of *Octopus ocellatus*. *Oceanologia et Limnologia Sinica, 44*(2), 477–481.

Gleadall, I. G. (2016). *Octopus sinensis* d'Orbigny, 1841 (Cephalopoda: Octopodidae): Valid species name for the commercially valuable East Asian common Octopus. *Species Diversity, 21*(1), 31–42. https://doi.org/10.12782/sd.21.1.031.

Gleadall, I. G. (1997). Hong Kong cephalopods: A brief review of current knowledge and identification of specimens collected in 1995. In B. Morton (Ed.). *The Marine Flora and Fauna of Hong Kong and Southern China*: IV. Hong Kong University Press https://www.jstor.org/stable/j.ctt2jc68s.37.

Gleadall, I. G., Moustahfid, H., Sauer, W. H. H., Ababouch, L., Arkhipkin, A. I., Bensbai, J., Elegbede, I., Faraj, A., Ferreiro-Velasco, P., González-Gómez, R., González-Vallés, C., Markaida, U., Morillo-Velarde, P. S., Pierce, G. J., Pirro, S., Pita, C., Roumbedakis, K., Sakurai, Y., Scheel, D., & Yamaguchi, T. (2024). Towards global traceability for sustainable cephalopod seafood. *Marine Biology, 171*(2), 14321793. https://doi.org/10.1007/s00227-023-04300-6.

Gleadall, I. G., & Naggs, F. C. (1991). The Asian ocellate octopuses II. The validity of *Octopus fangsiao* d'Orbigny. *Annals of Applied Information Sciences, 16*(2), 173–180.

Ignatius, B., Srinivasan, M., & Balakrishnan, S. (2011). Reproductive traits of sandbird octopus, *Amphioctopus aegina* (Gray, 1849) from Mandapam coastal water (Palk Bay), Southeast Coast of India. *Ocean Science Journal, 46*(3), 145–154. https://doi.org/10.1007/s12601-011-0013-z.

Jiang, D. H., Zheng, X. D., Qian, Y. S., & Zhang, Q. Q. (2020a). Development of *Amphioctopus fangsiao* (Mollusca: Cephalopoda) from eggs to hatchlings: indications for the embryonic developmental management. *Marine Life Science & Technology, 2*, 24–30. https://doi.org/10.1007/s42995-019-00009-y.

Jiang, D. H., Zheng, X. D., Qian, Y. S., & Zhang, Q. Q. (2020b). Embryonic development of *Amphioctopus fangsiao* under elevated temperatures: Implications for resource management and conservation. *Fisheries Research, 225*, 105479. https://doi.org/10.1016/j.fishres.2019.105479.

Li, J. H., Chen, S., Chen, W. D., Xie, S. W., Ni, X. P., & Zheng, X. D. (2022). Morphological and genetic diversity analysis of *Octopus sinensis* in Nanji Islands. [in Chinese]. *Oceanologia et Limnologia Sinica, 53*(2), 487–495. https://doi.org/10.11693/hyhz20210900225.

Lu, C. C., Zheng, X. D., & Lin, X. Z. (2012). Diversity of Cephalopoda from Chinese waters. In M. Lin, & C. G. Wang (Eds.), *Proceedings of the 1st Mainland and Taiwan symposium of marine biodiversity studies*. Ocean Press.

Mienis, H. K. (2004). New data concerning the presence of Lessepsian and other Indo-Pacific migrants among the molluscs in the Mediterranean Sea with emphasize on the situation in Israel. *Turkish Journal of Fisheries and Aquatic Sciences, 2*(2), 117–131. https://www.vliz.be/en/imis?module=ref&refid=130038.

Nabhitabhata, J. (2014). *Amphioctopus aegina*. In J. Iglesias, L. Fuentes, & R. Villanueva (Eds.), *Cephalopod Culture*. Springer https://doi.org/10.1007/978-94-017-8648-5_18.

Nabhitabhata, J., Petchkamnerd, J., Tuanapaya, S., Tongtherm, K., Yamrungrueng, A., & Hwai, V. T. S. (2024). *Amphioctopus aegina* sandbird octopus. In R. Rosa, I. G. Gleadall, G. J. Pierce, & R. Villanueva (Eds.), *Octopus Biology and Ecology*. Academic Press https://doi.org/10.1016/B978-0-12-820639-3.00004-2.

Nabhitabhata, J., Promboon, P., & Duengdee, T. (2010). *Embryonic and post-hatching development of marbled octopus,* Amphioctopus aegina *(Gray, 1849) (Cephalopoda: Octopodidae)*: 3. World Congress of Malacology.

Nakaoka, A. M. M., & Sugino, K. (2014). *Maturation and growth of* Octopus minor *in the bay of Fukuoka Japan Bulletin of Fukuoka Fisheries and Marine Technology Research Center, 24*, 41–47.

Norman, M. D. (1998). Octopodidae benthic octopuses. In K. E. Carpenter, & V. H. Niem (Eds.). *Cephalopods, crustaceans, holothurians and sharks. The living marine resources of the western central Pacific*: vol 2. Food and Agriculture Organization of the United Nations.

Osman, I. H., Gabr, H. R., El-Etreby, S. G., & Mohammed, S. Z. (2014). Morphometric variations and genetic analysis of Lessepsian migrant *Octopus aegina* (Cephalopoda: Octopodidae). *Journal of King Abdulaziz University: Marine Sciences, 25*(2), 23–40.

Öztürk, B., Doğan, A., Bitlis-Bakir, B., & Salman, A. (2014). Marine molluscs of the Turkish coasts: An updated checklist. *Turkish Journal of Zoology, 38*, 1–48. https://doi.org/10.3906/zoo-1405-78.

Pita, C., Roumbedakis, K., Fonseca, T., Matos, F. L., Pereira, J., Villasante, S., Pita, P., Bellido, J. M., Gonzalez, A. F., Tasende, M. G., Lefkaditou, E., Adamidou, A., Cuccu, D., Belcari, P., Moreno, A., & Pierce, G. J. (2021). Fisheries for common octopus in Europe: socioeconomic importance and management. *Fisheries Research., 235*, 105820. https://doi.org/10.1016/j.fishres.2020.105820.

Promboon, P., Nabhitabhata, J., & Duengdee, T. (2011). Life cycle of the marbled octopus, *Amphioctopus aegina* (Gray) (Cephalopoda: Octopodidae) reared in the laboratory. *Scientia Marina, 75*(4), 811–821. https://doi.org/10.3989/scimar.2011.75n4811.

Promwuthiporn, S. (2019). *Rearing of octopus (*Amphioctopus aegina *(Gray, 1849)) for restocking in Thai waters*. Department of Fisheries.

Qian, Y. S., Zheng, X. D., Liu, C., Wang, P. L., & Li, Q. (2013). Studies on the reproductive habit and embryonic development of *Octopus minor* under artificial conditions [in Chinese]. *Oceanologia et Limnologia Sinica, 44*(1), 165–170. https://doi.org/10.11693/hyhz201301024024.

Ren, J., & Zheng, X. D. (2022). *Aggregata sinensis* n. sp. (Apicomplexa: Aggregatidae), a new coccidian parasite from *Amphioctopus fangsiao* and *Octopus minor* (Mollusca: Octopodidae) in the Western Pacific Ocean [in Chinese]. *Parasitology Research, 121*, 373–381. https://doi.org/10.1007/s00436-021-07389-0.

Salman, A., Katagan, T., & Boletzky, S. (1999). New cephalopod molluscs in the Eastern Mediterranean previously unnoted species or *Lessepsian migrants*. *Vie Milieu, 49*(1), 11–17. https://api.semanticscholar.org/CorpusID:132674926.

Sauer, W. H. H., Gleadall, I. G., Downey-Breedt, N., Doubleday, Z., Gillespie, G., Haimovici, M., … Pecl, G. (2021). World Octopus Fisheries. *Reviews in Fisheries Science and Aquaculture, 29*(3), 279–429. https://doi.org/10.1080/23308249.2019.1680603.

Segawa, S., & Gleadall, I. G. (2024). *Amphioctopus fangsiao*, two-gold-ring octopus. In R. Rosa, I. G. Gleadall, G. J. Pierce, & R. Villanueva (Eds.), *Octopus Biology and Ecology*. Academic Press https://doi.org/10.1016/b978-0-12-820639-3.00002-9.

Song, M. P., Wang, J. H., & Zheng, X. D (2019). Prey preference of the common long-armed octopus *Octopus minor* (Cephalopoda: Octopodidae) on three different species of bivalves. *China Journal of Oceanology and Limnology, 37*(5), 1595–1603. https://doi.org/10.1007/s00343-019-8217-7.

Vidal, E. A. G., Villanueva, R., Andrade, J. P., Gleadall, I. G., Iglesias, J., Koueta, N., Rosas, C., Segawa, S., Grasse, B., Franco-Santos, R. M., Albertin, C. B., Caamal-Monsreal, C., Chimal, M. E., dsinger-Gonzales, E. E., Gallardo, P., Le Pabic, C., Pascual, C., Roumbedakis, K., & Wood, J. (2014). Cephalopod culture: current status of main biological models and research priorities. *Advances in Marine Biology, 67*, 1–98. https://doi.org/10.1016/B978-0-12-800287-2.00001-9.

Villanueva, R., Vidal, E. A. G., Fernàndez-Álvarez, F. Á., & Nabhitabhata, J. (2016). Early mode of life and hatchling size in cephalopod molluscs: Influence on the species distributional ranges. *PLoS One, 11*(11), e0165334. https://doi.org/10.1371/journal.pone.0165334.

Wang, W. J., Dong, G., Li, H. J., Liu, Z. C., Li, B., Feng, Y. W., & Yang, J. M. (2017). Feeding behavior of larval *Amphioctopus fangsiao* and the effect of different diets on its growth and survival. progress in fishery sciences [in Chinese]. *Progress in Fishery Sciences, 38*(6), 100–106. https://doi.org/10.11758/yykxjz.20160830001.

Zheng, X. D., Bo, Q. K., & Wang, J. H. (2023). *Biology of* Octopus minor *[in Chinese]*. Science Press.

Zheng, J., Tang, Y., Xu, R., Zhang, X. Y., & Zheng, X. D. (2023). Molecular phylogenetics and population demographic history of *Amphioctopus fangsiao*, inferred from mitochondrial and microsatellite DNA markers. *Acta Oceanologica Sinica, 42*(6), 39–48. https://doi.org/10.1007/s13131-022-2105-2.

Zheng, X. D., Bo, Q. K., Xing, D., Wang, J. H., Ren, J., & Tang, Y. (2024). *Octopus' minor*, long-armed octopus. In R. Rosa, I. G. Gleadall, G. J. Pierce, & R. Villanueva (Eds.), *Octopus Biology and Ecology*. Academic Press https://doi.org/10.1016/B978-0-12-820639-3.00005-4.

Zheng, X. D., Qian, Y. S., Liu, C., & Li, Q. (2014). Octopus minor. In J. Iglesias, L. Fuentes, & R. Villanueva (Eds.), *Cephalopod Culture*. Springer https://doi.org/10.1007/978-94-017-8648-5_22.

CHAPTER

20

Pathogens and diseases

Qinggang Xue[a], *Isabelle Arzul*[b], *Tristan Renault*[c] *and Haiyan Zhang*[a]

[a]Ninghai Institute of Mariculture Breeding and Seed Industry, Zhejiang Wanli University, Ningbo, China [b]ASIM Adaptation et Santé des Invertébrés Marins, Institut Francais de Recherche pour l'Exploitation de la Mer, La Tremblade, France [c]Département Ressources Biologiques et Environnement, Institut Francais de Recherche pour l'Exploitation de la Mer, Nantes, France

20.1 Introduction

Pathogens are biologically diverse agents that cause infectious diseases by invading host tissues, evading defenses, and producing structural or functional damage. Pathogenicity is determined by virulence that reflects the damage severity to the host and is commonly measured by the mortality in hosts such as mollusks, where disease-specific signs usually lack. Pathogenesis describes the process sequence of disease production from pathogen infection, which includes the steps of entry via appropriate portals, immune evasion, colonization and proliferation, host tissue damage, and transmission to new hosts mediated by complex molecular, cellular, and organismal interactions. At ecological scales, pathogen transmission and prevalence may produce epizootics affecting populations and ecosystems, while long-term host-pathogen interactions drive reciprocal evolutionary change in virulence and susceptibility. Environmental variables and microbial communities substantially modulate these processes by causing changes in pathogen traits and host defenses. Thus disease is the emergent outcome of pathogen-host interactions within environmental and microbiome contexts.

Molluscan pathogens encompass viruses, bacteria, fungi, unicellular protists, and metazoans; several have caused severe epizootics in cultured and wild populations (Arzul et al., 2017; Du et al., 2025; Gleason et al., 2017; Travers et al., 2015). Despite the diversity in etiological agents, most molluscan diseases lack characteristic pathology and clinical signs, and the diagnosis depends on pathogen detection. Molluscan disease ecology also differs from that of terrestrial animals in several respects. At the same time, many pathogens are persistent components of marine microbiota and interact continuously with diverse environmental microbiomes (Chen et al., 2024). In addition, the r-selected reproductive strategy of many mollusks, production of large numbers of vulnerable larvae, both maintains population persistence and generates genetic

Genetic Breeding and Aquaculture of Shellfish: A Collaborative Approach to a Sustainable Future
DOI: https://doi.org/10.1016/B978-0-443-33606-5.00020-5

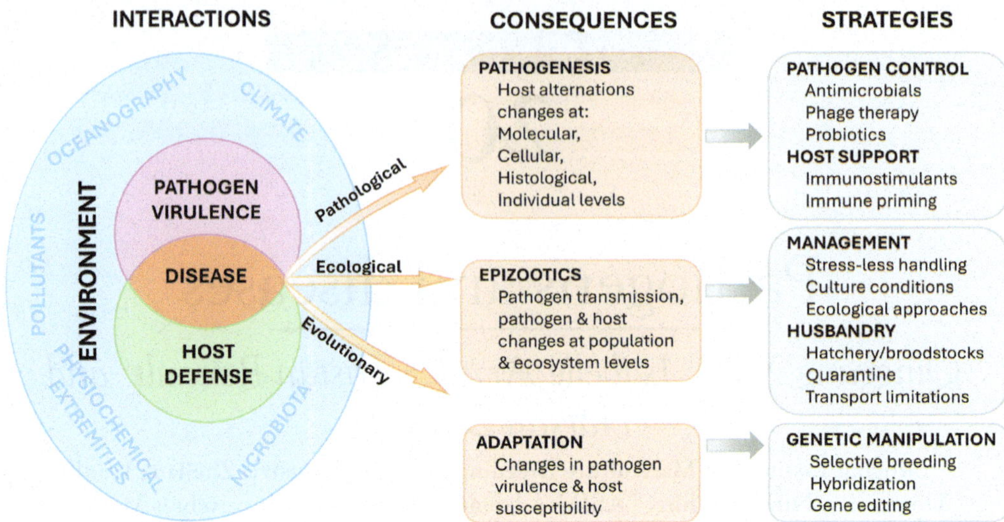

FIGURE 20.1 Framework for molluscan disease dynamics: Pathogen threats and multifactorial regulation.

variation that can fuel host-pathogen coevolution. Moreover, physical oceanography (currents, stratification, temperature, salinity, and oxygen) shapes pathogen survival, dispersal, and host susceptibility; anthropogenic transfers for aquaculture further facilitate geographic spread (Yisrael et al., 2024). These features establish a distinctive framework that governs molluscan disease dynamics at pathological, ecological, and evolutionary scales (Fig. 20.1). This chapter reviews principal pathogens, viruses, bacteria, and parasitic protists, and associated diseases that threaten molluscan aquaculture and fisheries.

20.2 Viral pathogens and viral diseases

Viruses are acellular infectious agents that replicate obligately in host cells; they consist of a DNA or RNA genome encased by a protein capsid and sometimes a host cell membrane-derived envelope. Viral infection starts with the virus entering the host cell via the mediation of specific receptors on the cell surface, followed by replication using the host cell machinery. Newly generated virions are released from the host cell through budding, exocytosis, or cell lysis, causing direct cytopathic effects and often eliciting host immune responses that contribute to pathology. While viruses are usually viewed as parasitic or pathogenic due to their intracellular living nature, many of them play a role in the maintenance of ecosystems via host population regulation and biogeochemical cycling. Viruses are often described according to their host organism and morphological characteristics. Systemic schemes classify viruses into seven groups in the Baltimore Classification, or into a hierarchical taxonomy consisting of Realm, Kingdom, Phylum, Class, Order, Family, Genus, and Species.

Molluscan virology has advanced substantially. A molluscan-specific viral family (family Malacoherpesviridae) is created to host herpesviruses that cause mortality-resulting infections in oysters and abalones (Davison et al., 2005, Davison et al., 2009). Numerous virus-like particles

(VLPs) are recognized in cultured and wild molluscan species, and virome surveys reveal complex viral assemblages associated with health and disease states (Jiang et al., 2023).

20.2.1 *Ostreid herpesvirus 1* and the infection in bivalves

Etiology: *Ostreid herpesvirus* type one (OsHV-1) is a herpesvirus belonging to the order Herpesvirales, family Malacoherpesviridae, and genus *Ostreavirus*, with a scientific species name of *Ostreavirus ostreidmalaco1*. OsHV-1 is an enveloped icosahedral virus of 100–180 nm in diameter (Le Deuff & Renault, 1999), with a reference genome of 207,439 bp DNA encompassing 132 open reading frames (ORFs) (Davison et al., 2005). Multiple genetic variants with microvariations in genome sequences have been reported in bivalves from different geographic areas, and some of them exhibit changes in virulence (Morga et al., 2021). Among them, the *Ostreid Herpesvirus* 1 μVar (OsHV-1 μVar) originally reported from dying juvenile French Pacific oysters in France in 2008 has gained wide distribution in France and Europe (Segarra et al., 2010). Other variants include OsHV-1-SB and the acute viral necrosis virus (AVNV), which are evolutionarily more distant to the "reference type" infecting different species (Ren et al., 2013; Zhang et al., 2025c).

Pathology: OsHV-1 usually shows tropism toward the fibroblastic-like cells of the connective tissues of all organs in oyster larvae and spat. Some variants (e.g., OsHV-1-SB) are more common in host hemocytes (Li et al., 2024). Infected cells exhibit nuclear margination, chromatin, and condensation. Nucleocapsids and empty capsids are visualized in the nuclei (Renault, 2024). There is also evidence for the viral infection suppressing host immunity (de Lorgeril et al., 2018). Infected larvae show reduced feeding, slowed growth, weak, circular, or no swimming. However, the most noticeable macroscopic signs are usually sudden increased mortality in all developmental stages.

Epizootiology: OsHV-1 and the microvariants infect oyster species, and larvae and juvenile Pacific oysters *Crassotrea gigas* are particularly impacted by epizootics associated mass mortalities. They have also been reported sporadically in Portugal oysters *Crassostrea angulata*, eastern oysters *Crassostrea virginica*, and Kumamoto oysters *Crassostrea sikamea*, and European flat oysters *Ostrea edulis* in addition to *C. gigas* (Burge et al., 2011; López Sanmartín et al., 2016). In addition, the viruses are detected in clams *Ruditapes decussatus* and *Ruditapes philippinarum*, giant scallops *Pecten maximus*, inducing high mortality among larvae (Arzul et al., 2001a, 2001b). In mussels *Mytilus edulis* and *Mytilus galloprovincialis*, pathological effects of the virus are, however, not confirmed (O' Reilly et al., 2018). OsHV-1-SB and AVNV are reported to impact broodstocks of ark clams, *Anadara broughtoni* and *Anadara kagoshimensis*, and the scallop *Chlamys farreri*, respectively (Bai et al., 2016; Ren et al., 2013; Xia et al., 2015; Zhang et al., 2025c). The geographic distribution of the OsHV-1 and its variants involves almost all the bivalve aquaculture areas in Europe, Asia, Australia, New Zealand, and the United States. The viruses in persistently infected hosts or carried by healthy hosts may serve as a source of transmission (Divilov et al., 2024; Evans et al., 2017). Studies reveal indications for coevolution between OsHV-1 and the host species in different environments (Pelletier et al., 2025). The virus is believed to transmit horizontally between host bivalves and vertically from broodstocks to the progenies (Barbosa-Solomieu et al., 2005; Lionel et al., 2013; López Sanmartín et al., 2016). Elevated water temperature appears to promote viral replication (Renault et al., 2014). Culture density and microbiome dynamics are also factors influencing viral prevalence (Pathirana et al., 2019).

Diagnosis: OsHV-1 infection diagnosis relies on cytopathological and histopathological analysis and virus direct detection by transmission electron microscopy. Molecular techniques that have been used for the routine diagnosis and virus identification include in situ hybridization, conventional PCR, real-time quantitative PCR, and loop-mediated isothermal amplification (LAMP) (Corbeil et al., 2015; Pepin et al., 2008; Ren et al., 2010). More specific variant genotyping is done by multiple loci or genome-wide analyses (Bai et al., 2019; Dotto-Maurel et al., 2025; Renault et al., 2012).

Control and management: Biosecurity, virus surveillance, stringent regulations, and husbandry operations to reduce host stress are some preventive measures in use (Hick et al., 2016; Whittington et al., 2019). Host immunity boosting using immunostimulants, probiotic bioregulators, or immune priming is assessed (de Kantzow, Hick, & Whittington, 2023). Selective breeding to generate viral infection more resistant stocks is being attempted (Divilov et al., 2023).

20.2.2 *Haliotid herpesvirus 1* and the abalone viral ganglioneuritis

Etiology: *Haliotid herpesvirus* 1 (HaHV-1) is another currently confirmed mollusk-specific herpesvirus in Family Malacoherpesviridae, belonging to Genus *Aurivirus* with a species name of *Aurivirus haliotidmalaco*1. HaHV-1 is an enveloped icosahedral virus about 150 nm with a nucleocapsid of 90–110 nm in diameter (Tan et al., 2008). It has a linear, double-stranded DNA genome of 212 kb encoding about 117 ORFs. Virus isolates are currently grouped into Australian genotypes, Taiwan genotypes, and South China genotypes, which show genetic variations of about 91.5% sequence similarity (Chen et al., 2016; Savin et al., 2010; Tan et al., 2008).

Pathology: HaHV-1 exhibits tropism toward host neural tissues and hemocytes. The Australian genotypes are typically neurotropic, invading and causing direct cytopathic changes in the central nervous system (Hooper et al., 2007). The Taiwan genotypes were initially observed to preferentially infect host ganglionic neural tissues, but isolates collected during chronic mortality events are also found in host hemocytes (Chen et al., 2016). The South China strains are tropistic to both hemocytes and neural ganglia (Bai et al., 2020). The virus likely reaches the target host tissues and cells via hemolymph after entering the host abalones (Wei et al., 2025). In abalone viral ganglioneuritis (AVG), cytopathology exhibits pyknosis, karyorrhexis, and the appearance of cytoplasmic and nucleic inclusion bodies in the ganglia. The infection causes necrosis and inflammation in pedal, cerebral, and pleuroviseral ganglia, with extensive neuronal degeneration, hemocyte infiltration, and foot muscle degeneration at the histopathological level (Hooper et al., 2007; Tan et al., 2008). Hemocyte degradation is observed in hemolymph and gill sinus (Bai et al., 2020; Wei et al., 2025). Infected abalones start to develop macroscopic signs such as weakness in adhering substrate, reduced mobility, mouth swelling, and excessive mucus around the foot and tentacles, typically 4–7 days after the virus exposure (Corbeil et al., 2012b). Mantle recessions and muscle atrophy can occur in chronic hemocyte infections (Chang et al., 2005; Chen et al., 2016), but neurotropic strain-infected AVA abalones often die before overt sign appearance, with the cumulative mortality >90% (Hooper et al., 2007).

Epizootiology: HaHV-1 was officially reported to cause acute and high mortalities in the farmed abalone *Haliotis diversicolor supertexta* in Taiwan in 2003 (Chang et al., 2005; Chen et al., 2016). Retrospective speculations, however, pinpoint the virus as the causative agent of massive *H. diversicolor* mortalities between 1999 and 2002 in Guangdong Province, China (Tan et al., 2008). The virus was then reported in 2007 as the cause of AVG under the name of *Abalone herpesvirus* (AbHV), which resulted in mortalities up to 90% in *H. rubra, H. laevigata*, and their hybrids in Victoria, Australia, between 2005 and 2006 (Hooper et al., 2007). HaHV-1 infects several abalone species of the genus *Haliotis*, particularly *H. laevigata, H. rubra, H. conicopora,* and *H. diversicolor supertexta*, at all ages (Tan et al., 2008), but *H. discus hannai* and *H. iris* are not susceptible in experimental infections (Bai et al., 2020). The virus remains infectious and highly pathogenic in 4–15°C seawater for at least one day and is believed to be transmitted horizontally via contaminated seawater (Corbeil et al., 2012a). Virus carriers in the environment, such as sediments, biofilms, and apparently healthy abalones, can be the virus sources for transmission.

Diagnosis: HaHV-1 infection diagnosis depends on histopathological and cytopathic analyses, and on specific virus detection by transmission electron microscopy for confirmation (Chang et al., 2005; Hooper et al., 2007). Conventional PCR, quantitative PCR, and LAMP are reported for detecting HaHV-1 in a variety of specimens (Chen et al., 2014; Corbeil et al., 2010). ISH has been used to visualize virus distribution and translocation in host tissues during the infection process (Bai et al., 2020).

Control and management: HaHV-1 infection is a molluscan disease listed in the World Organization for Animal Health (WOAH) Aquatic Code (https://www.woah.org/en/what-we-do/standards/codes-and-manuals/). Pathogen control and disease mitigation largely rely on biosecurity and virus surveillance-based prevention and husbandry management (Chang et al., 2005; Hooper et al., 2007). Virus infection resistance enhancements via antiviral immunity boosting and selective breeding are also being attempted.

20.2.3 Other virus-like particles reported from mollusks

VLPs often detected electron microscopically in mollusks collected in routine surveys or mortality investigations are reported (Meyers et al., 2009; Rolton et al., 2023). In most cases, these VLPs are tentatively characterized solely based on morphology, and there is normally insufficient information about the pathogenicity to establish disease causation. The following are a few examples.

Irido-like virus and infections in oysters: Irido-like viruses are reported to be associated with oyster diseases, including the gill necrosis virus (GNV) and hemocyte infection virus (HIV) in Portugal oysters *C. angulata* on the French Atlantic coasts in the late 1960s and early 1970s, and the oyster velar virus disease (OVVD) in larval Pacific oysters (Elston & Wilkinson, 1985). GNV invades *C. angulata,* typically resulting in yellow spots on the gills and labial palps, and sometimes on the adductor muscle and mantle, which eventually leads to the gill filament destruction and causes mortalities of up to 70% in an affected *C. angulata* population. HIV infects *C. angulata* hemocytes, causing acute infiltrations by virus-infected hemocytes, and the virus prevalence was speculated to be the most important factor resulting in the almost extinction of the Portuguese oyster populations on the French Atlantic coasts in the 1970s. In 1977, a similar type of HIV infection was observed in *C. gigas* in the Bay of Arcachon (France) associated with 15% mortality (Comps & Bonami, 1977). An irido-like virus similar to HIV was observed in *C. angulata* spat in France in 1984. Despite the difference in host tissue tropism, GNV and HIV are morphologically similar, being both icosahedral and 300–350 nm in diameter, and unclear whether representing different virus species or species genetic variants. OVVD was reported in *C. gigas* larvae from a hatchery in Washington State, USA, and is an icosahedral virus around 228 nm in diameter that forms intracytoplasmic inclusion bodies of 1.2–4 mm in diameter in the infected ciliated velar epithelium. The infection leads to the formation of characteristic blisters due to the detachment of epithelial cells from the velum. OVVD larvae lose mobility and start to die about 10 days after spawning when reaching 175–190 mm in shell height (Elston & Wilkinson, 1985).

Picorna-like VLPs in clams: VLPs morphologically resembling picornaviruses have been described in several bivalve species (Soni et al., 2024). VLPs of 27 nm in diameter, for example, were described in mussels, *Mytilus edulis*, from Denmark (Rasmussen, 1986). Electron-dense uncoated VLPs, ranging from 25 to 45 nm in diameter, were also associated with modified rough endoplasmic reticulum in farmed seed and adult green-lip mussels, *Perna canaliculus*, suffering mortality outbreaks and subsequently in *M. galloprovincialis* in New Zealand by transmission electron microscope examination (Jones et al., 1996). In addition, apparent replication of small VLPs of 22–30 nm in diameter was described to be present in an orderly array on the surfaces of the outer nuclear membrane and along the endoplasmic reticulum in digestive and secretory cells of scallops, *Pecten novaezelandiae*, and toheroa, *Paphies ventricosa*, collected during a mass mortality outbreak in New Zealand (Hine & Wesney, 1997). Moreover, a nonenveloped virus that is morphologically similar to those from *P. novaezelandiae*, 25–33 nm in diameter and named "Akoya-virus", was associated with necrosis and degeneration of muscle fibers in Japanese pearl oysters, *Pinctada fucata martensii*, collected during mass mortality events (Miyazaki et al., 1999). Finally, VLPs' icosahedral shape and 25–35 nm in diameter were described in wild and farmed Manila clams *R. philippinarum* and carpet shell clams *R. decussatus* in the United Kingdom and France, respectively, as belonging to the family Picornaviridae or the family Parvoviridae (Bateman et al., 2012; Dang et al., 2009).

Papilloma-like or polyoma-like VLPs and viral gametocytic hypertrophy: VLPs assigned as members of the family Papillomaviridae or family Polyomaviridae are linked to viral gametocytic hypertrophy (VGH) in *C. gigas, C. virginica*, and other bivalve species (Cheslett et al., 2009; Garcia et al., 2006). The VLPs are unenveloped icosahedrons of about 50 nm in diameter within basophilic inclusion bodies in the gamete nuclei, which results in hypertrophy in the affected cells. The particles are present in both male and female gonads. Similar basophilic inclusion bodies and VLPs are also reported in the digestive epithelia of razor clams *Ensis arcuatus* (Ruiz et al., 2011). Nevertheless, the presence of these VLPs is not associated with any pathological changes at histological and macroscopic levels.

Asfas-like virus and the abalone amyotrophia: An Asfas-like virus, named abalone asfa-like virus (AbALV), is reported as the etiological agent of abalone amyotrophia, an infectious disease of juvenile abalones *Haliotis gigantea, Haliotis discus discus, Haliotis discus hannai*, and *Haliotis diversicolor* in Japan since the 1980s (Matsuyama et al., 2020, 2021, 2023). The virus is currently assigned as a family Asfarviridae member, icosahedral in shape and about 200 nm in diameter, with a linear double-stranded DNA genome of 281 kb sharing 33.97% sequence identity with that of the African swine fever virus (ASFV) (Matsuyama et al., 2020, 2023). Virus infection-associated histopathology includes abnormal cell masses in the nerve trunks of the peripheral nerve surrounding the foot muscles. Macroscopic signs of the infection are typically related to foot pathology in the infected abalones and thus similar to AVG. Amyotrophia epizootics cause chronic mortalities of up to 80% susceptible juvenile abalones. The virus is believed to transmit horizontally through water, and infected abalones without showing disease signs can serve as virus reservoirs (Nakatsugawa et al., 2000).

20.3 Bacterial pathogens and bacterial diseases

Bacteria are prokaryotic microorganisms characterized by the absence of a membrane-bounded nucleus and organelles. They have a peptidoglycan cell wall protecting the plasma membrane, with a capsule or slime layer coating on the surface. Some bacteria have flagella for mobility and pili for adhesion and conjugation. Bacterial genomes are relatively small and structurally dynamic, ensuring genetic versatility for rapid evolutionary adaptation (Waters et al., 2025). Bacteria replicate by binary fission. They are classified into the taxonomic ranks based on a series of phenotypic traits, and new identifications increasingly depend on DNA-sequence analyses such as DNA-DNA hybridization (DDH), average nucleotide identity (ANI), multiple locus sequence analysis (MLSA), and 16S rRNA gene (Johnson et al., 2019). Over 25,000 species of bacteria have been described, and a tiny fraction of them are pathogenic to humans, animals, or plants (Soni et al., 2024). The genome, including plasmids, of pathogenic bacteria encodes a huge variety of virulence factors (VFs) that function in all steps of the pathogenesis. Unlike viruses and parasitic eukaryotes where VFs evolve principally by genetic mutation and selections, pathogenic bacteria acquire VFs very frequently and quickly from the environments or other bacteria via horizontal gene transfer (HGT) by plasmids or other transportable elements (Bakkeren et al., 2022), and the VF expression is coordinated at the bacterial population level by quorum sensing at the population level (Warrier et al., 2021). In line with the abundant composition of marine microbiomes, most currently described bacterial pathogens of aquacultural mollusks are Gram-negative bacteria (Danckert et al., 2021; Pierce & Ward, 2019).

20.3.1 Pathogenic vibrios and vibriosis in aquacultural mollusks

Etiology: Vibrios constitute one of the most heterotrophic and ecologically diverse genera of bacteria (Thompson et al., 2004). They are Gram-negative, facultatively anaerobic, curved or straight rods in shape, typically motile with a single polar flagellum, and predominantly halophilic or halotolerant. They grow on sodium-supplemented culture media such as marine agar, and on thiosulfate-citrate-bile salts-sucrose (TCBS) agar. MLSA on 8 housekeeping genes (i.e., *gapA, gyrB, ftsZ, mreB, pyrH, recA, rpoA,* and *topA*) grouped over 150 currently described *Vibrio* spp. into 51 evolutionary clades (Jiang et al., 2022). *Vibrio* genomes are highly plastic, consisting of two chromosomes and variable numbers of plasmids that encode traits for adaptation to different ecological roles, including pathogenesis in marine and estuarine environments (Huang et al., 2024). Vibrios encompass the most devastating bacterial pathogens for aquacultural mollusks and many other marine plants and animals (Vandeputte et al., 2024). More than 20 *Vibrio* spp. and subspecies involving 7 clades are currently reported as pathogens or opportunistic pathogens for marine bivalves and abalones (Table 20.1). In some cases, the possession of a pathogenicity island (e.g., the TCP-PAI for *V. mediterranei* and a specific *cuscop*-containing island for *V. aestuarianus*) or a regulatory gene (e.g., *varS* for *V. aestuarianus*) in the bacterial genome is found to render a genotypic characteristic distinguishing the pathogenic strains or subspecies from the nonpathogenic or low-virulence ones in a *Vibrio* sp. (Zhang et al., 2025b; Goudenège et al., 2015; Mesnil et al., 2023). Similar to many *Vibrio* spp. pathogenic to humans and land animals, *Vibrio* linages pathogenic to mollusks can also emerge and expand by clonal evolution (Mesnil et al., 2023).

Pathology: Mollusk-infecting vibrios have a huge array of VFs involving host adherence, hydrolysis of host macromolecules, cytolysis of host cells, nutrient acquisition, growth, and VF expression regulation, VF delivery, and mobility (Duperthuy et al., 2011; Fan et al., 2024; Galvis et al., 2021; Hasegawa et al., 2009; Islam et al., 2022; Morot et al., 2021; Vandeputte et al., 2024). There is research evidence for vibrios to acquire VFs via HGT (Bruto et al., 2017). Quantitative and qualitative differences in VFs are observed at the genetic and molecular levels,

TABLE 20.1 Major *Vibrio spp*. Infecting aquaculture bivalves and gastropods.

Clade	Species	Host and stage	Major VFs	Impact (mortality)	References
Anguillarum	*V. aestuarianus*	Oyster: *C. gigas* juvenile and adults (with high mortality rates), Edible Cockle: *Cerastoderma edule*	*hcp, rtx, vam, vpp, zot, aphA, aphB, luxU, luxO, toxR, toxS, ompU, djlA, mviN*, NlpI, T1SS, T2SS, T4SS, T6SS, pili, flagella/chem	Up to 100 % mortality of oysters in laboratory and up to 83% mortality of cockles	Mesnil et al., 2023), Lupo et al. (2019)
Coralliilyticus	*V. coralliilyticus*	Oysters: *C. gigas* larvae, *C. virginica* larvae, Mussel: *M. galloprovincialis* adult, Clams: *Tridacna crocea* adult, *M. mercenaria* larvae and juvenile	*toxR, ompU*, T6SS	Up to 100% mortality of oysters in laboratory; no mortality of mussel adults; up to 56.78% mortality of giant clam adults	Gómez-Chiarri et al., 2024
	V. neptunius	Oysters: *O. edulis* larvae, *C. gigas* larvae and juvenile, Clam: *Ruditapes philippinarum* larvae Mussel: *Perna canaliculus* larvae	VnpA, ColA, Piscibactin, Amphibactin	88% mortality of European flat oysters in laboratory; 80% mortality of Pacific oyster juveniles, 100% mortality of Manila clam; 75% mortality of Green shell mussel larvae	Gómez-Chiarri et al., 2024
	V. ostreicida	Oyster: *O. edulis* spat Mussel: *M. galloprovincialis*	-	Severe mortalities in the hatchery; up to 50% mortality of mussel	Lopez et al. (2025)

(*continued on next page*)

TABLE 20.1 Major *Vibrio spp*. Infecting aquaculture bivalves and gastropods—cont'd

Clade	Species	Host and stage	Major VFs	Impact (mortality)	References
Harveyi	*V. alginolyticus*	Oysters: *C. gigas* juveniles, *C.hongkongensis* Clam: *Ruditapes decussatus* larvae and juvenile Mussels: *Mytilus coruscus, Perna viridis,* Scallops: *Argopecten ventricosus* larvae, *Nodipecten subnodosus* larvae, Pen shell: *Atrina maura* larvae	Cytotoxicity, hemolysins, T3SS, ESBLs, AHL, *tcp, aspA,* Collagenase, *flaB, fur, ompW, tlh, toxR, vscC*	Up to 100% mortality of mussels, more than 60% mortality of oysters, 60% mortality of green mussel; up to 90% mortality of grooved carpet shell larvae; 100% mortality of Pacific calico scallop larvae; 100% mortality of giant lion's paw larvae; 100% mortality of Maura pen shell larvae	Vandeputte et al. (2024)
	V. jasicida	Oyster: *C. gigas*	-	Around 15% mortality of oyster	Gómez-Chiarri et al., 2024
	V. rotiferianus	Oyster: *C. gigas* larvae, Scallop: *Spondylus limbatus* larvae	T2SS, *tlh*	More than 80% mortality of scallop and oyster	Revilla et al. (2025)
	V. harveyi	Oyster: *C. gigas,* Abalones: *Haliotis tuberculata* adult, *H. discus hannai* adult, Snail: *Babylonia areolata,*	T3SS, *luxS, katA, katB, sodC, flgI, ompU,* Vibrioferrin	70% mortality of European abalone in 10 days, 80% mortality of ivory shell, 50% mortality of oyster	Morot et al. (2021), Lee et al. (2023)
	V. parahaemolyticus	Clam: *Meretrix meretrix* Abalone: *H. diversicolor supertexta* postlarvae	*tdh, trh,* serine protease, caseinase, lipase, gelatinase, vibrioferrin siderophore, Phospholipase	Up to 100% mortality of clam and 90% mortality of abalone in laboratory	Cai et al. (2007), Yue et al. (2010)

(*continued on next page*)

TABLE 20.1 Major *Vibrio spp*. Infecting aquaculture bivalves and gastropods—cont'd

Clade	Species	Host and stage	Major VFs	Impact (mortality)	References
Mediterranei	*V. mediterranei*	Oyster: *C. gigas* larvae Clam: *Sinonovacula constricta* larvae and juvenile Mussels: *M. galloprovincialis, Pinna nobilis* juvenile	*tcp, zot, ompU, rtx, vir, mshA, rpoS*	Up to 100% mortality of oyster larvae, razor clam larvae, and fan mussel juveniles	Gómez-Chiarri et al., 2024), Zhang et al. (2025a)
Orientalis	*V. bivalvicida*	Oyster: *Ostrea edulis* Scallop: *Argopecten purpuratus* larvae Clams: larval *R. philippinarum* and *R. decussatus, Donax trunculus*	Heat stable ciliostatic toxins and proteinases	Higher than 96% mortality of oyster and clam larvae	Rojas et al. (2019)
	V. europaeus	Oyster: juvenile *C. gigas,* Scallop: *A. purpuratus* Clam: *R. philliphinarum* larvae and juveniles	*vemA*	Up to 100% mortality of Manila clam larvae and juveniles; 88% mortality of Pacific oyster juveniles	Martinez et al. (2022)
	V. tubiashii	Oyster: *C. virginica* larvae Clams: larval *R. philippinarum* and *M. mercenaria* Snail: *B. areolate* juvenile	type I secretion outer membrane protein, TolC, lipoprotein, efflux system proteins, iron-regulated protein A, *vtpA, vtpB*	Up to 90 % mortality of ivory shell and 66% mortality of eastern oyster in laboratory, 81% mortality of manila clam larvae	Dubert et al. (2016), Wang et al. (2022)

(*continued on next page*)

TABLE 20.1 Major *Vibrio spp*. Infecting aquaculture bivalves and gastropods—cont'd

Clade	Species	Host and stage	Major VFs	Impact (mortality)	References
Pectenecida	*V. pectenecida*	Scallop: *Pecten maximus* larvae	VHKT	Up to 100% mortality of scallop	Sandlund, Torkildsen, Magnesen, Mortensen, & Bergh, 2006
Splendidus	*V. crassostreae*	Oyster: *C, gigas* Clam: *R. philippinarum* Mussel: *M. edulis* larvae Scallop: *A. purpuratus* juvenile	T6SS	73% mortality of blue mussel, 46.7% mortality of scallop juvenile by injection, and 100% mortality of clam, more than 50% mortality of oyster	Liu et al. (2025)
	V. splendidus	Oyster: *C. gigas* larvae Clams: *R. philippinarum* adult, *R. decussatus* spat Mussels: *P. canaliculus* larvae, *M. edulis* adult	*hop, fur, hfq, glide*	Up to 39% mortality of oyster and 100% mortality of clam in laboratory by injection; 83% mortality of greenshell mussel larvae, 48% mortality of carpet shell clam spats, 45% mortality of blue mussel adults	Vandeputte et al. (2024)
	V. chagasii	Scallop: *A. purpuratus* larvae	EPS type II secretion system, metallo-proteinase, and a capsular polysaccharide, thermo-labile hemolysin	Up to 65 % mortality of scallop in laboratory	Gómez-Chiarri et al., 2024
	V. cyclitrophicus	Oyster: *C. gigas* larvae Mussel: *M. coruscus*	-	Up to 75 % mortality of mussel, 25.5% mortality of oyster larvae	Oyanedel et al. (2023)

(*continued on next page*)

TABLE 20.1 Major *Vibrio spp*. Infecting aquaculture bivalves and gastropods—cont'd

Clade	Species	Host and stage	Major VFs	Impact (mortality)	References
	V. kanaloae	Oyster: *C. gigas* larvae Clam: *Scapharca broughtonii*	*iucA, iucC, hfq, lonA, luxO*, ABC transporter components, hemolysin, heme transport protein, and flagellins	Up to 98% mortality of clam in laboratory, 6% mortality of Pacific oyster larvae	Gómez-Chiarri et al., 2024
	V. tasmaniensis	Oyster: *C. gigas* spat Mussel: *M. edulis* larvae	*csrB, vsm, copA, AHL*	Up to 93% mortality of mussel, 60% mortality of Pacific oyster	Vandeputte et al. (2024), Islam et al. (2022)
Tapetis	*V. tapetis*	Clams: *R. philippinarum, R. decussatus*	T4SS, T6SS, haemolysins, transcriptional regulators, Type I restriction modification system, GGDEF domain proteins, several conjugative plasmids	Up to 100% of Manila clam mortality, up to 40% mortality of grooved carpet shell	Gómez-Chiarri et al., 2024

likely accounting for the intraspecific and interspecific variations in virulence between *Vibrio* spp. (Rubio et al., 2019; Zhang et al., 2025b). In addition, pathogenicity is usually a strain-specific trait with isolates of the same species differing significantly in virulence (Fan et al., 2023). Most vibrios exhibit a low host specificity, causing diseases and mortalities in a wide range of marine bivalves and gastropods (Destoumieux-Garzón et al., 2020; Worden et al., 2022). While most pathogenic *vibrios* usually impact more severely on larvae and juveniles than on adults in molluscan species, *V. aestuarianus* causes diseases mainly in adult Pacific oysters (Destoumieux-Garzón et al., 2020; Worden et al., 2022).

Vibrio infections cause diseases with systemic or localized pathological conditions collectively named vibriosis, which was once called bacillary necrosis in larval bivalves. Systemic vibriosis of bivalve larvae and spats likely starts with the bacteria being collected by the vellum and proliferating in the digestive glands and intestines, followed by rapid bacterial spread to the surrounding organs and the entire larvae (Dubert et al., 2016). Typical pathological changes involve massive hemocyte infiltration and necrosis in the digestive system and other affected tissues, and clinical signs include reduced or abnormal motility, velum detachment, bacterial swarming inside moribund or dead larvae, and acute mortalities of up to 100% (Dubert et al., 2017). In larval abalones, vibrios appear to establish an infection by entering the host through gills and the hypobranchial gland (Morot et al., 2025). Diseased animals show ulcer-like lesions with white patches at the foot margins, mantle detachment from the shell, epipodium discoloration, shrinkage of foot muscle and viscera, and up to 80% mortalities in severe cases (Cai et al., 2007).

Research on the infection of *V. aestuarianus* in Pacific oyster *C. gigas* reveals the septicemic characteristics in systemic vibriosis of adult mollusks, starting with bacterial colonization and proliferation in the hemolymph, followed by the spread of the pathogenic bacteria into all the other tissues in the host. Diseased and moribund animals can exhibit lysis of the sub-epithelial connective tissues in the mantle, epithelial atrophy and lumen dilation in the digestive diverticula, and hemocyte lysis, infiltration, or agglutination in the hemolymph vessels (Parizadeh et al., 2018). The disease caused a cumulative mortality of about 30% in market-size oysters at the end of the farming cycle and up to 75% in experimentally exposed animals (Parizadeh et al., 2018). Vibriosis caused by *V. harveyi* in juvenile and adult abalones may show lesions such as swelling and pustule formation in the foot muscles and thousands of affected abalones can die within a few days after the disease onset (Mérou et al., 2022; Sawabe et al., 2007). Systemic vibrioses in juvenile and adult bivalves are sometimes secondary to the infections by viruses or parasitic protozoa, or to the severe immune-compromise of the host due to extreme physiological and environmental stressors (Cowan et al., 2024; Yang et al., 2021). Examples of adult systemic vibriosis are the *Vibrio*-associated bacteremia and mortalities in Pacific oysters primarily infected by OsHV-1 (Petton et al., 2021) and the infections by strains of *V. splendidus* or *V. alginolyticus* in oyster summer mortality syndrome (Cowan et al., 2024; Yang et al., 2021). Inflammation and necrosis in connective tissues of the digestive gland are common, and bacteria are speculated to invade the host from the stomach in juvenile and adult mollusks. Infected adult bivalves can exhibit adductor muscle weakness, necrosis in the adductor muscle and connective tissues, and digestive diverticula atrophy, with mortalities up to 90% in severe epizootics (Travers et al., 2017).

Brown ring disease (BRD) of the Manila clam *R. philippinarum* and, to a less extent, the clam *R.decussatus* is an example of localized vibriosis. The disease is caused by the Tapetis clade *Vibrio* species, *V. tapetis*, which colonizes and disrupts the surface of the periostracal lamina, leading to anomalous deposition of periostracum and the formation of a brown ring around the mantle edge on the inner surface of the clam shells (Rodrigues et al., 2015). The infection also disturbs the normal calcification process in shell formation. *V. tapetis* can penetrate into extrapallial spaces and then into soft tissues through the mantle epithelium and eventually cause tissue damage and clam death. The pathogenesis is chronic, indicating active interactions between the bacterium and host defense at the histological and cytological levels. Under conditions favorable to the host, BRD can heal by nacreization or pearl formation. In addition to the characteristic "brown ring" on the inner shell surface, clinical signs may include reduced body weight, condition index, and glycogen contents, shrink of digestive gland and mantle, and even death (Rahmani et al., 2020).

Epizootiology: Larval vibriosis was first reported to impact bivalve larvae and juveniles in hatcheries in 1960s and remerges as a major threat to the seed productions of several species of oysters, clams, and scallops in hatcheries globally (Elston et al., 2008). In Europe, vibriosis caused by *V. aestuarianus* subspecies francensis has been posing a major threat to the cultured adult Pacific oysters since the disease reemerged in 2012 (Lupo et al., 2019; Mesnil et al., 2023). BRD epizootics are currently prevalent and posing a significant impact on clam aquaculture, mainly on the European Atlantic coasts.

Systemic vibriosis, particularly caused by *V. harveyi*, has also been reported to affect post-larval stages and juveniles of nearly all commercially explored abalone species in France,

Australia, and Japan (Cheng et al., 2008). Pathogenic *Vibrio* spp. that have been reported involve the clades of Anguillarum, Coralliuticus, Harveyi, Mediterranei, Orientalis, and Splendidus (Table 20.1). It generally remains poorly understood how vibriosis develops in wild and aquacultural molluscan populations. Results of research on *V. aestuarianus* infection in Pacific oysters using the combined experimental and modeling approaches indicate that the persistence of infected animals to provide highly virulent bacteria in the epidemic area is important for the transmission of the diseases; the three key parameters for the disease transmission include the half-infective dose defined by the LD50 of the bacterium, the bacterium shedding rate from the diseased oysters, and the initial dose of bacterial exposure, suggesting the significance of both virulence and concentration of the pathogenic *vibrios* for the development and transmission of vibriosis in the field of molluscan aquaculture (Mesnil et al., 2023). In addition, elevated water temperature is usually the most important environmental factor that triggers vibriosis associated with mortalities.

Diagnosis: *Vibrio* detection and characterization can start with bacterial isolation by culture on TCBS media, followed by basic phenotypic characterization (Thompson et al., 2004). Identification at the species or strain level is usually done using molecular methods such as conventional, quantitative, and multiplex PCR techniques (Costa et al., 2022; Zhang et al., 2025a). Initial characterization, particularly for microbiome analysis, can be carried out based on 16S rRNA gene sequencing (Johnson et al., 2019; Leunda-Esnaola et al., 2024). Confirmation with strain-level resolution can be realized using MLST or whole genome analysis (Jesser et al., 2019). On-site screening using LAMP and high-throughput identification using MALDI-TOF, DNA microarrays, and immunoassay have also been reported (Moussa et al., 2021; Rahman et al., 2022).

Control and management: Antimicrobial treatment can be effective, but is limited by environmental or human health concerns or regulatory restrictions. Husbandry practices such as optimizing water quality, reducing environmental stressors, monitoring total and pathogenic *vibrios*, and adjusting density have been applied for decreasing pathogen spread and mitigating disease impact (Dubert et al., 2017; Elston et al., 2008; Mackenzie et al., 2024). Immune priming using sympatric and allopatric *Vibrio* strains and microbial education using pathogen-free microbiota in early life stages likely lower host susceptibility to newly encountered pathogenic *vibrios* (Dantan et al., 2024; Fallet et al., 2022). New host-targeted strategies include selective breeding for infection-resistant stocks (Dégremont et al., 2020; Dietrich et al., 2022), and pathogen-targeted approaches, such as probiotics and bacteriophages, are also gaining progress.

20.3.2 *Aliiroseovarius crassostreae* and roseovarius oyster disease

Etiology: *Aliiroseovarius crassostreae*, originally named *Roseovarius crassostreae*, is an alphaproteobacterial bacterium belonging to the Phylum Pseudomonadota, which causes the roseovarius oyster disease (ROD) in eastern oysters *Crassostrea virginica*. The bacterium is Gram-negative, aerobic, and rod or ovoid in shape, and forms beige to pinkish-beige or greenish-yellow colonies on marine agar (Park et al., 2015). The type-strain (CV919-312) genome consists of a 3,706,831 bp chromosome and an 18,548 bp plasmid, encoding VFs involved in surface attachment, host hemocyte cytotoxins, and T4SS. Genetic variations are reported between different geographic isolates.

Pathology: *Aliiroseovarius crassostreae* infection in eastern oysters resembles the pathogenesis of BRD in clams. The disease causes deposition of the light to dark brown conchiolin on the inner shell surface around the mantle of the affected host oysters. Scanning electron microscopy reveals abundant colonization of live bacteria on the inner shell surface and in the conchiolin layer of the diseased oysters. Bacterial culture and 16S rRNA gene sequencing indicate numerical dominance of *A. crassostreae* in pallial fluid, soft tissues, and inner valve surfaces (Boardman et al., 2008). Lesions can appear on the mantle before visible deposition of conchiolin, which are highly correlated with subsequent conchiolin deposit and total mortality. In addition to the characteristic conchiolin deposition on the inner shell surface, mantle retraction can occur in juvenile oysters of 15–25 mm in shell length (Ford & Borrero, 2001).

Epizootiology: Originally reported as juvenile oyster disease (JOD) in the late 1980s, ROD was once speculated to be caused by a protistan parasite of 2–6 mm in size before *A.crassostreae* was characterized as the etiological agent (Ford & Borrero, 2001). ROD is known to impact the juvenile *C. virginica*. Disease-associated mortalities are consistently related to the host size. In oysters of <25 mm in shell length, mortality begins about 1 week after the conchiolin deposit and can reach 60%–90%. Oysters of <15 mm shell length may die directly without the occurrence of the typical symptom, whereas juveniles of 25–40 mm shell length show conchiolin deposit, but the mortalities do not exceed 30% (Ford & Borrero, 2001). ROD was first identified and is still limited geographically to the Atlantic coast north of New York in the United States, although a disease resembling ROD in clinical signs was reported in experimentally cultured *C. virginica* on the French Atlantic coast (Renault et al., 2002). The disease is reported only in *C. virginica* to date, and the bacterium likely exists in the seawater in epizootic areas and is transmitted horizontally between oysters and by contaminated water (Ford & Borrero, 2001). High-water temperature at 21–26°C and high salinity at 18–30 psu appear to facilitate pathogen transmission and disease development (Ford & Borrero, 2001).

Diagnosis: *Aliiroseovarius crassostreae* can be cultured on a seawater-tryptone (SWT) agar plate at room temperature (22–27°C) for 5–7 days, followed by identification based on colony growth and morphology (Boardman et al., 2008; Maloy et al., 2005). A PCR assay targeting the 16S–23S rDNA internal transcribed spacer (ITS) is reported to detect *A. crassostreae* in samples collected from the inner shell surface of the infected oysters (Maloy et al., 2005).

Control and management: No ROD-specific mitigation is available yet. Some recommended strategies for pathogen control and disease management include husbandry measures, earlier deployment of larger to avoid the high-water temperature (20°C), pathogen and disease monitoring, and development of resistant lines for culture (Rawson & Feindel, 2012). In addition, microbe-based approaches such as probiotics and bacteriophages are under exploration.

20.3.3 Abalone withering syndrome

Etiology: Withering syndrome (WS) is a chronic wasting disease of abalones, and a Rickettsiales-like organism (RLO) named *Candidatus* Xenohaliotis californiensis (CXc) has been reported as the etiological agent (Crosson et al., 2014). Rickettsiales are obligate intracellular, Gram-negative bacteria belonging to the class Alphaprotobacteria and encompassing symbionts of more than 60 molluscan species of aquaculture significance (Cruz-Flores & Cáceres-Martínez, 2020). Two types of RLOs, CXc and a stippled RLO (ST-RLO), are described to infect abalones in California, but only the CXc causes WS (Friedman et al., 2000). Multigenetic comparison places CXc as the basal lineage of the family Anaplasmataceae, and a genome assemblage of 91% completion has been reported recently (Kunselman et al., 2024). Some CXc cells contain a hyperparasite phage apparently close to Siphoviridae (Caudovirales), resulting in a distinctive morphology (Cruz-Flores & Cáceres-Martínez, 2016).

Pathology: CXc, sometimes called WS-RLO, infects abalone digestive epithelia, leading to severe histological abnormalities in the digestive glands (Friedman et al., 2000). Initial infections occur in the postesophagus and the intestine and subsequently in the digestive glands. Inclusion bodies form in the infected tissues, and the pathological alterations in the digestive glands result in physiological starvation. As a result, depletion of glycogen and energy reserves, metabolism reduction, pedal muscle absorption, soft tissue atrophy, and even death occur in advanced infections (Di et al., 2016). Changes in the gut and digestive gland microbiomes are reported from the infected abalones (Cicala et al., 2023). WS abalones can exhibit nonspecific clinical signs such as decreased feeding rates and feces production, and a severely shrunken body.

Epizootiology: WS was described from black abalones *Haliotis cracherodii* on the south shore of Santa Cruz Island, California, USA, in 1985, and is now reported to impact wild and cultured abalones on the Pacific coasts between Baja California, Mexico, and San Francisco County, California, USA (Lafferty & Ben-Horin, 2013). Bacteria that are morphologically and phylogenetically close to those prevalent in southern California, USA, and Baja California, Mexico, have, however, been detected in small abalones recently transferred to Thailand and China, indicating potentials of far-reached pathogen transmissions by transportation of infected abalones. The pathogen transmission is likely realized through the fecal–oral route but may also be through the contaminated seawater (Crosson et al., 2020). All abalone species of the genus *Haliotis* appear to be susceptible to WS-RLO. Interspecific differences in susceptibility are reported, and abalone species with cool water evolutionary histories appear to be most susceptible to WS (Crosson et al., 2014). Warm water temperature appears to enhance the disease severity, and recurrent epizootics can lead to decline and even collapse of the infected populations.

Diagnosis: WS diagnosis involves gross sign observation, histological examination, and pathogen detection. Histological examination of CXc cells and associated inclusion bodies in tissue squashes and histological sections is a rapid screening method. Molecular biology methods have been used in CXc-specific detection, which include in situ hybridization for the visualization of CXc cells in histopathological sections (Antonio et al., 2000; Friedman et al., 2014) and PCR techniques for measuring the bacterium in abalone tissues and water samples (Fuller et al., 2022).

Control and management: Health history and examinations before abalone movement are important for avoiding bacterial contact. Husbandry practices to reduce stress are also recommended. Oxytetracycline via oral or bath administration can be used to control the bacterium and to generate specific-pathogen-free abalone populations, but accumulations of the antibiotics in the abalone digestive glands may have undesired effects (Winkler et al., 2018). Research reveals that WS-RLO resistance is a phenotypic trait with moderate but significant heritability in abalones, implying the potential of selective breeding in WS control and management (Brokordt et al., 2017). In addition, phage therapy and abalone immune enhancement are also under investigation.

20.3.4 *Francisella halioticida* and francisellosis

Etiology: *Francisella halioticida* is a bacterial agent of francisellosis impacting abalones, scallops, and mussels (Brevik et al., 2011). The bacterium belongs to the genus *Francisella* of the phylum Pseudomonadota (proteobacteria) and the genus includes several species and subspecies that are pathogenic to humans, land animals, and fishes (Keim et al., 2007). The bacteria are Gram-negative, nonmotile, and pleomorphic coccobacilli. When cultured on modified Eugon agar (MEA), they typically form colonies of white to slightly yellow in color, with a smooth and very sticky surface (Kamaishi et al., 2010). The type-strain of the species, a strain isolated from the abalone in 2005 in Japan, has a circular chromosome genome of 2,197,430 bp. However, bacteria isolated from Yesso scallops, *Mizuhopecten yessoensis*, share a 99.84% genome sequence identity with the type-strain, and genotypic and phenotypic differences are observed between strains from some or different host species (Kawahara et al., 2021).

Pathology: Molluscan francisellosis typically manifests as an acute bacterial infection. The bacterium is visualized by in situ hybridization most commonly in the adductor muscle, followed by the digestive gland, gonad, and heart, and in a few specimens, the mantle and gills (Meyer et al., 2017). Infections result in histopathological alterations of muscle fiber lysis in the adductor muscle and intense hemocyte infiltration, granulocytoma, and hemocyte necrosis in connective tissues of different organs (Cano et al., 2022; Kawahara et al., 2021; Meyer et al., 2017). Gram-negative bacteria-like particles in the coccobacilli shape can be seen inside hemocytes (Kamaishi et al., 2010). Orange or pinkish lesions are observed in the adductor muscle of infected scallops, and mortalities of over 80% have been reported in the affected population (Kamaishi et al., 2010).

Epizootiology: *Francisella halioticida* was first described in farmed abalones *H. gigantea* from Shimane Prefecture, Japan, during a mass mortality event in 2005 (Brevik et al., 2011; Kamaishi et al., 2010). It is now reported to impact Yesso scallops, *Mizuhopecten yessoensis*, in Baynes Sound, British Columbia, Canada, and in southern Hokkaido, Japan (Meyer et al., 2017), and mussels *M. edulis* and *M. galloprovincialis* (Cano et al., 2022; Garcia et al., 2024). The bacterium is also reported to infect the abalone *Haliotis discus discus* in experimental challenges and to dominate the microbiota in diseased Yesso scallops collected in north China. In *M. yessoensis* populations, *F. halioticida* seems to be prevalent in low temperature seasons, but the associated mortalities are more common in high temperature seasons (Kawahara et al., 2024).

Diagnosis: Bacterial isolation can be carried out by culture on modified Eugon agar (MEA), and species identification can be done by 16S rRNA gene sequencing and bacteriological characterization (Kamaishi et al., 2010). In situ hybridization using species-specific probes was used to visualize bacteria in infected tissues and cells (Kamaishi et al., 2010; Meyer et al., 2017). PCR techniques have been applied in the detection of bacteria in field samples and in the identification of species-specific genes (Brevik et al., 2011; Charles et al., 2021; Kamaishi et al., 2010).

Control and management: No specific method is available yet. Management strategies to reduce environmental and handling stressors can be assessed.

20.4 Parasitic protists and parasitic diseases in mollusks

Protists refer to eukaryotic microorganisms, often unicellular or simple multicellular, that belong to various linages across the eukaryotic life tree and cover all groups of organisms not being classified as plants, animals, or fungi. They are grouped into clades, named supergroups, that are made up of lineages usually belonging to different phyla and even kingdoms in the Linnaean Taxonomy. Protists are found to be associated with mollusks in variable symbiotic relationships. In this chapter, parasitic protists indicate unicellular eukaryotes that are associated with a specific disease or mortality event in wild and cultured mollusk species. These mollusk–parasitic protists are covered in two of the three subclades (Alveolata and Rhizaria) of the supergroup SAR, and involve five genera (*Perkinsus, Marteilia, Haplosporidium, Bonamia*, and *Mikrocytos*) in two phyla (Myzozoa and Cerozoa) in the Linnaean Taxonomy (Table 20.2).

20.4.1 *Perkinsus* spp. and perkinsosis

Etiology: *Perkinsus* spp. are alveolate protists taxonomically classified in the phylum Myzozoa, class Perkinsea, order Perkinsida, family Perkinsidae, and genus *Perkinsus* (https://www.marinespecies.org/aphia.php?p=taxdetails&id=103344). Genus *Perkinsus* encompasses seven described and generally accepted species (Table 20.2) that cause diseases in different molluscan species collectively called perkinsosis, and the type species is *P. marinus*. *Perkinsus* life cycle involves different developmental stages, including trophozoites, hypnospores, zoosporangia, and zoospores (Villalba et al., 2004). Trophozoites are parasites under vegetative proliferation in host tissues and are released through host feces or upon tissue decay of dead hosts (Bushek et al., 2002). Observations from infected animals further indicate that both trophozoites and zoospores are infective to susceptible molluscan hosts (Bushek et al., 2002). In vitro studies and environmental observations show that trophozoites can develop into thick-walled dormant hypnospores or prezoosporangia in decomposing host tissues or after release into the environment (Delisle et al., 2025; Umeda et al., 2020), and some hypnospores further develop into zoosporangia that sporulate to produce and release biflagellate motile zoospores in seawater (Gajamange et al., 2020). The genomes of three species, *P. marinus* (GCF_000006405.1), *P. olsenii* (GCA_013115135.1), and *P. chesapeaki* (GCA_013115145.1), have recently been sequenced, revealing genomic divergences between strains and species (Bogema et al., 2021).

Pathology: *Perkinsus* spp. can invade molluscan hosts via gills, labial palps, mantle, or the digestive tract (Allam et al., 2013; Wang et al., 2018). The parasites proliferate to generate trophozoites intracellularly and extracellularly in host epithelial and connective tissues of different organs; an eastern oyster heavily infected by *P. marinus* can carry more than 10 million trophozoites (Remacha-Triviño et al., 2008), and a *P. olsenii*-infected Manila clam can contain up to 3.7×10^6 parasite cells per gram of tissue (Cho et al., 2025). Parasite proliferation disrupts host cells and tissues and provokes inflammatory responses, such as hemocyte infiltration. Pathological changes such as cellular and tissue necrosis and functional impairments are thus derived from direct disruption by parasite proliferation and indirect damage by host immune reactions, resulting in focal or systemic lesions and ultimate death of severely infected hosts. Parasite–host interactions involve numerous host defense elements and complex host immunity evasion (Hasanuzzaman et al., 2020; Remacha-Triviño et al., 2008). Host hemocytes likely function dually by eliminating invaded parasites through phagocytosis and by facilitating parasite spread in tissues (Lau et al., 2018). Macroscopic signs associated with the histopathology include pustules or abscesses varying in size (up to 10 mm) and in color (white, creamy, or yellowish), reduced condition index, impaired gametogenic development, severe emaciation, and retarded growth, notably in *P. olsenii*-infected clams (Subramaniam et al., 2025). In many cases, gradually increased mortalities are often the only observed sign in the field (Gaona-Hernández et al., 2025; Waki et al., 2018).

Epizootiology: Two *Perkinsus* species, *P. marinus* and *P. olsenii*, are considered major pathogens posing significant threats to global molluscan fisheries and thus are listed in the WOAH Aquatic Code. *Perkinsus marinus* was first reported as the cause of "dermo disease"

TABLE 20.2 Major mollusk–parasitic protists.

Genus	Species	Hosts	Effects	References
Bonamia	*B. exitiosa*	Oysters: *C. ariakensis, C. gigas, O. edulis, O. chilensis, O. angasi, O. puelchana, O. stentina, O. lurida, S. glomerata*	Haemocytes and connective tissues, up to 80% mortality	Arzul, Canier, Chollet, & Garcia, 2024
	B. ostreae	Oysters: *C. rivularis, O. edulis, O. chilensis*	Haemocyte, up to 100% mortality	Arzul, Canier, Chollet, & Garcia, 2024), Ronza et al. (2018)
	B. perspora	Oyster: *O. stentina*	Connective tissues	Arzul, Canier, Chollet, & Garcia, 2024
	B. roughleyi	Oysters: *S. glomerata*	Hemocytes and connective tissues	Arzul, Canier, Chollet, & Garcia, 2024
Haplosporidium	*H. pickfordi*	Snail: *Physella parkeri, L. stagnalis, Heliosoma companulatum*	Digestive gland	Reece et al. (2024a)
	H. armoricanum	Oyster: *O. edulis*	Heart	Reece et al. (2024a)
	H. costale	Oyster: *C. virginica, C. gigas*	Connective tissues of the digestive gland	Reece et al. (2024a), Wang et al. (2010)
	H. edule	Cockle: *C. edule*	Digestive gland	Reece et al. (2024a)
	H. hinei	Oyster: *P. maxima*	Connective tissue surrounding the digestive gland and mantle	Reece et al. (2024a)
	H. lusitanicum	Limpet: *Helcion pellucidus*	-	Reece et al. (2024a)
	H. montforti	Abalone: *H. tuberculata*	Connective tissue of gill, digestive gland and foot muscle	Reece et al. (2024a)
	H. nelsoni	Oysters: *O. edulis, C. gigas, C. virginica* Scallops: *P. yessoensis*	Epithelia of the digestive tubules	Reece et al. (2024a), Wang et al. (2012)
	H. patagon	False limpet: *Siphonaria lessonii, S. lateralis*	Digestive gland and hermaphroditic gonads	Reece et al. (2024a), Di Giorgio, Gilardoni, & Ituarte, 2014
	H. pickfordi	Snail: *Physella parkeri, L. stagnalis, Heliosoma companulatum*	Digestive gland	Reece et al. (2024a)
	H. pinnae	Clam: *Pinna nobilis*	Gills, mantle, and digestive gland, up to 93% mortality	Reece et al. (2024a)
	H. tumefacientis	Mussel: *M. californianus*	Digestive gland and kidney	Reece et al. (2024a)

(*continued on next page*)

TABLE 20.2 Major mollusk–parasitic protists—cont'd

Genus	Species	Hosts	Effects	References
Marteilia	*M. chriestenseni*	Clam*: Scrobicularia plana*	Epithelium of the digestive tubules	Comps (1983)
	M. cochilla	Oyster: *S. cucullata* Cockle: *C. edule*	Epithelial cells of the digestive tubules	Bower, 2024a, Stentiford et al. (2017)
	M. cocosarum	Cockles: *C. edule*	Gill, mantle, and connective tissues	Tidy et al., 2025
	M. lengehi	Oyster: *C. cucullata*	Digestive gland	Bower, 2024a
	M. octospora	Clam: *S. marginatus*	Digestive gland	Ruiz et al. (2016)
	M. pararefringens	Oysters: *O. edulis* Mussels: *M. edulis, M. trossulus, M. galloprovincialis*	Digestive system	Hooper et al. (2025)
	M. refringens	Oysters: *O. edulis, O. stentina* Mussels: *M. edulis, M. galloprovincialis* Clams: *Solen marginatus, Chamelea gallina*	Digestive system, with 87% mortality of *O. stentina*	Bower, 2024a
	M. sydneyi	Oysters: *C. commercialis, S. glomerata*	The hepatopancreas	Nenadic et al. (2025)
	M. tapetis	Clam: *R. philippinarum*	Digestive gland	Kang et al. (2019)
Mikrocytos	*M. boweri*	Oysters: *O. lurida, C. gigas*	-	Bower and Meyer, 2024b
	M. donaxi	Clam: *D. trunculusß*	Muscular and connective tissues, up to 80% mortality	Garcia et al., 2018
	M. mackini	Oysters: *C. gigas, C. virginica, O. edulis, O. lurida*	Mantle, labial palps, and adductor muscle	Bower and Meyer, 2024b
	M. mimicus	Oyster: *C. gigas*	-	Hartikainen et al. (2014)
	M. veneroïdes	Clam: *Donax trunculus*	Muscular and connective tissues, up to 80% mortality	Garcia et al., 2018

(*continued on next page*)

TABLE 20.2 Major mollusk–parasitic protists—cont'd

Genus	Species	Hosts	Effects	References
Perkinsus	*P. beihaiensis*	Oysters: *C. columbiensis, C. cortenziensis,C. hongkongensis, C. virginica, D. frons, S. prismata, I. alatus, O. angasi, O. lurida, O. stentina* Mussels: *M. coruscus, M. galloprovincialis*	Gill, mantle, and visceral mass	Gao et al. (2024)
	P. chesapeaki	Oysters: *C. virginica, C. cortenziensis, C. columbiensis, Dendostrea frons* Clam: *M. arenaria, M. balthica, R. philippinarum, Anadara trapezia, Tagelus plebeius*	Gill and visceral connective tissues, up to 100% mortality of *C. virginica*, 50% mortality of *M. balthica*, 30% mortality of *Mya arenaria*	Reece et al. (2024a)
	P. honshuensis	Clams: *R. variegatus, R. philippinarum*	Connective tissue of the digestive gland	Reece et al. (2024a)
	P. marinus	Oysters: *C. virginica, C. rhizophorae, C. gazar, C. cortenziensis, C. gigas, C. columbiensis, Striostrea prismatica, S. palmula* Clam: *M. mercenaria*	Digestive gland, up to 80% mortality of *C. virginica*	Reece et al. (2024a)
	P. mediterraneus	Oyster: *O. edulis* Clam: *Venus verrucosa, C. gallina, Arca noae* Scallop: *Mimachlamys varia*	Connective tissue	Reece et al. (2024a)
	P. olsenii	Abalones: *H. iris, H. laevigata, H. diversicolor* Oysters: *Isognomon alatus*, O. *angasi, Pinctada martensii, Pinctata imbricata* Mussels: *M. chilensis, Perna canali*culus Clams: *R. philippinarum, A. trapezia* Cockles: *Anadara kagoshimensis, Tegillarca granosa* Scallops: *Chlamys nobilis*	Hemal space of visceral mass connective tissue, up to 40% mortality of *H. laevigata*	Delisle et al. (2025)
	P. qugwadi	Scallop: *Patinopecten yessoensis* False oyster: *Pododesmus rudis*	Intercellular spaces and haemolymph sinuses	Garcia et al. (2018)

recurrent in eastern oysters *C. virginica* cultured in the Gulf of Mexico, United States, during the early to mid-1940s, under the name of *Dermocystidium marinum*, whereas *P. olsenii* was initially described as the agent causing mass mortalities in the abalone *Haliotis rubra* in Australia during the 1980s. Both species have subsequently been found to have broader geographic distribution and host species, with *P. olsenii* being detected in more than 50 bivalve species on nearly all major ocean coasts in the world and in *Haliotis* abalones (Table 20.2). Research also uncovers that two or perhaps more *Perkinsus* species can simultaneously infect the same molluscan species (Itoïz et al., 2023). Genetic differentiations indicative of parasite adaptation to new environments and hosts' coevolution are revealed between geographic strains, and some of the evolutionary adaptations may lead to the shifts of virulence phenotype in the parasites (Carnegie et al., 2021; Kim et al., 2025; Weatherup et al., 2024).

Perkinsus spp. transmits directly between host mollusks without involving intermediate hosts, and the parasites may spread several kilometers in seawater. Perkinsosis epizootics exhibit seasonal patterns that are associated with environmental dynamics. Elevated water temperature and high salinity are the most important known environmental factors that affect virulence and release, and thereby the prevalence and infection intensity of the parasites (Gignoux-Wolfsohn et al., 2021). The northward expansion of dermo prevalence in eastern oysters on the Atlantic coast, USA, was attributed to the global warming-associated temperature elevation and the lasting drought-caused salinity increase in seawater. Global transportation of the living mollusks is likely promoting the increased geographic distributions of *Perkinsus* spp. (Yisrael et al., 2024).

Diagnosis: Ray's fluid thioglycollate method (RFTM), which lets trophozoites enlarge into hypnospores in the fluid thioglycollate culture medium, followed by the microscopic observation after iodine-staining, has been a standard technique for *Perkinsus* detection and the assessment of parasite body burden for decades (Cho et al., 2025; Dungan & Bushek, 2015). Histological examination with standard hematoxylin and eosin-staining allows the visualization of infection-associated pathological changes. Immunoassays using polyclonal and monoclonal antibodies were reported to analyze the parasites at the species or developmental stage levels (Hanrio et al., 2021). In situ hybridization has been used to confirm *Perkinsus* spp. presence in histological sections, and several PCR-based methods are applied in parasite identification and routine diagnosis (Moss et al., 2007).

Control and management: Several chemicals, including antimicrobials, antiprotozoals, serine protease inhibitors, and coppers were shown to inhibit *P. marinus* propagation in vitro, but their application in the field to eradicate parasites not assessed (Elandalloussi et al., 2005; Xue et al., 2006). Management practices such as harvesting or moving oysters to low salinity areas before water temperature increases to 15–20°C and freshwater inflow control were tried to maintain the parasite infection intensity in eastern oyster populations below the lethal level (La Peyre et al., 2009). Development of disease-resistant or tolerant stocks for aquaculture is being under way through the identification of natural strains or selective breeding, and a genomic selection trial using a high-density SNP array has shown encouraging performance (Guo et al., 2023; Sambade et al., 2025; Wang et al., 2025).

20.4.2 *Marteilia* spp. and marteiliosis

Etiology: Nine formally described species make up the *Marteilia* genus that belongs to the phylum Cercozoa, class Ascetosporea, order Paramyxida, and family Marteiliidae, and two species, *M. refringens* and *M. sydneyi*, are particularly impactful on molluscan aquaculture (Table 20.2). The initial species of *M. refringens* was once divided into two types of "O" and "M" and has now been reclassified into two new species of *M. refringens* and *M. pararefringens* that respectively refer to the oyster-infecting type "O" and the mussel-infecting type "M" in the old classification (Bøgwald et al., 2024; Hooper et al., 2025; Kerr et al., 2018). Other species likely of aquacultural concerns include *M. cochilia* (Carballal et al., 2016). *Marteilia* life cycle includes proliferation in the bivalves, most often in the digestive gland, and development in intermediate hosts, at least for *M. refringens*, *M. pararefringens*, and *M. sydneyi*. The in-bivalve proliferation involves the paramyxids-characteristic cell-within-cell development in which the plasmodia-like primary cells and secondary cells inside them divide continuously as sporangia within host cells and tissues, and spores are generated and released into seawater after maturation in the sporangia (Darriba et al., 2020). Mature spores released from the host tissues may use copepods or polychaetes in local ecosystems as alternative or intermediate hosts for further development before infecting new molluscan hosts (Bøgwald et al., 2022; Boyer et al., 2013).

Pathology: Marteiliosis is sometimes called Aber disease in European flat oyster *O. edulis* or digestive gland disease (Berthe et al., 2004). The infections most commonly target the host digestive glands but have occasionally been described in the connective tissues of the mantle and gills, for example, in some mussels, *Mytilus* (Bøgwald et al., 2024), and in cockles (Tidy et al., 2025). Primary stages of *Marteilia refringens* and *Marteilia pararefringens* are observed in the epithelium of the labial palps, esophagus, and stomach, while more advanced stages occur in digestive gland tubules and ducts. *Marteilia sydneyi* enters the host Sydney rock oyster *Saccostrea glomerata* from the palps and gills and then undergoes an initial stage of extrasporogonic proliferation in the connective tissues and hemolymph before migrating to the digestive glands. Once inside the host digestive glands, parasite proliferation causes destruction and blockage of digestive tubules. *M. sydneyi* proliferation is reported to elicit hemocyte infiltration in the digestive tissues of *S. glomerate*, resulting in necrosis and granulocytoma formation and functional impairment in food digestion and glycogen storage. Hemocyte infiltration, epithelial hyperplasia, and hypertrophy can also occur in gills or gonadal follicles. Infected oysters may show changes in certain enzyme activities and protein expression levels (Butt & Raftos, 2008). Macroscopical signs such as emaciation, growth cessation, pale digestive glands, thin and watery flesh, mantle retraction, and increased mortalities can be seen in the affected oyster populations. Infection by *M. chungmuensis* in Pacific oysters *C. gigas* appears to start in the mantle, gills, and labial palps, and then the parasite sporulates in the oocytes.

Epizootiology: *Marteilia refringens* and *M. sydneyi* were first described in the 1970s as the causative agents of the digestive gland disease or Aber disease in the European flat oyster *O. edulis* in France and the QX disease in Sydney rock oysters in Australia, respectively. Both species have subsequently been reported in more host species, and geographic locations, and *M. refringens* is listed in the WOAH Aquatic Code. Newly described species and unidentified strains have also been reported ever since. Notably, three *Martejlia* spp., *M. refringens*, *M. sydneyi*, and *M. cochilia*, are of the most significant concerns for the molluscan fishery industry as they pose a major threat respectively to *O. edulis*, *S. glomerate*, and *C. edule* in different regions in the world (Carballal et al., 2016). It is not fully understood how the parasites transmit with the environmental dynamics in the field, but research reveals that elevated water temperature and low salinity likely facilitate the proliferation and transmission of *Marteilia* spp. A temperature above 17°C seems to be required for *M. refringens* sporulation and transmission, whereas high salinity and water renewal appeared extraneous to the parasite (Audemard et al., 2001). Spore survival is at least 20 days outside the host (Mérou et al., 2022). Field surveys of *M. cochilia* prevalence in *C. edule* uncovered evidence for the development of resistance or resilience to the parasites in the cockle populations, likely via natural selection (Villalba et al., 2023).

Diagnosis: Disease diagnosis depends on purposes. Examination of unstained tissue mounts and stained tissue imprints or smears for large cell-within-cell spherical bodies is used for rapid parasite screening. Histology with H&E-stained sections allows visualization of the proliferating parasite cells and the associated pathological changes in the infected host tissues. Electron microscopy is useful in studying the parasite structures at the subcellular level. Molecular approaches, including in situ hybridization and PCR techniques, have been reported for marteiliosis diagnosis and *Marteilia* species identification (Canier et al., 2020; Carrasco et al., 2017).

Control and management: Restriction of moving infected animals to non-epizootic areas and harvest or destruction of infected crops were recommended as management strategies. Growing oysters in high-salinity seawater was also attempted to mitigate *M. refringens* in *O.edulis* and *M. sydneyi* in *S. glomerata*. Given the possible presence of the parasite in the sediment, maintaining bivalves at a distance from the bottom may help reduce the number of infected animals (Mérou et al., 2023). In addition, selective breeding has shown encouraging results in developing QX-resistant *S. glomerata* stocks (Dove et al., 2013).

20.4.3 *Haplosporidium* spp. and multinucleate sphere unknown X and sea-side organism

Etiology: *Haplosporidium* spp. are haplosporidian protists belonging to Phylum Cercozoa, Class Ascetosporea, Order Haplosporida, Family Haplosporidiidea. The Genus *Haplosporidium* encompasses a large variety of species that are associated with different invertebrates (Davies et al., 2020; Reece et al., 2024a). Of particular importance for molluscan aquaculture are *H. nelsoni* and *H. costale*, which cause multinucleate sphere unknown X (MSX) disease and sea-side organism (SSO) disease in eastern oysters, respectively. Most of the other species were described based on histological detection in apparently normal mollusks. The complete life cycle of *Haplosporidium* spp. remains unknown. Two morphological forms, the multinucleate plasmodia and the uninucleate spores, have been observed in the infected host tissues. It is speculated that the plasmodia develop into sporocysts as spores are generated after the spore walls form around each nucleus, and spores are released when sporocysts rupture. The released spores appear not to be able to infect the oyster directly but instead require an intermediate host for the complete life cycle (Ford et al., 2018).

Pathology: Pathological changes are typically related to the damage of host cells and tissues pertaining to the parasite proliferation and the elicited host immune response (Burreson & Ford, 2004). In MSX, infections are acquired from the gill and mantle tissues and then rapidly spread throughout the oyster. Multinucleate plasmodia are initially visualized extracellularly in the connective tissues and epithelia, which multiply along the basal lamina in the gills and eventually break through into the circulatory system. Intense hemocyte infiltrations surrounding the plasmodial foci and necrotic areas are common in the infected host tissues. Sporulation, which rarely occurs in adults but is common in young spats, is limited in the epithelium of digestive glands without spores being seen in the connective tissues. Sporocysts inside the digestive epithelial cells may rupture, resulting in spore release into the digestive tubule lumen. Hemolymph enzyme activities, proteins, and tissue biochemicals may vary in the process

(Ford, 1986). Macroscopically, infected oysters show gametogenesis depression and emaciation, and may die without exhibiting visible disease signs in acute systemic infections. In SSO, the multinucleate plasmodia of *H. costale* also occur extracellularly in the connective tissues, similar as *H. nelsoni*. However, the sporulation takes place throughout the connective tissues of the digestive gland, mantle, and gonads, and spores are not present in the epithelia of the digestive tubules. Coinfections of MSX and SSO may occur.

Epizootiology: *Haplosporidium nelsoni* and *H. costale* were both initially described in eastern oysters *C. virginica* on the east coast of the United States in the 1960s, and reported subsequently in the Pacific oyster *C. gigas* in North America, Asia, and Europe without noticeable mortalities (Arzul et al., 2022; Lynch et al., 2020). There are speculations that *H. nelsoni* was natively present in *C. gigas* in Asia and became virulent to *C. virginica* after introduction to North America with its native hosts (Burreson et al., 2000). Up to date, MSX and SSO still limit to impact on eastern oysters along the Atlantic coast of North America (Batchelor et al., 2023; Marquis, Bishop, Record, Countway, & Fernández Robledo, 2020). Oyster infections occur from mid-May through October, with plasmodia being detected from mid-May to October and sporulation in late June through early July and in the autumn. Mortalities of 90%–95% in the period of July through October were reported in a cohort that had been planted for 2–3 years in the culture fields. Water temperature and salinity represent the most important environmental elements that influence MSX and SSO epizootics (Hofmann et al., 2001). *Haplosporidium nelsoni* requires a salinity of 15 psu for infection and 20 psu for rapid and high host mortality; *H. costale* does not occur in areas with a salinity less than 25 psu. Meanwhile, the parasites proliferate at water temperatures of 5–20°C, and oysters acquire infections and die at temperatures above 20°C. High temperature, however, is not sufficient to trigger an epizootic (Boeck et al., 2019). Research results indicate that long-term epizootic pressure may lead the molluscan host to the development of parasite resistance, which manifests as an ability to restrict infections and tolerate parasitism for prolonged periods (Ford & Bushek, 2012).

Diagnosis: Histopathological examination for parasites at different developmental stages (plasmodia, sporocysts, and spores) in the targeted tissues and cells of the infected oyster is a standard diagnostic method. Identifications at the species level, including distinction between *H. nelsoni* and *H. coastale* depends on sequence analysis on genes such as small subunit ribosomal DNA using molecular approaches; in situ hybridization and PCR-based methods have been reported for applications in the species-specific identifications, environmental parasite detection, and disease diagnosis (Arzul et al., 2022; Piesz et al., 2022).

Control and management: Husbandry measures such as using filtration or UV-treated water in hatcheries to produce healthy seeds, restriction of transporting oysters from epizootic areas to non-epizootic areas, holding oysters in low salinity, and early diagnosis-based harvest were proposed (Safi et al., 2025). Development of MSX and SSO-resistant oyster stocks for aquaculture has been attracting research investment in identifying natural or selectively bred resistant stocks (Hudock et al., 2025; Potts et al., 2021).

20.4.4 *Bonamia* spp. and bonamiosis

Etiology: *Bonamia* spp. are haplosporidian protists taxonomically classified in the Phylum Cercozoa, class Ascetosporea, Order Haplosporida, Family Haplosporidiidae. Genus *Bonamia* has four species, *B. ostreae*, *B. exitiosa*, *B. perspora*, and *B. roughleyi*, that are etiological agents of the disease collectively called bonamiosis in *Ostrea* oysters (Arzul, Canier, Chollet, & Garcia, 2024;

Engelsma et al., 2014). *Bonamia ostreae* and *B. exitiosa* are typically intracellular parasites, present and multiplying inside the host hemocytes and sometimes extracellularly, but *B. perspora* rarely appears inside any host cells (Arzul et al., 2011; Carnegie et al., 2006). *Bonamia roughleyi* was originally described as the cause of winter mortality (WM) in Sydney rock oysters *S. glomerata* under the name of *Mikrocytos roughleyi* (Cochennec-Laureau et al., 2003), and is proposed to be *Bonamia roughleyi* nomen dubium because of the inability in subsequent studies to validate its causation status in the *S. glomerata* WM and to distinguish it from *B. exitiosa* based on gene sequence comparison (Carnegie et al., 2014). *Bonamia* spp. typically appear in host cells as uninucleate "microcells" of 2–5 μm in diameter. They are believed to proliferate inside the host hemocytes via binary fission and reinitiate a new infection after release upon the lysis of the host cells. The parasites can also exist in host hemocytes or hemolymph in multinucleate plasmodia, which are common for *B. exitiosa* but rare for *B. ostreae*, and *B. perspora* was reported to be present in sporogonic forms in the host connective tissues (Carnegie et al., 2006).

Pathology: *Bonamia* spp. infect oysters by traversing the epithelial barriers and entering the subepithelial connective tissues and then the hemolymph circulation at the gills, mantle, or digestive tract. The parasites are phagocytosed by the hemocytes in hemolymph and proliferate intracellularly. Heavy hemocyte infiltrations are common in the connective tissues of the gills, mantle, and gonad during the pathogenesis (Kroeck, 2010). Alterations in hemolymph molecular and cellular parameters, and tissue granulomatous aggregations and necrosis are also reported in advanced infection (Longshaw et al., 2013). The infections can be subclinical at the individual level, but when stressed under environmental conditions like elevated water temperature, systemically infected oysters show reduced growth and condition index and sometimes impaired gametogenesis(Engelsma et al., 2010). Bonamiosis epizootics can cause more than 80% mortalities in the affected oyster populations (Buss et al., 2020a; Lallias et al., 2008). In addition, *Bonamia* spp. likely establish long-term endemic infections that manifest as chronic infection and parasite intracellular persistence in oyster stocks and wild populations.

Epizootiology: *Bonamia ostreae* was first described to cause mortalities in European flat oysters *O. edulis* in France in the late 1970s and represents one of the most important pathogens of *O. edulis* in the entire distribution range of the oyster species in Europe and North America (Engelsma et al., 2014; Marty et al., 2006). It has also been associated with mortalities in Chilean flat oysters *Ostrea chilensis* in New Zealand since 2015 (Lane et al., 2016) and detected in dead Suminoe oysters *Crassostrea rivularis* (syn. *ariakensis*) maintained in laboratory quarantine. *Bonamia exitiosa* was originally described as a new *Bonamia* species from the Chilean flat oyster, *O. chilensis*, in New Zealand in 2001, and is now known for its broad host species and geographic distribution, involving nearly 10 oyster species of the Genera *Ostrea* and *Crassostrea* on the coasts of Oceania, South and North America, Africa, and Europe (Bradley et al., 2025; Carrasco et al., 2012; Oehrens Kissner et al., 2014). *B. perspora* had, up to date, been detected only at low prevalence in the crested oyster *O. equestris* on the coast of the Atlantic in North Carolina, USA (Carnegie et al., 2006). *Bonamia exitiosa* and *B. ostreae* are both listed in the WOAH Aquatic Code. *Bonamia* spp. infect oysters at all developmental stages, but mortalities are more common in older flat oysters (Arzul et al., 2009; Lallias et al., 2008). Healthy oysters can be infected by cohabitation with infected animals, and the parasites are detected for a period in seawater, indicating a direct transmission of the parasites between the hosts at proximity (Buss et al., 2020a; Mérou et al., 2020). The infected larvae may also contribute to the parasite transmission during the planktonic life stage (Arzul et al., 2011). Oyster transportations may facilitate the long-distance spread of *Bonamia* spp., as close genetic connectivity was revealed between geographic strains of both *B. ostreae* and *B. exitiosa*, and the parasites were found to survive for at least 4 weeks in relocated living oysters (Helmer et al., 2020; Lane & Jones, 2020). Environmental and oyster physiological factors, such as warmer water temperature, high salinity, and gametogenesis and spawning, are speculated to influence *Bonamia* prevalence through impacting on parasite proliferation and transmission and multiplication, shaping the seasonal patterns of bonamiosis epizootics (Oehrens Kissner et al., 2014). Over time, recurrent epizootics lead to adaptive coevolution between the parasites and the oyster hosts (Arzul, Canier, Chollet, & Garcia, 2024; Engelsma et al., 2014).

Diagnosis: Parasite detection can be done by observation of the characteristic morphology of the parasite microcells in tissue imprints (heart smears) and in histological sections after proper staining. These morphology-based techniques are, however, not sensitive and cannot identify parasite species (Arzul, Canier, Chollet, & Garcia, 2024). Molecular approaches have been developed using conventional and real-time PCR methods and LAMP to detect *Bonamia* spp. at the species level (Canier et al., 2020; Cano et al., 2024). The application of the molecular techniques for diagnosis is facilitated by the development of new procedures with nonlethally sampled oyster tissues (Jørgensen et al., 2020; Mérou et al., 2020).

Control and management: There is no cure yet for bonamiosis in oysters. Restriction of oyster movement from *Bonamia* epizootic or status-unknown zones to the parasite-free areas is recommended by the WOAH Aquatic Animal Health Manual as a prevention for pathogen transmission. Treatments of equipment and facilities using chemical and physical approaches, such as UV-light radiation, may help with decreasing parasite density in the culture environments (Buss et al., 2020b; Fernández-Boo et al., 2021). Husbandry measures in oyster aquaculture, such as reducing handling stresses and lowering culture density, can be useful for lessening impacts from mortalities in the epizootic populations (Hilton et al., 2025). More recent approaches have involved selective breeding to generate oyster stocks with increased resilience to *Bonamia* infections (Sambade et al., 2022; Vera et al., 2019).

20.4.5 *Mikrocytos* spp. and associated disease in oysters

Etiology: *Mikrocytos* spp., together with *Bonamia* spp., are often called "microcell" parasites. They are taxonomically classified in the phylum Cercozoa, class Ascetosporea, order Mikrocytida, family Mikrocytiidae (Hartikainen et al., 2014). Genus *Mikrocytos* currently consists of six described species: *M. mackini, M. boweri, M. mimicus, M. veneroides, M. donaxi*, and *M. mytilicolid. M. mackini* represents a species of particular significance for molluscan aquaculture, infecting and causing mortalities in oysters. *Mikrocytos* spp. are intracellular microcells of 2–3 mm in diameter, typically present in the vesicular connective tissue cells and sometimes in the myocytes. Electron microscopy reveals three cell forms, quiescent cells (QC), vesicular cells (VC), and endosomal cells (EC), that are speculated to transform from one type to another during the development involving the cycle between vesicular connective tissue and myocytes with the transportation by hemocytes (Hartikainen et al., 2014). The entire life cycle of the parasites, however, remains unknown.

Pathology: *Mikrocytos mackini* infections cause focal hemocyte infiltration and tissue necrosis in the vesicular connective tissues and adductor muscle, resulting in visible green lesions in the mantle, adductor muscle, and labial palps (Bower and Meyer, 2024b). The histopathological changes resemble the oyster nocardiosis, a bacterial disease caused by the Gram-positive *Nocardia crassostreae*. About 10% *M. mackini*-infected oysters can recover from the infection.

Epizootiology: *Mikrocytos mackini* was originally reported as the causative agent of the Denman Island disease in Pacific oysters, *C. gigas*. It has subsequently been detected in several other naturally or experimentally infected oysters, including eastern oysters *C. virginica*, Kumamoto oysters *C. sikamea*, European flat oysters *O. edulis*, and Olympia oysters *O. lurida* (Elston et al., 2015; Gagné et al., 2008). Oysters show differences in susceptibility, and two clam species, *Panope abrupta* and *Venerupis (=Ruditapes) philippinarum*, are found to be resistant to *M. mackini* infection (Meyer et al., 2008). *M. mackini* is currently reported only in the cooler water zone on the northern Pacific coasts in North America, and the parasite isolates covering the entire distribution areas lack genetic divergencies, although an uncharacterized *Mikrocytos* spp. was detected at one sampling site (Abbott, Gilmore, Lowe, Meyer, & Bower, 2011). They are detectable histopathologically between March and June, and mortalities occur mainly in April and May among oysters 2 years and older. Transmission is likely realized by water, but the parasite loses infectivity rapidly when present extracellularly in seawater (Polinski et al., 2017). *Mikrocytos boweri* and *M. minicus* are up to date reported in *O. lurida* and *C. gigas* from southwestern British Columbia, Canada, and in *C. gigas* from the north Norfolk coast, UK, respectively (Abbott, Gilmore, Lowe, Meyer, & Bower, 2011; Hartikainen et al., 2014), and *M.* veneroides and *M. donaxi* are both observed by histology in the wedge clam *Donax trunculus* collected on the French Atlantic coasts during mortality events.

Diagnosis: Diagnosis of *Mikrocytos* spp. infections relies on histological and molecular techniques. Tissue imprints and histological sections can be observed for the presence of *Mikrocytos* microcell, but morphological characteristics, even at an ultrastructural level, cannot distinguish species. Molecular techniques, including ISH and PCR methods, have been developed for the detection and identification of *Mikrocytos* spp. at the species level (Polinski et al., 2021).

Control and management: Moving oysters from epizootic areas to non-epizootic areas, particularly in the cooler water zone, should be restricted as the parasite persists for a long time once entering the host. Parasite surveillance and management to harvest or relocate large oysters to the upper intertidal zone during winter or shortened production times were proposed.

20.5 Conclusion remarks

Health management is pivotal to aquaculture success and, consequently, to global food security, as pathogen- and disease-associated mortalities remain among the most serious threats to industry development. In marine molluscan aquaculture, viral outbreaks, bacterial infections, and protozoan parasitoses, often exacerbated by environmental stressors, regularly cause epizootics and mass mortalities across species, leading to substantial economic losses and ecological disruption. The challenge is expected to intensify under accelerating climate change and anthropogenic disturbances, which drive the emergence and evolution of novel pathogens and diseases.

Addressing these challenges requires new knowledge generated through innovative research approaches to enable more effective pathogen control and disease mitigation. Future studies on disease etiology and pathogen diversity should place greater emphasis on previously underexplored microbial groups, particularly viruses and pathogenic bacteria, as well as on polymicrobial infections and microbiome dysbiosis or the pathobiome concept (King et al., 2019). Studies on pathogen transmission in natural populations and wild ecosystems are also essential for improving disease forecasting and management in aquaculture systems.

Host defense mechanisms against biotic and abiotic stressors include species- or lineage-specific molecular and physiological components whose roles in infection pathogenesis merit deeper investigation (Lu et al., 2021; Ruan et al., 2022; Tan et al., 2025). Among host-targeted strategies, selective breeding and genetic manipulation to develop disease-resilient stocks represent promising, sustainable avenues for molluscan aquaculture (Jiang et al., 2024; Potts et al., 2021).

Ultimately, the study of molluscan pathogens and diseases will benefit greatly from systems biology approaches, integrating insights across pathology, ecology, and evolution to unravel complex host-pathogen interactions and support the design of innovative, ecologically grounded health management strategies.

References

Abbott, C. L., Gilmore, S. R., Lowe, G., Meyer, G., & Bower, S. (2011). Sequence homogeneity of internal transcribed spacer rDNA in Mikrocytos mackini and detection of *Mikrocytos* sp. in a new location. *Diseases of Aquatic Organisms, 93*(3), 243–250. doi:10.3354/dao02301.

Allam, B., Carden, W. E., Ward, J. E., Ralph, G., Winnicki, S., & Espinosa, E. (2013). Early host-pathogen interactions in marine bivalves: Evidence that the alveolate parasite *Perkinsus marinus* infects through the oyster mantle during rejection of pseudofeces. *Journal of Invertebrate Pathology, 113*(1), 26–34. doi:10.1016/j.jip.2012.12.011.

Antonio, D. B., Andree, K. B., Moore, J. D., Friedman, C. S., & Hedrick, R. P. (2000). Detection of Rickettsiales-like prokaryotes by *in situ* hybridization in black abalone, *Haliotis cracherodii*, with withering syndrome. *Journal of Invertebrate Pathology, 75*(2), 180–182. doi:10.1006/jipa.1999.4906.

Arzul, I., Nicolas, J. L., Davison, A. J., & Renault, T. (2001a). French scallops: A new host for ostreid herpesvirus-1. *Virology, 290*(2), 342–349. doi:10.1006/viro.2001.1186.

Arzul, I., Renault, T., & Lipart, C. (2001b). Experimental herpes-like viral infections in marine bivalves: Demonstration of interspecies transmission. *Diseases of Aquatic Organisms, 46*(1), 1–6. doi:10.3354/dao046001.

Arzul, I., Gagnaire, B., Bond, C., Chollet, B., Morga, B., Ferrand, S., Robert, M., & Renault, T. (2009). Effects of temperature and salinity on the survival of Bonamia ostreae, a parasite infecting flat oysters Ostrea edulis. *Diseases of Aquatic Organisms, 85*(1), 67–75. doi:10.3354/dao02047.

Arzul, I., Garcia, C., Chollet, B., Serpin, D., Lupo, C., Noyer, M., … Travers, M. (2022). First characterization of the parasite *Haplosporidium costale* in France and development of a real-time PCR assay for its rapid detection in the Pacific oyster, *Crassostrea gigas. Transboundary and Emerging Diseases, 69*(5), e2041–e2058. doi:10.1111/tbed.14541.

Arzul, I., Langlade, A., Chollet, B., Robert, M., Ferrand, S., Omnes, E., Lerond, S., Couraleau, Y., Joly, J., François, C., & Garcia, C. (2011). Can the protozoan parasite Bonamia ostreae infect larvae of flat oysters Ostrea edulis? *Veterinary Parasitology, 179*(1-3), 69–76. doi:10.1016/j.vetpar.2011.01.060.

Arzul, I., Canier, L., Chollet, B., & Garcia, C. (2024). *Bonamia* spp. infections of oysters. In *Diseases of bivalves: Historical and current perspectives* (pp. 1–14). Academic Press.

Arzul, I., Corbeil, S., Morga, B., & Renault, T. (2017). Viruses infecting marine molluscs. *Journal of Invertebrate Pathology, 147*, 118–135. doi:10.1016/j.jip.2017.01.009.

Audemard, C., Barnaud, A., Collins, C. M., Le Roux, F., Sauriau, P.-G., Coustau, C., … Berthe, F. C. J. (2001). Claire ponds as an experimental model for *Marteilia refringens* life-cycle studies: New perspectives. *Journal of Experimental Marine Biology and Ecology, 257*(1), 87–108. doi:10.1016/s0022-0981(00)00330-0.

Bai, C., Gao, W., Wang, C., Yu, T., Zhang, T., Qiu, Z., … Huang, J. (2016). Identification and characterization of ostreid herpesvirus 1 associated with massive mortalities of *Scapharca broughtonii* broodstocks in China. *Diseases of Aquatic Organisms, 118*(1), 65–75. doi:10.3354/dao02958.

Bai, C, Li, Y, Chang, P, Jiang, J, Xin, L, Li, C., … Wang, C. (2020). *In situ* hybridization revealed wide distribution of *Haliotid herpesvirus 1* in infected small abalone, *Haliotis diversicolor supertexta. Journal of Invertebrate Pathology, 173.* doi:10.1016/j.jip.2020.107356.

Bai, C., Morga, B., Rosani, U., Shi, J., Li, C., Xin, L., & Wang, C. (2019). Long-range PCR and high-throughput sequencing of *Ostreid herpesvirus* 1 indicate high genetic diversity and complex evolution process. *Virology, 526*, 81–90. doi:10.1016/j.virol.2018.09.026.

Bakkeren, E., Gül, E., Huisman, J. S., Steiger, Y., Rocker, A., Hardt, W. D., & Diard, M. (2022). Impact of horizontal gene transfer on emergence and stability of cooperative virulence in Salmonella Typhimurium. *Nature Communications, 13*(1), 1939. doi:10.1038/s41467-022-29597-7.

Barbosa-Solomieu, V., Dégremont, L., Vázquez-Juárez, R., Ascencio-Valle, F., Boudry, P., & Renault, T. (2005). Ostreid Herpesvirus 1 (OsHV-1) detection among three successive generations of Pacific oysters (*Crassostrea gigas*). *Virus Research, 107*(1), 47–56. doi:10.1016/j.virusres.2004.06.012.

Batchelor, S., Harrison, J. S., Greiman, S. E., Treible, L. M., & Carroll, J. M. (2023). Assessment of infection prevalence and intensity of disease-causing parasitic protozoans *Perkinsus marinus* and *Haplosporidium nelsoni* in Georgia oysters. *Microorganisms, 11*(7), 1808. doi:10.3390/microorganisms11071808.

Bateman, K. S., White, P., & Longshaw, M. (2012). Virus-like particles associated with mortalities of the Manila clam *Ruditapes philippinarum* in England. *Diseases of Aquatic Organisms, 99*(2), 163–167. doi:10.3354/dao02462.

Berthe, F. C. J., Le Roux, F., Adlard, R. D., & Figueras, A. (2004). Marteiliosis in molluscs: A review. *Aquatic Living Resources, 17*(4), 433–448. doi:10.1051/alr:2004051.

Boardman, C. L., Maloy, A. P., & Boettcher, K. J. (2008). Localization of the bacterial agent of juvenile oyster disease (*Roseovarius crassostreae*) within affected eastern oysters (*Crassostrea virginica*). *Journal of Invertebrate Pathology, 97*(2), 150–158. doi:10.1016/j.jip.2007.08.007.

Boeck, H. M., Laramore, S. E., Portnoy, D. S., & Scarpa, J. (2019). Survey of Texas Bay systems for the oyster parasites *Bonamia* spp., *Haplosporidium nelsoni*, and *Perkinsus marinus*. *Journal of Shellfish Research, 38*(3), 557–564. doi:10.2983/035.038.0306.

Bogema, D. R., Yam, J., Micallef, M. L., Gholipourkanani, H., Go, J., Jenkins, C., & Dang, C. (2021). Draft genomes of *Perkinsus olseni* and *Perkinsus chesapeaki* reveal polyploidy and regional differences in heterozygosity. *Genomics, 113*(1), 677–688. doi:10.1016/j.ygeno.2020.09.064.

Bøgwald, M, Skår, C. K., Karlsbakk, E., Alfjorden, A., Feist, S. W., Bass, D., & Mortensen, S. (2022). Infection cycle of *Marteilia pararefringens* in blue mussels *Mytilus edulis* in a heliothermic marine oyster lagoon in Norway. *Diseases of Aquatic Organisms, 148*, 153–166. doi:10.3354/dao03651.

Bøgwald, M, Larsson, S. B., & Mortensen, S. (2024). First detection of *Marteilia pararefringens* in *Mytilus trossulus*. *Diseases of Aquatic Organisms, 159*, 209–214. doi:10.3354/dao03818.

Bower, S. M. (2024a). *Marteilia* spp. infections in bivalves. In Roxanna Smolowitz (Ed.), *Diseases of bivalves. Historical and Current Perspectives*. Academic Press.

Bower, S. M., & Meyer, G. R. (2024b). *Mikrocytos* spp. infection in oysters and clams. In Roxanna Smolowitz (Ed.), *Diseases of bivalves. Historical and Current Perspectives* (pp. 133–142). Academic Press.

Boyer, S., Chollet, B., Bonnet, D., & Arzul, I. (2013). New evidence for the involvement of *Paracartia grani* (Copepoda, Calanoida) in the life cycle of *Marteilia refringens* (Paramyxea). *International Journal for Parasitology, 43*(14), 1089–1099. doi:10.1016/j.ijpara.2013.07.008.

Bradley, T. L., Mohr, P. G., Humphrey, J. D., Moody, N. J. G., Cummins, D., Slater, J., & Crane, M. S. J. (2025). *Bonamia exitiosa*: The cause of bonamiasis in native oysters *Ostrea angasi* in Australia in 2015. *Diseases of Aquatic Organisms, 162*, 99–113. doi:10.3354/dao03853.

Brevik, O. J., Ottem, K. F., Kamaishi, T., Watanabe, K., & Nylund, A. (2011). *Francisella halioticida* sp. nov., a pathogen of farmed giant abalone (*Haliotis gigantea*) in Japan. *Journal of Applied Microbiology, 111*(5), 1044–1056. doi:10.1111/j.1365-2672.2011.05133.xNorway.

Brokordt, K., González, R., Farías, W., Winkler, F. E., & Lohrmann, K. B. (2017). First insight into the heritable variation of the resistance to infection with the bacteria causing the withering syndrome disease in *Haliotis rufescens* abalone. *Journal of Invertebrate Pathology, 150*, 15–20. doi:10.1016/j.jip.2017.08.014.

Bruto, M., James, A., Petton, B., Labreuche, Y., Chenivesse, S., Alunno-Bruscia, M., … Le Roux, F. (2017). *Vibrio crassostreae*, a benign oyster colonizer turned into a pathogen after plasmid acquisition. *The ISME Journal, 11*(4), 1043–1052. doi:10.1038/ismej.2016.162.

Burge, C. A., Strenge, R. E., & Friedman, C. S. (2011). Detection of the oyster herpesvirus in commercial bivalves in northern California, USA: Conventional and quantitative PCR. *Diseases of Aquatic Organisms, 94*(2), 107–116. doi:10.3354/dao02314.

Burreson, E. M., Stokes, N. A., & Friedman, C. S. (2000). Increased virulence in an introduced pathogen: *Haplosporidium nelsoni* (MSX) in the eastern oyster *crassostrea virginica*. *Journal of Aquatic Animal Health, 12*(1), 1–8. doi:10.1577/1548-8667(2000)012⟨0001:IVIAIP⟩2.0.CO;2.

Burreson, E. M., & Ford, S. E. (2004). A review of recent information on the Haplosporidia, with special reference to *Haplosporidium nelsoni* (MSX disease). *Aquatic Living Resources, 17*(4), 499–517. doi:10.1051/alr:2004056.

Bushek, D., Ford, S. E., & Chintala, M. M. (2002). Comparison of in vitro-cultured and wild-type *Perkinsus marinus*. III. Fecal elimination and its role in transmission. *Diseases of Aquatic Organisms, 51*(3), 217–225. doi:10.3354/dao051217.

Buss, J. J., Harris, J. O., Tanner, J. E., Wiltshire, K. H., & Deveney, M. R. (2020a). Rapid transmission of *Bonamia exitiosa* by cohabitation causes mortality in *Ostrea angasi*. *Journal of Fish Diseases, 43*(2), 227–237. doi:10.1111/jfd.13116.

Buss, J. J., Wiltshire, K. H., Harris, J. O., & Deveney, M. R. (2020b). Decontamination of *Bonamia exitiosa*. *Aquaculture, 523*, 735210. doi:10.1016/j.aquaculture.2020.735210.

Butt, D., & Raftos, D. (2008). Phenoloxidase-associated cellular defence in the Sydney rock oyster, *Saccostrea glomerata*, provides resistance against QX disease infections. *Developmental & Comparative Immunology, 32*(3), 299–306. doi:10.1016/j.dci.2007.06.006.

Cai, J., Li, J., Thompson, K. D., Li, C., & Han, H. (2007). Isolation and characterization of pathogenic *Vibrio parahaemolyticus* from diseased post-larvae of abalone *Haliotis diversicolor supertexta*. *Journal of Basic Microbiology, 47*(1), 84–86. doi:10.1002/jobm.200610192.

Canier, L., Dubreuil, C., Noyer, M., Serpin, D., Chollet, B., Garcia, C., & Arzul, I. (2020). A new multiplex real-time PCR assay to improve the diagnosis of shellfish regulated parasites of the genus *Marteilia* and *Bonamia*. *Preventive Veterinary Medicine, 183*, 105126. doi:10.1016/j.prevetmed.2020.105126.

Cano, I., Parker, A., Ward, G. M., Green, M., Ross, S., Bignell, J., … Batista, F. M. (2022). First detection of *Francisella halioticida* infecting a wild population of blue mussels *Mytilus edulis* in the United Kingdom. *Pathogens, 11*(3), 329. doi:10.3390/pathogens11030329.

Cano, I., Wood, G., Stone, D., Noyer, M., Canier, L., & Arzul, I. (2024). Loop-mediated isothermal amplification for the fast detection of *Bonamia ostreae* and *Bonamia exitiosa* in flat oysters. *Pathogens, 13*(2), 132. doi:10.3390/pathogens13020132.

Carballal, M. J., Iglesias, D., Darriba, S., Cao, A., Mariño, J. C., Ramilo, A., … Villalba, A. (2016). Parasites, pathological conditions and resistance to *Marteilia cochillia* in lagoon cockle *Cerastoderma glaucum* from Galicia (NW Spain). *Diseases of Aquatic Organisms, 122*(2), 137–152. doi:10.3354/dao03070.

Carnegie, R. B., Ford, S. E., Crockett, R. K., Kingsley-Smith, P. R., Bienlien, L. M., Safi, L. S. L., … Burreson, E. M. (2021). A rapid phenotype change in the pathogen *Perkinsus marinus* was associated with a historically significant marine disease emergence in the eastern oyster. *Scientific Reports, 11*(1), 12872. doi:10.1038/s41598-021-92379-6.

Carnegie, R. B., Burreson, E. M., Hine, P. M., Stokes, N. A., Audemard, C., Bishop, M. J., & Peterson, C. H. (2006). *Bonamia perspora* n. sp. (Haplosporidia), a parasite of the oyster *Ostreola equestris*, is the first Bonamia species known to produce spores. *Journal of Eukaryotic Microbiology, 53*(4), 232–245. doi:10.1111/j.1550-7408.2006.00100.x.

Carnegie, R. B., Hill, K. M., Stokes, N. A., & Burreson, E. M. (2014). The haplosporidian *Bonamia exitiosa* is present in Australia, but the identity of the parasite described as *Bonamia* (formerly *Mikrocytos*) *roughleyi* is uncertain. *Journal of Invertebrate Pathology, 115*(1), 33–40. doi:10.1016/j.jip.2013.10.017.

Carrasco, N., Villalba, A., Andree, K. B., Engelsma, M. Y., Lacuesta, B., Ramilo, A., … Furones, M. D. (2012). *Bonamia exitiosa* (Haplosporidia) observed infecting the European flat oyster *Ostrea edulis* cultured on the Spanish Mediterranean coast. *Journal of Invertebrate Pathology, 110*(3), 307–313. doi:10.1016/j.jip.2012.03.015.

Carrasco, N., Voorbergen-Laarman, M., Lacuesta, B., Furones, D., & Engelsma, M. Y. (2017). Application of a competitive real time PCR for detection of *Marteilia refringens* genotype "O" and "M" in two geographical locations: The Ebro Delta, Spain and the Rhine-Meuse Delta, the Netherlands. *Journal of Invertebrate Pathology, 149*, 51–55. doi:10.1016/j.jip.2017.07.004.

Chang, P. H., Kuo, S. H., Lai, S. H., Yang, H. S., Ting, Y. Y., Hsu, C. L., & Chen, H. C. (2005). Herpes-like virus infection causing mortality of cultured abalone *Haliotis diversicolor supertexta* in Taiwan. *Diseases of Aquatic Organisms, 65*(1), 23–27. doi:10.3354/dao065023.

Charles, M., Quesnelle, Y., Glais, G., Trancart, S., & Houssin, M. (2021). Development of a semi-quantitative PCR assay for the detection of *Francisella halioticida* and its application to field samples. *Journal of Fish Diseases, 44*(8), 1169–1177. doi:10.1111/jfd.13377.

Chen, IW., Chang, P. H., Chen, M. S., Renault, T., Chen, M. M., Kuo, S. T., & Cheng, C. H. (2016). Exploring the chronic mortality affecting abalones in Taiwan: Differentiation of abalone herpes virus-associated acute infection from chronic mortality by PCR and *in situ* hybridization and histopathology. *Taiwan Veterinary Journal, 42*(01), 1–9. doi:10.1142/s1682648515500237.

Chen, J., Jia, Y., Sun, Y., Liu, K., Zhou, C., Liu, C., … Fan, G. (2024). Global marine microbial diversity and its potential in bioprospecting. *Nature, 633*(8029), 371–379. doi:10.1038/s41586-024-07891-2.

Chen, M. H., Kuo, S. T., Renault, T., & Chang, P. H. (2014). The development of a loop-mediated isothermal amplification assay for rapid and sensitive detection of abalone herpesvirus DNA. *Journal of Virological Methods, 196*, 199–203. doi:10.1016/j.jviromet.2013.11.011.

Cheng, L., Huang, J., Shi, C., Thompson, K. D., Mackey, B., & Cai, J. (2008). *Vibrio parahaemolyticus* associated with mass mortality of postlarval abalone, *Haliotis diversicolor supertexta* (L.), in Sanya, China. *Journal of the World Aquaculture Society, 39*(6), 746–757. doi:10.1111/j.1749-7345.2008.00210.x.

Cheslett, D., Mc Kiernan, F., Hickey, C., & Collins, E. (2009). Viral gametocytic hypertrophy of the Pacific oyster *Crassostrea gigas* in Ireland. *Diseases of Aquatic Organisms, 83*(3), 181–185. doi:10.3354/dao02020.

Cho, Y. G., Nobuhisa, K., Choi, H. J., Kwon, M. G., Yang, H. S., & Choi, K. S. (2025). Evaluation of the diagnostic reliability of the gill-based Ray's fluid thioglycollate medium assay for *Perkinsus olseni* infections in Manila clams (*Ruditapes philippinarum*) across Korean tidal flats. *Parasites, Hosts and Diseases, 63*(2), 157–167. doi:10.3347/phd.25014.

Cicala, F., Cisterna-Céliz, J. A., Paolinelli, M., Moore, J. D., Sevigny, J., & Rocha-Olivares, A. (2023). The role of diversity in mediating microbiota structural and functional differences in two sympatric species of abalone under stressed withering syndrome conditions. *Microbial Ecology, 85*(1), 277–287. doi:10.1007/s00248-022-01970-5.

Cochennec-Laureau, N., Reece, K. S., Berthe, F. C. J., & Hine, P. M. (2003). *Mikrocytos roughleyi* taxonomic affiliation leads to the genus *Bonamia* (Haplosporidia). *Diseases of Aquatic Organisms, 54*(3), 209–217. doi:10.3354/dao054209.

Comps, M., & Bonami, J. R. (1977). Viral infection associated with mortality in the oyster *Crassostrea gigas* Thunberg. *Comptes rendus hebdomadaires des seances de l'Academie des sciences. Serie D: Sciences naturelles, 285*(11), 1139–1140.

Comps, M. (1983). Etude morphologique de *Marteilia christenseni* sp. n. parasite du lavignon *Scrobicularia piperata* P. (mollusque pelecypode). *Revue des Travaux de l'Institut des Pêches Maritimes, 47*(1/ 2), 99–104.

Corbeil, S., Williams, L. M., Bergfeld, J., & Crane, M. S. J. (2012a). Abalone herpes virus stability in sea water and susceptibility to chemical disinfectants. *Aquaculture, 326-329*, 20–26. doi:10.1016/j.aquaculture.2011.11.031.

Corbeil, S., McColl, K. A., Williams, L. M., Mohammad, I., Hyatt, A. D., Crameri, S. G., Fegan, M., & Crane, M. S. J. (2012b). Abalone viral ganglioneuritis: Establishment and use of an experimental immersion challenge system for the study of abalone herpes virus infections in Australian abalone. *Virus Research, 165*(2), 207–213. doi:10.1016/j.virusres.2012.02.017.

Corbeil, S, Colling, A, Williams, L. M., Wong, F. Y. K., Savin, K., Warner, S., … Crane, M. S. J (2010). Development and validation of a TaqMan® PCR assay for the Australian abalone herpes-like virus. *Diseases of Aquatic Organisms, 92*(1), 1–10. doi:10.3354/dao02277.

Corbeil, S., Faury, N., Segarra, A., & Renault, T. (2015). Development of an *in situ* hybridization assay for the detection of ostreid herpesvirus type 1 mRNAs in the Pacific oyster, *Crassostrea gigas*. *Journal of Virological Methods, 211*, 43–50. doi:10.1016/j.jviromet.2014.10.007.

Costa, C., Ferreira, G. D., Simões, M., Silva, J. L., & Campos, M. J. (2022). Real-time PCR protocol for detection and quantification of three pathogenic members of the Vibrionaceae family. *Microorganisms, 10*(10), 2060. doi:10.3390/microorganisms10102060.

Cowan, M. W., Pearce, C. M., Green, T. J., Finston, T., Meyer, G. R., McAmmond, B., … de la Bastide, P. Y. (2024). Abundance of *Vibrio aestuarianus*, water temperature, and stocking density are associated with summer mortality of Pacific oysters in suspended culture. *Aquaculture International, 32*(4), 5045–5066. doi:10.1007/s10499-024-01415-5.

Crosson, L. M., Lottsfeldt, N. S., Weavil-Abueg, M. E., & Friedman, C. S. (2020). Abalone withering syndrome disease dynamics: Infectious dose and temporal stability in seawater. *Journal of Aquatic Animal Health, 32*(2), 83–92. doi:10.1002/aah.10102.

Crosson, L. M., Wight, N., VanBlaricom, G. R., Kiryu, I., Moore, J. D., & Friedman, C. S. (2014). Abalone withering syndrome: Distribution, impacts, current diagnostic methods and new findings. *Diseases of Aquatic Organisms, 108*(3), 261–270. doi:10.3354/dao02713.

Cruz-Flores, R., & Cáceres-Martínez, J. (2020). Rickettsiales-like organisms in bivalves and marine gastropods: A review. *Reviews in Aquaculture, 12*(4), 2010–2026. doi:10.1111/raq.12419.

Cruz-Flores, R., & Cáceres-Martínez, J. (2016). The hyperparasite of the rickettsiales-like prokaryote, *Candidatus* Xenohaliotis californiensis has morphological characteristics of a *Siphoviridae* (*Caudovirales*). *Journal of Invertebrate Pathology, 133*, 8–11. doi:10.1016/j.jip.2015.11.006.

Danckert, N. P., Wilson, N., Phan-Thien, K. Y., & Stone, D. A. J. (2021). The intestinal microbiome of Australian abalone, *Haliotis laevigata* and *Haliotis laevigata* × *Haliotis rubra*, over a 1-year period in aquaculture. *Aquaculture, 534*, 736245. doi:10.1016/j.aquaculture.2020.736245.

Dang, C., Gonzalez, P., Mesmer-Dudons, N., Bonami, J. R., Caill-Milly, N., & de Montaudouin, X. (2009). Virus-like particles associated with brown muscle disease in Manila clam, *Ruditapes philippinarum*, in Arcachon Bay (France). *Journal of Fish Diseases, 32*(7), 577–584. doi:10.1111/j.1365-2761.2009.01019.x.

Dantan, L., Carcassonne, P., Degrémont, L., Morga, B., Travers, M., Petton, B., ... Cosseau, C. (2024). Microbial education plays a crucial role in harnessing the beneficial properties of microbiota for infectious disease protection in *Crassostrea gigas*. *Scientific Reports, 14*(1), 26914. doi:10.1038/s41598-024-76096-4.

Darriba, S., Iglesias, D., & Carballal, M. J. (2020). *Marteilia cochillia* is released into seawater via cockle *Cerastoderma edule* faeces. *Journal of Invertebrate Pathology, 172*, 107364. doi:10.1016/j.jip.2020.107364.

Davies, C. E., Bass, D., Ward, G. M., Batista, F. M., Malkin, S. H., Thomas, J. E., ... Rowley, A. F. (2020). Diagnosis and prevalence of two new species of haplosporidians infecting shore crabs *Carcinus maenas*: *Haplosporidium carcini* n. sp., and *H. cranc* n. sp. *Parasitology, 147*(11), 1229–1237. doi:10.1017/s0031182020000980.

Davison, A. J., Eberle, R., Ehlers, B., Hayward, G. S., McGeoch, D. J., Minson, A. C., Pellett, P. E., Roizman, B., Studdert, M. J., & Thiry, E. (2009). The order Herpesvirales. *Archives of Virology, 154*(1), 171–177. doi:10.1007/s00705-008-0278-4.

Davison, A. J., Trus, B. L., Cheng, N., Steven, A., Watson, M. S., Cunningham, C., ... Renault, T. (2005). A novel class of herpesvirus with bivalve hosts. *Journal of General Virology, 86*(1), 41–53. doi:10.1099/vir.0.80382-0.

de Lorgeril, J., Lucasson, A., Petton, B., Toulza, E., Montagnani, C., Clerissi, C., ... Mitta, G. (2018). Immune-suppression by OsHV-1 viral infection causes fatal bacteraemia in Pacific oysters. *Nature Communications, 9*(1), 4215. doi:10.1038/s41467-018-06659-3.

Dégremont, L., Azéma, P., Maurouard, E., & Travers, M. A. (2020). Enhancing resistance to *Vibrio aestuarianus* in *Crassostrea gigas* by selection. *Aquaculture, 526*, 735429. doi:10.1016/j.aquaculture.2020.735429.

de Kantzow, M., Hick, P. M., & Whittington, R. J. (2023). Immune priming of Pacific oysters (*Crassostrea gigas*) to induce resistance to *ostreid herpesvirus 1*: Comparison of infectious and inactivated OsHV-1 with Poly I:C. *Viruses, 15*(9), 1943. doi:10.3390/v15091943.

Delisle, L., Bui, T., Copedo, J., Laroche, O., Von Ammon, U., Lane, H. S., & Huston, K. S. (2025). Isolation, culture, and optimal growth conditions for the shellfish protozoan parasite, *Perkinsus olseni*. *International Journal of Parasitology, 55*(7), 339–350.

Destoumieux-G. , D., Canesi, L., Oyanedel, D., Travers, M. A., Charrière, G. M., Pruzzo, C., & Vezzulli, L. (2020). Vibrio–bivalve interactions in health and disease. *Environmental Microbiology, 22*(10), 4323–4341. doi:10.1111/1462-2920.15055.

Di Giorgio, G., Gilardoni, C., & Ituarte, C. (2014). Pathology of *Haplosporidium patagon* affecting siphonariid gastropods in Patagonia. *Diseases of Aquatic Organisms, 112*(1), 59–67. doi:10.3354/dao02798.

Di, G., Kong, X., Zhu, G., Liu, S., Zhang, C., & Ke, C. (2016). Pathology and physiology of *Haliotis diversicolor* with withering syndrome. *Aquaculture, 453*, 1–9. doi:10.1016/j.aquaculture.2015.11.030.

Dietrich, J. P., Hicks, M. B. R., Hard, J. J., Nichols, K. M., Langdon, C. J., Divilov, K., ... Arkoosh, M. R. (2022). Heritability estimates of disease resistance to *Vibrio coralliiyticus* in Pacific oyster (*Crassostrea gigas*) larvae from a selective broodstock program. *Aquaculture, 560*, 738492. doi:10.1016/j.aquaculture.2022.738492.

Divilov, K., Wang, X., Swisher, A. E., Yeoman, P. C., Rintoul, M., Fleener, G. B., ... Jin, L. (2024). Ostreid herpesvirus 1 latent infection and reactivation in adult Pacific oysters, *Crassostrea gigas*. *Virus Research, 339*, 199245. doi:10.1016/j.virusres.2023.199245.

Divilov, K., Merz, N., Schoolfield, B., Green, T. J., & Langdon, C. (2023). Marker-assisted selection in a Pacific oyster population for an antiviral QTL conferring increased survival to OsHV-1 mortality events in Tomales Bay. *Aquaculture, 567*, 739291. doi:10.1016/j.aquaculture.2023.739291.

Dotto-Maurel, A., Pelletier, C., Degremont, L., Heurtebise, S., Arzul, I., Morga, B., & Chevignon, G. (2025). Evaluation of long-read sequencing for Ostreid herpesvirus type 1 genome characterization from *Magallana gigas* infected tissues. *Microbiology Spectrum, 13*(3), e0208224. doi:10.1128/spectrum.02082-24.

Dove, M. C., Nell, J. A., & O'Connor, W. A. (2013). Evaluation of the progeny of the fourth-generation Sydney rock oyster *Saccostrea glomerata* (Gould, 1850) breeding lines for resistance to QX disease (*Marteilia sydneyi*) and winter mortality (*Bonamia roughleyi*). *Aquaculture Research, 44*(11), 1791–1800. doi:10.1111/are.12012.

Du, X., Sun, J., Ju, H., Xu, Z., Tang, X., Fang, X., & Chang, M. S. (2025). Metazoan parasites associated with marine mollusks inhabiting the China Seas: A review. *Frontiers in Marine Science, 12*, 1488376. doi:10.3389/fmars.2025.1488376.

Dubert, J., Nelson, D. R., Spinard, E. J., Kessner, L., Gomez-Chiarri, M., Costa, F. D., … Barja, J. L. (2016). Following the infection process of vibriosis in Manila clam (*Ruditapes philippinarum*) larvae through GFP-tagged pathogenic Vibrio species. *Journal of Invertebrate Pathology, 133*, 27–33. doi:10.1016/j.jip.2015.11.008.

Dubert, J., Barja, J. L., & Romalde, J. L. (2017). New insights into pathogenic vibrios affecting bivalves in hatcheries: Present and future prospects. *Frontiers in Microbiology, 8*, 762. doi:10.3389/fmicb.2017.00762.

Dungan, C. F., & Bushek, D. (2015). Development and applications of Ray's fluid thioglycollate media for detection and manipulation of *Perkinsus* spp. pathogens of marine molluscs. *Journal of Invertebrate Pathology, 131*, 68–82. doi:10.1016/j.jip.2015.05.004.

Duperthuy, M., Schmitt, P., Garzón, E., Caro, A., Rosa, R. D., Le Roux, F., … Destoumieux-Garzón, D. (2011). Use of OmpU porins for attachment and invasion of *Crassostrea gigas* immune cells by the oyster pathogen *Vibrio splendidus*. *Proceedings of the National Academy of Sciences of the United States of America, 108*(7), 2993–2998. doi:10.1073/pnas.1015326108.

Degrémont, L., Guyader, T., Tourbiez, D., & Pépin, J. F. (2013). Is horizontal transmission of the Ostreid herpesvirus OsHV-1 in *Crassostrea gigas* affected by unselected or selected survival status in adults to juveniles? *Aquaculture, 408-409*, 51–57. doi:10.1016/j.aquaculture.2013.05.025.

Elandalloussi, L. M., Leite, R. B., Rodrigues, P. M., Afonso, R., Nunes, P. A., & Cancela, M. L. (2005). Effect of antiprotozoal drugs on the proliferation of the bivalve parasite *Perkinsus olseni*. *Aquaculture, 243*(1–4), 9–17. doi:10.1016/j.aquaculture.2004.09.038.

Elston, R., Friedman, C., Gustafson, L., Meyer, G., & Rogers, R. (2015). Denman Island disease in Washington State, USA: Distribution and prevalence in Pacific and Olympia oysters. *Diseases of Aquatic Organisms, 114*(2), 147–154. doi:10.3354/dao02862.

Elston, R. A., Hasegawa, H., Humphrey, K. L., Polyak, I. K., & Häse, C. C. (2008). Re-emergence of *Vibrio tubiashii* in bivalve shellfish aquaculture: severity, environmental drivers, geographic extent and management. *Diseases of Aquatic Organisms, 82*(2), 119–134. doi:10.3354/dao01982.

Elston, R. A., & Wilkinson, M. T. (1985). Pathology, management and diagnosis of oyster velar virus disease (OVVD). *Aquaculture, 48*(3-4), 189–210. doi:10.1016/0044-8486(85)90124-3.

Engelsma, M. Y., Culloty, S. C., Lynch, S. A., Arzul, I., & Carnegie, R. B. (2014). *Bonamia* parasites: A rapidly changing perspective on a genus of important mollusc pathogens. *Inter-Research, Netherlands Diseases of Aquatic Organisms, 110*(1–2), 5–23. doi:10.3354/dao02741.

Engelsma, M. Y., Kerkhoff, S., Roozenburg, I., Haenen, O. L. M, van Gool, A., Sistermans, W., … Hummel, H. (2010). Epidemiology of *Bonamia ostreae* infecting European flat oysters *Ostrea edulis* from Lake Grevelingen, The Netherlands. *Marine Ecology Progress Series, 409*, 131–142. doi:10.3354/meps08594.

Evans, O., Hick, P., & Whittington, R. J. (2017). Detection of Ostreid herpesvirus-1 microvariants in healthy *Crassostrea gigas* following disease events and their possible role as reservoirs of infection. *Journal of Invertebrate Pathology, 148*, 20–33. doi:10.1016/j.jip.2017.05.004.

Fan, C., Liu, S., Dai, W., He, L., Xu, H., Zhang, H., & Xue, Q. (2023). Characterization of Vibrio mediterranei isolates as causative agents of Vibriosis in marine bivalves. *Microbiology Spectrum, 11*(2). doi:10.1128/spectrum.04923-22.

Fallet, M., Montagnani, C., Petton, B., Dantan, L., de Lorgeril, J., Comarmond, S., … Cosseau, C. (2022). Early life microbial exposures shape the *Crassostrea gigas* immune system for lifelong and intergenerational disease protection. *Microbiome, 10*(1), 85. doi:10.1186/s40168-022-01280-5.

Fan, C., Dai, W., Zhang, H., Liu, S., Lin, Z., & Xue, Q. (2024). Genomic and proteomic analyses of extracellular products reveal major virulence factors likely accounting for differences in pathogenicity to bivalves between *Vibrio mediterranei* strains. *Animals, 14*(5), 692. doi:10.3390/ani14050692.

Fernández-Boo, S., Provot, C., Lecadet, C., Stavrakakis, C., Papin, M., Chollet, B., … Arzul, I. (2021). Inactivation of marine bivalve parasites using UV-C irradiation: Examples of *Perkinsus olseni* and *Bonamia ostreae*. *Aquaculture Reports, 21*, 100859. doi:10.1016/j.aqrep.2021.100859.

Ford, S. E., & Borrero, F. J. (2001). Epizootiology and pathology of juvenile oyster disease in the Eastern oyster, *Crassostrea virginica*. *Journal of Invertebrate Pathology, 78*(3), 141–154. doi:10.1006/jipa.2001.5052.

Ford, S. E., Stokes, N. A., Alcox, K. A., Kraus, B. S. F., Barber, R. D., Carnegie, R. B., & Burreson, E. M. (2018). Investigating the life cycle of *Haplosporidium Nelsoni* (MSX): A review. *Journal of Shellfish Research, 37*(4), 679–693. doi:10.2983/035.037.0402.

Ford, S. E. (1986). Comparison of hemolymph proteins from resistant and susceptible oysters, *Crassostrea virginica*, exposed to the parasite *Haplosporidium nelsoni* (MSX). *Journal of Invertebrate Pathology, 47*(3), 283–294. doi:10.1016/0022-2011(86)90098-4.

Ford, S. E., & Bushek, D. (2012). Development of resistance to an introduced marine pathogen by a native host. *Journal of Marine Research, 70*(2), 205–223. doi:10.1357/002224012802851922.

Friedman, C. S., Andree, K. B., Beauchamp, K. A., Moore, J. D., Robbins, T. T., Shields, J. D., & Hedrick, R. P. (2000). '*Candidatus* Xenohaliotis californiensis', a newly described pathogen of abalone, *Haliotis* spp., along the west coast of North America. *International Journal of Systematic and Evolutionary Microbiology, 50*(2), 847–855. doi:10.1099/00207713-50-2-847.

Friedman, C. S., Wight, N., Crosson, L. M., White, S. J., & Strenge, R. M. (2014). Validation of a quantitative PCR assay for detection and quantification of '*Candidatus* Xenohaliotis californiensis'. *Diseases of Aquatic Organisms, 108*(3), 251–259. doi:10.3354/dao02720.

Fuller, A. M., VanBlaricom, G. R., Neuman, M. J., Witting, D. A., & Friedman, C. S. (2022). A field sentinel study investigating withering syndrome transmission dynamics in California abalones. *Marine Environmental Research, 173*, 105540. doi:10.1016/j.marenvres.2021.105540.

Gagné, N., Cochennec, N., Stephenson, M., McGladdery, S., Meyer, G. R., & Bower, S. M. (2008). First report of a *Mikrocytos*-like parasite in European oysters *Ostrea edulis* from Canada after transport and quarantine in France. *Diseases of Aquatic Organisms, 80*(1), 27–35. doi:10.3354/dao01922.

Gajamange, D., Kim, S. H., Choi, K. S., Azevedo, C., & Park, K. I. (2020). Scanning electron microscopic observation of the *in* vitro cultured protozoan, *Perkinsus olseni*, isolated from the Manila clam, *Ruditapes philippinarum. BMC Microbiology, 20*(1), 238. doi:10.1186/s12866-020-01926-0.

Galvis, F., Ageitos, L., Rodríguez, J., Jiménez, C., Barja, J. L., Lemos, M. L., & Balado, M. (2021). *Vibrio neptunius* produces piscibactin and amphibactin and both siderophores contribute significantly to virulence for clams. *Frontiers in Cellular and Infection Microbiology, 11*, 750567. doi:10.3389/fcimb.2021.750567.

Gao, J., Yang, X., Zhai, J., Qi, P., Ren, Z., Zhu, D., & Fu, P. (2024). Survey on Perkinsus species in two economic mussels (*Mytilus coruscus* and *M. galloprovincialis*) along the coast of the East China Sea and the Yellow Sea. *Parasitology Research, 123*(7), 265. doi:10.1007/s00436-024-08287-x.

Gaona-Hernández, A., Jurgens, L. J., Pollack, J. B., Du, J., Jensen, C. C., & Quigg, A. (2025). Impact of *Perkinsus marinus* on Oyster Health: Survey findings after a regional data gap. *Journal of Shellfish Research, 44*(1), 125–137. doi:10.2983/035.044.0111.

Garcia, C., Charles, M., Chollet, B., Nadeau, A., Serpin, D., Quintric, L., … Lupo, C. (2024). Understanding the role of *Francisella halioticida* in mussel mortalities in France: An integrative approach. *Diseases of Aquatic Organisms, 158*, 81–99. doi:10.3354/dao03782.

Garcia, C., Haond, C., Chollet, B., Nerac, M., Omnes, E., Joly, J. P., … Arzul, I. (2018). Descriptions of *Mikrocytos veneroïdes* n. sp. and *Mikrocytos donaxi* n. sp. (Ascetosporea: Mikrocytida: Mikrocytiidae), detected during important mortality events of the wedge clam *Donax trunculus* Linnaeus (Veneroida: Donacidae), in France between 2008 and 2011. *Parasites & Vectors, 11*, 119.

Garcia, C., Robert, M., Arzul, I., Chollet, B., Joly, J. P., Miossec, L., … Berthe, F. (2006). Viral gametocytic hypertrophy of *Crassostrea gigas* in France: From occasional records to disease emergence? *Diseases of Aquatic Organisms, 70*(3), 193–199. doi:10.3354/dao070193.

Gignoux-Wolfsohn, S. A., Newcomb, M. S. R., Ruiz, G. M., & Pagenkopp Lohan, K. M. (2021). Environmental factors drive the release of *Perkinsus marinus* from infected oysters. *Parasitology, 148*(5), 532–538. doi:10.1017/S0031182020002383.

Gleason, F. H., Gadd, G. M., Pitt, J. I., & Larkum, A. W. D. (2017). The roles of endolithic fungi in bioerosion and disease in marine ecosystems. II. Potential facultatively parasitic anamorphic ascomycetes can cause disease in corals and molluscs. *Mycology, 8*(3), 216–227. doi:10.1080/21501203.2017.1371802.

Gómez-Chiarri, M., Andrews, J. S., Coppersmith, J., Guidry, M. E., Houtz, A., Mills, B., … Rowley, D. C. (2024). Vibriosis of bivalves. In *Diseases of bivalves. Historical and Current Perspectives* (pp. 143–156). Academic Press.

Goudenège, D., Travers, M. A., Lemire, A., Petton, B., Haffner, P., Labreuche, Y., … Le roux, F. (2015). A single regulatory gene is sufficient to alter *Vibrio aestuarianus* pathogenicity in oysters. *Environmental Microbiology, 17*(11), 4189–4199. doi:10.1111/1462-2920.12699.

Guo, X., Puritz, J. B., Wang, Z., Proestou, D., Allen, S., Small, J., … Wilbur, A. (2023). Development and evaluation of high-density SNP arrays for the Eastern Oyster *Crassostrea virginica*. *Marine Biotechnology, 25*(1), 174–191. doi:10.1007/s10126-022-10191-3.

Hanrio, E., Batley, J., Dungan, C. F., & Dang, C. (2021). Immunoassays and diagnostic antibodies for *Perkinsus* spp. pathogens of marine molluscs. *Diseases of Aquatic Organisms, 147*, 13–23. doi:10.3354/dao03635.

Hartikainen, H., Stentiford, G. D., Bateman, K. S., Berney, C., Feist, S. W., Longshaw, M., ... Bass, D. (2014). Mikrocytids are a broadly distributed and divergent radiation of parasites in aquatic invertebrates. *Current Biology, 24*(7), 807–812. doi:10.1016/j.cub.2014.02.033.

Hasanuzzaman, A. F. M., Cao, A., Ronza, P., Fernández-Boo, S., Rubiolo, J. A., Robledo, D., ... Martínez, P. (2020). New insights into the Manila clam – *Perkinsus olseni* interaction based on gene expression analysis of clam hemocytes and parasite trophozoites through in vitro challenges. *International Journal for Parasitology, 50*(3), 195–208. doi:10.1016/j.ijpara.2019.11.008.

Hasegawa, H., Gharaibeh, D. N., Lind, E. J., & Häse, C. C. (2009). Virulence of metalloproteases produced by *Vibrio species* on Pacific oyster *Crassostrea gigas* larvae. *Diseases of Aquatic Organisms, 85*(2), 123–131. doi:10.3354/dao02070.

Helmer, L., Hauton, C., Bean, T., Bass, D., Hendy, I., Harris-Scott, E., & Preston, J. (2020). Ephemeral detection of *Bonamia exitiosa* (Haplosporida) in adult and larval European flat oysters *Ostrea edulis* in the Solent, United Kingdom. *Journal of Invertebrate Pathology, 174*, 107421. doi:10.1016/j.jip.2020.107421.

Hick, P., Evans, O., Looi, R., English, C., & Whittington, R. J. (2016). Stability of ostreid herpesvirus-1 (OsHV-1) and assessment of disinfection of seawater and oyster tissues using a bioassay. *Aquaculture, 450*, 412–421. doi:10.1016/j.aquaculture.2015.08.025.

Hilton, Z., Rolton, A., Muznebin, F., Webb, S. C., Fidler, A., Elliot, A., ... Hutson, K. S. (2025). Reduced farming density of New Zealand flat oyster *Ostrea chilensis* is associated with decreased infection by *Bonamia ostreae* and other parasites. *Aquaculture, 607*, 742683. doi:10.1016/j.aquaculture.2025.742683.

Hine, P. M., & Wesney, B. (1997). Virus-like particles associated with cytopathology in the digestive gland epithelium of scallops *Pecten novaezelandiae* and toheroa *Paphies ventricosum*. *Diseases of Aquatic Organisms, 29*(3), 197–204. doi:10.3354/dao029197.

Hofmann, E., Ford, S., Powell, E., & Klinck, J. (2001). Modeling studies of the effect of climate variability on MSX disease in eastern oyster (*Crassostrea virginica*) populations. *Hydrobiologia, 460*, 195–212.

Hooper, C., Serpin, D., Arzul, I., Canier, L., Bogwald, J., Mortensen, S., ... Bass, D. (2025). Full ribosomal RNA gene arrays confirm *Marteilia refringens sensu stricto* and *Marteilia pararefringens* as separate species, and assess the validity of current diagnostic regions. *Parasitology, 152*, 1247–1262.

Hooper, C., Hardy-Smith, P., & Handlinger, J. (2007). Ganglioneuritis causing high mortalities in farmed Australian abalone (*Haliotis laevigata* and *Haliotis rubra*). *Australian Veterinary Journal, 85*(5), 188–193. doi:10.1111/j.1751-0813.2007.00155.x.

Huang, Z., Li, Y., Yu, K., Ma, L., Pang, B., Qin, Q., ... Kan, B. (2024). Genome-wide expanding of genetic evolution and potential pathogenicity in *Vibrio alginolyticus*. *Emerging Microbes and Infections, 13*(1), 2350164. doi:10.1080/22221751.2024.2350164.

Hudock, R. E., Campbell, B., Motz, N., Carlisle, A., & Hale, E. (2025). Performance of two selectively bred strains of eastern oyster, *Crassostrea virginica*, in Delaware Bay, USA with implications for living shoreline features. *Frontiers in Ecology and Evolution, 13*, 1608236. doi:10.3389/fevo.2025.1608236.

Islam, S. S., Zhang, S., Eggermont, M., Bruto, M., Le Roux, F., & Defoirdt, T. (2022). The impact of the multichannel quorum sensing systems of Vibrio tasmaniensis and *Vibrio crassostreae* on virulence towards blue mussel (*Mytilus edulis*) larvae. *Aquaculture, 547*, 737414. doi:10.1016/j.aquaculture.2021.737414.

Itoïz, S., Mouronvalle, C., Perennou, M., Chailler, E., Smits, M., Derelle, E., ... Chambouvet, A. (2023). Co-infection of two eukaryotic pathogens within clam populations in Arcachon Bay. *Frontiers in Microbiology, 14*, 1250947. doi:10.3389/fmicb.2023.1250947.

Jesser, K. J., Valdivia-Granda, W., Jones, J. L., & Noble, R. T. (2019). Clustering of *Vibrio parahaemolyticus* isolates using MLST and whole-genome phylogenetics and protein motif fingerprinting. *Frontiers in Public Health, 7*, 66. doi:10.3389/fpubh.2019.00066.

Jiang, C., Tanaka, M., Nishikawa, S., Mino, S., Romalde, J. L., Thompson, F. L., ... Sawabe, T. (2022). Vibrio Clade 3.0: New vibrionaceae evolutionary units using genome-based approach. *Current Microbiology, 79*(1), 10. doi:10.1007/s00284-021-02725-0.

Jiang, J. Z., Fang, Y. F., Wei, H. Y., Zhu, P., Liu, M., Yuan, W., ... Zhang, D. (2023). A remarkably diverse and well-organized virus community in a filter-feeding oyster. *Microbiome, 11*(1), 2. doi:10.1186/s40168-022-01431-8.

Jiang, K., Chen, C., Jiang, G., Chi, Y., Xu, C., Kong, L., ... Li, Q. (2024). Genetic improvement of oysters: Current status, challenges, and prospects. *Reviews in Aquaculture, 16*(2), 796–817. doi:10.1111/raq.12868.

Johnson, J. S., Spakowicz, D. J., Hong, B. Y., Petersen, L. M., Demkowicz, P., Chen, L., ... Weinstock, G. M. (2019). Evaluation of 16S rRNA gene sequencing for species and strain-level microbiome analysis. *Nature Communications, 10*(1), 5029. doi:10.1038/s41467-019-13036-1.

Jones, J. B., Scotti, P. D., Dearing, S. C., & Wesney, B. (1996). Virus-like particles associated with marine mussel mortalities in New Zealand. *Diseases of Aquatic Organisms, 25*(1-2), 143–149. doi:10.3354/dao025143.

Jørgensen, L. V. G., Nielsen, J. W., Villadsen, M. K., Vismann, B., Dalvin, S., Mathiessen, H., ... Buchmann, K. (2020). A non-lethal method for detection of *Bonamia ostreae* in flat oyster (*Ostrea edulis*) using environmental DNA. *Scientific Reports, 10*(1), 16143. doi:10.1038/s41598-020-72715-y.

Kamaishi, T., Miwa, S., Goto, E., Matsuyama, T., & Oseko, N. (2010). Mass mortality of giant abalone *Haliotis gigantea* caused by a *Francisella* sp. bacterium. *Diseases of Aquatic Organisms, 89*(2), 145–154. doi:10.3354/dao02188.

Kang, H. S., Itoh, N., Limpanont, Y., Lee, H. M., Whang, I., & Choi, K. S. (2019). A novel paramyxean parasite, *Marteilia tapetis* sp. nov. (Cercozoa) infecting the digestive gland of Manila clam *Ruditapes philippinarum* from the southeast coast of Korea. *Journal of Invertebrate Pathology, 163*, 86–93. doi:10.1016/j.jip.2019.03.006.

Kawahara, M., Yoshitake, K., Yoshinaga, T., & Itoh, N. (2021). Francisellosis of Yesso scallops *Mizuhopecten yessoensis* in Japan is caused by a novel type of *Francisella halioticida*. *Diseases of Aquatic Organisms, 144*, 9–19. doi:10.3354/dao03574.

Kawahara, M., Kanamori, M., Natsuike, M., Yoshinaga, T., & Itoh, N. (2024). Four-year field survey on francisellosis in Yesso scallops *Mizuhopecten yessoensis* cultured in southern Hokkaido, Japan. *Aquaculture, 578*, 740092. doi:10.1016/j.aquaculture.2023.740092.

Keim, P., Johansson, A., & Wagner, D. M. (2007). Molecular epidemiology, evolution, and ecology of *Francisella*. *Annals of the New York Academy of Sciences, 1105*, 30–66. doi:10.1196/annals.1409.011.

Kerr, R., Ward, G. M., Stentiford, G. D., Alfjorden, A., Mortensen, S., Bignell, J. P., ... Bass, D. (2018). *Marteilia refringens* and *Marteilia pararefringens* sp. nov. are distinct parasites of bivalves and have different European distributions. *Parasitology, 145*(11), 1483–1492. doi:10.1017/S003118201800063X.

Kim, S.-H., Kim, H. J., Bathige, S. D. N. K., Kim, S., & Park, K. I. (2025). Strain-specific virulence of Perkinsus marinus and related species in Eastern oysters: A comprehensive analysis of immune responses and mortality. *Fish & Shellfish Immunology, 157*, 110112. doi:10.1016/j.fsi.2025.110112.

King, W. L., Jenkins, C., Seymour, J. R., & Labbate, M. (2019). Oyster disease in a changing environment: Decrypting the link between pathogen, microbiome and environment. *Marine Environmental Research, 143*, 124–140. doi:10.1016/j.marenvres.2018.11.007.

Kroeck, M. A. (2010). Gross signs and histopathology of *Ostrea puelchana* infected by a *Bonamia exitiosa*-like parasite (Haplosporidia). *Diseases of Aquatic Organisms, 89*(3), 229–236. doi:10.3354/dao02186.

Kunselman, E., Allard, S., Burge, C., Marshman, B., Frederick, A., & Gilbert, J. (2024). Metagenome-assembled genome of withering syndrome causative agent, "*Candidatus* Xenohaliotis californiensis," from endangered white abalone (*Haliotis sorenseni*). *Microbiology Resource Announcements, 13*(6), e0008624. doi:10.1128/mra.00086-24.

La Peyre, M. K., Gossman, B., & La Peyre, J. F. (2009). Defining optimal freshwater flow for oyster production: Effects of freshet rate and magnitude of change and duration on Eastern Oysters and *Perkinsus marinus* infection. *Estuaries and Coasts, 32*(3), 522–534. doi:10.1007/s12237-009-9149-9.

Lafferty, K. D., & Ben-Horin, T. (2013). Abalone farm discharges the withering syndrome pathogen into the wild. *Frontiers in Microbiology, 4*, 373. doi:10.3389/fmicb.2013.00373.

Lallias, D., Arzul, I., Heurtebise, S., Ferrand, S., Chollet, B., Robert, M., ... Lapègue, S. (2008). *Bonamia ostreae*-induced mortalities in one-year old European flat oysters *Ostrea edulis*: Experimental infection by cohabitation challenge. *Aquatic Living Resources, 21*(4), 423–439. doi:10.1051/alr:2008053.

Lane, H. S., & Jones, J. B. (2020). Low internal transcribed spacer rDNA variation in New Zealand *Bonamia ostreae*: Evidence for a recent arrival. *Diseases of Aquatic Organisms, 139*, 121–130. doi:10.3354/dao03461.

Lane, H. S., Webb, S. C., & Duncan, J. (2016). *Bonamia ostreae* in the New Zealand oyster *Ostrea chilensis*: A new host and geographic record for this haplosporidian parasite. *Diseases of Aquatic Organisms, 118*(1), 55–63. doi:10.3354/dao02960.

Lau, Y. T., Gambino, L., Santos, B., Espinosa, E. P., & Allam, B. (2018). Transepithelial migration of mucosal hemocytes in *Crassostrea virginica* and potential role in *Perkinsus marinus* pathogenesis. *Journal of Invertebrate Pathology, 153*, 122–129. doi:10.1016/j.jip.2018.03.004.

Le Deuff, R. M., & Renault, T. (1999). Purification and partial genome characterization of a herpes-like virus infecting the Japanese oyster, *Crassostrea gigas*. *Journal of General Virology, 80*(5), 1317–1322. doi:10.1099/0022-1317-80-5-1317.

Lee, Y., Roh, H. J., Kim, A., Park, J., Lee, J. Y., Kim, Y. J., ... Kim, D. H. (2023). Molecular mechanisms underlying the vulnerability of Pacific abalone (*Haliotis discus hannai*) to *Vibrio harveyi* infection at higher water temperature. *Fish and Shellfish Immunology, 138*, 108844. doi:10.1016/j.fsi.2023.108844.

Leunda-Esnaola, A., Bunin, E., Arrufat, P., Pearman, P. B., & Kaberdin, V. R. (2024). Harnessing the intragenomic variability of rRNA operons to improve differentiation of Vibrio species. *Scientific Reports, 14*(1), 9908. doi:10.1038/s41598-024-60505-9.

Li, Y. N., Zhang, X., Huang, B. W., Xin, L. S., Wang, C. M., & Bai, C. M. (2024). Localization and tissue tropism of ostreid herpesvirus 1 in blood clam *Anadara broughtonii*. *Biology, 13*(9), 720. doi:10.3390/biology13090720.

Liu, Y., Liao, K., Zhu, Y., Zhang, Y., Huang, H., Ge, H., & Xu, J. (2025). Effects of dietary taurine on growth, taurine metabolism and *Vibrio crassostreae* resistance in juvenile clam *Ruditapes philippinarum*. *Fish & Shellfish Immunology, 161*, 110285. doi:10.1016/j.fsi.2025.110285.

Longshaw, M., Stone, D. M., Wood, G., Green, M. J., & White, P. (2013). Detection of *Bonamia exitiosa* (Haplosporidia) in European flat oysters *Ostrea edulis* cultivated in mainland Britain. *Diseases of Aquatic Organisms, 106*(2), 173–179. doi:10.3354/dao02643.

Lopez, Jose R., Leonessi, M., Fernandez-Tejedor, M., Auguste, M., Doni, L., Cruz-Quintana, Y., … Furones, D. (2025). First description of *Vibrio ostreicida* as putative causal agent of disease in Mediterranean mussel (*Mytilus galloprovincialis*) and development of a PCR protocol for its rapid diagnosis. *Journal of Invertebrate Pathology, 212*, 108399. doi:10.1016/j.jip.2025.108399.

López Sanmartín, M., Power, D. M., de la Herrán, R., Navas, J. I., & Batista, F. M. (2016). Evidence of vertical transmission of ostreid herpesvirus 1 in the Portuguese oyster *Crassostrea angulata*. *Journal of Invertebrate Pathology, 140*, 39–41. doi:10.1016/j.jip.2016.08.012.

Lu, J., Liu, S., Ruan, Z., Ye, J., Wu, Q., Wang, S., Lin, Z., & Xue, Q. (2021). Family I84 protease inhibitors likely constitute a Mollusca-specific protein family functioning in host defense. *Diseases of Aquatic Organisms, 145*, 89–100. doi:10.3354/dao03602.

Lupo, C., Travers, M. A., Tourbiez, D., Barthélémy, C. F., Beaunée, G., & Ezanno, P. (2019). Modeling the transmission of *Vibrio aestuarianus* in pacific oysters using experimental infection data. *Frontiers in Veterinary Science, 6*, 142. doi:10.3389/fvets.2019.00142.

Lynch, S. A., Lepée-Rivero, S., Kelly, R., Quinn, E., Coghlan, A., Bookelaar, B., Morgan, E., Finarelli, J. A., Carlsson, J., & Culloty, S. C. (2020). Detection of haplosporidian protistan parasites supports an increase to their known diversity, geographic range and bivalve host specificity. *Parasitology, 147*(5), 584–592. doi:10.1017/s0031182019001628.

Mackenzie, C. L., Raap, M. R., Leduc, S., Walker, C. Y. V., Green, T. J., Kim, E., … Pearce, C. M. (2024). Development of a nature-based solution for mitigation of Pacific oyster summer mortality: Use of the intertidal zone to improve resilience to environmental stressors. *Frontiers in Marine Science, 11*, 1345493. doi:10.3389/fmars.2024.1345493.

Maloy, A. P., Barber, B. J., & Boettcher, K. J. (2005). A PCR-based diagnostic assay for the detection of *Roseovarius crassostreae* in *Crassostrea virginica* affected by juvenile oyster disease (JOD). *Diseases of Aquatic Organisms, 67*(1–2), 155–162. doi:10.3354/dao067155.

Marquis, N. D., Bishop, T. J., Record, N. R., Countway, P. D., & Fernández Robledo, J. A. (2020). A qPCR-Based Survey of *Haplosporidium nelsoni* and *Perkinsus* spp. in the Eastern Oyster, *Crassostrea virginica* in Maine, USA. *Pathogens, 9*(4), 256. doi:10.3390/pathogens9040256.

Martinez, C., Rodriguez, S., Vences, A., Barja, J. L., Toranzo, A. E., & Dubert, J. (2022). Role of the vibriolysin VemA secreted by the emergent pathogen *Vibrio europaeus* in the colonization of Manila Clam mucus. *Microorganisms, 10*(12), 2475. doi:10.3390/microorganisms10122475.

Marty, G. D., Bower, S. M., Clarke, K. R., Meyer, G., Lowe, G., Osborn, A. L., … Robinson, J. H. (2006). Histopathology and a real-time PCR assay for detection of *Bonamia ostreae* in *Ostrea edulis* cultured in western Canada. *Aquaculture, 261*(1), 33–42. doi:10.1016/j.aquaculture.2006.07.024.

Matsuyama, T., Takano, T., Nishiki, I., Fujiwara, A., Kiryu, I., Inada, M., … Nakayasu, C. (2020). A novel Asfarvirus-like virus identified as a potential cause of mass mortality of abalone. *Scientific Reports, 10*(1), 4620. doi:10.1038/s41598-020-61492-3.

Matsuyama, T., Kiryu, I., Mekata, T., Takano, T., Umeda, K., & Matsuura, Y. (2023). Pathogenicity, genomic analysis and structure of abalone asfa-like virus: Evidence for classification in the family *Asfarviridae*. *Journal of General Virology, 104*(8), 001875. doi:10.1099/jgv.0.001875.

Matsuyama, T., Kiryu, I., Inada, M., Takano, T., Matsuura, Y., & Kamaishi, T. (2021). Susceptibility of four abalone species, *Haliotis gigantea, Haliotis discus discus, Haliotis discus hannai* and *Haliotis diversicolor*, to abalone asfa-like virus. *Viruses, 13*(11), 2315. doi:10.3390/v13112315.

Mérou, N., Lecadet, C., Pouvreau, S., & Arzul, I. (2020). An eDNA/eRNA-based approach to investigate the life cycle of non-cultivable shellfish micro-parasites: The case of *Bonamia ostreae*, a parasite of the European flat oyster *Ostrea edulis*. *Microbial Biotechnology, 13*(6), 1807–1818. doi:10.1111/1751-7915.13617.

Mérou, N., Lecadet, C., Ubertini, M., Pouvreau, S., & Arzul, I. (2023). Environmental distribution and seasonal dynamics of *Marteilia refringens* and *Bonamia ostreae*, two protozoan parasites of the European flat oyster, *Ostrea edulis*. *Frontiers in Cellular and Infection Microbiology, 13*, 1154484. doi:10.3389/fcimb.2023.1154484.

Mérou, N., Lecadet, C., Billon, T., Chollet, B., Pouvreau, S., & Arzul, I. (2022). Investigating the environmental survival of *Marteilia refringens*, a marine protozoan parasite of the flat Oyster *Ostrea edulis*, through an environmental DNA and microscopy-based approach. *Frontiers in Marine Science, 9*, 811284. doi:10.3389/fmars.2022.811284.

Mesnil, A., Jacquot, M., Garcia, C., Tourbiez, D., Canier, L., Bidois, A., … Travers, M. A. (2023). Emergence and clonal expansion of *Vibrio aestuarianus* lineages pathogenic for oysters in Europe. *Molecular Ecology, 32*(11), 2869–2883. doi:10.1111/mec.16910.

Meyer, G. R., Bower, S. M., Lowe, G., & Davies, S. (2008). Resistance of the Manila clam (*Venerupis philippinarum*) to infection with *Mikrocytos mackini*. *Journal of Invertebrate Pathology, 98*(1), 54–57. doi:10.1016/j.jip.2007.10.005.

Meyer, G. R., Lowe, G. J., Gilmore, S. R., & Bower, S. M. (2017). Disease and mortality among Yesso scallops *Patinopecten yessoensis* putatively caused by infection with *Francisella halioticida*. *Diseases of Aquatic Organisms, 125*(1), 79–84. doi:10.3354/dao03130.

Meyers, T. R., Burton, T., Evans, W., & Starkey, N. (2009). Detection of viruses and virus-like particles in four species of wild and farmed bivalve molluscs in Alaska, USA, from 1987 to 2009. *Diseases of Aquatic Organisms, 88*(1), 1–12. doi:10.3354/dao02154.

Miyazaki, T., Goto, K., Kobayashi, T., Kageyama, T., & Miyata, M. (1999). Mass mortalities associated with a virus disease in Japanese pearl oysters *Pinctada fucata* martensii. *Diseases of Aquatic Organisms, 37*, 1–12. doi:10.3354/dao037001.

Morga, B., Jacquot, M., Pelletier, C., Chevignon, G., Dégremont, L., Biétry, A., … Lamy, J. B. (2021). Genomic diversity of the Ostreid herpesvirus type 1 across time and location and among host species. *Frontiers in Microbiology, 12*, 711377. doi:10.3389/fmicb.2021.711377.

Morot, A., El Fekih, S., Bidault, A., Le Ferrand, A., Jouault, A., Kavousi, J., … Delavat, F. (2021). Virulence of *Vibrio harveyi* ORM4 towards the European abalone H*aliotis tuberculata* involves both quorum sensing and a type III secretion system. *Environmental Microbiology, 23*(9), 5273–5288. doi:10.1111/1462-2920.15592.

Morot, A., Lambert, C., Bidault, A., Dufour, A., Rodrigues, S., Delavat, F., & Paillard, C. (2025). *Vibrio harveyi* uses both type III secretion system and quorum sensing for the colonization of the European abalone. *Fish & Shellfish Immunology, 157*, 110103. doi:10.1016/j.fsi.2024.110103.

Moss, J. A., Burreson, E. M., Cordes, J. F., Dungan, C. F., Brown, G. D., Wang, A., … Reece, K. S. (2007). Pathogens in *Crassostrea ariakensis* and other Asian oyster species: Implications for non-native oyster introduction to Chesapeake Bay. *Diseases of Aquatic Organisms, 77*(3), 207–223. doi:10.3354/dao01829.

Moussa, M., Cauvin, E., Le Piouffle, A., Lucas, O., Bidault, A., Paillard, C., … Garcia, C. (2021). A MALDI-TOF MS database for fast identification of *Vibrio* spp. potentially pathogenic to marine mollusks. *Applied Microbiology and Biotechnology, 105*(6), 2527–2539. doi:10.1007/s00253-021-11141-0.

Nakatsugawa, T., Okabe, M., & Muroga, K. (2000). Horizontal transmission of amyotrophia in Japanese black abalone. *Fish Pathology, 35*(1), 11–14. doi:10.3147/jsfp.35.11.

Nenadic, N., Bar, I., & McDougall, C. (2025). Transcriptomic responses to *Marteilia sydneyi* infection in the Sydney rock oyster *Saccostrea glomerata*. *bioRxiv*, 111085. doi:10.1101/2025.02.02.636094.

O' Reilly, A. J., Laide, C., Maloy, A, Hutton, S., Bookelaar, B., O' Sullivan, K., … Culloty, S. C. (2018). The role of the mussel *Mytilus* spp. in the transmission of ostreid herpesvirus-1 microVar. *Parasitology, 145*(8), 1095–1104. doi:10.1017/s0031182017002244.

Oehrens K. , E. M., Doldan, M. S., Zaidman, P. C., Morsan, E. M., & Kroeck, M. A. (2014). Bonamiosis status in natural *Ostrea puelchana* beds in San Matías Gulf (Patagonia, Argentina), 14 years after an epizootic. *Diseases of Aquatic Organisms, 110*(1), 135–142. doi:10.3354/dao02707.

Oyanedel, D., Rojas, R., Brokordt, K., & Schmitt, P. (2023). *Crassostrea gigas* oysters from a non-intensive farming area naturally harbor potentially pathogenic vibrio strains. *Journal of Invertebrate Pathology, 196*, 107856. doi:10.1016/j.jip.2022.107856.

Parizadeh, L., Tourbiez, D., Garcia, C., Haffner, P., Dégremont, L., Le Roux, F., & Travers, M. A. (2018). Ecologically realistic model of infection for exploring the host damage caused by *Vibrio aestuarianus*. *Environmental Microbiology, 20*(12), 4343–4355. doi:10.1111/1462-2920.14350.

Park, S., Park, J. M., Kang, C. H., & Yoon, J. H. (2015). *Aliiroseovarius pelagivivens* gen. nov., sp. nov., isolated from seawater, and reclassification of three species of the genus *Roseovarius* as *Aliiroseovarius crassostreae* comb. nov.,

Aliiroseovarius halocynthiae comb. nov. and *Aliiroseovarius sediminilitoris* comb. nov. *International Journal of Systematic and Evolutionary Microbiology, 65*(8), 2646–2652. doi:10.1099/ijs.0.000315.

Pathirana, E., Fuhrmann, M., Whittington, R., & Hick, P. (2019). Influence of environment on the pathogenesis of *Ostreid herpesvirus-1* (OsHV-1) infections in Pacific oysters (*Crassostrea gigas*) through differential microbiome responses. *Heliyon, 5*(7), e02101. doi:10.1016/j.heliyon.2019.e02101.

Pelletier, C., Chevignon, G., Faury, N., Arzul, I., Garcia, C., Chollet, B., Renault, T., Morga, B., & Jacquot, M. (2025). Phylogenomic evidence for host specialization and genetic divergence in OsHV-1 infecting *Magallana gigas* and *Ostrea edulis*. *Infection, Genetics and Evolution, 134*, 105803. doi:10.1016/j.meegid.2025.105803.

Pepin, J. F., Riou, A., & Renault, T. (2008). Rapid and sensitive detection of ostreid herpesvirus 1 in oyster samples by real-time PCR. *Journal of Virological Methods, 149*(2), 269–276. doi:10.1016/j.jviromet.2008.01.022.

Petton, B., Destoumieux-Garzón, D., Pernet, F., Toulza, E., de Lorgeril, J., Degremont, L., & Mitta, G. (2021). The Pacific oyster mortality syndrome, a polymicrobial and multifactorial disease: State of knowledge and future directions. *Frontiers in Immunology, 12*, 630343. doi:10.3389/fimmu.2021.630343.

Pierce, M. L., & Ward, J. E. (2019). Gut Microbiomes of the Eastern Oyster (*Crassostrea virginica*) and the blue mussel (*Mytilus edulis*): Temporal variation and the influence of marine aggregate-associated microbial communities. *mSphere, 4*(6), e00730–19. doi:10.1128/msphere.00730-19.

Piesz, J. L., Scro, A. K., Corbett, R., Lundgren, K. M., Smolowitz, R., & Gomez-Chiarri, M. (2022). Development of a multiplex qPCR for the quantification of three protozoan parasites of the eastern oyster *Crassostrea virginica*. *Diseases of Aquatic Organisms, 151*, 111–121. doi:10.3354/dao03694.

Polinski, M. P., Laurin, E., Delphino, M. K. V. C., Lowe, G. J., Meyer, G. R., & Abbott, C. L. (2021). Evaluation of histopathology, PCR, and qPCR to detect *Mikrocytos mackini* in oysters *Crassostrea gigas* using Bayesian latent class analysis. *Diseases of Aquatic Organisms, 144*, 21–31. doi:10.3354/dao03566.

Polinski, M. P., Meyer, G. R., Lowe, G. J., & Abbott, C. L. (2017). Seawater detection and biological assessments regarding transmission of the oyster parasite *Mikrocytos mackini* using qPCR. *Diseases of Aquatic Organisms, 126*(2), 143–153. doi:10.3354/dao03167.

Potts, R. W. A., Gutierrez, A. P., Penaloza, C. S., Regan, T., Bean, T. P., & Houston, R. D. (2021). Potential of genomic technologies to improve disease resistance in molluscan aquaculture. *Philosophical Transactions of the Royal Society B: Biological Sciences, 376*(1825), 376(1825). doi:10.1098/rstb.2020.0168.

Rahman, A. M. A., Ransangan, J., & Subbiah, V. K. (2022). Improvements to the rapid detection of the marine pathogenic bacterium, *Vibrio harveyi*, using loop-mediated isothermal amplification (LAMP) in combination with SYBR green. *Microorganisms, 10*(12), 2346. doi:10.3390/microorganisms10122346.

Rahmani, A., Mathien, C., Bidault, A., Le Goïc, N., Paillard, C., & Pichereau, V. (2020). External pH modulation during the growth of *Vibrio tapetis*, the aetiological agent of brown ring disease. *Journal of Applied Microbiology, 129*(1), 3–16. doi:10.1111/jam.14699.

Rasmussen, L. P. D. (1986). Virus-associated granulocytomas in the marine mussel, *Mytilus edulis*, from three sites in Denmark. *Journal of Invertebrate Pathology, 48*(1), 117–123. doi:10.1016/0022-2011(86)90150-3.

Rawson, P., & Feindel, S. (2012). Growth and survival for genetically improved lines of Eastern oysters (*Crassostrea virginica*) and interline hybrids in Maine, USA. *Aquaculture, 326-329*, 61–67. doi:10.1016/j.aquaculture.2011.11.030.

Reece, K., Bushek, D., & Carnegie, R. (2024a). History and taxonomy of haplosporidians, and haplosporidiosis in bivalve mollusks. In *Diseases of bivalves: Historical and current perspectives* (pp. 95–116). Elsevier. doi:10.1016/B978-0-12-820339-2.00003-6.

Reece, K., Carnegie, R., & Bushek, D. (2024b). Perkinsosis (aka dermo) of bivalve mollusks. *Diseases of bivalves*. Elsevier. doi:10.1016/B978-0-12-820339-2.00020-6.

Remacha-Triviño, A., Borsay-Horowitz, D., Dungan, C., Gual-Arnau, X., Gómez-Leon, J., Villamil, L., & Gómez-Chiarri, M. (2008). Numerical quantification of *Perkinsus marinus* in the American oyster *Crassostrea virginica* (Gmelin, 1791), (Mollusca: Bivalvia) by modern stereology. *Journal of Parasitology, 94*(1), 125–136. doi:10.1645/ge-1148.1.

Ren, W., Chen, H., Renault, T., Cai, Y., Bai, C., Wang, C., & Huang, J. (2013). Complete genome sequence of acute viral necrosis virus associated with massive mortality outbreaks in the Chinese scallop, *Chlamys farreri*. *Virology Journal, 10*(1), 110. doi:10.1186/1743-422x-10-110.

Ren, W., Renault, T., Cai, Y., & Wang, C. (2010). Development of a loop-mediated isothermal amplification assay for rapid and sensitive detection of ostreid herpesvirus 1 DNA. *Journal of Virological Methods, 170*(1-2), 30–36. doi:10.1016/j.jviromet.2010.08.015.

Renault, T., Moreau, P., Faury, N., Pepin, J. F., Segarra, A., & Webb, S. (2012). Analysis of clinical ostreid herpesvirus 1 (*Malacoherpesviridae*) specimens by sequencing amplified fragments from three virus genome areas. *Journal of Virology, 86*(10), 5942–5947. doi:10.1128/JVI.06534-11France.

Renault, T. (2024). Herpesvirus infections in marine bivalves. *Diseases of Bivalves. Historical and Current Perspectives.* Academic Press https://doi.org/10.1016/B978-0-12-820339-2.00001-2.

Renault, T., Bouquet, A. L., Maurice, J. T., Lupo, C., & Blachier, P. (2014). Ostreid herpesvirus 1 infection among Pacific oyster (*Crassostrea gigas*) Spat: Relevance of water temperature to virus replication and circulation prior to the onset of mortality. *Applied and Environmental Microbiology, 80*(17), 5419–5426. doi:10.1128/AEM.00484-14.

Renault, T., Chollet, B., Cochennec, N., & Gerard, A. (2002). Shell disease in eastern oysters, *Crassostrea virginica*, reared in France. *Journal of Invertebrate Pathology, 79*(1), 1–6. doi:10.1016/s0022-2011(02)00009-5.

Revilla, J., Rodríguez-Rodríguez, S., Solórzano, R., Reyes, G., & Barja, J. L. (2025). First draft genome sequence of *Vibrio rotiferianus* isolated from diseased larvae of the spiny rock-scallop *Spondylus limbatus* during hatchery outbreaks. *Microbial Pathogenesis, 205*, 107617. doi:10.1016/j.micpath.2025.107617.

Rodrigues, S., Paillard, C., Pennec, G. L., Dufour, A., & Bazire, A. (2015). *Vibrio tapetis*, the causative agent of brown ring disease, forms biofilms with spherical components. *Frontiers in Microbiology, 6*, 1384. doi:10.3389/fmicb.2015.01384.

Rojas, R., Miranda, C. D., Romero, J., Barja, J. L., & Dubert, J. (2019). Isolation and pathogenic characterization of *Vibrio bivalvicida* associated with a massive larval mortality event in a commercial hatchery of scallop *Argopecten purpuratus* in Chile. *Frontiers in Microbiology, 10*, 855. doi:10.3389/fmicb.2019.00855.

Rolton, A., Webb, S. C., López-Sanmartín, M., & Hutson, K. S. (2023). Bivalve digestive epithelial virosis (DEV): A cause of disease or a natural process? *Journal of Invertebrate Pathology, 198*, 107924. doi:10.1016/j.jip.2023.107924.

Ronza, P., Cao, A., Robledo, D., Gómez-Tato, A., Álvarez-Dios, J. A., Hasanuzzaman, A. F. M., ... Martínez, P. (2018). Long-term affected flat oyster (*Ostrea edulis*) haemocytes show differential gene expression profiles from naïve oysters in response to *Bonamia ostreae*. *Genomics, 110*(6), 390–398. doi:10.1016/j.ygeno.2018.04.002.

Ruan, Z., Liu, Y., Chang, G., Lin, Z., & Xue, Q. (2022). Molecular characterization of two CuZn-SOD family proteins in the Pacific oyster *Crassostrea gigas*. *Comparative Biochemistry and Physiology Part B: Biochemistry and Molecular Biology, 260*, 110736. doi:10.1016/j.cbpb.2022.110736.

Rubio, T., Oyanedel, D., Labreuche, Y., Toulza, E., Luo, X., Bruto, M., Chaparro, C., Torres, M., de Lorgeril, J., Haffner, P., Vidal-Dupiol, J., Lagorce, A., Petton, B., Mitta, G., Jacq, A., Le Roux, F., Charrière, G. M., & Destoumieux-Garzón, D. (2019). Species-specific mechanisms of cytotoxicity toward immune cells determine the successful outcome of Vibrio infections. *Proceedings of the National Academy of Sciences, 116*(28), 14238–14247. doi:10.1073/pnas.1905747116.

Ruiz, M., López, C., Lee, R. S., Rodríguez, R., & Darriba, S. (2016). A novel paramyxean parasite, *Marteilia octospora* n. sp. (Cercozoa) infecting the Grooved Razor Shell clam *Solen marginatus* from Galicia (NW Spain). *Journal of Invertebrate Pathology, 135*, 34–42. doi:10.1016/j.jip.2016.02.002.

Ruiz, M., Darriba, S., Rodríguez, R., Iglesias, D., Lee, R., & López, C. (2011). *Viral basophilic* inclusions in the digestive gland of razor clams *Ensis arcuatus* (Pharidae) in Galicia (NW Spain). *Diseases of Aquatic Organisms, 94*(3), 239–241. doi:10.3354/dao02330.

Safi, L. S. L., McGurk, E. S., Carnegie, R. B., & Bushek, D. (2025). The biosecurity of bivalve seed transfers during different life stages with respect to their pathogens. *Aquaculture, 606*, 742582. doi:10.1016/j.aquaculture.2025.742582.

Sambade, I. M., Estêvão, J., Pampín, M., Cruz, A., Guévélou, E., Blanco, A., ... Martínez, P. (2025). Signatures of selection for resistance/tolerance to *Perkinsus olseni* in grooved carpet shell clam (*Ruditapes decussatus*) using a population genomics approach. *Evolutionary Applications, 18*(5), e70106. doi:10.1111/eva.70106.

Sambade, I. M., Casanova, A., Blanco, A., Gundappa, M. K., Bean, T. P., Macqueen, D. J., ... Martínez, P. (2022). A single genomic region involving a putative chromosome rearrangement in flat oyster (*Ostrea edulis*) is associated with differential host resilience to the parasite *Bonamia ostreae*. *Evolutionary Applications, 15*(9), 1408–1422. doi:10.1111/eva.13446.

Sandlund, N., Torkildsen, L., Magnesen, T., Mortensen, S., & Bergh, O. (2006). Immunohistochemistry of great scallop *Pecten maximus* larvae experimentally challenged with pathogenic bacteria. *Inter-Research, Norway Diseases of Aquatic Organisms, 69*(2–3), 163–173. doi:10.3354/dao069163.

Savin, K. W., Cocks, B. G., Wong, F., Sawbridge, T., Cogan, N., Savage, D., & Warner, S. (2010). A neurotropic herpesvirus infecting the gastropod, abalone, shares ancestry with oyster herpesvirus and a herpesvirus associated with the amphioxus genome. *Virology Journal, 7*, 308. doi:10.1186/1743-422X-7-308.

Sawabe, T., Kita-Tsukamoto, K., & Thompson, F. L. (2007). Inferring the evolutionary history of vibrios by means of multilocus sequence analysis. *Journal of Bacteriology, 189*(21), 7932–7936. doi:10.1128/JB.00693-07.

Segarra, A., Pépin, J. F., Arzul, I., Morga, B., Faury, N., & Renault, T. (2010). Detection and description of a particular Ostreid herpesvirus 1 genotype associated with massive mortality outbreaks of Pacific oysters, *Crassostrea gigas*, in France in 2008. *Virus Research, 153*(1), 92–99. doi:10.1016/j.virusres.2010.07.011.

Soni, J., Sinha, S., & Pandey, R. (2024). Understanding bacterial pathogenicity: A closer look at the journey of harmful microbes. *Frontiers in Microbiology, 15*, 1370818. doi:10.3389/fmicb.2024.1370818.

Stentiford, G. D., Ramilo, A., Abollo, E., Kerr, R., Bateman, K. S., Feist, S. W., … Villalba, A. (2017). *Hyperspora aquatica* n.gn., n.sp. (Microsporidia), hyperparasitic in *Marteilia cochillia* (Paramyxida), is closely related to crustacean-infecting microspordian taxa. *Parasitology, 144*(2), 186–199. doi:10.1017/S0031182016001633.

Subramaniam, T., Hong, H. K., Park, K. I., Cho, Y. G., & Choi, K. S. (2025). Effects of *Perkinsus olseni* parasitism and environmental conditions on the gonad maturation and reproductive effort of female Manila clam (*Ruditapes philippinarum*) on the west and south coasts of Korea. *Journal of Invertebrate Pathology, 212*, 108385. doi:10.1016/j.jip.2025.108385.

Tan, J., Lancaster, M., Hyatt, A., van Driel, R., Wong, F., & Warner, S. (2008). Purification of a herpes-like virus from abalone (*Haliotis* spp.) with ganglioneuritis and detection by transmission electron microscopy. *Journal of Virological Methods, 149*(2), 338–341. doi:10.1016/j.jviromet.2007.12.019.

Tan, L., Liu, Y., Sun, Y., Liu, S., Lin, Z., & Xue, Q. (2025). Copper only SOD repeat proteins likely act as an extracellular superoxide dismutase in oyster antioxidant defense. *Scientific Reports, 15*(1), 20465. doi:10.1038/s41598-025-06156-w.

Thompson, F. L., Iida, T., & Swings, J. (2004). Biodiversity of vibrios. *Microbiology and Molecular Biology Reviews, 68*(3), 403–431. doi:10.1128/MMBR.68.3.403-431.2004.

Tidy, A. M., Jessop, R., Ward, G. M., Green, M. J., Bateman, K. S., Bass, D., … Hooper, C. (2025). Characterisation of *Marteilia cocosarum* in the Wash Estuary, UK, linked to mass mortalities of cockles (*Cerastoderma edule*), and its relationship to closely related species. *bioRxiv*. doi:10.1101/2025.02.13.638039.

Travers, M. A., Boettcher Miller, K., Roque, A., & Friedman, C. S. (2015). Bacterial diseases in marine bivalves. *Journal of Invertebrate Pathology, 131*, 11–31. doi:10.1016/j.jip.2015.07.010.

Travers, M. A., Tourbiez, D., Parizadeh, L., Haffner, P., Kozic-Djellouli, A., Aboubaker, M., … Lupo, C. (2017). Several strains, one disease: Experimental investigation of *Vibrio aestuarianus* infection parameters in the Pacific oyster, *Crassostrea gigas*. *Veterinary Research, 48*(1), 32. doi:10.1186/s13567-017-0438-1.

Umeda, K., Yang, X., Waki, T., Yoshinaga, T., & Itoh, N. (2020). The effects of environmental and nutritional conditions on the development of *Perkinsus olseni* prezoosporangia. *Experimental Parasitology, 209*, 107827. doi:10.1016/j.exppara.2019.107827.

Vandeputte, M., Kashem, M. A., Bossier, P., & Vanrompay, D. (2024). Vibrio pathogens and their toxins in aquaculture: A comprehensive review. *Reviews in Aquaculture, 16*(4), 1858–1878. doi:10.1111/raq.12926.

Vera, M., Pardo, B. G., Cao, A., Vilas, R., Fernández, C., Blanco, A., … Martínez, P. (2019). Signatures of selection for bonamiosis resistance in European flat oyster (*Ostrea edulis*): New genomic tools for breeding programs and management of natural resources. *Evolutionary Applications, 12*(9), 1781–1796. doi:10.1111/eva.12832.

Villalba, A., Reece, K. S., Ordás, M. C., Casas, S. M., & Figueras, A. (2004). Perkinsosis in molluscs: A review. *Aquatic Living Resources, 17*(4), 411–432. doi:10.1051/alr:2004050.

Villalba, A., Coimbra, R. M., Pampín, M., Iglesias, D., Costas, D., Mariño, C., … Martínez, P. (2023). A common garden experiment supports a genetic component underlying the increased resilience of common cockle (*Cerastoderma edule*) to the parasite *Marteilia cochillia*. *Evolutionary Applications, 16*(11), 1789–1804. doi:10.1111/eva.13601.

Waki, T., Takahashi, M., Eki, T., Hiasa, M., Umeda, K., Karakawa, N., & Yoshinaga, T. (2018). Impact of *Perkinsus olseni* infection on a wild population of Manila clam *Ruditapes philippinarum* in Ariake Bay, Japan. *Journal of Invertebrate Pathology, 153*, 134–144. doi:10.1016/j.jip.2018.03.001.

Wang, R., Liu, X., Wang, J., Huizhu, X. L., Sun, J., Mou, H., … Ma, X. (2022). Proteomic differences between *Vibrio tubiashii* strains with high- or low-virulence levels isolated from diseased ivory snail *Babylonia areolata*. *Aquaculture International, 30*(5), 2579–2591. doi:10.1007/s10499-022-00919-2.

Wang, Y., Yoshinaga, T., & Itoh, N. (2018). New insights into the entrance of *Perkinsus olseni* in the Manila clam, *Ruditapes philippinarum*. *Journal of Invertebrate Pathology, 153*, 117–121. doi:10.1016/j.jip.2018.03.005.

Wang, Z., Casas, S., Peyre, J. L., Rikard, S., Williams, M. L., Tarnecki, A., … Guo, X. (2025). Genomic selection for dermo resistance in the Eastern Oyster *Crassostrea virginica*: Production and laboratory testing of F1 generation. *Journal of Shellfish Research, 44*(1), 75–87. doi:10.2983/035.044.0108.

Wang, Z, Lu, X, Liang, Y, & Wang, C (2010). *Haplosporidium nelsoni* and *H. costale* in the Pacific oyster *Crassostrea gigas* from China's coasts. *Diseases of Aquatic Organisms, 89*(3), 223–228. doi:10.3354/dao02196.

Wang, Z., Lu, X., & Liang, Y. (2012). *Haplosporidium nelsoni* (MSX) in Japanese scallops *Patinopecten yessoensis* (Jay, 1857) from Dalian along the northern coast of the Yellow Sea, China. *Parasitology Research, 110*(4), 1445–1451. doi:10.1007/s00436-011-2647-2.

Warrier, A., Satyamoorthy, K., & Murali, T. S. (2021). Quorum-sensing regulation of virulence factors in bacterial biofilm. *Future Microbiology, 16*(13), 1003–1021. doi:10.2217/fmb-2020-0301.

Waters, E. V., Cameron, S. K., Langridge, G. C., & Preston, A. (2025). Bacterial genome structural variation: Prevalence, mechanisms, and consequences. *Trends in Microbiology, 33*(8), 875–886. doi:10.1016/j.tim.2025.04.004.

Weatherup, E. F., Carnegie, R., Strand, A. E., & Sotka, E. E. (2024). Co-phylogeographic structure in a disease-causing parasite and its oyster host. *Parasitology, 151*(7), 671–678. doi:10.1017/S0031182024000611.

Wei, M., Li, Y., Wang, J., Ma, C., Kang, H., Li, P., … Bai, C. (2025). Mechanisms of HAHV-1 Interaction with Hemocytes in *Haliotis diversicolor supertexta*: An *in vitro* study. *Biology, 14*(2), 121. doi:10.3390/biology14020121.

Whittington, R. J., Liu, O., Hick, P. M., Dhand, N., & Rubio, A. (2019). Long-term temporal and spatial patterns of *Ostreid herpesvirus 1* (OsHV-1) infection and mortality in sentinel Pacific oyster spat (*Crassostrea gigas*) inform farm management. *Aquaculture, 513*, 734395. doi:10.1016/j.aquaculture.2019.734395.

Winkler, F. M., García, R., Valdivia, M. V., & Lohrmann, K. B. (2018). Assessment of oxytetracycline baths as therapeutic treatment for the control of the agent of withering syndrome (WS) in red abalone (*Haliotis rufescens*). *Journal of Invertebrate Pathology, 153*, 109–116. doi:10.1016/j.jip.2018.02.019.

Worden, P. J., Bogema, D. R., Micallef, M. L., Go, J., Deutscher, A. T., Labbate, M., … Jenkins, C. (2022). Phylogenomic diversity of *Vibrio* species and other Gammaproteobacteria isolated from Pacific oysters (*Crassostrea gigas*) during a summer mortality outbreak. *Microbial Genomics, 8*(12), mgen000883. doi:10.1099/mgen.0.000883.

Xia, J., Bai, C., Wang, C., Song, X., & Huang, J. (2015). Complete genome sequence of *Ostreid herpesvirus-1* associated with mortalities of *Scapharca broughtonii* broodstocks Herpes viruses. *Virology Journal, 12*(1), 110. doi:10.1186/s12985-015-0334-0.

Xue, Q., Waldrop, G. L., Schey, K. L., Itoh, N., Ogawa, M., Cooper, R. K., … La Peyre, J. F. (2006). A novel slow-tight binding serine protease inhibitor from eastern oyster (*Crassostrea virginica*) plasma inhibits perkinsin, the major extracellular protease of the oyster protozoan parasite *Perkinsus marinus*. *Comparative Biochemistry and Physiology Part B: Biochemistry and Molecular Biology, 145*(1), 16–26. doi:10.1016/j.cbpb.2006.05.010.

Yang, B., Zhai, S., Li, X., Tian, J., Li, Q., Shan, H., & Liu, S. (2021). Identification of *Vibrio alginolyticus* as a causative pathogen associated with mass summer mortality of the Pacific Oyster (*Crassostrea gigas*) in China. *Aquaculture, 535*, 736363. doi:10.1016/j.aquaculture.2021.736363.

Yisrael, L., DiMaria, R. A., Santos-Ciminera, P., Aguirre-Macedo, M. L., Vidal-Martínez, V. M., Ruiz, G. M., … Pagenkopp Lohan, K. M. (2024). Expanding the phylogeography and connectivity of *Perkinsus* species across North and Central America. *Diversity and Distributions, 30*(3), e13807. doi:10.1111/ddi.13807.

Yue, X., Liu, B., Xiang, J., & Jia, J. (2010). Identification and characterization of the pathogenic effect of a *Vibrio parahaemolyticus*-related bacterium isolated from clam *Meretrix meretrix* with mass mortality. *Journal of Invertebrate Pathology, 103*(2), 109–115. doi:10.1016/j.jip.2009.11.008.

Zhang, B., Qiu, Y., Shi, C., & Zhang, J. (2025a). Development of multiple real-time fluorescent quantitative PCR for *vibrio* pathogen detection in aquaculture. *Veterinary Sciences, 12*(4), 327. doi:10.3390/vetsci12040327.

Zhang, H., Zou, X., Hu, J., Liu, S., Fan, C., Dai, W., Lin, Z., & Xue, Q. (2025b). Genomic insights into mechanism underlying virulence variations between *Vibrio mediterranei* strains different in pathogenicity toward bivalves. *Aquaculture, 605*, 742524. doi:10.1016/j.aquaculture.2025.742524.

Zhang, X., Kang, H. G., Sun, T. Q., Huang, B. W., Wei, M. L., Wang, C. M., & Bai, C. M. (2025c). Pathogenic and genomic characterization of Ostreid herpesvirus 1 associated with mass mortalities of blood cockle (*Anadara kagoshimensis*). *Aquaculture, 595*. doi:10.1016/j.aquaculture.2024.741596.

CHAPTER

21

Carrying capacity assessment

Zengjie Jiang[a], Fan Lin[a], Antonio Agüera[b] and Jianguang Fang[a]

[a]Yellow Sea Fisheries Research Institute, Chinese Academy of Fishery Sciences, Qingdao, China [b]Benthic Ecology Research Group, Institute of Marine Research, Bergen, Norway

21.1 Carrying capacity in shellfish aquaculture

21.1.1 Development of the carrying capacity concept in aquaculture

The concept of carrying capacity (CC) has its roots in the broader field of ecology and resource management, and it was first introduced in the early 20th century by Aldo Leopold (Leopold, 1933) to describe the maximum population size of a species that a given habitat could support indefinitely. This early definition focused primarily on the availability of resources such as food and space. Leopold's work laid the foundation for understanding the limits of ecosystems and the need for sustainable management practices.

Over time, the concept of CC in aquaculture evolved to include a broader range of environmental and ecological factors. Researchers began to recognize that CC was not just about resource availability but also about the ability of the ecosystem to process waste products and maintain overall health. This led to the development of more sophisticated models and assessment methods that incorporated multiple parameters such as water quality, nutrient cycling, and habitat integrity.

The demand for exploring CC in aquaculture emerged in the middle-20th century as the industry began to expand rapidly. Early studies focused on the impacts of fish and shellfish farming on local ecosystems, particularly in terms of nutrient loading and benthic impacts. Since the 1960s, the application of CC in aquaculture began to take shape, with early studies focusing on the environmental and production limits of aquaculture developments. These studies highlighted the impacts of aquaculture on local ecosystems, particularly nutrient loading and benthic impacts (Dame, 1996). Over time, the understanding of CC in aquaculture has broadened to include a more comprehensive range of environmental and ecological factors, leading to an expanding system of sophisticated models and assessment methods.

Genetic Breeding and Aquaculture of Shellfish: A Collaborative Approach to a Sustainable Future
DOI: https://doi.org/10.1016/B978-0-443-33606-5.00001-1

21.1.2 Evolution of the methodology of carrying capacity assessment in shellfish aquaculture

The global shellfish aquaculture has experienced steady expansion in recent decades. FAO (2024) statistics indicate worldwide production reached 18.9 million metric tons in 2022, representing 20% of total aquaculture output. Shellfish aquaculture is defined by its filter-feeding nature, requiring no artificial feed as it directly extracts plankton and organic detritus from the water column. This generates significant ecosystem services beyond commercial production: water purification (removing suspended solids and nutrients), carbon sequestration (biodeposition and interaction with the planktonic community), and habitat creation ("oyster reefs"). However, its primary environmental impact centers on the artificially manipulated attributes: Manual stocking and harvesting processes will regulate the shellfish biomass in the ecosystem, intensively distributed culture facilities may alter the local current and accelerate the local biodeposition, potentially impose pressure on the benthic condition, and repeatedly shellfish aquaculture may trigger ecological community shifts and biodiversity loss. Management must balance these substantial services against localized ecological risks in the environment. Consequently, CC assessment has become essential for the sustainable operations of shellfish aquaculture.

Today, the concept of CC assessment in aquaculture is multifaceted, encompassing physical, ecological, production, economic, and even social dimensions. The requirement for sustainable aquaculture practices that balance ecological health with economic benefits has driven the development of advanced assessment tools and methodologies. Given shellfish's prominence in global marine aquaculture, they have remained a primary focus in CC methodology development since the concept's inception.

21.1.2.1 Mass balance methods and indexes

The early-stage methods started to focus on the environmental and production limits of aquaculture developments and helped to assess the potential impacts of shellfish farming on the benthic environment. The methodologies and insights from the works of Richard F. Dame and Peter M. J. Herman have significantly contributed to the understanding and estimation of CC in shellfish aquaculture (Dame, 1996; Heip et al., 1995; Herman, 1993). Their research, particularly on benthic-pelagic coupling and the role of turnover times, provides a robust framework for assessing the sustainability of shellfish populations in coastal and estuarine ecosystems. Most of these methods are mass balance based, which involve quantifying the sources and sinks of key parameters (usually the organic carbon) within the aquaculture system. Dame and Prins used a combination of empirical data analysis and modeling to estimate the CC of suspension-feeding bivalves in 11 coastal and estuarine ecosystems (Dame & Prins, 1997). Their approach focused on three key turnover times: water residence time (RT), primary production time (PPT), and bivalve clearance time (CT). These turnover times have become the foundation of constructing indexes, such as the Benthic Index (Dame/Herman), for estimating the environmental conditions and CC for shellfish. These indexes measure changes in benthic biomass and species diversity, providing a quantitative assessment of the health of benthic habitats (Heip et al., 1995).

Methods or indexes based on the mass balance have been widely used in aquaculture to assess the CC of shellfish aquaculture. For instance, Fang et al. (1996) were the first to study the CC of cultured scallop *Chlamys farreri* based on the estimation of individual consumption, local primary production, and estimated water exchange in Sanggou Bay, which is a pioneering work for CC assessment in the Chinese aquaculture scenario. In the Chesapeake Bay, mass balance methods were used to evaluate the impact of oyster farming on nutrient cycling. The results showed that oyster farms could significantly reduce nutrient loads in the water column, contributing to improved water quality (Newell et al., 2005).

Mass balance methods and indexes are the early developed tools for CC assessment of shellfish aquaculture. They provide a scientifically rigorous approach to understanding the ecological impacts and potential for growth in aquaculture systems based on a large-scale overview. However, the assumptions and the simplification of the dynamic ecosystems processes of the mass balance method have limited the application of the estimated CC, which is usually represented with biomass per unit area, being a spatial-temporal averaged indicator. In actual aquaculture practice, stakeholders or policymakers increasingly demand details of the interactions between aquaculture and the environment, thus seeking optimal productive efficiency and sustainability. This has, in turn, driven researchers to focus on the ecosystem processes of aquaculture under different scenarios of aquaculture layouts, stocking density, and the combination of culture species. Thereby promoting the further development of shellfish aquaculture CC assessment methods. Despite this, mass balance methods and indexes remain a crucial component in the sustainable management of shellfish aquaculture, providing insights that can inform general policy and management decisions.

21.1.2.2 Numerical modeling methods

As computer technology advanced, numerical models became an increasingly important tool in aquaculture CC assessment. Especially, the incorporation of hydrodynamic models brings in high temporal-spatial resolution information of the physical environment and fills the gaps in the observations. This enables researchers to study the interaction between the cultured shellfish population and the environment at selected temporal and spatial scales, and to further improve the CC assessment methodology.

Early applications of numerical models in CC assessments focused on exploring the impacts of aquaculture facilities on environmental processes in the farming area, thereby incorporating the temporal dynamics within the farming process. Jon Grant and Cedric Bacher (Grant and Bacher, 2001) have implemented a finite element circulation model to study the current modification from the culture structures for bivalves and seaweed in Sanggou Bay, which incorporates the effect of local hydrodynamics. The simulation concluded with a 41% decrease in the water exchange due to the presence of the aquaculture facilities, which may lead to an overestimation of particle renewal time, and as a result, the CC might be overestimated. Duarte et al. (2003) developed a 2D coupled physical–biogeochemical model that integrates hydrodynamics with ecological processes to simulate the interactions between the intensively cultured shellfish, kelp, and the local environment. Their model suggests that optimizing the spatial distribution and density of cultured species could significantly enhance production and highlight the importance of spatially resolved modeling in understanding and managing complex aquaculture systems sustainably.

One of the most influential works on the CC assessment of shellfish aquaculture with modeling tools was conducted by Inglis (Inglis et al., 2000) in New Zealand. The research group at NIWA has expressed a well-defined modeling framework for CC assessment: Interpreting the ecosystem process is fundamental for CC estimation, and the application of the CC results is contingent upon there being clear objectives to be achieved in the condition of the population

TABLE 21.1 Four types of carrying capacity defined by Inglis et al. (2000).

	Definition	Scale
Physical carrying capacity	Limits set by the physical space and conditions required for marine farms (size, situation, water depth, etc.)	Subembayment→Planning area
Production carrying capacity	The sustainable stocking density at which production levels are maximized	Farm → embayment → adjacent embayments → region
Ecological carrying capacity	Levels at which farm development causes significant changes in the ecosystem	Farm→embayment→region
Social carrying capacity	Levels at which farm development impairs or conflicts with other human uses	Embayment→region

or the resource (Inglis et al., 2000). They have defined four types of CC and the spatial scales at which they are most relevant, as presented in Table 21.1.

They have proposed a general technical roadmap for the development of the methodology on CC assessment: to enhance the spatial resolution of models to address smaller-scale processes related to individual farms and their cumulative impacts. This includes improving the representation of hydrodynamic processes and ecosystem interactions within the model concept. In addition, long-term monitoring and adaptive management are suggested to identify environmental changes associated with increased farming intensity at different scales. The development of indicators for regional ecosystem health and the use of historical data to establish baselines are also recommended. Research should also explore the social and cultural impacts of mussel farming through surveys and stakeholder engagement. Overall, future work should aim to integrate these diverse aspects into a cohesive approach that balances economic benefits with environmental sustainability and social acceptability.

Subsequent research on CC assessment has fallen within the methodological framework of Inglis (Inglis et al., 2000). Jon has studied the CC of suspended oyster aquaculture with an ecosystem model in Québec, Canada (Grant et al., 2007). An oyster individual growth model was integrated into the ecosystem model, and hydrodynamic model results were used to compute water exchange at the box boundaries. Guyondet et al. (2010) have coupled the individual growth model based on the dynamic energy budget (DEB) of cultured Pacific oysters to a nested fine resolution physical-biogeochemical model in the Gulf of Saint-Lawrence to study the interaction of oyster farms with the local coastal ecosystem. Shi et al. (2011) developed a three-dimensional physical-biological coupled model for Sanggou Bay, with the parameterized surface drag from culture structures. The kelp production, nitrogen budget, and current field were simulated for different culture densities to seek the optimum values for sustainable development. Ross and Ren have developed an ecosystem model for IMTA with multiple biological modules, including finfish, seaweed, shellfish, phytoplankton, zooplankton, carnivores, and benthic detritivores in Waihinau Bay and Pelorus Sound, New Zealand. Box models simplify the aquaculture system into discrete compartments to simulate the movement and transformation of substances (Ren et al., 2012; Ross et al., 2013).

In the above-mentioned studies, hydrodynamic models with finer temporal-spatial resolution were constructed and validated to provide better physical background information. Moreover, individual-based models (IBMs) simulate the growth, feeding, and reproductive behavior of shellfish individuals emerged as a significant advancement in the field. By coupling IBMs with ecosystem models, researchers can predict how the population level of the cultured shellfish, usually represented by culture density, affects the overall ecosystem. These numerical models above parameterized the interactions between cultured organisms and the ecosystem components and predicted individual growth and population state of targeted biological groups, as well as the significant environmental variables concerned. Ecosystem models often have such regional characteristics as the natural environment, ecosystem compositions, aquaculture facilities, and organism bio-physiology. Specific models should be implemented from general model frameworks with detailed localization, calibration, and validation. Then the results can provide suggestions for aquaculture management and details in environmental monitoring or physiological research.

21.1.2.3 *Incorporating social dimensions*

The previous methodologies mainly focus on the physical, production, and ecological CC. The social CC in aquaculture has attracted increasing research attention in the last 15 years, but it remains the least well-developed and understood component of CC (Weitzman & Filgueira, 2020). A major reason maybe is challenging to provide a universally agreed-upon definition for the social CC in aquaculture. It generally refers to the level of aquaculture development that can be supported without adverse social effects, but the specific criteria for "adverse effects" are not clearly defined. And the social CC assessments often focus on measuring stakeholder perceptions and conflicts. There are studies that have measured stakeholder attitudes toward different levels of aquaculture development, but these perceptions can vary widely among different groups (e.g., local communities, industry stakeholders, and environmental groups). Dalton uses normative evaluations to plan for and manage shellfish aquaculture development, focusing on stakeholder perceptions and social CC (Dalton et al., 2017). Kluger examines the use of media analysis to scope priorities in social CC assessments, providing a global perspective (Kluger et al., 2019). Ferreira derived four dimensions of CC for prospective fish aquaculture in African lakes and marine waters, with social acceptability gauged through prevailing legislative and management criteria (Ferreira, 2025). Pei et al. (2025) quantified the green-development performance of China's marine molluscan shellfish aquaculture; its social dimension was captured by three proxy indicators—employment generated by the sector, its contribution to food security, and the income uplift experienced by aquaculture practitioners. It is complex to assess due to the need to balance multiple stakeholders' interests and varieties of values. To develop robust indicators and frameworks for social CC, as well as the involvement of stakeholder/government engagement to define relevant thresholds, remains to be the major research task for integrating SCC into broader ecosystem-based management methodologies covering most aspects of the CC assessment of shellfish aquaculture. This requires interdisciplinary approaches and alignment with policy needs and is essential to provide scientific rigor and practical information to assist aquaculture decision-making.

21.2 Case studies of carrying capacity application

The primary focus of this chapter is on practical methods of CC assessment that can be directly applied in actual aquaculture production. These methods should be user-friendly and provide actionable insights for both industry stakeholders and government regulators. Practical CC assessment tools should integrate scientific rigor with operational feasibility, ensuring that they can be implemented with minimal disruption to existing aquaculture practices. For example, decision-support systems (DSS), which usually integrate environmental, biological, and social, and marketing information into logical system design, can help managers make informed decisions based on real-time data and predictive models (Chary et al., 2022; You et al., 2022).

CC assessment supports diverse management objectives, including physical, production, ecological, and social goals. While much of the research has focused on ecological and production capacities, recent studies have emphasized the need for a more holistic approach that considers the complex interactions between aquaculture and the broader ecosystem (Ross et al., 2013). The importance of CC in aquaculture management cannot be overstated. By providing a scientific basis for setting stocking densities and harvesting limits, CC assessments facilitate trade-off analysis between ecological and economic goals. This is crucial for the long-term viability of the aquaculture industry and the health of marine ecosystems. CC assessments have been successfully integrated into regulatory frameworks and industry guidelines in many countries, informing sustainable farming practices and leading to improved environmental conditions and stable production levels (Ferreira et al., 2013; Filgueira et al., 2014).

However, challenges remain in accurately predicting long-term impacts and ensuring compliance with CC-based regulations. Environmental variability and the complexity of ecosystem interactions can make it difficult to establish precise CC limits. In addition, socioeconomic factors, such as market demand and labor availability, can influence the feasibility of implementing CC-based management practices. Future research should focus on developing more robust and adaptable CC assessment methods that can account for these complexities and uncertainties (Weitzman & Filgueira, 2020). In this section, recent case studies on practical CC assessment with the latest technologies are summarized and discussed.

21.2.1 Canada: Shellfish production in Prince Edward Island and New Brunswick

Prince Edward Island (PEI) is Canada's leading shellfish producer. PEI produces 80% of Canadian mussels, equating to 20,000 tons. Oysters also play a significant role, with 3,000 tons harvested annually. Shellfish aquaculture in PEI has an important role in the economy of the province (Fisheries & Oceans Canada (DFO), 2024; Statistics Canada, 2023). An area of about 78 km^2 is dedicated to shellfish cultivation in PEI, divided in leases within small bays and estuaries around the island. Generally these bays are shallow, heavily influenced by a strong tidal exchange with the Gulf of St. Lawrence waters and different levels of riverine input with high summer phytoplankton availability. Mussels are cultivated in suspended lines, and oysters in both suspended and bottom systems. Concerns about the ecological impacts of shellfish production and uncertainty on CC prompted a moratorium in PEI for new shellfish production leases in the year 2000. This situation has prompted several assessments of CC being undertaken in some bays using different methods over the years. This is the case of Malpeque Bay, where a comprehensive assessment of CC overviewed the application of different methods of increasing complexity in the same system. Following that aim of a case study which compares different approaches with different complexity, we have added here the work done in the Richibucto estuary, NB. Although not in PEI, the Richibucto estuary has a significant oyster production and is especially interesting because of the application of advanced modeling studies and ecological complexity.

CC assessments in PEI and NB have evolved during the last 25 years from simple empirical and box-model approaches to spatially explicit couple hydrodynamic–biogeochemical–physiological models. The government of PEI runs a routine monitoring program that undertakes baseline surveys for the condition of the shellfish being produced, recruitment, physiological responses, and biomass produced across the leases (PEI Mussel Monitoring Program, 2024). This regular monitoring provides key background information such as production metrics, or performance of cultivated stocks, and is useful to detect when something goes wrong, and by itself has little predictive capacity and is not useful to base decisions on planning production at bay scale. However, the information collected in these surveys has been key in the validation and development of CC models (PEI Mussel Monitoring Program, 2024).

Malpeque Bay (PEI) is the second-largest bay in PEI, with about 204 km^2. It is very shallow (mean depth of 4m), and it is heavily influenced by the tidal exchange with the Gulf of St. Lawrence and fluvial inputs, but it is sheltered by a sand bar with several inlets. Malpeque Bay supports a large-scale and economically critical mussel industry (25% of PEI) and a growing oyster cultivation, yet it still faces a moratorium on new leases. Filgueira et al. (2015) approached the study of Malpeque Bay CC using different modeling approaches, with an increasing level of complexity, and analyzed the pros and cons of each approach.

At first approach, a box model was implemented, representing the area under study by a series of well-mixed boxes in which nutrient supply, phytoplankton production, and bivalves' feeding could be summed together. This approach is fast, transparent, and useful when data are limited; however, it lacks spatial resolution, and therefore it ignores spatial heterogeneity. The box model developed for Malpeque Bay had trouble assessing residence times, and therefore it could not identify areas and effects of localized depletion. Moreover is a lack of connectivity between leases. However, this study proved useful to inform posterior work with more complex models by identifying the areas of the bay where this need for complexity was warranted.

A 3D hydrodynamic + NPZD ecosystem model was the next step in the work by Filgueira (Filgueira et al., 2015). A circulation model forced by tides, wind, and riverine input was coupled with a biogeochemical module (NPZD) to simulate nutrient fields, primary production, and detritus pathways. This model was coupled with the feeding activity of the cultivated populations. The model approach was capable of capturing the transport of phytoplankton and detritus, and hence the connectivity between the different leases, the residence time, and the spatial patterns of phytoplankton depletion. The results of this study suggested that the load on the bay was not challenging the CC of the system for shellfish cultivation; moreover, the results suggested that almost doubling the area dedicated to shellfish production would not exceed the bay's ecological capacity. This new scenario will see that chlorophyll in the bay will decrease by ~18% and may produce a localized impact on mussel growth by 8%, which are within acceptable ranges. These results have prompted discussion about lifting the moratorium.

Richibucto estuary (NB): Located on the eastern coast of NB, it is a small (11.8 km^2) and shallow, mesotidal estuary influenced by both fluvial inputs from the Richibucto River and strong tidal exchange with the Northumberland Strait. The system supports productive shellfish grounds, particularly for Eastern oyster, alongside smaller-scale mussel culture. Its protected waters and high summer phytoplankton availability make it a priority site for aquaculture expansion in the province. Moreover, new leases are not subject to a moratorium, and expansion of activity is informed by available research on CC (https://www2.gnb.ca). Shellfish aquaculture in Richibucto is a cornerstone of the local rural economy. Oysters from the estuary supply both domestic markets and high-value export channels. The sector provides year-round employment for farm operators, seasonal processing jobs, and indirect benefits to service industries. The estuary's cultural heritage—rooted in both Acadian and Mi'kmaq communities—has long been intertwined with oyster harvesting (Aquaculture Industry Overview, 2022; Comea, 2013)

Richibucto estuary CC assessment was undertaken by Filgueira (Filgueira et al., 2014). The area was modeled using a fully coupled spatial 3D ecosystem model. Similar to the model applied in Malpeque, this model integrates a circulation model forced by tides, wind, and river input with an NPZD biogeochemical model that simulates nutrients, phytoplankton, zooplankton, and detritus dynamics. The main difference with the previous model is in how the cultivated stock is represented in the model; here, a DEB model that describes oyster growth, reproduction, and waste production under site-specific environmental conditions of resource availability and temperature is coupled to the model. More specifically, this DEB model is embedded within the NPZD and enables feedback between the oysters and the phytoplankton. Increasing the detail in the behavioral and physiological responses of the stock allows for the construction of highly realistic scenarios that could be tested to inform planning and analyze changes in spatial patterns. The Richibucto estuary CC assessment found that water residency is the main factor determining CC, with longer RT increasing the potential formation of depletion areas with negative effects on oyster growth. At the same time, the model allowed researchers to identify areas with high potential in the inner estuary where depletion thresholds are still within a safe margin. Still, scaling up oyster production could risk problems of food limitation in late summer when primary production decreases.

These case studies in eastern Canada can be used to draw some conclusions on the assessment of CC for shellfish aquaculture. First, how important is the local hydrodynamics for shellfish aquaculture and henceforth, how important is any intervention that causes changes in the dynamics of the system. Following this, the connectivity between the different leases or production plots is key and needed to be able to inform bay-scale planning to avoid localized overloading. CC happens at different scales. Which directly relates to the tradeoff between farm density and individual growth, increasing the density of the production may look like a way to boost production; however, it will be in detriment of individual growth and product quality. The assessments in Canada inform that complex models are best suited, with the needed spatial and temporal resolutions, and should be implemented in high-priority areas, such as Malpeque. Still, these models should not be used to the detriment of monitoring, which can identify changes in the system and be used for constant validation and recalibration of model approaches.

21.2.2 China: Shellfish production in Sanggou Bay

Sanggou Bay is a significant aquaculture region known for its large-scale integrated low-trophic-based aquaculture activities. It is a shallow semi-enclosed bay located at the east end of the Shandong Peninsula (37°01′-09′N, 122°24′-35′E) with an area of about 144 km^2. The favorable natural conditions, including calm waters, moderate climate, and nutrient-rich environment, support the intensive aquaculture of seaweed (kelp), shellfish (Pacific oysters, abalones), and sea cucumbers. Nearly two-thirds of the bay area have been occupied by bivalve and seaweed aquaculture, the annual production of kelp is estimated to be up to 84,500 tons in dry weight (Mao et al., 2018), and the annual production of shellfish is approximately 230,000 tons in wet weight (Shandong Province Fishery Statistical Yearbook, 2023). Aquaculture contributes significantly to the local economy by providing employment opportunities and generating substantial revenue, and is characterized by its commitment to sustainability, employing modern techniques and technologies to minimize environmental impact.

Sanggou Bay has long been a spot of scientific research regarding CC assessment since the 1990s. Fang and Zhang were the first to study the CC of cultured kelp with the nitrogen budget and estimated water exchange in Sanggou Bay (Fang & Zhang, 1996). The capacity was estimated to be about 54,000 MT in dry weight. The hydrodynamics in the bay with the effect of intensive longline rafts were also studied by Jon Grant and Cedric Bacher with a finite element circulation model (Grant and Bacher, 2001). The results concluded with a 41% decrease in the water exchange with the presence of the aquaculture facilities, which may lead to an overestimation of particle renewal time, thus overestimating the CC. Shi et al. (2011) have developed a three-dimensional physical-biological coupled model for Sanggou Bay, with parameterized surface drag from culture structures. The kelp production, nitrogen budget, and current field were simulated for different culture densities to seek the optimum values for sustainable development. Lin studied the CC of kelp and oysters with a physical-biological coupled ecosystem model (Fan et al., 2020). Sun et al. (2020) evaluated the growth potential of the same integrated culture system based on a numerical modelling method.

A recent case study for the CC assessment in Sanggou Bay is presented in this part. It is based on the research framework of an ecosystem model coupled with the hydrodynamic module, nutrient-plankton-zooplankton-detritus module, and the individual growth module based on the DEB theory.

Previous hydrodynamic simulations of Sanggou Bay mostly chose the bay entrance as the boundary and applied tides as the driving force, which may lose the hydrological information outside the bay. With boundary hydrological data of a validated hydrodynamic model established in the Yellow Sea and Bohai Sea (Lin et al., 2019), a hydrodynamic model with 200 m spatial resolution of Sanggou Bay has been developed in this case (Fig. 21.1). The hydrodynamic model was implemented using the Regional Ocean Modeling System (ROMS). Validated results provided hourly current fields, temperature, and salinity data. These outputs cover deep-water areas adjacent to Sanggou Bay, and this spatial coverage establishes an enhanced hydrodynamic background field for biological studies.

Based on the validated hydrodynamic results, the project team has integrated an Nutrient–Phytoplankton–Zooplankton–Detritus (NPZD) ecosystem model and individual growth models for cultured Pacific oysters, *Crassostrea gigas,* and kelp, *Saccharina japonica*. The model calculates the dynamic changes between dissolved inorganic nitrogen, phytoplankton, cultured oysters, kelp, and organic detritus on the hydrodynamic grid. The model incorporates ecological processes of oysters as filter-feeding populations and kelp as nutrient-consuming populations. Oysters are modeled using the individual growth model based on the DEB theory. They are attached to culture longlines at fixed locations in the model grid. The biological functions of oysters include filtering phytoplankton and detritus from the system and excreting nitrogen and feces. Cultured kelp and phytoplankton are modeled based on exponential growth with a maximum daily growth rate. Kelp acts as the primary dissolved inorganic nitrogen (DIN) absorber in the model, taking up inorganic nitrogen from the water column at the

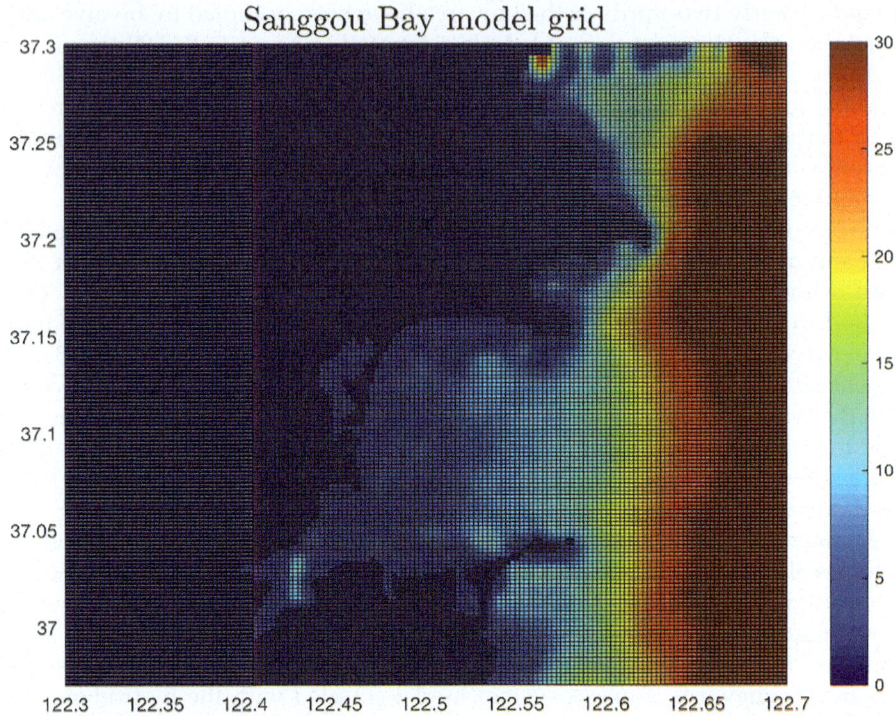

FIGURE 21.1 A ROMS model grid for Sanggou Bay, with a horizontal spatial resolution of about 200 m. *ROMS*, Regional Ocean Modeling System.

computational grid location. Phytoplankton are the main producers of the model, and the variation in phytoplankton biomass also includes exchange between adjacent computational grids. Growth of kelp and phytoplankton is limited by light, temperature, and nitrogen content. The model employs inorganic nitrogen as a currency to link the ecological processes of phytoplankton, kelp, and oysters. The particulate organic matter considered in the model is described by particulate organic carbon (POC) in detritus. POC mainly originates from oyster feces as suspended particles; it can be settled to the bottom or be advected between computational grids. Cultured oysters also feed on POC in the water. And at the end of each cultivation period, the biomass of the cultured organisms is removed from the ecosystem model as a harvest action.

The model determines whether oyster or kelp farming exists at different computational grid points by setting an aquaculture mask, as shown in Fig. 21.2. Oyster farming in Sanggou Bay is mainly concentrated in the central part of the bay entrance, while kelp farming covers a larger area, extending to the outside the bay for better water exchange. At every computational time step, each model grid point exchanges nutrients, phytoplankton, and POC with the surrounding waters and the external waters at the boundaries. When a computational grid point has oyster or kelp present, the corresponding physiological processes are added to complete the calculations of the ecosystem processes.

The ecosystem model was validated and calibrated against various observation data before application. Based on the actual aquaculture production in Sanggou Bay, oyster farming was set

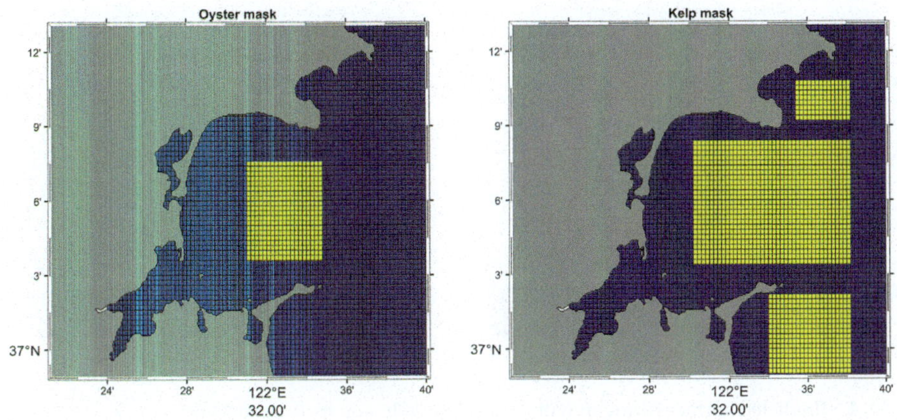

FIGURE 21.2 The aquaculture mask used in the ecosystem model to define whether to activate the computation of organism growth at each model grid.

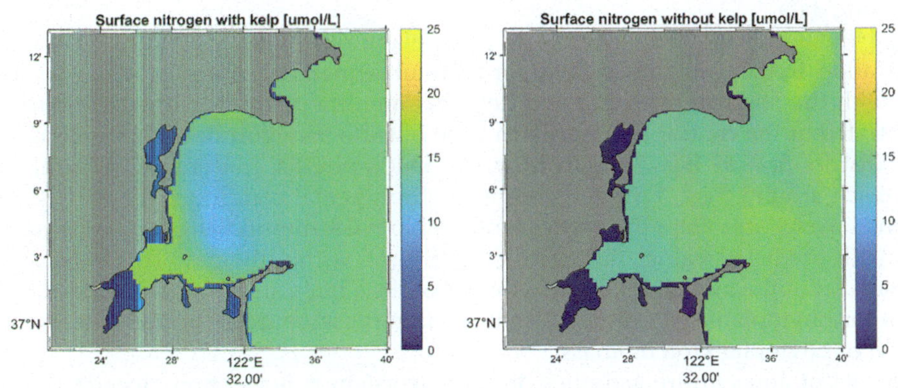

FIGURE 21.3 Predicted surface nitrogen with or without kelp aquaculture in May.

to start in June and last for one year; kelp farming began in November, and harvesting started in May of the following year. By conducting scenario simulation with different culture densities of oysters and kelp, the growth of cultured organisms and the feedback from the local environment are predicted with the model. Initial conditions for the ecosystem model as well as the individual growth model were taken from environmental and biological observations.

Fig. 21.3 shows the model-predicted surface distribution of DIN in the Sanggou Bay area under the scenario of the presence or absence of kelp farming in May. The cultured kelp reached harvest size and is being removed from the model area. The results demonstrated that DIN concentration in the area is significantly related to the kelp biomass. The major DIN source is replenished by the southward Yellow Sea Coastal Current outside the bay area. In the presented scenario, the kelp farm density was set at 5 per square meter. The results demonstrated that the nutrients captured by the growing kelp are not limited to the vicinity of the culture rafts but also extend within a local marine area, and lead to a significant decrease in nutrient levels inside

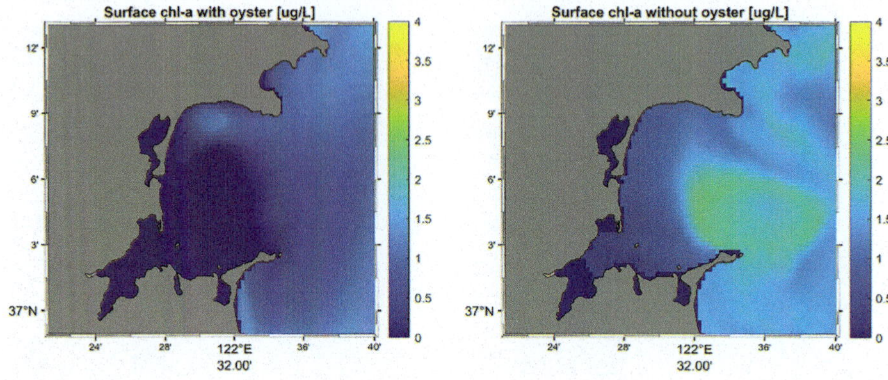

FIGURE 21.4 Predicted surface nitrogen with or without kelp aquaculture in May.

Sanggou Bay. The interaction between farmed organisms and the environment often occurs over certain temporal and spatial scales and needs to be discussed based on continuous observational and modeling data.

Fig. 21.4 Illustration of the distribution of phytoplankton biomass, as indicated by chlorophyll a concentration, in the Sanggou Bay with or without farmed oyster and kelp in May. Under the scenario with the absence of oyster and kelp farming, the predicted phytoplankton biomass is concentrated at the central and northern cape outside Sanggou Bay. This is consistent with the DIN distribution in Fig. 21.3. In May, the chlorophyll-a concentration within Sanggou Bay can reach around 3 μg/L under suitable environmental conditions. Under the projection with farmed oyster and kelp, the phytoplankton biomass around Sanggou Bay is significantly affected, with predicted values falling below 0.5 μg/L within the bay. This is because of two mechanisms set in the model: (1) Farmed kelp absorbs a large amount of DIN, causing nutrient limitation for phytoplankton growth; (2) the feeding pressure from farmed oysters suppresses the growth of phytoplankton biomass.

The impact of aquaculture activities on the ecosystem is not entirely localized. The model results reveal significant spatiotemporal differences in the changes of the aquatic environment. Therefore when assessing the environmental impact of aquaculture layout and planning, comprehensive consideration should be made in combination with observations and model results. Four sets of culture densities (30, 50, 70, and 90 oysters per square meter) were tested in the mariculture ecosystem model. The model results indicated that different culture density scenarios will have considerable variations in individual growth expectations and yield per unit area, as shown in Fig. 21.5A.

When the culture density is at 30 oysters/m^2, the growth of individual oysters is relatively even across the area and in good condition, with the final average wet weight of harvested oysters being around 12 g. When the culture density increases to 50 per square meter, the growth of individual oysters is somewhat affected, with the average wet weight of harvested individuals dropping to around 10 g. When the oyster culture density reaches 70 or 90 per square meter, the growth of oysters is significantly impacted and becomes more reliant on the supply of external phytoplankton. Oysters at the northern entrance of Sanggou Bay always grow better in each scenario, and in the interior of the bay, the individual quality of oysters is limited by the insufficient food supplement.

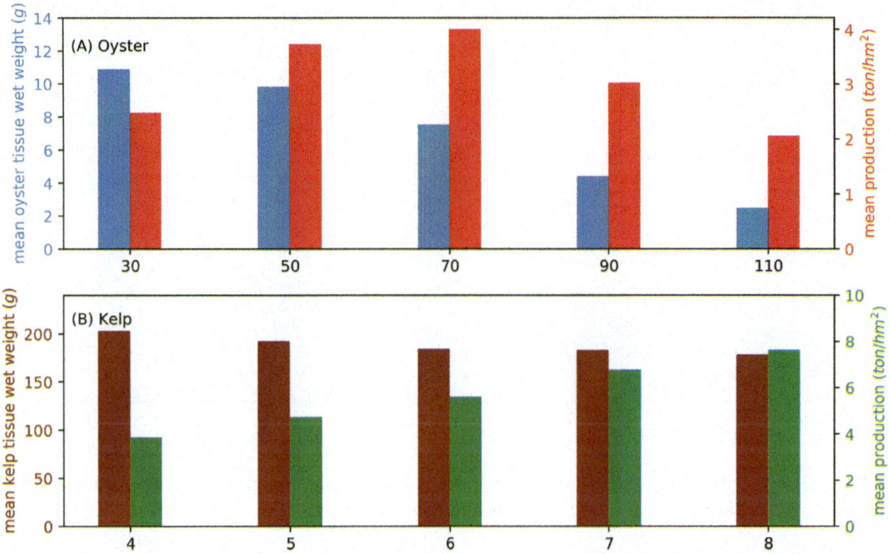

FIGURE 21.5 Model predicted individual growth and yield per unit area: (A) Pacific Oysters and (B) Kelp from the ecosystem model results. *From Fan et al. (2020).*

Statistics of oyster growth expectations under different culture densities from model results are shown in Fig. 21.5A. Under the relatively low culture density scenarios (below 70 ind./m^2), the soft tissue wet weight of harvested oysters is mostly expected to be in the range of 8–12 g. When 90 oysters were farmed together, the soft tissue wet weight of the oysters was expected to be less than 4 g, with only a few well-developed individuals. When the oyster culture density reaches 110 individuals, the expected soft tissue wet weight of most oysters at harvest is less than 2 g, indicating a very poor predicted growth situation. The harvestable oysters (based on soft tissue) can be estimated for each scenario, as shown in Fig. 21.6. There is a tradeoff between individual growth and yield from model predictions, and a balance can be sought for optimization of the aquaculture benefit. At a low density scenario (30–70 ind./m^2), the harvestable quantity (soft tissue weight without shell) is concentrated around 2–6 tons/ha. When the culture density exceeds 90 oysters, the spatial distribution of yield across different regions is significant. In suitable environmental areas, a higher yield (>8 tons/ha) can be achieved, but in most areas, due to the downregulation caused by high oyster biomass, the yield is within 2 tons, and the individual development is poor.

Kelp individual growth and yield predicted under different culture densities are shown in Fig. 21.5B. Whereas the statistics are presented in Fig. 21.7. Similar to the prediction with farmed oysters, kelp also achieves better growth with less seeding density. Individual kelp can reach an expectation of 240 g dry weight when the culture density is 5 kelp/m^2. As the culture density increases, the growth expectation of kelp begins to decline, but the decline is much milder compared to the oyster case. When the culture density exceeds 8 kelp/m^2, the harvestable individuals mostly weigh between 150 and 210 g. The simulated yields at harvest for different culture areas based on the simulation results are shown in Fig. 21.7B. As the culture density of kelp increases, the overall yield increases, resulting from the relatively sufficient supply of

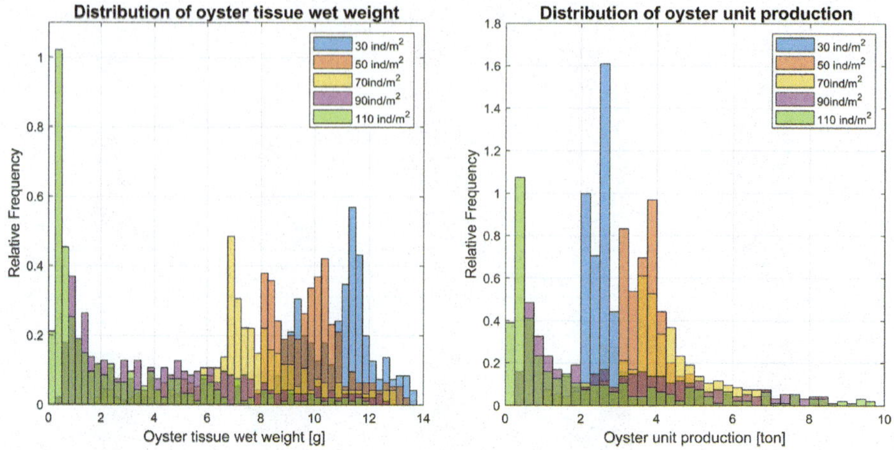

FIGURE 21.6 The distribution of individual oyster soft tissue wet weight at harvest under different culture densities and the corresponding yield of soft tissue wet weight per hectare.

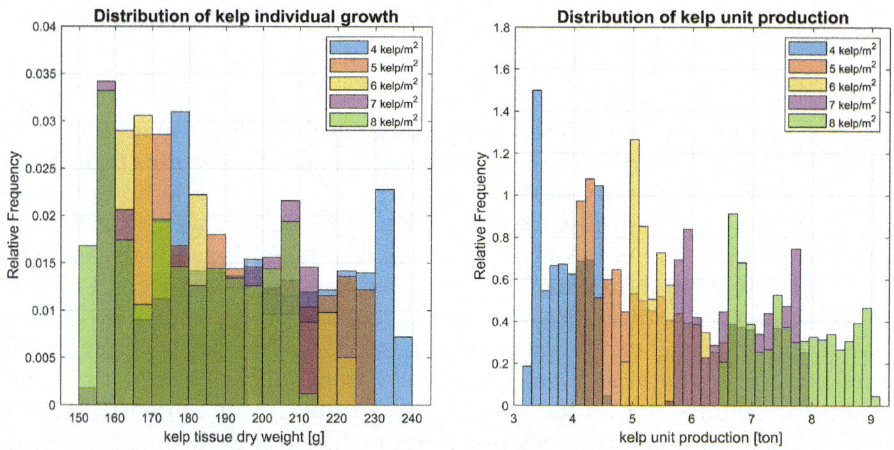

FIGURE 21.7 The distribution of individual kelp dry weight at harvest under different culture densities and the corresponding yield of wet weight per hectare.

nutrients in the model domain. With the increase in the number of harvestable individuals, the yield in most kelp farming areas of Sanggou Bay generally increases. However, an increase in kelp individuals means higher management and harvesting costs in actual production. And the physical space limitation might be the prioritized restrictions on the CC of kelp farming in Sanggou Bay.

Implementation of the ecosystem in Sanggou Bay for the oyster and kelp farming has offered a tool for studying the interaction between the cultured organisms and the environment. As the aquaculture fact in China, a baseline condition without cultured organisms is hard to assess, the effective construction and validation of such ecosystem models allows researchers and stakeholders to derive the management scenarios that cannot be carried out. For the Sanggou Bay case, the oyster aquaculture is projected to be limited by the primary production of phytoplankton, while the kelp has further yield potential with a relatively sufficient nutrient supply. The suggested culture densities, which combined the additional information with material cost, labor cost, and marketing, are 50–70 ind./m^2 for oyster farming and 5–6 ind./m^2 for kelp under the current model development.

21.3 Further development and application of carrying capacity assessment

Previously described case studies have demonstrated that the CC assessment activities have provided comprehensive information. This information interprets the interaction between aquaculture and the cultural environment under different aquaculture layouts, which can assist the stakeholders in making aquaculture planning decisions. However, the practical application of the CC results in aquaculture production still requiring further development in multiple aspects.

21.3.1 Carrying capacity and aquaculture management

A primary purpose of CC assessment is to help reduce the risk of ecosystem degradation under cultivation pressure while maximizing economic benefits. By providing a scientific basis for setting reasonable stocking densities and harvesting limits based on the production procedure, properly evaluating the CC of shellfish aquaculture helps to balance ecological and economic goals according to local laws/regulations. This is crucial for the long-term viability of the shellfish aquaculture industry and the marine ecosystem sustainability.

CC assessment results have been successfully integrated into regulatory frameworks and industry guidelines in many countries regarding the specific application of shellfish aquaculture. For example, CC assessments have informed sustainable farming practices in Canada, leading to acceptable environmental conditions and profitable production levels (Filgueira et al., 2012; Filgueira et al., 2015; Grant et al., 2007, 2008). In the Norwegian cases, interpretation of the interaction between farmed mussels and the regional environment has guided the development potential of low-trophic aquaculture, ensuring proper stocking densities under sustainable limits (Gatti et al., 2023). However, challenges remain in reasonably predicting long-term environmental impacts at the technical level and ensuring compliance with local socioeconomic conditions and regional regulations at the application level. As previously stated, interpreting the quantitative relationship based on ecosystem processes between the culture organism population and the surrounding environment is the kernel of CC estimation. Environmental variability and the complexity of ecosystem interactions can make it difficult to establish a precise description of the process within the aquaculture system, thus increasing the uncertainties of the estimated capacity. In addition, socioeconomic factors, such as market demand, labor cost, and seedling cost, can influence the feasibility of implementing capacity-based management practices. All these factors have resulted in the site-specific nature of actual CC assessment implementations in the shellfish aquaculture or any similar industry. While a scientifically based CC estimation framework forms the foundation of the application, successful implementation of capacity assessment results critically relies on in-depth collaboration with management authorities and stakeholders. Future research should focus on developing more robust and adaptable CC assessment methods or tools that can account for these complexities and uncertainties.

21.3.2 Toward the genetically bred species

In China, the rapid development of genetic breeding programs for shellfish has raised questions about the CC. Genetic improvements in growth rates and disease resistance may alter the metabolic demands and environmental impacts of shellfish populations. Genetically bred oysters and scallops may require higher nutrient inputs or have different feeding behaviors compared to natural strains. Large-scale aquaculture constitutes an artificially regulated activity. Within this context, promoting selectively bred shellfish varieties may impose specific impacts on the marine environment. This is exemplified in China's intensive shellfish aquaculture, where enhanced filtration capacity in selectively bred specimens can exacerbate preexisting bottom-up control effects. Consequently, such perturbations risk destabilizing marine ecosystem stability. Identifying and analyzing changes in these environmental processes constitutes the core of aquaculture capacity assessment and serves as a critical foundation for the sustainable development of the aquaculture industry. This could affect the carrying of the cultural environment and necessitate adjustments in management practices.

Future research should focus on conducting detailed life cycle assessments (LCAs) of genetically bred species to understand their specific CC requirements. LCAs can provide valuable insights into the environmental impacts of genetically bred species throughout their life cycle, from hatchery to harvest. This information can be used to develop tailored management strategies that ensure their sustainable integration into aquaculture systems. In addition to LCAs, advanced modeling techniques such as IBMs and ecosystem models can be used to simulate the impacts of genetically bred species on ecosystem dynamics. These models can incorporate genetic improvements in growth rates and disease resistance, providing more accurate predictions of sustainable stocking densities and harvesting limits.

21.3.3 Further development of methodology

Existing CC assessment methods can be enhanced by deeply incorporating advanced data collection techniques such as remote sensing, automated monitoring systems, and real-time data analytics. These techniques can provide more accurate and timely data on environmental conditions, shellfish growth rates, and ecosystem health. For example, remote sensing can be used to monitor changes in water quality and benthic habitats over large areas, while automated monitoring systems can provide real-time data on nutrient levels and oxygen concentrations.

CC assessment methods can also be improved by integrating multiple data sources, including environmental, biological, and socioeconomic data. This holistic approach can provide a more comprehensive understanding of the factors influencing CC and help develop more robust and adaptable management strategies. For example, integrating socioeconomic data on market demand and labor availability can help ensure that CC-based management practices are economically feasible and socially acceptable.

The emergence of new technologies such as machine learning (ML), large language models (LLMs), and artificial intelligence (AI) offers exciting opportunities for CC assessment. These technologies can be used to develop more sophisticated models that better capture the complexity of aquaculture ecosystems. ML algorithms can analyze large datasets to identify patterns and relationships that are difficult to detect through traditional methods. LLMs can be used to generate predictive models based on extensive knowledge bases, providing more accurate forecasts of CC limits. AI can optimize decision-making processes by simulating various scenarios and recommending the best management strategies.

21.3.3.1 *Integration of multiple data sources and advanced technologies*

Modern CC assessments increasingly integrate multiple data sources, including environmental, biological, and socioeconomic data. This holistic approach provides a more comprehensive understanding of the factors influencing CC and helps develop more robust and adaptable management strategies. For example, integrating remote sensing data, which can provide information on water quality, chlorophyll-a concentrations, and sea surface temperatures over large areas, with in-situ data from automated monitoring systems, which can measure parameters such as nutrient levels and oxygen concentrations in real - time, allows for a more accurate assessment of the environmental conditions affecting shellfish growth.

New technologies such as ML, LLMs, and AI are also being increasingly used in CC assessment. ML algorithms can analyze large datasets to identify patterns and relationships that are difficult to detect through traditional methods. For example, they can be used to predict shellfish growth rates based on a combination of environmental factors, providing more accurate forecasts of CC limits. LLMs can be used to generate predictive models based on extensive knowledge bases, while AI can optimize decision-making processes by simulating various scenarios and recommending the best management strategies.

21.3.3.2 *Field data collection techniques*

Accurate CC estimation begins with robust field data collection. Key parameters include water quality (e.g., temperature, salinity, dissolved oxygen, and nutrient levels), shellfish growth rates, and benthic community composition. Traditional methods involve manual sampling and laboratory analysis, which can be time-consuming and labor-intensive.

Recent advancements in automated monitoring systems have revolutionized data collection. These systems can provide continuous, real-time data on environmental conditions, significantly improving the accuracy and timeliness of CC assessments. For example, in situ sensors can measure water quality parameters at high frequencies, allowing for the detection of short-term variations that might be missed with manual sampling.

21.3.3.3 *Remote sensing and geographic information system applications*

Remote sensing and geographic information systems (GIS) are powerful tools for CC assessment. Remote sensing can provide large-scale data on water quality, chlorophyll-a concentrations, and sea surface temperatures, which are critical for understanding the environmental conditions affecting shellfish growth. GIS can integrate this data with other spatial information, such as bathymetry and land use, to create comprehensive models of aquaculture sites (McKindsey, 2013).

21.3.3.4 *Statistical and machine learning approaches*

Statistical methods, such as regression analysis and principal component analysis (PCA), are commonly used to identify relationships between environmental variables and shellfish growth.

These methods can help predict how changes in water quality or nutrient availability might impact CC.

ML techniques offer a more advanced approach to CC estimation. ML algorithms, such as random forests and neural networks, can analyze large datasets to identify complex patterns and make accurate predictions. For example, ML models have been used to predict oyster growth rates based on environmental data, providing valuable insights for CC assessment.

21.3.3.5 *Integrated modeling frameworks*

Integrated modeling frameworks combine multiple data sources and modeling approaches to provide a comprehensive assessment of CC. These frameworks often include hydrodynamic models, which simulate water flow and nutrient transport, coupled with ecological models that simulate shellfish growth and ecosystem dynamics.

21.4 Summary

CC assessment is a critical tool for sustainable shellfish aquaculture. It provides a scientific basis for setting stocking densities and harvesting limits, ensuring that farming activities do not exceed the ecosystem's ability to recover and regenerate resources. This chapter has explored the historical development of the CC concept, practical methods for its assessment, case studies from around the world, and future directions for research and application. As a crucial species in global aquaculture, shellfish significantly contribute to the provision of high-quality protein to the consumer market. The expansion of farming scales and the introduction of genetically bred new species have intensified the focus on the sustainable development of shellfish aquaculture. This dual emphasis necessitates a balanced approach, where farming activities must not only ensure basic economic returns for practitioners but also safeguard the sustainability of the local environment. Consequently, shellfish farming capacity assessments are required to deliver more informative results. These assessments must consider advancements in farming methods, species, models, and technologies, and provide actionable conclusions that genuinely support sustainable farming practices. The scientific rigor, practical applicability, and interdisciplinary integration of these assessment methods are becoming increasingly essential and specific for capacity assessments with practical significance. The continuous integration of new developments in physiological and ecological theories, computational technologies, and environmental innovations is fundamental to the ongoing evolution of CC assessment theories and methods for shellfish aquaculture. By integrating advanced data collection techniques, interdisciplinary approaches, and new technologies, practical CC assessments are expected to provide accurate and actionable insights for sustainable aquaculture management.

References

Aquaculture industry overview. Government of New Brunswick, & New Brunswick Department of Agriculture. (2022).

Chary, K., Brigolin, D., & Callier, M. D. (2022). Farm-scale models in fish aquaculture – An overview of methods and applications. *Reviews in Aquaculture, 14*(4), 2122–2157. doi:10.1111/raq.12695.

Comea, L. A. (2013). Suspended oyster culture in eastern Canada: Methods and production patterns. *Bulletin of the aquaculture association of Canada, 111*(2), 23–29.

Dalton, T., Di, J., Robert, T., & Allison, K. (2017). Using normative evaluations to plan for and manage shellfish aquaculture development in Rhode Island coastal waters. *Marine Policy, 83*, 194–203. doi:10.1016/j.marpol.2017.06.010.

Dame, R. F. (1996). The role of benthic-pelagic coupling in estuarine productivity and biogeochemical cycling. *Estuaries, 19*(2), 231–240.

Dame, R. F., & Prins, T. C. (1997). Bivalve carrying capacity in coastal ecosystems. *Aquatic Ecology, 31*(4), 409–421. doi:10.1023/A:1009997011583.

Duarte, P., Meneses, R., Hawkins, A. J. S., Zhu, M., Fang, J., & Grant, J. (2003). Mathematical modelling to assess the carrying capacity for multi-species culture within coastal waters. *Ecological Modelling, 168*(1-2), 109–143. doi:10.1016/s0304-3800(03)00205-9.

Fan, L. I. N., Meirong, D. U., Hui, L. I. U., Jianguang, F. A. N. G., Lars, A., & Zengjie, J. I. A. N. G. (2020). A physical-biological coupled ecosystem model for integrated aquaculture of bivalve and seaweed in Sanggou Bay. *Ecological Modelling, 431*, 109181. doi:10.1016/j.ecolmodel.2020.109181.

Fang, J. G., & Zhang, A. J. (1996). Study on the carrying capacity of *Chlamys farreri* aquaculture in Sanggou Bay. *Journal of Fisheries of China, 3*(4), 1–6.

FAO. (2024). The State of World Fisheries and Aquaculture 2024. Blue Transformation in action. Rome. 10.4060/cd0683en.

Ferreira, J. G. (2025). Aquaculture carrying capacity estimates show that major African lakes and marine waters could sustainably produce 10–11 Mt of fish per year. *Nature Food, 6*(5), 446–455. doi:10.1038/s43016-025-01114-1.

Ferreira, J. G., Grant, J, Verner-Jeffreys, D. W., & Taylor, N. G. H. (2013). Carrying capacity for aquaculture, modeling frameworks for determination of. In P. Christou, R. Savin, B. A. Costa-Pierce, I. Misztal, & C. B. A. Whitelaw (Eds.), *Sustainable food production*. doi:10.1007/978-1-4614-5797-8_904.

Filgueira, R., Grant, J., Bacher, C., & Carreau, M. (2012). A physical–biogeochemical coupling scheme for modeling marine coastal ecosystems. *Ecological Informatics, 7*(1), 71–80. doi:10.1016/j.ecoinf.2011.11.007.

Filgueira, R., Guyondet, T., Bacher, C., & Comeau, L. A. (2015). Informing marine spatial planning (MSP) with numerical modelling: A case-study on shellfish aquaculture in Malpeque Bay (Eastern Canada). *Marine Pollution Bulletin, 100*(1), 200–216. doi:10.1016/j.marpolbul.2015.08.048.

Filgueira, R., Guyondet, T., Comeau, L. A., & Grant, J. (2014). A fully-spatial ecosystem-DEB model of oyster (*Crassostrea virginica*) carrying capacity in the Richibucto Estuary, Eastern Canada. *Journal of Marine Systems, 136*(1), 42–54. doi:10.1016/j.jmarsys.2014.03.015.

Fisheries and Oceans Canada (DFO), Aquaculture production statistics – Prince Edward Island: Production volumes, acreage, and economic contribution for mussels and oysters in PEI. Government of Canada. (2024),

Gatti, P., Agüera, A., Gao, S., Strand, Ø., Strohmeier, T., & Skogen, M. D. (2023). Mussel farming production capacity and food web interactions in a mesotrophic environment. *Aquaculture Environment Interactions, 15*, 1–18. doi:10.3354/aei00448.

Grant, J., & Bacher, C. (2001). A numerical model of flow modification induced by suspended aquaculture in a Chinese bay. *Canadian Journal of Fisheries and Aquatic Sciences, 58*(5), 1003–1011. doi:10.1139/f01-027.

Grant, J., Bacher, C., Cranford, P. J., Guyondet, T., & Carreau, M. (2008). A spatially explicit ecosystem model of seston depletion in dense mussel culture. *Journal of Marine Systems, 73*(1-2), 155–168. doi:10.1016/j.jmarsys.2007.10.007.

Grant, J., Curran, K. J., Guyondet, T. L., Tita, G., Bacher, C., Koutitonsky, V., & Dowd, M. (2007). A box model of carrying capacity for suspended mussel aquaculture in Lagune de la Grande-Entrée, Iles-de-la-Madeleine, Québec. *Ecological Modelling, 200*(1-2), 193–206. doi:10.1016/j.ecolmodel.2006.07.026.

Guyondet, T., Roy, S., Koutitonsky, V. G., Grant, J., & Tita, G. (2010). Integrating multiple spatial scales in the carrying capacity assessment of a coastal ecosystem for bivalve aquaculture. *Journal of Sea Research, 64*(3), 341–359. doi:10.1016/j.seares.2010.05.003.

Heip, C. H. R., Goosen, N. K., Herman, P. M. J., Kromkamp, J., Middelburg, J. J., & Soetaert, K. (1995). In *Production and consumption of biological particles in temperate tidal estuaries oceanography and marine biology: An annual review: 33* (pp. 1–149). UCL Press, undefined.

Herman, P. M. J. (1993). *A set of models to investigate the role of benthic suspension feeders in estuarine ecosystems* (pp. 421–454). Springer Science and Business Media LLC. doi:10.1007/978-3-642-78353-1_12.

Inglis, G. J., Mallet, A. L., Ross, A. H. An overview of factors affecting the carrying capacity of coastal embayments for mussel culture. Aquaculture Research, (2000), 31, 39–53

Kluger, L. C., Filgueira, R., & Byron, C. J. (2019). Using media analysis to scope priorities in social carrying capacity assessments: A global perspective. *Marine Policy, 99*, 252–261. doi:10.1016/j.marpol.2018.10.042.

Leopold, A. (1933). *Game management*. University of Wisconsin Press.

Lin, F., Asplin, L., Budgell, W. P., Wei, H., & Fang, J. (2019). Currents on the northern shelf of the Yellow Sea. *Regional Studies in Marine Science, 32*, 100821. doi:10.1016/j.rsma.2019.100821.

Mao, Y., Li, J., Xue, S., Lin, F., Jiang, Z., Fang, J., & Tang, Q. N. (2018). Ecological functions of the kelp *Saccharina japonica* in integrated multi-trophic aquaculture, Sanggou Bay, China. *Shengtai Xuebao, 38*(9), 3230–3237. doi:10.5846/stxb201703160444.

McKindsey, C. W. (2013). Carrying capacity for sustainable bivalve aquaculture. In P. Christou, R. Savin, B. Costa-Pierce, I. Misztal, & B. Whitelaw (Eds.), *Sustainable food production* (pp. 449–466). Springer.

Newell, R. I. E., Fisher, T. R., Holyoke, R. R., & Cornwell, J. C. (2005). Influence of eastern oysters on nutrient and sediment dynamics: A review and future directions. *Aquatic Ecosystem Health & Management, 8*(4), 377–396.

PEI Mussel Monitoring Program, Technical Report # 283. Prince Edward Island Department of Fisheries, Tourism, Sport & Culture, Aquaculture Division. (2024)

Pei, R., Zhang, H., Liu, Z., Yu, C., Zhang, Y., & Mu, Y. (2025). Evaluating the green development level of marine molluscan shellfish aquaculture industry and analyzing obstacle factors that hinder its improvement: The case of China. *Aquaculture, 606*, 742603. doi:10.1016/j.aquaculture.2025.742603.

Ren, J. S., Stenton-Dozey, J., Plew, D. R., Fang, J., & Gall, M. (2012). An ecosystem model for optimising production in integrated multitrophic aquaculture systems. *Ecological Modelling, 246*(C), 34–46. doi:10.1016/j.ecolmodel.2012.07.020.

Ross, L. G., Telfer, T. C., Falconer, L., & Soto, D. (2013). Carrying capacities and site selection within the ecosystem approach to aquaculture. In L. G. Ross, T. C. Telfer, L. Falconer, D. Soto, & J. Aguilar-Manjarrez (Eds.), *Site selection and carrying capacities for inland and coastal aquaculture* (pp. 19–46). FAO/Institute of Aquaculture, University of Stirling, Expert Workshop, 6–8 December 2010.

Shandong Province fishery statistical yearbook, Shandong Provincial Department of Agriculture and Rural Affairs. (2023)

Shi, J., Wei, H., Zhao, L., Yuan, Y., Fang, J., & Zhang, J. (2011). A physical–biological coupled aquaculture model for a suspended aquaculture area of China. *Aquaculture, 318*(3-4), 412–424. doi:10.1016/j.aquaculture.2011.05.048.

Statistics Canada. (2023). *Aquaculture statistics*. ON. (National-level mussel production figures, with breakdown for PEI).

Sun, L.-Q., Liu, H., Gao, Y.-P., Jiang, Z.-J., Lin, F., Chang, L.-R., & Zhang, Y.-T. (2020). Food web structure and ecosystem attributes of integrated multi-trophic aquaculture waters in Sanggou Bay. *Aquaculture Reports, 16*, 100279. doi:10.1016/j.aqrep.2020.100279.

Weitzman, J., & Filgueira, R. (2020). The evolution and application of carrying capacity in aquaculture: Towards a research agenda. *Reviews in Aquaculture, 12*(3), 1297–1322. doi:10.1111/raq.12383.

You, J., Yu, L., Meillon, J., Gangnery, A., Bacher, C., Liu, H., & Strand, Ø. (2022). A set of web-based public decision support tools for integrated planning and management in aquaculture. *MethodsX, 9*, 101795. doi:10.1016/j.mex.2022.101795.

SECTION IV

Conclusion

CHAPTER

22

Future perspectives

Lisui Bao[a,b], *Shi Wang*[b] *and Zhenmin Bao*[b]

[a]Institute of Evolution & Marine Biodiversity, Ocean University of China, Qingdao, China [b]College of Marine Life Sciences, Ocean University of China, Qingdao, China

22.1 Introduction

The future of shellfish aquaculture stands at a critical juncture to tackle the growing challenges posed by climate change, increasing global populations, and evolving market demands. As the world's population is projected to exceed 9 billion by 2050, the demand for seafood, including shellfish, is expected to rise significantly (Costello et al., 2020). To meet these demands, breeding programs must focus on key traits such as growth rate, disease resistance, and nutritional content. Moreover, the shellfish industry is also positioning itself as a major contributor to the burgeoning market for bioactive materials, which have applications in pharmaceuticals, nutraceuticals, and functional foods (Chen, 2023).

In parallel, environmental pressures such as ocean warming, acidification, and biodiversity loss require aquaculture systems to adapt rapidly to maintain productivity and sustainability. As such, breeding for heat tolerance, acidification resistance, and disease resilience will be central to the future of shellfish aquaculture (Mackenzie et al., 2024). In addition, technological innovations in offshore and deep-sea aquaculture are expanding the boundaries of production, allowing for more sustainable operations that reduce coastal ecosystem stress (Gentry et al., 2017). Moreover, the next decade will likely witness shellfish playing an increasingly vital role in mitigating the impacts of climate change through ecological services such as carbon sequestration and water purification (van der Schatte Olivier et al., 2018).

The next generation of shellfish breeding will be centered on "breeding-by-design" as the overarching paradigm, characterized by the profound integration of advanced genomic tools, precision breeding technologies and ecosystem-based management practices. Achieving this goal depends first on the deep dissection of molecular basis underlying complex traits of economic and ecological importance. This functional understanding is being dramatically advanced by multi-omics platforms—encompassing genomics, transcriptomics, proteomics, and metabolomics—integrated with artificial intelligence (AI) and machine learning algorithms, which together illuminate the regulatory architecture and cryptic elements that have historically obscured trait mechanisms (Houston et al., 2020; Liu et al., 2025; Yu et al., 2025). Building on this

Genetic Breeding and Aquaculture of Shellfish: A Collaborative Approach to a Sustainable Future
DOI: https://doi.org/10.1016/B978-0-443-33606-5.00009-6

foundational knowledge, the next step involves genetic manipulation across the entire developmental trajectory of the organism, leveraging technologies such as genome-wide selection and precise gene editing via CRISPR/Cas9 to achieve multi-trait improvement with unprecedented speed and accuracy (Potts et al., 2021).

Within this molecular design breeding framework, stem-cell-based approaches are poised to play an instrumental supporting role. By isolating, rationally designing, and precisely editing stem cells, and subsequently directing their differentiation into functional germ cells for *in vitro* selection and reproduction, the laborious cycles of whole-organism breeding can be transferred to a more tractable cellular platform. This strategy—often conceptualized as "breeding in a tube"—offers the potential to dramatically compress generation intervals and streamline the breeding pipeline. Although such applications in shellfish remain prospective, recent advances in defining germline specification and gonadal niche biology (Gavery et al., 2025), alongside progress in *in vitro* gametogenesis and surrogate broodstock technologies in other aquaculture systems (Kleppe et al., 2025), suggest that germline stem-cell engineering may eventually enable faster, more precise shellfish breeding. This potential is further bolstered by the identification of conserved stem cell niches in invertebrate skeletons (Lian et al., 2025), underscoring the biological feasibility of manipulating stem cells to realize the ultimate vision of breeding-by-design in shellfish. To fully realize the biological potential of these superior, genetically improved lines, future aquaculture must co-evolve with automated, ecologically integrated offshore farming models. Simultaneously, the industry is trending toward full integration of the entire value chain—linking breeding, farming, processing, and marketing—to enhance quality control, traceability, and resource efficiency. This holistic approach helps ensure that advances in genetics and technology yield tangible benefits for both producers and consumers (Ababouch et al., 2023).

22.2 Future trait expectations of shellfish aquaculture

Shellfish breeding objectives are rapidly evolving to meet both market demands for nutritious seafood and the challenges imposed by climate change. Future breeding programs will need to integrate production traits (e.g., fast growth, high protein content, and desirable product quality) with resilience traits (e.g., tolerance to marine heatwaves, ocean acidification, and disease). Moreover, breeders are recognizing shellfish as not only commodities but also as ecosystem engineers that improve water quality and sequester carbon, so traits enhancing these ecosystem services are garnering attention.

22.2.1 Nutritional and market-driven traits

As the global population grows, shellfish are increasingly valued as a sustainable source of protein and nutrients. It is predicted that bivalves could contribute approximately 43% of the world's aquatic animal protein by 2050 (Costello et al., 2020). Future breeding will intensify focus on growth rate and feed conversion efficiency to boost production. Concurrently, product quality traits are rising in importance: Breeders aim to enhance the nutritional profile (e.g., improving amino acid composition and omega-3 content in scallop or oyster meat) and sensory properties (texture and flavor) to meet premium market standards. Multitrait selection is becoming feasible with advanced breeding-value estimation, allowing simultaneous improvement of yield and quality. In addition, traits conferring longer shelf-life and processing suitability may enter breeding objectives as the industry diversifies its product forms.

22.2.2 Bioactive compounds extraction

As the global demand for functional foods and bioactive materials grows, shellfish aquaculture is also poised to contribute significantly beyond traditional protein production. Marine mollusks, particularly bivalves and gastropods, are emerging as important sources of high-value secondary metabolites with diverse biological activities. These metabolites, including peptides, alkaloids, and terpenes, have shown promising pharmacological properties, such as anticancer, antiinflammatory, and antimicrobial effects (Chen, 2023). For instance, venom peptides from cone snails (conotoxins) are being tested as novel analgesics in clinical trials, exemplifying the pharmacological potential of mollusk-derived compounds (Sudewi et al., 2019). The growing interest in these compounds stems from their unique chemical structures and ecological roles, and this emerging sector offers exciting opportunities for aquaculture to expand its economic potential.

22.2.3 Climate resilience and environmental tolerance

With ocean conditions becoming more variable, climate-resilient phenotypes are a top priority. Notably, heat tolerance is critical as marine heatwaves and rising baseline temperatures have led to mass mortality events in bivalve farms. For instance, Pacific oyster farms suffered more than 70%–90% summer mortality globally during periods of warmer temperatures (Scanes et al., 2025). Temperature can significantly affect both the physiological processes of aquatic animals and the virulence of pathogenic bacteria (Chi et al., 2024; Li et al., 2023). Thus, breeding programs are responding by selecting for oysters that can withstand elevated temperatures, and breeding for resistance to pathogens complements breeding for environmental stress tolerance, since disease outbreaks often accompany climatic stressors. Relatedly, ocean acidification tolerance is sought, especially in larval and spat stages of calcifying shellfish that are sensitive to low pH (Goncalves et al., 2017). Hatchery-based experiments have demonstrated significant additive genetic variation for survival and growth under acidified conditions. Indeed, multigeneration breeding trials with Sydney rock oysters showed that families reared under elevated CO_2 had 1.3-fold higher survival than wild controls, indicating potential to improve tolerance via selection (Parker et al., 2010). Indeed, a recently bred Pacific oyster line exhibited higher survival and growth under experimental acidification stress than wild oysters, highlighting the feasibility of breeding for climate resilience (Jiang et al., 2025). Selective breeding programs for farmed species can create opportunities to stay ahead of changes in temperature and pH by raising more robust strains. In addition, environmental interventions such as intertidal exposure have been shown to mitigate summer mortality in oysters (Mackenzie et al., 2024).

22.2.4 Ecosystem service traits and climate mitigation

Importantly, the relationship between shellfish traits and climate is bidirectional. Shellfish confer ecological benefits that create positive feedback with climate-change mitigation. Future breeding may target traits that enhance these services. For instance, bivalves are filter feeders that remove particulates and nutrients from water, thereby improving water clarity and reducing eutrophication. Breeding for higher filtration rates could amplify water purification in farming areas, contributing to healthier coastal ecosystems. Similarly, shellfish contribute to blue carbon storage by forming shells of calcium carbonate, effectively locking away carbon in a stable form (Sea et al., 2022). This calcification process is analogous to the role of planktonic shell-formers in the open ocean, which have shown enhanced carbon export under elevated CO_2 conditions. Global assessments estimate that cultivated bivalves remove ~49,000 tons of nitrogen and 6,000 tons of phosphorus from coastal waters annually, services worth over $1.2 billion per year through nutrient mitigation (van der Schatte Olivier et al., 2018). They also facilitate carbon burial: constructed oyster farms in estuaries significantly accumulate organic carbon in sediments by trapping biodeposits and stabilizing shorelines (Fodrie et al., 2017; Sun et al., 2025). Notably, the carbon sequestration capacity of shellfish mariculture has been quantified in regions such as China, demonstrating a substantial contribution to coastal carbon sinks (Gu et al., 2022). Breeding programs may support these benefits by increasing shell biomass yield and stress tolerance to sustain large bivalve populations. Beyond water quality and carbon, shellfish provide habitat. Dense oyster reefs or mussel beds create complex structures that support fish and invertebrate biodiversity and protect coasts from erosion (van der Schatte Olivier et al., 2018). In short, the next generation of shellfish breeding will broaden the definition of "desirable traits" to encompass both production efficiency and ecological functionality. Future shellfish breeding aligns with principles of regenerative aquaculture: the traits we improve can, in turn, improve the environment, creating a virtuous cycle of climate adaptation.

22.3 Trait dissection through omics integration

Many genes in shellfish lack known homologs or functional annotation, which are often termed "dark genes" in genomics. Decoding complex traits and the genetic "dark matter" underlying them requires a holistic, systems-level approach. The past decade's revolution in omics technologies (high-throughput DNA, RNA, protein, and metabolite profiling) is now converging with powerful computational tools to accelerate trait discovery in shellfish. This section explores how enviromics, phenomics, genomics, transcriptomics, proteomics, and metabolomics can be integrated to dissect traits and identify key genetic factors for breeding. We highlight emerging strategies such as multiomics data integration, machine learning algorithms for pattern recognition, and international genome initiatives that together are illuminating previously "dark" genes, genomic elements, and cryptic biological functions in shellfish.

22.3.1 Phenomics and enviromics—High-resolution trait data

A persistent bottleneck in genetic improvement of shellfish is measuring complex traits accurately and affordably. Phenomics, the large-scale quantitative measurement of phenotypes, is now emerging in aquaculture. This can involve digital imaging of shellfish shape and color, sensor arrays for tracking growth and survival under different conditions, and automated recording of behavioral or physiological responses. For instance, new sensors can monitor a scallop's heart rate or the byssal thread attachment strength (the pull-off force), providing novel phenotypic traits. Coupled with these is enviromics—the tracking of detailed environmental parameters (temperature, salinity, oxygen, microalgal abundance, etc.) during field grow-out. By linking enviromic profiles to performance outcomes, breeders can identify genotype–environment interactions and include environment variables directly in genomic prediction models. Phenomics and enviromics thus enrich the trait dataset, moving beyond single-endpoint measurements to a more holistic view of phenotype, accounting for dynamic changes and environmental modulation, which is crucial for traits such as growth. To manage this influx, cloud databases and Internet of Things (IoT) sensor networks are being deployed on experimental farms. Real-time data streams from smart farms can feed into breeding databases, ensuring that genomic analyses consider the full context of trait expression.

22.3.2 Multiomics integration and systems biology

Historically, shellfish were genomically underrepresented, but this has changed dramatically with advances in sequencing technologies. High-quality reference genomes now exist for key shellfish species. However, telomere-to-telomere (T2T) genome assemblies and pangenomes are still lacking, especially given the high genetic variation within shellfish genomes. Genome-wide association studies (GWAS) and genomic selection in shellfish rely heavily on single-nucleotide polymorphism (SNP) markers, where any single marker type often explains only a limited fraction of genetic variation for traits such as growth or disease resistance. Thus combining different variant types (SNPs, InDels, structural variants, copy number variants, and epigenetic markers) is important to capture more heritable variation. In addition, integrating transcriptomics, proteomics, and metabolomics provides a more complete picture of biological mechanisms.

First, multiomics integration allows more precise identification of causal genes. Consider heat tolerance in mussels: Genomic scans might identify candidate variants in stress-response genes, but transcriptomics can validate whether those genes are upregulated under heat stress; proteomics can reveal if protective heat-shock proteins are produced in greater abundance in tolerant strains; and metabolomics might show changes in osmolytes or antioxidants. By overlaying these layers of information, one can filter out false-positive signals and focus on true functional genes—genes that not only vary in sequence but also drive downstream phenotypic differences.

Second, the holistic nature of multiomics integration is particularly powerful for uncovering cryptic functions, such as a gene of unknown function that actually plays a role in a trait. By observing that an unannotated gene's expression correlates strongly with a desirable phenotype and that it clusters in a coexpression network with known stress-response genes, researchers may discover a new functional gene. In shellfish, where a large fraction of genes has no assigned function, this approach is invaluable. The sheer volume of data does present challenges in distinguishing meaningful signals from noise, and this is where advanced bioinformatics and AI algorithms come into play. Large-scale comparative genomics across many mollusk species is also shedding light on lineage-specific "dark genes." Such analyses help flag novel genes that might underlie important adaptive traits. In short, the combination of expansive databases and cross-species genomics is illuminating the once-dark portions of the shellfish genome and opening them up to functional investigation.

Third, combining multiomics data enables a more comprehensive understanding of molecular mechanisms. Systems biology approaches, including gene coexpression network analysis and protein–protein interaction mapping, further help connect the dots between genotype and phenotype. These methods can highlight key regulatory modules and hub genes that orchestrate multiple pathways, providing excellent targets for marker-assisted breeding or gene editing. Multiomics does not just catalog molecular changes; it distills complex interactions into quantifiable factors associated with phenotypic variation, which can then directly inform breeding decisions. As multiomics studies accumulate, they feed back into improved genomic prediction models. Instead of using thousands of anonymous SNPs as predictors, breeders can weight markers by whether they lie in genes or regions with known functional impact, yielding more accurate and biologically informed predictions for selective breeding.

22.3.3 Artificial intelligence and the hunt for hidden heritability

The integration of diverse, large datasets can be achieved through classical statistical frameworks (e.g., correlation networks and Bayesian models) or increasingly through AI. Machine learning techniques ranging from random forests to neural networks can detect complex patterns in omics data that elude conventional statistical methods. For instance, AI can correlate epigenetic marks or chromatin features with environmental data to find stress-induced epigenetic changes that might be heritable and affect offspring performance. Deep learning models have already been tested in plant and livestock breeding to integrate image-based phenotypes with genomics for yield prediction, and similar applications are beginning to emerge in aquaculture. Importantly, AI can help crack the problem of epistasis, the nonlinear interactions among loci that contribute to complex traits. Traditional genomic selection models assume mostly additive gene effects, but in reality, combinations of alleles can have synergistic or antagonistic effects on shellfish phenotypes. Deep neural networks have shown improved prediction accuracy in genomic selection of crops by implicitly modeling such interactions. In an aquaculture context, a multimodal deep learning approach might simultaneously analyze genomic markers, gene expression profiles, and environmental parameters to predict growth or survival, thereby capturing latent interaction effects. While these models can behave as "black boxes," efforts are underway to improve their interpretability so that breeders can trust and understand the drivers of predictions.

22.3.4 Open data and collaborative platforms

Underlying all these omics advances is a trend toward open-access data sharing and collaboration. The complexity of multiomics research requires large, diverse datasets, which are difficult for any single institute to produce alone. International initiatives such as the Mollusk 10K Genome Project (M10K+) and databases such as MolluscDB are critical for this exchange. The launch of the M10K+ project, which aims to sequence 10,000 molluscan genomes, exemplifies the scale of current efforts. As part of this, the first comprehensive mollusk genomic database, MolluscDB, was constructed in 2021, integrating over 1,000 datasets (Liu et al., 2021). By 2025, MolluscDB 2.0 had expanded to cover >1,400 species and ~4,200 omics datasets, making it the largest molluscan genomics resource globally (Liu et al., 2025). These open-access platforms and consortium projects are rapidly filling gaps in molluscan genomic knowledge. Collaborative projects also foster standardization by establishing common protocols for phenotype measurement, environmental data logging, and baseline genetic panels, which makes integrative analysis more robust. The value of such coordination is recognized in programs such as the USDA-funded AquaFAANG (Functional Annotation of Animal Genomes for aquaculture), which applies standardized functional genomics assays (analogous to the livestock FAANG project) in species including oysters and mussels. Moreover, industry and researchers are adopting shared formats and metadata standards to enable seamless data merging. Ultimately, the multiomics era is transforming shellfish breeding from a trial-and-error endeavor into a data-driven science. In sum, trait dissection through omics is the compass guiding the next leaps in shellfish genetic improvement, ensuring that we understand why a certain broodstock is superior and how to reproducibly create even better ones.

22.4 Breakthroughs in breeding technologies

Armed with a better understanding of trait genetics, the shellfish aquaculture industry is entering an era of accelerated breeding powered by novel technologies. Traditional family-based selection and mass spawning are being augmented by approaches once confined to high-tech agriculture. This section surveys cutting-edge breeding technologies poised to revolutionize shellfish improvement: Genome-enabled selection with AI for capturing epistasis, automated genomic mating and gamete selection systems for high-throughput screening, digital twin simulations for in silico testing of breeding strategies, and molecular biotechnologies such as RNA interference and CRISPR/Cas9 for directly manipulating functional genes. These tools—many now proven in principle for shellfish—promise unprecedented precision and speed in developing superior cultivars.

22.4.1 Genomic selection and deep learning models

In recent years, genomic selection (GS) has become a game-changer for animal and plant breeding, and it is increasingly applied to aquaculture species. GS uses genome-wide marker information (such as thousands of SNPs) to predict the breeding values of individuals before phenotypes are measured, thus shortening generation intervals. For shellfish with high fecundity (e.g., a scallop can produce millions of larvae), GS can be extremely powerful: One can genotype a sample of larvae and forecast which will grow fastest or survive best, selecting those for further rearing. Early trials in scallops demonstrated the potential of GS, achieving moderate to high prediction accuracy for traits such as growth and disease resistance. However, one limitation of conventional GS methods is that they assume additive gene effects and may miss nonlinear interactions. This is where deep learning enters the picture. Researchers have begun applying deep neural networks to genomic prediction in aquaculture, following promising results in crop breeding. The complexity of shellfish genomes (many are highly polymorphic and some have polyploid or variable karyotypes) means that trait architecture can be complicated, with many loci of small effect and many interactions. Deep learning algorithms can model these interactions (epistasis) automatically. A comprehensive review by Montesinos-López et al. showed that across numerous GS studies, deep learning often matched or outperformed standard models, especially when genotype–environment interactions were present (Montesinos-López et al., 2021). Another advantage is the possibility of end-to-end selection: One can input raw data, such as DNA sequences or even image-derived phenotypes alongside genotypes, with no variant calling or imputation steps needed, and thus leverage traits that are hard to quantify by traditional means.

Although still experimental, these AI-driven genomic prediction models are expected to become part of the breeder's toolkit in the near future. In summary, genome-wide selection powered by advanced algorithms will enable breeders to capitalize on the troves of genomic data being generated, selecting superior broodstock more efficiently and with greater genetic gain per generation than traditional phenotype-based methods.

22.4.2 Automated genomic mating and gamete-level selection systems

A pivotal advancement in shellfish breeding is the integration of genomic mating strategies, particularly focusing on gamete-level selection to enhance genetic improvement. Traditional breeding methods, based largely on phenotypic evaluation, often overlook valuable genetic combinations due to random or uncontrolled mating. In contrast, genomic mating leverages genomic data at the gamete stage to optimize parentage and maximize genetic gains. In shellfish breeding programs, gamete-level selection involves screening and sorting gametes based on genomic information prior to fertilization. Recent technological breakthroughs, such as single-cell genotyping, now allow rapid genotypic assessment of individual gametes or early-stage embryos. This enables breeders to perform precise mate-pairing strategies that maximize genetic complementarity and heterozygosity, thereby significantly reducing inbreeding depression and enhancing desirable trait expression.

High-throughput automation is also being introduced at the earliest life stages of breeding. Microfluidic technologies and automated fluorescence-activated cell sorting (FACS) systems are being developed to apply fluorescent genetic or immunological markers to gametes and larvae. For example, fertilized eggs or larvae carrying a particular desirable genotype can be labeled using DNA probes or CRISPR-based reporters, then enriched via flow cytometry or imaging cell sorters from a mixed population. While these techniques are still experimental, their potential is immense: one could genotype larvae within days of fertilization and immediately isolate those with desired alleles for grow-out, dramatically speeding up breeding cycles and improving the accuracy of mass selection in shellfish hatcheries.

22.4.3 Precision gene editing and multitrait breeding

Perhaps the most direct way to improve a trait is to modify the key genes underlying it. Genome editing has swept through biomedical and agricultural research in the past decade, and shellfish are now part of that story. The CRISPR/Cas9 system has been successfully applied in several mollusks (Perry & Henry, 2015). Researchers further used CRISPR/Cas9 to knock-out a shell coiling gene in snails, producing mirror-image coiled offspring (Abe & Kuroda, 2019). In bivalves, different groups have delivered CRISPR/Cas9 components into fertilized oyster eggs via electroporation or knocked out a key ciliogenesis gene in a dwarf surf clam (Jin et al., 2021; Wang et al., 2022). This proved that CRISPR can penetrate tough egg membranes and make precise gene cuts in shellfish, although the timeline to practical application might be longer because of the complexity of producing stable edited lines. Nevertheless, we can foresee that in the next decade, gene editing will start to complement traditional breeding. Breeders will likely integrate editing with multitrait selection—using editing for big-effect genes and traits that are otherwise hard to improve, while continuing to apply GS for polygenic traits like general growth. Notably, the long-standing issue of toxin and heavy metal accumulation in shellfish may also be mitigated through genetic improvement—by editing genes involved in uptake or retention of harmful substances to enhance food safety. One can imagine future oyster strains combining a few targeted edits (e.g., alleles for disease resistance, faster growth via altered metabolism, heat tolerance, low-toxin and heavy-metal-free) on top of the polygenic gains accumulated by selection. The result could be a quantum leap in performance.

22.4.4 "Digital twin" simulations for in silico breeding

One of the most futuristic developments in aquaculture breeding is the concept of the digital twin—a virtual replica of a biological system that runs in parallel to the real world. Pioneering work has already demonstrated digital twins for biological systems in medical research, integrating real-time sensor data with AI and IoT to mirror cells, tissues, or organs such as the heart and brain (De Domenico et al., 2025; Osipov et al., 2024; Wang et al., 2024). Even at smaller scales, researchers are exploring similar simulations: Large-scale brain models illustrate the power of computational biology (Einevoll et al., 2019), and at the cellular level, a "virtual cell" can be constructed with AI to model complex interactions (Bunne et al., 2024). Extending this concept to shellfish, one can imagine creating a computational model that incorporates a shellfish's growth equations (using parameters such as temperature, food density, and metabolic rates), its genetic growth potential (estimated from multiomics regulatory networks and breeding values), and its environmental stress responses. By calibrating this model on real farm data, we obtain a "twin" that can then be experimented with virtually. For breeding, this means strategies could be evaluated in silico before implementation. For instance, a digital twin oyster farm could simulate multiyear production for each breeding line under various scenarios (e.g., projected warming or disease outbreaks) to predict which line yields the best outcomes. It could also simulate crossing those lines to see if an optimal combination emerges. This computational approach leverages the power of modeling to optimize decisions that traditionally would rely on lengthy field trials. Another use is virtual gene knock-out experiments: A digital twin would allow breeders to "knock-out" a hub gene in a heat-tolerance network in silico and observe the predicted impacts on farm performance across decades of climate projections. While these are still early days, the trajectory is clear: As computing and data integration improve, in silico testing will become a standard complement to wet trials, saving time and resources by narrowing the field of options to those most promising. The limitations lie in model accuracy, as biology is complex and models are simplifications. Unexpected factors can always throw off predictions (Potts et al., 2021). Therefore digital twin outputs are best used to inform rather than replace empirical testing.

22.5 Future aquaculture models: Toward ecological and intelligent farming

Improving the genetics of cultured shellfish is only part of the equation; the farming systems in which these animals grow must also evolve to realize their full potential. Key trends include a shift toward ecologically integrated farming, increased automation and real-time monitoring for responsive husbandry, and the expansion of cultivation to new frontiers, such as offshore and deep-sea environments, using specially designed infrastructure. These innovations aim to enhance the sustainability, yield, and climate resilience of shellfish aquaculture, creating synergy between the improved breeds and their cultivation environment.

22.5.1 Ecological and integrated farming systems

Traditional shellfish aquaculture already has a relatively light environmental footprint, and future models push this further by designing farms as components of the ecosystem, an approach epitomized by integrated multitrophic aquaculture (IMTA) (Ghosh et al., 2025). IMTA cocultures species at different trophic levels so that waste from one becomes input for another. A classic IMTA arrangement for shellfish is to combine a fed species, such as shrimp, with extractive species such as bivalves that consume particulates and plankton, alongside seaweeds that absorb dissolved nutrients. This creates a miniecosystem that recycles nutrients and increases overall efficiency. In the future, entire shellfish polycultures may be designed—for instance, mussels grown with oysters, with deposit-feeding sea cucumbers below to consume fallen detritus. Breeding programs might even coordinate with IMTA design by selecting shellfish genotypes that perform best in polyculture conditions. All these integrated approaches hinge on careful carrying-capacity management. Advancements in hydrodynamic and ecosystem modeling now allow real-time adjustments. For example, sensor networks can detect if chlorophyll levels are dropping in a bay due to overstocking of mussels and alert farmers to reduce stock or rotate harvest. Thus, breeding and farming models will co-evolve: the farm of the future might be custom-tailored to the stocks being cultivated, and vice versa.

22.5.2 Offshore aquaculture platforms

With coastal space often limited and increasingly affected by pollution and user conflicts, the next frontier is to move shellfish cultivation offshore—into deeper, more exposed but cleaner and more expansive waters (Peng et al., 2023). This poses engineering and operational challenges that are now being met with innovative designs. One approach is the development of large, self-contained offshore aquaculture vessels or platforms. China has been leading in this arena with prototypes such as "Guoxin-1" and "Wanqu Lingding." These are essentially floating fish-farm ships, though their concepts could extend to multispecies cultivation. The Wanqu Lingding platform, for example, spans 156 m in length and contains 12 independent aquaculture tanks, collectively holding ~80,000 m^3 of water, equivalent to roughly 330 ha of traditional fish ponds in capacity. Although initially geared toward finfish, its recirculating seawater system with open exchange could potentially be adapted to grow filter-feeding shellfish in contained conditions offshore. These vessels boast features such as automated feeding, intelligent monitoring, and even remote operation. Their mobility means they can follow plankton blooms or favorable water temperatures for optimal shellfish growth, essentially "chasing" the best environmental conditions.

For offshore shellfish farming, gear innovation is crucial: Submersible cages and robust longlines capable of withstanding open-ocean waves are in development (Aryai et al., 2021). Materials such as high-durability polymers and semi-rigid habitats that can be raised and lowered show promise. The environmental benefits of going offshore are multifold: Reduced coastal nutrient loading, freedom from nearshore contamination, potentially fewer disease issues due to distance from wild shellfish populations, and access to higher water quality and flow. Multiuse deep-sea vessels might even incorporate IMTA principles internally: An enclosed tank system could raise fish whose waste feeds algae that in turn feed shellfish in another section, all within one platform. Looking ahead, one could imagine fleets of aquaculture vessels roaming productive ocean gyres, such as combine harvesters of the sea; but instead of extracting wild fish, they would nurture and harvest farmed shellfish with minimal feed input. In summary, the aquaculture models of the future are poised to be more ecosystem-based, tech-driven, and spatially expanded than today's. It's the combination of better breeds and better "barns" that will elevate shellfish aquaculture to a new level of productivity and sustainability.

22.5.3 Automation, monitoring, and smart farming

Hand in hand with ecological integration is the rise of automation and "smart" technology in aquaculture operations. Much like precision agriculture on land, precision aquaculture seeks to continuously monitor key parameters and adjust management in real time (Wu et al., 2022). For shellfish, this means deploying arrays of sensors that track conditions such as temperature, salinity, dissolved oxygen, pH, and plankton density at the farm. Already, there are smart buoy systems that relay such data to farmers via mobile apps. The future will see more biosensors as well, including acoustic or accelerometer-based sensors on oyster cages that infer feeding rates from the sound of oysters snapping shut, or camera systems that detect gaping behavior indicative of stress. These data streams feed into farm management algorithms; for instance, if sensors detect a sudden drop in oxygen at the seabed at night, an automated upwelling system could be triggered to lift cages closer to the surface, or managers could be alerted to adjust stocking density. Remote monitoring extends beyond the environment to the stock themselves: New imaging techniques allow counting and size estimation of bivalves in cages or on longlines without manual sampling. Machine vision can assess fouling levels on oyster bags or mussel ropes. Feeding is not directly an issue for filter feeders, but in hatcheries, the feeding of microalgae to larvae can be automated with feedback control—optical sensors in larval tanks maintain algal cells at an optimal concentration by dosing from photobioreactors. Future technology might even include autonomous barges that traverse mussel longlines, mechanically harvesting the mussels, or underwater rovers that vacuum clams from seabed plots guided by GPS mapping of planting locations. Offshore and open-ocean farming will especially rely on automation, since regular human presence is more challenging in those environments. The integration of all these devices yields the "smart shellfish farm," where decisions are data-driven and many routine tasks are handled by machines. Notably, a digital twin approach combined with AIoT has been piloted in finfish aquaculture to enable intelligent farm management (Ubina et al., 2023), foreshadowing similar applications for shellfish farms. These improvements in farm practices complement genetic gains; for instance, a heat-tolerant oyster strain will reach its full potential only if the farm mitigates other stresses and avoids preventable losses such as low oxygen conditions.

22.6 Breeding–farming–market integration: Toward a holistic industry

The last piece of the puzzle is integration across the entire aquaculture value chain—linking breeding, farming, processing, and marketing into a cohesive system. Historically, these functions have often been fragmented, with independent hatcheries, separate grow-out farms, and different distributors typically operated in isolation. However, future models are trending toward vertical integration and full supply-chain coordination. In this section, we examine how large enterprises or cooperative alliances could adopt a "one-species, one-industry" approach, analogous to how some agribusiness firms coordinate crop genetics through to food product marketing. We explore the potential benefits of such integration for shellfish aquaculture, as well as considerations for sustainability and diversity. Ultimately, aligning breeding goals with farming realities and market demands through tight integration could accelerate innovation and ensure that improvements reach consumers effectively.

22.6.1 Vertical integration from hatchery to market

Vertical integration means a single entity controls multiple stages of production. In the context of shellfish, this might entail one company running the selective breeding program, the grow-out farms, the depuration and processing facilities, and even distribution to retailers. Early examples are already emerging: Some major oyster farming companies in the United States are vertically integrated—they operate their own broodstock hatcheries to produce seed, then raise and harvest the oysters, and even handle branding and direct sales. Taylor shellfish is a prime example, managing everything from seed to serving. Their tide-to-table operation has drawn comparisons to tech companies that control their supply chains to ensure quality.

The advantages of vertical integration include consistency and quality control: By managing seed production, a company can ensure only top-quality juveniles are used, leading to uniform crop performance. They can also time production to meet market demand. It reduces uncertainty in supply—farmers do not have to worry about sourcing quality seed from third parties if they produce it in-house. Moreover, integrated firms can capture more of the value-added; instead of selling bulk shellfish to a processor at low prices, they can process and brand the product themselves, capturing a premium from consumers. On a broader scale, if an industry adopts a coordinated "one-species" approach, it might resemble the salmon industry in Norway, which is dominated by a few firms that handle breeding, farming, and processing. This model has led to standardized products and significant R&D investment. Similarly, a consolidated shellfish industry could channel resources into advanced breeding at a scale that individual small farmers cannot. Cooperatives are another model: independent growers band together into a collective that operates like an integrated company for certain functions.

Another facet is data integration: With vertical integration, all performance data from farms can flow back into the breeding program's database seamlessly. This big-data feedback loop improves selection accuracy and allows rapid pinpointing of issues. Cooperatives could also ensure that small farmers have access to improved genetics without being locked out by proprietary monopolies. Traceability and safety also stand to gain. A fully integrated operation can implement measures such as unique tags for each harvest lot, linking it to broodstock. If an illness outbreak occurs, an integrated firm can quickly trace and recall products and adjust protocols, whereas in a fragmented industry, trace-backs can be slow and undermine consumer confidence. In marketing, being able to tell a story of integrated, sustainable production will resonate with consumers who value transparency.

Shellfish aquaculture of the future is likely to be more organized and consolidated, whether through corporate integration or tightly knit cooperative networks. This need not eliminate the small farmer—rather, small farmers might become part of cooperatives or serve as contract growers for larger brands. From a consumer perspective, a vertically integrated supply chain can deliver safer, more consistent shellfish with full traceability, addressing one of the public's main concerns with seafood. In essence, by bringing breeding, farming, and marketing under one roof, the industry can achieve both the transparency that diners crave and the control over quality that producers need. This holistic integration, combined with the scientific advances discussed earlier, completes the vision of a sustainable, high-tech shellfish aquaculture sector that can meet rising global seafood demand responsibly.

22.7 Conclusion

Shellfish aquaculture stands at an inflection point where advancements in genetics, technology, and industry organization converge to enable a more productive and sustainable future. By expecting more from the traits of cultured shellfish (not only faster growth and better nutrition but also resilience to climate change and provision of ecosystem services), we set ambitious targets that are increasingly attainable with modern science. Multiomics and AI are unraveling the complex genetic architecture of these traits, guiding breeders to make informed selections and even precise genomic edits. Meanwhile, breakthroughs in breeding technologies, from GS algorithms to CRISPR, are accelerating the development of superior stocks. Hand in hand with these biological improvements are evolving aquaculture models that are smarter, greener, and pushing into new spaces, such as the open ocean. Finally, a trend toward vertical integration promises to tightly connect the dots from broodstock to final product, ensuring that innovations in breeding translate to tangible benefits on the farm and value in the marketplace, under a regime of transparency and efficiency. Achieving this vision will require a careful balance, embracing innovation while safeguarding genetic diversity and ecosystem health. The coming decades will reveal whether these emerging perspectives become global realities, fulfilling the potential of shellfish as a cornerstone of sustainable "blue growth."

References

Ababouch, L., Nguyen, K. A. T., Castro de Souza, M., & Fernandez-Polanco, J. (2023). Value chains and market access for aquaculture products. *Journal of the World Aquaculture Society, 54*(2), 527–553. doi:10.1111/jwas.12964.

Abe, M., & Kuroda, R. (2019). The development of CRISPR for a mollusc establishes the formin Lsdia1 as the long-sought gene for snail dextral/sinistral coiling. *Development, 146*(9), dev175976. doi:10.1242/dev.175976.

Aryai, V., Abbassi, R., Abdussamie, N., Salehi, F., Garaniya, V., Asadnia, M., Baksh, A. A., Penesis, I., Karampour, H., Draper, S., Magee, A., Keng, A. K., Shearer, C., Sivandran, S., Yew, L. K., Cook, D., Underwood, M., Martini, A., Heasman, K., … Wang, C. M. (2021). Reliability of multi-purpose offshore-facilities: Present status and future direction in Australia. *Process Safety and Environmental Protection, 148*, 437–461. doi:10.1016/j.psep.2020.10.016.

Bunne, C., Roohani, Y., Rosen, Y., Gupta, A., Zhang, X., Roed, M., Alexandrov, T., AlQuraishi, M., Brennan, P., Burkhardt, D. B., Califano, A., Cool, J., Dernburg, A. F., Ewing, K., Fox, E. B., Haury, M., Herr, A. E., Horvitz, E., Hsu, P. D., … Quake, S. R. (2024). How to build the virtual cell with artificial intelligence: Priorities and opportunities. *Cell, 187*(25), 7045–7063. doi:10.1016/j.cell.2024.11.015.

Chen, Z.-H. (2023). Recent advances on marine mollusk-derived natural products: chemistry, chemical ecology and therapeutical potential. *Natural Product Reports, 40*(3), 509–556. doi:10.1039/D2NP00021K.

Chi, Y., Xu, C., & Li, Q. (2024). Evaluating survival after heat challenges and investigating their correlations with field summer survival for the juvenile Pacific oyster (Crassostrea gigas). *Aquaculture, 584*, 740643. doi:10.1016/j.aquaculture.2024.740643.

Costello, C., Cao, L., Gelcich, S., Cisneros-Mata, M., Free, C. M., Froehlich, H. E., Golden, C. D., Ishimura, G., Maier, J., Macadam-Somer, I., Mangin, T., Melnychuk, M. C., Miyahara, M., de Moor, C. L., Naylor, R., Nøstbakken, L., Ojea, E., O'Reilly, E., Parma, A. M., … Lubchenco, J. (2020). The future of food from the sea. *Nature, 588*(7836), 95–100. doi:10.1038/s41586-020-2616-y.

De Domenico, M., Allegri, L., Caldarelli, G., d'Andrea, V., Di Camillo, B., Rocha, L. M., … Zambelli, F. (2025). Challenges and opportunities for digital twins in precision medicine from a complex systems perspective. *npj Digital Medicine, 8*(1), 37. doi:10.1038/s41746-024-01402-3.

Einevoll, G. T., Destexhe, A., Diesmann, M., Grün, S., Jirsa, V., de Kamps, M., Migliore, M., Ness, T. V., Plesser, H. E., & Schürmann, F. (2019). The scientific case for brain simulations. *Neuron, 102*(4), 735–744. doi:10.1016/j.neuron.2019.03.027.

Fodrie, F. J., Rodriguez, A. B., Gittman, R. K., Grabowski, J. H., Lindquist, N. L., Peterson, C. H., ... Ridge, J. T. (2017). Oyster reefs as carbon sources and sinks. *Proceedings of the Royal Society B: Biological Sciences, 284*(1859), 20170891. doi:10.1098/rspb.2017.0891.

Gavery, M. R., Vandepas, L. E., Saunders, L. M., Vadopalas, B., Luckenbach, J. A., Trapnell, C., & Roberts, S. (2025). Primordial germ cell specification and early developmental cell states in Pacific oyster. *BMC Genomics, 26*(1), 951. doi:10.1186/s12864-025-12122-7.

Gentry, R. R., Froehlich, H. E., Grimm, D., Kareiva, P., Parke, M., Rust, M., Gaines, S. D., & Halpern, B. S. (2017). Mapping the global potential for marine aquaculture. *Nature Ecology & Evolution, 1*(9), 1317–1324. doi:10.1038/s41559-017-0257-9.

Ghosh, A. K., Hasanuzzaman, A. F. M., Islam, S. S., Sarower, M. G., Mistry, S. K., Arafat, S. T., & Huq, K. A. (2025). Integrated multi-trophic aquaculture (IMTA): Enhancing growth, production, immunological responses, and environmental management in aquaculture. *Aquaculture International, 33*(5), 336. doi:10.1007/s10499-025-02021-9.

Goncalves, P., Thompson, E. L., & Raftos, D. A. (2017). Contrasting impacts of ocean acidification and warming on the molecular responses of CO2-resilient oysters. *BMC Genomics, 18*(1), 431. doi:10.1186/s12864-017-3818-z.

Gu, Y., Lyu, S., Wang, L., Chen, Z., & Wang, X. (2022). Assessing the carbon sink capacity of coastal mariculture shellfish resources in China from 1981–2020. *Frontiers in Marine Science, 9*, 981569. doi:10.3389/fmars.2022.981569.

Houston, R. D., Bean, T. P., Macqueen, D. J., Gundappa, M. K., Jin, Y. H., Jenkins, T. L., Selly, S. L. C., Martin, S. A. M., Stevens, J. R., Santos, E. M., Davie, A., & Robledo, D. (2020). Harnessing genomics to fast-track genetic improvement in aquaculture. *Nature Reviews Genetics, 21*(7), 389–409. doi:10.1038/s41576-020-0227-y.

Jiang, X., Zhang, X., Guan, J., He, P., Wei, P., Zhang, L., Peng, J., & Zhao, L. (2025). Selective breeding boosts oyster resilience to ocean acidification via energy budget modulation. *Marine Environmental Research, 212*, 107544. doi:10.1016/j.marenvres.2025.107544.

Jin, K., Zhang, B., Jin, Q., Cai, Z., Wei, L., Wang, X., ... Wang, X. (2021). CRISPR/Cas9 system-mediated gene editing in the Fujian oysters (Crassostrea angulate) by electroporation. *Frontiers in Marine Science, 8*, 763470. doi:10.3389/fmars.2021.763470.

Kleppe, L., Crespo, D., Güralp, H. K., Lareyre, J.-J., Skaftnesmo, K. O., Ayllon, F., Fraser, T. W. K., Fjelldal, P. G., & Hansen, T. J. (2025). Production of donor-derived Atlantic salmon progeny using allogeneic surrogate broodstock technology. Aquaculture, 608, 742752. doi: 10.1016/j.aquaculture.2025.742752

Li, X., Shi, C., Yang, B., Li, Q. i, & Liu, S. (2023). High temperature aggravates mortalities of the Pacific oyster (Crassostrea gigas) infected with Vibrio: A perspective from homeostasis of digestive microbiota and immune response. *Aquaculture, 568*, 739309. doi:10.1016/j.aquaculture.2023.739309.

Lian, S., Hu, N., Chen, X., Dai, X., Zhu, X., Qiao, R., ... Wang, S. (2025). Widespread presence of bone marrow-like hematopoietic stem cell niche in invertebrate skeletons. *Science Advances, 11*(26), eadw0958. doi:10.1126/sciadv.adw0958.

Liu, F., Cai, B., Lian, S., Chang, X., Chen, D., Pu, Z., ... Li, Y. (2025). MolluscDB 2.0: A comprehensive functional and evolutionary genomics database for over 1400 molluscan species. *Nucleic Acids Research, 53*(D1), D1075–D1086. doi:10.1093/nar/gkae1026.

Liu, F., Li, Y., Yu, H., Zhang, L., Hu, J., Bao, Z., & Shi, W. (2021). MolluscDB: An integrated functional and evolutionary genomics database for the hyper-diverse animal phylum Mollusca. *Nucleic Acids Research, 49*(D1), D988–D997. doi:10.1093/nar/gkaa918.

Mackenzie, C. L., Raap, M. R., Leduc, S., Walker, C. Y. V., Green, T. J., Kim, E., ... Pearce, C. M. (2024). Development of a nature-based solution for mitigation of Pacific oyster summer mortality: use of the intertidal zone to improve resilience to environmental stressors. *Frontiers in Marine Science, 11*, 1345493. doi:10.3389/fmars.2024.1345493.

Montesinos-López, O. A., Montesinos-López, A., Pérez-Rodríguez, P., Barrón-López, J. A., Martini, J. W. R., Fajardo-Flores, S. B., Gaytan-Lugo, L. S., Santana-Mancilla, P. C., & Crossa, J. (2021). A review of deep learning applications for genomic selection. *BMC Genomics, 22*(1). doi:10.1186/s12864-020-07319-x.

Osipov, A., Nikolic, O., Gertych, A., Parker, S., Hendifar, A., Singh, P., Filippova, D., Dagliyan, G., Ferrone, C. R., Zheng, L., Moore, J. H., Tourtellotte, W., Van Eyk, J. E., & Theodorescu, D. (2024). The molecular twin artificial-intelligence platform integrates multi-omic data to predict outcomes for pancreatic adenocarcinoma patients. *Nature Cancer, 5*(2), 299–314. doi:10.1038/s43018-023-00697-7.

Parker, R., Ross, P. M, & O'Connor, W. A (2010). Populations of the Sydney rock oyster, Saccostrea glomerata, vary in response to ocean acidification. *Marine Biology, 158*(3), 689–697. doi:10.1007/s00227-010-1592-4.

Peng, D., Zhu, Y., & Chu, J. (2023). Strengthen management of offshore aquaculture. *Science, 381*(6661), 955. doi:10.1126/science.adj4352.

Perry, K. J., & Henry, J. Q. (2015). CRISPR/Cas9-mediated genome modification in the mollusc, Crepidula fornicata. *Genesis, 53*(2), 237–244. doi:10.1002/dvg.22843.

Potts, R. W. A., Gutierrez, A. P., Penaloza, C. S., Regan, T., Bean, T. P., & Houston, R. D. (2021). Potential of genomic technologies to improve disease resistance in molluscan aquaculture. *Philosophical Transactions of the Royal Society B, 376*(1825), 20200168. doi:10.1098/rstb.2020.0168.

Scanes, E., Siboni, N., Potts, J., Rao, S., Labbate, M., & Seymour, J. R. (2025). Temporal and spatial co-occurrence of Pacific oyster mortality and increased planktonic Vibrio abundance. *iScience, 28*(2), 111674. doi:10.1016/j.isci.2024.111674.

Sea, M. A, Hillman, J. R, & Thrush, S. F (2022). The influence of mussel restoration on coastal carbon cycling. *Global Change Biology, 28*(17), 5269–5282. doi:10.1111/gcb.16287.

Sudewi, A. A. R., Susilawathi, N. M., Mahardika, B. K., Mahendra, A. N., Pharmawati, M., Phuong, M. A., & Mahardika, G. N. (2019). Selecting potential neuronal drug leads from Conotoxins of various venomous marine cone snails in Bali, Indonesia. *ACS Omega., 4*(21), 19483–19490. doi:10.1021/acsomega.9b03122.

Sun, X., Filgueira, R., Sun, Y., Han, M., Tang, Q., & Sun, Y. (2025). Intensive oyster farming enhances carbon storage in sediments over decades. *Communications Earth & Environment, 6*(1), 383. doi:10.1038/s43247-025-02358-2.

Ubina, N. A., Lan, H. Y., Cheng, S. C., Chang, C. C., Lin, S. S., Zhang, K. X., Lu, H. Y., Cheng, C. Y., & Hsieh, Y. Z. (2023). Digital twin-based intelligent fish farming with Artificial Intelligence Internet of Things (AIoT). *Smart Agricultural Technology, 5*. doi:10.1016/j.atech.2023.100285.

van der Schatte Olivier, A., Jones, L., Vay, L. L., & Wilson, J. (2018). A global review of the ecosystem services provided by bivalve aquaculture. *Reviews in Aquaculture, 12*, 3–25. doi:10.1111/raq.12301.

Wang, H. E., Triebkorn, P., Breyton, M., Dollomaja, B., Lemarechal, J. D., Petkoski, S., ... Jirsa, V. K. (2024). Virtual brain twins: From basic neuroscience to clinical use. *National Science Review, 11*(5), nwae079. doi:10.1093/nsr/nwae079.

Wang, Y., Zhu, X., Lian, S., Li, Y., Hu, N., Hu, X., ... Shi, W. (2022). Functional characterization of Cfap206 for bivalve ciliogenesis by RNAi and CRISPR/Cas9 technologies. *Frontiers in Marine Science, 9*, 864037. doi:10.3389/fmars.2022.864037.

Wu, Y., Duan, Y., Wei, Y., An, D., & Liu, J. (2022). Application of intelligent and unmanned equipment in aquaculture: A review. *Computers and Electronics in Agriculture, 199*, 107201. doi:10.1016/j.compag.2022.107201.

Yu, X., Faggion, S., Liu, Y., Wang, B., Zeng, Q., Lu, C., Hu, J., Bargelloni, L., Fang, L., & Bao, Z. (2025). Role of multi-omics in aquaculture genetics and breeding: Current status and future perspective. *Science China Life Sciences, 68*(9), 2591–2604. doi:10.1007/s11427-024-2828-8.

Index

Page numbers followed by "*f*" and "*t*" indicate, figures and tables respectively.

D

E

F

G

H

Printed and bound by CPI Group (UK) Ltd, Croydon, CR0 4YY
14/09/2026
02216408-0019